Introduction to Quantum Field Theory with Applications to Quantum Gravity

Introduction to Quantum Field Theory with Applications to Quantum Gravity

Iosif L. Buchbinder

Department of Theoretical Physics, Tomsk State Pedagogical University, Tomsk, 634061, Russia

Ilya L. Shapiro

Departamento de Física – Instituto Ciências Exatas, Universidade Federal de Juiz de Fora, Juiz de Fora, CEP 36036-330, MG, Brazil

OXFORD
UNIVERSITY PRESS

OXFORD
UNIVERSITY PRESS
Great Clarendon Street, Oxford, OX2 6DP,
United Kingdom

Oxford University Press is a department of the University of Oxford.
It furthers the University's objective of excellence in research, scholarship,
and education by publishing worldwide. Oxford is a registered trade mark of
Oxford University Press in the UK and in certain other countries

First Edition published in 2021

Published in the United States of America by Oxford University Press
198 Madison Avenue, New York, NY 10016, United States of America

British Library Cataloguing in Publication
Data available

Library of Congress Control Number: 2020945719

ISBN 978–0–19–883831–9

DOI: 10.1093/oso/9780198838319.001.0001

Preface

For many decades, quantum field theory has played an important role in the successful description of the interactions of elementary particles. Besides, this area of theoretical physics has been always important due to the exchange of new ideas and methods with other branches of physics, such as statistical mechanics, condensed matter physics, gravitational physics, and cosmology. The last applications are becoming more important nowadays, especially because the amount of experimental and observational data demonstrates a fast growth and requires more detailed and reliable theoretical background. One of the most evident examples is the study of dark energy. Every few years, the estimates of its equation of state (EoS) become more precise and it can not be ruled out that, at some point, the EoS of the cosmological constant may be excluded from the list of phenomenologically acceptable possibilities. Does this necessarily mean that there is some special fluid (quintessence or alike) in the Universe? Or that the situation can be explained by the variable cosmological constant, e.g., some quantum effects? This is a phenomenologically relevant question, which should be answered at some point. On the other hand, this is a theoretical question, that can be answered only within a correctly formulated framework of quantum or semiclassical gravity.

In gravitational theory, general relativity represents a successful theory of relativistic gravitational phenomena, confirmed by various experiments in the laboratories and astronomical observations. Starting from the seventies and eighties, there has been a growing interest in the idea of the unification of all fundamental forces, including electroweak and strong interactions. Also, there is a general understanding that the final theory should also include gravitation. An important component of such unification is the demand for a quantum description of the gravitational field itself or, at least, a consistent formulation of the quantum theory of matter fields on the classical gravitational background, called semiclassical gravity.

The application of quantum field theory methods to gravitational physics, in both semiclassical and full quantum frameworks, requires a careful formulation of the fundamental base of quantum theory, with special attention to such important issues as renormalization, the quantum theory of gauge theories and especially effective action formalism. The existing literature on these subjects includes numerous review papers and also many books, e.g., [172, 56, 80, 150, 199, 240]. At the same time, the experience of the present authors, after giving many courses on the subject worldwide, shows that there is a real need to have a textbook with a more elementary introduction to the subject. This situation was one of the main motivations for writing this book which ended up being much longer than originally planned.

The textbook consists of two parts. Part I is based on the one-semester course given by I.B. in many places, including the Tomsk State Pedagogical University and the Federal University of Juiz de Fora. It includes a detailed introduction to the general methods of quantum field theory, which are relevant for quantum gravity, including its semiclassical part. Part II is mainly based on the one-semester course given regularly by I.Sh. in the Federal University of Juiz de Fora and on the numerous mini-courses in many countries. We did not pretend to do the impossible, that is, produce a comprehensive course of quantum field theory or quantum gravity. Instead, our purpose was to give a sufficiently detailed introduction to the fundamental, basic notions and methods, which would enable the interested reader to understand at least part of the current literature on the subject and, in some cases, start original research work.

It is a pleasure for us to acknowledge the collaborations on various subjects discussed in this book with M. Asorey, R. Balbinot, E.V. Gorbar, A. Fabbri, J.C. Fabris, J.-A. Helaël-Neto, P.M. Lavrov, T.P. Netto, S.D. Odintsov, F.O. Salles and A.A. Starobinsky. We would like also to thank many colleagues, especially A.O. Barvinsky, A.S. Belyaev, E.S. Fradkin, V.P. Frolov, S.J. Gates, E.A. Ivanov, D.I. Kazakov, S.M. Kuzenko, O. Lechnetfeld, H. Osborn, B.A. Ovrut, N.G. Pletnev, K. Stelle, A.A. Tseytlin, I.V. Tyutin, and G.A. Vilkovisky for fruitful discussions of the problems of quantum field theory.

We are grateful to Guilherme H.S. Camargo, Eduardo A. dos Reis, and especially to Wagno Cesar e Silva for communicating to us misprints and corrections; and also to Andreza R. Rodrigues and Yackelin Z. R. López for typing certain parts of the manuscript, and to Vadim Zyubanov for valuable technical assistance.

The main work on Part I of the book was done during the long-term visit of J.B. to the Federal University of Juiz de Fora (UFJF). The authors are grateful to UFJF and especially to the Physics Department for providing both kind hospitality and the conditions for productive work during this visit. Throughout the preparations of the manuscript, the work of the authors has been supported by a special project APQ-01205-16 from the Fundação de Amparo á Pesquisa de Minas Gerais (FAPEMIG). On the top of that, the scientific activity of I.Sh. was partially supported by the Conselho Nacional de Desenvolvimento Científico e Tecnológico (CNPq/Brazil). The authors are also grateful to the Russian Ministry of Science and High Education and Russian Foundation for Basic Research for their long-term support of the Center of Theoretical Physics at the Tomsk State Pedagogical University.

Contents

Part I

Introduction to Quantum Field Theory

1 Introduction

1.1 What is quantum field theory, and some preliminary notes

Quantum field theory (QFT) is part of the broader field of theoretical physics and is the study of quantum effects in continuous physical systems called fields. One can say that quantum field theory represents the unification of quantum mechanics and classical field theory. Since a natural and consistent description of fundamental interactions can be achieved in the framework of special relativity, it is also true to say that relativistic quantum field theory represents the unification of quantum mechanics and special relativity.

The main application of quantum field theory is the description of elementary particles and their interactions. However, QFT has also extensive applications in other areas of physics, including cosmology. Furthermore, quantum field theory plays an important role in the theoretical condensed matter physics, especially in the description of ensembles of a large number of interacting particles. The progress made in the theory of superconductivity, the theory of phase transitions and other areas of condensed matter physics is characterized by the consistent use of quantum field theory methods, and vice versa.

The first part of this book is devoted to the basic notions and fundamental elements of modern QFT formalism. In the second part, we present an introduction to the QFT in curved space and quantum gravity, which are less developed and essentially more complicated subjects.

The reader will note that the style of the two parts is different. In almost all of Part I and in most of Part II, we tried to give a detailed presentation, so that the reader could easily reproduce all calculations. However, following this approach for the whole topic of quantum gravity would enormously increase the size of the book and make it less readable. For this reason, in some places we avoided giving full technical details and, instead, just provided references of papers or preprints where the reader can find intermediate formulas. The same approach concerns the selection of the material. Since we intended to write an introductory textbook, in Part II we gave only the *need-to-know* information about quantum gravity. For this reason, many advanced subjects were not included. In addition, in some cases, only qualitative discussion and minimal references have been provided.

1.2 The notion of a quantized field

The field $\phi(x) = \phi(t, \mathbf{x})$ is defined as a function of time t and the space coordinates, that form a three-dimensional vector, $\mathbf{x}$. It is assumed that the values of the space

Introduction to Quantum Field Theory with Applications to Quantum Gravity. Iosif L. Buchbinder and Ilya L. Shapiro, Oxford University Press (2021). © Iosif L. Buchbinder and Ilya Shapiro.
DOI: 10.1093/oso/9780198838319.003.0001

coordinates correspond to a bounded or unbounded domain of the three-dimensional space. From the physical point of view, the field $\phi(t, \mathbf{x})$ can be treated as a dynamical object with an infinite number of degrees of freedom, marked by the three-dimensional vector index $\mathbf{x}$.

The notion of a field naturally arises in the framework of special relativity. Since there exists a maximal speed of propagation for any type of interaction, the physical bodies separated by space intervals can not affect each other instantly. Therefore, there should be a physical object responsible for transmitting perturbation from one body to another. Such an object is a field that fills the space between the bodies and carries perturbation from one body to another. The simplest example is an electromagnetic field that carries interaction between electrically charged bodies.

Taking into account quantum mechanical universality, it is natural to assume that fields should be quantized, like any other physical system. This means that quantum states are given by wave functions, while dynamical variables are given by operators acting on wave functions. Thus, in quantum theory, a field becomes an operator $\hat{\phi}(t, \mathbf{x})$, which is called a field operator.

As we have mentioned (and will discuss in more detail later on), a field is a system with an infinite number of degrees of freedom. However, it turns out that the state of the quantum field can be described in terms of either particles or fields. It turns out that the quantum field is a physical notion that is most suitable for the description of systems with an arbitrary number of particles.

It is well known that, in relativistic theory, there is a relation between momentum $\mathbf{p}$ and the energy $\varepsilon = \varepsilon(\mathbf{p})$ of a free particle,

$$\varepsilon^2 = m^2c^4 + c^2\mathbf{p}^2, \tag{1.1}$$

where c is the speed of light, and m is the mass of the particle. If the field can describe particles, it must take into account the relation (1.1) between energy and momentum. Let us try to clarify how the relation (1.1) can be implemented for the field. Let $\hat{\phi}(t, \mathbf{x})$ be the field operator, associated with the free particle. We can write the expansion as a Fourier integral,

$$\hat{\phi}(t, \mathbf{x}) = \int d^3p \, d\varepsilon \, e^{\frac{i}{\hbar}(\mathbf{p}\cdot\mathbf{x} - \varepsilon t)} \, \hat{\phi}(\varepsilon, \mathbf{p}) \,, \tag{1.2}$$

where $\hbar$ is the Planck constant. According to the standard interpretation, the vector $\mathbf{p}$ is treated as the momentum of a particle, and the quantity ε as the energy of the particle. Then, since, for each Fourier mode, ε and $\mathbf{p}$ are related by Eq. (1.1), the quantity ε under the integral (1.2) is not an independent variable but is a function of $\mathbf{p}$. In order to satisfy this condition, one can write

$$\hat{\phi}(\varepsilon, \mathbf{p}) = \delta(\varepsilon^2 - \varepsilon^2(\mathbf{p}))\hat{\phi}_*(\mathbf{p}), \tag{1.3}$$

where $\hat{\phi}_*(\mathbf{p})$ depends only on $\mathbf{p}$. As a result, we arrive at the representation

$$\hat{\phi}(t, \mathbf{x}) = \int d^3p \, d\varepsilon \, e^{\frac{i}{\hbar}(\mathbf{p}\mathbf{x} - \varepsilon t)} \delta(\varepsilon^2 - \varepsilon^2(\mathbf{p}))\hat{\phi}_*(\mathbf{p}). \tag{1.4}$$

Consider the following expression showing a d'Alembert operator acting on the field (1.2):

$$
\begin{aligned}
\Box\hat{\phi}(t,\mathbf{x}) &= \Big(\frac{1}{c^2}\frac{\partial^2}{\partial t^2} - \Delta\Big)\hat{\phi}(t,\mathbf{x}) \\
&= \int d^3p\, d\varepsilon\, e^{\frac{i}{\hbar}(\mathbf{p}\cdot\mathbf{x}-\varepsilon t)}\frac{1}{\hbar^2}\Big(\mathbf{p}^2 - \frac{1}{c^2}\varepsilon^2\Big)\delta(\varepsilon^2 - \varepsilon^2(\mathbf{p}))\hat{\phi}_*(\mathbf{p}) \\
&= \int d^3p\, e^{\frac{i}{\hbar}(\mathbf{p}\cdot\mathbf{x}-\varepsilon(\mathbf{p})t)}\frac{1}{\hbar^2}\Big[\mathbf{p}^2 - \frac{1}{c^2}\varepsilon^2(\mathbf{p})\Big]\,\delta(\varepsilon^2 - \varepsilon^2(\mathbf{p}))\hat{\phi}_*(\mathbf{p}) \\
&= \int d^3p\, e^{\frac{i}{\hbar}(\mathbf{p}\cdot\mathbf{x}-\varepsilon(\mathbf{p})t)}\frac{1}{\hbar^2}\Big[\mathbf{p}^2 - \frac{1}{c^2}(c^2\mathbf{p} + m^2c^4)\Big]\delta(\varepsilon^2 - \varepsilon^2(\mathbf{p}))\hat{\phi}_*(\mathbf{p}) \\
&= -\frac{m^2c^2}{\hbar^2}\,\hat{\phi}(t,\mathbf{x}).
\end{aligned}
$$

Thus, we find that the free field operator should satisfy

$$
\Big(\frac{1}{c^2}\frac{\partial^2}{\partial t^2} - \Delta + \frac{m^2c^2}{\hbar^2}\Big)\hat{\phi}(t,\mathbf{x}) \;=\; \Big(\Box + \frac{m^2c^2}{\hbar^2}\Big)\hat{\phi}(t,\mathbf{x}) = 0, \tag{1.5}
$$

the Klein–Gordon equation. Equation (1.5) is a direct consequence of the relativistic dispersion relation between the energy and the momentum of the particle.

If the field corresponds to a massless particle, the parameter m in Eq. (1.5) is zero. Therefore, the field operator of a free massless field satisfies the wave equation

$$
\Box\hat{\phi}(t,\mathbf{x}) \;=\; 0. \tag{1.6}
$$

Thus, any kind of a free relativistic quantum field is a spacetime-dependent operator satisfying the Klein–Gordon equation. In the case of interacting quantum fields, their dynamics is described by much more complicated equations which will be discussed in the following chapters.

1.3 Natural units, notations and conventions

It is evident that the units of measurements of physical quantities should correspond to the scales of phenomena where these units are used. For example, it is not reasonable to measure the masses of elementary particles in tons or grams, or the size of atomic nuclei in kilometers or centimeters.

When we consider the relativistic high-energy quantum phenomena in the fundamental quantum physics of elementary particles, it is natural to employ the units related to the fundamental constants of nature. This means that we have to choose the system of units where the speed of light is $c = 1$, and the Planck constant (which has the dimension of the action) is $\hbar = 1$. As a result, we obtain the natural system of units based only on the fundamental constants of nature. In these units, the action is dimensionless, the speed is dimensionless and the dimensions of energy and momentum coincide. As in quantum theory, there is a Planck formula, relating energy and frequency as $\varepsilon \sim \hbar\omega$, and $\omega \sim \frac{1}{t}$, where t is time, and the dimensions satisfy the relation

$$[\varepsilon] = [\mathbf{p}] = [m] = [l]^{-1} = [t]^{-1}. \tag{1.7}$$

Thus, we have only one remaining dimensional quantity, the unit of energy. Usually, the energy in high-energy physics is measured in electron-volts, such that the unit of energy is $1\,eV$, or $1\,GeV = 10^9\,eV$. The dimensions of length and time are identical. In what follows, we shall use this approach and assume the natural units of measurements described above, with $\hbar = c = 1$.

Other notations and conventions are as follows:

1) Minkowski space coordinates $x^\mu \equiv (x^0, \mathbf{x}) \equiv (t, \mathbf{x}) \equiv (x^0, x^i)$, where Greek letters represent the spacetime indices $\alpha, \dots, \mu = 0, 1, 2, 3$, while Latin letters are reserved for the space indices, $i, j, k, \dots = 1, 2, 3$.

2) Functions in Minkowski space are denoted as $\phi(x) \equiv \phi(x^0, x^i) \equiv \phi(x^0, \mathbf{x}) \equiv \phi(t, \mathbf{x})$.

3) The Minkowski metric is

$$\eta_{\mu\nu} = \begin{pmatrix} 1 & 0 & 0 & 0 \\ 0 & -1 & 0 & 0 \\ 0 & 0 & -1 & 0 \\ 0 & 0 & 0 & -1 \end{pmatrix} \equiv \mathrm{diag}(1, -1, -1, -1), \tag{1.8}$$

and the same is true for the inverse metric, $\eta^{\mu\nu} = \mathrm{diag}(1, -1, -1, -1)$. One can easily check the relations $\eta^{\mu\nu}\eta_{\nu\rho} = \delta^\mu{}_\rho$ and $\eta_{\mu\nu}\eta^{\nu\rho} = \delta_\mu{}^\rho$.

Furthermore, $\varepsilon^{\mu\nu\alpha\beta}$ is the four-dimensional, totally antisymmetric tensor. The sign convention is that $\varepsilon^{0123} = 1$ and hence $\varepsilon_{0123} = -1$.

4) Partial derivatives are denoted as

$$\frac{\partial}{\partial x^\mu} \equiv \partial_\mu, \qquad \frac{\partial^2}{\partial x^\mu \partial x^\nu} \equiv \partial_\mu \partial_\nu\,, \qquad \text{etc.} \tag{1.9}$$

5) Rising and lowering the indices looks like

$$A^\mu = \eta^{\mu\nu} A_\nu, \qquad A_\mu = \eta_{\mu\nu} A^\nu, \qquad \partial^\mu = \eta^{\mu\nu}\partial_\nu, \qquad \partial_\mu = \eta_{\mu\nu}\partial^\nu, \quad \text{etc.}$$

Let us note that these and some other rules will be changed in Part II, when we start to deal with curved spacetime.

6) The scalar product is as follows:

$$AB = A^\mu B_\mu = A^0 B_0 + A^i B_i = A_0 B_0 - A_i B_i.$$

In particular,

$$px = p_\mu x^\mu = p_0 x^0 + p_i x^i = p_0 x^0 - \mathbf{p} \cdot \mathbf{x},$$

where $p^\mu \equiv (p^0, \mathbf{p})$.

7) The integral over four-dimensional space is

$$\int d^4x = \int d^3x \int dx_0,$$

while the integral over three-dimensional space is $\int d^3x$.
8) Dirac's delta function in Minkowski space is

$$\delta^4(x - x') \equiv \delta(t - t')\delta(\mathbf{x} - \mathbf{x}') \equiv \delta(t - t')\delta(x^1 - x^{'1})\delta(x^2 - x^{'2})\delta(x^3 - x^{'3}). \quad (1.10)$$

In particular, this means $\int d^4x\, \delta^4(x - x')\phi(x') = \phi(x)$.
9) The d'Alambertian operator is

$$\Box = \partial_0^2 - \partial_1^2 - \partial_2^2 - \partial_3^2 = \eta^{\mu\nu}\partial_\mu\partial_\nu = \partial_0{}^2 - \Delta, \quad (1.11)$$

where the Laplace operator is

$$\Delta = \partial_0^2 + \partial_2^2 + \partial_3^2. \quad (1.12)$$

10) The convention is that repeated indices imply the summation in all cases, i.e.,

$$X_I Y_I = \sum_{I=1}^{N} X_I Y_I. \quad (1.13)$$

Comments

There are many books on quantum field theory that differ in their manner and level of presentation, targeting different audiences that range from beginners to more advanced readers. Let us present a short list of basic references, which is based on our preferences.

The standard textbooks covering the basic notions and methods are those by J.D. Bjorken and S.D. Drell [57], C. Itzykson and J.-B. Zuber [187], M.E. Peskin and D.V. Schroeder [250], M. Srednicki [304] and M.D. Schwartz [274].

A brief and self-contained introduction to modern quantum field theory can be found in the books by P. Ramond [256], M. Maggiore [215] and L. Alvarez-Gaume and M.A. Vazquez-Mozo [155].

Comprehensive monographs in modern quantum field theory, with extensive coverage but aimed for advanced readers are those by J. Zinn-Justin [356], S. Weinberg [345], B.S. DeWitt [106, 109] and W. Siegel [293].

There are also very useful lecture notes available online, e.g., those by H. Osborn [235]. For mathematical and axiomatical aspects and approaches to quantum field theory see, e.g., the book by N.N. Bogolubov, A.A. Logunov, A.I. Oksak and I. Todorov [60].

2
Relativistic symmetry

In this chapter, we briefly review special relativistic symmetry, which will be used in the rest of the book. In particular, we introduce basic notions of the Lorentz and Poincaré groups, which will be used in constructing classical and quantum fields.

In general, the principles of symmetry play a fundamental role in physics. One of the most universal symmetries of nature is the one that we can observe in the framework of special relativity.

2.1 Lorentz transformations

According to special relativity, a spacetime structure is determined by the following general principles:

1) Space and time are homogeneous.

2) Space is isotropic.

3) There exists a maximal speed of propagation of a physical signal. This maximal speed coincides with the speed of light. In all inertial reference frames the speed of light has the same value, c.

Let P_1 and P_2 be two infinitesimally separated events that are points in spacetime. In some inertial reference frame, the four-dimensional coordinates of these events are x^μ and $x^\mu + dx^\mu$. The interval between these two events is defined as

$$ds^2 = \eta_{\mu\nu} dx^\mu dx^\nu. \tag{2.1}$$

In another inertial reference frame, the same two events have the coordinates x'^μ and $x'^\mu + dx'^\mu$. The corresponding interval is

$$ds'^2 = \eta_{\mu\nu} dx'^\mu dx'^\nu. \tag{2.2}$$

The two intervals (2.1) and (2.2) are equal, that is, $ds'^2 = ds^2$, reflecting the independence of the speed of light on the choice of the inertial reference frame. Thus,

$$\eta_{\mu\nu} dx^\mu dx^\nu \;=\; \eta_{\alpha\beta}\, dx'^\alpha dx'^\beta. \tag{2.3}$$

Eq. (2.3) enables one to find the relation between the coordinates x'^α and x^μ. Let $x'^\alpha = f^\alpha(x)$, with some unknown function $f^\alpha(x)$. Substituting this relation into Eq. (2.3), one gets an equation for the function $f^\alpha(x)$ that can be solved in a general form. As a result,

$$x'^\alpha = \Lambda^\alpha{}_\mu x^\nu + a^\alpha, \tag{2.4}$$

Introduction to Quantum Field Theory with Applications to Quantum Gravity. Iosif L. Buchbinder and Ilya L. Shapiro, Oxford University Press (2021). © Iosif L. Buchbinder and Ilya Shapiro.
DOI: 10.1093/oso/9780198838319.003.0002

where $\Lambda \equiv (\Lambda^{\alpha}{}_{\mu})$ is a matrix with constant elements, and a^{α} is a constant four-vector. Substituting Eq. (2.4) into Eq. (2.3), we get

$$\eta_{\alpha\beta}\Lambda^{\alpha}{}_{\mu}\Lambda^{\beta}{}_{\nu} = \eta_{\mu\nu}. \tag{2.5}$$

The coordinate transformation (2.4) with the matrix $\Lambda^{\alpha}{}_{\mu}$, satisfying Eq. (2.5), is called the non-homogeneous Lorentz transformation. One can say that the non-homogeneous Lorentz transformation is the most general coordinate transformation preserving the form of the interval (2.1). If in Eq. (2.4) the vector $a^{\alpha} = 0$, the corresponding coordinate transformation is called the homogeneous Lorentz transformation, or simply the Lorentz transformation. Such a transformation has the form

$$x'^{\mu} = \Lambda^{\mu}{}_{\nu}x^{\nu} \tag{2.6}$$

with the matrix $\Lambda^{\mu}{}_{\nu}$ satisfying Eq. (2.5).

It is convenient to present the relation (2.5) in a matrix form. Let us introduce the matrices $\eta \equiv (\eta_{\alpha\beta})$ and $\Lambda \equiv (\Lambda^{\alpha}{}_{\mu})$. Then Eq. (2.5) can be written as

$$\Lambda^T \eta \Lambda = \eta, \tag{2.7}$$

where Λ^T is the transposed matrix with the elements $(\Lambda^T)_{\mu}{}^{\alpha} = \Lambda^{\alpha}{}_{\mu}$. One can regard Eq. (2.7) as a basic relation. Any homogeneous Lorentz transformation is characterized by the matrix Λ satisfying the basic relation, and vice versa. Therefore, the set of all homogeneous Lorentz transformations is equivalent to the set of all matrices Λ, satisfying (2.7).

Let us consider some important particular examples of Lorentz transformations:

1. Matrix Λ has the form

$$\Lambda = \begin{pmatrix} 1 & 0 \\ 0 & R^{i}{}_{j} \end{pmatrix} \tag{2.8}$$

where the matrix $R = (R^{i}{}_{j})$ transforms only space coordinates, $x'^{i} = R^{i}{}_{j}x^{j}$. Substituting eq.(2.8) into the basic relation (2.7), we obtain the orthogonality condition

$$R^T R = \mathbf{1_3}, \quad \text{or} \quad R^{i}{}_{k}\delta_{ij}R^{j}{}_{l} = \delta_{kl}, \tag{2.9}$$

where $\mathbf{1_3}$ is a three-dimensional unit matrix with elements δ_{ij}. Relation (2.9) defines the three-dimensional rotations

$$x'^{0} = x^{0}, \qquad x'^{i} = R^{i}{}_{j}x^{j}. \tag{2.10}$$

If matrix R satisfies Eq. (2.9), then the transformation (2.10) is the Lorentz transformation. Thus, the three-dimensional rotations represent a particular case of Lorentz transformation.

2. Consider a matrix Λ with the form

$$\Lambda = \begin{pmatrix} \frac{1}{\sqrt{1-\frac{v^2}{c^2}}} & 0 & 0 & \frac{\frac{v}{c}}{\sqrt{1-\frac{v^2}{c^2}}} \\ 0 & 1 & 0 & 0 \\ 0 & 0 & 1 & 0 \\ \frac{\frac{v}{c}}{\sqrt{1-\frac{v^2}{c^2}}} & 0 & 0 & \frac{1}{\sqrt{1-\frac{v^2}{c^2}}} \end{pmatrix}, \tag{2.11}$$

where $(v/c)^2 < 1$. It is easy to show that this matrix satisfies the basic relation. Therefore, this matrix describes a Lorentz transformation,

$$x'^0 = \frac{x^0 + \frac{v}{c}x^1}{\sqrt{1-\frac{v^2}{c^2}}}, \qquad x'^1 = x^1, \qquad x'^2 = x^2, \qquad x'^3 = \frac{x^3 + \frac{v}{c}x^0}{\sqrt{1-\frac{v^2}{c^2}}}. \tag{2.12}$$

This is the standard form of the Lorentz transformation for the case when one inertial frame moves with respect to another one in the x^3 direction. Indeed, one can construct a similar matrix describing relative motion in any other direction. Transformations of the type (2.12) are called boosts.

3. The matrix Λ corresponding to the time inversion, or T-transformation, is

$$\Lambda = \Lambda_T = \begin{pmatrix} -1 & 0 & 0 & 0 \\ 0 & 1 & 0 & 0 \\ 0 & 0 & 1 & 0 \\ 0 & 0 & 0 & 1 \end{pmatrix}.$$

This matrix corresponds to the Lorentz transformation

$$x'^0 = -x^0\,, \qquad x'^i = x^i. \tag{2.13}$$

4. Let the matrix Λ have the form

$$\Lambda = \Lambda_P = \begin{pmatrix} 1 & 0 & 0 & 0 \\ 0 & -1 & 0 & 0 \\ 0 & 0 & -1 & 0 \\ 0 & 0 & 0 & -1 \end{pmatrix}.$$

It is easy to check that the basic relation (2.7) is fulfilled in this case. This matrix corresponds to the following Lorentz transformation:

$$x'^0 = x^0 \qquad x'^i = -x^i, \tag{2.14}$$

which is called the space reflection or parity (P) transformation.

5. The matrix Λ with the form

$$\Lambda = \Lambda_{PT} = \Lambda_P\Lambda_T = \Lambda_T\Lambda_P = \begin{pmatrix} -1 & 0 & 0 & 0 \\ 0 & -1 & 0 & 0 \\ 0 & 0 & -1 & 0 \\ 0 & 0 & 0 & -1 \end{pmatrix}$$

corresponds to the following Lorentz transformation:

$$x'^\mu = -x^\mu, \tag{2.15}$$

which is called the full reflection.

Eqs. (2.13), (2.14), (2.15) are called discrete Lorentz transformations.

We will mainly need only the subclass of all Lorentz transformations that can be obtained by small deformations of the identical transformation. Let the transformation matrix have the form $\Lambda = I$, where I is the unit 4×4 matrix with elements $\delta^\mu{}_\nu$. Matrix I satisfies the basic relation (2.7). This matrix realizes the identical Lorentz transformation

$$x'^\mu = x^\mu.$$

Stipulating small deformations of identical transformations means that we consider matrices Λ of the form

$$\Lambda = I + \omega, \tag{2.16}$$

where ω is a matrix with infinitesimal elements $\omega^\mu{}_\nu$. Requiring that the matrix Λ from (2.16) correspond to a Lorentz transformation, we arrive at the relation

$$(I + \omega)^T \eta (I + \omega) = \eta.$$

Taking into account only the first-order terms in ω, one gets

$$\omega^T \eta + \eta\omega = 0.$$

Recovering the indices, we obtain

$$(\omega^T)_\mu{}^\alpha \eta_{\alpha\nu} + \eta_{\mu\alpha}\omega^\alpha{}_\nu = 0 \quad \Longrightarrow \quad \omega^\alpha{}_\mu \eta_{\alpha\nu} + \eta_{\mu\alpha}\omega^\alpha{}_\nu = \omega_{\mu\nu} + \omega_{\nu\mu} = 0. \tag{2.17}$$

One can see that the matrix ω is real and antisymmetric, and hence it has six independent elements. The matrix Λ (2.16) corresponds to the coordinate transformation

$$x'^\mu = x^\mu + \omega^\mu{}_\nu x^\nu,$$

which is called the infinitesimal Lorentz transformation.

2.2 Basic notions of group theory

Group theory is a branch of mathematics devoted to the study of the symmetries. In this subsection, we consider the basic notions of group theory that will be used in the rest of the book. It is worth noting that this section is not intended to replace a textbook on group theory. In what follows, we consequently omit rigorous definitions and proofs of the theorems and concentrate only on the main notions of our interest.

A set G of the elements g_1, g_2, $g_3, \ldots$, equipped with a law of composition (or product of elements, or multiplication rule, or composition law), e.g., $g_1 g_2$, is called a group if for each pair of elements $g_1, g_2 \in G$, the composition law satisfies the following set of conditions:

1) Closure, i.e., $\forall g_1,\, g_2 \in G$: $g_1 g_2 \in G$.
2) Associativity, i.e., $\forall g_1,\, g_2,\, g_3 \in G$, for the product $g_1(g_2 g_3) = (g_1 g_2) g_3$.
3) Existence of unit element, i.e., $\exists e \in G$, such that $\forall g \in G : ge = eg = g$.
4) Existence of inverse element, i.e., $\forall g \in G$, $\exists g^{-1} \in G$ such that $gg^{-1} = g^{-1}g = e$.

Using these conditions, one can prove the uniqueness of the unit and inverse elements.

A group is called Abelian or commutative if, $\forall g_1,\, g_2 \in G$, the product satisfies $g_1g_2 = g_2g_1$. In the opposite case, the group is called non-Abelian or non-commutative, i.e., $\exists g_1, g_2 \in G$ such that $g_1g_2 \neq g_2g_1$.

A subset $H \subset G$ is said to be a subgroup of group G if H itself is the group under the same multiplication rule as group G. In particular, this means if $h_1, h_2 \in H$, then $h_1h_2 \in H$. Also, $e \in H$, and if $h \in H$, then $h^{-1} \in H$.

A group consisting of a finite number of elements is called finite. In this case, it is possible to form a group table g_ig_j. A finite group is sometimes called a finite discrete group.

Let us consider a few examples:

1. Let G be a set of $n \times n$ real matrices M such that $\det M \neq 0$. It is evident that if $M_1,\, M_2 \in G$, then $\det M_1M_2 = \det M_1 \det M_2 \neq 0$ and hence $M_1,\, M_2 \in G$. Thus, this set forms a group under the usual matrix multiplication. The unit element is the unit matrix E, and the element inverse to the matrix M is the inverse matrix M^{-1}. We know that the multiplication of matrices is associative. Thus, all group conditions are fulfilled. This group is called a general linear n-dimensional real group and is denoted as $GL(n|\mathbb{R})$. Consider a subset $H \subset GL(n|\mathbb{R})$ consisting of matrices N that satisfy the condition $\det N = 1$. It is evident that $\det(N_1N_2) = \det N_1 \det N_2 = 1$. Hence

$$N_1,\, N_2 \in H \Longrightarrow N_1N_2 \in H.$$

Consider other properties of this group. It is evident that $E \in H$. On the top of this,

$$N \in G \quad \Longrightarrow \quad \det\left(N^{-1}\right) = (\det N)^{-1} = 1.$$

The last means $N^{-1} \in H$. Hence H is a subgroup of the group $GL(n|\mathbb{R})$. Group H is called a special linear n-dimensional real group and is denoted as $SL(n|\mathbb{R})$. In a similar way, one can introduce general and special complex groups $GL(n|\mathbb{C})$ and $SL(n|\mathbb{C})$, respectively, where $\mathbb{C}$ is a set of complex numbers.

2. Let G be a set of complex $n \times n$ matrices U such that $U^+U = UU^+ = E$, where E is the unit $n \times n$ matrix. Here, as usual, $(U^+)_{ab} = (U^*)_{ba}$ or $U^\dagger = (U^*)^T$, where $*$ means the operation of complex conjugation, and T means transposition. Evidently, $E \in G$ and, for any $U_1,\, U_2 \in G$, the following relations take place:

$$\begin{aligned}
(U_1U_2)^+(U_1U_2) &= U_2^+(U_1^+U_1)U_2 = U_2^+U_2 = E,\\
(U_1U_2)(U_1U_2)^+ &= U_1(U_2U_2^+)U_1^+ = U_1U_1^+ = E.
\end{aligned} \tag{2.18}$$

In addition, if $U \in G$, then $(U^{-1})^+U^{-1} = (UU^+)^{-1} = U^{-1}(U^{-1})^+ = (U^+U)^{-1} = E$. Therefore, if $U \in G$, then $U^{-1} \in G$ too. As a result, the set of matrices under consideration form a group. This group is called the n-dimensional unitary group $U(n)$.

The condition $U^+U = E$ leads to $|\det U|^2 = 1$. Hence $\det U = e^{i\alpha}$, where $\alpha \in \mathbb{R}$. One can also consider a subset of matrices $U \in U(n)$, that satisfy the relation $\det U = 1$. This subset forms a special unitary group and is denoted $SU(n)$.

Since the multiplication of matrices is, in general, a non-commutative operation, the matrix groups $GL(n,\mathbb{R})$, $SL(n,\mathbb{R})$, $U(n)$ and $SU(n)$ are, in general, non-Abelian.

A group G is called the Lie group if each of its element is a differentiable function of the finite number of parameters, and the product of any two group elements is a differentiable function of parameters of each of the factors. That is, consider, $\forall g \in G$, and for $g_1 = g(\xi_1^{(1)},\dots,\xi_N^{(1)})$ and $g_2 = g(\xi_1^{(2)},\dots,\xi_N^{(2)})$, $g = g_1 g_2 = g(\xi_1,\dots,\xi_N)$. Then

$$\xi_I = f_I(\xi_1^{(1)},\dots,\xi_N^{(1)},\xi_1^{(2)},\dots,\xi_N^{(2)}), \tag{2.19}$$

where $I = 1,2,\dots,N$ are the differentiable functions of the parameters $\xi_1^{(1)},\dots,\xi_N^{(1)},\xi_1^{(2)},\dots,\xi_N^{(2)}$. The Lie group is called compact if the parameters $\xi_1,\dots,\xi_N$ vary within a compact domain. One can prove that the parameters $\xi_1,\dots,\xi_N$ can be chosen in such a way that $g(0,\dots,0) = e$, where e is the unit element of the group.

All matrix groups described in the examples above are the Lie groups, where the role of parameters is played by independent matrix elements.

The two groups G and G' are called homomorphic if there exists a map f of the group G into the group G' such that, for any two elements $g_1, g_2 \in G$, the following conditions take place: $f(g_1g_2) = f(g_1)f(g_2)$, and if $f(g) = g'$, then $f(g^{-1}) = g'^{-1}$, where g'^{-1} is an inverse element in the group G'. Such a map is called homomorphism. One can prove that $f(e) = e'$, where e' is the unit element of the group G'. One-to-one homomorphism is called isomorphism, and the corresponding groups are called isomorphic. We will write, in this case, $G = G'$.

Let G be some group, and V be a real or complex linear space. Consider a map R such that, $\forall g \in G$, there exists an invertible operator $D_R(g)$ acting in the space V. Furthermore, let the operators $D_R(g)$ satisfy the following conditions:

1) $D_R(e) = I$, where I is a unit operator in the space V; and 2) $\forall g_1,\ g_2 \in G$, we have $D_R(g_1g_2) = D_R(g_1)D_R(g_2)$.

The map R is called a representation of the group G in the linear space V. Operators $D_R(g)$ are called the operators of representation, and the space V is called the space of the representation. One can prove that, $\forall g \in G$, there is $D_R(g^{-1}) = D_R^{-1}(g)$, where $D_R^{-1}(g)$ is the inverse operator for $D_R(g)$. Thus, the set of operators $D_R(g)$ forms a group where a multiplication rule is the usual operator product.

We will mainly concern ourselves with matrix representations, where the operators $D_R(g)$ are the $n \times n$ matrices $D_R(g)^i{}_j$, $\quad i,j = 1,2,\dots,n$. Let v be a vector in a space of representation with the coordinates $v^1, v^2,\dots,v^n$, in some basis. The matrices $D_R(g)^i{}_j$ generate the coordinate transformation of the form

$$v'^i = D_R(g)^i{}_j v^j.$$

Let R be a representation of the group G in the linear space V, and $\tilde{V}$ be a subspace in V, i.e., $\tilde{V} \subset V$. We assume that, for any vector $\tilde{v} \in \tilde{V}$ and for any operator $D_R(g)$, the condition $D_R(g)\tilde{v} \in \tilde{V}$ takes place. Then, the subspace $\tilde{V}$ is called the invariant subspace of the representation R. Any representation always has two invariant subspaces, which are called trivial. These are the subspace $\tilde{V} = V$, and the subspace $\tilde{V} = \{0\}$, which consists of a single zero element. All other invariant subspaces, if they exist, are called non-trivial. A representation R is called reducible if it has non-trivial invariant subspaces, and irreducible if it does not. In other words, the representation R is called irreducible if it has only trivial invariant subspaces. A representation is called completely irreducible if all representation matrices $D_R(g)$ have the block-diagonal form. This means that, in a certain basis,

$$D_R(g) = \begin{pmatrix} \boxed{D_1(g)} & 0 & 0 \;\ldots & 0 \\ 0 & \boxed{D_2(g)} & 0 \;\ldots & 0 \\ 0 & 0 & \ddots \;\ldots & 0 \\ 0 & 0 & 0 \;\ldots & \boxed{D_k(g)} \end{pmatrix}.$$

This situation means that the representation space has k non-trivial invariant subspaces. In each of such subspaces, one can define an irreducible representation, $D_k(g)$.

A given Lie group can have different representations, where the matrices $D_R(g)$ may have different forms. However, some properties are independent of the representation. Some of these properties can be formulated, e.g., in terms of Lie algebra. Let $D_R(g)$ be the operators of representation, and $g = g(\xi)$. Then, the operators $D_R(g)$ will be the functions of N parameters $\xi^1, \xi^2, \ldots, \xi^N$, i.e., $D_R(g) = D_R(\xi)$ and $D_R(\xi)|_{\xi^I=0} = D_R(e) = \mathbf{1}$, where $\mathbf{1}$ is a unit matrix in the given representation space. One can prove that, in an infinitesimal vicinity of the unit element, operators $D_R(\xi)$ can be presented in the form

$$D_R(\xi) = \mathbf{1} + i\xi^I T_{RI}, \quad \text{where} \quad T_{RI} = -i \left.\frac{\partial D_R(\xi)}{\partial \xi^I}\right|_{\xi=0}. \tag{2.20}$$

The operators T_{RI} are called the generators of the group G in the representation R. One can show that any operator $D_R(\xi)$ which is obtained by the continuous deformation from the unit element can be written as

$$D_R(\xi) = e^{i\xi^I T_{RI}}. \tag{2.21}$$

If the operator $D_R(\xi)$ is unitary, i.e., $D_R(\xi)D_R^+(\xi) = D_R^+(\xi)D_R(\xi) = \mathbf{1}$, then the generators T_{RI} are Hermitian, i.e., $T_{RI} = T_{RI}^+$. The generators of our interest satisfy the following relation in terms of commutators:

$$[T_{RI}, T_{RJ}] = i f_{IJ}{}^K T_{RK}, \tag{2.22}$$

where $f_{IJ}{}^K$ are the structure constants of the Lie group G. It is evident that $f_{IJ}{}^K = -f_{JI}{}^K$. In general, the form of the matrices T_{RI} depends on the representation. However, one can prove that the structure constants do not depend on the representation. Thus, these constants characterize the group G itself.

The group generators are closely related to the notion of Lie algebra. Let A be a real or complex linear space with the elements $a_1, a_2, \ldots$. A linear space A is called Lie algebra, if for each two elements $a_1, a_2 \in A$, there exists a composition law (also called multiplication or the Lie product) $[a_1, a_2]$, such that

$$
\begin{aligned}
&1)\ [a_1, a_2] \in A,\\
&2)\ [a_1, a_2] = -[a_2, a_1],\\
&3)\ [c_1 a_1 + c_2 a_2, a_3] = c_1[a_1, a_3] + c_2[a_2, a_3] and\\
&4)\ [a_1, [a_2, a_3]] + [a_2, [a_3, a_1]] + [a_3, [a_1, a_2]] = 0.
\end{aligned} \tag{2.23}
$$

Here, c_1 and c_2 are arbitrary real or complex numbers, and $a_3 \in A$. The composition law $[a_1, a_2]$ is called the Lie bracket, or the commutator. Property 4 is called the Jacobi identity.

It is easy to check that the commutator of the generators (2.22) of the representation of the Lie group G satisfies all properties of the composition law for the Lie algebra. Therefore, the generators T_{RI} form the Lie algebra which is called the Lie algebra associated with a given Lie group G. To be more precise, they form a representation of the Lie algebra. It means that one can define the map $T(a)$ of the Lie algebra into a linear space of operators such that

$$
T : a \longrightarrow T(a) \quad \text{and} \quad [a_1, a_2] \longrightarrow [T(a_1), T(a_2)]. \tag{2.24}
$$

A Lie algebra is called commutative, or Abelian, if, for any two elements $a_1, a_2 \in A$, $[a_1, a_2] = 0$. In the opposite case, the Lie algebra is called non-commutative, or non-Abelian. One can prove that the Lie algebra associated with an Abelian Lie group is Abelian.

2.3 The Lorentz and Poincaré groups

Consider the group properties of Lorentz transformations. The Lorentz transformation has been defined in the form

$$
x'^{\mu} = \Lambda^{\mu}{}_{\nu} x^{\nu},
$$

where the matrix Λ satisfies the basic relation (2.7). Let us show that Lorentz transformations form a group. Consider the set of all Lorentz transformations or, equivalently, the set of all matrices Λ satisfying $\Lambda^T \eta \Lambda = \eta$.

For the product of two matrices corresponding to the Lorentz transformations, Λ_1 and Λ_2, we have

$$
(\Lambda_1 \Lambda_2)^T \eta (\Lambda_1 \Lambda_2) = \Lambda_2{}^T \Lambda_1{}^T \eta \Lambda_1 \Lambda_2 = \Lambda_2{}^T \eta \Lambda_2 = \eta. \tag{2.25}
$$

Thus, the matrix product $\Lambda_1 \Lambda_2$ satisfies the basic relation (2.7), and hence two consequent Lorentz transformations are equivalent to another Lorentz transformation,

$$
x''^{\mu} = \Lambda_1{}^{\mu}{}_{\alpha} \Lambda_2{}^{\alpha}{}_{\nu} x^{\nu}. \tag{2.26}
$$

Let I be the unit 4×4 matrix with the elements $\delta^{\mu}{}_{\nu}$. It is evident that $I^T \eta I = \eta$, *i.e.*, the matrix I corresponds to a Lorentz transformation.

The next step is to check the existence of an inverse element. The basic relation (2.7) can be recast in the form $\Lambda^T \eta = \eta \Lambda^{-1}$ or, equivalently, $\eta = (\Lambda^T)^{-1} \eta \Lambda^{-1}$, or $(\Lambda^{-1})^T \eta \Lambda^{-1} = \eta$. Thus, matrix Λ^{-1} also corresponds to a Lorentz transformation.

Lorentz transformations form a group where the multiplication law is a standard product of matrices Λ. Such a group is called the Lorentz group. Let us explore it in more detail. From the basic relation $\Lambda^T\eta\Lambda = \eta$ follows

$$\det\Lambda^T \det\eta \det\Lambda = \det\eta \qquad \text{and} \qquad \det\eta = -1 \neq 0.$$

As a result, $(\det\Lambda)^2 = 1$, and hence $\det\Lambda = \pm 1$.

Starting from the relation $\Lambda^\alpha{}_\mu \eta_{\alpha\beta} \Lambda^\beta{}_\nu = \eta_{\mu\nu}$, and setting $\mu = 0$ and $\nu = 0$, one gets $1 = \Lambda^0{}_0\, \eta_{00}\, \Lambda^0{}_0 + \Lambda^i{}_0\, \eta_{ij}\, \Lambda^j{}_0$. Since $\eta_{ij} = -\delta_{ij}$ and $\eta_{00} = 1$, one obtains

$$1 = (\Lambda^0{}_0)^2 - \Lambda^i{}_0 \Lambda^i{}_0 \qquad \Longrightarrow \qquad (\Lambda^0{}_0)^2 = 1 + \Sigma_{i=1}^3 (\Lambda^i{}_0)^2.$$

Therefore, $(\Lambda^0{}_0)^2 \geq 1$, and hence $|\Lambda^0{}_0| \geq 1$. As a result, we have the two relations

$$(\det\Lambda)^2 = 1, \qquad |\Lambda^0{}_0| \geq 0. \tag{2.27}$$

Thus, the following cases are possible:

$$\begin{aligned} &\det\Lambda = 1, && \Lambda^0{}_0 > 0, \\ &\det\Lambda = -1, && \Lambda^0{}_0 > 0, \\ &\det\Lambda = 1, && \Lambda^0{}_0 < 0, \\ &\det\Lambda = -1, && \Lambda^0{}_0 < 0. \end{aligned}$$

It means that the set of all Lorentz transformation is separated into four subsets:

$$\begin{aligned} &L_+{}^\uparrow: && \text{the set of matrices } \Lambda \text{ such that } \det\Lambda = 1,\ \Lambda^0{}_0 > 0, \\ &L_-{}^\uparrow: && \text{the set of matrices } \Lambda \text{ such that } \det\Lambda = -1,\ \Lambda^0{}_0 > 0, \\ &L_+{}^\downarrow: && \text{the set of matrices } \Lambda \text{ such that } \det\Lambda = 1,\ \Lambda^0{}_0 < 0, \\ &L_-{}^\downarrow: && \text{the set of matrices } \Lambda \text{ such that } \det\Lambda = -1,\ \Lambda^0{}_0 < 0. \end{aligned} \tag{2.28}$$

It is easy to see that $I \in L_+{}^\uparrow$. One can show that the subset $L_+{}^\uparrow$ forms a group which is called the proper Lorentz group. It is evident that the proper Lorentz group is a subgroup of the Lorentz group. All other subsets do not form groups.

Remark. The infinitesimal Lorentz transformations are generated by the matrix $\Lambda = I + \omega$. Since $I \in L_+{}^\uparrow$, the matrices $\Lambda = I + \omega \in L_+{}^\uparrow$. Since the matrix ω has six real independent elements, the proper Lorentz group is a six-parametric real Lie group.

Now let us consider a set of non-homogeneous Lorentz transformations

$$x'^\mu = \Lambda^\mu{}_\nu x^\nu + a^\mu,$$

where the matrix Λ satisfies the basic relation, and a^μ is a constant four-vector. Applying two non-homogeneous Lorentz transformations, one after another, we get

$$x'^\alpha = \Lambda_1{}^\alpha{}_\nu x^\nu + a_1{}^\alpha, \qquad x''^\mu = \Lambda_2{}^\mu{}_\alpha x'^\alpha + a_2{}^\mu. \tag{2.29}$$

Substituting the first relation into the second one, we get

$$x''^{\mu} = \Lambda_2{}^{\mu}{}_{\alpha}(\Lambda_1{}^{\alpha}{}_{\nu}x^{\nu} + a_1{}^{\alpha}) + a_2{}^{\mu} \; = \; \Lambda_2{}^{\mu}{}_{\alpha}\Lambda_1{}^{\alpha}{}_{\nu}x^{\nu} + \Lambda_2{}^{\mu}{}_{\alpha}a_1{}^{\alpha} + a_2{}^{\mu}. \qquad (2.30)$$

According to what we proved before, the matrix $\Lambda_2{}^{\mu}{}_{\alpha}\Lambda_1{}^{\alpha}{}_{\nu} = (\Lambda_2\Lambda_1)^{\mu}{}_{\alpha}$ satisfies the basic relation. Furthermore, the quantity $\Lambda_2{}^{\mu}{}_{\alpha}a_1{}^{\alpha} + a_2{}^{\mu}$ is a constant four-vector. Let us denote

$$\Lambda^{\mu}{}_{\nu} = \Lambda_2{}^{\mu}{}_{\alpha}\Lambda_1{}^{\alpha}{}_{\nu}\,, \qquad a^{\mu} = \Lambda_2{}^{\mu}{}_{\alpha}a_1{}^{\alpha} + a_2{}^{\mu}.$$

Then the relation (2.30) becomes

$$x''^{\mu} = \Lambda^{\mu}{}_{\nu}x^{\nu} + a^{\mu}. \qquad (2.31)$$

Since the matrix $\Lambda^{\mu}{}_{\nu}$ satisfies the basic relation, Eq. (2.31) gives us again the non-homogeneous Lorentz transformation. We denote the non-homogeneous Lorentz transformation as (Λ, a), and define, as per (2.30), the multiplication rule on a set of all such transformations as follows:

$$(\Lambda_2, a_2)(\Lambda_1, a_1) = (\Lambda_2\Lambda_1, \Lambda_2 a_1 + a_2). \qquad (2.32)$$

Using the transformation $(I, 0)$, we arrive at the relations

$$(\Lambda, a)(I, 0) = (\Lambda I, \Lambda \cdot 0 + a) = (\Lambda, a) \; = \; (I, 0)(\Lambda, a).$$

It is clear that the transformation $(I, 0)$ plays the role of the identity transformation. Let (Λ, a) be a non-homogeneous Lorentz transformation, and consider the transformation $(\Lambda^{-1}, -\Lambda^{-1}a)$. We have

$$(\Lambda, a)(\Lambda^{-1}, -\Lambda^{-1}a) = (\Lambda\Lambda^{-1}, \, -\Lambda\Lambda^{-1}a + a) = (I, 0) = (\Lambda^{-1}, -\Lambda^{-1}a)(\Lambda, a).$$

Hence, the transformation $(\Lambda^{-1}, -\Lambda^{-1}a)$ is the inverse of (Λ, a). As a result, the set of all non-homogeneous Lorentz transformations forms a group, with the multiplication rule given by Eq. (2.32). This is the Poincaré group.

An infinitesimal non-homogeneous Lorentz transformation has the form

$$x'^{\mu} = x^{\mu} + \omega^{\mu}{}_{\nu}x^{\nu} + a^{\mu}, \qquad (2.33)$$

where $\omega^{\mu}{}_{\nu}$ is a matrix with infinitesimal elements, and a^{μ} is an infinitesimal constant four-vector. Since $\omega_{\mu\nu} = -\omega_{\nu\mu}$, the matrix $\omega_{\mu\nu}$ has six real independent elements. In the vicinity of the unit element $(I, 0)$, any element of the Poincaré group is determined by the real parameters $\omega_{\mu\nu}$ and a_{μ}. Thus, the Poincaré group is the ten-parametric Lie group.

2.4 Tensor representation

We defined a group representation as a map of the group into a group of matrices acting in a linear space. The Lorentz and Poincaré groups express the special relativity principles; therefore, the representations of these groups in the space of the fields define the types of the fields compatible with the principles of relativity. In this and the following sections, we consider the simplest representations of Lorentz and Poincaré groups. The general theory of the representations of these groups is well developed, but its detailed consideration is beyond the scope of this book.

Let us start with the linear space of the tensor fields. Consider a set of all coordinate systems related by the transformations

$$x'^{\mu} = \Lambda^{\mu}{}_{\nu}x^{\nu} + a^{\mu}, \tag{2.34}$$

with matrices Λ satisfying the basic relation (2.7). We will call all such coordinate systems admissible. Let us assume that, in some admissible coordinate system $\{x^{\mu}\}$, there is a set of 4^{m+n} functions $t^{\mu_1\cdots\mu_m}{}_{\nu_1\cdots\nu_n}(x)$, while, in another admissible coordinate system $\{x'^{\mu}\}$, there is a set of 4^{m+n} functions $t'^{\mu_1\cdots\mu_m}{}_{\nu_1\cdots\nu_n}(x')$. If these two sets are related to each other as

$$t'^{\mu_1\cdots\mu_m}{}_{\nu_1\cdots\nu_n}(x') = \frac{\partial x'^{\mu_1}}{\partial x^{\alpha_1}} \cdots \frac{\partial x'^{\mu_m}}{\partial x^{\alpha_m}} \frac{\partial x'^{\beta_1}}{\partial x^{\nu_1}} \cdots \frac{\partial x'^{\beta_n}}{\partial x^{\nu_n}} t^{\alpha_1\cdots\alpha_m}{}_{\beta_1\cdots\beta_n}(x), \tag{2.35}$$

then these functions form a tensor (or a tensor field) of the type (m, n). The numbers $t^{\mu_1\cdots\mu_n}{}_{\nu_1\cdots\nu_n}(x)$ are called the components of the tensor in the coordinate frame $\{x^{\mu}\}$.

Starting from the relation (2.34), we get

$$\frac{\partial x'^{\mu}}{\partial x^{\alpha}} = \Lambda^{\mu}{}_{\alpha}. \tag{2.36}$$

It is easy to see that

$$(\Lambda^{-1})^{\alpha}{}_{\mu}\Lambda^{\mu}{}_{\beta}\frac{\partial x^{\beta}}{\partial x'^{\nu}} = (\Lambda^{-1})^{\alpha}{}_{\mu}\delta^{\mu}{}_{\nu} \quad \Longrightarrow \quad \frac{\partial x^{\alpha}}{\partial x'^{\nu}} = (\Lambda^{-1})^{\alpha}{}_{\nu} = (\Lambda^{-1})^{T}{}_{\nu}{}^{\alpha}, \tag{2.37}$$

where T means a matrix transposition.

Then, the definition (2.35) can be rewritten as

$$\begin{aligned} t'^{\mu_t\cdots\mu_m}{}_{\nu_1\cdots\nu_n}(x') &= \Lambda^{\mu_1}{}_{\alpha_1}\ldots\Lambda^{\mu_m}{}_{\alpha_m} \\ &\times (\Lambda^{-1})^{T}{}_{\nu_1}{}^{\beta_1}\ldots(\Lambda^{-1})^{T}{}_{\nu_n}{}^{\beta_n} t^{\alpha_1\cdots\alpha_m}{}_{\beta_1\cdots\beta_n}(x). \end{aligned} \tag{2.38}$$

This relation means that, for any element of the Poincaré group (Λ, a), there is a tensor transformation (2.38). One can check that all conditions of the group representation are fulfilled for the transformations (2.38). Therefore, Eq. (2.38) defines a representation of the Poincaré group that is called the tensor representation.

Setting $a^{\mu} = 0$ in (2.38), we arrive at the tensor representation of the Lorentz group. There are special cases of tensor representations for either Poincaré or Lorentz groups. E.g., the tensor of rank zero, or type $(0, 0)$, is called a scalar. The relation (2.38) in this case is

$$\varphi'(x') = \varphi(x), \qquad x' = \Lambda x + a. \tag{2.39}$$

A tensor with components $t^{\mu}(x)$ is called a contravariant vector. The defining relation (2.38) in this case has the form

$$t'^{\mu}(\Lambda x + a) = \Lambda^{\mu}{}_{\nu}t^{\nu}(x). \tag{2.40}$$

A tensor with components t_μ is called a covariant vector. The defining relation (2.38) in this case looks like

$$t'_\mu(\Lambda x + a) = (\Lambda^{-1})^T{}_\mu{}^\nu t_\nu(x). \tag{2.41}$$

Remark. Using a metric, one can convert a covariant index into a contravariant, and vice versa. For example, for vectors, we have $t^\mu = \eta^{\mu\nu} t_\nu$ and $t_\mu = \eta_{\mu\nu} t^\nu$. Therefore, we can simply call the corresponding geometric object a vector, which may have covariant or contravariant components. Analogously, one can convert any upper tensor index into a lower tensor index, and vice versa.

Our main purpose in this section is to find an infinitesimal form of the Lorentz transformation of tensor components. Let us write

$$x'^\mu = x^\mu + \omega^\mu{}_\nu x^\nu + a^\mu, \tag{2.42}$$

where $\omega_{\mu\nu} = -\omega_{\nu\mu}$ is a matrix with infinitesimal elements, and a^μ is an infinitesimal vector. In what follows, we may omit the indices, when it is possible to do so without causing confusion.

Let us consider the transformation law for the scalar field,

$$\varphi'(x + \omega x + a) = \varphi(x) \quad \text{or} \quad \varphi'(x) + \partial_\mu \varphi(x)(\omega^\mu{}_\nu x^\nu + a^\mu) = \varphi(x). \tag{2.43}$$

Denote $\varphi'(x) - \varphi(x) = \delta\varphi(x)$, where $\delta\varphi(x)$ is a variation of a scalar field under the infinitesimal coordinate transformations. Let us introduce the operators P_μ and $J_{\alpha\beta}$ by the rule

$$P_\alpha = i\partial_\alpha\,, \qquad J_{\alpha\beta} = \eta_{\alpha\nu} x^\nu P_\beta - \eta_{\beta\nu} x^\nu P_\alpha. \tag{2.44}$$

P_μ and $J_{\alpha\beta}$ are called the generators of spacetime translations and the Lorentz rotations, respectively, in the scalar representation. Using these operators, the variation of the scalar field can be written as follows:

$$\delta\varphi(x) = \Big[ia^\alpha P_\alpha - \frac{i}{2}\,\omega^{\alpha\beta} J_{\alpha\beta}\Big]\varphi(x). \tag{2.45}$$

Let us consider the infinitesimal transformations of a vector field. We have

$$t'^\mu(x + \omega x + a) = (\delta^\mu{}_\nu + \omega^\mu{}_\nu) t^\nu(x),$$

which can be written as

$$t'^\mu(x) + \partial_\nu t^\mu(x)\big[\omega^\nu{}_\lambda x^\lambda + a^\nu\big] = t^\mu(x) + \omega^\mu{}_\nu t^\nu(x)$$

and, finally, as

$$\begin{aligned}
\delta t^\mu(x) &= -\partial_\nu t^\mu(x)\omega^\nu{}_\lambda x^\lambda - \partial_\nu t^\mu(x) a^\nu + \omega^\mu{}_\nu t^\nu(x) = ia^\alpha(i\partial_\alpha)\delta^\mu{}_\nu t^\nu(x) \\
&\quad + \frac{i}{2}\omega^{\alpha\beta}\Big[i(\delta^\mu{}_\beta \eta_{\alpha\nu} - \delta^\mu{}_\alpha \eta_{\beta\nu}) + i(\delta_\alpha{}^\gamma \eta_{\beta\lambda} x^\lambda - \delta_\beta{}^\gamma \eta_{\alpha\lambda} x^\lambda)\partial_\gamma\Big] t^\mu(x) \\
&= \Big[ia^\alpha (P_\alpha)^\mu{}_\nu - \frac{i}{2}\omega^{\alpha\beta}(J_{\alpha\beta})^\mu{}_\nu\Big] t^\nu(x),
\end{aligned} \tag{2.46}$$

where the following notations were used:

$$\begin{aligned}(P_\alpha)^\mu{}_\nu &= \delta^\mu{}_\nu(i\partial_\alpha),\\ (J_{\alpha\beta})^\mu{}_\nu &= (M_{\alpha\beta})^\mu{}_\nu + (S_{\alpha\beta})^\mu{}_\nu, \qquad (2.47)\\ (M_{\alpha\beta})^\mu{}_\nu &= \delta^\mu{}_\nu\big(x_\alpha P_\beta - x_\beta P_\alpha\big),\\ (S_{\alpha\beta})^\mu{}_\nu &= i\big(\delta^\mu{}_\alpha\eta_{\beta\nu} - \delta^\mu{}_\beta\eta_{\alpha\nu}\big). \qquad (2.48)\end{aligned}$$

The operator $(P_\alpha)^\mu{}_\nu$ is the generator of spacetime translations in the contravariant vector representations, and the operator $(J_{\alpha\beta})^\mu{}_\nu$ is the generator of Lorentz rotations in the covariant vector representation. Finally, P_α and $J_{\alpha\beta}$ are the Poincaré group generators in vector representations.

Similar considerations can be made for any tensor. For instance, in the particular case of the $(1,1)$-type tensor, the result can be written in the symbolic form

$$\delta t^{A'}{}_{B'}(x) = \Big[ia^\alpha(P_\alpha)^{A'}{}_{B'A}{}^B - \frac{i}{2}\omega^{\alpha\beta}(J_{\alpha\beta})^{A'}{}_{B'A}{}^B\Big]t^A{}_B(x),$$

where $A' \equiv (\mu'_1,\dots,\mu'_m)$, $B' \equiv (\nu'_1,\dots,\nu'_n)$, $A \equiv (\mu_1,\dots,\mu_m)$ and $B \equiv (\nu_1,\dots,\nu_n)$. As before, the operators P_α and $J_{\alpha\beta}$ are the generators of translations and the Lorentz rotations of the Poincaré group representation in the space of tensors $t^A{}_B$. The explicit form can be found in a way similar to that used in the case of a vector. If the vector $a^\alpha = 0$, one gets the tensor transformation law under the infinitesimal homogeneous Lorentz transformations.

2.5 Spinor representation

Along with tensors, there are other objects associated with the Lorentz group, called spinors. As we will see, in some sense, they are simpler than tensors.

Consider a set of 2×2 complex matrices N with unit determinants. Since N is not degenerate, there is an inverse matrix N^{-1}, and $\det N^{-1} = (\det N)^{-1} = 1$. For the two matrices N_1 and N_2 with $\det N_1 = \det N_2 = 1$, we have $\det(N_1N_2) = \det N_1 \times \det N_2 = 1$. The set of such matrices N forms a group which is called the two-dimensional special complex linear group and is denoted as $SL(2|\mathbb{C})$. We will show that there is a map of group $SL(2|\mathbb{C})$ into group $L_+{}^\uparrow$. Namely, for each matrix $N \in SL(2|\mathbb{C})$, there exists a matrix $\Lambda \in L_+{}^\uparrow$, and vice versa. Moreover,

$$\Lambda(N_1N_2) = \Lambda(N_1)\Lambda(N_2) \quad\text{and}\quad \Lambda(N_1) = \Lambda(N_2) \quad\Longrightarrow\quad N_1 = \pm N_2.$$

The construction of matrix $\Lambda(N)$ consists of several steps:

1. Consider a linear space of Hermitian 2×2 matrices X, $X = X^+$, where $X^+ = (X^*)^T$, where $*$ means the complex conjugation. A basis in this space can be taken as $\sigma_\mu = (\sigma_0,\sigma_j)$, where

$$\sigma_0 = \begin{pmatrix}1&0\\0&1\end{pmatrix}, \qquad \sigma_1 = \begin{pmatrix}0&1\\1&0\end{pmatrix}, \qquad \sigma_2 = \begin{pmatrix}0&-i\\i&0\end{pmatrix}, \qquad \sigma_3 = \begin{pmatrix}1&0\\0&-1\end{pmatrix}. \qquad (2.49)$$

Here σ_0 is the unit 2×2 matrix, and σ_j are the Pauli matrices. It is evident that all matrices σ_μ are Hermitian. Also, we introduce the matrices $\tilde{\sigma_\mu} = (\sigma_0,-\sigma_j)$. Then it is easy to check that the following relation takes place:

$$\operatorname{tr}\left(\tilde{\sigma_{\mu}}\sigma_{\nu}\right) = 2\eta_{\mu\nu}. \tag{2.50}$$

Let X be an arbitrary 2×2 Hermitian matrix, written in the basis σ_μ as

$$X = x^{\mu}\sigma_{\mu}. \tag{2.51}$$

Since the matrices X and σ_μ are Hermitian, the x^μ are real numbers which can be identified with coordinates in Minkowski space. The relations (2.50) and (2.51) lead to

$$x^{\mu} = \frac{1}{2}\operatorname{tr}\left(\tilde{\sigma}^{\mu}X\right). \tag{2.52}$$

2. Let $N \in SL(2|\mathbb{C})$. Consider the matrix

$$X' = NXN^{+}.$$

It is easy to see that X' is a Hermitian matrix. In addition, $\det X' = \det\left(NXN^{+}\right) = \det X$. Matrices X' and X can be expanded in the basis of σ_μ, providing $X' = x'^{\mu}\sigma_\mu$ and $X = x^\mu\sigma_\mu$. The coefficients x'^μ can be obtained according to (2.51) as

$$x'^{\mu} = \frac{1}{2}\operatorname{tr}\left(\tilde{\sigma}^{\mu}X'\right) = \frac{1}{2}\operatorname{tr}\left(\tilde{\sigma}^{\mu}N\sigma_{\nu}N^{+}\right)x^{\nu}.$$

Denoting

$$\Lambda^{\mu}{}_{\nu} = \frac{1}{2}\operatorname{tr}\left(\tilde{\sigma}^{\mu}N\sigma_{\nu}N^{+}\right) \equiv \Lambda^{\mu}{}_{\nu}(N), \tag{2.53}$$

we arrive at the transformation law

$$x'^{\mu} = \Lambda^{\mu}{}_{\nu}(N)x^{\nu}. \tag{2.54}$$

3. The $\det X$ can be exactly calculated using an explicit form of the matrices σ_μ,

$$\begin{aligned} X = x^{\mu}\sigma_{\mu} &= \begin{pmatrix} x^0 & 0 \\ 0 & x^0 \end{pmatrix} + \begin{pmatrix} 0 & x^1 \\ x^1 & 0 \end{pmatrix} + \begin{pmatrix} 0 & -ix^2 \\ ix^2 & -0 \end{pmatrix} + \begin{pmatrix} x^3 & 0 \\ 0 & -x^3 \end{pmatrix} \\ &= \begin{pmatrix} x^0 + x^3 & x^1 - ix^2 \\ x^1 + ix^2 & x^0 - x^3 \end{pmatrix}. \end{aligned}$$

Therefore,

$$\begin{aligned} \det X &= (x^0 + x^3)(x^0 - x^3) - (x^1 - ix^2)(x^1 + ix^2) \\ &= (x^0)^2 - (x^1)^2 - (x^2)^2 - (x^3)^2 = \eta_{\mu\nu}x^{\mu}x^{\nu}. \end{aligned}$$

Similarly, $\det X' = \eta_{\mu\nu}x'^{\mu}x'^{\nu}$. Hence, since the two determinants are equal, we get

$$\eta_{\mu\nu}x'^{\mu}x'^{\nu} = \eta_{\alpha\beta}x^{\alpha}x^{\beta}. \tag{2.55}$$

This is just the condition of preserving the interval (2.3).

4. Substitute (2.53) into (2.55) and get $\Lambda^T\eta\Lambda = \eta$. Therefore, the matrices $\Lambda(N)$ (2.53) satisfy the basic relation, and hence they realize the Lorentz transformations. Thus, the numbers x^μ from (2.52) can indeed be treated as the Minkowski space coordinates. Also, the matrix $\Lambda^\mu{}_\nu(N)$ (2.53) satisfies the relation $\det\Lambda(N) = 1$. In addition,

$$\Lambda^0{}_0(N) = \frac{1}{2}\,\mathrm{tr}\,\tilde{\sigma}^0 N\sigma_0 N^+ = \frac{1}{2}\,\mathrm{tr}\,(NN^+) > 0.$$

As a result, the matrices $\Lambda(N)$ (2.53) belong to the proper Lorentz group $L_+{}^\uparrow$.

5. We proved that, for each matrix $N \in SL(2|\mathbb{C})$, there is a matrix $\Lambda(N) \in L_+{}^\uparrow$. One can also prove the inverse statement. For each matrix $\Lambda(N) \in L_+{}^\uparrow$, there exists some matrix $N \in SL(2|\mathbb{C})$. Let

$$\Lambda(N_1) = \Lambda(N_2) \quad\Longleftrightarrow\quad \mathrm{tr}\,\tilde{\sigma}^\mu N_1\sigma_\nu N_1{}^+ = \mathrm{tr}\,\tilde{\sigma}^\mu N_2\sigma_\nu N_2{}^+ \quad\Longleftrightarrow\quad N_1 = \pm N_2.$$

There are two matrices N corresponding to a given matrix Λ. Performing two consequent transformations $X' = N_1 X N_1^+$ and $X'' = N_2 X' N_2^+$, one gets $x'^\alpha = \Lambda^\alpha{}_\nu(N_1)x^\nu$ and $x''^\mu = \Lambda^\mu{}_\alpha(N_2)x'^\alpha$. Therefore $x''^\mu = \Lambda^\mu{}_\alpha(N_2)\Lambda^\alpha{}_\nu(N_1)x^\nu$. On the other hand,

$$X'' = N_2 N_1 X N_1{}^+ N_2{}^+ = (N_2N_1)X(N_2N_1)^+.$$

Hence, $x''^\mu = \Lambda^\mu{}_\nu(N_2N_1)x^\nu$. As a result, $\Lambda^\mu{}_\nu(N_2N_1) = \Lambda^\mu{}_\alpha(N_2)\Lambda^\alpha{}_\nu(N_1)$. Thus, we have the mapping

$$\begin{aligned} SL(2|\mathbb{C}) \quad &\longrightarrow \quad L_+{}^\uparrow, \\ N_2N_1 \in SL(2|\mathbb{C}) \quad &\longrightarrow \quad \Lambda(N_2N_1) = \Lambda(N_2)\Lambda(N_1) \in L_+{}^\uparrow, \\ \Lambda_2 = \Lambda_1 \quad &\longrightarrow \quad N_2 = \pm N_1. \end{aligned}$$

We see that the group $SL(2|\mathbb{C})$ is closely related to the proper Lorentz group $L_+{}^\uparrow$. The coordinate transformations in Minkowski space, generated by the elements from the proper Lorentz group, create the transformations $X' = NXN^+$, which are generated by the matrices $N \in SL(2|\mathbb{C})$. Thus, we can consider the transformations $X' = NXN^+$ to be equal footing with the Lorentz transformations $x'^\mu = \Lambda^\mu{}_\nu x^\nu$.

The matrices N act in the two-dimensional complex space formed by the elements $\varphi \equiv \{\varphi_a\}$, with $a = 1, 2$. The rule looks like

$$\varphi'_a = N_a{}^b\varphi_b. \tag{2.56}$$

Since, for each matrix N, there is a matrix $\Lambda(N) \in L_+{}^\uparrow$, one can say that (2.56) is the transformation law of a complex two-component vector under the Lorentz transformation. The vectors $\varphi = \{\varphi_a\}$, transforming according to (2.56), are called left Weyl spinors, and a is called the spinor index. The representation of the $SL(2|\mathbb{C})$ group in the linear space of the left Weyl spinors is called the fundamental representation of the Lorentz group. Next, we introduce the matrix $\varepsilon = (\varepsilon_{ab})$ by the rule

$$\varepsilon = \begin{pmatrix} 0 & -1 \\ 1 & 0 \end{pmatrix} \quad\Longrightarrow\quad \varepsilon^{-1} = \begin{pmatrix} 0 & 1 \\ -1 & 0 \end{pmatrix} \tag{2.57}$$

and consider the expression

$$f_{ab} = N_a{}^c N_b{}^d \varepsilon_{cd} = N_a{}^1 N_b{}^2 \varepsilon_{12} + N_a{}^2 N_b{}^1 \varepsilon_{21} = N_a{}^2 N_b{}^1 - N_a{}^1 N_b{}^2.$$

It is evident that $f_{11} = f_{22} = 0$, $f_{12} = -f_{21}$ and we can calculate

$$f_{12} = N_1{}^2 N_2{}^1 - N_1{}^1 N_2{}^2 = -\det N = -1.$$

Then $f_{21} = 1$, and hence $f_{ab} = \varepsilon_{ab}$. As a result, one gets $N_a{}^c N_b{}^d \varepsilon_{cd} = \varepsilon_{ab}$, or $N\varepsilon N^T = \varepsilon$. This relation shows that the matrix ε (2.57) is an invariant matrix for group $SL(2|\mathbb{C})$.

Let us introduce the inverse matrix ε^{-1} with elements ε^{ab}, $\varepsilon^{ab}\varepsilon_{bc} = \delta^a{}_c$, $\varepsilon_{ab}\varepsilon^{bc} = \delta_a{}^c$. If ε in (2.57) is an invariant quantity, ε^{-1} is also invariant, and one can prove that

$$\varepsilon^{-1} = N^T \varepsilon^{-1} N. \tag{2.58}$$

The matrices ε and ε^{-1} can be used for raising and lowering the spinor indices

$$\varphi^a = \varepsilon^{ab}\varphi_b, \qquad \varphi_a = \varepsilon_{ab}\varphi^b. \tag{2.59}$$

Let us show that the expression $\varphi_1{}^a \varphi_{2a}$ is invariant. Indeed,

$$\begin{aligned}\varphi_1'{}^a \varphi_{2a}' &= \varepsilon^{ab}\varphi_{1b}'\varphi_{2a}' = \varepsilon^{ab} N_b{}^d \varphi_{1d} N_a{}^c \varphi_{2c} \\ &= (N^T)^c{}_a \varepsilon^{ab} N_b{}^d \varphi_{1d}\varphi_{2c} = \varepsilon^{cd}\varphi_{1d}\varphi_{2c} = \varphi_1{}^c \varphi_{2c}.\end{aligned}$$

Thus, we have learned how to construct a Lorentz invariant object from spinors.

Let N be an arbitrary matrix from $SL(2|\mathbb{C})$ and let the matrix N^* have complex conjugate elements. The elements of this matrix are denoted by definition as $N^*{}_{\dot a}{}^{\dot b}$; $\dot a, \dot b = \dot 1, \dot 2$. This matrix realizes the transformation

$$\chi_{\dot a}' = N^*{}_{\dot a}{}^{\dot b}\chi_{\dot b}. \tag{2.60}$$

The two-dimensional complex vector $\chi \equiv \{\chi_{\dot a}\}$, which transforms according to (2.60), is called the right Weyl spinor, and $\dot a$ is a spinor index. The representation of the $SL(2|\mathbb{C})$ group in the linear space of right Weyl spinors is called the conjugate representation.

It proves useful to introduce the matrices

$$\varepsilon_{\dot a\dot b} = \begin{pmatrix} 0 & -1 \\ 1 & 0 \end{pmatrix}, \qquad \varepsilon^{\dot a\dot b} = \begin{pmatrix} 0 & 1 \\ -1 & 0 \end{pmatrix},$$

which satisfy the relations

$$\varepsilon^{\dot a\dot b} = N^*{}_{\dot a}{}^{\dot c} N^*{}_{\dot b}{}^{\dot d}\varepsilon_{\dot c\dot d}, \qquad \varepsilon^{\dot a\dot b} = N^*{}_{\dot c}{}^{\dot a} N^*{}_{\dot d}{}^{\dot b}\varepsilon^{\dot c\dot d}. \tag{2.61}$$

This means that $\varepsilon_{\dot a\dot b}$ and $\varepsilon^{\dot a\dot b}$ are invariant tensors of the group $SL(2|\mathbb{C})$. These matrices can be used for raising and lowering the dotted spinor indices:

$$\chi^{\dot a} = \varepsilon^{\dot a\dot b}\chi_{\dot b}, \qquad \chi_{\dot a} = \varepsilon_{\dot a\dot b}\chi^{\dot b}.$$

It is evident that the expression $\chi_{1\dot a}\chi_2{}^{\dot a}$ is Lorentz invariant; hence, we obtain, once again, a recipe for how to construct the Lorentz invariants from spinors.

Generalization of the undotted φ_a and dotted $\chi_{\dot{a}}$ spinors is a general spin tensor $\varphi_{a_1,\ldots,a_m,\dot{a}_1,\ldots,\dot{a}_n}$, defined by the transformation law under the Lorentz transformation as follows:

$$\varphi'_{a_1,\ldots,a_m,\dot{a}_1,\ldots,\dot{a}_n}(x') = N_{a_1}{}^{b_1}\ldots N_{a_m}{}^{b_m}N^*{}_{\dot{a}_1}{}^{\dot{b}_1}\ldots N^*{}_{\dot{a}_n}{}^{\dot{b}_n}\varphi_{b_1,\ldots,b_m,\dot{b}_1,\ldots,\dot{b}_n}(x). \quad (2.62)$$

Consider the matrices $X' = x'^{\mu}\sigma_{\mu}$ and $X = x^{\mu}\sigma_{\mu}$, where $X' = NXN^+$ and $x'^{\mu} = \Lambda^{\mu}{}_{\nu}(N)x^{\nu}$. This means $\Lambda^{\mu}{}_{\nu}(N)x^{\nu}\sigma_{\mu} = N\sigma_{\nu}N^+x^{\nu}$. Therefore,

$$\sigma_{\mu} = (\Lambda^T)^{-1}{}_{\mu}{}^{\nu}N\sigma_{\nu}N^+. \quad (2.63)$$

This relation shows that the matrices σ_{μ} form invariant objects of the group $SL(2|\mathbb{C})$. The matrix elements of σ_{μ} are denoted as $(\sigma_{\mu})_{a\dot{a}}$. Rising the spinor indices, one gets

$$(\sigma_{\mu})^{a\dot{a}} = \varepsilon^{ab}\varepsilon^{\dot{a}\dot{b}}(\sigma_{\mu})_{b\dot{b}} \equiv (\tilde{\sigma}_{\mu})^{\dot{a}a}. \quad (2.64)$$

Using the explicit form of $\sigma_{\mu} = (\sigma_0, \sigma_j)$ and that of ε^{ab}, $\varepsilon^{\dot{a}\dot{b}}$, one can find $\tilde{\sigma}_{\mu} = (\sigma_0, -\sigma_j)$. This matrix has already been introduced, at the beginning of section 2.5.

Starting from the spinors φ_a, $\chi_{\dot{a}}$, the matrices σ_{μ} and the relations (2.56), (2.60) and (2.63), we can construct (complex) vectors under Lorentz transformations, e.g.,

$$\begin{aligned}\varphi'^{a}(\sigma^{\mu})_{a\dot{a}}\chi'^{\dot{a}} &= \varepsilon^{ab}\varphi'_b(\sigma^{\mu})_{a\dot{a}}\varepsilon^{\dot{a}\dot{b}}\chi'_{\dot{b}} \\ &= \Lambda^{\mu}{}_{\nu}\varepsilon^{ab}N_b{}^c\varphi_c(N\sigma^{\nu}N^{\dagger})_{a\dot{a}}\varepsilon^{\dot{a}\dot{b}}(N^*)_{\dot{b}}{}^{\dot{c}}\chi_{\dot{c}} = \Lambda^{\mu}{}_{\nu}\varphi^d(\sigma^{\nu})_{d\dot{d}}\chi^{\dot{d}}.\end{aligned} \quad (2.65)$$

Thus, we have obtained the transformation law for the contravariant vector under the Lorentz transformation. Analogously, one can prove that the expression $\chi_{\dot{a}}(\tilde{\sigma}^{\mu})^{\dot{a}a}\varphi_a$ is also the contravariant vector. As a result, we arrive at a prescription for how to construct Lorentz vectors from spinors and σ-matrices.

The matrices σ_{μ}, $\tilde{\sigma}_{\mu}$ possess many useful properties that can be established by direct calculations, e.g.,

$$\begin{aligned}&(\sigma_{\mu}\tilde{\sigma}_{\nu} + \sigma_{\nu}\tilde{\sigma}_{\mu})_a{}^b = 2\eta_{\mu\nu}\delta_a{}^b, && (\tilde{\sigma}_{\mu}\sigma_{\nu} + \tilde{\sigma}_{\nu}\sigma_{\mu})^{\dot{a}}{}_{\dot{b}} = 2\eta_{\mu\nu}\delta^{\dot{a}}{}_{\dot{b}} \\ &\mathrm{tr}\,\sigma_{\mu}\tilde{\sigma}_{\nu} = 2\eta_{\mu\nu}, && \sigma^{\mu}{}_{a\dot{a}}\tilde{\sigma}_{\mu}{}^{\dot{b}b} = 2\delta_a{}^b\delta_{\dot{a}}{}^{\dot{b}}.\end{aligned} \quad (2.66)$$

The matrices σ_{μ}, $\tilde{\sigma}_{\mu}$ make it possible to convert the vector indices into a pair of spinor indices, and vice versa. For example, if we have a vector t_{μ}, we can construct the object $t_{a\dot{a}} \sim (\sigma^{\mu})_{a\dot{a}}t_{\mu}$. Consider an arbitrary spin tensor with an equal number of dotted and undotted indices, $\varphi_{a_1\ldots a_n\dot{a}_1\ldots\dot{a}_n}$. Then we can construct a tensor of rank n as

$$\varphi_{\mu_1\ldots\mu_n} = \sigma_{\mu_1}{}^{a_1\dot{a}_1}\sigma_{\mu_2}{}^{a_2\dot{a}_2}\ldots\sigma_{\mu_n}{}^{a_n\dot{a}_n}\varphi_{a_1\ldots a_n\dot{a}_1\ldots\dot{a}_n} \quad (2.67)$$

and vice versa,

$$\varphi_{a_1\ldots a_n\dot{a}_1\ldots\dot{a}_n} = \frac{1}{2^n}\tilde{\sigma}^{\mu_1}{}_{\dot{a}_1a_1}\ldots\tilde{\sigma}^{\mu_n}{}_{\dot{a}_na_n}\varphi_{\mu_1\ldots\mu_n}. \quad (2.68)$$

For the spin tensors $\varphi_{a_1 \ldots a_n a \dot{a}_1 \ldots \dot{a}_n}$ and $\chi_{a_1 \ldots a_n \dot{a}_1 \ldots \dot{a}_n}{}^{\dot{a}}$, one can construct the quantities

$$\varphi_{\mu_1 \ldots \mu_n a} = \sigma_{\mu_1}{}^{a_1 \dot{a}_1} \ldots \sigma_{\mu_n}{}^{a_n \dot{a}_n} \varphi_{a_1 \ldots a_n a \dot{a}_1 \ldots \dot{a}_n},$$
$$\chi_{\mu_1 \ldots \mu_n^{\dot{a}}} = \sigma_{\mu_1}{}^{a_1 \dot{a}_1} \ldots \sigma_{\mu_n}{}^{a_n \dot{a}_n} \chi_{a_1 \ldots a_n \dot{a}_1 \ldots \dot{a}_n}{}^{\dot{a}}. \tag{2.69}$$

A tensor with the spinor components

$$\psi_{\mu_1 \ldots \mu_n} = \begin{pmatrix} \varphi_{\mu_1 \ldots \mu_n} \\ \chi_{\mu_1 \ldots \mu_n} \end{pmatrix} \tag{2.70}$$

is called the Dirac tensor spinor.

The relations (2.56) and (2.60) make it possible to derive the generators of the Lorentz group in the fundamental and conjugate representations. In the vicinity of the unit element, the matrices $N \in SL(2|\mathbb{C})$ have the form $N = E+T$, or $N_a{}^b = \delta_a{}^b + T_a{}^b$, where E is the unit 2×2 matrix, and $T_a{}^b$ is a 2×2 matrix with infinitesimal elements. It is known that $\det N = 1 + \operatorname{tr} T$ is linear in T; therefore, $\operatorname{tr} T = 0$, since $\det N = 1$. The matrix T can be expanded in the basis σ_μ; however, since T is traceless, we can write $T = z_i \sigma_i$, where z_i are the tree complex numbers. Thus, in the vicinity of the unit element, each matrix N is parametrized by six real numbers, as it should be for the Lorentz transformations.

On the other hand, the elements of the Lorentz group representation in the vicinity of the unit element have the form $e^{-\frac{i}{2}\omega^{\alpha\beta} J_{\alpha\beta}}$, where $\omega^{\alpha\beta} = -\omega^{\beta\alpha}$ and $J_{\alpha\beta}$ are the Lorentz group generators in the given representation. We will need these generators in the fundamental and conjugate representations. For the fundamental representation in the infinitesimal vicinity of the unit element, we have $E + z_i \sigma_i = \mathbf{1} - \frac{i}{2}\omega^{\alpha\beta} J_{\alpha\beta}$, where, in the case under consideration, $\mathbf{1} = E$. Let us parameterize the complex numbers z_i as follows:

$$z_1 = -\frac{1}{2}(\omega^{01} + i\omega^{23}), \quad z_2 = -\frac{1}{2}(\omega^{02} + i\omega^{31}), \quad z_3 = -\frac{1}{2}(\omega^{03} + i\omega^{12}). \tag{2.71}$$

In this notation, $z_i \sigma_i = -\frac{i}{2}\omega^{\alpha\beta}(i\sigma_{\alpha\beta})$, where

$$(\sigma_{\alpha\beta})_a{}^b = \frac{1}{4}(\sigma_\alpha \tilde{\sigma}_\beta - \sigma_\beta \tilde{\sigma}_\alpha)_a{}^b\,. \tag{2.72}$$

Therefore, the generators of the Lorentz group in the fundamental representation are

$$J^{(F)}_{\alpha\beta} = i\sigma_{\alpha\beta}. \tag{2.73}$$

Here, (F) labels the fundamental representation. Similar consideration shows that the generators of the Lorentz group in the conjugate representation have the form

$$J^{(\bar{F})}_{\alpha\beta} = i\tilde{\sigma}_{\alpha\beta}, \tag{2.74}$$

where $$(\tilde{\sigma}_{\alpha\beta})^{\dot{a}}{}_{\dot{b}} = \frac{1}{4}(\tilde{\sigma}_\alpha \sigma_\beta - \tilde{\sigma}_\beta \sigma_\alpha)^{\dot{a}}{}_{\dot{b}}. \tag{2.75}$$

As a result, the transformations laws under infinitesimal Lorentz transformation for two-component spinors have the form

$$\delta\varphi = -\frac{i}{2}\omega^{\alpha\beta}(i\sigma_{\alpha\beta})\varphi, \qquad \delta\chi = -\frac{i}{2}\omega^{\alpha\beta}(i\tilde{\sigma}_{\alpha\beta})\chi. \tag{2.76}$$

The matrices $\sigma_{\alpha\beta}$ and $\tilde{\sigma}_{\alpha\beta}$ possesses useful properties:

$$\begin{aligned} &\sigma_{\alpha\beta} = -\sigma_{\beta\alpha}, \qquad &\tilde{\sigma}_{\alpha\beta} = -\tilde{\sigma}_{\beta\alpha}, \\ &(\sigma_{\alpha\beta})^{ab} = (\sigma_{\alpha\beta})^{ba}, \qquad &(\tilde{\sigma}_{\alpha\beta})^{\dot{a}\dot{b}} = (\tilde{\sigma}_{\alpha\beta})^{\dot{b}\dot{a}}. \end{aligned} \tag{2.77}$$

The generators of the Lorentz group that is the representation of arbitrary tensor spinors are found quite analogously by using the transformation law (2.62) and considering the infinitesimal forms of the matrices N and N^*. Also, we can derive the transformations of tensor spinors under non-homogeneous Lorentz transformations. We will not discuss this in detail but only formulate the final form. Consider an arbitrary spin tensor $\varphi_{a_1\ldots a_m\dot{a}_1\ldots\dot{a}_n} \equiv \varphi_{a(m)\dot{a}(n)}$. The variation of $\varphi_{a(m)\dot{a}(n)}(x)$ under the infinitesimal transformations $x'^\mu = x^\mu + \omega^\mu{}_\nu x^\nu + a^\mu$ is written as follows:

$$\delta\varphi_{a(m)\dot{a}(n)} = \Big[ia^\alpha(P_\alpha)_{a(m)\dot{a}(n)}{}^{b(m)\dot{b}(n)} - \frac{i}{2}\omega^{\alpha\beta}(J_{\alpha\beta})_{a(m)\dot{a}(n)}{}^{b(m)\dot{b}(n)}\Big]\varphi_{b(m)\dot{b}(n)}, \tag{2.78}$$

with Poincaré group generators $(P_\alpha)_{a(m)\dot{a}(n)}{}^{b(m)\dot{b}(n)}$ and $(J_{\alpha\beta})_{a(m)\dot{a}(n)}{}^{b(m)\dot{b}(n)}$, which can be found in an explicit form in the same way as we derived (2.76).

2.6 Irreducible representations of the Poincaré group

Irreducible representations of the Poincaré group determine relativistic physical systems with given mass and spin. These systems can be called elementary and are associated with elementary particles. The aim of this subsection is to formulate the basic notions and results concerning these systems. Let us note that we do not pretend to consider a complete description of the representations of the Poincaré group, which can be found in more specialized literature.

In section 2.3 we denoted the elements of the Poincaré group as (Λ, a). The multiplication rule in this group has the form (2.32), where the matrix Λ satisfies the basic relation (2.7). Let us denote the operators of some representation of the Poincaré group as $\mathcal{U}(\Lambda, a)$. According to the definition of a representation, we have

$$\mathcal{U}(\Lambda_1, a_1)\mathcal{U}(\Lambda_2, a_2) = \mathcal{U}((\Lambda_1, a_1)\cdot(\Lambda_2, a_2)) = \mathcal{U}(\Lambda_1\Lambda_2, \Lambda_1 a_2 + a_1). \tag{2.79}$$

As we know, the infinitesimal non-homogeneous Lorentz transformation is (2.33), with the parameters a^μ and $\omega^\mu{}_\nu$. Denoting the generators of the Poincaré group as P_α and $J_{\alpha\beta}$, in the vicinity of the unit element, one can write

$$\mathcal{U}(\Lambda, a) = e^{ia^\alpha P_\alpha - \frac{i}{2}\omega^{\alpha\beta}J_{\alpha\beta}}. \tag{2.80}$$

The operator P_α is the generator of spacetime translations, and the operator $J_{\alpha\beta}$ is the generator of Lorentz rotations. Examples of these generators for scalar and vector representations were given in section 2.3. Eqs. (2.79) and (2.80) allow one to derive, in a purely algebraic way, the commutation relations

$$[P_\alpha, P_\beta] = 0,$$
$$[P_\mu, J_{\alpha\beta}] = i(\eta_{\mu\alpha}P_\beta - \eta_{\mu\beta}P_\alpha),$$
$$[J_{\mu\nu}, J_{\alpha\beta}] = i(\eta_{\mu\alpha}J_{\nu\beta} + \eta_{\nu\beta}J_{\mu\alpha} - \eta_{\mu\beta}J_{\nu\alpha} - \eta_{\nu\alpha}J_{\mu\beta}). \tag{2.81}$$

These equations define the Lie algebra of the Poincaré group, which is sometimes called the Poincaré algebra. The last line defines the Lie algebra of the Lorentz group.

One can show that the algebra (2.81) has two operators, C_1, C_2, which commute with all generators. These are the Casimir operators

$$C_1 = P^\mu P_\mu \quad \text{and} \quad C_2 = W^\mu W_\mu\,, \tag{2.82}$$

where W^μ is the Lubanski-Pauli vector

$$W^\mu = -\frac{1}{2}\varepsilon^{\mu\nu\alpha\beta}P_\nu J_{\alpha\beta}. \tag{2.83}$$

One can check that the invariants (2.82) satisfy the relations

$$[P_\mu, C_1] = [P_\mu, C_2] = 0, \qquad [J_{\alpha\beta}, C_1] = [J_{\alpha\beta}, C_2] = 0, \qquad [C_1, C_2] = 0. \tag{2.84}$$

One can distinguish two kinds of physically acceptable irreducible representations of the Poincaré group: massive and massless. The basis vectors $|p, m, s\rangle$ of a massive irreducible representation are defined by the following equations:

$$P_\mu|p, m, s\rangle = p_\mu|p, m, s\rangle,$$
$$P^\mu P_\mu|p, m, s\rangle = m^2|p, m, s\rangle,$$
$$W^\mu W_\mu|p, m, s\rangle = -m^2 s(s+1)|p, m, s\rangle. \tag{2.85}$$

The first two equations mean that $p^2 = p_0^2 - \mathbf{p}^2 = m^2$, which is the well-known relation (1.1). Thus, massive irreducible representations of Poincaré group are naturally associated with free massive relativistic particles. The parameter m plays a role in mass and should be positive. One can prove that the parameter s takes the values $s = 0, \frac{1}{2}, 1, \frac{3}{2}, 2, \ldots$. For each fixed s, there are $2s+1$ different vectors $|p, m, s\rangle$, and s is called the spin of a massive relativistic particle.

For the massless irreducible representation, the basis vectors of the state $|p, \lambda\rangle$ are defined by the equations

$$P_\mu|p, \lambda\rangle = p_\mu|p, \lambda\rangle, \qquad P^\mu P_\mu|p, \lambda\rangle = 0, \qquad W^\mu W_\mu|p, \lambda\rangle = 0. \tag{2.86}$$

The first two of these equations mean that $p^2 = p_0^2 - \mathbf{p}^2 = 0$. This is the relation between the energy and the three-dimensional momentum of a free massless relativistic particle. Thus, the massless irreducible representations of the Poincaré group are naturally associated with free massless relativistic particles. One can prove that, in the case under consideration, $W_\mu = \lambda P_\mu$, where λ takes the values $\lambda = 0, \pm\frac{1}{2}, \pm 1, \pm\frac{3}{2}, \pm 2, \ldots$.

This parameter is called the helicity of the massless elementary particle. Sometimes the value $|\lambda|$ is called the spin of a massless particle.

One can show that the Poincaré algebra can be realized in the linear space of the tensor fields $\varphi_{\mu_1\dots\mu_n}$ or the tensor spinor fields $\psi_{\mu_1\dots\mu_n}$, as defined by Eq. (2.70). The fields $\varphi_{\mu_1\dots\mu_n}$ or $\psi_{\mu_1\dots\mu_n}$, corresponding to the massive irreducible representations of the Poincaré algebra, are characterized by their masses and spin. The relativistic field with the given mass m and an integer spin $s = n$, is defined by the system of equations

$$\begin{aligned} &\varphi_{\mu_1\dots\mu_s}(x) = \varphi_{(\mu_1\dots\mu_n)}(x), && (\Box + m^2)\varphi_{\mu_1\dots\mu_s}(x) = 0, \\ &\partial^{\mu_1}\varphi_{\mu_1\mu_2\dots\mu_s}(x) = 0, && \eta^{\mu_1\mu_2}\varphi_{\mu_1\mu_2\mu_3\dots\mu_s}(x) = 0. \end{aligned} \tag{2.87}$$

Consider the following few examples:

$$1)\ s = 0 \quad (\Box + m^2)\varphi = 0, \tag{2.88}$$

$$2)\ s = 1 \quad (\Box + m^2)\varphi_\mu = 0, \quad \partial^\mu\varphi_\mu = 0, \tag{2.89}$$

$$3)\ s = 2 \quad (\Box + m^2)\varphi_{\mu\nu} = 0, \quad \partial^\mu\varphi_{\mu\nu} = 0, \ \ \eta^{\mu\nu}\varphi_{\mu\nu} = 0, \ \ \varphi_{\mu\nu}(x) = \varphi_{(\mu\nu)}. \tag{2.90}$$

In the last formula, $\varphi_{(\mu_1\dots\mu_n)}$ means a total symmetrization of the indices.

Eq. (2.88) is the Klein–Gordon equation (1.5), corresponding to a free massive scalar field. Eqs. (2.89) define the free massive vector field equations of motion. Eqs. (2.90) define the equation of motion of a massive symmetric second-rank tensor field. Equations for massless relativistic fields with integer spin can be obtained from Eqs. (2.87) at $m = 0$.

The relativistic fields with given mass m and given half-integer spin $s = n + \frac{1}{2}$ are defined in terms of the Dirac tensor spinors $\psi_{\mu_1\dots\mu_n}$ (2.70) by the system of equations

$$\begin{aligned} &\psi_{\mu_1\dots\mu_n} = \psi_{(\mu_1\dots\mu_n)}, && (i\gamma^\mu\partial_\mu - m)\psi_{\mu_1\dots\mu_n} = 0, \\ &\partial^{\mu_1}\psi_{\mu_1\mu_2\dots\mu_n} = 0, && \gamma^{\mu_1}\psi_{\mu_1\mu_2\dots\mu_n} = 0, \end{aligned} \tag{2.91}$$

where Dirac matrices γ^μ are defined in terms of the matrices σ^μ and $\tilde{\sigma}^\mu$, as defined in section 2.5,

$$\gamma^\mu = \begin{pmatrix} 0 & \sigma^\mu \\ \tilde{\sigma}^\mu & 0 \end{pmatrix}, \tag{2.92}$$

satisfying the basic relation (also called the Clifford algebra)

$$\begin{aligned} \gamma^\mu\gamma^\nu + \gamma^\nu\gamma^\mu &= \begin{pmatrix} 0 & \sigma^\mu \\ \tilde{\sigma}^\mu & 0 \end{pmatrix}\begin{pmatrix} 0 & \sigma^\nu \\ \tilde{\sigma}^\nu & 0 \end{pmatrix} + \begin{pmatrix} 0 & \sigma^\nu \\ \tilde{\sigma}^\nu & 0 \end{pmatrix}\begin{pmatrix} 0 & \sigma^\mu \\ \tilde{\sigma}^\mu & 0 \end{pmatrix} \\ &= \begin{pmatrix} \sigma^\mu\tilde{\sigma}^\nu + \sigma^\nu\tilde{\sigma}^\mu & 0 \\ 0 & \tilde{\sigma}^\mu\sigma^\nu + \tilde{\sigma}^\nu\sigma^\mu \end{pmatrix} = \begin{pmatrix} 2\eta^{\mu\nu} & 0 \\ 0 & 2\eta^{\mu\nu} \end{pmatrix} = 2\eta^{\mu\nu} I, \end{aligned} \tag{2.93}$$

where I is the four-dimensional unit matrix.

Some examples of the construction described above, are

$$s = \frac{1}{2}, \quad (i\gamma^\mu\partial_\mu - m)\psi(x) = 0, \tag{2.94}$$

$$s = \frac{3}{2}, \quad (i\gamma^\mu\partial_\mu - m)\psi_\nu(x) = 0, \quad \text{with } \partial^\nu\psi_\nu(x) = 0, \ \ \gamma^\nu\psi_\nu(x) = 0. \tag{2.95}$$

Equation (2.94) is a free, massive, spin-$\frac{1}{2}$ equation of motion. It is called the Dirac equation. The four-component field ψ is called the Dirac spinor field, or the Dirac fermion. The relations (2.95) define the massive spin-$\frac{3}{2}$ field equation of motion, which is called the Rarita-Schwinger equation. Massless relativistic fields with half-integer spins are described by Eqs. (2.91) in the limit $m = 0$.

Exercises

2.1. Let G be the group of $n \times n$ matrices O such that $O^T O = E$, where T means transposition, and E is the unit matrix. Prove that this set forms a group where the multiplication law is the matrix product. This group is called the rotation group and is denoted $O(n)$. Consider a subset of the matrices from $O(n)$ with a unit determinant. Show that this subset is a subgroup of $O(n)$, called $SO(n)$.

2.2 Prove that the groups $SO(2)$ and $U(1)$ are isomorphic.

2.3. Consider the coordinate transformation $x'^\mu = f^\mu(x)$. Prove that the Eq. (2.3) leads to $f^\mu(x) = \Lambda^\mu{}_\nu x^\nu + a^\mu$, where the matrix $\Lambda^\mu{}_\nu$ satisfies the basic relation (2.7).

2.4. Let Λ be a matrix realizing the Lorentz transformation. Show that $\Lambda^{-1} = \eta^{-1}\Lambda^T\eta$.

2.5. Consider a set 5×5 of matrices of the form

$$\begin{pmatrix} \Lambda^\mu{}_\nu & a^\mu \\ 0 & 1 \end{pmatrix}.$$

Show that the group of these matrices is a realization of the Poincaré group.

2.6. Consider a matrix $A = E + \alpha X$, where α is an infinitesimal parameter, and $\det A = 1$. Show that $\operatorname{tr} X = 0$.

2.7. Let $A \in SO(n)$ and $A = E + \alpha X$, where α is an infinitesimal real parameter. Show that $X^T = -X$.

2.8. Let $A \in SU(n)$ and $A = E + i\alpha X$, where α is an infinitesimal real parameter. Show that $X^\dagger = X$.

2.9. Consider the Lie algebra with generators T_i, where $[T_i, T_j] = i f_{ijk} T_k$ and the structure constants are totally antisymmetric. Prove that $C = T_i T_i$ is the Casimir operator.

2.10. Prove that the vector product in the three-dimensional Euclidean space possesses all of the properties of the multiplication law in the Lie algebra.

2.11. Consider a phase space of some dynamical system with phase coordinates q^i, p_i. Assuming that $f(q,p)$ and $g(q,p)$ are functions on the phase space, prove that the Poisson bracket

$$\{f, g\} = \sum_{i=1}^{n} \left(\frac{\partial f}{\partial q^i}\frac{\partial g}{\partial p_i} - \frac{\partial f}{\partial p_i}\frac{\partial g}{\partial q^i} \right)$$

possesses all of the properties of the multiplication law in the Lie algebra.

2.12. Prove that the commutator of the operators possesses all of the properties of the multiplication law in the Lie algebra.

2.13. Let A be an $n \times n$ matrix and $\det A \neq 0$. Prove that $\det A = e^{\mathrm{tr}\,\log A}$.

2.14. Let $J^{\alpha\beta} = x^\alpha P^\beta - x^\beta P^\alpha + S^{\alpha\beta}$. Show that $W_\alpha = -\frac{1}{2}\varepsilon_{\alpha\beta\gamma\delta}P^\beta S^{\gamma\delta}$.

2.15. Calculate the commutator $[J_{\alpha\beta}, J_{\mu\nu}]$ in the scalar representation.

2.16. Calculate the commutator $[S_{\alpha\beta}, S_{\mu\nu}]$, where $S_{\alpha\beta}$ are defined by (2.48).

2.17. Show that all matrices σ_μ are Hermitian.

2.18. Prove the identity $\sigma_i\sigma_j = \sigma_0\delta_{ij} + i\varepsilon_{ijk}\sigma_k$.

2.19. Prove the identity $(\sigma\mathbf{n_1})(\sigma\mathbf{n_2}) = (\mathbf{n_1}\mathbf{n_2}) + i(\mathbf{n_1} \times \mathbf{n_2})\sigma$.

2.20. Prove the identity $\mathrm{tr}\,(\tilde{\sigma}_\mu\sigma_\nu) = 2\eta_{\mu\nu}$.

2.21. Prove that the matrices $i\sigma_{\mu\nu}$ satisfy the commutation relation for the generators of Lorentz rotations.

2.22. Let $v_{a\dot{a}} = (\sigma^\mu)_{a\dot{a}}v_\mu$. Find the v_μ from this relation.

2.23. Prove the relation $W^2 = -\frac{1}{2}\big(S_{\beta\gamma}S^{\beta\gamma}P^2 + S_{\beta\gamma}S^{\alpha\beta}P_\alpha P^\gamma + S_{\beta\gamma}S^{\gamma\alpha}P_\alpha P\beta\big)$.

2.24. Consider $W^2\varphi^\mu(x)$, where $\varphi^\mu(x)$ is an arbitrary vector field. Using the result of the previous exercise and the relation (2.48), formulate the conditions for $W^2\varphi^\mu(x) = -2m^2\varphi^\mu(x)$. Explain on the basis of the last relation how the spin of a massive vector field is equal to 1.

Comments

There are many excellent books on special relativity, e.g., the eminent book by L.D. Landau and E.M. Lifshitz [202] as well as the ones by W. Rindler [264], P.M. Schwarz and J.H. Schwarz [275] and G.L. Naber [228]. Many details of special relativity are usually considered in books on general relativity and gravitation, e.g., in the books by S. Weinberg [340], C.W. Misner, K.S. Thorne and J.A. Wheeler [219] and J.B. Hartle [175].

Group theory for physicists is considered in many books, e.g., in those by W. K. Tung [320], A.O. Barut, R. Raczka [29] and P. Ramond [257]. There are also excellent lecture notes on group theory for physicists available on-line (see e.g., the notes by H. Osborn [236]).

Representations of Lorentz and Poincaré groups are considered with different levels of detail in, e.g., the books by I.M. Gelfand, R.A. Minlos and Z.Ya. Shapiro [156], W. K. Tung [320], A.O. Barut and R. Raczka [29], I.L. Buchbinder and S.M. Kuzenko [81].

Group theory is considered in the physical context in books on quantum field theory, e.g., S. Gasiorowicz [152], S. Schweber [276] and S. Weinberg [345]. Our considerations here followed, in a simplified form, those in [81].

3

Lagrange formalism in field theory

In this chapter, we briefly present the minimal amount of information required about classical fields for the subsequent treatment of quantum theory in the rest of the book.

3.1 The principle of least action, and the equations of motion

Consider Minkowski space with the coordinates $x = x^\mu$. As we already know, the function of coordinates $\phi \equiv \phi(x)$ defined in Minkowski space is called a field. The field can be real or complex, one component or multi component and it can be a scalar, a tensor or a spinor. In particular, this means that the field can have various indices. Then it can be written as $\phi^i(x)$, where i is a set of all indices (tensor, spinor or any other). Usually, we will not use indices if there is no special reason to do so.

It is supposed that the dynamics of the field ϕ is described in terms of the action functional $S = S[\phi]$. It is postulated that the action has the following form:

$$S = \int_\Omega d^4x \, \mathcal{L} \, . \tag{3.1}$$

Here Ω is a domain in Minkowski space bounded by two space-like hypersurfaces, $\sigma(x) = \sigma_1$ and $\sigma(x) = \sigma_2$, as shown in the figure below. Remember that a hypersurface is called space-like if its normal vector $n_\mu(x) = \frac{\partial \sigma(x)}{\partial x^\mu}$ is time-like at any point x^μ, i.e., $n_\mu n^\mu > 0$. In this case, there is an inertial reference frame such that these two hypersurfaces are written as $t = t_1$, and $t = t_2$, where t is a time coordinate.

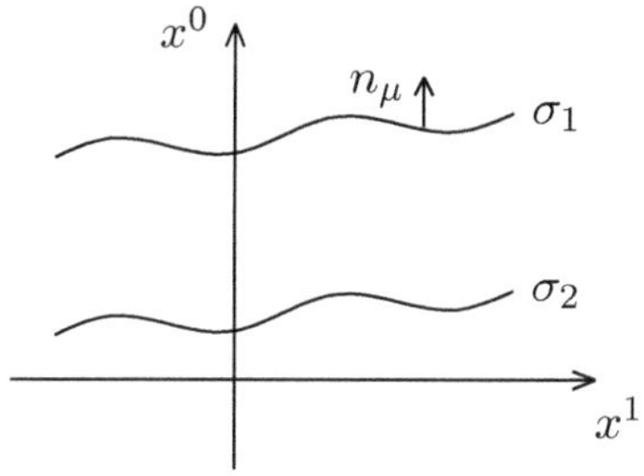

Usually, it is assumed that the domain Ω coincides with the whole Minkowski space, meaning $t_1 \to -\infty$ and $t_2 \to \infty$. It is postulated that the function $\mathcal{L}$ in (3.1) is a real scalar field under the Lorentz transformations. This guarantees that the action $S[\phi]$ is a real Lorenz invariant.

It is generally assumed that the model of field theory is defined when the set of fields $\phi^i(x)$ and the function $\mathcal{L}$ are specified.

Introduction to Quantum Field Theory with Applications to Quantum Gravity. Iosif L. Buchbinder and Ilya L. Shapiro, Oxford University Press (2021). © Iosif L. Buchbinder and Ilya Shapiro.
DOI: 10.1093/oso/9780198838319.003.0003

The function $\mathcal{L}$ is called Lagrangian. In the special reference frame described above, the action (3.1) can be presented in the form

$$S = \int_{t_1}^{t_2} dt \int d^3x\mathcal{L} = \int_{t_1}^{t_2} dt\, L,$$

where $L = \int d^3x\mathcal{L}$ is the Lagrange function, similar to the one in classical mechanics. In this framework, the Lagrangian is nothing else but the density of the Lagrange function.

Let us discuss the analogy with classical mechanics. The field $\phi(x) = \phi^i(x)$ can be regarded as $\phi^i(t, \mathbf{x}) \equiv \phi^i{}_{\mathbf{x}}(t)$, with space coordinates $\mathbf{x}$ playing the role of indices. Thus, one can understand a relativistic field as a mechanical system with generalized coordinates $\phi^i{}_{\mathbf{x}}(t)$, characterized by discrete indices i and by the three-dimensional vector $\mathbf{x}$. This means that we can consider a field as a system with an infinite (continuous) number of degrees of freedom.

The Lagrangian is postulated to be a real function of the field and of its spacetime derivatives taken at the same spacetime point x^μ,

$$\mathcal{L} = \mathcal{L}(\phi(x), \partial_\mu\phi(x),\, \partial_{\mu_1}\partial_{\mu_2}\phi(x),\, \dots, \partial_{\mu_1}\partial_{\mu_2}\dots\partial_{\mu_n}\phi(x)). \tag{3.2}$$

Usually, it is assumed that the Lagrangian includes only first derivatives of the field. However, there are models containing derivatives of order higher than first. Such models are called higher-derivatives theories. We will mainly consider the models with only the first derivatives of the fields in Lagrangian, at least in Part I of this book.

The integral in (3.1) is convergent if it requires that $\mathcal{L} \to 0$ at the space infinity, when $|\mathbf{x}| \to \pm\infty$. In most cases, it is sufficient to consider that $\phi(x) \to 0$ at $|\mathbf{x}| \to \pm\infty$. To get the convergent integral in (3.1), when Ω coincides with the whole Minkowski space, we demand that $\phi(x) \to 0$ at $t \to \pm\infty$. As a result, one gets the standard boundary conditions $\phi(x) \to \infty$ at $x^\mu \to \pm\infty$. However, in some cases, we need to deal with theories of fields that are defined on some domains in Minkowski space, which are bounded in the space directions. In this case, the boundary conditions for the field require a special consideration.

Field dynamics is defined by the least action principle: physically admissible configurations correspond to the minimum of the action. The mathematical formulation of this principle is as follows. Let $\phi^i(x)$ be some field and $\phi'^i(x)$ be another field with the same set of indices. The difference $\delta\phi^i(x) = \phi'^i(x) - \phi^i(x)$ is called a field variation. We assume that the difference $S[\phi + \delta\phi] - S[\phi]$ can be represented in the form

$$S[\phi + \delta\phi] - S[\phi] = \int_\Omega d^4x\, A(x)\, \delta\phi(x) + \dots\,,$$

where the dots mean the terms with higher than the first power of $\delta\phi$. The expression

$$\delta S[\phi] = \int_\Omega d^4x A(x)\, \delta\phi(x) \tag{3.3}$$

is called a variation of the functional $S[\phi]$. The function $A(x)$ is called a variational or functional derivative and it is denoted $\frac{\delta S[\phi]}{\delta\phi(x)}$. Hence, we get

$$\delta S[\phi] = \int_\Omega d^4x \, \frac{\delta S[\phi]}{\delta \phi^i(x)} \, \delta\phi^i(x). \tag{3.4}$$

Let us use the following theorem from the variational calculus: if the field $\phi(x)$ corresponds to an extremum of the functional $S[\phi]$, then the corresponding variation $\delta S[\phi] = 0$ for any $\delta\phi^i(x)$. Since $\delta\phi^i(x)$ is arbitrary, Eq. (3.4) leads to

$$\frac{\delta S[\phi]}{\delta \phi^i(x)} = 0. \tag{3.5}$$

Eq. (3.5) is called a classical equation of motion or simply the field equation. The solutions of this equation determine the physically admissible field configurations $\phi^i(x)$.

Consider the calculation of the variational derivative of the functional $S[\phi]$ (3.1), assuming that $\mathcal{L} = \mathcal{L}(\phi, \partial_\mu\phi)$. In this case, we have

$$S[\phi] = \int_\Omega d^4x \, \mathcal{L}\big(\phi(x), \partial_\mu\phi(x)\big). \tag{3.6}$$

Here Ω is a domain in Minkowski space, bounded by $\partial\Omega$, that consists of the space-like hypersurfaces σ_1 and σ_2, where the hypersurface σ_2 lies in the future, relative to the hypersurface σ_1. Let $\phi(x)\Big|_{\sigma_1} = \phi_1(\mathbf{x})$, and $\phi(x)\Big|_{\sigma_2} = \phi_2(\mathbf{x})$, where $\phi(x)\Big|_{\sigma} = \phi(x)\Big|_{x\in\sigma}$, $\sigma = \sigma(x)$ is a space-like hypersurface and $\mathbf{x}$ is the vector formed by independent three-dimensional coordinates on this hypersurface. Let $\phi(x)$ be the field corresponding to an extremum of the functional $S[\phi]$, and $\phi'(x)$ an arbitrary field. We assume that both $\phi(x)$ and $\phi'(x)$ satisfy the same boundary conditions. Then, for the variation $\delta\phi(x)$=$\phi'(x)$-$\phi(x)$, we get $\delta\phi(x)\big|_{\sigma_1} = \delta\phi(x)\big|_{\sigma_2} = 0$. On the top of that, we assume $\delta\phi \to 0$ at $x^i \to \pm\infty$.

Consider

$$S[\phi + \delta\phi] - S[\phi] = \int_\Omega d^4x \{ \mathcal{L}(\phi + \delta\phi, \partial_\mu\phi + \partial_\mu\delta\phi) - \mathcal{L}(\phi, \partial_\mu\phi) \}. \tag{3.7}$$

Since the field variation is not related to the change of coordinates, $\partial_\mu\delta\phi = \partial_\mu\phi' - \partial_\mu\phi = \delta\partial_\mu\phi(x)$. Expanding the integral in (3.7) in the Taylor series in $\delta\phi$ up to the first order, we arrive at

$$\begin{aligned}
\delta S[\phi] &= \int_\Omega d^4x \Big\{ \frac{\partial \mathcal{L}}{\partial \phi}\delta\phi + \frac{\partial \mathcal{L}}{\partial\partial_\mu\phi}\partial_\mu\delta\phi \Big\} \\
&= \int_\Omega d^4x \Big\{ \frac{\partial \mathcal{L}}{\partial \phi}\delta\phi + \partial_\mu\Big(\frac{\partial \mathcal{L}}{\partial\partial_\mu\phi}\delta\phi\Big) - \Big(\partial_\mu\frac{\partial \mathcal{L}}{\partial\partial_\mu\phi}\Big)\delta\phi \Big\} \\
&= \int_\Omega d^4x \Big[\frac{\partial \mathcal{L}}{\partial \phi} - \partial_\mu\Big(\frac{\partial \mathcal{L}}{\partial\partial_\mu\phi}\Big)\Big]\delta\phi + \int_{\partial\Omega} d\sigma_\mu \Big(\frac{\partial \mathcal{L}}{\partial\partial_\mu\phi}\Big)\delta\phi.
\end{aligned} \tag{3.8}$$

Here $d\sigma_\mu$ is an element of the surface $\partial\Omega$, and the Gauss theorem has been used. According to the boundary conditions, $\delta\phi \to 0$ on $\partial\Omega$. Therefore,

$$\delta S[\phi] = \int_\Omega d^4x \Big[\frac{\partial \mathcal{L}}{\partial \phi} - \partial_\mu\Big(\frac{\partial \mathcal{L}}{\partial\partial_\mu\phi}\Big)\Big]\delta\phi(x) = 0, \tag{3.9}$$

and hence the variation of the action has the form (3.4). Thus, the functional derivative of the action is

$$\frac{\delta S[\phi]}{\delta \phi(x)} = \frac{\partial \mathcal{L}}{\partial \phi} - \partial_\mu \Big(\frac{\partial \mathcal{L}}{\partial \partial_\mu \phi} \Big), \tag{3.10}$$

and the equations of motion take the form of the Lagrange equations for the field ϕ,

$$\frac{\partial \mathcal{L}}{\partial \phi} - \partial_\mu \Big(\frac{\partial \mathcal{L}}{\partial \partial_\mu \phi} \Big) = 0. \tag{3.11}$$

Two observations are in order. First of all, the Lagrangian $\mathcal{L}$ is not uniquely defined. For instance, the two Lagrangians $\mathcal{L}$ and $\mathcal{L} + \partial_\mu R^\mu(\phi)$ lead to the same equations of motion (3.11).

The second observation is that, in some field models, we need to define a field ϕ in the space with boundaries, assuming that at least some of the space coordinates x^i take their values in the finite domains. Then the equations of motion (3.11) take place under the boundary conditions of the modified form. Usually these boundary conditions are defined by the requirement $\delta\phi\big|_{\partial\Omega} = 0$, and then the variation $\delta S[\phi]$ has the standard form (3.3).

3.2 Global symmetries

Following the analogy with classical mechanics, let us explore global symmetries of the Lagrangian approach for the fields, and its relation with the conservation laws.

Consider a theory of the fields $\phi = \phi^i(x)$ with the action (3.6). The infinitesimal transformations of coordinates and fields

$$x'^\mu = x^\mu + \delta x^\mu, \tag{3.12}$$

$$\phi'^i(x') = \phi^i(x) + \Delta\phi^i(x) \tag{3.13}$$

are symmetry transformations if they leave the action invariant, i.e.,

$$S[\phi] = S'[\phi'], \tag{3.14}$$

or, in the detailed form,

$$\int_\Omega d^4x \, \mathcal{L}(\phi(x), \partial_\mu \phi(x)) = \int_{\Omega'} d^4x' \, \mathcal{L}(\phi'(x'), \partial'_\mu \phi'(x')), \tag{3.15}$$

where Ω' is a domain of integration in terms of coordinates x'^μ. The last means that the equations defining Ω' are obtained from the equations defining Ω by the coordinate transformations $x^\mu = x'^\mu - \delta x^\mu$.

The transformations (3.13) can be reformulated as follows. Rewriting Eq. (3.13) as

$$\phi'^i(x + \delta x) = \phi'^i(x) + \partial_\mu \phi^i(x)\delta x^\mu = \phi^i(x) + \Delta\phi^i(x), \tag{3.16}$$

we arrive at

$$\Delta\phi^i(x) = \phi'^i(x) - \phi^i(x) + \partial_\mu\phi^i\delta x^\mu.$$

The quantity $\phi'^i(x) - \phi^i(x) = \delta\phi^i(x)$, where $\delta\phi^i(x)$ is a field variation, separated from the variation of the independent coordinates. Therefore,

$$\Delta\phi^i(x) = \delta\phi^i(x) + \partial_\mu\phi^i(x)\delta x^\mu. \tag{3.17}$$

One can assume that the transformations (3.13) are characterized by a finite set of parameters $\xi^1, \xi^2, \ldots, \xi^N$, such that

$$\begin{aligned}\delta x^\mu &= X^\mu{}_I(x)\xi^I,\\ \delta\phi^i(x) &= Y^i{}_I\big(x, \phi(x), \partial_\mu\phi(x)\big)\xi^I,\end{aligned} \tag{3.18}$$

where $I = 1, 2, ..., N$ and there is summation over the index I. The transformations (3.18) are called the N-parametric global transformations. The term global means that the parameters ξ^I are coordinate-independent. In the opposite case, the transformations are called local, as it is the case for the gauge transformations. For now, we consider only global transformations.

Taking into account Eq. (3.18), one can rewrite Eq. (3.13) as follows,

$$\begin{aligned}\phi'^i(x') &= \phi^i(x) + \Delta\phi^i(x), \qquad \text{where}\\ \Delta\phi^i(x) &= \big[Y^i{}_I\big(x, \phi(x), \partial_\mu\phi(x)\big) + \partial_\mu\phi^i(x)X^\mu{}_I(x)\big]\xi^I.\end{aligned} \tag{3.19}$$

Thus, in order to specify the global symmetry transformations, one should identify a field model and define the functions $X^\mu{}_I(x)$ and $Y^i{}_I\big(x, \phi(x), \partial_\mu\phi(x)\big)$.

The global symmetry transformations can be classified into spacetime transformations and internal symmetry transformations. In the last case, $\delta x^\mu = 0$ and $Y^i{}_I = Y^i{}_I\big(\phi(x)\big)$. This means that the internal symmetry transformations are transformations of the fields with fixed coordinates.

Spacetime symmetry transformations. Consider this type of symmetry transformations by dealing with two important examples.

As we already mentioned above, the Lagrangian should be a scalar under the Lorentz transformations, i.e., the action must be Lorentz invariant. The infinitesimal Lorentz transformations of coordinates have the form

$$\delta x^\mu = \omega^\mu{}_\nu x^\nu, \tag{3.20}$$

where $\omega_{\mu\nu} = -\omega_{\nu\mu}$ are the transformation parameters.

Usually, it is assumed that the fields are tensors or spinors under the Lorentz transformations. The reason for this is that, for such types of fields, it is easy to control the transformation rules, including constructing the Lagrangian as a Lorentz scalar. In the most general case, one can assume that all the fields are spin tensors $\phi_A(x)$, where A is a collection of tensor and spinor indices. Then, under Lorentz transformations, the fields ϕ_A transform as follows:

$$\delta\phi_A = -\frac{i}{2}\omega^{\alpha\beta}(J_{\alpha\beta})_A{}^B\phi_B, \tag{3.21}$$

with the generators of the Lorentz transformations $(J_{\alpha\beta})_A{}^B$ in the corresponding representation. The transformations, (3.20) and (3.21), represent an example of transformations related to the spacetime symmetry.

Another example is the symmetry under spacetime translations $x'^\mu = x^\mu + a^\mu$, which is equivalent to a shift of the reference frame as a whole. Since all points of the Minkowski space are physically equivalent, a transformed field in a transformed point must coincide with the initial field in the initial point, which is $\phi'^i(x') = \phi^i(x)$, and the action $S[\phi]$ should be invariant. Then $\phi'^i(x)+\partial_\mu\phi^i a^\mu = \phi'(x)$ or $\delta\phi^i(x) = -\partial_\mu\phi^i(x)a^\mu$. This transformation, together with $x'^\mu = x^\mu + a^\mu$, gives us another example of a spacetime symmetry transformation.

Internal symmetries. Consider the field $\phi^r{}_A(x)$, where A is a set of all Lorentz indices (tensor or spinor) and $r = 1, 2, \ldots, n$. The field $\phi^r{}_A(x)$ can be treated as a vector in some n-dimensional linear space. Let this vector transform according to

$$\begin{aligned} \phi'^r{}_A(x) &= \phi^r{}_A(x) + \delta\phi^r{}_A(x), \\ \delta\phi^r{}_A(x) &= i(T^I)^r{}_s\phi^s{}_A(x)\xi^I, \end{aligned} \tag{3.22}$$

where ξ^I are constant parameters, $I = 1, 2, \ldots, N$. The matrices $(T^I)^r{}_s$ are assumed to satisfy the condition

$$[T^I, T^J] = if^{IJ}{}_K T^K, \tag{3.23}$$

where $f^{IJ}{}_K = -f^{JI}{}_K$, and $f^{IJ}{}_K$ are constants. The relations (3.22) and (3.23) mean that we have a representation of some Lie group, in the linear space of vectors $\phi^r{}_A$. The matrices $(T^I)^r{}_s$ are generators of the group, and the quantities $f^{IJ}{}_K$ are the corresponding structure constants, while $\xi^1, \xi^2, \ldots, \xi^N$ are the group parameters. One can compare this relation to Eq. (2.22).

If the action $S[\phi]$ is invariant under the transformations (3.22), the last correspond to internal symmetry. Later on, we shall discuss the concrete examples of internal symmetries.

3.3 Noether's theorem

Noether's theorem gives a general method for finding the conserved quantities corresponding to the symmetries of the theory. Additive conserving quantities are also called dynamical invariants.

Theorem. Each N-parameters continuous symmetry transformation corresponds to N dynamical invariants.

Proof. Consider the transformations (3.12), (3.13) in the form (3.18), and assume the action is invariant, $\delta S[\Phi] = 0$. The variation of action $\delta S[\phi] = S'[\phi'] - S[\phi]$ follows from

$$S[\phi] = \int\limits_\Omega d^4x \mathcal{L}(\phi(x), \partial_\mu\phi(x)) \quad \text{and} \quad S'[\phi'] = \int\limits_{\Omega'} d^4x' \mathcal{L}(\phi'(x'), \partial'_\mu\phi'(x')). \tag{3.24}$$

Making the change of variables $x'^\mu = x^\mu + \delta x^\mu$ in the *r.h.s.* of the expression for $S'[\phi']$, the domain Ω' is transformed into the domain Ω. In the linear order in δx^μ, the Jacobian of this change of variables is

$$\det\left(\frac{\partial x'^{\mu}}{\partial x^{\nu}}\right) = \det\left(\delta^{\mu}{}_{\nu} + \partial_{\nu}\delta x^{\mu}\right) = 1 + \frac{\partial\delta x^{\mu}}{\partial x^{\mu}}. \tag{3.25}$$

Here we have taken into account that all the terms out of the main diagonal of the matrix in the *l.h.s.* enter the determinant in at least second order in ξ^I, and therefore these terms are irrelevant for the first-order expression. Thus, we get

$$S'[\phi'] = \int_{\Omega} d^4x\,(1 + \partial_{\mu}\delta x^{\mu})\,\mathcal{L}\big(\phi'(x+\delta x),\, \frac{\partial x^{\nu}}{\partial x'^{\mu}}\partial_{\nu}\phi'(x+\delta x)\big). \tag{3.26}$$

According to the definition of symmetry transformations, $\phi'(x+\delta x) = \phi(x) + \Delta\phi(x)$. Furthermore, in the lowest order,

$$\frac{\partial x^{\nu}}{\partial x'^{\mu}} = \delta^{\nu}{}_{\mu} - \partial_{\mu}\delta x^{\nu} \qquad \text{and} \qquad \frac{\partial x'^{\mu}}{\partial x^{\alpha}} = \delta^{\mu}{}_{\alpha} + \partial_{\alpha}\delta x^{\mu}. \tag{3.27}$$

Finally,

$$S'[\phi'] = \int\limits_{\Omega} d^4x\,(1 + \partial_{\mu}\delta x^{\mu})\,\mathcal{L}\big(\phi + \Delta\phi, (\delta^{\nu}{}_{\mu} - \partial_{\mu}\delta x^{\nu})(\partial_{\nu}\phi + \partial_{\nu}\Delta\phi)\big). \tag{3.28}$$

The Lagrangian in the *r.h.s.* of the last expression can be expanded into the power series up to the linear terms in $\Delta\phi$ and δx^{ν}, to give

$$\begin{aligned}
S'[\phi'] &= \int\limits_{\Omega} d^4x\Big\{\mathcal{L}(\phi, \partial_{\mu}\phi)\,[1 + \partial_{\mu}\delta x^{\mu}] + \frac{\partial\mathcal{L}}{\partial\phi^i}\Delta\phi^i + \frac{\partial\mathcal{L}}{\partial\partial_{\mu}\phi^i}\left(\partial_{\mu}\Delta\phi^i - \partial_{\nu}\phi^i\partial_{\mu}\delta x^{\nu}\right)\Big\} \\
&= \int\limits_{\Omega} d^4x\Big\{\mathcal{L} + \mathcal{L}\partial_{\mu}\delta x^{\mu} + \frac{\partial\mathcal{L}}{\partial\phi^i}\delta\phi^i + \frac{\partial\mathcal{L}}{\partial\partial_{\mu}\phi^i}\partial_{\mu}\delta\phi^i\,\frac{\partial\mathcal{L}}{\partial\phi^i}\partial_{\mu}\phi^i\delta x^{\mu} + \frac{\partial\mathcal{L}}{\partial\partial_{\nu}\phi^i}\partial_{\nu}\partial_{\mu}\phi^i\delta x^{\mu}\Big\} \\
&+ \int\limits_{\Omega} d^4x\,\Big\{\mathcal{L} + \mathcal{L}(\partial_{\mu}\delta x^{\mu}) + (\partial_{\mu}\mathcal{L})\delta x^{\mu} + \frac{\partial\mathcal{L}}{\partial\phi^i}\delta\phi^i + \frac{\partial\mathcal{L}}{\partial\partial_{\mu}\phi^i}\partial_{\mu}\delta\phi^i\Big\},
\end{aligned} \tag{3.29}$$

where we used the feature of coordinate-independent variations, $\delta\partial_{\mu}\phi^i = \partial_{\mu}\delta\phi^i$. Using the equations of motion (3.11) in the expression (3.29), we obtain

$$S'[\phi'] = \int\limits_{\Omega} d^4x\,\Big\{\mathcal{L} + \partial_{\mu}(\mathcal{L}\delta x^{\mu}) + \partial_{\mu}\left(\frac{\partial\mathcal{L}}{\partial\partial_{\mu}\phi^i}\delta\phi^i\right)\Big\}. \tag{3.30}$$

Next, substituting (3.18) and (3.19) into the last expression yields, for the first variation of the action,

$$\begin{aligned}
\delta S[\phi] = S'[\phi'] - S|\phi| &= \int\limits_{\Omega} d^4x\partial_{\mu}\left\{\frac{\partial\mathcal{L}}{\partial\partial_{\mu}\phi^i}\delta\phi^i + \mathcal{L}\delta x^{\mu}\right\} \\
&= \int\limits_{\Omega} d^4x\;\partial_{\mu}\Big\{\frac{\partial\mathcal{L}}{\partial\partial_{\mu}\phi^i}Y^i{}_I + \mathcal{L}X^{\mu}{}_I\Big\}\xi^I = 0,
\end{aligned} \tag{3.31}$$

since (3.18) corresponds to the symmetry transformations. Then, taking into account that the parameters ξ^I are linear independent and an arbitrariness of the domain Ω, one gets

$$\partial_\mu J^\mu_I = 0, \quad \text{for} \quad I = 1, 2, \dots, N\,, \tag{3.32}$$

where we have introduced the notation for the Noether's current, or generalized current,

$$J^\mu_I = -\Big(\frac{\partial \mathcal{L}}{\partial\, \partial_\mu \phi^i}\, Y^i{}_I + \mathcal{L}\, X^\mu{}_I\Big). \tag{3.33}$$

The relation (3.32) is the local conservation law of generalized current. Starting from this identity, we can get the integrated form of the conservation law. Using the Gauss theorem, one gets

$$\int_{\partial\Omega} d\sigma_\mu\, J^\mu_I = 0. \tag{3.34}$$

Since the fields $\phi^i(x)$ are vanishing at the space infinity, the identity (3.34) implies that

$$\int_{\sigma_2} d\sigma_\mu\, J^\mu_I - \int_{\sigma_1} d\sigma_\mu\, J^\mu_I \;=\; 0, \tag{3.35}$$

where the change of sign is stipulated by the change of direction of the vector n_μ normal to the hypersurface $\sigma(x) = \sigma_1$.

One can introduce the functionals depending on the hypersurface σ,

$$C_I[\sigma] = \int_\sigma d\sigma_\mu\, J^\mu_I, \qquad\qquad I = 1, 2, \dots, N. \tag{3.36}$$

Then Eq. (3.35) gives, for the two space-like hypersurfaces σ_1 and σ_1,

$$C_I[\sigma_1] \;=\; C_I[\sigma_2], \qquad I = 1, 2, \dots, N. \tag{3.37}$$

Thus, the functionals $C_I[\sigma]$ do not depend on choice of the hypersurface σ, such that $C_I[\sigma] = const$. Choosing a constant time hypersurface, $\sigma(x) = t$, we arrive at

$$C_I[t_2] \;=\; C_I[t_1], \qquad \text{where} \qquad C_I[t] \;=\; \int d^3x\, J^0_I(x). \tag{3.38}$$

The conditions $C_I[\sigma] = const$ mean that the functionals (3.36) are conserved quantities. Thus, we have shown that there is a conservation law for each continuous symmetry transformation with a fixed parameter ξ^I. Since the functionals $C_I[\sigma]$ are given by the integrals over hypersurfaces, they are additive quantities, which completes the proof.

Remark 1. Since Eq. (3.37) was derived using the equations of motion, the quantities $C_I[\sigma]$ in (3.37) are conserved only *on shell* (or on the mass shell), when the fields $\phi^i(x)$ are solutions to the equations of motion.

Remark 2. It s important that the generalized current is defined in a non-unique way. For instance, let $J^\mu{}_I$ be a generalized current. One can introduce the quantity

$$\tilde{J}^\mu_I = J^\mu_I + \partial_\nu f^{\mu\nu}{}_I, \tag{3.39}$$

where $f^{\mu\nu}{}_I = -f^{\nu\mu}{}_I$ is an arbitrary function of the fields and their derivatives, that is antisymmetric in the indices μ and ν. Obviously,

$$\partial_\nu \tilde{J}^\mu_I = \partial_\mu J^\mu_I + \partial_\mu \partial_\nu f^{\mu\nu}{}_I = \partial_\mu J^\mu_I. \tag{3.40}$$

In other words, if $\partial_\mu J^\mu{}_I = 0$, then $\partial_\mu \tilde{J}^\mu{}_I = 0$ too. Thus, the local conservation law for the current (3.32) does not change under the modification of the current (3.39). Consequently, the dynamical invariants $C_I[\sigma]$ remain conserved quantities under the same operation. This arbitrariness can be used to impose additional conditions on the generalized current.

As an application of the general Noether's theorem, consider the conservation law corresponding to internal symmetries, when $\delta x^\mu = 0$. According to (3.18) and (3.22),

$$X^\mu_I = 0, \qquad Y^r_I = i(T^I)^r{}_s \phi^s. \tag{3.41}$$

Then, the Noether's current is

$$J^\mu_I = -i\,\frac{\partial \mathcal{L}}{\partial\, \partial_\mu \phi^r}\,(T^I)^r{}_s \phi^s\,. \tag{3.42}$$

The conserving quantities associated with internal symmetries are called charges and are denoted as Q^I. Using (3.36) and (3.42), one gets

$$Q^I = \int d^3x\, J^0{}_I = -i \int d^3x\, \frac{\partial \mathcal{L}}{\partial\, \dot{\phi}^r}\,(T^I)^r{}_s \phi^s. \tag{3.43}$$

3.4 The energy-momentum tensor

One of the most important conservation laws is related to the invariance under the spacetime translations $x'^\mu = x^\mu + a^\mu$, where a^μ is an arbitrary constant four-vector. It is clear that if the Lagrangian does not depend explicitly on the coordinates, the spacetime translations are the symmetry transformations. In this case,

$$\delta x^\mu = a^\mu = \delta^\mu{}_\nu a^\nu, \tag{3.44}$$

i.e., $X^\mu_I = \delta^\mu_\nu$ in the general relation $\delta x^\mu = X^\mu_I \xi^I$, and a^μ plays the role of parameters ξ^I. As we already know, the field transforms under translation as

$$\delta\phi^i = -\partial_\mu \phi^i a^\mu = -\delta^\mu{}_\nu \partial_\mu \phi^i a^\nu. \tag{3.45}$$

It means that the role of the function $Y^i{}_I$ in the general relation $\delta\phi^i = Y^i{}_I \xi^I$ is played by $-\delta^\mu{}_\nu \partial_\mu \phi^i$. The generalized current, corresponding to the symmetry under the spacetime translations described above, is called the canonical energy-momentum tensor and is denoted as $T^\mu{}_\nu$. Let us note in passing that in Part II we shall introduce another (dynamical) definition of the energy-momentum tensor. The two definitions are equivalent in all known cases, but the general proof of this fact is not known yet.

The expression for generalized current (3.33), in the case of (3.44), yields

$$T^\mu{}_\nu = \frac{\partial \mathcal{L}}{\partial \partial_\mu \phi^i} \partial_\nu \phi^i - \mathcal{L} \delta^\mu{}_\nu, \tag{3.46}$$

while the local conservation law has the form

$$\partial_\mu T^\mu{}_\nu = 0. \tag{3.47}$$

The dynamical invariants corresponding to the symmetry under the spacetime translations are denoted P_ν. According to (3.36), they have the form

$$P_\nu = \int_\sigma d\sigma_\mu \, T^\mu{}_\nu. \tag{3.48}$$

Consider the expression (3.48) in more detail. Let the hypersurface σ in (3.48) be a surface of a constant time t. In this case,

$$P_\nu = \int d^3x T^0{}_\nu = \int d^3x \Big(\frac{\partial \mathcal{L}}{\partial \partial_0 \phi^i} \partial_\nu \phi^i - \mathcal{L} \delta^0{}_\nu \Big). \tag{3.49}$$

In particular, the component P_0 has the form

$$P_0 = \int d^3x \Big(\frac{\partial \mathcal{L}}{\partial \dot{\phi}^i} \dot{\phi}^i - \mathcal{L} \Big). \tag{3.50}$$

By analogy with classical mechanics, one defines the momenta $\pi_i = \frac{\partial \mathcal{L}}{\partial \dot{\Phi}^i}$, canonically conjugate to the fields ϕ^i. Then, the component P_0 becomes

$$P_0 = \int d^3x (\pi_i \dot{\phi}^i - \mathcal{L}) = H. \tag{3.51}$$

The expression H is analogous to the classical Hamilton function, or energy. Thus, the component P_0 of the vector P_ν is energy. Then, due to relativistic covariance, P_ν is the energy-momentum vector.

Exercises

3.1. Consider the higher-derivative theory with the Lagrangian depending on higher derivatives of the fields,

$$\mathcal{L} = \mathcal{L}(\phi, \partial_\mu \phi, \partial_{\mu_1} \partial_{\mu_2} \phi, \dots \partial_{\mu_1} \partial_{\mu_2} \dots, \partial_{\mu_n} \phi). \tag{3.52}$$

Formulate the boundary conditions for the variations of the field and its derivatives, which enable one to derive the Lagrange equations from the least action principle. Calculate the variational derivative of the action and obtain the equations of motion.

3.2. Prove, without taking a variational derivative of the actions, that the Lagrangians $\mathcal{L}(\phi, \partial_\alpha \phi)$ and $\mathcal{L}(\phi, \partial_\alpha \phi) + \partial_\mu R^\mu(\phi)$ lead to the same equations of motion.

3.3. Prove that the Lagrangians $\mathcal{L}(\phi, \partial_\mu \phi)$ and $\tilde{\mathcal{L}} = \mathcal{L}(\phi, \pi_\mu) + \varrho^\mu(\pi_\mu - \partial_\mu \phi)$ lead to the same equations of motion. Here $\pi_\mu = \pi_\mu(x)$ and $\varrho^\mu = \varrho^\mu(x)$ are arbitrary vector functions.

3.4. Formulate the conditions under which the equations of motion for a theory with the Lagrangian $\mathcal{L}(\phi^i, \partial_\mu \phi^i)$ have the form

$$A_{ij}^{\mu\nu} \partial_\mu \partial_\nu \phi^j + B_{ij}^\mu \partial_\mu \phi^j + C_i = 0$$

with constant coefficients, and find the explicit form for the $A_{ij}^{\mu\nu}$, B_{ij}^μ, C_i in this case.

3.5. Let the symmetry transformations be $\delta x^\mu = X^\mu(x, \phi, \partial_\alpha \phi)$. Construct the proof of Noether's theorem in this case.

3.6. Let the condition of invariance (3.15) have the alternative form

$$\int_{\Omega'} d^4x' \, \mathcal{L}(\phi'(x'), \partial'_\mu \phi'(x')) = \int_\Omega d^4x \left[\mathcal{L}(\phi(x), \partial_\mu \phi(x) + \partial_\mu R^\mu(\phi) \right],$$

with an arbitrary function $R^\mu(\phi)$. Construct a proof of Noether's theorem in this case. Explore whether the conserved charges depend on $R^\mu(\phi)$.

Comments

Different aspects of Lagrange formalism in field theory are considered in practically all books on relativistic field theory, e.g., in [187], [250], [304], [274], [256], [215], [155], [345], [105] [109], [293], [276], [59].

4
Field models

In this chapter, we consider the constructions of Lagrangians for various field models and discuss the basic properties of these models.

4.1 Basic assumptions about the structure of Lagrangians

Consider an arbitrary theory with a set of fields $\phi^i(x)$ and with the action (3.6). In what follows, we do not need to specify the choice of Ω, and deal with the action

$$S[\phi] = \int d^4x\, \mathcal{L}\big(\phi(x), \partial_\mu \phi(x)\big). \tag{4.1}$$

The choice of the field model is related by the specification of the set of fields and the Lagrangian $\mathcal{L}$. Usually, it is assumed that fields ϕ^i are the spin tensors, e.g., we will explore scalar, vector and spinor field models, higher-rank tensors, etc. Models with different types of fields in the same Lagrangian are also possible.

As to the choice of Lagrangian, it is assumed that it should be a function of fields and their derivatives, being taken in the same point x^μ (this is called an assumption of locality), that can be always divided into the sum of the two terms

$$\mathcal{L} = \mathcal{L}_0 + \mathcal{L}_{int}, \tag{4.2}$$

where $\mathcal{L}_0$ is bilinear in fields and their derivatives, while $\mathcal{L}_{int}$ contains powers of the fields and the derivatives higher than the second. The part $\mathcal{L}_0$ is called the free Lagrangian, and $\mathcal{L}_{int}$ is called the interaction Lagrangian.

The equations of motion for the theory with the Lagrangian $\mathcal{L}$ can be written in the form

$$\frac{\partial \mathcal{L}_0}{\partial \phi^i} - \partial_\mu \frac{\partial \mathcal{L}_0}{\partial(\partial_\mu \phi^i)} = - \Big[\frac{\partial \mathcal{L}_{int}}{\partial \phi^i} - \partial_\mu \frac{\partial \mathcal{L}_{int}}{\partial(\partial_\mu \phi^i)}\Big]. \tag{4.3}$$

Since the Lagrangian $\mathcal{L}_0$ is quadratic in fields and their derivatives, the *l.h.s.* of the equations (4.3) is a linear equation, containing no more than two derivatives of fields. If $\mathcal{L}_{int} = 0$, the corresponding equations of motion will be linear partial differential equations, typically of the order not higher than the second. If $\mathcal{L}_{int} \neq 0$, then the equations of motion will be non-linear. This feature explains the terms “free Lagrangian” and “interacting Lagrangian.”

Equations to the free Lagrangian are called free equations of motion, which are linear partial differential equations for spin tensors. Requiring that the fields transform under irreducible spin-tensor representations of the Poincaré group, the corresponding

Introduction to Quantum Field Theory with Applications to Quantum Gravity. Iosif L. Buchbinder and Ilya L. Shapiro, Oxford University Press (2021).
DOI: 10.1093/oso/9780198838319.003.0004

equations of motion must be compatible with the relations (2.87) and (2.91) or with their massless versions, defining the irreducible representations of the Poincaré group in the linear space of fields. Due to the Lorentz covariance, the Lagrangian can be constructed from fields and their derivatives, and other covariant objects of the Lorentz group, such as $\eta_{\mu\nu}$ and spinor quantities, such as $(\sigma^\mu)_{a\dot{a}}$, ε_{ab} and so on.

In principle, the free Lagrangian for any kind of spin-tensor fields can be restored based on the above relations. The main problem in the construction of the Lagrangian $\mathcal{L}$ consists of finding the $\mathcal{L}_{int}$. There is no general prescription for this. The construction of an interacting Lagrangian for a concrete field model is based on the use of additional physical and mathematical assumptions, including the arguments based on the quantum consistency of the theory.

4.2 Scalar field models

Consider the simplest example of a field model, namely, a scalar field.

4.2.1 Real scalar fields

According to (2.87), the real scalar field φ describes the massive or massless irreducible representation of the Poincaré group with spin $s = 0$ under the Klein–Gordon equation

$$(\Box + m^2)\varphi = 0. \tag{4.4}$$

The free Lagrangian $\mathcal{L}_0$ for the field φ is constructed as follows. Since the equation (4.4) is linear, the corresponding Lagrangian should be quadratic in φ and $\partial_\mu\varphi$. Hence the most general expression for $\mathcal{L}_0$ is

$$\mathcal{L}_0 = \frac{1}{2}c_1\eta^{\mu\nu}\partial_\mu\varphi\partial_\nu\varphi + \frac{1}{2}c_2 m^2\varphi^2, \tag{4.5}$$

where c_1, c_2 are some arbitrary numerical coefficients. The term $\varphi\partial_\mu\varphi$ is ruled out by Lorentz covariance. Derive the equations of motion for the Lagrangian (4.5) gives

$$\frac{\partial\mathcal{L}_0}{\partial\varphi} - \partial_\mu\frac{\partial\mathcal{L}_0}{\partial(\partial_\mu\varphi)} = c_2 m^2\varphi - c_1\Box\varphi = 0. \tag{4.6}$$

The comparison of this equation with Eq. (4.4) shows that $c_2 = -c_1$, and we obtain the Lagrangian $\mathcal{L}_0$ in the form

$$\mathcal{L}_0 = \frac{c_1}{2}\Big(\eta^{\mu\nu}\partial_\mu\varphi\partial_\nu\varphi - m^2\varphi^2\Big). \tag{4.7}$$

To fix the coefficient c_1, one has to derive the energy (3.50) corresponding to the Lagrangian (4.7),

$$E = P_0 = \int d^3x\left(\frac{\partial\mathcal{L}}{\partial\dot{\varphi}^i}\dot{\varphi}^i - \mathcal{L}\right) = \frac{c_1}{2}\int d^3x\left(\dot{\varphi}^2 + \partial_j\varphi\partial_j\varphi + m^2\varphi^2\right), \tag{4.8}$$

where $j = 1, 2, 3$. Requiring that the energy is positively defined, we arrive at a positive value of c_1. The absolute value of this constant can be modified by rescaling $\varphi \to k\varphi$

with $k = const$, therefore it can be fixed by convenience. The standard choice is $c_1 = 1$, bringing the free Lagrangian of the real scalar field to be

$$\mathcal{L}_0 = \frac{1}{2}\eta^{\mu\nu}\partial_\mu\varphi\partial_\nu\varphi - \frac{1}{2}m^2\varphi^2, \tag{4.9}$$

The term $\frac{1}{2}\eta^{\mu\nu}\partial_\mu\varphi\partial_\nu\varphi$ is called kinetic, and the term $\frac{1}{2}m^2\varphi^2$ is called massive.

To include interaction, one has to add to (4.9) a potential term, $-V(\varphi)$, with an arbitrary real function of φ. Thus, the full Lagrangian of a real scalar field is

$$\mathcal{L} = \frac{1}{2}\eta^{\mu\nu}\partial_\mu\varphi\partial_\nu\varphi - \frac{1}{2}m^2\varphi^2 - V(\varphi). \tag{4.10}$$

It is easy to see that the Lagrangian $\mathcal{L}$ (4.10) has the form (4.2), where $\mathcal{L}_{int} = -V(\varphi)$. The expression $U(\varphi) = \frac{1}{2}m^2\varphi^2 + V(\varphi)$ is called the scalar potential. The interaction potential is usually chosen as a finite order polynomial in φ,

$$V(\varphi) = \sum_{k=3}^{N}\frac{f_k}{k!}\varphi^k, \tag{4.11}$$

where real parameters f_k are called coupling constants, or just couplings.

The quantities φ, m and f_k (except f_4) are dimensional. To find the dimensions, one has to remember that the action is dimensionless in the natural units with $c = 1$ and $\hbar = 1$. In the same units, the mass dimensions of coordinates are $[x^\mu] = -1$, and the dimensions of the derivatives are $[\partial_\mu] = 1$. Since the action is the integral of the Lagrangian over the four-volume, the dimensions of all terms in the Lagrangian are $[\mathcal{L}] = 4$. For the kinetic term in the Lagrangian, this gives $4 = 2[\partial_\mu] + 2[\varphi]$. Therefore, the dimension of the scalar field is $[\varphi] = 1$. The massive term has dimension $m^2\varphi^2$; therefore, the dimension of mass is $[m] = 1$, as we should expect. The interaction potential has dimension $[V(\varphi)] = 4$. Using Eq. (4.11), one gets $[f_k] = 4 - k$. Then, the coupling constant f_4 is dimensionless. In what follows, we denote $f_4 = f$. Thus, the simplest interaction potential with the dimensionless coupling constant has the form

$$V(\varphi) = \frac{f}{4!}\varphi^4. \tag{4.12}$$

One can formulate a multi-scalar model with the fields φ^i, $i = 1, 2, \ldots, n$. The corresponding Lagrangian is a natural generalization of Eq. (4.10),

$$\mathcal{L} = \frac{1}{2}\delta_{ij}\partial^\mu\varphi^i\partial_\mu\varphi^j - \frac{1}{2}m^2_{ij}\varphi^i\varphi^j - V(\varphi^i). \tag{4.13}$$

Assuming that the interaction potential $V(\varphi^i)$ is a polynomial containing only third and fourth powers in the field φ^i, one gets

$$V(\varphi^i) = \frac{1}{3!}f_{ijk}\,\varphi^i\varphi^j\varphi^k + \frac{1}{4!}f_{ijkl}\,\varphi^i\varphi^j\varphi^k\varphi^l. \tag{4.14}$$

The quantities f_{ijk} and f_{ijkl} are called scalar coupling constants, and m^2_{ij} is called the mass matrix.

Observation 1. Later on, we shall see that the potentials of the (4.14) type, without the couplings with *negative* mass dimensions, are very special. Only these couplings lead to theories that are consistent at all energy scales in the quantum theory. Indeed, this is the main reason why we usually prefer to restrict the considerations by using these types of potentials. On the other hand, all powers of scalars can be used for constructing the effective field theory models, which are used in the restricted energy domains.

Observation 2. One should not mix the multi-component scalar field φ^i (where $i = 1, \dots, n$) with the generic type of field ϕ^i, where i includes all discrete indices (Lorentz indices, the ones of internal symmetries, etc,). In the present section, we discuss only scalar fields.

4.2.2 Complex scalar fields

Consider a complex scalar field

$$\varphi = \sqrt{2}(\varphi_1 + i\varphi_2) \quad \text{and} \quad \varphi^* = \sqrt{2}(\varphi_1 - i\varphi_2), \tag{4.15}$$

where φ_1 and φ_2 are the real scalar functions. Since the Lagrangian for a real scalar field is given by (4.10), in the case under consideration we have, assuming equal masses,

$$\mathcal{L} = \frac{1}{2}\eta^{\mu\nu}\partial_\mu\varphi_1\partial_\nu\varphi_1 - \frac{1}{2}m^2{\varphi_1}^2 + \frac{1}{2}\eta^{\mu\nu}\partial_\mu\varphi_2\partial_\nu\varphi_2 - \frac{1}{2}m^2{\varphi_2}^2 - V(\varphi_1, \varphi_2). \tag{4.16}$$

Rewriting the Lagrangian (4.16) in terms of the fields (4.15) yields

$$\mathcal{L} = \eta^{\mu\nu}\partial_\mu\varphi^*\partial_\nu\varphi - m^2\varphi^*\varphi - V(\varphi^*, \varphi), \tag{4.17}$$

where the potentials are related by $V(\varphi^*, \varphi) = V(\varphi_1, \varphi_2)$, with (4.15) defining the relation between the two sets of variables. The free part of the Lagrangian (4.17) is invariant under the transformations

$$\varphi' = e^{ie\,\xi}\varphi, \qquad \varphi'^* = e^{-ie\,\xi}\varphi^*, \tag{4.18}$$

where ξ is a real constant parameter. Requiring the potential $V(\varphi^*, \varphi)$ to be invariant under these transformations, we provide that $V(\varphi^*, \varphi) = V(\varphi^*, \varphi)$.

4.2.3 Sigma model

The sigma model is a more complicated kind of a multi-scalar theory, which has many applications ranging from pion physics to string theory.

This field model is formulated as follows. Let us require that, in the Lagrangian (4.13), the masses and interaction terms vanish, $m^2_{ij} = 0$ and $V(\varphi^i) = 0$. Replacing δ_{ij} with the field-dependent matrix $g_{ij}(\varphi)$, where $\varphi = (\varphi^i;\ i = 1, 2, \dots, n)$, we arrive at the Lagrangian

$$\mathcal{L} = \frac{1}{2}\eta^{\mu\nu}g_{ij}\,(\varphi)\partial_\mu\varphi^i\partial_\nu\varphi^j. \tag{4.19}$$

By construction, $g_{ij}(\varphi) = g_{ji}(\varphi)$. Let us demand the Lagrangian (4.19) to be invariant under an arbitrary reparametrization of the fields, $\phi'^i = f^i(\varphi)$. This implies

$$\mathcal{L} = \frac{1}{2}\,g_{ij}(\varphi)\frac{\partial\varphi^i}{\partial\varphi'^k}\frac{\partial\varphi^j}{\partial\varphi'^l}\,\partial^\mu\varphi'^k\partial_\mu\varphi'^l = \frac{1}{2}g'_{kl}(\varphi')\partial^\mu\varphi'^k\partial_\mu\varphi'^l.$$

The invariance means that

$$g'_{kl}(\varphi') = \frac{\partial\varphi^i}{\partial\varphi'^k}\frac{\partial\varphi^j}{\partial\varphi'^l}\,g_{ij}(\varphi). \tag{4.20}$$

This relation is a transformation law for the components of the covariant second-rank tensor field under the coordinate transformations. The relation (4.20) admits the following interpretation. The fields φ^i are treated as the coordinates on some manifolds, and the $g_{ij}(\varphi)$ should be regarded as the Riemann metric on this manifold. As a result, the Lagrangian of the sigma model (4.19) is formulated in the geometric terms, using the Minkowski spacetime and the space of the scalar fields.

4.3 Spinor field models

In this section, we discuss the Lagrangian for the model of the field with spin $s = \frac{1}{2}$.

4.3.1 Equations of motion for a free spinor field

According to Eq. (2.91), the equation of motion for massive spin $s = \frac{1}{2}$ field is the Dirac equation (2.94),

$$(i\gamma^\mu\partial_\mu - m)\psi = 0, \tag{4.21}$$

where [see Eq. (2.92)]

$$\gamma^\mu = \begin{pmatrix} 0 & \sigma^\mu \\ \tilde{\sigma}^\mu & 0 \end{pmatrix} \quad \text{and} \quad \psi = \begin{pmatrix} \varphi_a \\ \chi^{\dot{a}} \end{pmatrix} \tag{4.22}$$

is the four-component Dirac spinor constructed in terms of the two-component left and right Weyl spinors φ_a and $\chi^{\dot{a}}$. The basic relation for the Dirac matrices γ^μ has the form of the Clifford algebra (2.93), $\gamma^\mu\gamma^\nu + \gamma^\nu\gamma^\mu = 2\eta^{\mu\nu}I$.

It proves useful to start by formulating the massless equation for the spinor field φ_a. According to relativistic symmetry, such an equation can be constructed using only $\partial_\mu\varphi_a$ and other covariant quantities including σ^μ, $\tilde{\sigma}^\mu$, $\eta_{\mu\nu}$. The simplest possibility for such an equation looks like

$$i(\tilde{\sigma}^\mu)^{\dot{a}a}\,\partial_\mu\varphi_a = 0. \tag{4.23}$$

Furthermore, each component of the field φ_a has to satisfy the wave equation. Let us check that the equation (4.23) is compatible with the wave equation. Acting with the operator $-i\sigma^\nu{}_{b\dot{a}}\partial_\nu$ on Eq. (4.23), one gets

$$(\sigma^\nu\tilde{\sigma}^\mu)_b{}^a\,\partial_\nu\partial_\mu\varphi_a = \frac{1}{2}\big(\sigma^\nu\tilde{\sigma}^\mu + \sigma^\mu\tilde{\sigma}^\nu\big)_b{}^a\,\partial_\nu\partial_\mu\varphi_a = \Box\varphi_b = 0, \tag{4.24}$$

where we used the relation

$$(\sigma^\nu\tilde{\sigma}^\mu + \sigma^\mu\tilde{\sigma}^\nu)_b{}^a = 2\eta^{\mu\nu}\delta_b{}^a. \tag{4.25}$$

The basic requirements defining the irreducible representation are fulfilled, and thus eq. (4.23) is admissible as the equation of motion for the spinor field φ_a. Using the analysis carried out in section 2.6, one can derive that, in the case under consideration, the Lubansky-Pauli vector W_μ (2.83) is related to the momentum operator P_μ according to $W_\mu = \frac{1}{2}P_\mu$. Therefore, the field φ_a satisfying Eq. (4.23) corresponds to an irreducible massless representation of the Poincaré group with helicity $\lambda = \frac{1}{2}$. Eq. (4.23) is called the Weyl equation.

In a similar way, one can show that the free equation of motion for a spinor field $\chi^{\dot{a}}(x)$ is written as

$$i(\sigma^\mu)_{a\dot{a}}\, \partial_\mu \chi^{\dot{a}} = 0, \tag{4.26}$$

which is also called Weyl equation. The field satisfying this equation corresponds to the irreducible massless representation of the Poincaré group with the helicity $\lambda = -\frac{1}{2}$.

Let us now turn to the equation of motion for the massive field. Due to relativistic symmetry, such an equation can be constructed only from the partial derivatives ∂_μ and from the invariant quantities of the Lorentz group. Let us begin with the Weyl equations (4.23) and (4.26) for φ_a and $\chi_{\dot{a}}$, and introduce a massive term. It is easy to note that Eq. (4.23) contains a free index $\dot{a}$; however, the field φ_a does not have an index. Thus, the equation of motion for the massive spinor field cannot be constructed only from φ_a. The same consideration applies to $\chi_{\dot{a}}$ and Eq. (4.26). Thus, we have to use both spinors, so let us combine both equations in the form

$$\begin{aligned} i(\tilde{\sigma}^\mu)^{\dot{a}a}\partial_\mu \varphi_a - m\chi^{\dot{a}} &= 0, \\ i(\sigma^\mu)_{a\dot{a}}\partial_\mu \chi^{\dot{a}} - m\varphi_a &= 0. \end{aligned} \tag{4.27}$$

The consistency of this system of equations includes the proof that both components φ and χ satisfy wave equations $(\Box + m^2)\varphi_a = 0$ and $(\Box + m^2)\chi^{\dot{a}} = 0$.

Acting by the operator $i(\sigma^\nu)_{b\dot{a}}\partial_\nu$ on the first of Eq. (4.27), one gets

$$-(\sigma^\nu \tilde{\sigma}^\mu)_b{}^a\, \partial_\mu \partial_\nu \varphi_a - im(\sigma^\nu)_{b\dot{a}}\partial_\nu \chi^{\dot{a}} \;=\; 0.$$

The expression $i(\sigma^\nu)_{b\dot{a}}\partial_\nu \chi^{\dot{a}}$ can be found from the second of Eq. (4.27). Then, we get

$$-\,(\sigma^\nu \tilde{\sigma}^\mu)_b{}^a \partial_\mu \partial_\nu \varphi_a - m^2 \varphi_b = 0.$$

Due to the commutativity of partial derivatives, this equation can be rewritten as

$$-\frac{1}{2}(\sigma^\nu \tilde{\sigma}^\mu + \sigma^\mu \tilde{\sigma}^\nu)_b{}^a \partial_\mu \partial_\nu \varphi_a - m^2 \varphi_b \;=\; (\Box + m^2)\varphi_b \;=\; 0, \tag{4.28}$$

where we used (4.25). In a similar way one can derive from Eq. (4.27)

$$(\Box + m^2)\chi^{\dot{a}} = 0. \tag{4.29}$$

Thus, Eq. (4.27) are compatible with the relativistic dispersion relation between energy and the three-dimensional momentum of a relativistic particle. Therefore, these equations are consistent candidates to be the equations of motion for massive spinor fields. Writing (4.27) in the matrix form, we meet

$$\left\{ i \begin{pmatrix} 0 & \sigma^{\mu} \\ \tilde{\sigma}^{\mu} & 0 \end{pmatrix} \partial_{\mu} - m \begin{pmatrix} \tilde{\sigma}^{0} & 0 \\ 0 & \sigma^{0} \end{pmatrix} \right\} \begin{pmatrix} \varphi \\ \chi \end{pmatrix} = 0 , \tag{4.30}$$

where $\tilde{\sigma}_0 = \sigma_0 = I_2$. Introducing the four-component column (4.22), it is easy to show that Eq. (4.30) can be recast in terms of the four-dimensional gamma matrices defined in (4.22), as the Dirac equation (4.21), for the four-component Dirac spinor field ψ defined in (4.22).

Let us elaborate further on the Dirac equation. According to the definition (4.22), the Dirac spinor ψ is a column. The four-component line

$$\bar{\psi} = \psi^{\dagger}\gamma^{0} \tag{4.31}$$

is called the Dirac conjugate spinor. The expression for $\bar{\psi}$ in terms of the two-component spinors can be found on the base of relations (4.22),

$$\bar{\psi} = (\bar{\chi}^{a}, \bar{\varphi}_{\dot{a}}). \tag{4.32}$$

Performing Hermitian conjugation on Eq. (4.21), one gets

$$-i\partial_{\mu}\psi^{\dagger}(\gamma^{\mu})^{\dagger} - m\psi^{\dagger} = 0. \tag{4.33}$$

Next, since $\bar{\psi} = \psi^{\dagger}\gamma^{0}$, we have $\bar{\psi}\gamma^{0} = \psi^{\dagger}(\gamma^{0})^{2}$. The Clifford algebra (2.93) leads to the relation $(\gamma^{0})^{2} = I$. Thus, $\psi^{\dagger} = \bar{\psi}\gamma^{0}$, and we end up with the equation

$$i\partial_{\mu}\bar{\psi}\,\gamma^{0}(\gamma^{\mu})^{\dagger} + m\bar{\psi}\gamma^{0} = 0 \qquad \Longrightarrow \qquad i\partial_{\mu}\bar{\psi}\,\gamma^{0}(\gamma^{\mu})^{\dagger}\,\gamma^{0} + m\bar{\psi} = 0. \tag{4.34}$$

Using Eq. (2.93), one can easily prove that $\gamma^{0}(\gamma^{\mu})^{\dagger}\,\gamma^{0} = \gamma^{\mu}$. As a result, we arrive at

$$i\partial_{\mu}\bar{\psi}\,\gamma^{\mu} + m\bar{\psi} = 0, \tag{4.35}$$

which is the equation of motion for a free Dirac conjugate field.

4.3.2 Properties of the Dirac spinors and gamma matrices

The four-dimensional Dirac spinor ψ is defined by the relation (4.22), in terms of the two-component spinors φ_a and $\chi_{\dot{a}}$. The transformation law of the Dirac spinor under infinitesimal Lorentz transformations can be obtained from the transformation laws for the two-component spinors (2.76),

$$\delta\psi \;=\; -\frac{i}{2}\omega^{\alpha\beta}\Sigma_{\alpha\beta}\psi, \tag{4.36}$$

where the generators of Lorentz transformations $\Sigma_{\alpha\beta}$ in the four-component spinor representation are

$$\Sigma_{\alpha\beta} = \begin{pmatrix} i\sigma_{\alpha\beta} & 0 \\ 0 & i\tilde{\sigma}_{\alpha\beta} \end{pmatrix} = \frac{i}{4}(\gamma_{\alpha}\gamma_{\beta} - \gamma_{\beta}\gamma_{\alpha}). \tag{4.37}$$

In the previous section, we observed the difference between four-component Dirac spinor and the two-component Weyl spinors. In particular, we saw that the Weyl spinors are necessary massless. Let us consider an alternative way to define the spinor with smaller amount of degrees of freedom.

Starting from the general definition of the four-component Dirac spinor (4.22), one can define the Majorana spinor ψ_M as follows:

$$\psi_M = \begin{pmatrix} \varphi_a \\ \bar{\varphi}^{\dot{a}} \end{pmatrix}. \tag{4.38}$$

This spinor has two lower components, which are conjugate to the two upper components. Therefore, the Majorana spinor ψ_M has half of the independent components, compared to an arbitrary Dirac spinor ψ.

To compare the Weil and Majorana spinors, we note that the left and right Weyl spinors can be written in the four-component form as

$$\psi_W^{(l)} = \begin{pmatrix} \varphi_a \\ 0 \end{pmatrix} \qquad \text{and} \qquad \psi_W^{(r)} = \begin{pmatrix} 0 \\ \chi^{\dot{a}} \end{pmatrix}. \tag{4.39}$$

Therefore, an arbitrary Dirac spinor can be written as $\psi = \psi_W^{(l)} + \psi_W^{(r)}$. Other special features of the Weyl spinors can be revealed if we introduce one more gamma matrix $\gamma_5 = \gamma^5$, according to the definition

$$\gamma_5 = \frac{i}{4!}\,\varepsilon_{\alpha\beta\mu\nu}\,\gamma^\alpha\gamma^\beta\gamma^\mu\gamma^\nu = -i\gamma_0\gamma_1\gamma_2\gamma_3. \tag{4.40}$$

The expressions (4.22) for the gamma matrices leads to the expression for γ_5 in terms of the two-component unit matrix,

$$\gamma_5 = \begin{pmatrix} \sigma_0 & 0 \\ 0 & -\sigma_0 \end{pmatrix}. \tag{4.41}$$

Taking into account that σ_0 is a 2×2 unit matrix, we obtain immediately that $(\gamma_5)^2 = I$, where I is the 4×4 unit matrix.

The basic relation (2.93) for the gamma matrices is invariant under the transformation

$$\gamma'_\mu = M\gamma_\mu M^\dagger, \tag{4.42}$$

where M is an arbitrary unitary matrix. It is easy to see that the Dirac equation (2.94) is invariant under the transformation (4.42), if we also fulfill the transformation of the Dirac spinor as follows:

$$\psi' = M\psi. \tag{4.43}$$

Using unitary matrices M, one can change the components of the Dirac spinors and the representations of the gamma matrices. However, even after this change, the relations obtained on the basis of the two-component spinors and 2×2 matrices σ_μ will remain the same. In particular, the spinors obtained through a matrix M from the spinors $\psi_W^{(l)}$ or $\psi_W^{(r)}$ (4.39), will still depend only on the two-component spinor φ_a from the fundamental representation of the $SL(2|\mathbb{C})$ group, or only on the two-component spinor $\chi_{\dot{a}}$ from the conjugate representation. Analogously, for any choice of the matrix M, the Majorana spinor ψ_M from Eq. (4.38) will include both two-component spinors φ_a and $\bar{\varphi}_{\dot{a}}$. Thus, the unitary transformation is not able to transform Weyl spinors into Majorana spinors, or vice versa.

As we already know, the Weyl equations (4.23) and (4.26) can be written in the four-component form, like the Dirac equation. How can we know which type of spinor we are dealing with?

Let us define the matrices P_+ and P_- as

$$P_+ = \frac{1}{2}(I + \gamma_5) \qquad \text{and} \qquad P_- = \frac{1}{2}(I - \gamma_5). \tag{4.44}$$

Using the relation $(\gamma_5)^2 = I$, it is easy to check the following relations:

$$P_+ + P_- = I, \quad P_+P_- = P_-P_+ = 0, \quad P_+^2 = P_+, \quad P_-^2 = P_-, \tag{4.45}$$

indicating that P_+ and P_- matrices are projection operators. In terms of two-component σ-matrices, these projectors have the form

$$P_+ = \begin{pmatrix} \sigma_0 & 0 \\ 0 & 0 \end{pmatrix} \qquad \text{and} \qquad P_- = \begin{pmatrix} 0 & 0 \\ 0 & \sigma_0 \end{pmatrix}. \tag{4.46}$$

It is evident that acting by these projectors on an arbitrary Dirac spinor results in Weyl spinors,

$$\psi_W^{(l)} = P_+\psi \qquad \text{and} \qquad \psi^{(r)} = P_-\psi. \tag{4.47}$$

These relations, together with the definition (4.44) and the properties of the matrices $P_\pm$, remain unchanged under the transformations (4.42) and (4.43).

The last observation is that one can rewrite the Weyl equations (4.23) and (4.26) in terms of the projection operators P_+ and P_-. The result is

$$i\gamma^\mu P_+ \partial_\mu \psi = 0\,, \qquad i\gamma^\mu P_- \partial_\mu \psi = 0. \tag{4.48}$$

4.3.3 The Lagrangian for a spinor field

A Lagrangian for a free field should provide that the corresponding Lagrange equations reproduce (2.87) or (2.91), depending on the irreducible representation of the Poincaré group. Let us show that the Lagrangian

$$\mathcal{L}_0 = \bar{\psi}(i\gamma^\mu \partial_\mu \psi - m\psi) \tag{4.49}$$

reproduces the Dirac equation (4.21). Taking derivatives, we obtain

$$\frac{\partial \mathcal{L}_0}{\partial \bar{\psi}} = i\gamma^\mu \partial_\mu \psi - m\psi \qquad \text{and} \qquad \frac{\partial \mathcal{L}_0}{\partial \partial_\mu \bar{\psi}} = 0, \tag{4.50}$$

and the Lagrange equations in the form

$$\frac{\partial \mathcal{L}_0}{\partial \bar{\psi}} - \partial_\mu \Big(\frac{\partial \mathcal{L}_0}{\partial \partial_\mu \bar{\psi}} \Big) = 0,$$

yielding Eq. (4.21). Analogously,

$$\frac{\partial \mathcal{L}_0}{\partial \psi} = -m\bar{\psi}, \qquad \frac{\partial \mathcal{L}_0}{\partial \partial_\mu \psi} = \bar{\psi}\gamma^\mu, \tag{4.51}$$

and we get the equation for the conjugate Dirac spinor (4.35).

The next step is to ensure that the Lagrangian (4.49) is invariant under the Lorentz transformations. The formalism of two-component spinors provides a simple way to do this. First of all, one can note that

$$\bar{\psi}\psi = \big(\bar{\chi}^a \; \bar{\varphi}_{\dot{a}} \big) \begin{pmatrix} \varphi_a \\ \chi^{\dot{a}} \end{pmatrix} = \bar{\chi}^a \varphi_a + \bar{\varphi}_{\dot{a}} \chi^{\dot{a}}.$$

It was shown in section 2.5 that $\chi^a \varphi_a$ and $\bar{\varphi}_{\dot{a}} \chi^{\dot{a}}$ are Lorentz scalars. Moreover, it is easy to verify that their sum is real. The second step is to calculate

$$\begin{aligned} \bar{\psi}\gamma^\mu \partial_\mu \psi &= \big(\bar{\chi}^a \; \bar{\varphi}_{\dot{a}} \big) \begin{pmatrix} 0 & \sigma^\mu \partial_\mu \\ \tilde{\sigma}^\mu \partial_\mu & 0 \end{pmatrix} \begin{pmatrix} \varphi_a \\ \chi^{\dot{a}} \end{pmatrix} \\ &= \big(\bar{\chi}^a \; \bar{\varphi}_{\dot{a}} \big) \begin{pmatrix} (\sigma^\mu)_{a\dot{a}} \, \partial_\mu \chi^{\dot{a}} \\ (\tilde{\sigma}^\mu)^{\dot{a}a} \, \partial_\mu \varphi_a \end{pmatrix} = \bar{\chi}^a (\sigma^\mu)_{a\dot{a}} \, \partial_\mu \chi^{\dot{a}} + \bar{\varphi}_{\dot{a}} (\tilde{\sigma}^\mu)^{\dot{a}a} \partial_\mu \varphi_a. \end{aligned} \tag{4.52}$$

It was also shown in section 2.5, that both these expressions are Lorentz scalars and each of them is real, up to a total divergence. Thus, the Lagrangian (4.49) is a real Lorentz scalar.

To construct an interaction between spinors, one can follow the same strategy as in scalar field theory, adding potential to the free Lagrangian. According to this idea,

$$\mathcal{L} = \bar{\psi}(i\gamma^\mu \partial_\mu \, \psi - m\psi) - V(\bar{\psi}, \psi), \tag{4.53}$$

where the potential $V(\bar{\psi}, \psi)$ is a real Lorentz scalar function. The simplest choice is

$$V(\bar{\psi}, \psi) = V(\bar{\psi}\psi) = \lambda(\bar{\psi}\psi)^2. \tag{4.54}$$

Other possible interaction terms that turn out to be useful in particle physics are

$$\begin{aligned} V_1(\bar{\psi}, \psi) &= \lambda_1 (\bar{\psi}\gamma^5 \psi)^2, \\ V_2(\bar{\psi}, \psi) &= \lambda_2 (\bar{\psi}\gamma^\mu \psi)(\bar{\psi}\gamma_\mu \psi), \\ V_3(\bar{\psi}, \psi) &= \lambda_3 (\bar{\psi}\gamma^5 \gamma^\mu \psi)(\bar{\psi}\gamma^5 \gamma_\mu \psi). \end{aligned} \tag{4.55}$$

Let us discuss the dimensions of the field and parameters in the Lagrangian (4.53). As usual, the Lagrangian dimension is $[\mathcal{L}] = 4$. Therefore, we have, for the dimension of the Dirac spinor $[\psi]$, the equations $4 = 2[\psi] + [\partial_\mu] = 2[\psi] + [m]$. Since $[\partial_\mu] = 1$, one gets $[\psi] = \frac{3}{2}$, $[m] = 1$. The dimension of the coupling constant in the potential (4.54) is defined from the relation $4 = [\lambda] + 4[\psi]$. This gives $[\lambda] = -2$. Thus, the dimension of the coupling constant in the potential (4.54) is of the inverse mass-squared. It is easy to check that the dimensions of other terms (4.55) in the self-interacting potential of the spinors are the same, $[\lambda_{1,2,3}] = -2$.

The combinations of the potentials (4.54) and (4.55) are typical in the effective models of weak and strong interactions, such as the Fermi model, or phenomenological theories of quark interactions. These models cannot be regarded as fundamental exactly because the coupling constants have inverse-mass dimensions. As we shall learn in the next chapters, this leads to problems at the quantum level.

If we restrict the consideration by the interactions involving only couplings with zero and positive mass dimensions, it is necessary to couple spinors to scalars or vectors. In the theory with spinor and scalar fields, following the requirement formulated above, we arrive at the Lagrangian

$$\mathcal{L} = \frac{1}{2}\eta^{\mu\nu}\partial_\mu\varphi\,\partial_\nu\varphi - \frac{1}{2}m_1^2\varphi^2 - \frac{f}{4!}\varphi^4 + i\bar{\psi}\gamma^\mu\partial_\mu\psi - m_2\bar{\psi}\psi - h\bar{\psi}\psi\varphi. \tag{4.56}$$

Here m_1 and m_2 are the masses of scalar and spinor fields, respectively. The Lagrangian (4.56), the potential

$$V(\bar{\psi}, \psi, \varphi) = h\bar{\psi}\psi\varphi \tag{4.57}$$

and the dimensionless constant h are called the Yukawa model, interaction and coupling constant, correspondingly.

4.4 Models of free vector fields

In this section, we construct the Lagrangians for free massive and massless vector fields. The interaction between vectors and fermions (spinors) and scalars is one of the most important subjects for particle physics applications. We shall deal with these interactions in the following subsections.

4.4.1 Massive vector field

According to relations (2.87), the irreducible massive representation of the spin $s = 1$ is described by the vector field φ_μ, under the conditions

$$(\Box + m^2)\varphi_\mu = 0, \tag{4.58}$$

$$\partial_\mu\varphi^\mu = 0. \tag{4.59}$$

Our first purpose will be to find the Lagrangian reproducing these equations. It is easy to see that such a Lagrangian should be special enough. The Lagrange equations for a vector field,

$$\frac{\partial\mathcal{L}}{\partial\varphi_\mu} - \partial_\nu\frac{\partial\mathcal{L}}{\partial\partial_\nu\varphi_\mu} = 0, \tag{4.60}$$

consist of the four equations. But the number of Eq. (4.58) is five.

In order to construct such a Lagrangian, we introduce the antisymmetric second rank tensor field $F_{\mu\nu}$,

$$F_{\mu\nu} = \partial_\mu \varphi_\nu - \partial_\nu \varphi_\mu \tag{4.61}$$

and consider the equation

$$\partial_\mu F^{\mu\nu} + m^2 \varphi^\nu = 0. \tag{4.62}$$

Substituting Eq. (4.61) into Eq. (4.62), one gets

$$\Box \varphi^\nu - \partial^\nu \partial_\mu \varphi^\mu + m^2 \varphi^\nu = 0. \tag{4.63}$$

On the other hand, acting by ∂_ν on Eq. (4.62), we obtain

$$\partial_\nu \partial_\mu F^{\mu\nu} + m^2 \partial_\nu \varphi^\nu = 0.$$

Since $\partial_\nu \partial_\mu F^{\mu\nu} \equiv 0$, we get Eq. (4.59) as a consequence of Eq. (4.62). Next, we use Eq. (4.59) in Eq. (4.63), to arrive at Eq. (4.58). All in all, the single Eq. (4.62) leads to both Eqs. (4.58) and (4.59). Thus, Eq. (4.62), called the Proca equation, is appropriate to describe the dynamics of a free massive vector field.

Let us show that the Lagrangian leading to Eq. (4.62) has the form

$$\mathcal{L}_P = -\frac{1}{4} F_{\mu\nu} F^{\mu\nu} + \frac{1}{2} m^2 \varphi_\mu \varphi^\mu. \tag{4.64}$$

For this, we calculate the variation of the action, corresponding to Lagrangian (4.64),

$$\begin{aligned} \delta S|\varphi| &= \int_\Omega d^4x \left\{ -\frac{1}{2} F^{\mu\nu} \delta F_{\mu\nu} + m^2 \varphi^\mu \delta \varphi_\mu \right\} = \int_\Omega d^4x \left\{ F^{\mu\nu} \partial_\nu \delta \varphi_\mu + m^2 \varphi^\mu \delta \varphi_\mu \right\} \\ &= \int_\Omega d^4x \left\{ -\partial_\nu F^{\mu\nu} + m^2 \varphi^\mu \right\} \delta \varphi_\mu, \end{aligned}$$

where the total divergence was omitted due to standard boundary condition for the variation. The result is

$$\frac{\delta S[\varphi]}{\delta \varphi_\mu} = \partial_\nu F^{\nu\mu} + m^2 \varphi^\mu, \tag{4.65}$$

and the equations of motion are exactly Eq. (4.62). Thus, Eq. (4.64) is a Lagrangian of a free massive vector field: the Proca model.

The dimension of the vector field is $[\varphi_\mu] = 1$. It can be found in the same way as the one for the scalar field. We leave the details as an exercise for the reader.

4.4.2 Massless vector fields

A massless vector field is usually denoted as A_μ and is defined by the conditions

$$\Box A_\mu = 0, \tag{4.66}$$

$$\partial_\mu A^\mu = 0. \tag{4.67}$$

This field is mostly used to describe electromagnetic interactions.

Let us show that the conditions (4.66) and (4.67) can be obtained from Eq. (4.62) at $m = 0$. Such an equation has the form

$$\partial_\mu F^{\mu\nu} = 0, \tag{4.68}$$

where

$$F_{\mu\nu} = \partial_\mu A_\nu - \partial_\nu A_\mu. \tag{4.69}$$

Substituting Eq. (4.69) into Eq. (4.68), we obtain

$$\Box A^\nu - \partial^\nu(\partial_\mu A^\mu) = 0. \tag{4.70}$$

Eq. (4.70) does not coincide with (4.66), since one needs also (4.67). However, let us pay attention to the fact that the field A_μ is not defined uniquely. Let A_μ be a field satisfying Eq. (4.68) and A'_μ results from the *gauge transformation*,

$$A'_\mu(x) = A_\mu(x) + \partial_\mu \xi(x), \tag{4.71}$$

where $\xi(x)$ is an arbitrary scalar field. It is evident that

$$F'_{\mu\nu} \;=\; \partial_\mu A'_\nu - \partial_\nu A'_\mu \;=\; \partial_\mu A_\nu - \partial_\nu A_\mu + \partial_\mu\partial_\nu\xi - \partial_\mu\partial_\nu\xi \;=\; F_{\mu\nu}, \tag{4.72}$$

which means the tensor $F_{\mu\nu}$ is invariant under the transformation (4.71). Hence, Eq. (4.68) has the same form for any fields A'_μ and A_μ which are related by Eq. (4.71). As a consequence, we can construct a special field $A'_\mu(x)$ by using a special choice of $\xi(x)$, while Eq. (4.68) would be the same for any $\xi(x)$. Let A_μ be such a special field, satisfying the condition (4.67). Then, the second term in Eq. (4.70) vanishes, and this equation takes the form of the condition (4.66).

It remains to demonstrate that the arbitrariness in the choice of ξ in (4.71) is sufficient to impose the condition (4.67) on the field A_μ for an arbitrary field A'_μ . Suppose that $\partial_\mu A'^\mu = f(x) \neq 0$ and consider

$$\partial_\mu A'^\mu = \partial_\mu A^\mu + \partial_\mu \partial^\mu \xi \qquad \Longrightarrow \qquad \partial_\mu A^\mu = -\Box\xi(x) + f(x). \tag{4.73}$$

Let $\xi(x)$ satisfy the equation $\Box\xi(x) = f(x)$. According to the well-known mathematical theorem, this equation has a solution. Then, from (4.73) follows $\partial_\mu A^\mu = 0$, meaning that the desired condition (4.67) can be imposed. If the initial field A_μ does not satisfy the condition (4.67), one can perform the transformation (4.71), with ξ satisfying the equation $\Box\xi = \partial_\mu A^\mu$, and obtain the new field A_μ satisfying (4.67).

The scalar field $\xi(x)$ is called the gauge parameter, and the condition (4.67) is called the Lorentz gauge-fixing condition, or Lorentz gauge. The tensor $F_{\mu\nu}$ (4.69) is called the electromagnetic strength tensor. The property $F'_{\mu\nu} = F_{\mu\nu}$ means that the strength tensor is gauge invariant. The equations (4.68) are part of the Maxwell equations in empty space without electric charges and currents.

The Lagrangian for a free massless vector field can be obtained from the Lagrangian (4.64) at $m = 0$ and has the form

$$\mathcal{L} = -\frac{1}{4} F_{\mu\nu} F^{\mu\nu}. \tag{4.74}$$

The equations of motion corresponding to this Lagrangian are (4.68). The Lagrangian (4.74) is gauge invariant, since it depends only on the invariant strength tensor $F_{\mu\nu}$.

The dimension of the massless vector field is the same as that of the massive one, $[A_\mu] = 1$.

4.4.3 The gauge-invariant form of the Proca Lagrangian and Stüeckelberg fields

The equations of motion for a massive vector field (4.63) consists of two terms. The first one is based on the field strength $F_{\mu\nu}$ and hence is automatically gauge invariant. It is easy to see that the second term is not gauge-invariant. However, there is a special procedure to construct a gauge invariant description of a massive vector field.

Instead of using the Lagrangian (4.64), let us consider another Lagrangian,

$$\mathcal{L} = -\frac{1}{4}F_{\mu\nu}F^{\mu\nu} + \frac{1}{2}m^2(\varphi_\mu - \partial_\mu\varphi)(\varphi^\mu - \partial^\mu\varphi), \tag{4.75}$$

which includes, along with the vector field φ_μ, an additional scalar field φ and describes a dynamics of the two fields, φ_μ ad φ. On the other hand, this Lagrangian is invariant under the gauge transformations

$$\varphi'_\mu = \varphi_\mu + \partial_\mu\xi, \qquad \varphi' = \varphi + \xi. \tag{4.76}$$

It is worth noticing that the gauge transformation for φ does not contain derivatives. Taking into account the gauge freedom in the choice of the parameter $\xi(x)$, one can impose the gauge-fixing condition. The simplest version of such a condition is $\varphi = 0$. After that, the gauge parameter is fixed, gauge symmetry breaks down and we arrive at the massive vector field theory with the usual Lagrangian (4.64). In this sense, theories with the Lagrangians (4.64) and (4.75) are equivalent. At the same time, one can use other gauge-fixing conditions, for example, $\partial_\mu\varphi^\mu = 0$.

The scalar field φ with the gauge transformation without derivatives is called the Stüeckelberg field. The procedure of obtaining the gauge-invariant Lagrangian for non-gauge-invariant massive theory is called the Stüeckelberg mechanism.

It is useful to see how the the Stüeckelberg mechanism works in the equations of motion. The ones corresponding to the Lagrangian (4.75) have the form

$$\begin{aligned} &\partial_\mu F^{\mu\nu} + m^2(\varphi^\nu - \partial^\nu\varphi) = 0, \\ &\Box\varphi - \partial_\mu\varphi^\mu = 0. \end{aligned} \tag{4.77}$$

The equations (4.77) are invariant under the gauge transformations (4.76), and hence we can impose the gauge-fixing conditions. In the gauge $\varphi = 0$, one gets the equations

$$\partial_\mu F^{\mu\nu} + m^2\varphi^\nu = 0, \qquad \partial_\mu\varphi^\mu = 0, \tag{4.78}$$

where the second equation is a consequence of the first one. As a result, we arrive at the known equations of motion for the free massive vector field (4.62).

4.5 Scalar and spinor filelds interacting with an electromagnetic field

Let us consider a model of a complex scalar field with the Lagrangian

$$\mathcal{L} = \eta^{\mu\nu}\partial_\mu\varphi^*\partial_\nu\varphi - m^2\varphi^*\varphi - V(\varphi^*\varphi). \tag{4.79}$$

This Lagrangian is invariant under the global transformations (4.18). The infinitesimal form of these transformations is

$$\delta\varphi = ie\xi\varphi, \qquad \delta\varphi^* = -ie\xi\varphi^*. \tag{4.80}$$

According to Noether's theorem, the invariance under the transformations (4.80) leads to the current

$$j^\mu = -\Big(\frac{\partial\mathcal{L}}{\partial\partial_\mu\varphi}\,ie\varphi - \frac{\partial\mathcal{L}}{\partial\partial_\mu\varphi^*}\,ie\varphi^*\Big) = ie(\varphi^*\partial^\mu\varphi - \varphi\partial^\mu\varphi^*). \tag{4.81}$$

Furthermore, the local conservation law $\partial_\mu j^\mu = 0$ defines the conserved charge

$$Q = \int d^3x\, j^0 = ie\int d^3x\,\varphi^*\overleftrightarrow{\partial_0}\varphi, \tag{4.82}$$

where $\varphi^*\overleftrightarrow{\partial_0}\varphi = \varphi^*\partial_0\varphi - \partial_0\varphi^*\varphi$. The expression (4.82) is called the electric charge of the complex scalar field, and the constant e is the electric charge of the particles corresponding to this field.

Now let us consider the spinor field Lagrangian

$$\mathcal{L}_0 = i\bar{\psi}\gamma^\mu\partial_\mu\psi - m\bar{\psi}\psi. \tag{4.83}$$

This Lagrangian is invariant under the global transformations

$$\psi' = e^{ie\,\xi}\psi, \qquad \bar{\psi}' = e^{-ie\,\xi}\bar{\psi} \tag{4.84}$$

or, in the infinitesimal form,

$$\delta\psi' = ie\xi\psi, \qquad \delta\bar{\psi}' = -ie\xi\bar{\psi}. \tag{4.85}$$

Using Noether's theorem, one gets the current

$$j^\mu = -\frac{\partial\mathcal{L}}{\partial\partial_\mu\psi}\,ie\psi = e\bar{\psi}\gamma^\mu\psi, \tag{4.86}$$

satisfying the local conservation law $\partial_\mu j^\mu = 0$. The conserved charge stipulated by the current (4.86) is

$$Q = \int d^3x\, j^0 = e\int d^3x\bar{\psi}\gamma^0\psi = e\int d^3x\psi^\dagger\psi. \tag{4.87}$$

The quantity Q in (4.87) is called the electric charge of the spinor field. The constant e is the electric charge of particles corresponding to this field.

Our next purpose is to construct the interacting Lagrangians of scalars and spinor fields coupled to the electromagnetic field. As we pointed out above, the electromagnetic field is described by a massless vector field $A_\mu(x)$, which is defined up to the gauge transformation $A'_\mu = A_\mu + \partial_\mu \xi(x)$. We know that the Lagrangian of a free electromagnetic field is invariant under this transformation, with the parameter $\xi = \xi(x)$ being a function of coordinates. It is natural to require that the full Lagrangian is also invariant under the gauge transformation with an arbitrary scalar field $\xi(x)$.

To achieve this, we introduce a new kind of derivative,

$$\mathcal{D}_\mu \varphi = \partial_\mu \varphi - ieA_\mu \varphi, \qquad \text{such that} \qquad (\mathcal{D}_\mu \varphi)^* = \partial_\mu \varphi^* + ieA_\mu \varphi^*, \tag{4.88}$$

and demonstrate that they are covariant with respect to the local gauge transformation. The gauge transformations of a complex scalar field are

$$\varphi' = e^{ie\,\xi(x)} \varphi \qquad \Longrightarrow \qquad \varphi'^* = e^{-ie\,\xi(x)} \varphi^* . \tag{4.89}$$

Then,

$$\begin{aligned}
\mathcal{D}'_\mu \varphi' &= \partial_\mu \varphi' - ieA'_\mu \varphi' = \partial_\mu (e^{ie\,\xi(x)} \varphi) - ieA_\mu e^{ie\,\xi(x)} \varphi - ie\,\partial_\mu \xi(x) e^{ie\,\xi(x)} \varphi \\
&= e^{ie\,\xi(x)} \big[\partial_\mu \varphi - ieA_\mu \varphi + ie\,\partial_\mu \xi(x) \varphi - ie \partial_\mu \xi(x) \varphi \big] \\
&= e^{ie\xi(x)} (\partial_\mu - ieA_\mu) \varphi = e^{ie\xi(x)} \mathcal{D}_\mu \varphi,
\end{aligned} \tag{4.90}$$

and, similarly,

$$(\mathcal{D}'_\mu \varphi')^* = e^{-ie\,\xi(x)} (\mathcal{D}_\mu \varphi)^* . \tag{4.91}$$

According to these relations, (4.88) defines the covariant derivatives. It is evident that $(\mathcal{D}_\mu \varphi)^* (\mathcal{D}^\mu \varphi) = (\mathcal{D}'_\mu \varphi')^* (\mathcal{D}'^\mu \varphi')$. Thus, replacing the partial derivatives by the covariant derivatives $\mathcal{D}_\mu \varphi$ and $(\mathcal{D}_\mu \varphi)^*$ in the Lagrangian of the free complex scalar field (4.79), we obtain the gauge-invariant Lagrangian

$$\mathcal{L} = \eta^{\mu\nu} (\mathcal{D}_\mu \varphi)^* (\mathcal{D}_\nu \varphi) - m\varphi^* \varphi - V(\varphi^* \varphi). \tag{4.92}$$

Let us now consider the spinor field theory. The coupling between the spinor ψ and the vector field A_μ, without introducing parameters with inverse-mass dimensions, can be defined as

$$\mathcal{L}_{int} = j^\mu A_\mu = e\bar{\psi} \gamma^\mu \psi A_\mu . \tag{4.93}$$

Adding this interaction Lagrangian to the free spinor field expression gives

$$\mathcal{L} = i\bar{\psi} \gamma^\mu (\partial_\mu - ieA_\mu) \psi - m\bar{\psi} \psi . \tag{4.94}$$

It is easy to show that the last Lagrangian is gauge invariant. Using the gauge transformation for the vector field, $A_\mu = A'_\mu - \partial_\mu \xi$, one gets

$$\mathcal{L} = i\bar{\psi} \gamma^\mu (\partial_\mu - ieA'_\mu + ie\,\partial_\mu \xi) \psi - m\bar{\psi} \psi . \tag{4.95}$$

Introducing the gauge transformations of the spinor field,

$$\psi' = e^{ie\,\xi(x)} \psi, \qquad \bar{\psi}' = e^{-ie\,\xi(x)} \bar{\psi}, \tag{4.96}$$

These transformations look similar to (4.84), with the constant parameter ξ replaced by a scalar field $\xi(x)$. Substituting the transformations (4.96) into (4.95), one gets

$$\begin{aligned}\mathcal{L} &= \bar{\psi}' e^{ie\xi(x)} i\gamma^\mu \big[\partial_\mu (e^{-ie\xi(x)}\psi') - ieA'_\mu e^{-ie\xi(x)}\psi' + ie\,\partial_\mu \xi e^{-ie\xi(x)}\psi'\big] \\ &- m\bar{\psi}' e^{ie\xi(x)} e^{-ie\xi(x)}\psi' = i\bar{\psi}'\gamma^\mu(\partial_\mu - ieA'_\mu)\psi' - m\bar{\psi}'\psi' = \mathcal{L}',\end{aligned} \tag{4.97}$$

where $\mathcal{L}'$ is the same Lagrangian (4.94) but expressed in terms of $A'_\mu, \psi', \bar{\psi}'$. As a result, the Lagrangian (4.94) is invariant under the transformations $A'_\mu = A_\mu + \partial_\mu \xi$ completed by the transformations (4.96). These transformations are called the gauge transformations of the spinor field. Thus, we proved that the Lagrangian (4.94) is gauge invariant.

Since the derivative ∂_μ enters the Lagrangian (4.94) only in the combination

$$\mathcal{D}_\mu = \partial_\mu - ieA_\mu, \tag{4.98}$$

let us consider the transformation rule of this derivative applied to a spinor,

$$\begin{aligned}\mathcal{D}'_\mu \psi' &= (\partial_\mu - ieA'_\mu)\psi' = (\partial_\mu - ieA_\mu - ie\,\partial_\mu \xi)\, e^{ie\xi(x)}\psi \\ &= e^{ie\xi(x)}(\partial_\mu - ieA_\mu)\psi = e^{ie\xi(x)}\mathcal{D}_\mu \psi.\end{aligned} \tag{4.99}$$

It is evident that the expression $\mathcal{D}_\mu \psi$ transforms exactly like ψ. This property allows us to call the expression $\mathcal{D}_\mu$ (4.98) the covariant derivative.

We can conclude that the gauge-invariant Lagrangian (4.95) is obtained from the free spinor field Lagrangian (4.83) by means of replacing the partial derivative ∂_μ with the covariant derivative $\mathcal{D}_\mu$. Such a procedure can be called "minimal" for including the interaction with the vector field.

The Lagrangians (4.94) and (4.92) include the vector field A_μ but do not contain its derivatives. Therefore, these Lagrangians cannot describe the dynamics of the vector field. To provide such dynamics, one has to introduce the Lagrangian of the free gauge field, which we already know, namely, (4.74). After this, the Lagrangians describing the scalar and spinor fields coupled with the electromagnetic field have the form

$$\mathcal{L} = -\frac{1}{4}F_{\mu\nu}F^{\mu\nu} + \eta^{\mu\nu}(\mathcal{D}_\mu \varphi)^* \mathcal{D}_\nu \varphi - m\varphi^*\varphi - V(\varphi^*\varphi) \tag{4.100}$$

and

$$\mathcal{L} = -\frac{1}{4}F_{\mu\nu}F^{\mu\nu} + i\bar{\psi}\gamma^\mu \mathcal{D}_\nu \psi - m\bar{\psi}\psi. \tag{4.101}$$

The field model with the Lagrangian (4.101) describes spinor electrodynamics, which is basis for quantum electrodynamics (QED). The field model with the Lagrangian (4.100) defines scalar electrodynamics (SQED).

To conclude, we note that the set of transformation $z' = e^{i\alpha}z$, with the real parameter α, forms an Abelian group $U(1)$. Therefore, our considerations have shown that the Lagrangians (4.100) and (4.101) are invariant under the local group $U(1)$.

4.6 The Yang-Mills field

The construction of the Lagrangians describing the interactions of scalar and spinor fields with the gauge vector taught us a useful lesson. A field model that is invariant under the global transformations from the $U(1)$ group can be modified in such a way that its new version is invariant under the same transformations but with local parameters of the transformation. Such a modification involves interaction with a vector field. This vector field enters the matter (scalar or spinor) Lagrangians through the covariant derivative $\mathcal{D}_\mu = \partial_\mu - ie\,A_\mu$. The commutator of covariant derivatives, $[\mathcal{D}_\mu, \mathcal{D}_\nu] = -ieF_{\mu\nu}$, defines a strength tensor $F_{\mu\nu}$ of the vector field. In this section, we extend this construction to the generic case, when the original global symmetry transformations form a non-Abelian Lie group.

To establish the terminology, let us agree that the transformations of fields with spacetime-dependent parameters are called gauge transformations. If the action of a field model is invariant under gauge transformations, such a model is called a gauge theory or a gauge model.

Consider a field model with the action invariant under a set of global field transformations, forming a Lie group. In this case, there exists a general method to include interaction with a vector field, which is called a gauge principle. This approach works as follows: a field model invariant under a Lie group of global transformations can be reformulated by introducing a vector field in such a way that it becomes invariant under the same transformation group but with local parameters.

The interaction of scalars or spinors with a vector field is introduced by replacing the partial derivatives in a Lagrangian by the covariant derivatives, similar to what we discussed for the Abelian case in the previous section. In general, the vector field that is introduced on the basis of the gauge principle is called a Yang-Mills, or non-Abelian gauge field. Let us note that the Yang-Mills field is a necessary element in the description of fundamental forces, including the standard model of the electroweak and strong interactions. The main difference between the Abelian models and the Yang-Mills theory is the possibility of introducing self-interaction between the gauge vector fields. In what follows, we describe how this interaction can be consistently introduced.

Consider the Lagrangian $\mathcal{L}(\phi^i{}_A, \partial_\mu\phi^i{}_A)$, depending on a set of fields $\phi^i{}_A$, where A is a set of all possible Lorentz indices. In what follows, we omit the indices A for brevity and indicate only i, which is an internal symmetry index. We assume that the fields ϕ^i transform under a representation of some compact Lie group with global parameters ξ^a, where $a = 1, 2, \dots, N$. The Lagrangian $\mathcal{L}(\phi^i, \partial_\mu\phi^i)$ is assumed to be invariant under this transformation group, $\mathcal{L}(\phi, \partial_\mu\phi) = \mathcal{L}(\phi', \partial_\mu\phi')$. Here

$$\phi'^i(x) = h^i_j\,(\xi)\phi^j(x), \qquad \text{where} \qquad \xi = (\xi^1, \xi^2, ..., \xi^N) \tag{4.102}$$

and all ξ^a are coordinate independent. In this case, the transformation group is called global. In Eq. (4.102), $h^i_j\,(\xi)$ are matrices of representation of a Lie group. Then, we can consider $h(\xi) = e^{ig\xi^a T^a}$, where the matrices $(T^a)^i_j$ are representations of the generators of the corresponding Lie algebra, satisfying the relation

$$[T^a, T^b] = if^{ab}{}_c T_c, \tag{4.103}$$

where $f^{ab}{}_c$ are the structure constants. We will call T^a generators. Finally, the parameter g is a gauge-coupling constant.

Now we consider the local version of the field transformations,

$$\phi'^i(x) = h^i_j(\xi(x))\,\phi^j(x), \tag{4.104}$$

where $\xi(x) = \xi^a(x) = \xi^1(x), \xi^2(x), \dots, \xi^N(x)$. However, unlike the case for Eq. (4.102), in this case replaced the parameters ξ^a with the scalar fields $\xi^a(x)$. For constant parameters ξ^a, there is a symmetry

$$\mathcal{L}(\phi', \partial_\mu\phi') \;=\; \mathcal{L}(h\phi, \partial_\mu h\phi) \;=\; \mathcal{L}(h\phi, h\partial_\mu\phi) \;=\; \mathcal{L}(\phi, \partial_\mu\phi). \tag{4.105}$$

However, in the case of $\xi^a = \xi^a(x)$, this identity does not take place, because

$$\partial_\mu\phi' \;=\; \partial_\mu(h\phi) \;=\; h\big(\partial_\mu\phi + h^{-1}(\partial_\mu h)\phi\big), \tag{4.106}$$

where h^{-1} is a group element inverse to h. As result, $\mathcal{L}(\phi', \partial_\mu\phi') \neq \mathcal{L}(\phi, \partial_\mu\phi)$; hence, the Lagrangian is not invariant under the transformations (4.104). The gauge invariance of the Lagrangian requires $\partial_\mu(h\phi) = h\partial_\mu\phi$, which is possible only for $h = const$.

Let us introduce an expression similar to (4.98),

$$\mathcal{D}_\mu = \partial_\mu - igA_\mu. \tag{4.107}$$

By definition, the vector field A_μ takes the values in the representation of the Lie algebra, corresponding to the transformations (4.104). Let T^a be the generators of the representation of the Lie algebra. In this case,

$$A_\mu(x) = A_\mu{}^a(x)\,T^a, \tag{4.108}$$

where $A^a_\mu(x)$ is a set of ordinary vector fields, with $a = 1, 2, \dots, N$. Thus, the number of vector fields coincides with the number of gauge parameters. Both $A_\mu(x)$ and $A^a_\mu(x)$ are called Yang-Mills fields.

According to our previous experience, the $\mathcal{D}_\mu$ (4.107) is a covariant derivative, if

$$\mathcal{D}'_\mu\phi' = h\mathcal{D}_\mu\phi, \tag{4.109}$$

resolving the problem of Eq. (4.106). The condition (4.109) defines the transformation law for the vector field. Namely, we need to have $\partial_\mu\phi' - igA'_\mu\phi' = h(\partial_\mu\phi - igA_\mu\phi)$, or

$$h\,\partial_\mu\phi + \partial_\mu h\,\phi - igA'_\mu h\phi = h\,\partial_\mu\phi - ighA_\mu\phi.$$

Since ϕ is an arbitrary field, we require $\partial_\mu h - ig\,A'_\mu h = -igh\,A_\mu$; hence,

$$A'_\mu \;=\; h\,A_\mu h^{-1} - \frac{i}{g}(\partial_\mu h)^{-1} \;=\; h\,A_\mu h^{-1} + \frac{i}{g}h\,\partial_\mu h^{-1}, \tag{4.110}$$

which defines the gauge-transformation rule of the Yang-Mills field.

It is not difficult to see that transformations (4.110) form a group. Assume that

$$A'_\mu = h_1 A_\mu h_1^{-1} + \frac{i}{g} h_1 \partial_\mu h_1^{-1}, \quad \text{and} \quad A''_\mu = h_2 A'_\mu h_2^{-1} + \frac{i}{g} h_2 \partial_\mu h_2^{-1},$$

where h_1 and h_2 are two group elements. Then,

$$\begin{aligned} A''_\mu &= h_2 \Big(h_1 A_\mu h_1^{-1} + \frac{i}{g} h_1 \partial_\mu h_1^{-1} \Big) h_2^{-1} + \frac{i}{g} h_2 \partial_\mu h_2^{-1} = h_2 h_1 A_\mu (h_2 h_1)^{-1} \\ &+ \frac{i}{g} h_2 h_1 (\partial_\mu h_1^{-1}) h_2^{-1} + \frac{i}{g} h_2 \partial_\mu h_2^{-1} = h_2 h_1 A_\mu (h_2 h_1)^{-1} + \frac{i}{g} h_2 h_1 \partial_\mu (h_2 h_1)^{-1}, \end{aligned} \tag{4.111}$$

meaning two consecutive gauge transformations are equivalent to the third one.

The next problem is to construct a Lagrangian that is invariant under the gauge transformations. To do this, one replaces, in the initial Lagrangian $\mathcal{L}(\phi, \partial_\mu \phi)$, the ordinary partial derivative $\partial_\mu \phi$ with the covariant derivative $\mathcal{D}_\mu \phi$. As a result, one gets a new Lagrangian $\mathcal{L}(\phi, \mathcal{D}_\mu \phi)$, which satisfies the identity

$$\mathcal{L}(h\phi, \mathcal{D}_\mu h\phi) = \mathcal{L}(h\phi, h\mathcal{D}_\mu \phi) = \mathcal{L}(\phi, \mathcal{D}_\mu \phi). \tag{4.112}$$

Thus, we arrived at the Lagrangian invariant under the joint transformations of (4.104) and (4.110), with the vector field A_μ entering the Lagrangian through the covariant derivative $\mathcal{D}_\mu$.

Let us find the infinitesimal form of the gauge transformation of the Yang-Mills field. For the infinitesimal parameters ξ^a, the relation $\phi'^i = (e^{ig\xi^a T^a})^i_j \phi^j$ takes the form $\delta\phi^i = ig(T^a)^i_j \phi^j \xi^a$; hence, $h = 1 + igT^a\xi^a$ and $h^{-1} = 1 - igT^a\xi^a$. Therefore,

$$A'_\mu{}^a T^a = (1 + igT^b\xi^b) A_\mu{}^d T^d (1 - igT^c\xi^c) + \frac{i}{g}(1 + igT^b\xi^b)(-ig) T^a \partial_\mu \xi^a, \tag{4.113}$$

or

$$A'_\mu{}^a T^a = A_\mu{}^a T^a + igA_\mu{}^c [T^b, T^c] \xi^b + T^a \partial_\mu \xi^a. \tag{4.114}$$

Let us remember that the generators T^b and T^c satisfy the commutation relation (4.103). For the sake of simplicity, in what follows we consider only a special class of Lie groups, the semisimple Lie group. In this case, the structure constants are totally antisymmetric in all indices, $f^{bc}{}_a = f^{bca}$. Then, Eq. (4.113) gives

$$A'_\mu{}^a = A_\mu{}^a + \partial_\mu \xi^a + gf^{acb} A^c{}_\mu \xi^b. \tag{4.115}$$

Using the covariant derivative

$$\mathcal{D}_\mu{}^{ab} = \delta^{ab} \partial_\mu + gf^{acb} A_\mu{}^c, \tag{4.116}$$

one gets the infinitesimal gauge transformations of the Yang-Mills field in the useful covariant form,

$$\delta A_\mu{}^a = \mathcal{D}_\mu{}^{ab} \xi^b. \tag{4.117}$$

The Lagrangian $\mathcal{L}(\phi, \mathcal{D}_\mu\phi)$ describes an interaction of the fields ϕ and A_μ, but not the dynamics of the field A_μ. To describe such dynamics, we have to introduce an addition Lagrangian, including the derivatives of A_μ. This can be done analogously to the case for Abelian theory considered in section 4.5. Let us calculate the commutator of the covariant derivatives,

$$[\mathcal{D}_\mu, \mathcal{D}_\nu] = [\partial_\mu - igA_\mu, \partial_\nu - igA_\nu] \;=\; -igG_{\mu\nu}, \tag{4.118}$$

where we introduced the antisymmetric second-rank tensor $G_{\mu\nu}$, called the Yang-Mills strength tensor,

$$G_{\mu\nu} = \partial_\mu A_\nu - \partial_\nu A_\mu - ig[A_\mu, A_\nu]. \tag{4.119}$$

Consider the transformation law for $G_{\mu\nu}$ under the gauge transformations. As we know, $\mathcal{D}'_\mu\phi' = \mathcal{D}'_\mu h\phi = h\mathcal{D}_\mu\phi$. Since the field ϕ is arbitrary, this implies

$$\mathcal{D}'_\mu = h\mathcal{D}_\mu h^{-1}. \tag{4.120}$$

Then,

$$\begin{aligned} G'_{\mu\nu} &= \frac{i}{g}[\mathcal{D}'_\mu, \mathcal{D}'_\nu] \;=\; \frac{i}{g}(h\,\mathcal{D}_\mu h^{-1}h\mathcal{D}_\nu h^{-1} - h\mathcal{D}_\nu h^{-1}h\mathcal{D}_\mu h^{-1}) \\ &= \frac{i}{g}h[\mathcal{D}_\mu, \mathcal{D}_\nu]h^{-1} = h\,G_{\mu\nu}h^{-1}, \end{aligned} \tag{4.121}$$

meaning the Yang-Mills strength transforms homogenously. Using the definition of $G_{\mu\nu}$ (4.119), we can rewrite the Yang-Mill strength in the form

$$\begin{aligned} G_{\mu\nu} &= \partial_\mu A_\nu{}^a T^a - \partial_\nu A_\mu{}^a T^a - ig\,A_\mu{}^b A_\nu{}^c [T^b, T^c] \\ &= (\partial_\mu A_\nu{}^a - \partial_\nu A_\mu{}^a)T^a - igA_\mu{}^b A_\nu{}^c (ig) f^{bca} T^a \;=\; G^a{}_{\mu\nu}T^a, \end{aligned} \tag{4.122}$$

where

$$G^a{}_{\mu\nu} = \partial_\mu A_\nu{}^a - \partial_\nu A_\mu{}^a + gf^{abc}A_\mu{}^b A_\nu{}^c. \tag{4.123}$$

We can see that $G_{\mu\nu}$ takes the values in the Lie algebra, just as A_μ does.

For the trace over matrix indices, we meet

$$\operatorname{tr} G'_{\mu\nu}G'^{\mu\nu} \;=\; \operatorname{tr} h\,G_{\mu\nu}h^{-1}h\,G^{\mu\nu}h^{-1} \;=\; \operatorname{tr} h\,G_{\mu\nu}G^{\mu\nu}h^{-1} \;=\; \operatorname{tr} G_{\mu\nu}G^{\mu\nu}.$$

According to this relation, the quantity $\operatorname{tr} G_{\mu\nu}G^{\mu\nu}$ is gauge invariant. On top of this, $\operatorname{tr} G_{\mu\nu}G^{\mu\nu} = G^a{}_{\mu\nu}G^{b\mu\nu} \operatorname{tr} T^aT^b$. For semisimple Lie algebras, the generators satisfy the normalization condition $\operatorname{tr} T^aT^b \sim \delta^{ab}$. This is a relevant relation, since the quantity $G^a{}_{\mu\nu}G^{a\mu\nu}$ is the Lagrangian of the pure Yang-Mills field,

$$\mathcal{L}_{YM} \;=\; -\frac{1}{4}\,G^a{}_{\mu\nu}G^{a\mu\nu} \;=\; -\frac{1}{4}\operatorname{tr} G_{\mu\nu}G^{\mu\nu}. \tag{4.124}$$

The overall gauge-invariant Lagrangian for interacting fields ϕ and A_μ has the form

$$\mathcal{L}_{total}(\phi, \partial_\mu\phi, A_\nu, \partial_\mu A_\nu) \;=\; \mathcal{L}(\phi, \mathcal{D}_\mu\phi) + \mathcal{L}_{YM}(A_\nu, \partial_\mu A_\nu). \tag{4.125}$$

In the context of gauge theories, the fields ϕ (spinor and scalar) are usually called matter, or matter fields. The Lagrangian $\mathcal{L}(\phi, \mathcal{D}_\mu\phi)$ is called the matter Lagrangian and is sometimes denoted as $\mathcal{L}_m(\phi, \mathcal{D}_\mu\phi)$. The equations of motion for the theory with the Lagrangian (4.125) can be written as follows:

$$\frac{\partial \mathcal{L}_{YM}}{\partial A_\mu{}^a} - \partial_\nu \frac{\partial \mathcal{L}_{YM}}{\partial\left(\partial_\nu A_\mu{}^a\right)} + \frac{\partial \mathcal{L}_m}{\partial A_\mu{}^a} = 0, \tag{4.126}$$

$$\frac{\partial \mathcal{L}_m}{\partial \phi} - \partial_\nu \frac{\partial \mathcal{L}_m}{\partial(\partial_\nu \phi)} = 0. \tag{4.127}$$

The reduced version of (4.126), without matter fields, is

$$\frac{\partial \mathcal{L}_{YM}}{\partial A_\mu{}^a} - \partial_\nu \frac{\partial \mathcal{L}_{YM}}{\partial\left(\partial_\nu A_\mu{}^a\right)} = 0, \tag{4.128}$$

and represents the equation of motion for the pure Yang-Mills field. Let us calculate the explicit form of this equation. We leave the direct derivation of the variation of the action as an exercise and use the Lagrange equation approach. Then, we need

$$\begin{aligned}\frac{\partial \mathcal{L}_{YM}}{\partial A_\mu{}^a} &= -\frac{1}{2} G^{b\alpha\beta} \frac{\partial G^b{}_{\alpha\beta}}{\partial A_\mu{}^a} = -\frac{1}{2} G^{b\alpha\beta}\, g f^{bcd}\left(\delta^\mu_\alpha \delta^{ac} A_\beta{}^d + A_\alpha{}^c \delta^{ad} \delta^\mu_\beta\right) \\ &= -\frac{1}{2} g\left(G^{b\mu\beta} f^{bad} A_\beta{}^d + G^{b\alpha\mu} f^{bca} A_\alpha{}^c\right) = g f^{abc} G^{b\mu\alpha} A_\alpha^{\;c}\end{aligned} \tag{4.129}$$

and

$$\partial_\nu \frac{\partial \mathcal{L}_{YM}}{\partial\left(\partial_\nu A_\mu{}^a\right)} = -\frac{1}{2} G^{b\alpha\beta} \frac{\partial G^b{}_{\alpha\beta}}{\partial\, \partial_\nu A_\mu{}^a} = -\frac{1}{2}\, G^{b\alpha\beta}\left(\delta^\alpha_\nu \delta^b_\mu \delta^{ab} - \delta^\beta_\nu \delta^\alpha_\mu \delta^{ab}\right) = G^{a\mu\nu}.$$

Substituting these expressions into Lagrange equations (4.128), we get

$$\partial_\nu G^{a\mu\nu} + g f^{acb} G^{b\mu\nu} A_\nu{}^c = \left(\delta^{ab} \partial_\nu + g f^{acb} A_\nu{}^c\right) G^{b\mu\nu} = \mathcal{D}_\mu{}^{ab} G^{b\mu\nu} = 0, \tag{4.130}$$

where we used the covariant derivative $\mathcal{D}_\mu{}^{ab}$ defined in Eq. (4.116). The equation of motion for the pure Yang-Mills field (4.130) is obviously non-linear, indicating that the Yang-Mills field is self-interacting.

The gauge invariance provides another piece of important information about the the equations of motion in the theory. Let $S[\phi, A]$ be the full action of the system, including both Yang-Mills and matter fields,

$$S[\phi, A] = S_{YM}[A] + S_m[\phi, A] = \int_\Omega d^4x\, \mathcal{L}_{YM} + \int_\Omega d^4x\, \mathcal{L}_m. \tag{4.131}$$

The gauge invariance of such an action means that

$$S[\phi + \delta\phi, A + \delta A] = S[\phi, A], \tag{4.132}$$

where $\delta\phi$ and δA_μ are gauge variations of the matter and gauge fields, respectively. In the first order in gauge parameters, the last relation leads to the equation

$$\int_\Omega d^4x \Big(\frac{\delta S[\phi, A]}{\delta \phi^i} \delta\phi^i + \frac{\delta S[\phi, A]}{\delta A_\mu{}^a} \delta A_\mu{}^a \Big) = 0. \tag{4.133}$$

If matter fields satisfy the equation of motion, $\frac{\delta S[\phi,A]}{\delta\phi^i}=0$, the first term in (4.133) vanishes. Using the transformation rule (4.117), we assume that the gauge parameters $\xi^a(x)=0$ on the boundary of the domain Ω. Then, the relation (4.133) boils down to

$$\int_\Omega dx\,\frac{\delta S[\Phi,A]}{\delta A_\mu^{\ a}}\,\mathcal{D}_\mu{}^{ab}\xi^b = 0. \tag{4.134}$$

In the explicit form, this gives, after integrating by parts,

$$\begin{aligned}
&\int d^4x\,\frac{\delta S[\Phi,A]}{\delta A_\mu{}^a}\,\big(\partial_\mu\xi^a + gf^{acb}A_\mu{}^c\xi^b\big)\\
&= \int d^4x\left(-\partial_\mu\frac{\delta S[\Phi,A]}{\delta A_\mu{}^a}\xi^a + gf^{bca}A^c{}_\mu\frac{\delta S[\Phi,A]}{\delta A_\mu{}^b}\xi^a\right)\\
&= -\int d^4x\Big(\mathcal{D}_\mu{}^{ab}\frac{\delta S[\Phi,A]}{\delta A_\mu{}^b}\Big)\xi^a(x) = 0.
\end{aligned} \tag{4.135}$$

Here we have taken into account that the surface terms are absent because the parameters $\xi^a(x)$ vanish at the boundary of Ω. Since the parameters $\xi^a(x)$ are arbitrary, we arrive at Noether's identities,

$$\mathcal{D}_\mu{}^{ab}\,\frac{\delta S[\Phi,A]}{\delta A_\mu{}^b} = 0. \tag{4.136}$$

The identities (4.136) lead to important consequences. Since $a=1,\dots,N$, the number of the Yang-Mills equations of motion $\frac{\delta S[\Phi,A]}{\delta A_\mu{}^a}=0$ is equal to $4N$. However, since these equations satisfy N identities (4.136), only $3N$ of these equations are functionally independent.

Due to the identity (4.136), the Yang-Mills equations of motion do not allow one to define uniquely all $4N$ components of the vector field $A_\mu^{\ a}(x)$, because N of these components are arbitrary. Indeed, such a situation is not surprising. Imagine that we found all $4N$ component of the field $A_\mu^{\ a}(x)$, and then perform a gauge transformation $A'^{\ a}_\mu(x) = A_\mu^{\ a}(x)+\mathcal{D}_\mu{}^{ab}\xi^b(x)$, depending on N arbitrary gauge parameters $\xi^a(x)$. Due to the gauge invariance, if the field $A_\mu^{\ a}(x)$ is a solution to the equations of motion, then $A'^{\ a}_\mu(x)$ is also a solution, for any choice of the parameters $\xi^a(x)$. Therefore, N components of the field $A_\mu^{\ a}(x)$, which can be found from the equations of motion uniquely, are still arbitrary, and we arrive at the contradiction.

To fix completely all components of the field $A_\mu^{\ a}(x)$, one has to supplement the equations of motion by N additional gauge non-invariant equations. These extra equations are called gauge conditions, gauge fixing, or simply gauges. An example of gauge fixing for the Yang-Mills field is the Lorentz gauge $\partial^\mu A_\mu^{\ a}(x)=0$.

To conclude this section, consider a particular example of Yang-Mill theory together with matter, which can be regarded as a simplified version of quantum chromodynamics, the theory describing strong interactions at the fundamental level. The model of our interest includes the set of spinor fields ψ_k with $k=1,2,\dots,n_f$, belonging to the fundamental representation of the $SU(3)$ group. These fields are associated with quarks, with n_f being the number of flavors. The matrices of the generators in the fundamental representation of the $SU(3)$ group are denoted as $(\lambda_a)^I{}_J$ and are called the Gell-Mann matrices. The theory is defined by the action

$$S[\psi, A] = \int d^4x \Big\{ -\frac{1}{4} G^a{}_{\mu\nu} G^{a\mu\nu} + \sum_{k=1}^{n_f} [i\bar{\psi}^I{}_k (\delta^I{}_J \partial_\mu - igA^a{}_\mu (\lambda_a)^I{}_J) \psi_k{}^J - m_k \bar{\psi}_k{}^I \psi_k{}^I] \Big\}. \tag{4.137}$$

Here the indices $I, J = 1, 2, 3$, while the index a takes the values $1, 2, \ldots, 8$. The gauge fields $A_\mu^a(x)$ are associated with gluons. I and J are called the color indices, and the invariance under local $SU(3)$ group is called color symmetry.

Exercises

4.1. The inner product in the space of solutions to the Klein-Gordon equation is defined by the expression

$$(\varphi_1, \varphi_2) = i \int d^3x \, \big(\varphi_1^*(x) \overleftrightarrow{\partial_0} \varphi_2(x)\big).$$

Here $\varphi_1(x)$ and $\varphi_2(x)$ are two (generally speaking, complex) solutions of the Klein-Gordon equations, and $\varphi_1^*(x) \overleftrightarrow{\partial_0} \varphi_2(x) = \varphi_1^*(x)(\partial_0 \varphi_2(x)) - (\partial_0 \varphi_1^*(x))\varphi_2(x)$. Prove that this product does not depend on time.

4.2. The Lagrangian of a real scalar field has the form (4.10). Calculate the canonical energy-momentum tensor and show that it satisfies the local conservation law.

4.3. The Lagrangian of a complex scalar field has the form (4.17). Consider the parametrization $\varphi(x) = \varrho(x)e^{i\chi(x)}$, where ϱ and χ are the two real scalar fields. Derive the equations of motion for the fields $\varrho(x)$ and $\chi(x)$.

4.4. Show that the equations of motion in the sigma model (4.19) have the form

$$\Box \varphi^i + \eta^{\mu\nu} \Gamma^i{}_{jk} \partial_\mu \varphi^j \partial_\nu \varphi^k,$$

where $\Gamma^i{}_{jk}(\varphi)$ are the Christoffel symbols constructed with the metric $g_{ij}(\varphi)$. Discuss the possibility of formulating these equations in a covariant form.

4.5. A model of a free multi-component scalar field $\varphi^i(x)$, with $i = 1, 2, \ldots, n$, is described by the Lagrangian $\mathcal{L} = \frac{1}{2}\big(\eta^{\mu\nu}\partial_\mu\varphi^i\partial_\nu\varphi^i - m^2\varphi^i\varphi^i\big)$. Prove that the Lagrangian is invariant under the infinitesimal transformations $\delta\varphi^i = \xi^{ij}\varphi^j$, where $\xi^{ij} = -\xi^{ji}$, and find the corresponding Noether's current.

4.6. The Lagrangian of a spinor field has the form (4.53). Show that the Lagrangian $\tilde{\mathcal{L}} = \frac{1}{2}\big(\bar{\psi} i\gamma^\mu\partial_\mu\psi - i\partial^\mu\bar{\psi}\gamma^\mu\psi\big) - m\bar{\psi}\psi - V(\bar{\psi}\psi)$ leads to the same equations of motion.

4.7. Find the energy-momentum tensor for the theory (4.53) and show that it satisfies the local conservation law.

4.8. The inner product in space of the solutions to Dirac equation is given by the following expression:

$$(\psi_1, \psi_2) = \int_t d^3x\, \bar{\psi}_1(x)\gamma^0\psi_2(x), \tag{4.138}$$

where ψ_1 and ψ_2 are solutions to the Dirac equation. Prove that this inner product does not depend on time.

4.9. Prove the relation $\partial_\mu j^\mu = 0$, where $j^\mu = \bar{\psi}(x)\gamma^\mu\psi(x)$ and $\psi(x)$ satisfies the Dirac equation.

4.10. a) By using only definition (4.40) and the basic relation (2.93), show that $\gamma_5^2 = I$. b) The projectors P_+ and P_- were defined in Eq. (4.44). Prove that $P_+^2 = P_+$, $P_-^2 = P_-$ and $P_+P_- = P_-P_+ = 0$. c) Let ψ be a Dirac spinor. Show that the equations $i\gamma^\mu\partial_\mu P_+\psi = 0$ and $i\gamma^\mu\partial_\mu P_-\psi = 0$ coincide with the Weyl equations.

4.11. Show that the matrices $\gamma_\mu' = (a - b\gamma_5)\gamma_\mu$ satisfy the basic relation for Dirac matrices (2.93), provided that $a^2 - b^2 = 1$.

4.12. Prove that the Lagrangian of the massless Dirac spinor field is invariant under the chiral transformation $\psi' = e^{i\xi\gamma_5}\psi$, where ξ is a constant parameter.

4.13. Using only the basic relation (Clifford algebra) for the gamma matrices, verify the following formulas, which hold in the four-dimensional spacetime:

$$\begin{aligned} &\gamma_\mu\gamma^\mu = 4, \qquad \gamma_\mu\gamma_\nu\gamma^\mu = -2\gamma_\nu, \qquad \gamma_\mu\gamma^\alpha\gamma^\beta\gamma^\mu = 4\eta^{\alpha\beta}, \\ &\gamma_\mu\gamma^\alpha\gamma^\beta\gamma^\lambda\gamma^\mu = -2\gamma^\lambda\gamma^\beta\gamma^\alpha, \qquad \gamma_\mu\gamma^\alpha\gamma^\beta\gamma^\lambda\gamma^\tau\gamma^\mu = 2\big(\gamma^\tau\gamma^\alpha\gamma^\beta\gamma^\lambda + \gamma^\lambda\gamma^\beta\gamma^\alpha\gamma^\tau\big). \end{aligned} \tag{4.139}$$

4.14. a) Using only the basic relation for the gamma matrices, prove the identities $\operatorname{tr}\gamma^\mu = 0$, $\operatorname{tr}\gamma_5 = 0$ and $\operatorname{tr}\gamma^\alpha\gamma^\beta\gamma^\gamma = 0$. After solving the next exercise, **4.15**, explain how the last identities can be proved without any calculations.
b) Prove the relations

$$\operatorname{tr}\gamma^\mu\gamma^\nu = 4\eta^{\mu\nu}, \quad \operatorname{tr}\gamma^\mu\gamma^\nu\gamma^\alpha\gamma^\beta = 4\big(\eta^{\mu\nu}\eta^{\alpha\beta} - \eta^{\mu\alpha}\eta^{\nu\beta} + \eta^{\mu\beta}\eta^{\nu\alpha}\big). \tag{4.140}$$

4.15. Consider the set of sixteen matrices Γ, consisting of

$$\begin{aligned} &\Gamma_S = I, \quad \Gamma_V^\mu = \gamma^\mu, \quad \Gamma_P = \gamma^5, \quad \Gamma_A^\mu = \gamma^5\gamma^\mu, \\ &\Gamma_T^{\mu\nu} = \sigma^{\mu\nu} = \frac{i}{2}\big(\gamma^\mu\gamma^\nu - \gamma^\nu\gamma^\mu\big). \end{aligned} \tag{4.141}$$

a) Prove that these matrices form a linearly independent set. b) Explore the properties of γ^μ and of all matrices (4.141) under Hermitian conjugation (derived from transposition and complex conjugation). c) Using the previous two points, prove that the matrices (4.141) form a complete set in the space of the matrices that can be obtained by multiplying any finite number of gamma matrices. d) In order to check how the previous part of the exercise works, as an example, construct an expansion of the product $\gamma^\alpha\gamma^\beta\gamma^\lambda$ into the basis of the matrices (4.141).
Hint. Remember that the coefficients of this expansion can be only *numerical* matrices, which means only $\eta_{\alpha\beta}$ and $\varepsilon_{\alpha\beta\lambda\tau}$ can be used as building blocks. The results of the previous exercise may be very useful in finding the coefficients.

4.16. Calculate the commutator $[\gamma^\alpha\gamma^\beta, \gamma^\mu\gamma^\nu]$, using only the basic relation for the gamma matrices.

4.17. a) The Lagrangian of a real massive vector field has the form (4.64). Find the energy-momentum tensor in this model and show that it satisfies the local conservation law.
b) The Lagrangian of a massive complex vector field has the form

$$\mathcal{L} = -\frac{1}{2}F^*_{\mu\nu}F^{\mu\nu} + m^2\varphi^*_\mu\varphi^\mu. \tag{4.142}$$

Verify the following symmetry transformation: $\varphi'_\mu = e^{i\xi}\varphi_\mu$, $\varphi'^*_\mu = e^{-i\xi}\varphi^*_\mu$, where ξ ia a constant real parameter. Find the Noether's current corresponding to this symmetry.

4.18. Derive the equations of motion for the fields A_μ, ψ, $\bar{\psi}$ in spinor electrodynamics, Eq. (4.101), and the fields A_μ, ϕ, ϕ^* in scalar electrodynamics, Eq. (4.100). Explain why in both cases the equations for the couples of conjugate matter variables ϕ and ϕ^*, or ψ and $\bar{\psi}$, give equivalent results.

4.19. Consider the massless vector field theory with self-interaction that does not violate the gauge symmetry. The Lagrangian of such a theory has the form

$$\mathcal{L} = -\frac{1}{4}F_{\mu\nu}F^{\mu\nu} + \frac{\lambda}{4!}(F_{\mu\nu}F^{\mu\nu})^2. \tag{4.143}$$

a) Find the dimension of the coupling constant λ. b) Using both Lagrange equations and making a direct variation of the action, derive the equations of motion. c) Calculate the energy-momentum tensor of this model. d) Verify the gauge invariance of the action, and obtain the corresponding Noether's current.

4.20. Let $F_{\mu\nu}$ be the Abelian vector field strength tensor and $\tilde{F}_{\mu\nu} = \frac{1}{2}\varepsilon_{\mu\nu\alpha\beta}F^{\alpha\beta}$. Prove that the expression $\tilde{F}_{\mu\nu}F^{\mu\nu}$ is a total derivative. Consider the self-interaction term of the form $(\tilde{F}_{\mu\nu}F^{\mu\nu})^2$ and discuss its relation to the interaction term from the previous exercise.
Hint. Try to solve this exercise with and without explicit calculations.

4.21. Prove the Jacobi identities for the covariant derivatives and show, on this basis, that the Yang-Mills strength tensor satisfies the Bianchi identity $\varepsilon^{\alpha\beta\gamma\delta}\mathcal{D}_\beta{}^{ab}G^b{}_{\gamma\delta} = 0$.

4.22. Prove that the expression $\varepsilon^{\alpha\beta\gamma\delta}G^a{}_{\alpha\beta}G^a{}_{\gamma\delta}$ is a total derivative.

4.23. Check that $A_\mu = -\frac{i}{g}h^{-1}\partial_\mu h$, where $h = e^{ig\xi^a(x)T^a}$, T^a are generators of the gauge group and $\xi^a(x)$ arbitrary scalar fields, is a solution to the equation $G_{\mu\nu} = 0$.

4.24. Consider the field model with a set of real scalar fields φ^I, where $I = 1, 2, \dots, n$, and spinor and vector fields, with the total Lagrangian

$$\mathcal{L} = \frac{1}{2}\partial_\mu\varphi^I\partial^\mu\varphi^I - U(\varphi) + \dots.$$

Here $U(\varphi)$ is a scalar potential and the dots mean the terms containing the spinor and vector fields. Find the Lorentz covariant conditions for the field configuration, which provides the minimum for the full energy.

4.25. Consider the real scalar field model with the Lagrangian

$$\mathcal{L} = \frac{1}{2}\partial_\alpha\varphi\,\partial^\alpha\varphi - U(\varphi).$$

Let the scalar potential have the form $U(\varphi) = -\frac{1}{2}\mu^2\varphi^2 + \frac{\lambda}{4!}\varphi^4$, where the μ and λ are the real parameters and $\lambda > 0$.
a) Find the constant value of the field v where the scalar potential $U(\varphi)$ has a minimum, and the value of the potential at this minimal point.
b) Let $\varphi(x) = v + \sigma(x)$. Rewrite the Lagrangian in terms of the field $\sigma(x)$. Show that, although the initial Lagrangian is invariant under the transformation $\varphi \to -\varphi$, the Lagrangian in terms of the field σ is not invariant under $\sigma \to -\sigma$. Such a phenomenon is called spontaneous symmetry breaking.

4.26. The irreducible massive representation of the Poincaré group with spin $s = 2$ is defined by the conditions

$$\varphi_{\mu\nu}(x) = \varphi_{\nu\mu}(x), \quad (\Box + m^2)\varphi_{\mu\nu}(x) = 0, \quad \varphi^\mu{}_\mu(x) = 0, \quad \partial^\mu\varphi_{\mu\nu}(x) = 0. \quad (4.144)$$

Prove that the most general linear equation compatible with these conditions has the form of the Pauli-Fiertz equation [134],

$$(\Box + m^2)\varphi_{\mu\nu} - \partial_\mu\partial^\alpha\varphi_{\nu\alpha} - \partial_\nu\partial^\alpha\varphi_{\mu\alpha} + \eta_{\mu\nu}\partial^\alpha\partial^\beta\varphi_{\alpha\beta} + \partial_\mu\partial_\nu\varphi - \eta_{\mu\nu}(\Box + m^2)\varphi = 0,$$

where $\varphi = \eta^{\mu\nu}\varphi_{\mu\nu}$. At $m = 0$, this equation coincides with linearized Einstein equations, which we shall discuss in Part II. Derive the Lagrangian for the field $\varphi_{\mu\nu}$.

4.27. Consider the theory of an antisymmetric tensor field called the Ogievetsky-Polubarinov model [233]. Let $B_{\alpha\beta}(x)$ be the antisymmetric second-rank tensor field, and let the strength of this field be $F_{\alpha\beta\gamma} = \partial_\alpha B_{\beta\gamma} + \partial_\beta B_{\gamma\alpha} + \partial_\gamma B_{\alpha\beta}$. Let the Lagrangian of the theory have the form

$$\mathcal{L} = -\frac{1}{12}F_{\alpha\beta\gamma}F^{\alpha\beta\gamma}.$$

a) Show the invariance of this Lagrangian under the gauge transformation $\delta B_{\alpha\beta} = \partial_\alpha\xi_\beta - \partial_\beta\xi_\alpha$, where $\xi_\alpha(x)$ is an arbitrary vector field. b) Show that, at $\xi_\alpha(x) = \partial_\alpha\xi(x)$, there is $\delta B_{\alpha\beta\gamma} = 0$, which means the gauge transformations are degenerate. c) Derive the equations of motion for the theory under consideration. d) Prove that these equations are equivalent to the equations of motion in the massless free scalar field theory.

4.28. A free massless spin $s = \frac{3}{2}$ field ψ_α is described by the vector-spinor field ψ_μ, where the Dirac spinor index is omitted. The Lagrangian of this field model has been constructed by Rarita and Schwinger [258], and has the form

$$\mathcal{L} = \varepsilon^{\mu\nu\alpha\beta}\bar{\psi}_\mu\gamma_5\gamma_\nu\partial_\alpha\psi_\beta.$$

Prove that this Lagrangian is invariant under the gauge transformations $\delta\psi_\mu = \partial_\mu\xi(x)$, where $\xi(x)$ is an arbitrary local spinor parameter (field).

Comments

The models of scalar, spinor, and vector fields are discussed in the textbooks on classical and quantum field theory, e.g., in [187], [250], [304], [274], which we already mentioned above, and also in the book by V.A. Rubakov [267]. The model of spin $s = 2$ field was constructed in a paper by M. Fierz and W. Pauli [134], the model of spin $s = \frac{3}{2}$ was introduced in a paper by W. Rarita and J. Schwinger [258], and the theory of the antisymmetric second rank tensor field was proposed by V.I. Ogievetsky and I.V. Polubarinov [233].

5

Canonical quantization of free fields

In this chapter, we briefly consider canonical quantization in field theory and show how the notion of a particle arises within the frameworks of the concept of a field.

5.1 Principles of canonical quantization

Canonical quantization is the process of constructing a quantum theory on the basis of a classical theory. In what follows, we briefly consider the main elements of this procedure, starting from its simplest version in the framework of classical mechanics.

Let us start from a Lagrange function $L(q, \dot{q})$ of a classical system with generalized coordinates $q = q_i$, where $i = 1, 2, \ldots, n$. The action is

$$S[q] = \int_{t_1}^{t_2} L(q, \dot{q})\, dt, \tag{5.1}$$

and the corresponding equations of motion have the form

$$\frac{\partial L}{\partial q_i} - \frac{d}{dt}\frac{\partial L}{\partial \dot{q}_i} = 0. \tag{5.2}$$

The procedure of canonical quantization is based on the canonical, or Hamiltonian formulation of classical theory. This formulation consists of the following three steps:

1) Introduce the generalized momenta p_i corresponding to the coordinate q_i,

$$p_i = \frac{\partial L}{\partial \dot{q}_i}. \tag{5.3}$$

2) It is assumed that the equations (5.3) make it possible to express all velocities $\dot{q}_i$ through the coordinates q_i and momenta p_i in the form $\dot{q}_i = f_i(q, p)$. It can be done under the condition $\det \frac{\partial^2 L}{\partial \dot{q}_i \partial \dot{q}_j} \neq 0$, which is assumed to be fulfilled.

3) Introduce the Hamiltonian function $H(q, p)$ according to the rule

$$H(q, p) = \Big\{ \sum_{i=1}^{n} p_i \dot{q}_i - L(q, \dot{q}) \Big\}\Big|_{\dot{q}_i = f_i(q,p)}. \tag{5.4}$$

The Hamiltonian canonical equations,

$$\dot{q}_i = \frac{\partial H}{\partial p_i}, \qquad \dot{p}_i = -\frac{\partial H}{\partial q_i}, \tag{5.5}$$

Introduction to Quantum Field Theory with Applications to Quantum Gravity. Iosif L. Buchbinder and Ilya L. Shapiro, Oxford University Press (2021). © Iosif L. Buchbinder and Ilya Shapiro.
DOI: 10.1093/oso/9780198838319.003.0005

are dynamically equivalent to Eq. (5.2), and the action becomes

$$S_H[q,p] = \int_{t_1}^{t_2} dt \left\{ \sum_{i=1}^{n} p_i \dot{q}_i - H(q,p) \right\}, \tag{5.6}$$

with the boundary conditions $\delta q_i(t)\big|_{t_1,t_2} = 0$. No conditions for $\delta p_i(t)$ are assumed.

The Hamiltonian equations (5.5) can be written in the symmetric form using the notion of the Poisson bracket. For $A(q,p)$ and $B(q,p)$ being arbitrary functions of q and p, the Poisson bracket of A and B is defined as

$$\{A,B\} = \sum_{i=1}^{n} \Big(\frac{\partial A}{\partial q_i}\frac{\partial B}{\partial p_i} - \frac{\partial A}{\partial p_i}\frac{\partial B}{\partial q_i} \Big). \tag{5.7}$$

In terms of the Poisson brackets, the equations (5.5) are written in the form

$$\dot{q}_i = \{q_i, H\}, \qquad \dot{p}_i = \{p_i, H\}.$$

The canonical quantization is based on the following postulates:

Postulate 1. *The pure quantum state of the system is given by a normalized vector* $|\Psi\rangle$ *of a Hilbert space, with* $\langle\Psi|\Psi\rangle = 1$.

Postulate 2. *The coordinates* q_i *and* p_i *correspond to the Hermitian operators* $\hat{q}_i$ *and* $\hat{p}_i$*, acting in this Hilbert space and satisfying the commutation relations*

$$[\hat{q}_i, \hat{q}_j] = 0, \qquad [\hat{p}_i, \hat{p}_j] = 0, \qquad [\hat{q}_i, \hat{p}_j] = i\hbar\, \delta_{ij}\, \hat{1}. \tag{5.8}$$

Each classical observable $A(q,p)$ *corresponds to a Hermitian operator*

$$\hat{A} = A(q,p)\big|_{q_i=\hat{q}_i, p_i=\hat{p}_i} \tag{5.9}$$

acting in this Hilbert space. The set of all possible values of an observable A *coincides with the spectrum of eigenvalues of the corresponding operator* $\hat{A}$.

Postulate 3. *The dynamics of the quantum state is described by the Schrödinger equation*

$$i\hbar \frac{d\,|\Psi\rangle^t}{dt} = \hat{H}|\Psi\rangle^t, \tag{5.10}$$

where $\hat{H} = H(q,p)\big|_{q_i=\hat{q}_i, p_i=\hat{p}_i}$ *is the Hamiltonian operator, also called the Hamiltonian. The subscript* t *indicates time dependence.*

Postulate 4. *The expectation value of the observable* A *in the state* $|\Psi\rangle^t$ *is given by the expression*

$$\langle A\rangle_\Psi^t = {}^t\langle\Psi|\hat{A}|\Psi\rangle^t. \tag{5.11}$$

The probability of observing a system in state $|\Psi_2\rangle$ *under the condition that the system was prepared in the state* $|\Psi_1\rangle$ *is given by*

$$|\langle\Psi_2|\Psi_1\rangle|^2. \tag{5.12}$$

In particular, this means that the probability amplitude of observing a system at the time instant t_2 in the state $|\Psi_2\rangle$, under the condition that at the instant t_1 a system was prepared in the state $|\Psi_1\rangle$, is given by

$$\langle\Psi_2|\Psi_1\rangle^{t_2} = \langle\Psi_2|\,\hat{U}(t_2,t_1)\,|\Psi_1\rangle, \tag{5.13}$$

where $U(t,t')$ is an evolution operator satisfying the equation

$$i\hbar\,\frac{\partial\hat{U}(t,t')}{\partial t} = \hat{H}\hat{U}(t,t') \tag{5.14}$$

with the boundary condition $\hat{U}(t,t')\big|_{t=t'} = \hat{1}$. Here $\hat{1}$ is the unit operator in the Hilbert space under consideration.

Postulate 5. *The state of a system of identical particles can be either completely symmetric or purely antisymmetric under the permutation of particles.*

The postulates listed above define a basic procedure of canonical quantization. Before going further, let us make some observations.

Postulate 5 allows the classification of the particles as bosons or fermions. In field theory, it is more convenient to replace this postulate with another equivalent statement. We saw in section 2.6 that the irreducible representations of the Poincaré group are characterized by spins (or helicities) taking either integer or half-integer values. The formulation of this postulate, which we shall follow in the canonical quantization in field theory can be formulated as follows.

Postulate 5'. *Fields with integer and half-integer spin are considered to be classical commuting and anticommuting functions. These fields are quantized using commutators and anticommutators, respectively.*

In particular, scalar and vector fields, are quantized with commutators and spinor fields are quantized with the help of anticommutators.

This postulate means that the particles corresponding to integer spin fields are bosons, while the particles corresponding to half-integer spin fields are fermions. The definition and properties of anticommuting variables will be given later on.

Remark 1. We formulated the postulates of canonical quantization for pure quantum states. However, the most general quantum state is not pure but mixed and cannot be described by a vector in Hilbert space. In general, the mixed quantum state corresponds to a statistical operator (density matrix) $\hat{\varrho}$, acting in the Hilbert space. This is important, for instance, for the description of the multi-particle systems in statistical mechanics. In this book, we will consider only pure quantum states and do not use the statistical operator.

Remark 2. Let $A(q,p)$ be some classical observable. How can we construct the corresponding operator $\hat{A}$? In the definition given above, we wrote the formal expression $\hat{A} = A(q,p)\big|_{q=\hat{q},p=\hat{p}}$. Certainly, the classical quantities q_i and p_j commute, and hence their products do not depend on the order. However, the quantum operators $\hat{p}_i$ and $\hat{q}_j$ do not commute, and hence the products of our interest may strongly depend on the ordering of these operators.

Thus, we face a serious question about the order in which the operators $\hat{p}_i$ and $\hat{q}_j$ should be placed in constructing an operator $\hat{A}$ for the classical observable A. There is no answer to this question in the system of quantum-mechanical postulates listed above. One can assume that the operators $\hat{p}_i$ and $\hat{q}_j$ should be ordered in such a way that the operator $\hat{A}$ would be Hermitian, and all physical quantities computed by means of this operator would correspond to the experimental results. From the formal point of view, this means that there is some freedom in the choice of ordering for a product of the non-commuting operators $\hat{p}_i$ and $\hat{q}_j$. This freedom can be used to impose additional conditions when constructing the quantum operator for a given classical observable.

In what follows, we shall consider only time-independent Hamiltonians $\hat{H}$. Then the equation (5.14), together with the boundary condition $\hat{U}(t,t')\big|_{t=t'} = \hat{1}$, has the formal solution

$$\hat{U}(t,t') = e^{-\frac{i}{\hbar}(t-t')\hat{H}}. \tag{5.15}$$

Each concrete realization of Hilbert space and commutation relations (5.8) is called a representation. Let us recall two representations that are conventionally used in quantum mechanics:

a. Coordinate representation. The Hilbert space is realized as a complex linear space of functions of n variables, $\Psi(q_1, \dots, q_n)$, with the scalar product

$$\langle \Psi_1 | \Psi_2 \rangle = \int dq_1 \dots dq_n \, \Psi_1^*(q_1, \dots, q_n) \Psi_2(q_1, \dots, q_n).$$

$\Psi(q_1, \dots, q_n)$ are called wave functions in the coordinate representation. The operators $\hat{q}_i$ and $\hat{p}_i$ act on the wave function as follows:

$$\begin{aligned} \hat{q}_i \Psi(q_1, \dots, q_n), &= q_i \Psi(q_1, \dots, q_n) \\ \hat{p}_i \Psi(q_1, \; dots, q_n) &= -i\hbar \, \frac{\partial \Psi(q_1, \dots, q_n)}{\partial q_i}. \end{aligned} \tag{5.16}$$

b. Momentum representation. The Hilbert space is realized as a complex linear space of the functions of n variables $\tilde{\Psi}(p_1, \dots, p_n)$, with the scalar product defined as

$$\langle \Psi_1 | \Psi_2 \rangle = \int dp_1 \dots dp_n \, \tilde{\Psi}_1^*(p_1, \dots, p_n) \tilde{\Psi}_2(p_1, \dots, p_n). \tag{5.17}$$

The operators $\hat{q}_i$ and $\hat{p}_j$ are realized as follows:

$$\begin{aligned} \hat{q}_i \tilde{\Psi}(p_1, \dots, p_n) &= i\hbar \, \frac{\partial \tilde{\Psi}(p_1, \dots, p_n)}{\partial p_i} \\ \hat{p}_i \tilde{\Psi}(p_1, \dots, p_n) &= p_i \tilde{\Psi}(p_1, \dots, p_n). \end{aligned}$$

The functions $\tilde{\Psi}(p_1, \dots, p_n)$ are called the wave functions in momentum representation.

The wave functions $\Psi(q_1, \dots, q_n)$ and $\tilde{\Psi}(p_1, \dots, p_n)$ of the same pure state $|\Psi\rangle$ are related as

$$\begin{aligned} \Psi(q_1, \dots, q_n) &= \int \frac{dp_1 \dots dp_n}{(2\pi\hbar)^{n/2}} \, e^{\frac{i}{\hbar}(q_1 p_1 + \dots q_n p_n)} \tilde{\Psi}(p_1, \dots, p_n), \\ \tilde{\Psi}(p_1, \dots, p_n) &= \int \frac{dq_1 \dots dq_n}{(2\pi\hbar)^{n/2}} \, e^{-\frac{i}{\hbar}(q_1 p_1 + \dots q_n p_n)} \Psi(q_1, \dots, q_n). \end{aligned} \tag{5.18}$$

The procedure of canonical quantization defines quantum theory in the so-called Schrödinger dynamical picture. In this case, the quantum states (wave functions) depend on time, while operators do not depend on time, if the corresponding classical observable is time independent.

On the other hand, the quantum dynamics can be equivalently reformulated in an alternative way called the Heisenberg dynamical picture. In this case, states are time independent and the operators are time dependent. There is a simple relation between these two pictures, as shown in the table below:

Schrödinger picture	Heisenberg picture
$\lvert\Psi\rangle^t$	$\lvert\Psi\rangle$
$\hat{A}$	$\hat{A}(\text{t})$
$i\hbar\frac{d\lvert\Psi\rangle^t}{dt} = \hat{H}\lvert\Psi\rangle^t$	$i\hbar\frac{d\hat{A}(t)}{dt} = [\hat{A}(t), \hat{H}]$

Expectation values of the observables are the same in both pictures,

$$ {}^t\langle\Psi\big|\hat{A}\big|\Psi\rangle^t = \langle\Psi\big|\hat{A}(t)\big|\Psi\rangle, \tag{5.19}$$

while the states are related as

$$ \lvert\Psi\rangle^t = \hat{U}(t,t_0)\lvert\Psi\rangle \tag{5.20}$$

and the operators as

$$ \hat{A}(t) = \hat{U}^\dagger(t,t_0)\,\hat{A}\,\hat{U}(t,t_0). \tag{5.21}$$

Let us consider the equal-time commutators for the operators of coordinates and momenta in the Heisenberg dynamical picture. According to (5.21),

$$ \hat{q}_i(t) = \hat{U}^\dagger(t,t_0)\,\hat{q}_i\,\hat{U}(t,t_0) \qquad \text{and} \qquad \hat{p}_i(t) = \hat{U}^\dagger(t,t_0)\,\hat{p}_i\,\hat{U}(t,t_0). \tag{5.22}$$

Therefore,

$$ \begin{aligned} [\hat{q}_i(t),\hat{p}_j(t)] &= [\hat{U}^\dagger(t,t_0)\,\hat{q}_i\,\hat{U}(t,t_0), \hat{U}^\dagger(t,t_0)\,\hat{p}_j\,\hat{U}(t,t_0)] \\ &= \hat{U}^\dagger(t,t_0)[\hat{q}_i,\hat{p}_j]\hat{U}(t,t_0) \;=\; i\hbar\,\delta_{ij}\,\hat{1}. \end{aligned} \tag{5.23}$$

In a similar way, one can derive the relations

$$ [\hat{q}_i(t),\hat{q}_j(t)] = 0, \qquad [\hat{p}_i(t),\hat{p}_j(t)] = 0. \tag{5.24}$$

Equal-time commutators of Heisenberg operators of coordinates and momenta satisfy the same canonical commutation relations as the corresponding Schrödinger operators.

5.2 Canonical quantization in field theory

Consider the field model with a set of fields $\phi^i(x)$ and the action

$$ S[\phi] = \int d^4x\,\mathcal{L}(\phi,\partial_\mu\phi). $$

We intend to discuss how, in principle, the postulates of canonical quantization can be applied in field theory.

Let us start the discussion of the quantization from the simplest theory of a single scalar field. Thus, as a field $\phi^i(x)$, we use a scalar, $\varphi(x)$. After that, we briefly consider the specific details related to the quantization of spinor and vector fields. The Lagrangian of a scalar field is

$$\mathcal{L} = \frac{1}{2}\eta^{\mu\nu}\,\partial_\mu\varphi\,\partial_\nu\varphi - U(\varphi), \tag{5.25}$$

where $U(\varphi) = \frac{1}{2}m^2\varphi^2 + V(\varphi)$. We already discussed earlier that the field $\varphi(x)$ can be regarded as generalized coordinates, indexed by a three-dimensional vector $\mathbf{x}$. The useful notation in this case is $\varphi(x) = \varphi_{\mathbf{x}}(t)$. Then the Lagrangian (5.25) becomes

$$\mathcal{L} = \frac{1}{2}\dot{\varphi}^2_{\mathbf{x}}(t) - \frac{1}{2}\partial_j\varphi_{\mathbf{x}}(t)\,\partial_j\varphi_{\mathbf{x}}(t) - U\big(\varphi_{\mathbf{x}}(t)\big). \tag{5.26}$$

As in classical mechanics, we introduce the field momentum $\pi_{\mathbf{x}}(t)$ in the form

$$\pi_{\mathbf{x}}(t) = \frac{\partial\mathcal{L}}{\partial\dot{\varphi}_{\mathbf{x}}(t)} = \dot{\varphi}_{\mathbf{x}}(t) \equiv \pi(x). \tag{5.27}$$

Now we are in a position to construct the Hamiltonian function

$$\begin{aligned} H &= \int d^3x\{\pi(x)\dot{\varphi}(x) - \mathcal{L}\}\Big|_{\dot{\varphi}\pi} \\ &= \int d^3x\left\{\frac{1}{2}\pi^2 + \frac{1}{2}\partial_j\varphi\,\partial_j\varphi + U(\varphi)\right\} = \int d^3x\,\mathcal{H}. \end{aligned} \tag{5.28}$$

The corresponding Hamiltonian action is

$$S_H[\varphi,\pi] = \int dt\,d^3x\,\big(\pi\dot{\varphi} - \mathcal{H}\big) = \int d^4x\Big[\pi\dot{\varphi} - \frac{1}{2}\pi^2 - \frac{1}{2}\partial_j\varphi\,\partial_j\varphi - U(\varphi)\Big], \tag{5.29}$$

and the Hamiltonian equations of motion are

$$\frac{\delta S_H}{\delta\varphi} = 0, \qquad \frac{\delta S_H}{\delta\pi} = 0. \tag{5.30}$$

Direct calculation shows that

$$\frac{\delta S_H}{\delta\varphi} = -\dot{\pi} + \Delta\varphi - \frac{dU}{d\varphi}, \qquad \text{and} \qquad \frac{\delta S_H}{\delta\pi} = \dot{\varphi} - \pi. \tag{5.31}$$

Thus, the canonical equations of motion are

$$\dot{\varphi} = \pi\,, \qquad \dot{\pi} = \Delta\phi - \frac{dU}{d\varphi}, \tag{5.32}$$

where Δ is the Laplace operator (1.12). Substituting the first equation into the second one, we get

$$\Box\phi + \frac{dU}{d\phi} = 0, \tag{5.33}$$

which is exactly the Lagrange equation for the Lagrangian (5.25).

Now, since the theory was successfully formulated in the Hamiltonian form, we can apply the postulates of canonical quantization.

1) The pure state of the quantum field is represented by the normalized on the unit vector $|\Psi\rangle$ of some Hilbert space.

2) The field $\varphi(\mathbf{x})$ and momentum $\pi(\mathbf{x})$ at initial time moment $t = 0$ are represented by the Hermitian operators $\hat{\varphi}(\mathbf{x})$ and $\hat{\pi}(\mathbf{x})$ satisfying the commutation relation

$$[\hat{\varphi}(\mathbf{x}), \hat{\varphi}(\mathbf{y})] = 0, \qquad [\hat{\pi}(\mathbf{x}), \hat{\pi}(\mathbf{y})] = 0, \qquad [\hat{\varphi}(\mathbf{x}), \hat{\pi}(\mathbf{y})] = i\hbar\,\delta(\mathbf{x}-\mathbf{y})\,\hat{1}. \tag{5.34}$$

3) Field dynamics is defined by the Schrödinger equation

$$i\hbar\frac{d|\Psi\rangle^t}{dt} = \hat{H}|\Psi\rangle^t,$$

where

$$\hat{H} = H(\varphi,\pi)\big|_{\varphi=\hat{\varphi},\pi=\hat{\pi}} = \int d^3x\Big[\frac{1}{2}\hat{\pi}^2 + \frac{1}{2}\partial_j\hat{\varphi}\,\partial_j\hat{\varphi} + U(\hat{\varphi})\Big]. \tag{5.35}$$

We see that the procedure is completely analogous to one for a system with the finite number of degrees of freedom.

In the coordinate representation, a pure state of the field $|\Psi\rangle$ is represented by the wave functional $\Psi[\varphi]$, where $\varphi \equiv \varphi(\mathbf{x})$. Then $\hat{\varphi}(\mathbf{x})$ is the multiplication operator

$$\hat{\varphi}(\mathbf{x})\Psi[\varphi] = \varphi(\mathbf{x})\Psi[\varphi] \tag{5.36}$$

and the operator $\hat{\pi}(\mathbf{x})$ acts as follows:

$$\hat{\pi}(\mathbf{x})\Psi[\varphi] = -i\hbar\,\frac{\delta\Psi[\varphi]}{\delta\phi(\mathbf{x})}. \tag{5.37}$$

The Schrödinger equation looks like

$$i\hbar\,\frac{\partial\Psi[\varphi,t]}{\partial t} = \int d^3x\Big\{-\frac{\hbar^2}{2}\frac{\delta}{\delta\varphi(\mathbf{x})}\frac{\delta}{\delta\varphi(\mathbf{x})} + \partial_j\varphi\,\partial_j\varphi + U(\varphi)\Big\}\Psi[\varphi,t]. \tag{5.38}$$

Thus, the procedure of quantization works perfectly well for the scalar field model.

As we shall see in brief, the quantization scheme described above meets essential difficulties when applied to spinor and vector fields. Let us now consider these models.

a) The Lagrangian of the spinor field is

$$\mathcal{L} = \bar{\psi}i\gamma^\mu\partial_\mu\psi - m\bar{\psi}\psi = \bar{\psi}i\gamma^0\dot{\psi} + \bar{\psi}i\gamma^j\partial_j\psi - m\bar{\psi}\psi. \tag{5.39}$$

Then,

$$\pi_\psi = \frac{\partial\mathcal{L}}{\partial\dot{\psi}} = \bar{\psi}i\gamma^0, \qquad \pi_{\bar{\psi}} = \frac{\partial\mathcal{L}}{\partial\dot{\bar{\psi}}} = 0.$$

It is easy to see that these equations do not allow one to express the generalized velocities $\dot{\psi}$ and $\dot{\bar{\psi}}$ through π_ψ and $\pi_{\bar{\psi}}$. Instead, we got a constraint $\pi_{\bar{\psi}} = 0$ imposed on the phase variable. Thus, the standard canonical procedure does not work in this case.

b) In the theory of a massless vector field, the Lagrangian can be transformed as

$$\mathcal{L} = -\frac{1}{4}F_{\mu\nu}F^{\mu\nu} = -\frac{1}{4}(F_{00}F^{00} + 2\,F_{0i}F^{oi} + F_{ij}F^{ij}) = \frac{1}{2}F_{0i}F_{0i} - \frac{1}{4}F_{ij}F_{ij}$$
$$= \frac{1}{2}(\partial_0 A_i - \partial_i A_0)^2 + \frac{1}{4}F_{ij}F_{ij} = \frac{1}{2}(\dot{A}_i - \partial_i A_0)(\dot{A}_i - \partial_i A_0) - \frac{1}{4}F_{ij}F_{ij}. \tag{5.40}$$

Therefore,

$$\pi_0 = \frac{\partial \mathcal{L}}{\partial \dot{A}_0} = 0, \qquad \pi_i = \frac{\partial \mathcal{L}}{\partial \dot{A}_i} = \dot{A}_i - \partial_i A_0. \tag{5.41}$$

We see that these equations do not enable one to express the velocity $\dot{A}_0$ through π_0 and π_i. Once again, we met a constraint on the phase variables, such that the standard canonical procedure does not work.

Let us conclude this section by stating that the standard, classical, canonical procedure is not applicable to spin 1/2 and massless vector fields. It turns out that the same problem holds for both Yang-Mills theory and gravity. One can say that such a situation is typical in field theory. Indeed, there is a very general and powerful approach to canonical quantization, originated by Dirac, that works even in the case when not all velocities can be expressed through the canonical momenta. On the other hand, for the relatively simple models, there are essentially simpler indirect methods. These methods will be used below, since the discussion of the general Dirac approach is beyond the scope of this book. Let us mention that the correctness of these methods in the cases of our interest was confirmed within the Dirac approach.

5.3 Canonical quantization of a free real scalar field

Let us consider the quantization for a scalar field in more detail than it was done in section 5.2. Our main purpose is to quantize the free theory, but, for the sake of generality, until some point we will consider also an interaction term. In this section and further (if this is not indicated explicitly), we will use the natural units system and set $\hbar = 1$.

The Lagrangian of a real scalar field has the form (4.10), and the corresponding classical equation of motion is

$$\Box\varphi(x) + m^2\varphi(x) + V'(\varphi) = 0. \tag{5.42}$$

The canonical quantization, considered in section 5.2, leads to operators $\hat{\varphi}(\mathbf{x})$ and $\hat{\pi}(\mathbf{x})$ and to the Hamilton function

$$\hat{H} = \int d^3x \left\{\frac{1}{2}\hat{\pi}^2(\mathbf{x}) + \frac{1}{2}\partial_j\hat{\varphi}(\mathbf{x})\,\partial_j\hat{\varphi}(\mathbf{x}) + \frac{1}{2}m^2\hat{\varphi}^2(\mathbf{x}) + V(\hat{\varphi}^2(\mathbf{x}))\right\}. \tag{5.43}$$

Introducing field operators in the Heisenberg picture

$$\hat{\varphi}(x) = e^{it\hat{H}}\,\hat{\varphi}(\mathbf{x})\,e^{-it\hat{H}}, \qquad \hat{\pi}(x) = e^{it\hat{H}}\,\hat{\pi}(\mathbf{x})\,e^{-it\hat{H}}, \tag{5.44}$$

one obtains the operator equations of motion

$$i\dot{\hat{\varphi}}(x) = [\hat{\varphi}(x), \hat{H}], \qquad i\dot{\hat{\pi}}(x) = [\hat{\pi}(x), \hat{H}]. \tag{5.45}$$

The calculation of the commutators in the *r.h.s.* of Eq. (5.45), using canonical commutation relations, gives

$$[\hat{\varphi}(x), \hat{H}] = i\hat{\pi}(x), \qquad [\hat{\pi}(x), \hat{H}] = i\Delta\hat{\varphi}(x) - im^2\hat{\varphi}(x) - iV'\big(\hat{\varphi}(x)\big). \tag{5.46}$$

Eq. (5.45) and (5.46) lead to

$$\dot{\hat{\varphi}}(x) = \hat{\pi}(x), \qquad \dot{\hat{\pi}}(x) = \Delta\hat{\varphi}(x) - m^2\hat{\varphi}(x) - V\big(\hat{\varphi}(x)\big). \tag{5.47}$$

Substituting the first equation into the second one, we get

$$\Box\hat{\varphi}(x) + m^2\hat{\varphi}(x) + V'\big(\hat{\varphi}(x)\big) = 0. \tag{5.48}$$

Thus, the field operator $\hat{\varphi}(x)$ satisfies the same equation (5.42) as the classical field.

Starting from this point, we consider the free theory. The important reason is that we will need an exact solution of the classical equations of motion. This is something one cannot achieve for the theory with interaction, especially with an undefined potential $V'(\varphi)$.

In the free-field case, the equation of motion for the field operator has the form

$$\Box\hat{\varphi}(x) + m^2\hat{\varphi}(x) = 0. \tag{5.49}$$

The solution to Eq. (5.49) can be written as a linear combination of flat waves,

$$\hat{\varphi}(x) = \int \frac{d^3p}{\sqrt{2(2\pi)^3\varepsilon(\mathbf{p})}} \Big[e^{-ipx}\hat{a}(\mathbf{p}) + e^{ipx}\hat{a}^\dagger(\mathbf{p})\Big]. \tag{5.50}$$

Taking into account that $\hat{\pi} = \dot{\hat{\varphi}}$, one gets

$$\hat{\pi}(x) = -i\int \frac{d^3p\,\varepsilon(\mathbf{p})}{\sqrt{2(2\pi)^3\varepsilon(\mathbf{p})}} \Big[e^{-ipx}\hat{a}(\mathbf{p}) - e^{-ipx}\hat{a}^\dagger(\mathbf{p})\Big], \tag{5.51}$$

where we used short notations $p_0 = \varepsilon(\mathbf{p})$ and $px = p_0t - \mathbf{px}$. Using the relations (5.50) and (5.51), one can express the operators $\hat{a}(\mathbf{p})$ and $\hat{a}^\dagger(\mathbf{p})$ through the field operators $\hat{\varphi}(x)$ and $\dot{\hat{\varphi}}(x)$.

As a next step, consider a linear space of solutions to the Klein–Gordon equation, and define an inner product in this space as follows. Let $\varphi_1(x)$ and $\varphi_2(x)$ be the two (generally speaking, complex) solutions of this equation. The inner product (φ_1, φ_2) of the solutions $\varphi_1(x)$ and $\varphi_2(x)$ is given by

$$(\varphi_1, \varphi_2) = i\int_t d^3x\, \varphi_1^* \overset{\leftrightarrow}{\partial_0} \varphi_2. \tag{5.52}$$

Here we used the notation similar to that defined in section 4.5,

$$\varphi_1^* \overset{\leftrightarrow}{\partial_0} \varphi_2 = \varphi_1^*(\partial_0\varphi_2) - (\partial_0\varphi_1^*)\varphi_2. \tag{5.53}$$

The integral in the *r.h.s.* of Eq. (5.52) is taken in a certain instant of time t, and therefore the *l.h.s.* formally can depend on time. However, differentiating the relation

(5.52) with respect t and taking into account that the fields $\varphi_1(x)$ and $\varphi_2(x)$ satisfy Klein–Gordon equation, it is easy to show that this derivative is zero. Therefore, the inner product (5.52) in the space of solutions of the Klein–Gordon equation is time independent and the integral in the *r.h.s.* can be calculated at any t. Let us denote

$$\varphi_{\mathbf{p}}(x) = \frac{1}{\sqrt{2(2\pi)^3\varepsilon(\mathbf{p})}} e^{-ipx}, \tag{5.54}$$

where, as before, $p_0 = \varepsilon(\mathbf{p})$ and $px = p_0 t - \mathbf{px}$. Using the definition of the scalar product (5.52), it is easy to see that the functions (5.54) satisfy the normalization conditions

$$(\varphi_{\mathbf{p}}, \varphi_{\mathbf{q}}) = \delta(\mathbf{p} - \mathbf{q}), \qquad (\varphi^*_{\mathbf{p}}, \varphi_{\mathbf{q}}) = 0. \tag{5.55}$$

The relations (5.50), (5.52) and (5.54) normalization conditions (5.55) lead to the following expressions for the coefficients of the expansion in flat waves (5.50),

$$\hat{a}(\mathbf{p}) = (\varphi_{\mathbf{p}}, \hat{\varphi}), \qquad \hat{a}^\dagger(\mathbf{p}) = (\hat{\varphi}, \varphi_{\mathbf{p}}). \tag{5.56}$$

Imposing the canonical, equal-time, commutation relations (5.8) on the operators $\hat{\varphi}(t, \mathbf{x})$ and $\hat{\pi}(t, \mathbf{x})$, one can find the commutation relations for the operators $\hat{a}(\mathbf{p})$ and $\hat{a}^\dagger(\mathbf{p})$, (5.56). To do this, one can use the commutators for the operators (5.56) and the relations (5.8). The result has the form

$$[\hat{a}(\mathbf{p}), \hat{a}(\mathbf{p}')] = 0, \qquad [\hat{a}^\dagger(\mathbf{p}), \hat{a}^\dagger(\mathbf{p}')] = 0, \qquad [\hat{a}(\mathbf{p}), \hat{a}^\dagger(\mathbf{p}')] = \delta(\mathbf{p} - \mathbf{p}'). \tag{5.57}$$

We see that $\hat{a}(\mathbf{p})$ and $\hat{a}^\dagger(\mathbf{p})$ satisfy the same commutation relations that hold for the annihilation and creation operators in the quantized harmonic oscillator in quantum mechanics. Thus, we can consider the scalar field as the infinite set of quantized harmonic oscillators and regard $\hat{a}(\mathbf{p})$ and $\hat{a}^\dagger(\mathbf{p})$ as annihilation and creation operators of the states with momentum $\mathbf{p}$.

The construction of the Hamiltonian and momentum operators is based on the expressions for classical Hamiltonian function and three-dimensional momentum. Based on Noether's theorem, we get

$$\begin{aligned} H &= \frac{1}{2}\int d^3x\big(\pi^2 + \partial_j\varphi\,\partial_j\varphi + m^2\varphi^2\big), \\ P_j &= -\int d^3x\,\frac{\partial\mathcal{L}}{\partial\dot{\varphi}}\,\partial_j\varphi = -\int d^3x\,\pi\,\partial_j\varphi. \end{aligned} \tag{5.58}$$

Therefore, the corresponding Hermitian operators are

$$\hat{H} = \frac{1}{2}\int d^3x\big(\hat{\pi}^2 + \partial_j\hat{\varphi}\,\partial_j\hat{\varphi} + m^2\hat{\varphi}^2\big), \quad \hat{P}_j = -\frac{1}{2}\int d^3x\,\big(\hat{\pi}\,\partial_j\hat{\varphi} + \partial_j\hat{\varphi}\,\hat{\pi}\big). \tag{5.59}$$

Substituting (5.50) and (5.51) into (5.59), one gets

$$\begin{aligned} \hat{H} &= \frac{1}{2}\int d^3p\;\varepsilon(\mathbf{p})\big[\hat{a}^\dagger(\mathbf{p})\hat{a}(\mathbf{p}) + \hat{a}(\mathbf{p})\hat{a}^\dagger(\mathbf{p})\big], \\ \hat{P}^j &= \frac{1}{2}\int d^3p\;p^j\,\big[\hat{a}^\dagger(\mathbf{p})\hat{a}(\mathbf{p}) + \hat{a}(\mathbf{p})\hat{a}^\dagger(\mathbf{p})\big]. \end{aligned} \tag{5.60}$$

The two expressions (5.60) can be written in an alternative form using the last of the commutation relations (5.57),

$$\hat{H} = \int d^3p\, \varepsilon(\mathbf{p})\, \hat{a}^\dagger(\mathbf{p})\hat{a}(\mathbf{p}) + c^0, \qquad \hat{P}^j = \int d^3p\, p^j\, \hat{a}^\dagger(\mathbf{p})\hat{a}(\mathbf{p}) + c^j, \tag{5.61}$$

where

$$c^0 = \delta^{(3)}(\mathbf{p})\Big|_{\mathbf{p}=0} \times \int \frac{d^3p}{2}\, \varepsilon(\mathbf{p}) \quad \text{and} \quad c^j = \delta^{(3)}(\mathbf{p})\Big|_{\mathbf{p}=0} \times \int \frac{d^3p}{2}\, p^j. \tag{5.62}$$

It is easy to see that the constants c^0 and c^j are ill-defined, which is one of the simplest examples of divergences in quantum field theory. Thus, these expressions require an additional prescription of how they should be understood. Such a prescription is related to the arbitrariness in the definition of the operators corresponding to classical observables. One can use such an arbitrariness as follows.

a) Start from the classical observables H and P^j, according to (5.58).

b) Construct the corresponding Hermitian operators according to (5.59).

c) Substitute Eq. (5.50) and (5.51) into (5.59).

d) Define the final operators $\hat{H}$ and $\hat{P}^j$ in such a way that all operators $\hat{a}^\dagger(\mathbf{p})$ are placed on the left from all operators $\hat{a}(\mathbf{p})$. Such a prescription is called normal ordering.

Using the definition described above, we finally arrive at the new expressions

$$\hat{H} = \int d^3p\, \varepsilon(\mathbf{p})\, \hat{n}(\mathbf{p}) \qquad \text{and} \qquad \hat{P}^j = \int d^3p\, p^j\, \hat{n}(\mathbf{p}), \tag{5.63}$$

where $\varepsilon(\mathbf{p})$ is the energy of a relativistic particle with momentum $\mathbf{p}$ and

$$\hat{n}(\mathbf{p}) = \hat{a}^\dagger(\mathbf{p})\hat{a}(\mathbf{p}). \tag{5.64}$$

The relations (5.63) should be taken as the definitions of the Hamiltonian operator and the operator of three-dimensional momentum of a free real scalar field.

Consider the problem of the energy spectrum for the quantized free real scalar field. It is important that the operators $\hat{H}$ and $\hat{P}_j$ commute and therefore they have the same system of eigenvectors. Let us introduce the operator

$$\hat{N} = \int d^3p\, \hat{n}(\mathbf{p}) = \int d^3p\, \hat{a}^\dagger(\mathbf{p})\hat{a}(\mathbf{p}), \tag{5.65}$$

which also commutes with $\hat{H}$ and $\hat{P}^j$. Thus, the eigenvectors of $\hat{N}$ will be the eigenvectors for $\hat{H}$ and $\hat{P}^j$ too.

As we know from quantum mechanics, the eigenvectors of the operator $\hat{N}$ are

$$\big|\mathbf{p}_1, \dots, \mathbf{p}_n\big\rangle = \frac{1}{\sqrt{n!}}\, \hat{a}^\dagger(\mathbf{p}_1) \dots \hat{a}^\dagger(\mathbf{p}_n)\big|0\big\rangle, \tag{5.66}$$

where $n = 0, 1, 2, \dots$ and the vector $|0\rangle$ is defined as a solution to the equation

$$\hat{a}(\mathbf{p})|0\rangle = 0, \qquad \forall\, \mathbf{p}. \tag{5.67}$$

Since the operators $\hat{a}^\dagger(\mathbf{p})$ and $\hat{a}^\dagger(\mathbf{q})$ commute, the state (5.66) is totally symmetric under the permutation of the particles. This fact completely corresponds to the quantum postulate of the symmetry for integer-spin particles.

Direct computations with the help of commutations relation (5.57) lead to

$$\hat{N}\big|\mathbf{p}_1,\ldots,\mathbf{p}_n\big\rangle = n\big|\mathbf{p}_1,\ldots,\mathbf{p}_n\big\rangle, \tag{5.68}$$

where $n = 0, 1, 2, \ldots$. We see that the eigenvalues of the operator $\hat{N}$ are integer numbers. Using the commutation relations (5.57), one gets

$$\hat{\mathbf{P}}|\mathbf{p}_1,\ldots,\mathbf{p}_n\rangle = (\mathbf{p}_1 + \mathbf{p}_2 + \ldots + \mathbf{p}_n)|\mathbf{p}_1,\ldots,\mathbf{p}_n\rangle \tag{5.69}$$

and

$$\hat{H}\big|\mathbf{p}_1,\ldots,\mathbf{p}_n\big\rangle = \Big(\varepsilon(\mathbf{p}_1) + \ldots + \varepsilon(\mathbf{p}_n)\Big)\big|\mathbf{p}_1,\ldots,\mathbf{p}_n\big\rangle. \tag{5.70}$$

The relations (5.68), (5.69) and (5.70) serve as the base for the interpretation of the states of quantized free real scalar field in terms of particles. The state (5.68) is called the n-particle state. The operator $\hat{N}$ (5.65) is called the operator of particle number, and its eigenvalues are the possible numbers of particle in the states $|\mathbf{p}_1,\ldots,\mathbf{p}_n\rangle$. Eq. (5.69) shows that the n-particle state has a certain momentum $\mathbf{p}_1 + \ldots + \mathbf{p}_n$ and Eq. (5.70) shows that the n-particle state has a certain energy $\varepsilon(\mathbf{p}_1) + \ldots + \varepsilon(\mathbf{p}_n)$. The full three-dimensional momentum is a sum of momenta of individual particles, and the full energy is a sum of energies of individual particles.

Eq. (5.69) and (5.70) show that

$$\begin{aligned}
&\hat{H}|0\rangle = 0, \qquad \hat{\mathbf{P}}\,|0\rangle = 0, \qquad \hat{N}\,|0\rangle = 0,\\
&\hat{H}|\mathbf{p}\rangle = \varepsilon(\mathbf{p})\,|\mathbf{p}\rangle, \qquad \hat{\mathbf{P}}\,|\mathbf{p}\rangle = \mathbf{p}\,|\mathbf{p}\rangle, \qquad \hat{N}\,|\mathbf{p}\rangle = |\mathbf{p}\rangle,\\
&\hat{H}|\mathbf{p}_1,\ldots,\mathbf{p}_n\rangle = \{\varepsilon(\mathbf{p}_1) + \ldots + \varepsilon(\mathbf{p}_n)\}|\mathbf{p}_1,\ldots,\mathbf{p}_n\rangle.
\end{aligned} \tag{5.71}$$

We see that the energy spectrum is bounded from below and there is a state $|0\rangle$ corresponding to the minimal energy. Such a state is called the vacuum. The particle number of the vacuum is zero, and the same is true for the three-dimensional momentum and the energy of the vacuum. As we have seen above, the last two conclusions are due to the normal ordering prescription.

The vectors $|\mathbf{p}_1,\ldots,\mathbf{p}_n\rangle$ form a basis in the space of quantum states of a real scalar field. This is called the Fock basis. Any state $\big|\Phi\big\rangle$ of a real scalar field can be expanded in this basis as

$$|\Phi\rangle = \sum_{n=0}^{\infty}\int d^3p_1 \ldots d^3p_n\, \Phi_n(\mathbf{p}_1 \ldots \mathbf{p}_n)|\mathbf{p}_1,\ldots,\mathbf{p}_n\rangle, \tag{5.72}$$

where $\Phi_n(\mathbf{p}_1 \ldots \mathbf{p}_n)$ is a totally symmetric function. The set of states (5.72) is called the Fock space.

Finally, we can conclude that the states of a quantized scalar field are described in terms of particles. The specific feature of quantum field theory is that there are states with an arbitrarily large numbers of particles.

5.4 Canonical quantization of a free complex scalar field

As we already know, the Lagrangian of a free complex scalar field has the form

$$\mathcal{L} = \eta^{\mu\nu}\,\partial_\mu\varphi^*\,\partial_\nu\varphi - m^2\varphi^*\varphi. \tag{5.73}$$

Since $\mathcal{L}$ depends on the two fields φ and φ^*, there are two conjugated momenta

$$\pi_\varphi = \frac{\partial\mathcal{L}}{\partial\dot{\varphi}} = \dot{\varphi}^* \qquad \text{and} \qquad \pi_{\varphi^*} = \frac{\partial\mathcal{L}}{\partial\dot{\varphi}^*} = \dot{\varphi}. \tag{5.74}$$

Thus, $\pi_{\varphi^*} = \pi_\varphi^*$. The Hamiltonian function is constructed in a standard way,

$$\begin{aligned} H &= \int d^3x\,\big(\pi_\varphi\dot{\varphi} + \pi_{\varphi^*}\dot{\varphi}^* - \mathcal{L}\big)\Big|_{\dot{\varphi}=\pi_{\varphi^*},\,\dot{\varphi}^*=\pi_\varphi} \\ &= \int d^3x\big(\pi_\varphi^*\pi_\varphi + \partial_j\varphi^*\,\partial_j\varphi + m^2\varphi^*\varphi\big). \end{aligned} \tag{5.75}$$

To quantize the theory, one introduces the operators $\hat{\varphi}(\mathbf{x})$, $\hat{\varphi}^\dagger(\mathbf{x})$, $\hat{\pi}(\mathbf{x})$ and $\hat{\pi}^\dagger(\mathbf{x})$, satisfying the commutation relations

$$[\hat{\varphi}(\mathbf{x}), \hat{\pi}(\mathbf{x}')] = i\delta(\mathbf{x}-\mathbf{x}'), \qquad [\hat{\varphi}^\dagger(\mathbf{x}), \hat{\pi}^\dagger(\mathbf{x}')] = i\delta(\mathbf{x}-\mathbf{x}'). \tag{5.76}$$

All other commutators vanish.

The Hamiltonian operator corresponding to the classical Hamiltonian function is

$$\hat{H} = \int d^3x\{\hat{\pi}(\mathbf{x})\hat{\pi}^\dagger(\mathbf{x}) + \partial_j\hat{\varphi}^\dagger(\mathbf{x})\,\partial_j\hat{\varphi}(\mathbf{x}) + m^2\hat{\varphi}^\dagger(\mathbf{x})\hat{\varphi}(\mathbf{x})\}. \tag{5.77}$$

Now we introduce the Heisenberg operators $\hat{\varphi}(x)$, $\hat{\varphi}^\dagger(x)$, $\hat{\pi}(x)$ and $\hat{\pi}^\dagger(x)$. Using the Heisenberg equations of motion, one gets

$$\Box\hat{\varphi}(x) + m^2\hat{\varphi}(x) = 0, \qquad \Box\hat{\varphi}^\dagger(x) + m^2\hat{\varphi}^\dagger(x) = 0. \tag{5.78}$$

The solutions to Eq. (5.78) are

$$\begin{aligned} \hat{\varphi}(x) &= \int \frac{d^3p}{\sqrt{2(2\pi)^3\varepsilon(\mathbf{p})}}\Big[\hat{a}(\mathbf{p})e^{-ip\,x} + \hat{b}^\dagger(\mathbf{p})e^{ip\,x}\Big], \\ \hat{\varphi}^\dagger(x) &= \int \frac{d^3p}{\sqrt{2(2\pi)^3\varepsilon(\mathbf{p})}}\Big[\hat{a}^\dagger(\mathbf{p})e^{ip\,x} + \hat{b}(\mathbf{p})e^{-ip\,x}\Big], \end{aligned} \tag{5.79}$$

where $p_0 = \varepsilon(\mathbf{p}) = \sqrt{\mathbf{p}^2 + m^2}$. In this case, we have two sets of bosonic annihilation and creation operators. Taking into account that $\hat{\pi} = \dot{\hat{\varphi}}^\dagger$ and $\hat{\pi}^\dagger = \dot{\hat{\varphi}}$, and substituting (5.79) into (5.76), one obtains

$$\begin{aligned} [\hat{a}(\mathbf{p}), \hat{a}^\dagger(\mathbf{p}')] &= \delta(\mathbf{p}-\mathbf{p}'), \\ [\hat{b}(\mathbf{p}), \hat{b}^\dagger(\mathbf{p}')] &= \delta(\mathbf{p}-\mathbf{p}'). \end{aligned} \tag{5.80}$$

All other commutation relations among $\hat{a}$, $\hat{a^\dagger}$, $\hat{b}$ and $\hat{b^\dagger}$ vanish.

Using the Lagrangian (5.73), on the basis of Noether's theorem, we can construct the three-dimensional momentum and electric charge of the field,

$$P^j = -\int d^3x \,(\pi\partial_j\varphi + \pi^*\partial_j\varphi^*),$$
$$Q = ie\int d^3x\,(\varphi^*\pi^* - \varphi\pi). \tag{5.81}$$

Let us note that the existence of the charge Q is the characteristic feature of the complex scalar field, as a consequence of the invariance of the Lagrangian under $U(1)$ transformation group of the fields φ and φ^*.

The next step is the construction of the operators corresponding to the classical observables H, P^j and Q. We shall follow the following prescription:

a) The operators corresponding to P^j and Q are constructed with the rule leading to the Hermitian operators, namely,

$$\pi\partial_j\varphi \longrightarrow \frac{1}{2}(\hat\pi\partial_j\hat\varphi + \partial_j\hat\varphi\,\hat\pi), \qquad \varphi\pi \longrightarrow \frac{1}{2}(\hat\varphi\,\hat\pi + \hat\pi\hat\varphi),$$
$$\pi^*\partial_j\varphi^* \longrightarrow \frac{1}{2}(\hat\pi^\dagger\partial_j\hat\varphi^\dagger + \partial_j\hat\varphi^\dagger\,\hat\pi^\dagger), \qquad \varphi^*\pi^* \longrightarrow \frac{1}{2}(\hat\varphi^\dagger\,\hat\pi^\dagger + \hat\pi^\dagger\hat\varphi^\dagger). \tag{5.82}$$

As a result, we arrive at the Hermitian (by construction) operators

$$\hat H = \int d^3x(\hat\pi^\dagger\hat\pi + \partial_j\hat\varphi^\dagger\,\partial_j\hat\varphi^\dagger + m^2\hat\varphi^\dagger\hat\varphi),$$
$$\hat P^j = -\frac{1}{2}\int d^3x(\hat\pi\partial_j\hat\varphi + \partial_j\hat\varphi\,\hat\pi + \hat\pi^\dagger\partial_j\hat\varphi^\dagger + \partial_j\hat\varphi^\dagger\hat\pi^\dagger),$$
$$\hat Q = \frac{ie}{2}\int d^3x(\hat\varphi^\dagger\hat\pi^\dagger + \hat\pi^\dagger\hat\varphi^\dagger - \hat\varphi\,\hat\pi - \hat\pi\hat\varphi). \tag{5.83}$$

b) Using Eq. (5.80), express the operators $\hat\varphi$, $\hat\varphi^\dagger$, $\hat\pi$, $\hat\pi^\dagger$ in terms of $\hat a$, $\hat a^\dagger$, $\hat b$, $\hat b^\dagger$.

c) Put all operators $\hat a^\dagger$ and $\hat b^\dagger$ on the left of all operators $\hat a$ and $\hat b$, that is, follow normal ordering. As a result, one gets

$$\hat H = \int d^3p\,\varepsilon(\mathbf{p})\big[\hat a^\dagger(\mathbf{p})\hat a(\mathbf{p}) + \hat b^\dagger(\mathbf{p})\hat b(\mathbf{p})\big],$$
$$\hat{\mathbf{P}} = \int d^3p\,\mathbf{p}\big[\hat a^\dagger(\mathbf{p})\hat a(\mathbf{p}) + \hat b^\dagger(\mathbf{p})\hat b(\mathbf{p})\big],$$
$$\hat Q = e\int d^3p\,\big[\hat a^\dagger(\mathbf{p})\hat a(\mathbf{p}) - \hat b^\dagger(\mathbf{p})\hat b(\mathbf{p})\big]. \tag{5.84}$$

All the operators in the *r.h.s.* are Hermitian. The expressions (5.84) are taken as the definitions for the Hamiltonian operator, tree dimensional momentum and the operator of electric charge in quantized, free, complex scalar field theory.

As in section 5.3, our aim is to develop an interpretation of quantum states in terms of particles. It is convenient to introduce the operators

$$\hat{N}_+ = \int d^3p\, \hat{a}^\dagger(\mathbf{p})\hat{a}(\mathbf{p}), \quad \hat{N}_- = \int d^3p\, \hat{b}^\dagger(\mathbf{p})\hat{b}(\mathbf{p}), \quad \hat{N} = \hat{N}_+ + \hat{N}_-. \tag{5.85}$$

The operators $\hat{H}$, $\hat{\mathbf{P}}$, $\hat{N}_+$ and $\hat{N}_-$ commute and therefore have a common system of eigenvectors. For example, if we find the eigenvectors of the operator $\hat{N}$, they automatically will be the eigenvectors for the Hamiltonian.

The eigenvectors of the operator $\hat{N}$ are the products of eigenvectors of the operators $\hat{N}_+$ and $\hat{N}_-$. The last vectors are the same as in the theory for a real scalar field. Therefore, the eigenvectors for the operator $\hat{N}$ are

$$|\mathbf{p}_1, \dots, \mathbf{p}_n, \mathbf{q}_1, \dots, \mathbf{q}_m\rangle = \frac{1}{\sqrt{n!\,m!}}\hat{a}^\dagger(\mathbf{p}_1)\dots\hat{a}^\dagger(\mathbf{p}_n)\hat{b}^\dagger(\mathbf{q}_1)\dots\hat{b}^\dagger(\mathbf{q}_m)\,|0\rangle, \tag{5.86}$$

where the vacuum state vector $|0\rangle$ is defined by the conditions

$$\hat{a}(\mathbf{p})|0\rangle = 0, \qquad \hat{b}(\mathbf{q})|0\rangle = 0, \qquad \forall\, \mathbf{p}, \mathbf{q}. \tag{5.87}$$

Direct computations, using commutation relations (5.80), give

$$\begin{aligned}
&\hat{H}|\mathbf{p_1}, \dots, \mathbf{p_n}, \mathbf{q_1}, \dots, \mathbf{q_m}\rangle = \big[\varepsilon(\mathbf{p_1}) + \dots + \varepsilon(\mathbf{p_n}) + \varepsilon(\mathbf{q_1}) + \dots + \varepsilon(\mathbf{q_m})\big] \\
&\qquad \times |\mathbf{p_1}, \dots, \mathbf{p_n}, \mathbf{q_1}, \dots, \mathbf{q_m}\rangle, \\
&\hat{\mathbf{P}}|\mathbf{p_1}, \dots, \mathbf{p_n}, \mathbf{q_1}, \dots, \mathbf{q_m}\rangle = \big[\mathbf{p_1} + \dots + \mathbf{p_n} + \mathbf{q_1} + \dots + \mathbf{q_m}\big] \\
&\qquad \times |\mathbf{p_1}, \dots, \mathbf{p_n}, \mathbf{q_1}, \dots, \mathbf{q_m}\rangle, \\
&\hat{N}_+|\mathbf{p_1}, \dots, \mathbf{p_n}, \mathbf{q_1}, \dots, \mathbf{q_m}\rangle = n|\mathbf{p_1}, \dots, \mathbf{p_n}, \mathbf{q_1}, \dots, \mathbf{q_m}\rangle, \\
&\hat{N}_-|\mathbf{p_1}, \dots, \mathbf{p_n}, \mathbf{q_1}, \dots, \mathbf{q_m}\rangle = m|\mathbf{p_1}, \dots, \mathbf{p_n}, \mathbf{q_1}, \dots, \mathbf{q_m}\rangle, \\
&\hat{Q}|\mathbf{p_1}, \dots, \mathbf{p_n}, \mathbf{q_1}, \dots, \mathbf{q_m}\rangle = e(n-m)|\mathbf{p_1}, \dots, \mathbf{p_n}, \mathbf{q_1}, \dots, \mathbf{q_m}\rangle.
\end{aligned} \tag{5.88}$$

The relations (5.88) constitute a base for interpreting the states of a quantized free complex scalar field in terms of particles. We see that these states have a certain energy, certain three-dimensional momentum and certain electric charge. Full energy and momentum are sums of energies and momenta of individual particles. Moreover, there are two types of particles, namely, n particles have charge e, while m particles $-e$. The full charge is a sum of charges of individual particles. The situation completely corresponds to the intuitive understanding of the properties of a system of charged non-interacting particles. Particles with charge e can be identified as proper particles, and particles with the opposite charge $-e$ as antiparticles. Both particles and antiparticles are real particles with positive energy.

The state $|0\rangle$ has the minimal energy $E_0 = 0$, and the energy spectrum is bounded from below. The three-dimensional momentum of this state is also zero. In addition, $\hat{N}|0\rangle = 0 = 0|0\rangle$ and $\hat{Q}|0\rangle = 0 = 0|0\rangle$. We can see that vacuum is the state with minimal energy, zero momentum and electric charge. The number of particles of each type in the vacuum state is also zero.

The vectors $|\mathbf{p}_1, \dots, \mathbf{p}_n, \mathbf{q}_1, \dots, \mathbf{q}_m\rangle$, where $n, m = 0, 1, 2, \dots$ form the Fock basis in the space of quantum states of a complex scalar field. An arbitrary vector $|\Phi\rangle$ is expanded in this basis in the form

$$|\Phi\rangle = \sum_{n,m=0}^{\infty} \int d^3p_1 \dots d^3p_n \, d^3q_1 \dots d^3q_m \, \times$$
$$\times \, \Phi_{nm}(\mathbf{p}_1, \dots, \mathbf{p}_n, \mathbf{q}_1, \dots, \mathbf{q}_m) |\mathbf{p}_1, \dots, \mathbf{p}_n, \mathbf{q}_1, \dots, \mathbf{q}_m\rangle. \tag{5.89}$$

The set of vectors (5.89) defines the Fock space of the complex scalar field. As in the case of a real scalar field, the quantized field states are described in terms of particles.

5.5 Quantization of a free spinor field

The Lagrangian of the free spinor field has the form (4.49). For the quantization of this theory, we will try to follow the same scheme as in the case of the complex scalar theory. However, we have to replace the basic commutation relations by anticommutators. Our main purpose is to construct n-particle states, to treat the field quantum states in terms of particles.

As we have explained in section 5.2, the direct canonical formulation of the spinor field theory does not work. Therefore, we will follow an indirect approach.

Let us define the energy-momentum tensor on the basis of Noether's theorem,

$$\begin{aligned} T^{\mu}_{\nu} &= \frac{\partial \mathcal{L}}{\partial \partial_\mu \psi} \partial_\nu \psi + \frac{\partial \mathcal{L}}{\partial \partial_\mu \bar{\psi}} \partial_\nu \bar{\psi} - \mathcal{L} \, \delta^{\mu}{}_{\nu} = i\bar{\psi}\gamma^\mu \partial_\nu \psi - \delta^\mu_\nu (i\bar{\psi}\gamma^\alpha \partial_\alpha \psi - m\bar{\psi}\psi) \\ &= i\bar{\psi}\gamma^\mu \partial_\nu \psi - \delta^\mu_\nu \bar{\psi}(i\gamma^\alpha \partial_\alpha \psi - m\psi). \end{aligned}$$

The second term in the last expression is proportional to the equation of motion (4.21). Since the energy-momentum tensor is defined for the fields satisfying the equation of motion, we get

$$T^{\mu}{}_{\nu} = i\bar{\psi}\gamma^\mu \partial_\nu \psi. \tag{5.90}$$

In particular, the four-vector of energy-momentum is

$$P^\mu = \int d^3x \, T^{0\mu} = \int d^3x \, \bar{\psi} i\gamma^0 \partial^\mu \psi. \tag{5.91}$$

According to the standard interpretation, P^0 is the full energy H, and P^j is the three-dimensional momentum. Hence, the Hamiltonian function of the spinor field is

$$H = i \int d^3x \, \bar{\psi}\gamma^0 \dot{\psi}. \tag{5.92}$$

Introducing the momentum π_ψ corresponding to the field ψ,

$$\pi_\psi = \frac{\partial \mathcal{L}}{\partial \dot{\psi}} = i\bar{\psi}\gamma^0, \tag{5.93}$$

the Hamiltonian function, the three-dimensional field momentum and the electric charge of the field become

$$H = \int d^3x\, \pi_\psi \dot{\psi}, \tag{5.94}$$

$$P^j = -\int d^3x\, \bar{\psi} i\gamma^0 \partial_j \psi = -\int d^3x\, \pi_\psi \partial_j \psi, \tag{5.95}$$

$$Q = e \int d^3x\, \bar{\psi}\gamma^0\psi = e\int d^3x\, \psi^\dagger \psi. \tag{5.96}$$

The three quantities H, P_j and Q are the main dynamical invariants for the free classical spinor field.

The equation of motion for the spinor field is the Dirac equation (4.21). The solution to this equation can be written as

$$\psi(x) = \sum_{\lambda=\pm 1/2} \int \frac{d^3p}{\sqrt{(2\pi)^3 \frac{\varepsilon(\mathbf{p})}{m}}} \big\{ a(\mathbf{p},\lambda) u_{\mathbf{p},\lambda}\, e^{-ipx} + b^*(\mathbf{p},\lambda) v_{\mathbf{p},\lambda}\, e^{ipx} \big\}, \tag{5.97}$$

where $p_0 = \varepsilon(\mathbf{p})$ and the spinors $u_{\mathbf{p},\lambda}$ and $v_{\mathbf{p},\lambda}$ satisfy the following system of linear algebraic equations:

$$(\gamma^\mu p_\mu - m)u_{\mathbf{p},\lambda} = 0, \qquad (\gamma^\mu p_\mu + m)v_{\mathbf{p},\lambda} = 0. \tag{5.98}$$

These equations can be solved exactly, but we are not going to do it here (see, e.g., the standard textbook [57]). For our purposes, the most essential are the following properties of the solutions:

$$\begin{aligned} &\bar{u}_{\mathbf{p},\lambda}\, \gamma^0\, v_{-\mathbf{p},\lambda'} = 0, && \bar{v}_{\mathbf{p},\lambda}\, \gamma^0\, u_{-\mathbf{p},\lambda'} = 0, \\ &\bar{u}_{\mathbf{p},\lambda}\, \gamma^0\, u_{\mathbf{p},\lambda'} = \frac{\varepsilon(\mathbf{p})}{m}\delta_{\lambda\lambda'}, && \bar{v}_{\mathbf{p},\lambda}\, \gamma^0\, v_{\mathbf{p},\lambda'} = \frac{\varepsilon(\mathbf{p})}{m}\delta_{\lambda\lambda'}. \end{aligned} \tag{5.99}$$

In quantum theory, we introduce the operators corresponding to the classical fields $\psi(x)$ and $\bar{\psi}(x)$ and obtain the operators $\hat{\psi}(x)$ and $\hat{\bar{\psi}}(x)$. According to Remark 1 of section 5.2, the quantum relations for spinor field operators should be imposed in terms of anticommutators.

Before using the anticommutation relations, let us remember that, in the free theory with linear equations of motion, the field operators satisfy the same equations of motion as the corresponding classical fields. This property was proved for free real and complex scalar fields, but it can be easily extended for other fields, including the Dirac spinor. Thus, in our case, the field the operator $\hat{\psi}(x)$ satisfies is the equation (4.21). The solution to this operator equation is obtained from the solution (5.97) by replacing the classical amplitudes $a(\mathbf{p},\lambda), b(\mathbf{p},\lambda)$ and their conjugate by the operators $\hat{a}(\mathbf{p},\lambda), \hat{b}(\mathbf{p},\lambda)$ and their Hermitian conjugate. In this way, we arrive at the expansions

$$\begin{aligned} \hat{\psi}(x) &= \sum_{\lambda=\pm 1/2} \int \frac{d^3p}{\sqrt{(2\pi)^3 \frac{\varepsilon(\mathbf{p})}{m}}} \Big\{ \hat{a}(\mathbf{p},\lambda) u_{\mathbf{p},\lambda}\, e^{-ipx} + \hat{b}^\dagger(\mathbf{p},\lambda) v_{\mathbf{p},\lambda}\, e^{ipx} \Big\}, \\ \hat{\bar{\psi}}(x) &= \sum_{\lambda=\pm 1/2} \int \frac{d^3p}{\sqrt{(2\pi)^3 \frac{\varepsilon(\mathbf{p})}{m}}} \big\{ \hat{a}^\dagger(\mathbf{p},\lambda) \bar{u}_{\mathbf{p},\lambda}\, e^{ipx} + \hat{b}(\mathbf{p},\lambda) \bar{v}_{\mathbf{p},\lambda}\, e^{-ipx} \big\}. \end{aligned} \tag{5.100}$$

Since $\hat{\psi}(x)$ and $\hat{\bar{\psi}}(x)$ are operators, the objects $\hat{a}(\mathbf{p},\lambda)$, $\hat{a}^\dagger(\mathbf{p},\lambda)$, $\hat{b}(\mathbf{p},\lambda)$ and $\hat{b}^\dagger(\mathbf{p},\lambda)$ (5.100) should be operators too.

As the canonical quantization does not work directly for spinor field theory, we still do not know which operators we have to impose the anticommutator relations for. To resolve this issue, we can impose some physically motivated conditions. Let us suppose that the desired anticommutation relations provide a positively defined, Hermitian Hamiltonian.

The Hamiltonian operator is constructed in correspondence to the classical Hamilton function H (5.92). That is,

$$\hat{H} \;=\; H\big|_{\psi\to\hat{\psi},\bar{\psi}\to\hat{\bar{\psi}}}. \tag{5.101}$$

Substituting (5.100) into this expression and using the relations (5.98), we get

$$\hat{H} = \sum_{\lambda=\pm 1/2} \int d^3p\,\varepsilon(\mathbf{p})\,\big[\hat{a}^\dagger(\mathbf{p},\lambda)\,\hat{a}(\mathbf{p},\lambda) - \hat{b}(\mathbf{p},\lambda)\,\hat{b}^\dagger(\mathbf{p},\lambda)\big]. \tag{5.102}$$

This expression may look similar to the Hamiltonian of the complex scalar field (5.84), but there is an essential difference, such that a, $a^\dagger$ and b, $b^\dagger$ cannot be the two sets of bosonic annihilation and creation operators. The point is that the integrand of the expression (5.102) has the negative sign in the second term. Therefore, assuming that the operators $\hat{a}$, $\hat{b}$, $\hat{a}^\dagger$ and $\hat{b}^\dagger$ are bosonic annihilation and creation operators, the Hamiltonian (5.102) is not positively defined. For this reason, the quantum relations for the operators $\hat{a}$, $\hat{b}$, $\hat{a}^\dagger$ and $\hat{b}^\dagger$ should be given in terms of the anticommutators.

Taking into account this important detail, we postulate the following anticommutation relations:

$$\begin{aligned}\big\{\hat{a}(\mathbf{p},\lambda),\hat{a}^\dagger(\mathbf{p}',\lambda')\big\} &= \delta_{\lambda,\lambda'}\,\delta(\mathbf{p}-\mathbf{p}'),\\ \big\{\hat{b}(\mathbf{p},\lambda),\hat{b}^\dagger(\mathbf{p}',\lambda')\big\} &= \delta_{\lambda,\lambda'}\,\delta(\mathbf{p}-\mathbf{p}').\end{aligned} \tag{5.103}$$

Here $\{\hat{A},\hat{B}\} = \hat{A}\hat{B} + \hat{B}\hat{A}$ is the notation for the anticommutator. All other anticommutators of the operators $\hat{a}$, $\hat{a}^\dagger$, $\hat{b}$ and $\hat{b}^\dagger$, besides (5.103), vanish. As a result, we obtain the two types of fermionic annihilation and creation operators.

Starting from the relations (5.103), we can establish the anticommutators for the field operators $\hat{\psi}(x)$ and $\hat{\pi}_\psi(x) = i\hat{\bar{\psi}}\gamma^0$. To do that, we need the completeness relations for the flat waves,

$$\sum_{\lambda=\pm 1/2} \Big[u_{\mathbf{p},\lambda}\bar{u}_{\mathbf{p},\lambda}\gamma^0 + v_{-\mathbf{p},\lambda}\bar{v}_{-\mathbf{p},\lambda}\gamma^0\Big] \;=\; \frac{\varepsilon(\mathbf{p})}{m}. \tag{5.104}$$

The anticommutation relations (5.103), together with the formula (5.104), lead to the simultaneous anticommutation relations for the fields,

$$\{\hat{\psi}(x),\hat{\psi}(x')\}_{t'=t} = 0, \quad \{\hat{\pi}_\psi(x),\hat{\pi}_\psi(x')\}_{t'=t} = 0, \quad \{\hat{\psi}(x),\hat{\pi}_\psi(x')\}_{t'=t} \;=\; i\delta(x-x'). \tag{5.105}$$

These anticommutators represent the basic relations for the quantization of the spinor field. The crucial difference with the complex scalar field is that, for the spinor fields, these relations are given by anticommutators. This is a realization of the general quantization postulate discussed in section 5.2, namely, that the fields with the integer spin are quantized in terms of commutators, while the fields with a half-integer spin are quantized in terms of anticommutators.

Substituting $\hat{\pi}_\psi(x) = i\hat{\bar{\psi}}\gamma^0 = i\hat{\psi}^\dagger$ into Eq. (5.103), one gets

$$\{\hat{\psi}(t,\mathbf{x}),\, \hat{\psi}^\dagger(t,\mathbf{x}')\} \;=\; \delta(\mathbf{x}-\mathbf{x}'). \tag{5.106}$$

In the classical limit, all the operators should be replaced by the corresponding classical fields. Then the relations (5.105) and (5.106) mean, according to the standard quantization postulate, that the classical spinor fields should be anticommuting.

It is useful to define a normal form of the product of the fermionic annihilation and creation operators. Such a normal ordering means that all the creation operators are located on the left from all annihilation operators. Under the normal ordering, the permutations of fermionic annihilation and creation operators are performed by the anticommutation rules. In particular, the normal product

$$N(\hat{b}\hat{b}^\dagger) \;=\; -\,N(\hat{b}^\dagger\hat{b}) \;=\; -\,\hat{b}^\dagger\hat{b}. \tag{5.107}$$

Thus, transforming the Hamiltonian (5.102) to the normal form, one gets the positively defined expression defined

$$\hat{H} \;=\; \sum_{\lambda=\pm 1/2}\int d^3p\,\varepsilon(\mathbf{p})\,\big[\hat{a}^\dagger(\mathbf{p},\lambda)\,\hat{a}(\mathbf{p},\lambda) + \hat{b}^\dagger(\mathbf{p},\lambda)\hat{b}(\mathbf{p},\lambda)\,\big]. \tag{5.108}$$

Now let us express the operators of electric charge $\hat{Q}$ and three-dimensional momentum $\hat{\mathbf{P}}$ in terms of $\hat{a}$, $\hat{a}^\dagger$, $\hat{b}$, $\hat{b}^\dagger$. The operator $\hat{Q}$ is obtained from the classical dynamical invariant Q defined in Eq. (5.96), by replacing classical fields with field operators, using the expansions (5.100) and the normal ordering prescription for fermionic operators. The result is

$$\hat{Q} = e\int d^3x\hat{\bar{\psi}}\gamma^0\hat{\psi} \;=\; e\sum_{\lambda=\pm 1/2}\int d^3p\Big[\hat{a}^\dagger(\mathbf{p},\lambda),\hat{a}(\mathbf{p},\lambda) - \hat{b}^\dagger(\mathbf{p},\lambda),\hat{b}(\mathbf{p},\lambda)\Big]. \tag{5.109}$$

One can note both Hamiltonian (5.108) and charge operator (5.109) are written in normal form. This ordering is considered to be an element of definition for operators $\hat{H}$ and $\hat{Q}$. The operator of the three-dimensional momentum $\hat{\mathbf{P}}$ is constructed analogously to $\hat{H}$ and $\hat{Q}$. Using the normal ordering for fermionic operators, one gets

$$\hat{\mathbf{P}} \;=\; \sum_{\lambda=\pm 1/2}\int d^3p\,\mathbf{p}\big[\hat{a}^\dagger(\mathbf{p},\lambda)\,\hat{a}(\mathbf{p},\lambda) + \hat{b}^\dagger(\mathbf{p},\lambda)\,\hat{b}(\mathbf{p},\lambda)\big]. \tag{5.110}$$

The $\hat{H}, \hat{\mathbf{P}}, \hat{Q}$ are the operators of basic dynamical invariants in the quantized free spinor field theory. To describe the quantum states of free spinor field theory in terms of particles, we consider an eigenvalue problem for the Hamiltonian and find the energy spectrum of the model. To this end, we introduce the operators

$$\hat{N}_+ = \sum_{\lambda=\pm 1/2} \int d^3p\, \hat{a}^\dagger(\mathbf{p},\lambda)\hat{a}(\mathbf{p},\lambda), \qquad \hat{N}_- = \sum_{\lambda=\pm 1/2} \int d^3p\, \hat{b}^\dagger(\mathbf{p},\lambda)\hat{b}(\mathbf{p},\lambda), \quad (5.111)$$

with $\hat{N} = \hat{N}_+ + \hat{N}_-$. Using these operators, the charge operator $\hat{Q}$ can be cast in the form

$$\hat{Q} = e(\hat{N}_+ - \hat{N}_-). \quad (5.112)$$

It is not difficult to prove that the operators $\hat{H}$, $\hat{\mathbf{P}}$, $\hat{Q}$, $\hat{N}_+$ and $\hat{N}_-$ commute. Therefore, these operators have the same system of eigenvectors. The eigenvectors can be easily found for the operators $\hat{N}_+$ and $\hat{N}_-$, and have the form

$$\begin{aligned} &|\mathbf{p}_1,\lambda_1;\mathbf{p}_2,\lambda_2;\dots;\mathbf{p}_n,\lambda_n;\mathbf{p}'_1,\lambda'_1;\mathbf{p}'_2,\lambda'_2;\dots;\mathbf{p}'_m,\lambda'_m;\rangle \\ &= \hat{a}^\dagger(\mathbf{p}_1,\lambda_1)\dots\hat{a}^\dagger(\mathbf{p}_n,\lambda_n)\hat{b}^\dagger(\mathbf{p}'_1,\lambda'_1)\dots\hat{b}^\dagger(\mathbf{p}'_m,\lambda'_m)|0\rangle = |\Psi_{n,m}\rangle, \end{aligned} \quad (5.113)$$

where the vector $|0\rangle$ is defined by the conditions

$$\hat{a}(\mathbf{p},\lambda)|0\rangle = 0 \quad \text{and} \quad \hat{b}(\mathbf{p},\lambda)|0\rangle = 0, \quad \forall\,\mathbf{p},\ \forall\,\lambda. \quad (5.114)$$

Since the operators $\hat{a}^\dagger(\mathbf{p},\lambda)$ and $\hat{b}^\dagger(\mathbf{p},\lambda)$ anticommute, the state (5.113) is totally antisymmetric under permutation of the particles. This property perfectly corresponds to the antisymmetrization quantum postulate for the particles with a half-integer spin. In particular, $[\hat{a}^\dagger(\mathbf{p},\lambda)]^2 = [\hat{b}^\dagger(\mathbf{p},\lambda)]^2 = 0$. This relation is a mathematical expression of the Pauli principle in quantum mechanics.

Using some algebra, one can obtain the relation

$$\begin{aligned} \hat{H}|\Psi_{n,m}\rangle &= \left[\varepsilon(\mathbf{p}_1) + \dots + \varepsilon(\mathbf{p}_n) + \dots + \varepsilon(\mathbf{p}'_1) + \dots + \varepsilon(\mathbf{p}'_m)\right]|\Psi_{n,m}\rangle, \\ \hat{\mathbf{P}}|\Psi_{n,m}\rangle &= \left(\mathbf{p}_1 + \dots \mathbf{p}_n + \mathbf{p}'_1 + \dots + \mathbf{p}'_m\right)|\Psi_{n,m}\rangle, \\ \hat{N}_+|\Psi_{n,m}\rangle &= n|\Psi_{n,m}\rangle, \qquad \hat{N}_-|\Psi_{n,m}\rangle = m|\Psi_{n,m}\rangle, \\ \hat{Q}|\Psi_{n,m}\rangle &= e(n-m)|\Psi_{n,m}\rangle. \end{aligned} \quad (5.115)$$

The relations (5.115) serve as the basis for interpreting the states of the quantized free spinor field in terms of particles. Vectors $|\Psi_{n,m}\rangle$ (5.113) describe the states of a quantized spinor field with n particles of charge e and m particles with charge $-e$. The energy of the field in this state is given by the sum

$$\varepsilon(\mathbf{p}_1) + \dots + \varepsilon(\mathbf{p}_n) + \dots + \varepsilon(\mathbf{p}'_1) + \dots + \varepsilon(\mathbf{p}'_m)$$

and the momentum of the field is

$$\mathbf{p}_1 + \dots + \mathbf{p}_n + \mathbf{p}'_1 + \dots + \mathbf{p}'_m.$$

Thus, $\hat{a}(\mathbf{p},\lambda)$ and $\hat{a}^\dagger(\mathbf{p},\lambda)$ are the fermionic annihilation and creation operators, respectively, of particles with charge e, and $\hat{b}(\mathbf{p},\lambda)$ and $\hat{b}^\dagger(\mathbf{p},\lambda)$ are the fermionic annihilation and creation operators, respectively, of particles with charge $-e$. If e is the electron charge, then particles with charge e are electrons, and particles with charge $-e$ are positrons. In this sense, a positron is the antiparticle of the electron.

The vector $|0\rangle$ possesses the properties $\hat{H}|0\rangle = 0$, $\hat{\mathbf{P}}|0\rangle = 0$, $\hat{Q}|0\rangle = 0$. That is, the vector $|0\rangle$ is a state with a minimal (zero in normal ordering) energy, with zero momentum, zero charge and no particles or antiparticles. This vector is called the vacuum state of the spinor field.

The vectors $|\Psi_{n,m}\rangle$ (5.113) form a basis in the space of states of a quantized spinor field. This is the Fock basis for fermions. An arbitrary state vector $|\Psi\rangle$ can be expanded in this basis in the form

$$
\begin{aligned}
|\Psi\rangle = \sum_{n,m=0}^{\infty} \sum_{\lambda_1,..,\lambda_n=\pm 1/2} \sum_{\lambda'_1,...,\lambda'_m=\pm 1/2} \int d^3p_1 \dots d^3p_n, d^3p'_1 \dots d^3p'_m \\
\times \Psi_{n,m}(\mathbf{p}_1, \lambda_1; \dots ; \mathbf{p}_n, \lambda_m; \mathbf{p}'_1, \lambda'_1; \dots ; \mathbf{p}'_m, \lambda'_m) \\
\times |\mathbf{p}_1, \lambda_1; \dots ; \mathbf{p}_n, \lambda_n; \mathbf{p}'_1, \lambda'_1; \dots ; \mathbf{p}'_m, \lambda'_m\rangle,
\end{aligned}
\tag{5.116}
$$

where $\Psi_{n,m}$ are numerical coefficients. The set of vectors $|\Psi\rangle$ form the Fock space of the quantized spinor field.

Remark, The use of commutators or anticommutators for the quantization of a field means that we choose statistics for the corresponding particles. We have formulated a postulate according to which scalar field should be quantized in terms of commutators and spinor field in terms of anticommutators.

In principle, there is another way to achieve the same result, without introducing a special postulate. Instead, it is necessary to request the quantum theory to be Lorentz invariant and causal and to have a positive energy. It is possible to prove that, under these assumptions, the integer spin fields should be quantized as bosons, and the half-integer spin fields should be quantized as fermions. This statement is called the spin-statistics theorem (see, e.g., [250]).

5.6 Quantization of a free electromagnetic field

The action of the electromagnetic field has the form

$$
S[A] = -\frac{1}{4} \int d^4x \, F_{\mu\nu} F^{\mu\nu},
\tag{5.117}
$$

where $F_{\mu\nu} = \partial_\mu A_\nu - \partial_\nu A_\mu$. As we already know, this action is invariant under the gauge transformation (4.71), that is, $A'_\mu = A_\mu + \partial_\mu \xi(x)$, where $\xi(x)$ is an arbitrary scalar field.

As we already pointed out, the canonical formalism in its original formulation does not work directly in this theory. Therefore, we will use a modified approach. First of all, let us reformulate the model by rewriting

$$
\begin{aligned}
F_{\mu\nu}F^{\mu\nu} &= (\partial_\mu A_\nu - \partial_\nu A_\mu)(\partial^\mu A^\nu - \partial^\nu A^\mu) = 2\,\partial_\mu A_\nu(\partial^\mu A^\nu - \partial^\nu A^\mu) \\
&= -2\,A_\nu(\Box A^\nu - \partial^\nu \partial_\mu A^\mu) + 2\,\partial_\mu[A_\nu(\partial^\mu A^\nu - \partial^\nu A^\mu)].
\end{aligned}
$$

Omitting the total divergence term, we recast the action in the form

$$S[A] = \frac{1}{2}\int d^4x\, A_\mu(\eta^{\mu\nu}\Box - \partial^\mu\partial^\nu)A_\nu = \frac{1}{2}\int d^4x\, A_\mu \Box\Big(\eta^{\mu\nu} - \frac{\partial^\mu\partial^\nu}{\Box}\Big)A_\nu. \quad (5.118)$$

Here the operator $\frac{1}{\Box}$ is formally defined by its action on an arbitrary function $f(x)$ by the rule

$$\frac{1}{\Box} f(x) = \int \frac{d^4k}{(2\pi)^4}\, e^{ikx}\Big(-\frac{1}{k^2}\Big)\tilde{f}(k), \quad (5.119)$$

where the function $\tilde{f}(k)$ is defined through the Fourier transformation

$$f(x) = \int \frac{d^4k}{(2\pi)^4}\, e^{ikx}\tilde{f}(k) \quad (5.120)$$

and $kx = k_0t - \mathbf{kx}$ and $k^2 = \eta^{\mu\nu}k_\mu k_\mu = k_0^2 - \mathbf{k}^2$, as usual.

Introducing the operators

$$\theta^\mu{}_\nu = \delta^\mu{}_\nu - \frac{\partial^\mu\partial_\nu}{\Box}, \qquad \omega^\mu{}_\nu = \frac{\partial^\mu\partial_\nu}{\Box}, \quad (5.121)$$

it is easy to show that

$$\begin{aligned} &\theta^\mu{}_\alpha\,\theta^\alpha{}_\nu = \theta^\mu{}_\nu, \qquad \theta^\mu{}_\alpha\,\omega^\alpha{}_\nu = 0, \qquad \omega^\mu{}_\alpha\,\omega^\alpha{}_\nu = \omega^\mu{}_\nu, \\ &\omega^\mu{}_\alpha\,\theta^\alpha{}_\nu = 0, \qquad \theta^\mu{}_\alpha + \omega^\mu{}_\alpha = \delta^\mu{}_\nu. \end{aligned} \quad (5.122)$$

Therefore, the operators $\theta^\mu{}_\nu$ and $\omega^\mu{}_\nu$ are projection operators, or projectors.

Let us denote

$$A_\mu{}^\perp = \theta^\mu_\nu A^\nu \qquad \text{and} \qquad A_\mu^{\|} = \omega^\mu{}_\nu A^\nu. \quad (5.123)$$

It is easy to see that

$$\partial_\mu A^{\perp\,\mu} = 0, \qquad \theta_\mu^\nu A_\nu^{\|} = 0 \qquad \text{and} \qquad A_\mu = A_\mu^\perp + A_\mu^{\|}. \quad (5.124)$$

The field $A_\mu^{\|}$ can be always regarded as a divergence of some scalar field. Indeed,

$$A_\mu^{\|} = \omega_\mu{}^\nu A_\nu = \frac{\partial_\mu\partial^\nu}{\Box} A_\nu = \partial_\mu\varphi,$$

where $\varphi = \Box^{-1}\,\partial^\nu A_\nu$. As a result, $A_\mu = A_\mu^\perp + \partial_\mu\varphi$. $A_\mu^\perp$ is the transverse component of the vector field A_μ and $A_\mu^{\|}$ is the longitudinal component of the field.

Taking into account the definition $A_\mu^\perp$, we can rewrite the action (5.117) as follows:

$$\begin{aligned} S[A] &= \frac{1}{2}\int d^4x\, A_\mu \Box\theta^\mu{}_\nu A^\nu = \frac{1}{2}\int d^4x\, A_\mu\theta^\mu{}_\alpha\Box\theta^\alpha{}_\nu A^\nu \\ &= \frac{1}{2}\int d^4x\, A_\mu{}^\perp\Box A^{\perp\mu} = S[A^\perp]. \end{aligned} \quad (5.125)$$

Thus, the action of electromagnetic field depends only on the component $A_\mu^\perp$, and there is no $A_\mu^{\|}$-dependence. The equations of motion of a free field are $\partial^\mu F_{\mu\nu} = 0$, or

$$\Box A_\nu - \partial_\nu\partial^\mu A_\mu = \Box\theta_\nu{}^\alpha A_\alpha = \Box A_\nu^\perp = 0. \quad (5.126)$$

Due to the properties (5.124), under the gauge transformation $A'_\mu = A_\mu + \partial_\mu\partial_\mu\xi$, we have $A'^\perp_\mu = A_\mu^\perp$ and $\varphi' = \varphi + \xi$. It is clear that the component $A_\mu^\perp$ is gauge invariant. On the other hand, the component $A_\mu^{\|}$ is gauge dependent and can be changed arbitrarily by the gauge transformation. But since this component does not enter into the equations of motion, this is irrelevant for us.

Since the action does not depend on $A_\mu^{\|}$, we can consider this component arbitrary and fix it in a way we like. Fixing $A_\mu^{\|}$ is called gauge fixing, similar to what we considered in the classical treatment of the massless vector field. The simplest gauge fixing is $A_\mu^{\|} = 0$ or, equivalently, $\varphi = 0$, or simply the Lorentz gauge (4.67).

Although the action $S[A_\mu^{\perp}]$ (5.125) is written in terms of the gauge-invariant field $A_\mu^{\perp}$, this field is non-local. Therefore, it is more convenient to work in term of the initial gauge-dependent field A_μ but to impose Lorentz gauge fixing (4.67). In this gauge, we have $A_\mu^{\perp} = A_\mu$. Therefore, the action (5.125) becomes

$$S = \frac{1}{2}\int d^4x\, A_\mu \Box A^\mu, \tag{5.127}$$

with the equations of motion

$$\Box A_\mu = 0. \tag{5.128}$$

This equation must be supplemented by the gauge condition (4.67). This theory can be canonically quantized using the standard method. However, we have to remember that all consideration corresponds to the given gauge condition (4.67). The dependence on the choice of the gauge condition can be kept under control, as will be discussed below.

The Lagrangian of the action (5.127) can be transformed as follows:

$$\mathcal{L} = \frac{1}{2}A_\mu \Box A^\mu = -\frac{1}{2}\partial_\mu A_\nu \partial^\mu A^\nu + \text{total divergence}.$$

Omitting the total divergence, one obtains

$$\mathcal{L} = -\frac{1}{2}\eta^{\mu\alpha}\eta^{\nu\beta}\partial_\mu A_\nu \partial_\alpha A_\beta = \frac{1}{2}\big(\dot{A}_i\dot{A}_i - \partial_j A_i\,\partial_j A_i\big) - \frac{1}{2}\big(\dot{A}_0^2 - \partial_i A_0 \partial_i A_0\big). \tag{5.129}$$

Introducing the momenta

$$\pi_i = \frac{\partial \mathcal{L}}{\partial \dot{A}_i} = \dot{A}_i, \qquad \pi_0 = \frac{\partial \mathcal{L}}{\partial \dot{A}_0} = -\dot{A}_0, \tag{5.130}$$

after a small algebra we can construct the Hamiltonian function

$$\begin{aligned} H &= \int d^3x\big(\pi_i \dot{A}_i + \pi_0 \dot{A}_0 - \mathcal{L}\big)\Big|_{\dot{A}_i=\pi_i,\,\dot{A}_0=-\pi_0} \\ &= \frac{1}{2}\int d^3x\Big\{\big(\pi_i\pi_i + \partial_j A_i\,\partial_j A_i\big) - \big(\pi_0^2 + \partial_i A_0 \partial_i A_0\big)\Big\}. \end{aligned} \tag{5.131}$$

It is clear that we face a problem of the positiveness of energy of the classical theory in this case.

To quantize the theory, we introduce the operators $\hat{A}_i(\mathbf{x})$, $\hat{A}_0(\mathbf{x})$, $\hat{\pi}_i(\mathbf{x})$ and $\hat{\pi}_0(\mathbf{x})$. The canonical commutation relations are imposed in a standard way,

$$[\hat{A}_i(\mathbf{x}), \hat{\pi}_j(\mathbf{x}')] = i\delta(\mathbf{x}-\mathbf{x}')\,\delta_{ij} \quad \text{and} \quad [\hat{A}_0(\mathbf{x}), \hat{\pi}_0(\mathbf{x}')] = i\delta(\mathbf{x}-\mathbf{x}'), \tag{5.132}$$

while other commutators vanish. Let us remember that we follow the canonical quantization scheme in the Schrödinger picture, where the operators are time independent.

The Hamiltonian is constructed by the standard rule,

$$\hat{H} = \frac{1}{2}\int d^3x \Big\{ \big(\hat{\pi}_i\hat{\pi}_i + \partial_j\hat{A}_i\,\partial_j\hat{A}_i\big) - \big(\hat{\pi}_0^2 + \partial_i\hat{A}_0\partial_i\hat{A}_0\big)\Big\}. \tag{5.133}$$

The operators in the Heisenberg picture are defined as

$$\hat{A}_\mu(x) = e^{it\hat{H}}\,\hat{A}_\mu(\mathbf{x})\,e^{-it\hat{H}}, \qquad \hat{\pi}_\mu(x) = e^{it\hat{H}}\,\hat{\pi}_\mu(\mathbf{x})\,e^{-it\hat{H}}. \tag{5.134}$$

Then the Heisenberg equations of motion are

$$i\dot{\hat{A}}_\mu(x) = [\hat{A}_\mu, \hat{H}], \qquad \dot{\hat{\pi}}_\mu(x) = [\hat{\pi}_\mu, \hat{H}]. \tag{5.135}$$

Using the Hamiltonian operator (5.133) and canonical commutation relations, one gets the operator equation of motion

$$\Box\hat{A}_\mu(x) = 0, \tag{5.136}$$

which is the same as Eq. (5.128) for a classical field $A_\mu(x)$.

The solution to Eq. (5.136) can be written as a Fourier expansion in flat waves,

$$\hat{A}_\mu(x) = \int \frac{d^3k}{\sqrt{2(2\pi)^3\omega(\mathbf{k})}}\, e_\mu^{(\lambda)}(\mathbf{k})\,\Big[e^{-ikx}\hat{c}_\lambda(\mathbf{k}) + e^{ikx}\hat{c}_\lambda^\dagger(\mathbf{k})\Big], \tag{5.137}$$

where $e_\mu^{(\lambda)}(\mathbf{k})$, with $\lambda = 0, 1, 2, 3$, are polarization vectors, satisfying the conditions

$$k_\mu e^{\mu(1,2)} = 0, \qquad k_\mu e^{\mu(0)} = \omega(\mathbf{k}), \qquad k_\mu e^{\mu(3)} = -\omega(\mathbf{k}) \tag{5.138}$$

and also the normalization condition

$$\eta^{\mu\nu}\, e_\mu^{(\lambda_1)}(\mathbf{k})\, e_\nu^{(\lambda_2)}(\mathbf{k}) = \eta^{\lambda_1\lambda_2}. \tag{5.139}$$

It is possible to find an explicit form of the vectors $e_\mu^{(\lambda)}(\mathbf{k})$, but we do not need to do so. Since the *l.h.s.* of (5.137) is an operator, $\hat{c}_\lambda(\mathbf{k})$ and $\hat{c}_\lambda^\dagger(\mathbf{k})$ in the *r.h.s.* are operators too. To find the commutation relations for these operators, one can use commutators (5.132). It is convenient to rewrite the relations (5.132) in a form that is little bit different, substituting $\hat{\pi}_j = \dot{\hat{A}}_j$ and $\hat{\pi}_0 = -\dot{\hat{A}}_0$. This procedure leads to

$$\begin{aligned} &[\hat{A}_\mu(x), \hat{A}_\nu(x')]\big|_{t'=t} = 0, \qquad [\dot{\hat{A}}_\mu(x), \dot{\hat{A}}_\nu(x')]\big|_{t'=t} = 0, \\ &[\hat{A}_\mu(x), \dot{\hat{A}}_\nu(x')]\big|_{t'=t} = -i\eta_{\mu\nu}\,\delta(\mathbf{x}-\mathbf{x}'). \end{aligned} \tag{5.140}$$

These relations enable one to find the commutation relations for the operators $\hat{c}_\lambda(\mathbf{k})$ and $\hat{c}^\dagger_\lambda(\mathbf{k})$. For this, we substitute $\hat{A}_\mu(x)$ from Eq. (5.137) and the corresponding $\dot{\hat{A}}_\mu(x)$ into (5.140) and use the polarization vectors properties (5.138). As a result, we get

$$
\begin{aligned}
&[\hat{c}_\lambda(\mathbf{k}), \hat{c}_{\lambda'}(\mathbf{k}')] = 0, \qquad [\hat{c}^\dagger_\lambda(\mathbf{k}), \hat{c}^\dagger_{\lambda'}(\mathbf{k}')] = 0,\\
&[\hat{c}_\lambda(\mathbf{k}), \hat{c}^\dagger_{\lambda'}(\mathbf{k}')] = -\eta_{\lambda\lambda'}\delta(\mathbf{k}-\mathbf{k}').
\end{aligned}
\tag{5.141}
$$

Consider the structure of the last commutators in (5.141) separately for $\lambda = i,\ \lambda' = j$ and $\lambda = \lambda' = 0$:

$$
[\hat{c}_i(\mathbf{k}), \hat{c}^\dagger_j(\mathbf{k}')] = \delta_{ij}\delta(\mathbf{k}-\mathbf{k}') \qquad \text{and} \qquad [\hat{c}_0(\mathbf{k}), \hat{c}^\dagger_0(\mathbf{k}')] = -\delta(\mathbf{k}-\mathbf{k}'). \tag{5.142}
$$

Thus, the operators $\hat{c}_i(\mathbf{k})$ and $\hat{c}^\dagger_i(\mathbf{k})$ obey the standard commutation relations for the bosonic annihilation and creation operators. The main difference with the scalar case is that the operators $\hat{c}_0(\mathbf{k})$ and $\hat{c}^\dagger_0(\mathbf{k})$ have the opposite sign in the commutator.

Expressing the Hamiltonian operator (5.133) in terms of the operators $\hat{c}_\lambda(\mathbf{k})$ and $\hat{c}^\dagger_\lambda(\mathbf{k})$ can be done using the explicit form (5.137) of the field operator $\hat{A}_\mu(x)$. Using the normalization condition (5.139) and the normal ordering, we arrive at

$$
\hat{H} = -\int d^3k\, \omega(\mathbf{k})\eta^{\lambda\lambda'}\hat{c}^\dagger_\lambda(\mathbf{k})\hat{c}_{\lambda'}(\mathbf{k}) = \int d^3k\, \omega(\mathbf{k})\big[\hat{c}^\dagger_i(\mathbf{k})\hat{c}_i(\mathbf{k}) - \hat{c}^\dagger_0(\mathbf{k})\,\hat{c}_0(\mathbf{k})\big], \tag{5.143}
$$

where $\omega(\mathbf{k}) = |k^0| = \sqrt{\mathbf{k}^2}$. It is evident that the Hamiltonian (5.143) is not positively defined, and, therefore, one can expect a problem with the positiveness of energy.

The Fock basis is constructed analogously to the scalar field theory. The vacuum $|0\rangle$ is defined by the equations

$$
\hat{c}_\lambda(\mathbf{k})|0\rangle = 0, \qquad \forall\,\mathbf{k},\ \forall\lambda. \tag{5.144}
$$

Then the basis vectors can be built by acting the operators $\hat{c}^\dagger_\lambda(\mathbf{k})$ on the vacuum state. However, in the present case, this construction yields a problem. Let us consider the one-particle state $|\varphi\rangle = \int d^3k\, \varphi(\mathbf{k})\, \hat{c}^\dagger_0(\mathbf{k})|0\rangle$ and calculate

$$
\begin{aligned}
\langle\varphi|\varphi\rangle &= \int d^3k\, d^3k'\varphi^*(\mathbf{k})\,\varphi(\mathbf{k}')\,\langle 0|\hat{c}_0(\mathbf{k})\hat{c}^\dagger_0(\mathbf{k}')0\rangle\\
&= -\int d^3k\, d^3k'\varphi^*(\mathbf{k})\,\varphi(\mathbf{k}')\delta(\mathbf{k}-\mathbf{k}') = -\int d^3k\, |\varphi(\mathbf{k})|^2 < 0.
\end{aligned}
\tag{5.145}
$$

The last relation means that the corresponding Fock space contains a vector with the negative norm, leading to a problem with the probability interpretation in quantum theory. This is a price one has to pay for considering the modified Lagrangian $\frac{1}{2}A_\mu\Box A^\mu$ instead of the true Lagrangian $-\frac{1}{4}F_{\mu\nu}F^{\mu\nu}$.

Indeed, both of the problems with the positiveness of energy and the norm are due to that fact that, until now, we did not take into account the Lorentz gauge fixing condition (4.67), that is, $\partial_\mu A^\mu = 0$. Indeed, it is only under this condition that the equations of motion in the modified theory are equivalent to the ones in the initial theory. Thus, let us discuss how the condition can be used in quantum theory. The first "natural" idea would be to implement the operator condition $\partial_\mu \hat{A}^\mu(x) = 0$; however, this approach leads to a contradiction. The aforementioned relation means $\hat{\pi}_0 + \partial_i \hat{A}_i = 0$. Then,

$$0 = [\hat{A}_0(x), \partial_\mu \hat{A}^\mu(x)]\Big|_{t=t'} = -[\hat{A}_0(\mathbf{x}'), \hat{\pi}_0(\mathbf{x}) + \partial_i \hat{A}_i(\mathbf{x})]\Big|_{t=t'} = -i\delta(\mathbf{x} - \mathbf{x}') \neq 0.$$

Thus, imposing the operator equality $\partial_\mu \hat{A}^\mu(x) = 0$ is inconsistent.

Let us relax operator equality by constructing a subset of Fock space such that all matrix elements of the operator $\partial_\mu \hat{A}^\mu(x)$ over the states from this subset vanish. We will call these *the physical states* and denote them $|phys\rangle$. By definition, for any two such states,

$$\langle phys'|\partial_\mu \hat{A}^\mu(x)|phys\rangle = 0. \tag{5.146}$$

The solution of (5.137) enables us to write $\hat{A}_\mu(x) = \hat{A}_\mu^{(+)}(x) + \hat{A}_\mu^{(-)}(x)$, where

$$\hat{A}_\mu^{(+)}(x) = \int \frac{d^3k}{\sqrt{2(2\pi)^3\omega(\mathbf{k})}} e_\mu^{(\lambda)}(\mathbf{k})\, e^{-ikx}\, \hat{c}_\lambda(\mathbf{k}), \tag{5.147}$$

and $\hat{A}_\mu^{(-)}(x) = \big[\hat{A}_\mu^{(+)}(x)\big]^\dagger$. Then the condition (5.146) is cast in the form

$$\langle phys'|\partial_\mu \hat{A}^{(+)\mu}(x) + \partial_\mu \hat{A}^{(-)\mu}(x)|phys\rangle = 0.$$

To satisfy this condition, it is sufficient to require

$$\partial_\mu \hat{A}^{(+)\mu}(x)|phys\rangle = 0, \qquad \forall\, |phys\rangle. \tag{5.148}$$

Eq. (5.148) should be regarded as conditions to find the physical states $|phys\rangle$. Let us explore these conditions in more detail. Using (5.147), one gets

$$\partial_\mu \hat{A}^{(+)\mu}(x) = -i \int \frac{d^3k}{\sqrt{2(2\pi)^3\omega(\mathbf{k})}}\, k^\mu\, e_\mu^{(\lambda)}(\mathbf{k})\, e^{-ikx}\, \hat{c}_\lambda(\mathbf{k}). \tag{5.149}$$

Next, we use the properties of polarization vectors (5.138) and get

$$\partial_\mu \hat{A}^{(+)\mu}(x)|phys\rangle = -i \int \frac{d^3k}{\sqrt{2(2\pi)^3\omega(\mathbf{k})}}\, \omega(\mathbf{k})\, e^{-ikx} \Big[\hat{c}_0(\mathbf{k}) - \hat{c}_3(\mathbf{k})\Big]|phys\rangle.$$

Thus, the equation for the physical states has the form

$$\big[\hat{c}_0(\mathbf{k}) - \hat{c}_3(\mathbf{k})\big]|phys\rangle = 0, \qquad \forall\, |phys\rangle \quad \text{and} \quad \forall\, \mathbf{k}. \tag{5.150}$$

Let us calculate the matrix element of the Hamiltonian between the two physical states

$$\langle phys'|\hat{H}|phys\rangle = \int d^3k\; \omega(\mathbf{k}) \Big\{ \langle phys'|\, \hat{c}_1^\dagger(\mathbf{k})\hat{c}_1(\mathbf{k}) + \hat{c}_2^\dagger(\mathbf{k})\hat{c}_2(\mathbf{k})|phys\rangle + \langle phys'|\, \hat{c}_3^\dagger(\mathbf{k})\hat{c}_3(\mathbf{k}) - \hat{c}_0^\dagger(\mathbf{k})\hat{c}_0(\mathbf{k})|phys\rangle \Big\}. \tag{5.151}$$

On the other hand, $\hat{c}_0(\mathbf{k})|phys\rangle = \hat{c}_3(\mathbf{k})|phys\rangle$, $\forall\ |phys\rangle$. Taking this into account,

$$\langle phys'|\hat{H}|phys\rangle = \int d^3k\,\omega(\mathbf{k})\ \langle phys'|\hat{c}_1^\dagger(\mathbf{k})\hat{c}_1(\mathbf{k}) + \hat{c}_2^\dagger(\mathbf{k})\hat{c}_2(\mathbf{k})|phys\rangle. \qquad (5.152)$$

Therefore, in the subspace of physical states, the Hamiltonian is reduced to

$$\hat{H} = \int d^3k\,\omega(\mathbf{k})\,\big[\hat{c}_1^\dagger(\mathbf{k})\hat{c}_1(\mathbf{k}) + \hat{c}_2^\dagger(\mathbf{k})\hat{c}_2(\mathbf{k})\big]. \qquad (5.153)$$

We can see that, in this physical subspace, the energy is automatically non-negative.

Usually, the following terminology is used. The particles described by the annihilation and creation operators $\hat{c}_\lambda(\mathbf{k})$ and $\hat{c}_\lambda^\dagger(\mathbf{k})$ are called photons. The photons with $\lambda = 1, 2$ are called physical. Namely, these are the photons which can be detected in experiments. This identification is in complete agreement with the fact that the electromagnetic wave has two transverse polarizations. Furthermore, the photons with $\lambda = 3$ are called longitudinal states, and the photons with $\lambda = 0$ are called temporal states.

The longitudinal and temporal photons are non-physical, and their role is to provide the relativistic symmetry of the theory. The equation (5.148) for finding the physical state shows that such a state can contain longitudinal and temporal photons, but only in a very special combination. The described way of quantizing the electromagnetic field is called the Gupta-Bleuler quantization. To complete the story, let us mention that there exists a general solution to Eq. (5.148), which we did not consider since, for our purposes, it was not necessary.

Exercises

5.1. Assume that the operators $\hat{a}$ and $\hat{a}^\dagger$ satisfy the commutation relation $[\hat{a}, \hat{a}^\dagger] = \hat{1}$. a) Calculate $[\hat{a}, (\hat{a}^\dagger)^n]$. b) Let the vector $|0\rangle$ satisfy the condition $\hat{a}|0\rangle = 0$. Consider the vectors $|n\rangle = \frac{1}{\sqrt{n!}}(\hat{a}^\dagger)^n|0\rangle$ and the operator $\hat{n} = \hat{a}^\dagger\hat{a}$. Prove that $\hat{n}|n\rangle = n|n\rangle$, where $n = 0, 1, 2, \dots$. c) Prove that $\hat{n}\hat{a}^\dagger|n\rangle = (n+1)\hat{a}^\dagger|n\rangle$ and $\hat{n}\hat{a}|n\rangle = (n-1)\hat{a}|n\rangle$. d) Let $\langle 0|0\rangle = 1$. Show that $\langle m|n\rangle = \delta_{mn}$.

5.2. Let the operators $\hat{b}$ and $\hat{b}^\dagger$ satisfy the anticommutation relations $\{\hat{b}, \hat{b}^\dagger\} = \hat{1}$ and $\{\hat{b}^\dagger, \hat{b}^\dagger\} = \{\hat{b}, \hat{b}\} = 0$. a) Let the vector $|0\rangle$ be such that $\hat{b}|0\rangle = 0$. Consider the vectors $|n\rangle = \frac{1}{\sqrt{n!}}(\hat{b}^\dagger)^n|0\rangle$. Denote $\hat{n} = \hat{b}^\dagger\hat{b}$ and prove that $\hat{n}|n\rangle = n|n\rangle$. Which are the possible values of n? b) Prove that $\hat{n}\hat{b}^\dagger|n\rangle = (n+1)\hat{b}^\dagger|n\rangle$ and $\hat{n}\hat{b}|n\rangle = (n-1)\hat{b}|n\rangle$. c) Assuming the normalization of the vacuum state $\langle 0|0\rangle = 1$, show that $\langle m|n\rangle = \delta_{mn}$.

5.3. Consider $\hat{\varphi}(x)$ to be the operator of a free real scalar field. Calculate the commutator function $\Delta^{(0)}(x, x') = [\hat{\varphi}(x), \hat{\varphi}(x')]$ and find

$$\Delta(x, x')\Big|_{t'=t} \qquad \text{and} \qquad \frac{\partial\Delta(x, x')}{\partial t}\bigg|_{t'=t}.$$

Explain how the last two quantities can be found without special calculations.

5.4. The Heaviside step function is defined as

$$\theta(t) = \begin{cases} 0\ , & t < 0 \\ 1/2\ , & t = 0\ . \\ 1\ , & t > 0 \end{cases} \tag{5.154}$$

Prove the integral representation

$$\theta(t) = \frac{1}{2\pi i}\int_{-\infty}^{\infty} \frac{d\omega}{\omega - i\varepsilon} e^{i\omega t}, \qquad \text{where} \qquad \varepsilon \to +0.$$

5.5. Denote

$$D_{ret}(x,x') = -i\theta(t-t')\Delta^{(0)}(x,x'), \qquad D_{adv}(x,x') = i\theta(t'-t)\Delta^{(0)}(x,x'),$$
$$D(x,x') = -i(\theta(t-t')\langle 0|\hat{\varphi}(x)\hat{\varphi}(x')|0\rangle + \theta(t'-t)\langle 0|\hat{\varphi}(x')\hat{\varphi}(x)|0\rangle).$$

Functions $D_{ret}(x,x')$, $D_{adv}(x,x'')$, $andD(x,x')$ are, respectively, the retarded, advanced and causal Green functions of the Klein–Gordon equation. Function $D(x,x')$ is also called the Feynman propagator of the scalar field.
a) Derive the equations of motion for the functions $D_{ret}(x,x')$, $D_{adv}(x,x')$, $andD(x,x')$.
b) Calculate the Fourier transform of the functions $D_{ret}(x,x')$, $D_{adv}(x,x')$, $andD(x,x')$.

5.6. Consider the theory of the free real scalar field with the Lagrangian

$$\mathcal{L} = \frac{1}{2}\partial^\mu\varphi\partial_\mu\varphi - \frac{1}{2}m^2\varphi^2 + \varphi(x)J(\mathbf{x}).$$

Here $J(\mathbf{x})$ is the time-independent external field. Carry out the canonical quantization of this model, introduce annihilation and creation operators, construct the Hamiltonian with a normal ordering and find the ground (vacuum) state energy.

5.7 Let the operator of the energy-momentum in the real scalar field theory be

$$\hat{P}_\mu = \int d^3p\ p_\mu\, \hat{a}^\dagger(\mathbf{p})\hat{a}(\mathbf{p}),$$

where $p_0 = \varepsilon(\mathbf{p})$. Calculate $[\hat{P}_\mu, \hat{\varphi}(x)]$.

5.8. Verify the solution (5.97) and the relations (5.99).

5.9. Derive the relations (5.104).

5.10. Let $\hat{\psi}(x)$ and $\hat{\bar{\psi}}(x)$ be the operators of a free spinor field. Calculate the anticommutator function $\Delta^{(\frac{1}{2})}(x,x') = \{\hat{\psi}(x), \hat{\bar{\psi}}(x')\}$, and find $\Delta^{(\frac{1}{2})}(x,x')|_{t'=t}$. Explain why the last quantity can be written down without special calculations.

5.11. Denote

$$\begin{aligned} S_{ret}(x,x') &= -i\theta(t-t')\Delta^{(\frac{1}{2})}(x,x'), \\ S_{adv}(x,x') &= -i\theta(t'-t)\Delta^{(\frac{1}{2})}(x,x'), \\ S(x,x') &= -i(\theta(t-t')\langle 0|\hat{\psi}(x)\hat{\bar{\psi}}(x')|0\rangle - \theta(t'-t)\langle 0|\hat{\bar{\psi}}(x')\hat{\psi}(x)|0\rangle). \end{aligned} \tag{5.155}$$

Functions $S_{ret}(x,x')$, $S_{adv}(x,x')$, $andS(x,x')$ are called the retarded, advanced and causal Green functions of the Dirac equation, respectively. The function $S(x,x')$ is

also called the Feynman propagator of the spinor field.
a) Derive the equations of motion for the functions $S_{ret}(x,x')$, $S_{adv}(x,x')$, $and S(x,x')$. b) Calculate the Fourier transform of the functions $S_{ret}(x,x')$, $S_{adv}(x,x')$, $and S(x,x')$. c) Prove that

$$S(x,x') = (i\gamma^\mu \partial_\mu + m)D(x,x'), \tag{5.156}$$

where $D(x,x')$ is the causal Green function of the Klein–Gordon equation.

5.12 Verify the solution (5.137) with the conditions (5.138) and (5.139).

5.13. Let $\hat{A}_\mu(x)$ be the operator of a free electromagnetic field. Calculate the commutator function $\Delta^{(1)}_{\mu\nu}(x,x') = [\hat{A}_\mu(x), \hat{A}_\nu(x')]$, and discuss its relation with a similar commutator for a real scalar model.

5.14. The function

$$\Delta_{\mu\nu}(x,x') = -i(\theta(t-t')\langle 0|\hat{A}_\mu(x)\hat{A}_\nu(x')|0\rangle + \theta(t'-t)\langle 0|\hat{A}_\nu(x')\hat{A}_\mu(x)|0\rangle) \tag{5.157}$$

is called the Feynman propagator of the electromagnetic field. Derive the equations of motion for $\Delta_{\mu\nu}(x,x')$ in the Lorentz gauge, and calculate the Fourier transform of this function.

5.15. Consider the equations of motion of free electromagnetic field in the Coulomb gauge $\partial_i A_i = 0$. Show that the physical components of the potential $A_\mu(x)$ are $A_i^{tr}(x)$, where the three-dimensional transverse vector satisfies the condition $\partial_i A_i^{tr} \equiv 0$. Carry out the canonical quantization, and construct the Hamiltonian in terms of the creation and annihilation operators of the physical photons.

Comments

Quantization of free scalar, spinor, and electromagnetic fields and interpretation of quantum states in terms of particles is discussed in full technical detail in many textbooks (see, e.g., [187], [250], [304], [274], [152], [276]). Here we discussed only the general scheme of quantization, and refer the reader to look for the further details in these books.

The Gupta-Bleuler quantization of electromagnetic fields, including the general solution to the equation (5.148), is described in detail, e.g., in the book by S. Schweber [276].

As we pointed out, the standard canonical quantization scheme does not apply in its literal form to spinor and electromagnetic fields. The same is true for the Yang-Mills and gravitational fields. From the formal point of view, these theories are characterized by constraints in the phase space which should not occur in the standard canonical construction. The Hamiltonian formalism for theories with constraints in phase space was developed by Dirac (see, e.g., the book by P.A.M. Dirac, [111]). The modern and more general approach to the Hamiltonian formalism for theories with constraints in phase space and its application to quantum field theory models are considered in the book by D.M. Gitman and I.V. Tyutin [160].

6
The scattering matrix and the Green functions

In the previous chapter, we considered the quantization of free fields and discussed how the quantum states of free fields are described in terms of particles. Starting from this chapter, we develop a perturbative formalism to describe the interactions of quantized fields and, in particular, the interactions of particles in terms of their quantum fields.

Two general observations are in order. First of all, the main reason why we cannot deal with the interacting theory in the same way as we did with the free theory is that the interacting theories are more complicated. From the practical side, this means that, in interacting theories, one cannot write the general solution of the equation for a field operator as a simple linear combination of the flat waves, such as we did, for instance, in Eq. (5.50) for a scalar field. Consequently, it is not possible to construct the annihilation and creation operators in a simple way, identify the vacuum state in a simple way, etc. All that is simple in a free theory gets complicated when interactions are switched on. The second question is, why do we need perturbation theory? The reason is that perturbation theory is sufficient in many cases, because many interactions of practical interest are weak. And in other interesting cases, e.g., in quantum chromodynamics (QCD), describing strong interactions, there are well-developed non-perturbative approaches. However, the fact that they are considered "well-developed" does not mean that they are as perturbative quantum theory is. All in all, the perturbative approach remains the cornerstone and the solid base of modern quantum field theory, and we will concentrate on it in this book.

6.1 Particle interactions and asymptotic states

The main applications of the quantum field theory are related to elementary particle physics. Schematically, the experiments in particle physics are organized as follows. Beams composed of a finite number of initially free particles with certain energies, three-dimensional momenta, spin projections, masses, electric charges and other individual characteristics collide with each other or with some targets, interact and then fly away, becoming free again. In the process of interaction, some of the particles may disappear, some new particles may emerge and the parameters characterizing the particles may change. The experimentally measurable quantities are the scattering cross-sections, related to the transition probabilities. Since the particles are described in terms of quantized fields, the methods of calculating the cross-sections are formulated in the framework of quantum field theory.

Introduction to Quantum Field Theory with Applications to Quantum Gravity. Iosif L. Buchbinder and Ilya L. Shapiro, Oxford University Press (2021). © Iosif L. Buchbinder and Ilya Shapiro.
DOI: 10.1093/oso/9780198838319.003.0006

From the formal point of view, the calculation of the cross-sections means the following quantum mechanical problem. At some initial time moment t_1, there is certain quantum state $|\Psi_1\rangle$ of a system of initially free particles. What is the probability w of detecting a system of final free particles in a certain quantum state $|\Psi_2\rangle$, at the time instant t_2? According to the postulates of quantum mechanics, the answer is given by the expression

$$w = \big|\langle\Psi_2|\hat{U}(t_2,t_1)|\Psi_1\rangle\big|^2 = \big|\langle\Psi_2,t_2|\Psi_1,t_1\rangle\big|^2. \tag{6.1}$$

Here $\hat{U}(t_2,t_1) = e^{-i(t_2-t_1)\hat{H}}$ is the evolution operator, $|\Psi,t\rangle = e^{it\hat{H}}|\Psi\rangle$ is the quantum state in the Heisenberg dynamical picture, and the states $|\Psi_1\rangle$ and $|\Psi_2\rangle$ are defined in the Schrödinger dynamical picture. It is important to note that usually the time-independent Heisenberg states are obtained from the time-dependent Schrödinger states; see Eq. (5.20). However, in (6.1) we perform an inverse transformation of the expression for the probability in terms of the time-independent Schrödinger states, into the time-dependent Heisenberg states.

From the formal point of view, the initial and final particles cannot be treated as free while they are located at finite distances one from another, since the fields associated with the particles fill the whole space and, hence, interactions are not switched off at any finite distances. Therefore, we should only consider initial and final particles that are infinitely far away from each other. But particles can only go infinitely far away from each other when there are infinite time intervals. This means that we have to assume that the time difference $t_2 - t_1$ tends to infinity for asymptotic free states. For example, we can set $t_1 \to -\infty$ and $t_2 \to \infty$. According to this assumption, the correct expression for the probability (6.1) is as follows:

$$w = \lim_{t_1\to-\infty,t_2\to\infty} \big|\langle\Psi_2,t_2|\Psi_1,t_1\rangle\big|^2. \tag{6.2}$$

We denote the probability amplitude (or transition amplitude) from initial to the final state in expression (6.2) as

$$\mathcal{A}_{in\to out} = \langle\Psi_2,out|\Psi_1,in\rangle, \tag{6.3}$$

and call the states $|\Psi_1,in\rangle$ and $|\Psi_1,out\rangle$ the asymptotic states of the initial and final particles, respectively.

For simplicity and definiteness, further consideration will be carried out for the interacting real scalar field theory. The generalizations for other theories are mostly straightforward. For example, the spinor and vector field theories are technically more complicated, but the general approach is the same.

Let us clarify a form of the states $|\Psi_1,in\rangle$ and $|\Psi_2,out\rangle$ in Eq. (6.3). Consider $\hat{\varphi}(x)$ to be the Hermitian scalar field operator in the Heisenberg's dynamical picture. We assume that, in the distant past, at $t\to-\infty$, there was a system of initially free identical particles with masses m.

The free particles are described by the solution for a free quantized field, which we denote $\hat{\varphi}_{in}(x)$. The free field $\hat{\varphi}_{in}(x)$ satisfies the Klein-Gordon equation

$$(\Box + m^2)\hat{\varphi}_{in}(x) = 0,$$

and, as was shown in section 5.3, this field can be presented as

$$\hat{\varphi}_{in}(x) = \int \frac{d^3p}{\sqrt{2(2\pi)^3\varepsilon(\mathbf{p})}} \Big[\hat{a}_{in}(\mathbf{p})e^{-ip\,x} + \hat{a}^\dagger_{in}(\mathbf{p})e^{ip\,x}\Big]. \tag{6.4}$$

Here $\hat{a}_{in}(\mathbf{p})$ and $\hat{a}^\dagger_{in}(\mathbf{p})$ are the annihilation and creation operators of the free initial particles with momentum $\mathbf{p}$, respectively. It is assumed that, in the distant past, all observables for the system are constructed in terms of the initial free field $\hat{\varphi}_{in}(x)$ or, equivalently, in terms of the operators $\hat{a}_{in}(\mathbf{p})$ and $\hat{a}^\dagger_{in}(\mathbf{p})$. On the other hand, after quantization, all observables in the theory are given by operators constructed from the field operator $\hat{\varphi}(x)$ and the corresponding momentum operator $\hat{\pi}(x)$. Therefore, we are assuming that

$$t \to -\infty, \qquad \hat{\varphi}(x) \to \hat{\varphi}_{in}(x). \tag{6.5}$$

Thus, at $t \to -\infty$ all of the effects of interaction in the operator equations of motion for $\hat{\varphi}(x)$ adiabatically vanish. Regardless of the clear physical sense of the relation (6.5), it turns out to be extremely non-trivial, as will be discussed in brief.

Taking into account relations (6.5) and (6.4), we can construct the Fock space of the free initial particles with the basis

$$|\mathbf{q}_1, \dots, \mathbf{q}_N\rangle_{in} \;=\; \hat{a}^\dagger_{in}(\mathbf{q}_1)\dots\hat{a}^\dagger_{in}(\mathbf{q}_N)|0\rangle_{in}, \qquad N = 0, 1, 2 \dots . \tag{6.6}$$

Here $|0\rangle_{in}$ is the vacuum of the initial free particles. The vectors (6.6) describe the quantum state with N free particles, before being scattered by interaction forces.

Analogously, we assume that, in the distant future, at $t \to \infty$, there will be a system of free final particles with observable masses m. These free particles are described by the free field operator, which we denote $\hat{\varphi}_{out}(x)$. This operator satisfies the Klein-Gordon equation

$$(\Box + m^2)\hat{\varphi}_{in}(x) = 0 \tag{6.7}$$

and can be written in the form

$$\hat{\varphi}_{out}(x) = \int \frac{d^3p}{\sqrt{2(2\pi)^3\varepsilon(\mathbf{p})}} \big[\hat{a}_{out}(\mathbf{p})e^{-ip\,x} + \hat{a}^\dagger_{out}(\mathbf{p})e^{ip\,x}\big]. \tag{6.8}$$

Here $\hat{a}_{out}(\mathbf{p})$ and $\hat{a}^\dagger_{out}(\mathbf{p})$ are, respectively, the annihilation and creation operators of the free final particles. It is assumed that, in the distant future, all observables for the system under consideration are constructed in terms of the final free field $\hat{\varphi}_{out}(x)$ or, which is the same, in terms of the operators $\hat{a}_{out}(\mathbf{p})$ and $\hat{a}^\dagger_{out}(\mathbf{p})$. On the other hand, after quantization, all observables in the theory are given by operators constructed in terms of the field operator $\hat{\varphi}(x)$ and the corresponding momentum operator $\hat{\pi}(x)$. Therefore, it is assumed that

$$t \to \infty, \qquad \hat{\varphi}(x) \to \hat{\varphi}_{out}(x). \tag{6.9}$$

This relation means that, at $t \to \infty$, respectively, the effects of interaction in the operator equations of motion for $\hat{\varphi}(x)$ adiabatically vanish, and thus we get the free equations of motion. The assumption that the interaction terms in the equations of motion efficiently vanish in the distant past and the distant future is called the adiabatic hypothesis. The mathematical expression of the adiabatic hypothesis are relations (6.5) and (6.9).

Taking into account the relations (6.9) and (6.8), we construct the Fock space of the free final particles with the basis

$$|\mathbf{p}_1, \ldots, \mathbf{p}_M\rangle_{out} = \hat{a}^\dagger_{out}(\mathbf{p}_1) \ldots \hat{a}^\dagger_{out}(\mathbf{p}_M)|0\rangle_{out}. \tag{6.10}$$

Here $|0\rangle_{out}$ is the vacuum of the free final particles, and $M = 0, 1, 2 \ldots$. The vectors (6.10) describe the quantum state with M particles after collision.

It is clear that the vectors $|N\rangle_{in} = |\mathbf{q}_1, \ldots, \mathbf{q}_N\rangle_{in}$ and $|M\rangle_{out} = |\mathbf{p}_1, \ldots, \mathbf{p}_M\rangle_{out}$ should be used in Eq. (6.3) as the asymptotic states $|\Psi_1, in\rangle$ and $|\Psi_2, out\rangle$ of the initial and final states, respectively. As a result, the transition amplitude (6.3) has the form

$$\mathcal{S}_{MN} = \mathcal{S}_{N\to M} = {}_{out}\langle M|N\rangle_{in}. \tag{6.11}$$

The infinite-dimensional matrix $\mathcal{S}_{MN}$ is called the scattering matrix or $\mathcal{S}$-matrix. The $\mathcal{S}$-matrix is one of the main objects of interest in quantum field theory.

Let $\hat{H}$ be an exact Hamiltonian, constructed using canonical quantization in terms of the operators $\hat{\varphi}(x)$ and $\hat{\pi}(x)$. According to the hypothesis formulated above, one can write

$$\begin{aligned} t \to -\infty, &\qquad \hat{H} \to \hat{H}_{in}; \\ t \to \infty, &\qquad \hat{H} \to \hat{H}_{out}. \end{aligned} \tag{6.12}$$

Here $\hat{H}_{in}$ and $\hat{H}_{out}$ are the Hamiltonians of the free initial and final particles, correspondingly. The spectrum of the free Hamiltonian has been described in section 5.3, while the exact spectrum of the Hamiltonian $\hat{H}$ is unknown. However, physical arguments allow us to make reasonable assumptions about some of the eigenvectors of $\hat{H}$.

Consider an empty three-dimensional space without initial particles and external fields. The corresponding quantum state is the initial vacuum $|0\rangle_{in}$. Since in this space there are no particles, there is no interaction that could change this state. This means that the vacuum state is stable and unique,

$$|0\rangle_{in} = |0\rangle_{out} = |0\rangle, \tag{6.13}$$

where $|0\rangle$ is the ground state of the full Hamiltonian. Without loss of generality, we can consider the corresponding eigenvalue to be zero. Therefore, the state $|0\rangle$ is the vacuum for the full interacting theory. Physically, this assumption means that the particles can not be spontaneously created from the vacuum. Now, consider a space with a single initial particle in a state with a certain three-dimensional momentum $|\mathbf{p}\rangle_{in}$. Since there are no other particles in this space the external fields are absent, there are no interactions capable of changing this state. Therefore, this one-particle state is stable and unique,

$$|\mathbf{p}\rangle_{in} = |\mathbf{p}\rangle_{out} = |\mathbf{p}\rangle, \tag{6.14}$$

where $|\mathbf{p}\rangle$ is an exact one-particle eigenstate of the full Hamiltonian with a certain three-dimensional momentum. However, if we consider a system with two or more particles, the situation changes drastically. The particles can collide, interact, affect each other and exchange their individual energies, momenta and so on. As a result, their states can be time dependent. One can say that the vacuum and the one-particle state correspond to the discrete spectrum of the full Hamiltonian $\hat{H}$. All other eigenvectors of $\hat{H}$ lie in the continuum spectrum. The detailed discussion of the spectrum of the Hamiltonian in quantum field theory, on the basis of relativistic symmetry, is discussed, e.g., in the books [57, 303, 152].

Remark 1. Although the relations (6.5) and (6.9) look physically evident, they generate many questions from the mathematical point of view. First of all, in what sense could these expressions to be understood? The operators $\hat{\varphi}(x)$, $\hat{\varphi}_{in}(x)$ and $\hat{\varphi}_{out}(x)$ are ones that act in the infinite-dimensional space of functions. For such operators, there may be different and non-equivalent definitions of the limits.

One can show that the limit in relations (6.5) and (6.9) is understood in the weak sense. The definition of this notion is as follows. Suppose $\hat{A}(\tau)$ and $\hat{B}$ are two operators acting in an infinite-dimensional space and that the first of them depends on the parameter τ. It is said that the operator $\hat{B}$ is a weak limit of the operator $\hat{A}(\tau)$ if

$$\lim_{\tau\to\tau_0} \langle\Phi|\hat{A}(\tau)|\Psi\rangle \;=\; \langle\Phi|\hat{B}|\Psi\rangle, \tag{6.15}$$

where $|\Phi\rangle$ and $|\Psi\rangle$ are two arbitrary vectors in the space under consideration. The detailed discussion of this point can be found, e.g., in the books [57, 303, 152].

Remark 2. Consider the equation of motion for the operator $\hat{\varphi}(x)$,

$$(\Box + m^2)\hat{\varphi}(x) = \hat{\jmath}(x), \tag{6.16}$$

where $\hat{\jmath}(x) = -\frac{dV(\varphi)}{d\varphi}\big|_{\varphi=\hat{\varphi}(x)}$ and $V(\varphi)$ is an interaction potential. The solution to Eq. (6.16) with the initial condition $\hat{\varphi}(x) \to \hat{\varphi}_{in}(x)$ at $t \to -\infty$ has the form

$$\hat{\varphi}(x) \;=\; \hat{\varphi}_{in}(x) + \int d^4x' D_{ret}(x-x')\hat{\jmath}(x'), \tag{6.17}$$

where $D_{ret}(x-x')$ is the retarded Green function of the Klein–Gordon equation. This function is defined by the conditions

$$(\Box_x + m^2)D_{ret}(x-x') = \delta^4(x-x') \qquad \text{and} \qquad D_{ret}(x-x') = 0$$

at $(x-x')^2 > 0$ and $x^0 - x'^0 < 0$. Applying the iteration procedure to Eq. (6.17), one gets an expression for $\hat{\varphi}(x)$ in terms of the initial field $\hat{\varphi}_{in}(x)$. Schematically, such an expression looks like $\hat{\varphi}(x) = \hat{\varphi}_{in}(x) + \hat{F}(x|\hat{\varphi}_{in})$, where $\hat{F}(x|\hat{\varphi}_{in})$ is stipulated by the iteration procedure and depends on the interaction potential. Thus, the above hypothesis means that the weak limit of $\hat{F}(x|\hat{\varphi}_{in})$ vanishes, even though this is not mathematically obvious.

The more rigorous analysis based on the Eq. (5.50) and the canonical commutation relations shows that the asymptotic relations (6.5) and (6.9) should be rewritten as

$$\begin{aligned} t \to -\infty, &\qquad \hat{\varphi}(x) \to Z^{\frac{1}{2}}\hat{\varphi}_{in}(x), \\ t \to \infty, &\qquad \hat{\varphi}(x) \to Z^{\frac{1}{2}}\hat{\varphi}_{out}(x), \end{aligned} \tag{6.18}$$

where the constant Z must satisfy the inequalities $0 \leq Z < 1$. However, the limits (6.18) and inequalities for Z have been obtained under the assumption that the matrix elements of the products of the field operators at the same point x^μ are well defined. Unfortunately, these expressions are ill-defined, being divergent in the high-energy limit. Thus, the constant Z is infinite. For this reason, the relations (6.46) need additional definitions. Here we consider the relations (6.5) and (6.9) just as a kind of motivation for the adiabatic hypothesis, allowing us to define the $\mathcal{S}$-matrix; hence, we will not write the constant Z.

Let us note that the problem of divergences and the technique of dealing with infinities in quantum field theory will be discussed separately in Chapter 9.

6.2 Reduction of the $\mathcal{S}$-matrix to Green functions

In this section, we will show that the $\mathcal{S}$-matrix can be reduced to the Green functions, constructed from the Heisenberg field operators. As in the previous section, the considerations will be performed for a real scalar field theory.

Consider the $\mathcal{S}$-matrix

$$\mathcal{S}_{MN} = {}_{out}\langle M|N\rangle_{in} = {}_{out}\langle \mathbf{p}_1, \dots, \mathbf{p}_M|\mathbf{q}_1, \dots, \mathbf{q}_N\rangle_{in}, \tag{6.19}$$

where the vectors $|M\rangle_{out}$ and $|N\rangle_{in}$ are given by (6.6) and (6.10), respectively. For the sake of simplicity, we assume that all the momenta in the final M-particle state differ from any of the momenta in the initial N-particle state. If some momentum $\mathbf{q}_i$ coincides with some momentum $\mathbf{p}_j$, this should be interpreted as meaning that one of the particles did not interact with other particles. Then, it is sufficient to consider only other particles, leaving the special one aside.

The annihilation and creation operators can be written according to Eq. (5.56) as

$$\begin{aligned}
\hat{a}_{in}(\mathbf{q}) &= i\int_{t\to-\infty} d^3x\, \varphi^*_{\mathbf{q}}(x)\, \overleftrightarrow{\partial_0}\, \hat{\varphi}_{in}(x), \\
\hat{a}^\dagger_{in}(\mathbf{q}) &= i\int_{t\to-\infty} d^3x\, \hat{\varphi}_{in}(x)\, \overleftrightarrow{\partial_0}\, \varphi_{\mathbf{q}}(x),
\end{aligned} \tag{6.20}$$

$$\begin{aligned}
\hat{a}_{out}(\mathbf{p}) &= i\int_{t\to\infty} d^3x\, \varphi^*_{\mathbf{p}}(x)\, \overleftrightarrow{\partial_0}\, \hat{\varphi}_{out}(x), \\
\hat{a}^\dagger_{out}(\mathbf{p}) &= i\int_{t\to\infty} d^3x\, \hat{\varphi}_{out}(x)\, \overleftrightarrow{\partial_0}\, \varphi_{\mathbf{p}}(x).
\end{aligned} \tag{6.21}$$

Taking into account the asymptotic conditions, we can replace the $\hat{\varphi}_{in}(x)$ in (6.20) with $\hat{\varphi}(x)$, and in (6.21), replace $\hat{\varphi}_{out}(x)$ with $\hat{\varphi}(x)$. Besides, as was pointed out in section 5.3, the integrals in the *r.h.s.* of Eqs. (6.20) and (6.21) do not depend on time and hence can be calculated at any instant t.

Let us write the $\mathcal{S}$-matrix (6.19) in the form

$$\mathcal{S}_{MN} = {}_{out}\langle M|N\rangle_{in} = {}_{out}\langle M, -\mathbf{p}|\hat{a}_{out}(\mathbf{p})|N\rangle_{in}, \tag{6.22}$$

where the symbolic notation ${}_{out}\langle M, -\mathbf{p}|$ means

$$
{}_{out}\langle \mathbf{p}_1, \dots, \mathbf{p}_M | \,=\, {}_{out}\langle M, -\mathbf{p} | \hat{a}_{out}(\mathbf{p}), \tag{6.23}
$$

with $\mathbf{p}$ being one of the momenta $\mathbf{p}_1, \dots, \mathbf{p}_M$. Indeed, ${}_{out}\langle M, -\mathbf{p}|$ is an $(M-1)$-particle state where the particle with one of the momenta, $\mathbf{p}$, is absent.

Using the first of the relations (6.21), one gets

$$
\mathcal{S}_{MN} \,=\, i \int_{x^0 \to \infty} d^3x \varphi^*_{\mathbf{p}}(x) \overset{\leftrightarrow}{\partial_0} {}_{out}\langle M, -\mathbf{p} | \hat{\varphi}_{out}(x) | N \rangle_{in}. \tag{6.24}
$$

Applying the asymptotic condition (6.9), one can rewrite this expression as

$$
\mathcal{S}_{MN} = i \int_{x^0 \to \infty} d^3x \; \varphi^*_{\mathbf{p}}(x) \overset{\leftrightarrow}{\partial_0} {}_{out}\langle M, -\mathbf{p} | \hat{\varphi}(x) | N \rangle_{in}. \tag{6.25}
$$

The *r.h.s.* of the last formula can be transformed into

$$
\begin{aligned}
& i \int\limits_{x^0 \to -\infty} d^3x \varphi^*_{\mathbf{p}}(x) \overset{\leftrightarrow}{\partial_0} {}_{out}\langle M, -\mathbf{p} | \hat{\varphi}(x) | N \rangle_{in} \\
& \quad + \int\limits_{-\infty}^{\infty} dx^0 \, \partial_0 \Big[i \int d^3x \, \varphi^*_{\mathbf{p}}(x) \overset{\leftrightarrow}{\partial_0} {}_{out}\langle M, -\mathbf{p} | \hat{\varphi}(x) | N \rangle_{in} \Big].
\end{aligned} \tag{6.26}
$$

In the first term, we used the asymptotic condition, which enables one to replace $\hat{\varphi}(x)$ by $\hat{\varphi}_{in}(x)$. After taking into account Eq. (6.20), this term takes the form

$$
\langle M, -\mathbf{p} | \hat{a}_{in}(\mathbf{p}) | N \rangle_{in} \,=\, 0, \tag{6.27}
$$

assuming that the momentum $\mathbf{p}$ does not coincide with any of the momenta $\mathbf{q}$.

Now we consider the second term in the *r.h.s.* of Eq. (6.26). There is a relation

$$
\begin{aligned}
& \partial_0 \big(\varphi^*_{\mathbf{p}}(x) \overset{\leftrightarrow}{\partial_0} {}_{out}\langle M, -\mathbf{p} | \hat{\varphi}(x) | N \rangle_{in} \big) \\
& = \; \varphi^*_{\mathbf{p}}(x) \partial_0^2 \big({}_{out}\langle M, -\mathbf{p} | \hat{\varphi}(x) | N \rangle_{in} \big) - \big(\Delta - m^2 \big) \varphi^*_{\mathbf{p}}(x) {}_{out}\langle M, -\mathbf{p} | \hat{\varphi}(x) | N \rangle_{in},
\end{aligned} \tag{6.28}
$$

where we have taken into account that the function $\varphi_{\mathbf{p}}$ satisfies the Klein–Gordon equation. The above expression should be integrated over d^3x; hence, we can integrate by parts and transfer Δ onto the second factor. Thus,

$$
\mathcal{S}_{MN} \,=\, i \int d^4x \; \varphi^*_{\mathbf{p}}(x) \, \big(\Box + m^2 \big) {}_{out}\langle M, -\mathbf{p} | \hat{\varphi}(x) | N \rangle_{in}. \tag{6.29}
$$

We conclude that the first step in the desired reduction has been performed.

Let us start the second part of the reduction. The state $|N\rangle_{in}$ can be written as $a^\dagger_{in}(\mathbf{q}) |N, -\mathbf{q}\rangle_{in}$. The symbol $|N, -\mathbf{q}\rangle_{in}$ means that the momentum $\mathbf{q}$ is removed from the set of momenta defining the state $|N\rangle_{in}$. Using the second of the relations (6.21), one gets

$$
\begin{aligned}
{}_{out}\langle M,-\mathbf{p}|\hat{\varphi}(x)|N\rangle_{in} &= -i\int\limits_{y^0\to-\infty} d^3y\;\varphi_{\mathbf{q}}(y)\;\overleftrightarrow{\partial}_{y^0}\;{}_{out}\langle M,-\mathbf{p}|\hat{\varphi}(x)\hat{\varphi}_{in}(y)|N,-\mathbf{q}\rangle_{in} \\
&= -i\int\limits_{y^0\to-\infty} d^3y\;\varphi_{\mathbf{q}}(y)\;\overleftrightarrow{\partial}_{y^0}\;{}_{out}\langle M,-\mathbf{p}|\hat{\varphi}(x)\hat{\varphi}(y)|N,-\mathbf{q}\rangle_{in}.
\end{aligned}
\tag{6.30}
$$

Here the asymptotic condition (6.5) has been used. It is worth noticing that, in the expression (6.30), the time arguments satisfy the inequality $x^0 > y^0$.

In deriving relation (6.29), we have used the relation (6.26). This allowed us to transform the operator $\hat{a}_{out}(\mathbf{p})$, which acts on ${}_{out}\langle M,-\mathbf{p}|$, to the operator $\hat{a}_{in}(\mathbf{p})$, which acts on $|N\rangle_{in}$ and vanish. Let us try to apply the same trick to the expression (6.30). It means that we intend to transform the operator $\hat{a}^\dagger_{in}(\mathbf{q})$, which acts on the state $|N,-\mathbf{q}\rangle_{in}$, to the operator $\hat{a}^\dagger_{out}(\mathbf{q})$, which acts on ${}_{out}\langle M,-\mathbf{p}|$ and then vanishes. However, this same trick will not work in the case under consideration. The point is that, in order to transform an in-operator into an out-operator, the field $\hat{\varphi}(y)$ should be to the left of the field $\hat{\varphi}(x)$. But, in the expression (6.30), these operators are in the opposite order. To solve this problem, we define a chronological product, or T-product, of the time-dependent operators, following the rule

$$
T(\hat{\varphi}(x)\hat{\varphi}(y)) = \theta(x^0-y^0)\hat{\varphi}(x)\hat{\varphi}(y) + \theta(y^0-x^0)\hat{\varphi}(y)\hat{\varphi}(x),
\tag{6.31}
$$

where $\theta(x^0-y^0)$ is a step function. According to this definition, a chronological product arranges the operators in order of decreasing time from the left to the right. Therefore, under the sign of the T-product, the field operators behave as classical commuting functions. With the help of this definition, the expression (6.30) can be rewritten as follows:

$$
\begin{aligned}
&{}_{out}\langle M,-\mathbf{p}|\hat{\varphi}(x)|N\rangle_{in} \\
&\quad = -i\int_{y^0\to-\infty} d^3y\;\varphi_{\mathbf{q}}(y)\;\overleftrightarrow{\partial}_{y^0}\;{}_{out}\langle M,-\mathbf{p}|T(\hat{\varphi}(x)\hat{\varphi}(y))|N,-\mathbf{q}\rangle_{in}.
\end{aligned}
\tag{6.32}
$$

Since $x^0 > y^0$, expression (6.32) coincides with (6.30).

The *r.h.s.* of (6.32) is rewritten in the form

$$
\begin{aligned}
&-i\int_{y^0\to-\infty} d^3y\;\varphi_{\mathbf{q}}(y)\;\overleftrightarrow{\partial}_{y^0}\;{}_{out}\langle M,-\mathbf{p}|T(\hat{\varphi}(x)\hat{\varphi}(y))|N,-\mathbf{q}\rangle_{in} \\
&= -i\int_{y^0\to\infty} d^3y\;\varphi_{\mathbf{q}}(y)\;\overleftrightarrow{\partial}_{y^0}\;{}_{out}\langle M,-\mathbf{p}|T(\hat{\varphi}(x)\hat{\varphi}(y))|N,-\mathbf{q}\rangle_{in} \\
&\quad + i\int d^4y\;\partial_{y^0}\big(\varphi_{\mathbf{q}}(y)\;\overleftrightarrow{\partial}_{y^0}\;{}_{out}\langle M,-\mathbf{p}|T(\hat{\varphi}(x)\hat{\varphi}(y))|N,-\mathbf{q}\rangle_{in}\big).
\end{aligned}
\tag{6.33}
$$

In the first term in the *r.h.s.*, $y^0 > x^0$ and, owing to the T-product, one can put $\hat{\varphi}(y)$ to the left of $\hat{\varphi}(x)$. After that, this term includes the expression

${}_{out}\langle M, -\mathbf{p}|\hat{a}^{\dagger}_{out}(\mathbf{q})\hat{\varphi}(x)|N, -\mathbf{q}\rangle_{in} = 0$, since we assumed that none of the momenta of $\mathbf{q}$ coincides with any of $\mathbf{p}$. The second term in (6.33) is transformed in the same way as in (6.28). As a result, one gets

$$\begin{aligned} &{}_{out}\langle M, -\mathbf{p}|\hat{\varphi}(x)|N\rangle_{in} \\ &\quad = i\int d^4y\varphi_{\mathbf{q}}(y)\big(\Box_y + m^2\big)\,{}_{out}\langle M, -\mathbf{p}|T(\hat{\varphi}(x)\hat{\varphi}(y))|N, -\mathbf{q}\rangle_{in}. \end{aligned} \tag{6.34}$$

Now let us go back to Eq. (6.29). Substituting (6.34) into (6.29), we obtain

$$\begin{aligned} \mathcal{S}_{MN} = i^2 \int d^4x d^4y\; &\varphi^*_{\mathbf{p}}(x)\big(\overrightarrow{\Box}_x + m^2\big)\,{}_{out}\langle M, -\mathbf{p}|T\big(\hat{\varphi}(x)\hat{\varphi}(y)\big)|N, -\mathbf{q}\rangle_{in} \\ &\times \big(\overleftarrow{\Box}_y + m^2\big)\varphi_{\mathbf{q}}(y). \end{aligned} \tag{6.35}$$

The relation (6.35) shows that removing a particle from the states ${}_{out}\langle M|$ and $|N\rangle_{in}$ is equivalent to inserting the field operator $\hat{\varphi}$ into the matrix element with one less particle. The index x in the $\Box_x$ shows that the operator acts only on the argument x, and the arrows indicate whether the operator acts on the right or on the left.

At the next stage, starting from ${}_{out}\langle M, -\mathbf{p}|T\big(\hat{\varphi}(x)\hat{\varphi}(y)\big)|N, -\mathbf{q}\rangle_{in}$ in (6.35), one has to repeat the same procedure until all particles are reduced and we arrive at the vacuum state. The result is

$$\begin{aligned} \mathcal{S}_{MN} = (i)^{M+N} \int d^4x_1 \dots d^4x_M\, d^4y_1 \dots d^4y_N \prod_{k=1}^{M} \Big[\varphi^*_{\mathbf{p}_k}(x_k)\big(\overrightarrow{\Box}_{x_k} + m^2\big)\Big] \\ \times\, G_{N+M}(x_1, \dots, x_M, y_1, \dots, y_N) \prod_{l=1}^{N} \Big[\big(\overleftarrow{\Box}_{y_l} + m^2\big)\varphi_{\mathbf{q}_l}(y_l)\Big]. \end{aligned} \tag{6.36}$$

Here we introduced the function $G_{N+M}(x_1, \dots, x_M, y_1, \dots, y_N)$, which is called the $(M + N)$-point Green function of the Heisenberg operators

$$G_n(x_1, \dots, x_n) = \langle 0|T(\hat{\varphi}(x_1) \dots \hat{\varphi}(x_n)|0\rangle. \tag{6.37}$$

The T-product of n time-dependent operators is defined as

$$T\{\hat{q}(t_1), \dots \hat{q}(t_n)\} = \sum_P \theta(t_{\alpha_1} - t_{\alpha_2}) \dots \theta(t_{\alpha_{n-1}} - t_{\alpha_n})\hat{q}(t_{\alpha_1}) \dots \hat{q}(t_{\alpha_n}), \tag{6.38}$$

where the sum over P means the sum over all permutations $\{\alpha_1, \dots, \alpha_n\}$ of the given n numbers. It is easy to see that the T-product operation acts on the product of time-depending operators by rearranging them in order of decreasing time from the left to the right. Thus, under the T-product, all operators commute. As a result, the Green function (6.37) is a symmetric function of its arguments.

All in all, we see that the calculation of the $\mathcal{S}$-matrix can be reduced to the calculation of the Green functions (6.37).

The expression (6.37) gets simplified in the momentum representation. The Fourier transform of the Green function is defined as

$$G_n(x_1, x_2, \dots, x_n) = \int \frac{d^4p_1}{(2\pi)^4} \cdots \frac{d^4p_n}{(2\pi)^4} \, e^{-i(p_1x_1+\cdots+p_nx_n)} \, \tilde{G}_n(p_1, p_2, \dots, p_n). \tag{6.39}$$

Substituting Eq. (6.39) into (6.36) and taking into account that

$$\varphi_{\mathbf{p}}(x) = \frac{e^{-ipx}}{\sqrt{2(2\pi)^3\varepsilon(\mathbf{p})}}, \tag{6.40}$$

with $p_0 = \varepsilon(\mathbf{p})$, one gets

$$\mathcal{S}_{MN}(\mathbf{p}_1, \mathbf{p}_2, \dots, \mathbf{p}_M, \mathbf{q}_1, \mathbf{q}_2, \dots, \mathbf{q}_N) \tag{6.41}$$

$$= \left[- \frac{i}{2(2\pi)^3} \right]^{M+N} \frac{(p_1^2 - m^2)\dots(p_M^2 - m^2)(q_1^2 - m^2)\dots(q_N^2 - m^2)}{\sqrt{\varepsilon(\mathbf{p}_1)\,\dots\,\varepsilon(\mathbf{p}_M)\;\varepsilon(\mathbf{q}_1)\,\dots\,\varepsilon(\mathbf{q}_N)}}$$

$$\times \; \tilde{G}_{M+N}(p_1, \dots, p_M, -q_1 \dots, -q_N). \tag{6.42}$$

Here the arguments $\tilde{G}_{M+N}(p_1, \dots, p_M, -q_1 \dots, -q_N)$ satisfy the relations $p_k^2 = q_l^2 = m^2$, with $k = 1, 2, \dots, M$ and $l = 1, 2, \dots, N$. The equation $p^2 = m^2$ defines a special surface in the momentum space, which is called the *mass shell*. The relation (6.41) shows that the $\mathcal{S}$-matrix element is obtained from the Fourier transform of the Green function when all the momentum arguments are settled on the mass shell. In this case, it is said that the Green function is taken *on shell*.

Remark 1. On first sight, it looks like, owing to the mass-shell relations, the expression (6.42) vanishes on shell. However, this is not true. Later on, we shall see that the function $\tilde{G}_{M+N}(p_1, \dots, p_M, -q_1 \dots, -q_N)$ has its poles at $p_k^2 = m^2$ and $q_l^2 = m^2$. These poles cancel the terms $p_k^2 - m^2$ and $q_l^2 - m^2$ in the numerator, and the final result is non-zero.

Remark 2. The translational invariance imposes an additional restriction on the momentum dependence of the $\mathcal{S}$-matrix. Owing to the homogeneity of spacetime, the Green function $G_n(x_1, x_2, \dots, x_n)$ is invariant under the shift of all coordinates on a constant four-vector a^μ, which means the symmetry

$$G_n(x_1, x_2, \dots, x_n) = G_n(x_1 + a_1, x_2 + a_2, \dots, x_n + a_n). \tag{6.43}$$

Rewriting this relation in terms of momenta, one arrives at the conclusion that

$$\tilde{G}_n(p_1, \dots, p_n) \sim \delta^4(p_1 + \dots + p_n). \tag{6.44}$$

This property means that an element of the scattering matrix

$$\mathcal{S}_{MN}(\mathbf{p}_1, \dots, \mathbf{p}_M, \mathbf{q}_1 \dots, \mathbf{q}_N) \tag{6.45}$$

contains the factor of the delta function $\delta^4(p_1 + p_2 + \dots, p_M - q_1 - q_2 - \dots - q_N)$. This is the mathematical expression of the conservation law that relates initial and final overall energies and momenta, namely, $q_1 + q_2 + \dots + q_N = p_1 + p_2 + \dots + p_M$.

To conclude this section, we note that the formula for reducing the $\mathcal{S}$-matrix elements to the on-shell Green functions was originally derived by H. Lehmann, K. Symanzik and W. Zimmermann [212] and is called the LSZ reduction formula.

6.3 Generating functionals of Green functions and the $\mathcal{S}$-matrix

In this section, we continue to study different aspects of the $\mathcal{S}$-matrix for the real scalar field model. Our main purpose is to formulate the main statements in a more formal way.

Consider the n-point Green function $G_n(x_1, x_2, \dots, x_n)$, with $n = 0, 1, 2, \dots$, and introduce an arbitrary external scalar *source* field $J(x)$. Consider the functional

$$Z[J] = \sum_{n=0}^{\infty} \frac{i^n}{n!} \int d^4x_1 d^4x_2 \dots d^4x_n \, G_n(x_1, x_2, \dots, x_n) \, J(x_1)J(x_2)\dots J(x_n). \quad (6.46)$$

Taking n functional derivatives of the functional $Z[J]$ at the vanishing source, one gets

$$\frac{\delta^n Z[J]}{\delta iJ(x_1)\delta iJ(x_2)\dots\delta iJ(x_n)}\bigg|_{J=0} = G_n(x_1, x_2, \dots, x_n). \quad (6.47)$$

Thus, the derivatives of $Z[J]$ at the vanishing source generate the Green functions. For this reason, the functional $Z[J]$ is called the generating functional of Green functions.

The expression for $Z[J]$ in Eq. (6.46) can be rewritten in a compact form as follows. Substituting the expression for the Green function (6.37) to (6.46), one gets

$$\begin{aligned} Z[J] &= \sum_{n=0}^{\infty} \frac{i^n}{n!} \int d^4x_1 d^4x_2 \dots d^4x_n \, \langle 0|\, T(\hat{\varphi}(x_1)\varphi(x_2)\dots\hat{\varphi}(x_n))\,|0\rangle \, J(x_1)\dots J(x_n) \\ &= \sum_{n=0}^{\infty} \frac{i^n}{n!} \langle 0|\, T\Big\{ \int d^4x_1 \hat{\varphi}(x_1)J(x_1) \times \dots \times \int d^4x_n \hat{\varphi}(x_n)J(x_n)\Big\}\, |0\rangle \\ &= \langle 0|\, T \sum_{n=0}^{\infty} \frac{i^n}{n!} \Big\{ i \int d^4x \hat{\varphi}(x)J(x)\Big\}^n \,|0\rangle \;=\; \langle 0|Te^{i\int d^4x\hat{\varphi}(x)J(x)}|0\rangle. \end{aligned} \quad (6.48)$$

The expression (6.48) for the generating functional $Z[J]$ contains information about all the Green functions $G_n(x_1, x_2, \dots, x_n)$. For this reason, the generating functional $Z[J]$ represents a very convenient form for the analysis of the general properties of the Green functions and related objects, including the $\mathcal{S}$-matrix.

The generating functional of the Green functions satisfies some equation of motion. Consider

$$\begin{aligned} \frac{\delta Z[J]}{\delta iJ(x)} &= \langle 0|T(\hat{\varphi}(x)e^{i\int d^4x\, \hat{\varphi}(x)J(x)})|0\rangle \\ &= \langle 0|\Big\{Te^{i\int\limits_{-\infty}^{x^0} d^4x\, \hat{\varphi}(x)J(x)}\Big\}\, \hat{\varphi}(x)\, \Big\{Te^{i\int\limits_{x^0}^{\infty} d^4x\hat{\varphi}(x)J(x)}\Big\}\, |0\rangle, \end{aligned} \quad (6.49)$$

where we have denoted $\int\limits_{-\infty}^{x^0} d^4x = \int\limits_{-\infty}^{x^0} dt \int d^3x$ and $\int\limits_{x^0}^{\infty} d^4x = \int\limits_{x^0}^{\infty} dt \int d^3x$. Therefore,

$$\frac{\partial}{\partial x^0}\frac{\delta Z[J]}{\delta iJ(x)} = i\int d^3y\langle 0|\Big\{Te^{i\int_{-\infty}^{x^0} d^4y\hat{\varphi}(y)J(y)}\Big\}\left[\hat{\varphi}(x^0,\mathbf{y}),\hat{\varphi}(x^0,\mathbf{x})\right] \qquad (6.50)$$
$$\times\Big\{Te^{i\int_{x^0}^{\infty} d^4x\hat{\varphi}(x)J(x)})|0\rangle J(x^0,\mathbf{y}\Big\} + \langle 0|\{Te^{i\int d^4x\hat{\varphi}(x)J(x)})\partial_{x^0}\hat{\varphi}(x)\}|0\rangle.$$

Since the equal-time commutators of the field operators vanish, the first term in the *r.h.s.* of the last relation is zero. The next step is to consider

$$\frac{\partial^2}{\partial(x^0)^2}\frac{\delta Z[J]}{\delta iJ(x)} = i\int d^3y\,\langle 0|\Big\{Te^{i\int_{-\infty}^{x^0} d^4y\hat{\varphi}(y)J(y)}\Big\}\left[\hat{\varphi}(x^0,\mathbf{y}),\partial_{x^0}\hat{\varphi}(x^0,\mathbf{x})\right]$$
$$\times\Big\{Te^{i\int_{x^0}^{\infty} d^4x\hat{\varphi}(x)J(x)}\Big\}|0\rangle J(x^0,\mathbf{y}) + \langle 0|\Big\{Te^{i\int d^4x\hat{\varphi}(x)J(x)}\Big\}\frac{\partial^2\hat{\varphi}(x)}{\partial(x^0)^2}\Big\}|0\rangle.$$

In the real scalar field theory, $\partial_{x^0}\varphi(x) = \pi(x)$, where $\pi(x)$ is the conjugate momentum to $\varphi(x)$. Therefore, according to the canonical commutation relations

$$[\hat{\varphi}(x^0,\mathbf{y}),\partial_{x^0}\hat{\varphi}(x^0,\mathbf{x})] = i\delta^3(\mathbf{y}-\mathbf{x}).$$

The operator equation of motion for the Heisenberg field operator $\hat{\varphi}(x)$ leads to

$$\partial^2_{x^0}\hat{\varphi}(x) = (\Delta - m^2)\hat{\varphi}(x) - V'(\hat{\varphi}(x)), \qquad (6.51)$$

where $V'(\hat{\varphi}(x))$ is the operator corresponding to $\frac{dV}{d\varphi(x)}$ and V is a classical interaction potential. Substituting the commutation relation and the operator equation of motion into Eq. (6.51), we also take into account that the *r.h.s.* of the equation of motion (6.51) has no time derivatives. After that, we obtain

$$(\Box + m^2)\frac{\delta Z[J]}{\delta iJ(x)} = -J(x)Z[J] - \langle 0|T\big(e^{i\int d^4x\hat{\varphi}(x)J(x)}V'(\hat{\varphi}(x))\big)|0\rangle. \qquad (6.52)$$

Let us derive an identity which will prove useful in simplifying the last term in Eq. (6.52). Consider the set of functions $F_n(x_1,\dots,x_n)$, and define

$$F[\varphi] = \sum_{n=1}^{\infty}\frac{1}{n!}\int d^4x_1\dots d^4x_n\,F_n(x_1,\dots,x_n)\,\varphi(x_1)\dots\varphi(x_n).$$

The identity of interest is

$$F[\varphi]\,e^{i\int d^4x\varphi(x)J(x)} = F[\varphi]\Big|_{\varphi=\frac{\delta}{\delta iJ}} e^{i\int d^4x\varphi(x)J(x)}. \qquad (6.53)$$

To prove it, we note that the *l.h.s.* of this relation can be cast into the form

$$\sum_{n=1}^{\infty}\frac{1}{n!}\int d^4x_1\dots d^4x_n\,F_n(x_1,\dots,x_n)\,\varphi(x_1)\dots\varphi(x_n)\,e^{i\int d^4x\varphi(x)J(x)}$$
$$= \sum_{n=1}^{\infty}\frac{1}{n!}\int d^4x_1\dots d^4x_n\,F_n(x_1,\dots,x_n)\,\frac{\delta}{\delta iJ(x_1)}\cdots\frac{\delta}{\delta iJ(x_n)}\,e^{i\int d^4x\varphi(x)J(x)},$$

which coincides with the *r.h.s.* of (6.53).

Now, let us consider the last term in the *r.h.s.* of (6.52). Under the sign of the T-product, all the operators commute. Therefore, one can write

$$\langle 0|T\Big\{e^{i\int d^4x\hat{\varphi}(x)J(x)}\,V'(\hat{\varphi}(x)\Big\}|0\rangle \;=\; \langle 0|T\Big\{V'(\hat{\varphi}(x))e^{i\int d^4x\hat{\varphi}(x)J(x)}\Big\}|0\rangle$$

$$= \langle 0|T\Big\{V'(\varphi)\Big|_{\varphi(x)=\frac{\delta}{\delta iJ(x)}} e^{i\int d^4x\hat{\varphi}(x)J(x)}\Big\}|0\rangle \;=\; V'(\varphi)\Big|_{\varphi(x)=\frac{\delta}{\delta iJ(x)}} Z[J].$$

Substituting this relation into (6.52) one gets, finally

$$(\Box + m^2)\frac{\delta Z[J]}{\delta iJ(x)} + V'(\varphi)\Big|_{\varphi(x)\,=\,\frac{\delta}{\delta iJ(x)}} Z[J] = -J(x)Z[J]. \tag{6.54}$$

This is the equation of motion for the generating functional of Green functions; it is called the Schwinger-Dyson equation.

Let us now go back to the $\mathcal{S}$-matrix (6.36) and show how it can be related to the generating functional of Green functions. Introduce the functional

$$S[\alpha,\alpha^*] \;=\; \langle 0|Te^{i\int d^4x\hat{\varphi}(x)(\overleftarrow{\Box}+m^2)f(x|\alpha,\alpha^*)}|0\rangle. \tag{6.55}$$

Here the function $f(x|\alpha,\alpha^*)$ is defined as

$$f(x|\alpha,\alpha^*) \;=\; \int d^3p\{\varphi_{\mathbf{p}}(x)\alpha(\mathbf{p}) + \varphi^*_{\mathbf{p}}(x)\alpha^*(\mathbf{p})\}, \tag{6.56}$$

where $\varphi_{\mathbf{p}}(x)$ is defined by Eq. (5.54), while $\alpha(\mathbf{p})$ and $\alpha^*(\mathbf{p})$ are arbitrary complex functions of the three-dimensional momentum. It is worth noting that the expression (6.55) is obtained from the generating functional of the Green function $Z[J]$, where, formally, the source $J(x)$ is taken as

$$J(x) = (\overleftarrow{\Box} + m^2)f(x|\alpha,\alpha^*). \tag{6.57}$$

Consider the derivatives

$$\frac{\delta S[\alpha,\alpha^*]}{\delta\alpha(\mathbf{p})} \;=\; \int d^4x\frac{\delta Z[J]}{\delta iJ(x)}\frac{\delta iJ(x)}{\delta\alpha(\mathbf{p})} = i\int d^4x\frac{\delta Z[J]}{\delta iJ(x)}(\overleftarrow{\Box} + m^2)\varphi_{\mathbf{p}}(x) \tag{6.58}$$

and

$$\frac{\delta S[\alpha,\alpha^*]}{\delta\alpha^*(\mathbf{p})} \;=\; \int d^4x\frac{\delta Z[J]}{\delta iJ(x)}\frac{\delta iJ(x)}{\delta\alpha^*(\mathbf{p})} \;=\; i\int d^4x\varphi^*_{\mathbf{p}}(x)(\overrightarrow{\Box} + m^2)\frac{\delta Z[J]}{\delta iJ(x)}. \tag{6.59}$$

Using the last relations (6.58) and (6.59) several times, one gets

$$\frac{\delta^{M+N}S[\alpha,\alpha^*]}{\delta\alpha^*(\mathbf{p}_1)\ldots\alpha^*(\mathbf{p}_M)\delta\alpha(\mathbf{q}_1)\ldots\alpha(\mathbf{q}_N)}\bigg|_{\alpha=\alpha^*=0} \;=\; \mathcal{S}_{MN}(\mathbf{p}_1,\ldots,\mathbf{p}_M,\mathbf{q}_1,\ldots,\mathbf{q}_N), \tag{6.60}$$

where the expression $\mathcal{S}_{MN}(\mathbf{p}_1,\ldots,\mathbf{p}_M,\mathbf{q}_1,\ldots,\mathbf{p}_N)$ is the $\mathcal{S}$-matrix (6.36). Thus, we see that the derivatives of the functional $S[\alpha,\alpha^*]$ with respect to α and α^*, with subsequently vanishing α and α^*, generate the elements of the $\mathcal{S}$-matrix. For this reason, the functional (6.55) is called the generating functional of the $\mathcal{S}$-matrix.

6.4 The $\mathcal{S}$-matrix and the Green functions for spinor fields

In this section, we consider the definition of the $\mathcal{S}$-matrix in spinor field theory.

6.4.1 Definition of the spinor field $\mathcal{S}$-matrix and Green functions

To describe the scattering of spinor particles, we introduce the asymptotic initial and final states, similar to what was done for scalar field theory. In the presence of spinor fields, the asymptotic conditions are formulated as follows:

$$\begin{aligned} t \to -\infty, \quad & \hat{\psi}(x) \to \hat{\psi}_{in}(x), \qquad \hat{\bar{\psi}}(x) \to \hat{\bar{\psi}}_{in}(x), \\ t \to \infty, \quad & \hat{\psi}(x) \to \hat{\psi}_{out}(x), \qquad \hat{\bar{\psi}}(x) \to \hat{\bar{\psi}}_{out}(x). \end{aligned} \tag{6.61}$$

Here $\hat{\psi}(x)$ is the Heisenberg spinor field operator, and $\hat{\psi}_{in}(x)$ and $\hat{\psi}_{out}(x)$ are the corresponding free field operators describing the initial and final particles and satisfying the Dirac equation with observable mass.

Consider some properties of the solutions to the Dirac equation (4.21). For the two solutions, $\psi_1(x)$ and $\psi_2(x)$, there is a quantity (4.138) which is time independent and is called the invariant inner product in the linear space of solutions to the Dirac equation.

The free field operators $\hat{\psi}^{(0)}(x)$ and $\hat{\bar{\psi}}^{(0)}(x)$, satisfying the Dirac equation, are written according to (5.100), as

$$\begin{aligned} \hat{\psi}^{(0)}(x) &= \sum_{\lambda=\pm1/2} \int d^3p\{\psi_{\mathbf{p},\lambda}(x)\hat{a}(\mathbf{p},\lambda) + \chi_{\mathbf{p},\lambda}(x)\hat{b}^{\dagger}(\mathbf{p},\lambda)\} \\ \hat{\bar{\psi}}^{(0)}(x) &= \sum_{\lambda=\pm1/2} \int d^3p\,\{\bar{\psi}_{\mathbf{p},\lambda}(x)\hat{a}^{\dagger}(\mathbf{p},\lambda) + \bar{\chi}_{\mathbf{p},\lambda}(x)\hat{b}(\mathbf{p},\lambda)\}, \end{aligned} \tag{6.62}$$

where

$$\psi_{\mathbf{p},\lambda}(x) = \frac{1}{(2\pi)^{3/2}}\sqrt{\frac{m}{\varepsilon(\mathbf{p})}}u_{\mathbf{p},\lambda}e^{-ipx}, \quad \chi_{\mathbf{p},\lambda}(x) = \frac{1}{(2\pi)^{3/2}}\sqrt{\frac{m}{\varepsilon(\mathbf{p})}}v_{\mathbf{p},\lambda}e^{ipx} \tag{6.63}$$

and the spinors $u_{\mathbf{p},\lambda}$ and $v_{\mathbf{p},\lambda}$ have been introduced in section 5.5. Using relations (5.99), one can show that the functions $\psi_{\mathbf{p},\lambda}(x)$ and $\chi_{\mathbf{p},\lambda}(x)$ satisfy the following orthogonality conditions:

$$\begin{aligned} (\psi_{\mathbf{p}_1,\lambda_1}, \psi_{\mathbf{p}_2,\lambda_2}) &= \delta_{\lambda_1,\lambda_2}\delta(\mathbf{p}_1 - \mathbf{p}_2), \qquad (\psi_{\mathbf{p}_1,\lambda_1}, \chi_{\mathbf{p}_2,\lambda_2}) = 0, \\ (\chi_{\mathbf{p}_1,\lambda_1}, \chi_{\mathbf{p}_2,\lambda_2}) &= \delta_{\lambda_1,\lambda_2}\delta(\mathbf{p}_1 - \mathbf{p}_2), \qquad (\chi_{\mathbf{p}_1,\lambda_1}, \psi_{\mathbf{p}_2,\lambda_2}) = 0. \end{aligned} \tag{6.64}$$

The relations (6.62) and (6.64) enable us to express the annihilation and creation operators in terms of the free field operators

$$\begin{aligned} \hat{a}(\mathbf{p},\lambda) &= (\psi_{\mathbf{p},\lambda}, \hat{\psi}^{(0)}), \qquad \hat{a}^{\dagger}(\mathbf{p},\lambda) = (\hat{\psi}^{(0)}, \psi_{\mathbf{p},\lambda}), \\ \hat{b}(\mathbf{p},\lambda) &= (\hat{\psi}^{(0)}, \chi_{\mathbf{p},\lambda}), \qquad \hat{b}^{\dagger}(\mathbf{p},\lambda) = (\chi_{\mathbf{p},\lambda}, \hat{\psi}^{(0)}). \end{aligned} \tag{6.65}$$

Since the initial particles are free, they are described by the free field operators $\hat{\psi}_{in}(x)$ and $\hat{\bar{\psi}}_{in}(x)$, satisfying the Dirac equation and being linear combinations of the annihilation and creation operators $\hat{a}_{in}(\mathbf{p},\lambda)$, $\hat{b}_{in}(\mathbf{p},\lambda)$, $\hat{a}^{\dagger}_{in}(\mathbf{p},\lambda)$ and $\hat{b}^{\dagger}_{in}(\mathbf{p},\lambda)$. With these operators, one can construct the state with N_1 initial particles that are type a, and N_2 initial particles that are type b:

$$|N_1,N_2\rangle_{in} = \hat{a}^{\dagger}_{in}(\mathbf{p}_1,\lambda_1)\dots,\hat{a}^{\dagger}_{in}(\mathbf{p}_{N_1},\lambda_{N_1})\hat{b}^{\dagger}_{in}(\mathbf{p}'_1,\lambda'_1)\,\dots\,\hat{b}^{\dagger}_{in}(\mathbf{p}'_{N_2},\lambda'_{N_2})|0\rangle. \quad (6.66)$$

In a similar way, one can introduce the field operators of the final free particles $\hat{\psi}_{out}(x)$ and $\hat{\bar{\psi}}_{out}(x)$ and the corresponding annihilation and creation operators, $\hat{a}_{out}(\mathbf{p},\lambda)$, $\hat{b}_{out}(\mathbf{p},\lambda)$, $\hat{a}^{\dagger}_{out}(\mathbf{p},\lambda)$ and $\hat{b}^{\dagger}_{out}(\mathbf{p},\lambda)$. Using these operators, we construct the state with M_1 final particles of type a and M_2 final particles of type b:

$$|M_1,M_2\rangle_{out} = \hat{a}^{\dagger}_{out}(\mathbf{p}_1,\lambda_1)\,\dots\,\hat{a}^{\dagger}_{out}(\mathbf{p}_{M_1},\lambda_{M_1})\hat{b}^{\dagger}_{out}(\mathbf{p}'_1,\lambda'_1)\,\dots\,\hat{b}^{\dagger}_{out}(\mathbf{p}'_{N_2},\lambda'_{M_2})|0\rangle. \quad (6.67)$$

The vacuum $|0\rangle$ is the same in Eqs. (6.66) and (6.67) since, according to the adiabatic hypothesis, the vacuum is a stable state.

The $\mathcal{S}$-matrix of the spinor particles is defined as

$$\mathcal{S}_{M_1,M_2;N_1,N_2} = {}_{out}\langle M_1,M_2|N_1,N_2\rangle_{in}, \quad (6.68)$$

where the states $|M_1,M_2\rangle_{out}$ and $|N_1,N_2\rangle_{in}$ are defined by Eqs. (6.67) and (6.66), respectively.

The reduction of the $\mathcal{S}$-matrix to Green functions is carried out according to the same scheme as used for scalar field theory. We use the relations (6.65), allowing us to express the in and out annihilation and creation operators in terms of in and out free field operators. Taking into account the time independence of the inner products and using the asymptotic conditions, we replace the free field operators by the Heisenberg field operators $\hat{\psi}(x)$ and $\hat{\bar{\psi}}(x)$. As in scalar field theory, for simplicity, it is assumed that none of the initial particles' momenta coincides with any of the final particles' momenta.

All other calculations are the same as in section 6.2. The final result has the form

$$\begin{aligned}\mathcal{S}_{M_1,M_2;N_1,N_2} &= (-i)^{M_1+N_2}(i)^{M_2+N_1}\\ &\times \int d^4x_1\dots d^4x_{M_1}d^4x'_1\dots d^4x'_{M_2}d^4y_1\dots d^4y_{N_1}d^4y'_1\dots d^4y'_{N_2} \quad (6.69)\\ &\times \overrightarrow{O}^{(1)}_{M_1N_2}\, G_{M_1+M_2+N_1+N_2}(x_1,..,x_{M_1}x'_1,..,x'_{M_2},y_1,\dots,y_{N_1},..,y'_1,..,y'_{N_2})\,\overleftarrow{O}^{(2)}_{M_2N_1},\end{aligned}$$

where the operators $\overrightarrow{O}^{(1)}_{M_1N_2}$ and $\overleftarrow{O}^{(2)}_{N_2M_1}$ are defined as

$$\overrightarrow{O}^{(1)}_{M_1N_2} = \prod_{k=1}^{M_1}\bar{\psi}_{\mathbf{p}_k,\lambda_k}(x_k)(i\gamma^{\mu_k}\overrightarrow{\partial}_{\mu_k}-m)\prod_{k'=1}^{N_2}\chi_{\mathbf{q}'_{k'},\lambda'_{k'}}(y_{k'})(-i\gamma^{\mu'_k}\overrightarrow{\partial}_{\mu_{k'}}-m), \quad (6.70)$$

$$\overleftarrow{O}^{(2)}_{M_2N_1} = \prod_{l=1}^{M_2}(i\gamma^{\nu_l}\overleftarrow{\partial}_{\nu_l}-m)\psi_{\mathbf{p}'_l,\lambda'_l}(y_l)\prod_{l'=1}^{N_1}(-i\gamma^{\nu_{l'}}\overleftarrow{\partial}_{\nu_{l'}}-m)\bar{\chi}_{\mathbf{p}'_{l'},\lambda'_{l'}}(x'_{l'}). \quad (6.71)$$

Here, in Eq. (6.70), the derivatives $\overrightarrow{\partial}_{\mu_k}$ act on the coordinates x^{μ_k}, and the derivatives $\overrightarrow{\partial}_{\mu_{k'}}$ act on the coordinates $y^{\mu_{k'}}$. In (6.71) the derivatives $\overleftarrow{\partial}_{\nu_l}$ act on the coordinates x'^{ν_l}, and the derivatives $\overleftarrow{\partial}_{\nu_{l'}}$ act on the coordinates $y'^{\nu_{l'}}$.

In Eq. (6.69), $G_{M_1+M_2+N_1+N_2}$ is the Green function of the Heisenberg spinor field operators

$$G_{m+n}(x_1,\dots,x_m,y_1,\dots,y_n) = \langle 0|T\{\hat{\psi}(x_1)\dots\hat{\psi}(x_m)\hat{\bar{\psi}}(y_1)\dots\hat{\bar{\psi}}(y_n)\}|0\rangle, \quad (6.72)$$

where $T\{\dots\}$ is the T-product of the time-dependent fermionic operators. By definition, the T-product rearranges the operators in order of decreasing time from left to the right with the sign factor $(-1)^P$, where P is the parity of the permutation from an initial arrangement of operators to the chronological ordering. The last feature is specific for the T-product of spinor field operators. Under the sign of the T-product, the operators behave like classical functions. But, as we already emphasized above, the classical spinor fields are anticommuting, defining the presence of the sign factor in the T-product for spinor field operators. In particular,

$$T(\hat{\psi}(x)\hat{\bar{\psi}}(y)) = -T(\hat{\bar{\psi}}(y)\hat{\psi}(x)). \quad (6.73)$$

6.4.2 Generating functional of spinor field Green functions

In general, the generating functional of Green functions can be constructed analogously to the scalar case. However, since, in the spinor case, the sign of the Green function changes under the permutation of neighboring operators, the corresponding sources should be anticommuting. Consider the Green functions

$$G_{m+n}(x_1,\dots,x_m,y_1,\dots,y_n) = \langle 0|T(\hat{\psi}(x_1)\dots\hat{\psi}(x_m)\hat{\bar{\psi}}(y_1)\dots\hat{\bar{\psi}}(y_n)|0\rangle \quad (6.74)$$

and $\eta(x)$, $\bar{\eta}(x)$, the classical source Dirac fields. The generating functional of the spinor Green functions is defined as

$$Z[\bar{\eta},\eta] = \sum_{m,n=1}^{\infty}\frac{i^{m+n}}{m!\,n!}\int d^4x_1\dots d^4x_m\,d^4y_1\dots d^4y_n\,\bar{\eta}(x_m)\dots\bar{\eta}(x_1)$$
$$\times\, G_{m+n}(x_1,\dots,x_m,y_1,\dots,y_n)\,\eta(y_n)\dots\eta(y_1). \quad (6.75)$$

As we know, the T-product of spinor field operators may change its sign under the permutation of neighboring operators. Therefore, the Green function (6.72) should change its sign under the permutation of neighboring arguments. Then, since the *l.h.s.* of Eq. (6.75) should not change under the permutation of integrating variables, the spinor fields $\eta(x)$ and $\bar{\eta}(x)$ should be anticommuting sources.

There are two types of functional (variational) derivatives with respect to the anticommuting functions, namely, the left derivative $\frac{\overrightarrow{\delta}}{\delta\bar{\eta}_C(z)}$ and the right one, $\frac{\overleftarrow{\delta}}{\delta\eta_C(z)}$. These two operations are defined as follows:

$$\textbf{1.}\quad \frac{\overrightarrow{\delta}}{\delta\bar{\eta}_{C_1}(z_1)}\frac{\overrightarrow{\delta}}{\delta\bar{\eta}_{C_2}(z_2)} = -\frac{\overrightarrow{\delta}}{\delta\bar{\eta}_{C_2}(z_2)}\frac{\overrightarrow{\delta}}{\delta\bar{\eta}_{C_1}(z_1)},$$
$$\frac{\overleftarrow{\delta}}{\delta\eta_{C_1}(z_1)}\frac{\overleftarrow{\delta}}{\delta\eta_{C_2}(z_2)} = -\frac{\overleftarrow{\delta}}{\delta\eta_{C_2}(z_2)}\frac{\overleftarrow{\delta}}{\delta\eta_{C_1}(z_1)},$$
$$\frac{\overrightarrow{\delta}}{\delta\bar{\eta}_{C_1}(z_1)}\frac{\overleftarrow{\delta}}{\delta\eta_{C_2}(z_2)} = -\frac{\overleftarrow{\delta}}{\delta\eta_{C_2}(z_2)}\frac{\overrightarrow{\delta}}{\delta\bar{\eta}_{C_1}(z_1)}, \quad (6.76)$$

2. $$\frac{\overrightarrow{\delta}}{\delta\bar{\eta}_{C_1}(z_1)}\bar{\eta}_{C_2}(z_2) = \delta_{C_1C_2}\delta^4(z_1 - z_2),$$

$$\eta_{C_2}(z_2)\frac{\overleftarrow{\delta}}{\delta\eta_{C_1}(z_1)} = \delta_{C_1C_2}\delta^4(z_1 - z_2),$$

$$\frac{\overrightarrow{\delta}}{\delta\bar{\eta}_{C_1}(z_1)}\eta_{C_2}(z_2) = 0, \qquad \bar{\eta}_{C_2}(z_2)\frac{\overleftarrow{\delta}}{\delta\eta_{C_1}(z_1)} = 0. \tag{6.77}$$

Here the indices C_1 and C_2 are the components of the four-component spinors. The rules (6.76) and (6.77) allow us to take the functional derivative of any expressions constructed from the products $\bar{\eta}(x_m)\ldots\bar{\eta}(x_1)\eta(y_n)\ldots\eta(y_1)$.

Applying the rules (6.76) and (6.77) to $Z[\bar{\eta},\eta]$ in (6.75), one gets

$$G_{m+n}(x_1,\ldots,x_m,y_1,\ldots,y_n)$$
$$= \left\{\frac{\overrightarrow{\delta}}{\delta i\bar{\eta}(x_m)}\cdots\frac{\overrightarrow{\delta}}{\delta i\bar{\eta}(x_1)}Z[\bar{\eta},\eta]\frac{\overleftarrow{\delta}}{\delta i\eta(y_n)}\cdots\frac{\overleftarrow{\delta}}{\delta i\eta(y_1)}\right\}\bigg|_{\eta=\bar{\eta}=0}. \tag{6.78}$$

We see that the derivatives of the functional $Z[\bar{\eta},\eta]$ with respect to the sources, at vanishing η and $\bar{\eta}$, generate the Green functions, as it should be for the generating functional.

Substituting the expression for the Green function (6.72) into (6.75) leads to

$$Z[\bar{\eta},\eta] = \langle 0|Te^{i\int d^4x(\bar{\eta}(x)\hat{\psi}(x)+\hat{\bar{\psi}}(x)\eta(x))}|0\rangle. \tag{6.79}$$

This is a convenient expression for exploring the general properties of the generating functional of spinor Green functions.

Exercises

6.1. Let $f(x)$ be a function that is expandable into a power series and let $\hat{A}$ and $\hat{B}$ be some operators that are, in general, non-commuting. Prove the identity $e^{\hat{A}}f(\hat{B})e^{-\hat{A}} = f(e^{\hat{A}}Be^{-\hat{A}})$.
6.2. Using the relation $|\Psi\rangle^{t_2} = \hat{U}(t_2,t_1)|\Psi\rangle^{t_1}$, prove that $\hat{U}(t_3,t_2) = \hat{U}(t_3,t_2)\hat{U}(t_2,t_1)$.
6.3. Consider the quantum scalar field model in the Schrödinger picture, with the Hamiltonian $\hat{H} = \hat{H}_0 + \hat{V}(\hat{\varphi},\hat{\pi})$, where $\hat{\varphi} = \hat{\varphi}(\mathbf{x})$, $\hat{\pi} = \hat{\pi}(\mathbf{x})$. Define the operator $\hat{S}(t,t_0)$ by the relation $\hat{U}(t,t_0) = e^{-i(t-t_0)\hat{H}_0}\hat{S}(t,t_0)$, where t_0 is an arbitrary initial time moment. Derive the equation of motion and the initial condition for the operator $\hat{S}(t,t_0)$, and show that its dynamics is ruled by the operator $\hat{V}(\hat{\varphi}_{t_0}(t),\hat{\pi}_{t_0}(t))$. Find out how the operators $\hat{\varphi}_{t_0}(t)$ and $\hat{\pi}_{t_0}(t)$ are related to the operators $\hat{\varphi}$ and $\hat{\pi}$.
6.4 Prove that the operator $\hat{S}(t,t_0)$ satisfies the relation $\hat{S}(t_3,t_2) = \hat{S}(t_3,t_2)\hat{S}(t_2,t_1)$.
6.5 Prove that the operator $\hat{S}(t,t_0)$ is unitary.
6.6. Show that the operator $\hat{S}(t,-\infty)$ satisfies the integral equation

$$\hat{S}(t,-\infty) = \hat{1} - i\int_{-\infty}^{t} dt'\,\hat{V}_{in}(t')\,\hat{S}(t',-\infty),$$

where $\hat{V}_{in}(t) = \hat{V}(\hat{\varphi}_{in}(t), \hat{\pi}_{in}(t))$. Prove that the solution to the above equation can be written in the form (the Dyson formula)

$$\hat{S}(t, -\infty) = Te^{-i\int_{-\infty}^{t} dt\, \hat{V}_{in}(t)}.$$

6.7. Show that the Green function of the Heisenberg field operators (6.37) can be written as

$$G(x_1, x_2, \ldots, x_n) = \frac{\langle 0|T(\hat{\varphi}_{in}(x_1), \hat{\varphi}_{in}(x_2), \ldots, \varphi_{in}(x_n)\hat{S}(\infty, -\infty))|0\rangle}{\langle 0|\hat{S}(\infty, -\infty)|0\rangle}.$$

Comments

The LSZ formulas reducing the calculations of $\mathcal{S}$-matrix elements to calculating the Green functions are discussed in much detail in the books [57], [187], [250], and [152].

7
Functional integrals

In this chapter, we consider an alternative approach to the quantization of fields. This approach will be critically important to the development of quantum field theory in curved space, which is the subject of Part II.

7.1 Representation of the evolution operator by a functional integral

We start the consideration of functional integrals from the simplest quantum-mechanical models and then go to the more complex case of field theory.

7.1.1 The functional integral for the one-dimensional quantum system

Consider a quantum system with one degree of freedom. This means that there is a single coordinate operator $\hat{q}$ and a single momentum operator $\hat{p}$. The system is described by the Hamiltonian $\hat{H} = H(q,p)|_{q\to\hat{q},\ p\to\hat{p}}$, which is constructed on the basis of the classical Hamiltonian function $H(q,p)$ with a certain ordering prescription for products of the non-commuting operators $\hat{q}$ and $\hat{p}$. The dynamics of the system is described by the evolution operator

$$\hat{U}(t'',t') = e^{-\frac{i}{\hbar}(t''-t')\hat{H}}. \tag{7.1}$$

The purpose of this section is to construct a special representation of the matrix elements of the evolution operator, one which is very convenient in quantum field theory as well as other areas of physics. Let us note that, in this part of the book, we shall write the Planck constant $\hbar$ explicitly; also, in this subsection, x is a one-dimensional coordinate.

Let $|q\rangle$ be a state with a certain coordinate, $\hat{q}|q\rangle = q|q\rangle$, and let $|p\rangle$ be a state with a certain momentum, $\hat{p}|p\rangle = p|p\rangle$. In the coordinate representation, the states $|q\rangle$ and $|p\rangle$ are given by the wave functions

$$\psi_q(x) = \langle x|q\rangle = \delta(q-x) \qquad \text{and} \qquad \psi_p(x) = \langle x|p\rangle = \frac{1}{\sqrt{2\pi\hbar}}e^{\frac{ipx}{\hbar}}. \tag{7.2}$$

We know that

$$\langle p|q\rangle = \int_{-\infty}^{\infty} dx\, \psi_p^*(x)\psi_q(x) = \frac{1}{\sqrt{2\pi\hbar}}\int_{-\infty}^{\infty} dx e^{-\frac{ipx}{\hbar}}\delta(q-x) = \frac{1}{\sqrt{2\pi\hbar}}e^{-\frac{ipq}{\hbar}}. \tag{7.3}$$

Let us consider the matrix element of the evolution operator of the form

Introduction to Quantum Field Theory with Applications to Quantum Gravity. Iosif L. Buchbinder and Ilya L. Shapiro, Oxford University Press (2021). © Iosif L. Buchbinder and Ilya Shapiro.
DOI: 10·1093/oso/9780198838319·003·0007

$$\langle q''|\hat{U}(t'',t')|q'\rangle = \langle q''|e^{-\frac{i}{\hbar}(t''-t')\hat{H}}|q'\rangle, \tag{7.4}$$

where $|q'\rangle$ and $|q''\rangle$ are states with certain values of the coordinates q' and q''.

Consider a matrix element of the Hamiltonian. As was pointed out above, $\hat{H}$ is constructed assuming some ordering prescription for the non-commuting operators $\hat{q}$ and $\hat{p}$. Thus, we assume that some initial ordering is fixed. Using the commutation relation $[\hat{q},\hat{p}] = i\hbar$, one can commute the operators $\hat{q}$ and $\hat{p}$ in the expression for $\hat{H}$ in such a way that all operators $\hat{p}$ will be to the left of all operators $\hat{q}$. Then the matrix element $\langle p|\hat{H}|q\rangle$ can be obtained by replacing the operator $\hat{p}$ by its eigenvalue p, and the operator $\hat{q}$ by its eigenvalue q:

$$\langle p|\hat{H}|q\rangle = H_{pq}\langle p|q\rangle = \frac{1}{\sqrt{2\pi\hbar}}e^{-\frac{ipq}{\hbar}}H_{pq}(q,p). \tag{7.5}$$

The indices pq in $H_{pq}(q,p)$ mean that we wrote $\hat{H}$ in the form where all $\hat{p}$ are to the left of all $\hat{q}$. It is evident that $H_{pq}(q,p)$ is a c-valued function, called the pq-symbol of the operator $\hat{H}$. If $\hbar \to 0$, then $H_{pq}(q,p) \to H(q,p)$, where $H(q,p)$ is a classical Hamilton function. It is important to emphasize that, in general, the function $H_{pq}(q,p)$ depends on $\hbar$.

The matrix element (7.4) can be written as

$$\langle q''|\hat{U}(t'',t')|q'\rangle = \int dp\langle q''|p\rangle\langle p|\hat{U}(t'',t')|q'\rangle. \tag{7.6}$$

If the time interval $t''-t'$ is very small, we can use the linear approximation,

$$\hat{U}(t'',t') = e^{-\frac{i}{\hbar}(t''-t')\hat{H}} \approx \hat{1} - \frac{i}{\hbar}(t''-t')\hat{H}.$$

Then, in the same approximation,

$$\begin{aligned}
\langle p|\hat{U}(t'',t')|q'\rangle &= \langle p|q'\rangle - \frac{i}{\hbar}(t''-t')\langle p|\hat{H}|q'\rangle \\
&= \langle p|q'\rangle - \frac{i}{\hbar}(t''-t')H_{pq}(q',p)\langle p|q'\rangle \;=\; \langle p|q'\rangle\Big[1 - \frac{i}{\hbar}(t''-t')H_{pq}(q',p)\Big] \\
&= \langle p|q'\rangle\, e^{-\frac{i}{\hbar}(t''-t')H_{pq}(q',p)} \;=\; \frac{1}{\sqrt{2\pi\hbar}}e^{-\frac{i}{\hbar}pq' - \frac{i}{\hbar}(t''-t')H_{pq}(q',p)}.
\end{aligned}$$

Therefore,

$$\begin{aligned}
\langle q''|\hat{U}(t'',t')|q'\rangle &\approx \int \frac{dp}{\sqrt{2\pi\hbar}}e^{\frac{i}{\hbar}pq''}\frac{1}{\sqrt{2\pi\hbar}}e^{-\frac{i}{\hbar}pq' - \frac{i}{\hbar}(t''-t')H_{pq}(q',p)} \\
&= \int \frac{dp}{2\pi\hbar}e^{\frac{i}{\hbar}\left[p(q''-q') - (t''-t')H_{pq}(q',p)\right]} \\
&= \int \frac{dp}{2\pi\hbar}e^{\frac{i}{\hbar}(t''-t')\left[p\frac{q''-q'}{t''-t'} - H_{pq}(q',p)\right]}.
\end{aligned} \tag{7.7}$$

For a finite time interval $t''-t'$, we can go beyond the linear approximation as follows. Let us divide the time interval $t''-t'$ into N small intervals Δt by the points

$t_1, t_2, \ldots, t_{N-1}$, where $t_i - t_{i-1} \equiv \Delta t$; $i = 1, \ldots, N$; $t' = t_0$ and $t'' = t_N$. Using the property of the evolution operator

$$\hat{U}(t'', t') = \hat{U}(t'', t_{N-1})\, \hat{U}(t_{N-1}, t_{N-2}) \ldots \hat{U}(t_2, t_1)\, \hat{U}(t_1, t'), \tag{7.8}$$

we get

$$\begin{aligned} \langle q''|\hat{U}(t'', t')|q'\rangle &= \int dq_{N-1} \ldots dq_1\, \langle q''|\hat{U}(t'', t_{N-1})|q_{N-1}\rangle \\ &\times \langle q_{N-1}|\hat{U}(t_{N-1}, t_{N-2})|q_{N-2}\rangle \ldots \langle q_2|\hat{U}(t_2, t_1)|q_1\rangle \langle q_1|\hat{U}(t_1, t')|q'\rangle. \end{aligned} \tag{7.9}$$

Since all the intervals are equal to the same small quantity Δt, the relation (7.7) can be applied to each of the matrix elements in Eq. (7.9). Thus,

$$\begin{aligned} \langle q''|\hat{U}(t'', t')|q'\rangle &\approx \int \frac{dq_1 dp_1}{2\pi\hbar} \frac{dq_2 dp_2}{2\pi\hbar} \ldots \frac{dq_{N-1} dp_{N-1}}{2\pi\hbar} \frac{dp_N}{2\pi\hbar} \exp \frac{i}{\hbar} \Big\{ p_N (q'' - q_{N-1}) \\ &+ p_{N-1}(q_{N-1} - q_{N-2}) + \cdots + p_1(q_1 - q') - \Delta t \Big[H_{pq}(q_{N-1}, p_N) \\ &+ H_{pq}(q_{N-2}, p_{N-1}) + \cdots + H_{pq}(q_1, p_2) + H_{pq}(q', p_1) \Big] \Big\} \\ &= \int \frac{dq_1 dp_1}{2\pi\hbar} \ldots \frac{dq_{N-1} dp_{N-1}}{2\pi\hbar} \frac{dp_N}{2\pi\hbar} \exp \frac{i}{\hbar} \Delta t \Big\{ p_N \frac{q'' - q_{N-1}}{\Delta t} \\ &+ p_{N-1} \frac{q_{N-1} - q_{N-2}}{\Delta t} + \ldots + p_2 \frac{q_2 - q_1}{\Delta t} + p_1 \frac{q_1 - q'}{\Delta t} \\ &- \Big[H_{pq}(q_{N-1}, p_N) + \cdots + H_{pq}(q_1, p_2) + H_{pq}(q', p_1) \Big] \Big\}. \end{aligned} \tag{7.10}$$

In order to obtain exact equality, consider the limit $\Delta t \to 0$ and $N \to \infty$, but keeping $\Delta t N \equiv t'' - t'$. The expression in the exponential of (7.10) is an integral sum for

$$\int_{t'}^{t''} dt \{ p(t)\dot{q}(t) - H_{pq}(q(t), p(t)) \} = S_H[q, p], \tag{7.11}$$

where $S_H[q, p]$ is the action in the classical canonical formalism, corresponding to the Hamiltonian function $H_{pq}(q, p)$ (depending on $\hbar$, in general). When $\Delta t \to 0$ and $N \to \infty$, the number of integrals in (7.10) tends to infinity. We can consider this limit as an integral over all functions $q(t)$, where $t' < t < t''$ (there are no integrals over q', q'') and over all functions $p(t)$, where $t' < t \leq t''$. Here $q(t') = q'$, $q(t'') = q''$ with fixed q' and q''. The limit $\Delta t \to 0$, $N \to \infty$ of (7.10) is denoted

$$\int \mathcal{D}q\, \mathcal{D}p\; e^{\frac{i}{\hbar} S_H[q, p]}, \tag{7.12}$$

which is called the functional integral, or the path integral, over the phase space. This integral is taken over all possible trajectories $q(t)$, $p(t)$ starting at $t = t'$ and ending at $t = t''$. The time instants t' and t'' define the expression for the action $S_H[q, p]$ (7.11). The values of the functions $q(t)$ at $t = t'$ and $t = t''$ are fixed, $q(t') = q'$, $q(t'') = q''$ and there are no boundary conditions for the momenta $p(t)$. As a result, we get the expression for the matrix elements of the evolution operator in the form of the functional integral over the phase space,

$$\langle q''|\hat{U}(t'',t')|q'\rangle = \int \mathcal{D}q\mathcal{D}p \; e^{\frac{i}{\hbar}S_H[q,p]}. \tag{7.13}$$

Consider an example when the classical Hamiltonian function is

$$H(q,p) = \frac{p^2}{2} + V(q).$$

Then $\hat{H} = \frac{\hat{p}^2}{2} + V(\hat{q})$ and $H_{pq}(q,p) = H(q,p)$, where H(q, p) is a classical Hamiltonian function. Therefore, in the case under consideration, one gets

$$\langle q''|\hat{U}(t'',t')|q'\rangle = \int \mathcal{D}q\,\mathcal{D}p\, e^{\frac{i}{\hbar}\int_{t'}^{t''} dt\left[p(t)\dot{q}(t)-\frac{p^2(t)}{2}-V(q(t))\right]}.$$

Let us perform the change of variables $p(t) \to p(t) + \dot{q}(t)$ in the last integral. It is natural to expect that the integration measure $\mathcal{D}q\,\mathcal{D}p$ does not change, since we shift the function $p(t)$ on the function $\dot{q}(t)$, which is independent of $p(t)$. Since there is no boundary condition for $p(t)$, this shift does not affect the conditions $q(t') = q'$, $q(t'') = q''$. In this way, we get

$$\begin{aligned}\langle q''|\hat{U}(t'',t')|q'\rangle &= \int \mathcal{D}q\mathcal{D}p \exp\frac{i}{\hbar}\int_{t'}^{t''} dt\Big\{p\dot{q} + \dot{q}^2 - \frac{p^2}{2} - p\dot{q} - \frac{\dot{q}^2}{2} - V(q)\Big\} \\ &= \int \mathcal{D}p e^{\frac{i}{\hbar}\int_{t'}^{t''} dt\frac{p^2(t)}{2}} \int \mathcal{D}q e^{\frac{i}{\hbar}\int_{t'}^{t''} dt\left[\frac{\dot{q}^2(t)}{2}-V(q(t))\right]}.\end{aligned} \tag{7.14}$$

Obviously,

$$\int_{t'}^{t''} dt\left\{\frac{\dot{q}^2(t)}{2} - V(q(t))\right\} = S[q], \tag{7.15}$$

where $S[q]$ is the action in the Lagrange formalism. We denote

$$\int \mathcal{D}p \exp\left\{-\frac{i}{\hbar}\int_{t'}^{t''} dt\,\frac{p^2(t)}{2}\right\} = N^{-1}(t'',t'), \tag{7.16}$$

which is a function that does not depend on q' and q''. Therefore, (7.14) becomes

$$\langle q''|\hat{U}(t'',t')|q'\rangle = \frac{1}{N(t'',t')}\int \mathcal{D}q\; e^{\frac{i}{\hbar}S[q]}. \tag{7.17}$$

In principle, the quantity $N^{-1}(t'',t')$ can be included into the integration measure $\mathcal{D}q$. The integral in (7.17) is taken over all possible paths or trajectories in configuration space that start at the time instant t' at the point $q(t') = q'$ and end at t'' at the point $q(t'') = q''$. Eq. (7.17) is a representation of the matrix element of the evolution operator by a functional integral over configuration space.

Indeed, the representation (7.13) was derived only on the basis of canonical quantization, and hence it is more fundamental, compared to (7.17), requiring an assumption about the form of the Hamiltonian operator.

7.1.2 The functional integral for quantum system with a finite number of degrees of freedom

Consider a system described by n generalized coordinates $q \equiv (q_1, q_2, \dots, q_n)$. In such a system, we denote $q' \equiv (q'_1, q'_2, \dots, q'_n)$ and $q'' \equiv (q''_1, q''_2, \dots, q''_n)$. The matrix element of the evolution operator $\langle q''|\hat{U}(t'', t')|q'\rangle$ can be worked out by employing the same method as for the one-dimensional case described above. As a result, we obtain

$$\langle q''|\hat{U}(t'', t')|q'\rangle = \int \prod_{i=1}^{n} \mathcal{D}q_i \mathcal{D}p_i e^{\frac{i}{\hbar} S_{\mathcal{H}}[q,p]}, \tag{7.18}$$

where

$$S_H[q,p] = \int_{t'}^{t''} dt \left\{ \sum_{i=1}^{n} p_i(t)\dot{q}_i(t) - H_{pq}(q(t), p(t)) \right\}$$

is the action in the classical canonical formalism with the Hamilton function $H_{pq}(q,p)$, the pq-symbol of the Hamiltonian $\hat{H}$. In the case when the Hamiltonian has the form

$$\hat{H} = \sum_{i=1}^{n} \frac{\hat{p}_i^2}{2} + V(\hat{q}_i, \dots, \hat{q}_n),$$

(7.18) can be transformed to

$$\langle q''|\hat{U}(t'', t')|q'\rangle = \frac{1}{N(t'', t')} \int \prod_{i=1}^{n} \mathcal{D}q_i \, e^{\frac{i}{\hbar} S[q]}. \tag{7.19}$$

Here $S[q]$ is the action in the Lagrangian formalism, and

$$N^{-1}(t'', t') = \prod_{i=1}^{n} \int \mathcal{D}p_i \, \exp\left\{ -\frac{i}{\hbar} \int_{t'}^{t''} dt \, \frac{\hat{p}_i^2(t)}{2} \right\}. \tag{7.20}$$

In what follows, in most of the cases, the expression $N^{-1}(t'', t') \prod_{i=1}^{n} \mathcal{D}q_i$ is denoted as $\mathcal{D}q$, i.e.,, the *normalization factor* $N^{-1}(t'', t')$ is included in the integration measure.

7.1.3 The functional integral in scalar field theory

Now we consider a scalar field model with the Lagrangian $\mathcal{L}(\varphi, \partial_\mu \varphi)$. As we pointed out above, the field $\varphi(x)$ can be treated as $\varphi_{\mathbf{x}}(t)$, that is, as generalized time-dependent coordinates indexed by the vector $\mathbf{x}$. In the coordinate representation, the field state is given by the functional $\Psi[\varphi]$, where $\varphi \equiv \varphi(\mathbf{x})$ and the operators $\hat{\varphi}(\mathbf{x})$ and $\hat{\pi}(\mathbf{x})$ are defined by the relations following from (5.36) and (5.37),

$$\hat{\varphi}(\mathbf{x})\Psi[\varphi] = \varphi(\mathbf{x})\Psi[\varphi], \qquad \hat{\pi}(\mathbf{x})\Psi[\varphi] = -i\hbar \frac{\delta \Psi[\varphi]}{\delta \varphi(\mathbf{x})}. \tag{7.21}$$

Consider the matrix element of the evolution operator

$$\langle \varphi''|\hat{U}(t'', t')|\varphi'\rangle = \langle \varphi''|e^{-\frac{i}{\hbar}(t''-t')\hat{H}}|\varphi'\rangle = \langle \varphi'', t''|\varphi', t'\rangle. \tag{7.22}$$

This is the probability amplitude for finding a field configuration $\varphi''(\mathbf{x})$ at the time instant t'' under the condition that, at the instant t', the field was in the configuration

$\varphi'(\mathbf{x})$. The quantity $\langle\varphi'', t''|\varphi', t'\rangle$ is also treated as a transition amplitude from the field configuration $\varphi'(\mathbf{x})$ to the field configuration $\varphi''(\mathbf{x})$ in the time interval $(t',\, t'')$.

To find the matrix element of the evolution operator $\langle\varphi'', t''|\varphi', t'\rangle$, one can repeat the consideration for the finite-dimensional systems. The result is written in the form

$$\langle\varphi'', t''|\varphi', t'\rangle = \int \mathcal{D}\varphi\mathcal{D}\pi\, e^{\frac{i}{\hbar}S_H[\varphi,\pi]}, \tag{7.23}$$

where $S_H[\varphi, \pi]$ is the action in the canonical formalism

$$S_H[\varphi,\pi] = \int_{t'}^{t''} dt\, \Big\{ \int d^3x\, \pi(t,\mathbf{x})\, \dot{\varphi}(t,\mathbf{x}) \,-\, \mathcal{H}_{\pi\varphi}\big(\varphi(\mathbf{x}), \partial_j\varphi(\mathbf{x}), \pi(\mathbf{x})\big)\Big\}. \tag{7.24}$$

Here $\mathcal{H}_{\pi\varphi}(\varphi, \partial_j\varphi, \pi)$ is the $\pi\varphi$-symbol of the Hamiltonian $\hat{\mathcal{H}}$. We remind the reader that the Hamiltonian operator is $\hat{H} = \int d^3x\, \hat{\mathcal{H}}$. The integral in (7.23) is taken over all trajectories or paths in the phase space of scalar field theory, parameterized by the functions $\varphi = \varphi(t,\mathbf{x})$, $\pi = \pi(t,\mathbf{x})$ with the fixed ends $\varphi(t'',\mathbf{x}) = \varphi''(\mathbf{x})$ and $\varphi(t',\mathbf{x}) = \varphi'(\mathbf{x})$.

If the Hamiltonian has the special form

$$\hat{H} = \int d^3x\, \Big\{\frac{1}{2}\hat{\pi}^2(\mathbf{x}) + V\big(\hat{\varphi}(\mathbf{x}), \partial_j\hat{\varphi}(\mathbf{x})\big)\Big\}, \tag{7.25}$$

we can repeat the previous consideration for the finite-dimensional case and obtain

$$\langle\varphi'', t''|\varphi', t'\rangle = \int \mathcal{D}\varphi\, e^{\frac{i}{\hbar}S[\varphi]}, \tag{7.26}$$

where

$$\mathcal{S}[\varphi] = \int_{t'}^{t''} dt \int d^3x\, \mathcal{L}(\varphi, \partial_\mu\varphi) \tag{7.27}$$

is the action in Lagrange formalism. The integral (7.26) is taken over all trajectories $\varphi = \varphi(t,\mathbf{x})$ in the configuration space under the conditions $\varphi(t'',\mathbf{x}) = \varphi''(\mathbf{x})$ and $\varphi(t',\mathbf{x}) = \varphi'(\mathbf{x})$. The fields $\varphi'(\mathbf{x})$, $\varphi''(\mathbf{x})$ are arguments of the matrix element $\langle\varphi'', t''|\varphi', t'\rangle$.

The *r.h.s.* of (7.27) is not a Lorentz scalar, since it includes manifestly a definite integral between the time instants t' and t''. On the other hand, when we discussed the $\mathcal{S}$-matrix in chapter 6, it was pointed out that the scattering problem in quantum field theory is well defined only when $t' \to -\infty$ and $t'' \to \infty$. Taking these limits resolves the problem of Lorentz invariance, since, in this case, the integral in the expression (7.27) is taken over the whole Minkowski space. This integration will be assumed in what follows, if the opposite is not indicated explicitly. Denoting

$$\lim_{t'\to-\infty, t''\to\infty} \langle\varphi'', t''|\varphi', t'\rangle = \langle\varphi'', \infty|\varphi', -\infty\rangle, \tag{7.28}$$

one gets

$$\langle\varphi'',\infty|\varphi',-\infty\rangle \;=\; \int\mathcal{D}\,\varphi e^{\frac{i}{\hbar}S[\varphi]}\,. \tag{7.29}$$

We define a classical asymptotic field in quantum theory as an expectation value of the free field operator on some state in the corresponding Fock state, that is,

$$\varphi(\mathbf{x}) \;=\; \langle\Psi|\hat{\varphi}(\mathbf{x})|\Psi\rangle, \tag{7.30}$$

where $\hat{\varphi}(\mathbf{x})$ is given by Eq. (5.50). The state $|\Psi\rangle$ can be the vacuum state $|0\rangle$. Then the classical field in the vacuum is

$$\varphi(\mathbf{x}) = \langle 0|\hat{\varphi}(\mathbf{x})|0\rangle = \int\frac{d^3p}{\sqrt{2(2\pi^3)\varepsilon(\mathbf{p})}}\Big[e^{\frac{i}{\hbar}\mathbf{px}}\langle 0|\hat{a}(\mathbf{p})|0\rangle + e^{-\frac{i}{\hbar}\mathbf{px}}\langle 0|\hat{a}^{\dagger}(\mathbf{p})|0\rangle\Big] = 0. \tag{7.31}$$

The matrix element $\langle\varphi=0,\infty|\varphi=0,-\infty\rangle$ is defined as

$$\langle\varphi=0,\infty|\varphi=0,-\infty\rangle \;=\; \lim_{t'\to-\infty,\,t''\to\infty}\langle\varphi'',t''|\varphi',t'\rangle\Big|_{\varphi'=\varphi''=0}. \tag{7.32}$$

The expression $\langle\varphi=0,\infty|\varphi=0,-\infty\rangle$ is the probability amplitude of finding a field in a vacuum state in a distant future under the condition that, in a distant past, a field was also in a vacuum state. Therefore, one can write

$$\langle\varphi=0,\infty|\varphi=0,-\infty\rangle \;=\; \int\mathcal{D}\varphi\, e^{\frac{i}{\hbar}S[\varphi]}. \tag{7.33}$$

Let us stress that the functional integral in (7.33) is taken over all fields $\varphi(x)$, satisfying the conditions $\varphi(x)|_{t=-\infty} = 0$ and $\varphi(x)|_{t=\infty} = 0$. Eqs. (7.29) and (7.33) will be the basic relations for the subsequent analysis.

7.2 Functional representation of Green functions

Let us now turn our attention to the Green functions.

It is useful to start, once again, with the one-dimensional theory, which is quantum mechanics. Our target is the matrix element $\langle q'',t''|\hat{q}(t)|q',t'\rangle$, where $t'<t<t''$ and $\hat{q}(t)$ is the coordinate operator in the Heisenberg picture. For the calculation, we use the same approach, as in the previous section:

$$\begin{aligned}
\langle q'',t''|\hat{q}(t)|q',t'\rangle &= \langle q''|e^{-\frac{i}{\hbar}t''\hat{H}}e^{\frac{i}{\hbar}t\hat{H}}\,\hat{q}\,e^{-\frac{i}{\hbar}t\hat{H}}e^{\frac{i}{\hbar}t'\hat{H}}|q'\rangle \\
&= \langle q''|e^{-\frac{i}{\hbar}(t''-t)\hat{H}}\,\hat{q}\,e^{-\frac{i}{\hbar}(t-t')\hat{H}}|q'\rangle \\
&= \langle q''|\hat{U}(t'',t)\hat{q}\hat{U}(t,t')|q'\rangle \\
&= \int dq\langle q''|\hat{U}(t'',t)\,\hat{q}\,|q\rangle\langle q|\hat{U}(t,t')|q'\rangle \\
&= \int dq\langle q''|\hat{U}(t'',t)|q\rangle\, q\,\langle q|\hat{U}(t,t')|q'\rangle.
\end{aligned} \tag{7.34}$$

For the Hamiltonian quadratic in the momentum operators, each of the matrix elements $\langle q''|\hat{U}(t'',t')|q\rangle$ and $\langle q|\hat{U}(t,t')|q'\rangle$ is represented by a functional integral over configuration space. Therefore, we can write

$$\langle q'', t''|\hat{q}(t)|q', t'\rangle = \int dq(t)\Big(\int \mathcal{D}q e^{\frac{i}{\hbar}S(t'',t)}\Big)\, q(t)\,\Big(\int \mathcal{D}q e^{\frac{i}{\hbar}S(t,t')}\Big).$$
$$= \int \mathcal{D}q q(t) e^{\frac{i}{\hbar}S(t'',t')}. \tag{7.35}$$

Here $S(t,t') = \int_{t'}^{t} dt L(q,\dot{q})$ is the action between t' and t in the Lagrange formalism. In addition, we replaced the integration variable q with $q(t)$. The integral is taken over all possible trajectories in configuration space, which start at t' at the point q' and end at t'' at the point q''.

For a more complicated matrix element, we have, with $t' < t_2 < t_1 < t''$,

$$\langle q'', t''|\hat{q}(t_1)\hat{q}(t_2)|q', t'\rangle = \langle q''|e^{-\frac{i}{\hbar}t''\hat{H}} e^{\frac{i}{\hbar}t_1\hat{H}} \hat{q} e^{-\frac{i}{\hbar}t_1\hat{H}} e^{\frac{i}{\hbar}t_2\hat{H}} \hat{q} e^{-\frac{i}{\hbar}t_2\hat{H}} e^{\frac{i}{\hbar}t'\hat{H}}|q'\rangle$$
$$= \int dq_1 dq_2 \langle q''|\hat{U}(t'',t_1)\hat{q}|q_1\rangle\langle q_1|\hat{U}(t_1,t_2)\hat{q}|q_2\rangle\langle q_2|\hat{U}(t_2,t')|q'\rangle$$
$$= \int dq_1 dq_2 \langle q''|\hat{U}(t'',t_1)q_1\rangle q_1 \langle q_1|\hat{U}(t_1,t_2)|q_2\rangle q_2 \langle q_2|\hat{U}(t_2,t_1)|q'\rangle. \tag{7.36}$$

Each of the matrix elements of the evolution operator in (7.36) can be represented by a functional integral. In this way, we get

$$\langle q'', t''|\hat{q}(t_1)\hat{q}(t_2)|q', t'\rangle$$
$$= \int dq_1 dq_2 \Big(\int \mathcal{D}q e^{\frac{i}{\hbar}S(t'',t_1)}\Big) q_1 \Big(\int \mathcal{D}q e^{\frac{i}{\hbar}S(t_1,t_2)}\Big) q_2 \Big(\int \mathcal{D}q e^{\frac{i}{\hbar}S(t_2,t')}\Big)$$
$$= \int dq(t_1) dq(t_2) \Big(\int \mathcal{D}q e^{\frac{i}{\hbar}S(t'',t_1)}\Big) q(t_1) \Big(\int \mathcal{D}q e^{\frac{i}{\hbar}S(t_1,t_2)}\Big) q(t_2) \Big(\int \mathcal{D}q e^{\frac{i}{\hbar}S(t_2,t')}\Big)$$
$$= \int \mathcal{D}q\; q(t_1) q(t_2) e^{\frac{i}{\hbar}S(t'',t')}. \tag{7.37}$$

The last path integral is taken over all possible trajectories $q(t)$, which start at t' at the point q' and end at t'' st the point q''.

The next step is to consider the matrix element $\langle q'', t''|\hat{q}(t_2)\hat{q}(t_1)|q', t'\rangle$, where $t' < t_1 < t_2 < t''$. After the same transformations as in the previous case, we find

$$\langle q'', t''|\hat{q}(t_2)\hat{q}(t_1)|q', t'\rangle = \int \mathcal{D}q\, q(t_2) q(t_1) e^{\frac{i}{\hbar}S(t'',t')}. \tag{7.38}$$

As a result,

$$\int \mathcal{D}q q(t_1) q(t_2) e^{\frac{i}{\hbar}S(t'',t')} = \begin{cases} \langle q'', t''|\hat{q}(t_1)\hat{q}(t_2)|q', t'\rangle, & t_1 > t_2, \\ \langle q'', t''|\hat{q}(t_2)\hat{q}(t_1)|q', t'\rangle, & t_1 < t_2 \end{cases} \tag{7.39}$$
$$= \theta(t_1 - t_2)\langle q'', t''|\hat{q}(t_1)\hat{q}(t_2)|q', t'\rangle + \theta(t_2 - t_1)\langle q'', t''|\hat{q}(t_2)\hat{q}(t_1)|q', t'\rangle.$$

Using the T-ordering introduced in Eq. (6.31) Chapter 6, we can write

$$\langle q'', t''|T(\hat{q}(t_1)\hat{q}(t_2))|q', t'\rangle = \int \mathcal{D}q\, q(t_1) q(t_2) e^{\frac{i}{\hbar}S[q]}. \tag{7.40}$$

In the general case, similar considerations yield

$$\langle q'',t''|T(\hat{q}(t_1)\dots\hat{q}(t_n))|q',t'\rangle = \int \mathcal{D}q\, q(t_1)\dots q(t_n)e^{\frac{i}{\hbar}S[q]}. \tag{7.41}$$

Here all times $t_1,\dots,t_n$ are smaller then t'' and greater then t'.

Let us now pass to scalar field theory. In this case, we obtain

$$\langle \varphi'',t''|T(\hat{\varphi}(x_1)\dots\hat{\varphi}(x_n))|\varphi',t'\rangle = \int \mathcal{D}\varphi\, \varphi(x_1)\dots\varphi(x_n)\, e^{\frac{i}{\hbar}S[\varphi]}, \tag{7.42}$$

where the T-product is defined according to (6.31) and the time integral in the action $S[\varphi]$ is taken from t' to t''. Taking the limits $t'\to-\infty$ and $t''\to\infty$ in Eq. (7.42), we also assume $\varphi'=0$ and $\varphi''=0$. Then, in the *l.h.s.* we have

$$\langle \varphi=0,\infty|T(\hat{\varphi}(x_1)\dots\hat{\varphi}(x_n))|\varphi=0,-\infty\rangle, \tag{7.43}$$

where the time arguments are already unrestricted and the action contains the integral over the full Minkowski space. Therefore,

$$\langle \varphi=0,\infty|T(\hat{\varphi}(x_1)\dots\hat{\varphi}(x_n))|\varphi=0,-\infty\rangle = \int \mathcal{D}\varphi\, \varphi(x_1)\dots\varphi(x_n)\, e^{\frac{i}{\hbar}S[\varphi]}. \tag{7.44}$$

We can analyse the situation from a slightly different perspective. Let $\hat{H}$ be the full Hamiltonian of the scalar field theory. As was pointed out in section 6.1, under physically reasonable assumptions, one can show that the vacuum $|0\rangle$ of the free theory is an exact eigenvector of $\hat{H}$, with the corresponding eigenvalue equal to zero.

We can denote the eigenvectors of the Hamiltonian as $|l\rangle$, which means

$$\hat{H}|l\rangle = E_l|l\rangle, \qquad \text{where} \qquad l=0,1,2,\dots, \qquad \text{and} \qquad E_0=0. \tag{7.45}$$

Some of the eigenvalues correspond to the continuous spectrum. In this case, the sums over these eigenvalues should be replaced by integrals. For the mean values, we get

$$\begin{aligned}
&\langle \varphi=0,t''|T(\hat{\varphi}(x_1)\dots\hat{\varphi}(x_n))|\varphi=0,t'\rangle \\
&= \langle \varphi=0|e^{-\frac{i}{\hbar}t''\hat{H}}T(\hat{\varphi}(x_1)\dots\hat{\varphi}(x_n))e^{-\frac{i}{\hbar}t'\hat{H}}|\varphi=0\rangle \\
&= \langle \varphi=0|0\rangle\langle 0|T(\hat{\varphi}(x_1)\dots\hat{\varphi}(x_n))|0\rangle\langle 0|\varphi=0\rangle \\
&\quad + \sum_{l=1}^{\infty}\langle \varphi=0|l\rangle\langle l|T(\hat{\varphi}(x_1)\dots\hat{\varphi}(x_n))|0\rangle\langle 0|\varphi=0\rangle e^{-\frac{i}{\hbar}t''E_l} \\
&\quad + \sum_{l'=1}^{\infty}\langle \varphi=0|0\rangle\langle 0|T(\hat{\varphi}(x_1)\dots\hat{\varphi}(x_n))|l'\rangle\langle l'|\varphi=0\rangle e^{\frac{i}{\hbar}t'E_{l'}} \\
&\quad + \sum_{l,l'=1}^{\infty}\langle \varphi=0|l\rangle\langle l|T(\hat{\varphi}(x_1)\dots\hat{\varphi}(x_n))|l'\rangle\langle l'|\varphi=0\rangle e^{-\frac{i}{\hbar}t''E_l+\frac{i}{\hbar}t'E_{l'}}.
\end{aligned} \tag{7.46}$$

In the last expression, $|\varphi=0\rangle = |\varphi=0,t\rangle_{t=0}$. Remember that, in our notation, $|0\rangle$ is a vacuum and $\hat{H}|0\rangle=0$. On the other hand, the state $|\varphi=0\rangle$ is not a vacuum, but $|\varphi=0\rangle = |\varphi,t\rangle|_{t=0,\ \varphi=0}$.

Taking the limits $t' \to -\infty$ and $t'' \to \infty$ in Eq. (7.46), we obtain in the *l.h.s.*

$$\langle\varphi = 0, \infty|T(\hat{\varphi}(x_1)\ldots\hat{\varphi}(x_n))|\varphi = 0, -\infty\rangle. \tag{7.47}$$

On the other hand, in the *r.h.s.*, (7.46), all time-dependent terms with the exponential factors oscillate very fast and hence vanish in the limit under consideration. This argument can be made more precise by replacing t' with $t'(1+i\epsilon)$, and t'' with $t''(1-i\epsilon)$, where $\epsilon \to +0$. We leave it to the reader to elaborate this argument in more detail.

In the end, in the *r.h.s.* of (7.46) there remains only the single term

$$\langle\varphi = 0|0\rangle\langle 0|T(\hat{\varphi}(x_1)\ldots\hat{\varphi}(x_n))|0\rangle\langle 0|\varphi = 0\rangle, \tag{7.48}$$

which is proportional to the vacuum average of the T-product of the field operators. However, remember that

$$\langle 0|T\hat{\varphi}(x_1)\ldots\hat{\varphi}(x_n)|0\rangle \;=\; G_n(x_1,\ldots,x_n) \tag{7.49}$$

is the n-point Green function.

Thus, the relations (7.47), (7.48) and (7.49) enable us to write the n-point Green function in the form

$$G_n(x_1,\ldots,x_n) \;=\; \frac{\langle\varphi = 0, \infty|T(\hat{\varphi}(x_1)\ldots\hat{\varphi}(x_n))|\varphi = 0, -\infty\rangle}{\langle\varphi = 0|0\rangle\langle 0|\varphi = 0\rangle}. \tag{7.50}$$

The numerator of this formula is the expression (7.44). Thus, the expression (7.47) differs from the Green function only by a constant factor of $\langle\varphi = 0|0\rangle\langle 0|\varphi = 0\rangle$.

In order to calculate this constant factor, consider the relation (7.26) and rewrite its *l.h.s.* using the basis of the eigenvectors $|n\rangle$. This leads to the following relation:

$$\begin{aligned}\langle\varphi'', t''|\varphi', t'\rangle &= \sum_{l=0}^{\infty}\langle\varphi''|l\rangle\langle l|e^{-\frac{i}{\hbar}(t''-t')\hat{H}}|\varphi'\rangle \\ &= \langle\varphi''|0\rangle\langle 0|\varphi'\rangle + \sum_{l=1}^{\infty}\langle\varphi''|l\rangle\langle l|\varphi'\rangle e^{-\frac{i}{\hbar}(t''-t')E_l}\,.\end{aligned} \tag{7.51}$$

Setting $\varphi' = 0$, $\varphi'' = 0$, we take the limits $t' \to -\infty$ and $t'' \to \infty$. Since the time-dependent terms vanish, we arrive at

$$\langle\varphi = 0, \infty|\varphi = 0, -\infty\rangle \;=\; \langle\varphi = 0|0\rangle\langle 0|\varphi = 0\rangle. \tag{7.52}$$

Substituting this result into relation (7.26), the constant of our interest has the form

$$\langle\varphi = 0|0\rangle\langle 0|\varphi = 0\rangle = \int \mathcal{D}\varphi e^{\frac{i}{\hbar}S[\varphi]}. \tag{7.53}$$

Finally, using the relations (7.50) and (7.53), we get

$$G_n(x_1,\ldots,x_n) \;=\; \frac{\int\mathcal{D}\varphi\,\varphi(x_1)\ldots\varphi(x_n)e^{\frac{i}{\hbar}S[\varphi]}}{\int\mathcal{D}\varphi\, e^{\frac{i}{\hbar}S[\varphi]}}, \tag{7.54}$$

which is the final representation of the Green function by the functional integral.

7.3 Functional representation of generating functionals

Starting from this section and until it is useful for us to stop, we omit the $\hbar$, setting it to unit. The generating functional of the Green functions $Z[J]$ has been defined by Eq. (6.46), and its relation with the Green functions by Eq. (6.47).

Our first purpose is to show that the generating functional $Z[J]$ can be represented by a functional integral. To do this we substitute Eq. (7.54) into Eq. (6.47), to obtain

$$\begin{aligned}
Z[J] \textstyle\int \mathcal{D}\varphi\, e^{S[\varphi]} &= \sum_{n=0}^{\infty} \frac{i^n}{n!} \textstyle\int d^4x_1 \dots d^4x_n \Big(\textstyle\int \mathcal{D}\varphi\, \varphi(x_1) \dots \varphi(x_n) e^{iS[\varphi]} \Big) J(x_1) \dots J(x_n) \\
&= \textstyle\int \mathcal{D}\varphi \sum_{n=0}^{\infty} \frac{i^n}{n!} \Big(\textstyle\int d^4x_1 \varphi(x_1) J(x_1) \Big) \dots \Big(\textstyle\int d^4x_n \varphi(x_n) J(x_n) \Big) e^{iS[\varphi]} \qquad (7.55) \\
&= \textstyle\int \mathcal{D}\varphi e^{iS[\varphi]} \Big(\sum_{n=0}^{\infty} \frac{i^n}{n!} \textstyle\int d^4x \varphi(x) J(x) \Big)^n = \textstyle\int \mathcal{D}\varphi e^{i\left\{ S[\varphi] + \int d^4x \varphi(x) J(x) \right\}}.
\end{aligned}$$

As a result, we can state that

$$Z[J] = \frac{\int \mathcal{D}\varphi e^{i\left\{ S[\varphi] + \int d^4x \varphi(x) J(x) \right\}}}{\int \mathcal{D}\varphi\, e^{S[\varphi]}}. \qquad (7.56)$$

Indeed, this is a standard representation of the generating functional of Green functions by functional integral.

It proves useful to denote

$$S_J[\varphi] = S[\varphi] + \int d^4x \varphi(x) J(x).$$

The equation of motion for theory with the action $S_J[\varphi]$ has the form

$$\frac{\delta S[\varphi]}{\delta \varphi(x)} + J(x) = 0. \qquad (7.57)$$

Incorporating this equation into the functional integral, we obtain

$$\int \mathcal{D}\varphi \Big[\frac{\delta S[\varphi]}{\delta \varphi(x)} + J(x) \Big] e^{i\left\{ S[\varphi] + i \int d^4x \varphi(x) J(x) \right\}} = 0, \qquad (7.58)$$

which means

$$\int \mathcal{D}\varphi \frac{\delta}{\delta \varphi(x)} e^{i\left(S[\varphi] + \int d^4x \varphi(x) J(x) \right)} = 0. \qquad (7.59)$$

The relation (7.59) expresses an important property of the functional integral, namely, that the functional integral of the total variational derivative vanishes.

The identity (7.59) leads to an interesting consequence. Let us insert the identity (6.53) into the integral in (7.59). The result is

$$\Big\{ \frac{\delta S[\varphi]}{\delta \varphi(x)} + J(x) \Big\} \Big|_{\varphi = \frac{\delta}{\delta i J(x)}} Z[J] = 0, \qquad (7.60)$$

which can be regarded as the equation for calculating the generating functional of the Green functions $Z[J]$.

Substituting the action of scalar field theory into Eq. (7.60), we obtain the equation of motion for the generating functional of Green functions (6.54). This demonstrates the equivalence between the operatorial methods described in Chapter 6 and the functional integral approach. Since the constant in the denominator of (7.56) does not depend on the source $J(x)$, the presence of this constant does not change the derivation of Eq. (7.60) on the basis of Eq. (7.58).

Finally, consider the functional representation of the $\mathcal{S}$-matrix. It was shown that the generating functional of the $\mathcal{S}$-matrix $S[\alpha, \alpha^*]$ in Eq. (6.55) can be obtained from the generating functional of the Green function by substituting an independent source $J(x)$ function with the expression $J(x) = (\overleftarrow{\Box} + m^2) f(x|\alpha, \alpha^*)$, where the function $f(x|\alpha, \alpha^*)$ is given by Eq. (6.56). Thus, we can write the relation

$$\mathcal{S}[\alpha, \alpha^*] = \frac{\int \mathcal{D}\varphi\, e^{i\left(S[\varphi] + \int d^4x \varphi(x)(\overleftarrow{\Box} + m^2) f(x|\alpha,\alpha^*)\right)}}{\int \mathcal{D}\varphi\, e^{iS[\varphi]}}. \tag{7.61}$$

It is clear that the functional derivatives of $S[\alpha, \alpha^*]$ with respect to $\alpha(\mathbf{p})$ and $\alpha^*(\mathbf{p})$ at $\alpha = \alpha^* = 0$ generate the elements of the $\mathcal{S}$-matrix.

7.4 Functional integrals for fermionic theories

In Chapter 6 we argued that classical spinor fields must be anticommuting and used some evident properties of anticommuting fields. However, the integration over anticommuting variables requires a more detailed mathematical approach, which will be developed in this section.

7.4.1 Grassmann algebra and anticommuting variables

To construct the functional integral over anticommuting fields, we need a mathematical notion called Grassmann algebra. The exposition of these issues will mainly follow the Berezin approach [49], in a slightly simplified form.

Consider a real or complex linear space L with the elements $a, b, c, \ldots$. As we already know from section 2.2, the linear space L is called an algebra if some multiplication operation is defined for any two of its elements such that

1. $\forall a, b \in L: \; ab \in L$.

2. $\forall a, b, c, \forall \alpha, \beta \in \mathbb{K}: \; a(\alpha b + \beta c) = \alpha ab + \beta ac$; $(\alpha b + \beta c)a = \alpha ba + \beta ca$, where $\mathbb{K}$ is a set of real $\mathbb{R}$ or complex $\mathbb{C}$ numbers. In the first case, the algebra is called real and, in the second case, it is called complex.

An algebra L is called commutative if $\forall a, b \in L: \; ab = ba$. The algebra is called associative if $\forall a, b, c \in L: \; a(bc) = (ab)c$. The algebra is called finite-dimensional if the corresponding linear space has finite dimension. In the opposite case, it is called infinite-dimensional. The algebra is called an algebra with unit if $\exists e \in L$ and $\forall a \in L$, $ae = ea = a$. Then e is called the unit element of the algebra.

Examples.

1. A set of $n \times n$ matrices is an associative algebra with unit, where the multiplication law is given by the standard matrix product.

2. Any Lie algebra is a non-associative algebra without unit.

Let N be a subset of L and let L be an algebra with unit. Let $F(N)$ be the set of all polynomials constructed from the elements of N. In other words, $F(N)$ is the set of all of the finite sums of the form

$$f = f^{(0)}e + \sum_{k=1} f^{(k)}{}_{i_1 \dots i_k}\, a^{i_1} \dots a^{i_k}, \tag{7.62}$$

where $a^i \in N$, $f^{(k)}_{i_1 \dots i_k} \in \mathbb{K}$. One can prove that $F(N)$ is also an algebra, with the same multiplication law as for the algebra L. If there exists such a set $N \subset L$ that $F(N) = L$, then this set N is called the generating set (or a system of generating elements) of the algebra L.

An associative algebra with unit is called a Grassmann algebra if there exists such a generating set $\{\theta^i;\ i = 1, \dots, n\}$ that the following conditions take place:

1. *Basic property:* $\theta^i\theta^j + \theta^j\theta^i = 0$. In particular, this means $(\theta^i)^2 = 0$.

2. Any relation for the generating elements θ^i follows from the basic property.

A Grassmann algebra with n generating elements is denoted Λ_n. Here and in what follows, we assume that the unit element e coincides with the number 1. According to the basic property, an arbitrary element of the Grassmann algebra can be written in the form

$$f^{(0)}1 + f^{(1)}{}_i\theta^i + f^{(2)}{}_{i_1 i_2}\theta^{i_1}\theta^{i_2} + \dots + f^{(n)}{}_{i_1 i_2 \dots i_n}\theta^{i_1}\theta^{i_2} \dots \theta^{i_n}. \tag{7.63}$$

Here we took into account that $\theta^{i_1}\theta^{i_2} \dots \theta^{i_n}\theta^{i_{n+1}} \equiv 0$. Eq. (7.63) can be considered an expansion of an arbitrary element $f \in \Lambda_n$ in the basis formed by the monomials

$$1, \quad \theta^i, \quad \theta^{i_1}\theta^{i_2}, \quad \dots, \quad \theta^{i_1}\,\theta^{i_2} \dots \theta^{i_n}. \tag{7.64}$$

The dimension of the algebra Λ_n is 2^n. The expansion (7.63) can be written as

$$f(\theta) = \sum_{k=0}^{n} f^{(k)}{}_{i_1 i_2 \dots i_k}\, \theta^{i_1}\theta^{i_2} \dots \theta^{i_k}, \tag{7.65}$$

where $\theta = \{\theta^1, \theta^2, \dots, \theta^n\}$ and the coefficients $f^{(k)}{}_{i_1 i_2 \dots i_k}$ are antisymmetric in all their indices. The expression (7.65) is called the function of the anticommuting variables.

The relation (7.63) shows that any element $f \in \Lambda_n$ can be represented in the form $f = f_{(0)} + f_{(1)}$, where $f_{(0)}$, $f_{(1)}$ are linear combinations of the monomials $\theta^{i_1}\theta^{i_2} \dots \theta^{i_k}$, with even and odd k, respectively. $f_{(0)}$ are called the even elements and, $f_{(1)}$ are the odd elements. The elements θ^i by themselves are odd by definition.

Let $\Lambda^{(0)}_n$ be the set of all even elements of Λ_n and let $\Lambda^{(1)}_n$ be the set of all odd elements of Λ_n. It is easy to see that the sets $\Lambda^{(0)}_n$ and $\Lambda^{(1)}_n$ are linear spaces, and $\Lambda_n = \Lambda^{(0)}_n \oplus \Lambda^{(1)}_n$.

It proves useful to introduce the function $\varepsilon(f)$, called the Grassmann parity of the element $f \in \Lambda_n$,

$$\varepsilon(f) = \begin{cases} 0, & \text{for} \quad f \in \Lambda^{(0)}_n, \\ 1, & \text{for} \quad f \in \Lambda^{(1)}_n. \end{cases} \tag{7.66}$$

Indeed, for $f = f_{(0)} + f_{(1)}$, with $f_{(0)} \neq 0$ and $f_{(1)} \neq 0$, the function $\varepsilon(f)$ is not defined.

The main properties of the Grassmann parity are given by the following relations:
1. $\varepsilon(fg) = \big[\varepsilon(f) + \varepsilon(g)\big] \bmod 2$.
2. For any two $f, g \in \Lambda_n$ with definite $\varepsilon(f)$ and $\varepsilon(g)$, $fg = (-1)^{\varepsilon(f)+\varepsilon(g)} gf$.

We will need to take derivatives with respect to the anticommuting variables θ^i. There are two types of such derivatives, namely, the left ones and the right ones. Let $f(\theta)$ be the function (7.65). Then the left derivative is denoted as $\frac{\vec{\partial}}{\partial\theta^i}$ and is defined by the following two conditions:

1. The left derivative is a linear operation and Leibnitz's rule holds. Therefore, it is sufficient to define the derivative of the monomials $\theta^1\theta^2 \dots \theta^k$.

2. If there is no element θ^i in the monomials composing a given function, then the left derivative of this function with respect to θ^i is zero. If the monomial has the element θ^i, one has to move it to the left position, using the basic property $\theta^i\theta^j = -\theta^j\theta^i$, and then strike it out.

Both these conditions are expressed by the rules

$$\frac{\vec{\partial}}{\partial\theta^i}\theta^j = \frac{\vec{\partial}\,\theta^j}{\partial\theta^i} = \delta_i{}^j, \qquad \frac{\vec{\partial}}{\partial\theta^i}\big[\theta^j f(\theta)\big] = \frac{\vec{\partial}\,\theta^j}{\partial\theta^i} f(\theta) - \theta^j \frac{\vec{\partial}}{\partial\theta^i} f(\theta). \tag{7.67}$$

It is evident that the derivative with respect to θ anticommutes with the proper θ. The right derivative $\frac{\overleftarrow{\partial}}{\partial\theta^i}$ can be defined in a similar way; however, the element θ^i must be moved to the right position and then eliminated.

By using these definitions, one can formulate the following differentiation rules:

$$\textbf{1.}\quad \frac{\vec{\partial}}{\partial\theta^i}\frac{\vec{\partial}}{\partial\theta^j} = -\frac{\vec{\partial}}{\partial\theta^j}\frac{\vec{\partial}}{\partial\theta^i}, \qquad \frac{\overleftarrow{\partial}}{\partial\theta^i}\frac{\overleftarrow{\partial}}{\partial\theta^j} = -\frac{\overleftarrow{\partial}}{\partial\theta^j}\frac{\overleftarrow{\partial}}{\partial\theta^i}. \tag{7.68}$$

$$\textbf{2.}\quad \frac{\vec{\partial}}{\partial\theta^i}\frac{\overleftarrow{\partial}}{\partial\theta^j} = \frac{\overleftarrow{\partial}}{\partial\theta^j}\frac{\vec{\partial}}{\partial\theta^i}. \tag{7.69}$$

In particular, from (7.68), it follows that $\left(\frac{\vec{\partial}}{\partial\theta^i}\right)^2 = \left(\frac{\overleftarrow{\partial}}{\partial\theta^i}\right)^2 = 0$. In what follows, we will mainly use the left derivative.

The next step is to introduce integration over the anticommuting variables. Since $(\frac{\partial}{\partial\theta^i})^2 = 0$, we cannot define an integral as an operation inverse to differentiation. The definition of the integral over anticommuting variables has been given by F. Berezin. This definition consists from the following set of properties:

1. Integration is a linear operation. Therefore, it is sufficient to define the integral of a monomial $\theta^1\theta^2 \dots \theta^k$.

2. Consider the Grassmann algebra Λ_{2n} with the generating elements θ^i and $d\theta_j$, where the last is an independent differential of θ^j. Owing to the main feature of Grassmann variables, $\theta^i d\theta_j = -d\theta_j\theta^i$ and $d\theta_i d\theta_j = -d\theta_j d\theta_i$. Then we get

$$\int d\theta_i\, c = 0, \qquad \int d\theta_i \theta^j = \delta_i{}^j, \tag{7.70}$$

where c is an arbitrary constant.

3. The multiple integral is defined as a repeatedly taken simple integral.

The three properties listed above form the rule of integrating any function of anticommuting variables. It is interesting to note the following property of the Berezin's integral:

$$\int d\theta_i \theta^j = \frac{\partial}{\partial \theta^i}\theta^j, \tag{7.71}$$

that is, the integral is, in this case, equivalent to the derivative.

The definition of the Berezin integral leads to important and useful properties of the integral.

1. Calculating the integral. Let $f(\theta)$ be an arbitrary function of anticommuting variables and $d\theta = d\theta_n \dots d\theta_2 d\theta_1$. Consider

$$\int d\theta f(\theta) = \sum_{k=0}^{n} f^{(k)}_{\ i_1 i_2 \dots i_k} \int d\theta_n \dots d\theta_2 d\theta_1 \theta^{i_1}\theta^{i_2} \dots \theta^{i_k}.$$

It is evident that, for $k < n$, this integral is zero, because it contains n derivatives $\frac{\partial}{\partial \theta^i}$ of the monomial $\theta^{i_1}\theta^{i_2}\dots\theta^{i_k}$ with $k < n$. Therefore,

$$\int d\theta f(\theta) = f^{(n)}_{\ i_1 i_2 \dots i_n} \int d\theta_n \dots d\theta_2 d\theta_1\, \theta^{i_1}\theta^{i_2} \dots \theta^{i_n} = n!\, f^{(n)}_{\ 12\dots n}, \tag{7.72}$$

where we used the total antisymmetry of the coefficients $f^{(k)}_{\ i_1 i_2 \dots i_k}$.

2. Integration by parts. Suppose $f(\theta)$ and $g(\theta)$ are two functions of anticommuting variables. Then

$$\int d\theta f(\theta)\Big[\frac{\overrightarrow{\partial}}{\partial \theta^i} g(\theta)\Big] = \int d\theta \Big[f(\theta)\frac{\overleftarrow{\partial}}{\partial \theta^i}\Big] g(\theta). \tag{7.73}$$

3. Invariance under the shift of the integration variable. Consider the Grassmann algebra Λ_{2n} with the generating elements θ^i and η^i. Then

$$\int d\theta f(\theta + \eta) = \int d\theta f(\theta), \tag{7.74}$$

where $\theta + \eta = \{\theta^i + \eta^i;\ i = 1, 2, \dots, n\}$. The proof is based on the direct calculations of the *l.h.s.* and *r.h.s.* of the equation, using (7.72).

4. Linear change of anticommuting variables. Let $f(\theta)$ be an arbitrary function of anticommuting variables with the generating elements θ^i, and θ'^i be another system of such elements, with $\theta^i = A^i_{\ j}\theta'^j$, $\det A \neq 0$ and $f'(\theta') = f(\theta)$. Then

$$\int d\theta f(\theta) = \big(\det A\big)^{-1} \int d\theta' f'(\theta'). \tag{7.75}$$

To prove this relation, we just need to note that

$$\int d\theta f(\theta) = n!\, f^{(n)}_{\ 12\dots n} \quad \text{and} \quad \int d\theta' f'(\theta') = n!\, f^{(n)}_{\ 12\dots n} \det A,$$

where we used Eq. (7.72).

5. Gaussian integral over fermionic variable. As an example, consider the derivation of the Gaussian integral over fermionic variables, which is a Grassmannian analog of the well-known Gaussian integral over bosonic variables. Let us note that the systematic derivation of the functional integral over both types of fields will be dealt with in section 7.6.

Let θ^i be generating elements of the Grassmann algebra Λ_{2n}, η_i be independent Grassmannian sources and b_{ij} be an antisymmetric matrix with $\det(b_{ij}) \neq 0$. The expression for the Gaussian integral over anticommuting variables is

$$I(\eta) = \int d\theta e^{\frac{1}{2}\theta^i b_{ij}\theta^j + \eta_i \theta^i}. \tag{7.76}$$

Since $b_{ij} = -b_{ji}$ and $\det(b_{ij}) \neq 0$, the dimension n is necessarily even. Thus, we can restrict our attention to the cases of even n.

Calculating the integral (7.76) consists of several steps. First, let us make a shift of variables $\theta^i \to \theta^i + b^{ij}\eta_j$, where b^{ij} is an inverse matrix, $b^{ij}b_{jk} = \delta^i_{\ k}$. Then, the relation (7.76) becomes

$$I(\eta) = e^{\frac{1}{2}\eta_i b^{ij}\eta_j} I(0). \tag{7.77}$$

The second step is the calculation of $I(0)$. To contemplate this, consider a variation of $I(0)$ with respect to b_{ij},

$$\delta I(0) = \frac{1}{2}\delta b_{ij} \int d\theta\, \theta^i\theta^j\, e^{\frac{1}{2}\theta^k b_{kl}\theta^l}. \tag{7.78}$$

This last integral can be written in the form

$$\int d\theta\, \theta^i\theta^j e^{\frac{1}{2}\theta^k b_{kl}\theta^l} = -\frac{\partial}{\partial\eta_j}\frac{\partial}{\partial\eta_i} I(\eta)\Big|_{\eta=0}, \tag{7.79}$$

where the negative sign in the *r.h.s.* originates from the relation $\frac{\partial}{\partial\eta_j}\theta^i = -\eta^i\frac{\partial}{\partial\eta_j}$. Using the relation (7.77), we obtain

$$\int d\theta\, \theta^i\theta^j\, e^{\frac{1}{2}\theta^k b_{kl}\theta^l} = -\frac{\partial}{\partial\eta_j}\frac{\partial}{\partial\eta_i} e^{\frac{1}{2}\eta_k b^{kl}\eta_l} I(0) = -b^{ij}\, I(0).$$

Therefore,

$$\delta I(0) = -\frac{1}{2}\delta b_{ij}\, b^{ij}\, I(0) = \frac{1}{2}\delta b_{ij}\, b^{ji}\, I(0) = \frac{1}{2}\delta\,\{\mathrm{tr}\,\log(b_{ij})\}\, I(0). \tag{7.80}$$

Solving this differential equation, we get

$$I(0) = C\big[\det(b_{ij})\big]^{1/2}, \tag{7.81}$$

where C is an arbitrary constant that does not depend on b_{ij}. As a result, we have

$$I(\eta) = C\big[\det(b_{ij})\big]^{\frac{1}{2}}\, e^{\frac{1}{2}\eta_i b^{ij}\eta_j}. \tag{7.82}$$

To find the constant C, we choose the matrix b_{ij} in the form $b_{ij} = \epsilon_{ij}$, with

$$\epsilon = \begin{pmatrix} \boxed{D_1} & 0 & 0 & \dots 0 \\ 0 & \boxed{D_2} & 0 & \dots 0 \\ 0 & 0 & \ddots & 0 \\ 0 & 0 & 0 \;\dots & \boxed{D_{\frac{n}{2}}} \end{pmatrix},$$

where

$$D_k = \begin{pmatrix} 0 & 1 \\ -1 & 0 \end{pmatrix}; \qquad k = 1, 2, \dots, \frac{n}{2}$$

and $\det(\epsilon_{ij}) = 1$. In this case, one gets

$$I(0) = \int d\theta e^{\theta^1\theta^2 + \theta^3\theta^4 + \dots + \theta^{n-1}\theta^n} = 1.$$

Thus, $C = 1$, and we finally get

$$I(\eta) = \int d\,\theta e^{\frac{1}{2}\theta^i b_{ij}\theta^j + \eta_i\theta^i} = \big[\det(b_{ij})\big]^{1/2} e^{\frac{1}{2}\eta_i b^{ij}\eta_j}. \tag{7.83}$$

It is instructive to compare two Gauss integrals, one over anticommuting variables (7.83) and another over n-component commuting ones,

$$\int d^n x\, e^{-\frac{1}{2}x^i a_{ij} x^j + x^i c_i} = \pi^{n/2}\big[\det(a_{ij})\big]^{-1/2} e^{\frac{1}{2}c_i (a^{-1})^{ij} c_j}. \tag{7.84}$$

It is easy to see that the powers of the determinants have opposite signs.

Now we consider the Grassmann algebra with an infinite amount of generating elements. To be more precise, we will assume that the generating elements are functions of the spacetime coordinates, which take values in the Grassmann algebra $\theta = \theta(x)$. It is assumed that these functions satisfy the basic anticommuting relation

$$\theta(x)\theta(x') + \theta(x')\theta(x) = 0. \tag{7.85}$$

An arbitrary element of such an algebra is written as follows:

$$f[\theta] = \sum_{k=0}^{\infty} \int d^4x_1 d^4x_2 \dots d^4x_k\, f^{(k)}(x_1, x_2, \dots, x_k)\, \theta(x_1)\theta(x_2)\dots\theta(x_k). \tag{7.86}$$

The object $f[\theta]$ can be treated as a functional depending on the functions $\theta(x)$. The functions $f^{(k)}(x_1, x_2, \dots, x_k)$ can be considered totally antisymmetric in their arguments. By default, we assume left functional derivatives, $\frac{\vec{\delta}}{\delta\theta(x)}$. All the properties of these derivatives are analogous to the finite-dimensional case, with the additional rules

$$\frac{\delta}{\delta\theta(x)}\big(\theta(x')f[\theta]\big) = \frac{\delta\theta(x')}{\delta\theta(x)} f[\theta] - \theta(x')\frac{\delta}{\delta\theta(x)} f[\theta], \quad \frac{\delta\theta(x')}{\delta\theta(x)} = \delta^4(x - x'). \tag{7.87}$$

These relations will prove useful for developing the functional methods for fermionic theories.

7.4.2 Generating functional of spinor Green functions

Consider the functional representation of the generating functional of spinor Green functions. These functions have been defined by the relation (6.72), and the corresponding generating functional $Z[\bar{\eta}, \eta]$ is given by (6.75). The Green functions can be derived using the relations (6.78), in terms of the left functional derivatives with respect to $\bar{\eta}(x)$, and right functional derivatives with respect to $\eta(x)$.

Representation of the spinor Green functions in terms of the functional integral is constructed in the same way as for scalar field theory in sections 7.2 and 7.3. However, according to the corresponding discussion, the functional integral should be taken over the anticommuting fields $\psi(x)$ and $\bar{\psi}(x)$. The basic properties of the anticommuting fields are given by the relations

$$\begin{aligned} &\psi(x)\psi(x') = -\psi(x')\psi(x), \qquad \bar{\psi}(x)\bar{\psi}(x') = -\bar{\psi}(x')\bar{\psi}(x), \\ &\psi(x)\bar{\psi}(x') = -\bar{\psi}(x')\psi(x). \end{aligned} \tag{7.88}$$

Repeating the same consideration as for the derivation of Eq. (7.56), one gets

$$Z[\bar{\eta}, \eta] = \frac{\int \mathcal{D}\bar{\psi}\mathcal{D}\psi \, e^{iS[\bar{\psi},\psi]+i\int d^4x\{\bar{\eta}(x)\psi(x)+\bar{\psi}(x)\eta(x)\}}}{\int \mathcal{D}\psi\mathcal{D}\bar{\psi}e^{iS[\bar{\psi},\psi]}}, \tag{7.89}$$

where $S[\bar{\psi}, \psi]$ is the action of some spinor field theory.

As a more complicated example, consider the model of interacting scalar and spinor fields with the action $S[\varphi, \bar{\psi}, \psi]$ and the Green functions

$$\begin{aligned} G_{kmn} \; &(x_1, \dots, x_k, y_1, \dots, y_m, z_1, \dots, z_n) \\ &= \langle 0|T\big(\hat{\varphi}(x_1)\dots\hat{\varphi}(x_k)\hat{\psi}(y_1)\dots\hat{\psi}(y_m)\hat{\bar{\psi}}(z_1)\dots\hat{\bar{\psi}}(z_n)\big)|0\rangle. \end{aligned} \tag{7.90}$$

The generating functional for these Green functions has the form

$$Z[J, \bar{\eta}, \eta] = \langle 0|e^{i\int d^4x\{\hat{\varphi}(x)J(x)+\bar{\eta}(x)\hat{\psi}(x)+\hat{\bar{\psi}}(x)\eta(x)\}}|0\rangle, \tag{7.91}$$

and its functional integral representation is written as follows:

$$Z[J, \bar{\eta}, \eta] = \frac{\int \mathcal{D}\varphi\mathcal{D}\bar{\psi}\mathcal{D}\psi \, e^{i\{S[\varphi,\bar{\psi},\psi]+\int d^4x(\varphi(x)J(x)+\bar{\eta}(x)\psi(x)+\bar{\psi}(x)\eta(x))\}}}{\int \mathcal{D}\varphi\mathcal{D}\bar{\psi}\mathcal{D}\psi \, e^{iS[\varphi,\bar{\psi},\psi]}}. \tag{7.92}$$

The functional integral in Eq. (7.92) contains integration over the commuting field $\varphi(x)$ and over the anticommuting fields $\psi(x), \bar{\psi}(x)$. The same pattern can be applied to other models with mixed (boson and fermion) fields.

7.5 Perturbative calculation of generating functionals

From this poin,t we start to explore interacting theories and discuss perturbation theory in terms of functional integrals.

Let us come back to scalar field theory and consider the generating functional of Green functions. This functional is given by the expression (7.56), and our purpose is to develop a perturbation theory with respect to the interaction term. For this sake, we write the action in the form

$$S[\varphi] = S_0[\varphi] + S_{int}[\varphi], \tag{7.93}$$

where $S_0[\varphi]$ is a free action and $S_{int}[\varphi]$ is the interaction term. Applying the identity (6.53), we arrive at the expression

$$Z[J] = \frac{e^{S_{int}[\frac{\delta}{\delta iJ}]} Z_0[J]}{e^{S_{int}[\frac{\delta}{\delta iJ}]} Z_0[J]\big|_{J=0}}, \tag{7.94}$$

where

$$Z_0[J] = \int \mathcal{D}\varphi e^{i\left\{S_0[\varphi]+\int d^4x\varphi(x)J(x)\right\}} \tag{7.95}$$

is the generating functional in the free theory, or of the free Green functions.

Expanding $e^{S_{int}[\frac{\delta}{\delta iJ}}$ in Eq. (7.94) into a power series, we get

$$Z[J] = \frac{\{1+\sum_{n=1}^{\infty}\frac{i^n}{n!}\big(S_{int}[\frac{\delta}{\delta iJ}]\big)^n\}Z_0[J]}{\{1+\sum_{n=1}^{\infty}\frac{i^n}{n!}\big(S_{int}[\frac{\delta}{\delta iJ}]\big)^n\}Z_0[J]\big|_{J=0}}. \tag{7.96}$$

It is also easy to see that

$$Z_0[J]\Big|_{J=0} = \langle 0|Te^{i\int d^4x\hat{\varphi}_{in}(x)J(x)}|0\rangle\Big|_{J=0} = \langle 0|0\rangle = 1. \tag{7.97}$$

It is easy to see that (7.96) has the following general structure:

$$Z[J] = \frac{Z_0[J] + \text{terms containing } S_{int}}{1 + \text{terms containing } S_{int}} = Z_0$$
$$+ \text{ terms generated by interaction.} \tag{7.98}$$

All the terms generated by interactions are stipulated by expressions of the type $\big(S_{int}[\frac{\delta}{\delta iJ}]\big)^n Z_0[J]$, with various numbers n. Thus, the relation (7.96) determines the perturbation series for the calculation of the generating functional in the full theory with interaction. The problem of the perturbative calculation of $Z[J]$ is reduced to taking variational derivatives of $Z_0[J]$. At this point, the main unknown element of this scheme is the generating functional of free Green functions.

Thus, we proceed to calculate $Z_0[J]$ using the important property of the functional integral (7.59). In the case under consideration, this formula yields

$$\int \mathcal{D}\varphi \,\frac{\delta}{\delta\varphi(x)}\, e^{i\left\{S_0[\varphi]+\int d^4x\varphi(x)J(x)\right\}} = 0. \tag{7.99}$$

After integration by parts, the action S_0 is written as

$$S_0[\varphi] = -\frac{1}{2}\int d^4x\,\varphi(x)\{\Box + m^2\}\varphi(x). \tag{7.100}$$

To provide the convergence of the path integral (7.95), we modify it by introducing in the exponential the infinitesimal term $-\frac{1}{2}\int d^4x\epsilon\varphi^2(x)$. The limit $\epsilon \to +0$ is assumed in all final physical expressions. As a result, we arrive at the formula

$$\int \mathcal{D}\varphi \frac{\delta}{\delta\varphi(x)} \exp\Big\{ -\frac{i}{2}\int d^4x\varphi(x)(\Box + m^2 - i\epsilon)\varphi(x) + i\int d^4x\varphi(x)J(x)\Big\} = 0.$$

This relation leads to

$$(\Box + m^2 - i\epsilon)\frac{\delta Z_0[J]}{\delta iJ(x)} = J(x)Z_0[J], \tag{7.101}$$

and we get the closed equation for $Z_0[J]$.

To solve Eq. (7.101), we introduce $K(x,y)$, the kernel of the operator $\Box + m^2 - i\epsilon$:

$$K(x,y) = (\Box_x + m^2 - i\epsilon)\delta^4(x-y). \tag{7.102}$$

Let $K^{-1}(x,y)$ be the kernel of the inverse operator, which is defined by the equation

$$\int d^4z\, K(x,z)K^{-1}(z,y) = \int d^4z\, K^{-1}(x,z)K(z,y) = \delta^4(x-y). \tag{7.103}$$

Eq. (7.101) can be rewritten as

$$\int d^4y\, K(z,y)\frac{\delta Z_0[J]}{\delta iJ(y)} = J(z)Z_0[J]. \tag{7.104}$$

Multiplying this expression by $K^{-1}(x,z)$ and integrating over d^4z, one gets

$$\frac{\delta Z_0[J]}{\delta iJ(x)} = \int d^4\, yK^{-1}(x,y)J(y)\, Z_0[J]. \tag{7.105}$$

The solution to this equation, with the initial condition $Z_0[0] = \langle 0|0\rangle = 1$, has the form

$$Z_0[J] = e^{-\frac{i}{2}\int d^4x d^4y J(x)D(x,y)J(y)}, \tag{7.106}$$

where the function $D(x,y)$ is defined as

$$D(x,y) = -K^{-1}(x,y). \tag{7.107}$$

The function $D(x,y)$ is called the Feynman propagator of the real scalar field. Expression (7.106) is the final form of the generating functional of free Green functions.

To summarize, we have constructed a general scheme for perturbative calculations of the generating functional of Green functions. The calculations are based on the relation (7.96), where the functional $Z_0[J]$ is given by Eq. (7.106).

The perturbative calculations of the generating functional of the spinor Green functions are organized following the same scheme. The corresponding generating functional of the free spinor Green functions has the form

$$Z_0[\bar{\eta}, \eta] = e^{-i \int d^4x d^4y \bar{\eta}(x) S(x,y) \eta(y)}, \tag{7.108}$$

where the function $S(x, y)$ is defined by the equation

$$(i\gamma^\mu \partial_\mu - m + i\epsilon) S(x, y) = \delta^4(x - y) \tag{7.109}$$

and the derivatives are taken with respect to x. This function is called the Feynman propagator of the spinor field. One can prove that spinor and scalar propagators are related as

$$S(x, y) = (i\gamma^\mu \partial_\mu + m) D(x, y). \tag{7.110}$$

It is useful to obtain the propagators in the momentum representation. Due to the translational invariance, the propagators depend only on the difference of the arguments, $D(x, y) = D(x-y)$ and $S(x, y) = S(x-y)$. Consider the Fourier transform of the scalar propagator,

$$D(x - y) = \int \frac{d^4p}{(2\pi)^4} e^{-ip(x-y)} \tilde{D}(p). \tag{7.111}$$

Substituting the relation (7.111) into the equation for the propagator $D(x, y)$,

$$(\Box_x + m^2 - i\epsilon) D(x, y) = -\delta^4(x - y), \tag{7.112}$$

one gets

$$\tilde{D}(p) = \frac{1}{p^2 - m^2 + i\epsilon}. \tag{7.113}$$

Similarly, one can show that, for the fermion propagator,

$$\tilde{S}(p) = \frac{\gamma^\mu p_\mu + m}{p^2 - m^2 + i\epsilon}. \tag{7.114}$$

These propagators will be intensively used in the rest of the book. The infinitesimal correction $i\epsilon$ in the propagators (7.113) and (7.114) defines the rules of integrating the expression with the pole terms in the momentum integrals. One can find the detailed discussion of this important aspect in many books, e.g., in [57].

7.6 Properties of functional integrals

As we pointed out above, the generating functional of the Green functions is formulated in terms of the functional integral

$$\int \mathcal{D}\varphi e^{i\{S_0[\varphi] + S_{int}[\varphi] + \int d^4x \varphi(x) J(x)\}}.$$

Expanding the $e^{iS_{int}[\varphi]}$ into a power series under the integral, we get functional integrals of the form

$$\int \mathcal{D}\varphi e^{iS_0[\varphi]}\varphi(x_1)\ldots\varphi(x_n), \qquad n = 1, 2, \ldots .$$

All such integrals can be calculated on the basis of the Gaussian functional integral over bosonic fields, which is defined as

$$G[J] = \int \mathcal{D}\phi e^{-\frac{i}{2}\phi K\varphi + i\phi J}, \tag{7.115}$$

where ϕ is a set of bosonic fields including the scalar field φ, the vector field A_μ, the second-rank tensor field $h_{\mu\nu}$, etc. The external field J is the source for the field ϕ. For brevity and simplicity, we will use the condensed notations

$$\phi K \phi = \int d^4x d^4y\, \phi(x)K(x,y)\phi(y), \qquad \phi J = \int d^4x\, \phi(x)J(x). \tag{7.116}$$

In this expression, $K(x,y)$ is the kernel of some symmetric differential operator K acting on the fields ϕ. We assume that the inverse operator K^{-1} has the kernel $K^{-1}(x,y)$,

$$\int d^4z\, K(x,z)K^{-1}(z,y) = \int d^4z\, K^1(x,z)K(z,y) = \delta^4(x-y). \tag{7.117}$$

The explicit form of the operator K is defined by a concrete field model and cannot be found in a general form. In this section, we will describe a calculation of the integral (7.115), assuming that the field ϕ is a real scalar field φ. Calculations for any other bosonic field are quite analogous.

1. *Finite-dimensional illustration of calculating the Gaussian functional integral.* The finite-dimensional Gaussian integral has the form

$$I(a,j) = \int_{-\infty}^{\infty} \prod_{i=1}^{n} dx^i e^{-\frac{1}{2}x^i a_{ij} x^j + x^i j_i} = \Big(\det \frac{a}{2\pi}\Big)^{-\frac{1}{2}} e^{j_k a^{kl} j_l}, \tag{7.118}$$

where x^i with $i = 1, 2, \ldots, n$ are real variables and $a = \|a_{ij}\|$ is a non-singular, symmetric, positively defined $n \times n$ matrix. $\|a^{kl}\|$ is the inverse matrix, i.e., $a^{ik}a_{kj} = \delta^i_{\ j}$.

Consider the derivation of the Gaussian integral formula. This method illustrates how to find the dependence of the integral (7.118) on the matrix a_{ij} and can be generalized to the functional integral (7.115) for the dependence on $K(x,y)$.

The first operation is the shift of the integration variables $x^i \to x^i + a^{ik}j_k$. After that, one gets $I(a,j) = e^{\frac{1}{2}j_k a^{kl} j_l} I(a,0)$. Next, consider the differential related to the independent variation of the matrix a_{ij},

$$\begin{aligned} dI(a,0) &= -\frac{1}{2}da_{il}\int \prod_{k=1}^{n} dx^k\, x^i x^l\, e^{-\frac{1}{2}x^k a_{kl} x^l} = -\frac{1}{2}\frac{\partial^2}{\partial j_i \partial j_l} I(a,j)\Big|_{j=0} da_{il} \\ &= -\frac{I(a,0)}{2}\, a^{il}\, da_{il} = -\frac{1}{2}d\big[\mathrm{tr}\,\log(a_{il})\big]\, I(a,0). \end{aligned} \tag{7.119}$$

Hence, $I(a,0) = C\det^{-\frac{1}{2}}(a_{ij})$, where C is a constant. In order to find this constant, we take $a_{ij} = \delta_{ij}$. It gives $C = (2\pi)^{\frac{n}{2}}$, and we thus obtain the relation (7.118).

The integral (7.118) possesses some remarkable properties. In particular, this integral is invariant under the shift $x^i \to x^i + C^i$ with constant parameters C^i.

2. *Calculation of the Gaussian functional integral.*

As for the finite-dimensional case, we assume that the functional integral (7.115) is invariant under the functional shift $\varphi(x) \to \varphi(x) + \Phi(x)$, where $\Phi(x)$ is a given field belonging to the same functional space as $\varphi(x)$. After this shift, we get

$$G[J] = \int \mathcal{D}\varphi e^{-\frac{i}{2}\varphi K\varphi - i\varphi K\Phi - \frac{i}{2}\Phi K\Phi + i\varphi J + i\Phi J}.$$

It is useful to choose the field Φ to eliminate the terms linear in φ. Solving the equation of motion in the theory with the action $S[\Phi] = -\frac{1}{2}\Phi K\Phi + \Phi J$ yields $\Phi = K^{-1}J$. After that, the integral (7.115) becomes

$$G[J] = e^{\frac{i}{2}JK^{-1}J}G[0], \qquad \text{where} \qquad G[0] = \int \mathcal{D}\varphi e^{-\frac{i}{2}\varphi K\varphi}. \tag{7.120}$$

Taking a variation of $G[0]$ with respect to the operator K,

$$\delta G[0] = -\frac{i}{2}\int d^4x d^4y\, \delta K(x,y) \int \mathcal{D}\varphi\; \varphi(x)\varphi(y)\, e^{-\frac{i}{2}\varphi K\varphi}, \tag{7.121}$$

the remaining functional integral in (7.121) is

$$\int \mathcal{D}\varphi\, \varphi(x)\varphi(y)\, e^{-\frac{i}{2}\varphi K\varphi} = \frac{\delta}{\delta iJ(x)}\frac{\delta}{\delta iJ(y)}G[J]\Big|_{J=0} = -iK^{-1}(x,y)G[0]. \tag{7.122}$$

Thus, as in the finite-dimensional case, we have

$$\delta G[0] = -\frac{1}{2}\delta\big[\operatorname{Tr}\log K\big]\,G[0]. \tag{7.123}$$

Here Tr means a functional trace of the operator, defined as

$$\operatorname{Tr} A = \int d^4x A(x,y)\Big|_{y=x}, \tag{7.124}$$

and $A(x,y)$ is a kernel of the operator A. In this way, we arrive at

$$G[0] = C(\operatorname{Det} K)^{-\frac{1}{2}}, \tag{7.125}$$

where $\operatorname{Det} K$ is the functional determinant of the operator K, defined by the relation $\operatorname{Det} K = e^{\operatorname{Tr}\log K}$, and C is the integration constant. Let us note that the scheme of calculating functional determinants is briefly discussed in section 7.7. The final result for the Gaussian functional integral has the form

$$\int \mathcal{D}\varphi e^{-\frac{i}{2}\varphi K\varphi + i\varphi J} = \big(\operatorname{Det} K\big)^{-\frac{1}{2}}\, e^{\frac{i}{2}JK^{-1}J}, \tag{7.126}$$

where the inessential constant C is included in the functional measure $\mathcal{D}\varphi$.

3. *Elimination of the functional integral of the functional derivative.* Starting from the relation (7.115), we denote

$$\mathcal{F}[\varphi] = e^{-\frac{i}{2}\varphi K\varphi + i\varphi J}. \tag{7.127}$$

The condition of shift invariance of the Gaussian functional integral has the form

$$\int \mathcal{D}\varphi \mathcal{F}[\varphi] = \int \mathcal{D}\varphi \mathcal{F}[\varphi + \Phi].$$

Let us replace the field Φ with $\varepsilon\Phi(x)$. In the first order in ε, one gets

$$\int \mathcal{D}\varphi \mathcal{F}[\varphi] = \int \mathcal{D}\varphi \Big\{ \mathcal{F}[\varphi] + \varepsilon \int d^4x\, \Phi(x) \frac{\delta}{\delta\varphi(x)} \mathcal{F}[\varphi] \Big\}. \tag{7.128}$$

As far as $\Phi(x)$ is arbitrary, we obtain

$$\int \mathcal{D}\varphi \frac{\delta\, \mathcal{F}[\varphi]}{\delta\varphi(x)} = 0. \tag{7.129}$$

This general property of functional integrals was derived earlier via another method for a particular form of functional $\mathcal{F}[\varphi]$.

4. *Integration by parts in the functional integral.* Consider Eq. (7.129) with the functional being the product $\mathcal{F}[\varphi] = \mathcal{F}_1[\varphi]\mathcal{F}_2[\varphi]$:

$$\int \mathcal{D}\varphi \Big\{ \frac{\delta}{\delta\varphi(x)} \mathcal{F}_1[\varphi] \Big\} \mathcal{F}_2[\varphi] = - \int \mathcal{D}\varphi\, \mathcal{F}_1[\varphi] \Big\{ \frac{\delta}{\delta\varphi(x)} \mathcal{F}_2[\varphi] \Big\}. \tag{7.130}$$

5. *Demonstration of the functional delta function.*

Let Q be an invertible operator with the kernel $Q(x,y)$. The functional delta function, $\delta[f - Q\varphi]$, is defined by the following functional integral relation valid for any $f(x)$:

$$\delta[f - Q\varphi] = \int \mathcal{D}\Psi\, e^{i \int d^4x \Psi(x) \big[f(x) - \int d^4y (Q(x,y)\varphi(y) \big]}. \tag{7.131}$$

To clarify the property of the functional delta function, let us consider

$$\int \mathcal{D}\varphi e^{-\frac{i}{2}\varphi K\varphi + i\varphi J} \delta[f - Q\varphi] = \int \mathcal{D}\varphi \mathcal{D}\Psi e^{-\frac{i}{2}\varphi K\varphi + i\Psi(f - Q\varphi)}$$
$$= \int \mathcal{D}\Psi e^{i\Psi f} \int \mathcal{D}\varphi\, e^{-\frac{i}{2}\varphi K\varphi + i\varphi(J - Q^T\Psi)}, \tag{7.132}$$

where we used the condensed notations (7.116). The transposed operator Q^T has the kernel $Q^T(x,y) = Q(y,x)$. Thus, we obtained the Gaussian functional integral (7.132), which is calculated according to Eq. (7.126). After that, one gets the integral

$$\big(\operatorname{Det} K\big)^{-\frac{1}{2}}\, e^{-\frac{i}{2}JK^{-1}J} \int \mathcal{D}\Psi e^{-\frac{i}{2}\Psi QK^{-1}Q^T\Psi + i\Psi(f + QK^{-1}J)}, \tag{7.133}$$

that is, again, the Gaussian functional integral. Its calculation leads to

$$\big(\operatorname{Det} K\big)^{-\frac{1}{2}} e^{\frac{i}{2}JK^{-1}J}\big[\operatorname{Det}\big(QK^{-1}Q^T\big)\big]^{\frac{1}{2}} e^{\frac{i}{2}(f-QK^{-1}J)(-QK^{-1}Q^T)^{-1}(f-K^{-1}J)}.$$

After a small algebra, we arrive at

$$\int \mathcal{D}\varphi\, e^{-\frac{i}{2}\varphi K\varphi+i\varphi J}\delta[f-Q\varphi] \,=\, \big(\operatorname{Det} Q\big)^{-1} e^{-\frac{i}{2}\varphi K\varphi+i\varphi J}\Big|_{\varphi=Q^{-1}f}. \tag{7.134}$$

Thus, under the Gaussian functional integral, the functional delta function (7.131) possesses the standard property of the ordinary delta function.

6. *Change of variables in the functional integral.*

Consider the Gaussian functional integral $G[J]$ from Eq. (7.115), and define

$$Q(x|\tilde{\varphi}) = q(x) + \sum_{n=1}^{\infty}\frac{1}{n!}\int d^4x_1 d^4x_2\dots d^4x_n\, q_n(x;x_1,x_2,\dots,x_n)$$
$$\times\ \tilde{\varphi}(x_1)\tilde{\varphi}(x_2)\dots\tilde{\varphi}(x_n), \tag{7.135}$$

where $q(x)$ and $q_k(x)$ are some scalar functions, and $\tilde{\varphi}(x)$ a scalar field. Our intention is to make a change of variables:

$$\varphi(x) = Q(x|\tilde{\varphi}) \equiv Q(\tilde{\varphi}). \tag{7.136}$$

The statement is that

$$G[J] \,=\, \int \mathcal{D}\tilde{\varphi}\ \operatorname{Det}\Big(\frac{\delta Q}{\delta\tilde{\varphi}}\Big)\, e^{-\frac{i}{2}\varphi K\varphi+iJ\varphi}\Big|_{\varphi=Q(\tilde{\varphi})}, \tag{7.137}$$

$$\text{where}\qquad \frac{\delta Q}{\delta\tilde{\varphi}} = \frac{\delta Q(x|\tilde{\varphi})}{\delta\tilde{\varphi}(y)} \qquad\text{and}\qquad \operatorname{Det}\Big(\frac{\delta Q}{\delta\tilde{\varphi}}\Big) \tag{7.138}$$

is called the functional Jacobian. The calculation of functional determinants, including of this type, will be discussed in section 7.7.

The derivation of (7.137) starts from the functional Fourier transform of $G[J]$,

$$\int \mathcal{D}J\, G[J]\, e^{-iJ\Phi} = \int \mathcal{D}\varphi\,\mathcal{D}J\, e^{-\frac{i}{2}\varphi K\varphi+iJ(\varphi-\Phi)}$$
$$= \int \mathcal{D}\varphi\, e^{-\frac{i}{2}\varphi K\varphi}\,\delta[\phi-\Phi] \,=\, e^{-\frac{i}{2}\Phi K\Phi}. \tag{7.139}$$

On the other hand, consider

$$\int\ \mathcal{D}J\mathcal{D}\tilde{\varphi}\ \operatorname{Det}\Big(\frac{\delta Q}{\delta\tilde{\varphi}}\Big)\, e^{-\frac{i}{2}Q(\tilde{\varphi})KQ(\tilde{\varphi})+iJ(Q\tilde{\varphi}-\Phi)}$$
$$= \int \mathcal{D}\tilde{\varphi}\ \operatorname{Det}\Big(\frac{\delta Q}{\delta\tilde{\varphi}}\Big)\, e^{-\frac{i}{2}Q(\tilde{\varphi})KQ(\tilde{\varphi})}\,\delta\big(Q\tilde{\varphi}-\Psi\big)$$
$$= \operatorname{Det}\Big(\frac{\delta Q}{\delta\tilde{\varphi}}\Big)\Big|_{Q\tilde{\varphi}=\Psi}\Big(\operatorname{Det}\frac{\delta Q}{\delta\tilde{\varphi}}\Big)^{-1} e^{-\frac{i}{2}\Psi K\Psi} \,=\, e^{-\frac{i}{2}\Psi K\Psi}. \tag{7.140}$$

We see that the expressions (7.139) and (7.140) coincide. Performing an inverse Fourier transform, one gets the relation (7.137).

The functional integral over anticommuting or fermionic fields is considered in a complete analogy. The properties of the functional integrals over bosonic fields, such as the functional shift invariance, the vanishing of the integral of the derivative, and integration by parts, can be proved to hold also for the functional integral over fermionic fields, as the reader can ensure as an exercise.

Indeed, only the derivation of the Gaussian functional integral requires special consideration. Consider the Gaussian functional integral over fermionic fields,

$$G_F[\eta] = \int \mathcal{D}\psi \, e^{\frac{i}{2}\psi K_F \psi + i\eta\psi}, \qquad (7.141)$$

where $\psi(x)$ are the anticommuting fields, $\eta(x)$ are external anticommuting fields (anticommuting sources) and K_F is an invertible operator with the kernel $K_F(x, y)$, where

$$K_F(x, y) = -K_F(y, x), \qquad (7.142)$$

because the fields $\psi(x)$ are anticommuting. In expression (7.141), we used the notations

$$\psi K_F \psi = \int d^4x d^4y \psi(x) \, K_F(x, y)\psi(y) \qquad \text{and} \qquad \eta\psi = \int d^4x \, \eta(x)\psi(x).$$

The derivation of the integral (7.141) is performed according to the same scheme used for (7.126). Consider the integral (7.141) and fulfill the shift of variables, $\psi \to \psi + \Psi$. Then one gets

$$G_F[\eta] = \int \mathcal{D}\psi e^{\frac{i}{2}\psi K_F \psi + i\Psi K_F \psi + \frac{i}{2}\Psi K_F \Psi + i\eta\psi + i\eta\Psi}, \qquad (7.143)$$

where the relation (7.142) has been used. The external field Ψ is a solution of the condition $\Psi K_F + \eta = 0$, that gives $\Psi = K_F^{-1}\eta$. Substituting the last relation into Eq. (7.143), we obtain

$$G_F[\eta] = e^{\frac{i}{2}\eta K_F^{-1}\eta} \, G_F[0], \qquad \text{where} \qquad G_F[0] = \int \mathcal{D}\psi e^{\frac{i}{2}\psi K_F \psi}. \qquad (7.144)$$

Consider a variation of $G[0]$ in Eq. (7.144) with respect to the parameters of the operator K_F,

$$\delta G_F[0] = \frac{i}{2} \int d^4x d^4y \, \delta K_F(x, y) \int \mathcal{D}\psi \, \psi(x)\psi(y) \, e^{\frac{i}{2}\psi K_F \psi}. \qquad (7.145)$$

The functional integral in the relation (7.145) is calculated using (7.142),

$$\int \mathcal{D}\psi\psi(x)\psi(y)e^{\frac{i}{2}\psi K_F \psi} = -\frac{\delta}{\delta i\eta(y)} \frac{\delta}{\delta i\eta(x)} G_F[\eta]\Big|_{\eta=0} = iK_F^{-1}(x, y)G_F[0].$$

Then

$$\delta G_F[0] = -\frac{1}{2} \int d^4x d^4y \, \delta K_F(x, y) \, K_F^{-1}(x, y) \, G_F[0].$$

Therefore,

$$\delta \log G_F[0] = \frac{1}{2}\delta \,\mathrm{Tr}\, \log K_F = \delta\big(\log\, \mathrm{Det}\, K_F\big)^{\frac{1}{2}}. \tag{7.146}$$

Thus, we arrive at the relation $G_F[0] = C(\,\mathrm{Det}\, K_F)^{\frac{1}{2}}$, where C is an arbitrary constant, which is the same for all Gaussian-type integrals. This constant can be included in the definition of the functional measure $\mathcal{D}\psi$, so we do not need to write it explicitly.

Finally, the Gaussian functional integral over fermionic fields has the form

$$G_F[\eta] = (\,\mathrm{Det}\, K_F)^{\frac{1}{2}}\, e^{\frac{i}{2}\eta K_F^{-1}\eta}. \tag{7.147}$$

7.7 Techniques for calculating functional determinants

In the previous section, we saw that the functional integrals for free theories reduce to the functional determinants. Let us consider these determinants in more detail.

Functional determinants are the generalization of the determinants of ordinary, finite-dimensional, square matrices. Consider a finite-dimensional determinant of a non-degenerate matrix M_{ij}, assuming λ_n are its eigenvalues. Then,

$$\det M = \prod_n \lambda_n = e^{\sum_n \log \lambda_n} = e^{\,\mathrm{tr}\, \log M}. \tag{7.148}$$

Similarly, an invertible symmetric differential operator K has an infinite-dimensional matrix representation and the kernel $K(x, y)$. By definition,

$$\mathrm{Det}\, K = e^{\,\mathrm{Tr}\, \log K}. \tag{7.149}$$

The functional trace of the operator acting on a scalar is defined as

$$\mathrm{Tr}\, K = \int d^4x K(x, y)\big|_{y=x}. \tag{7.150}$$

Many differential operators of our interest have the form $K = \Box + X$, or can be transformed into this form. The operator X contains the derivatives of no higher than first order and depends on background fields and parameters such as masses and coupling constants. The kernel of such an operator can be written in the form

$$K(x, y) = (\Box + m^2 + X)\delta^4(x - y). \tag{7.151}$$

In this case, the Tr $\log K$ is well defined perturbatively as the series

$$\mathrm{Tr}\, \log K \;=\; \mathrm{Tr}\, \log\,(\Box + m^2) + \sum_{n=1}^{\infty} \frac{(-1)^{n-1}}{n}\, \mathrm{Tr}\, \Big(\frac{1}{\Box + m^2}\, X\Big)^n, \tag{7.152}$$

where the first term Tr $\log(\Box + m^2)$ is a constant. In principle, every term in this series can be calculated. However, it is useful to formulate the general methods of calculating the functional determinants, which are not related formally with an expansion into series. We will consider two such methods in this book. The Fock-Schwinger proper-time method will be considered as part of quantum field theory in curved space, in Part II, where it is called the Schwinger-DeWitt, or heat-kernel technique. Now we briefly review another approach.

Calculation of functional determinants using a generalized ζ-function.

Let $\{\lambda_n\}$ be the eigenvalues of the operator K, and assume that $\lambda_n \neq 0$. The generalized ζ-function associated with operator K is defined as

$$\zeta(z|K) = \sum_n \frac{1}{\lambda_n{}^z} = \mathrm{Tr}\, K^{-z}, \tag{7.153}$$

where z is a complex variable. One can show that the last expression is well defined for sufficiently large values of $\mathrm{Re}\, z$ and can be analytically continued to the complex plane of z. It is easy to see that Eq. (7.153) leads to the relation

$$\zeta'(z|K)\Big|_{z=0} = -\,\mathrm{Tr}\,\log K = -\,\mathrm{Det}\, K. \tag{7.154}$$

Therefore,

$$\mathrm{Det}\, K = e^{-\zeta'(0|K)}. \tag{7.155}$$

The relation (7.155) can be taken as a formal definition of a functional determinant. As a result, calculating the functional determinant of the operator is reduced to finding the ζ-function associated with this operator. The generalized ζ-function associated with the operator K can be related to the Schwinger kernel of this operator (see, e.g., [121, 324]).

Exercises

7.1. Consider model with the Lagrangian $L = \frac{1}{2}\, m_{ij}(q)\dot{q}^i\dot{q}^j$; $i,j = 1,2\ldots,n$, where $q = q^i$ are real generalized coordinates and $\det\|m_{ij}(q)\| \neq 0$. Construct the Hamilton function, carry out canonical quantization, define the Hamiltonian and find its p,q-symbol.

7.2. Using (7.10), calculate matrix elements of the evolution operator for the one-dimensional harmonic oscillator.

7.3. Calculate the following integrals:

1) $$Z(j) = \int_{-\infty}^{\infty} dx\, e^{-\frac{1}{2}x^2 + xj}, \quad \text{where } j \text{ is a parameter.}$$

2) $$\int_{-\infty}^{\infty} dx\; x^n e^{-\frac{1}{2}x^2 + xj}, \quad \text{where } n \text{ is a natural number.}$$

7.4. Calculate the following functional integral for a real scalar field:

$$\int \mathcal{D}\varphi\; \varphi(x_1)\varphi(x_2)\varphi(x_3)\varphi(x_4)\, e^{iS_0[\varphi]}.$$

7.5. Let θ_i be anticommuting variables, with $i = 1, 2, 3, 4, 5$. Calculate the following derivatives

$$\frac{\vec{\partial}}{\partial\theta_2}\theta_1\theta_3\theta_2\theta_4\theta_5, \qquad \theta_1\theta_3\theta_2\theta_4\theta_5\frac{\overleftarrow{\partial}}{\partial\theta_1}. \tag{7.156}$$

7.6. Let $\theta(x)$ be the anticommuting function $\theta(x)\theta(x') = -\theta(x')\theta(x)$. Calculate the functional derivatives

$$\frac{\vec{\delta}}{\delta\theta(x)}\theta(x')\theta(x'')\theta(y)\theta(y'), \quad \theta(x')\theta(x'')\theta(y)\theta(y')\frac{\overleftarrow{\delta}}{\delta\theta(x)}.$$

7.7. Calculate the functional derivatives

$$\frac{\vec{\delta}}{\delta\psi(x)}e^{i\int d^4y\big(\bar{\psi}(y)\eta(y)+\bar{\eta}(y)\psi(y)\big)}, \qquad e^{i\int d^4y\big(\bar{\psi}(y)\eta(y)+\bar{\eta}(y)\psi(y)\big)}\frac{\overleftarrow{\delta}}{\delta\bar{\eta}(x)}. \tag{7.157}$$

7.8. Consider the two real scalar fields φ_1 and φ_2 with the Lagrangian

$$\mathcal{L} = \frac{1}{2}\big(\partial_\mu\varphi_1\partial^\mu\varphi_1 - {m_1}^2{\varphi_1}^2\big) + \frac{1}{2}\big(\partial_\mu\varphi_2\partial^\mu\varphi_2 - {m_2}^2{\varphi_2}^2\big) - \lambda\varphi_1\varphi_2.$$

In addition, consider the functional integral

$$Z[J_1, J_2] = \int \mathcal{D}\varphi_1\mathcal{D}\varphi_2 e^{iS[\varphi_2,\varphi_2]+i\int d^4x(\varphi_1 J_1+\varphi_2 J_2)}.$$

Find the matrix K for this case and calculate this functional integral in terms of K and J_1, J_2.

7.9 Consider a complex scalar field with the action

$$S[\varphi^\dagger, \varphi] = \int d^4x\Big\{\varphi^\dagger(x)K_{12}\varphi(x) + \frac{1}{2}\varphi^\dagger(x)K_{11}\varphi^\dagger(x) + \frac{1}{2}\varphi(x)K_{22}\varphi(x)\Big\},$$

where K_{12}, K_{11}, K_{22} are differential operators, with action on the functions of $x = x^\mu$. Introduce a line $(\varphi^\dagger, \varphi)$. Show that the above action can be written in the form

$$S[\varphi^\dagger, \varphi] = \frac{1}{2}\int d^4x\ (\varphi^\dagger, \varphi)\,\hat{K}\,(\varphi^\dagger, \varphi)^T,$$

where T means transposition and $\hat{K}$ is a matrix differential operator. Write down the matrix kernel $\hat{K}(x, y)$ in terms of the kernels of K_{12}, K_{11} and K_{22}. Find the restrictions on the matrix kernel $\hat{K}(x, y)$, that ensure the action is a real-valued quantity.

7.10. Consider the electromagnetic field action

$$S[A] = -\frac{1}{4}\int d^4x\, F_{\mu\nu}F^{\mu\nu},$$

where $F_{\mu\nu} = \partial_\mu A_\nu = \partial_\nu A_\mu$. Show that this action can be rewritten in the form

$$S[A] = \frac{1}{2}\int d^4x\ A^\mu K_{\mu\nu}A^\nu$$

and find an explicit form of the matrix kernel $K_{\mu\nu}(x, y)$. Calculate the determinant of $K_{\mu\nu}$ in the momentum representation, and prove that it is zero. Try to solve the last exercise in two distinct ways.

7.11. Calculate the Gaussian functional integral over fermionic fields

$$\int \mathcal{D}\psi\, e^{\frac{i}{2}\psi_l K_{lj}\psi_j + i\psi_l\eta_l},$$

where

$$\psi_l K_{lj}\psi_j = \int d^4x d^4y\ \psi_l(x) K_{lj}(x,y)\psi_j(y), \qquad \psi_l\eta_l = \int d^4x\, \psi_l(x)\eta_l(x), \tag{7.158}$$

and $K_{lj}(x,y) = -K_{jl}(y,x)$. Write the result in terms of the matrix K and the column η.

Comments

The functional integral (also called the path integral or continual integral) was introduced to quantum mechanics and quantum field theory by R.P. Feynman [131]; see also the book by R.P. Feynman and A.R. Hibbs [133].

The formulation of quantum field theory in terms of functional integrals is considered in detail in all quantum field theory books published after 1980, see, e.g., [187], [250], [304], [274], [256], and [345].

The Grassmann algebra and properties of the anticommuting variables, including differentiation and integration, are considered in the most mathematically rigorous way in the book by F.A. Berezin [48]; see also the books by B.S. DeWitt [108] and by I.L. Buchbinder and S.M. Kuzenko [81].

The mathematically rigorous definition of Gaussian functional integrals over commuting and anticommuting fields is given in the aforementioned book, [48]. The properties of functional integrals are formulated in the framework of perturbation theory in the paper by A.A. Slavnov [298]; see also the books by L.D. Faddeev and A.A. Slavnov [127] and A.N. Vasiliev [323].

The proper-time representation of the Green function was introduced by V.A. Fock (see, e.g., [136]). The Proper time technique for calculating functional determinants was initiated by J. Schwinger [277] and presented in the generally covariant form by B.S. DeWitt [105]. We discuss the details of the Schwinger-DeWitt technique in Part II of this book.

The first implementation of the generalized ζ-function for the calculation of functional determinants in quantum field theory was given by J.S. Dowker and R. Critchley [115] and S.W Hawking [177]. For a more detailed consideration of the generalized ζ-function and its application to the calculation of the functional determinants in quantum field theory see, e.g., the books by E. Elizalde, S.D. Odintsov, A, Romeo and A.A. Bytsentko [122, 121] and D. Fursaev and D. Vassilevich [151].

8
Perturbation theory

In this chapter, we show that each term of the perturbative expansion for a Green function can be presented by specific graphs called Feynman diagrams. We will illustrate the construction of Feynman diagrams for the model interacting real scalar field. The generalizations for the other field models are mainly evident, but we will briefly consider the Feynman diagrams for the theory of interacting scalar and spinor fields with Yukawa coupling and for spinor electrodynamics.

8.1 Perturbation theory in terms of Feynman diagrams

As was shown in chapter 6, the reduction formulas (6.36) express the elements of the $\mathcal{S}$-matrix through the Green functions of the Heisenberg field operators. In turn, the n-point Green functions $G_n(x_1, x_2, \dots, x_n)$ are obtained by taking variational derivatives of the generating functional of Green functions $Z[J]$, according to Eq. (6.47). In the framework of perturbation theory, the functional $Z[J]$ is given by the expression (7.96), where the generating functional of free Green functions has the form (7.106). Thus, we get the expression for the n-point Green function as a power series in the interactions,

$$G_n(x_1, x_2, \dots, x_n) = \frac{\frac{\delta}{\delta iJ(x_1)} \cdots \frac{\delta}{\delta iJ(x_n)} \big\{1 + \sum_{m=1}^{\infty} \frac{i^m}{m!} (S_{int}[\frac{\delta}{\delta iJ}])^m \big\} Z_0[J] \big|_{J=0}}{\big\{1 + \sum_{m=1}^{\infty} \frac{i^m}{m!} (S_{int}[\frac{\delta}{\delta iJ}])^m \big\} Z_0[J] \big|_{J=0}}, \tag{8.1}$$

and we assume the result (7.106), which can be written as $Z_0[J] = \exp\big\{ -\frac{i}{2} JDJ \big\}$.

Let us investigate the structure of the relation (8.1). First of all, the expression for $G_n(x_1, x_2, \dots, x_n)$ has the form

$$G_n(x_1, x_2, \dots, x_n) = G_n^{(0)}(x_1, x_2, \dots, x_n) + \sum_{m=1}^{\infty} G_n^{(m)}(x_1, x_2, \dots, x_n), \tag{8.2}$$

where $G_n^{(m)}(x_1, x_2, \dots, x_n)$ with $m = 1, 2, \dots$ is a contribution to the Green function in the m-th order in interaction, and the zero-order Green function is

$$G_n^{(0)}(x_1, x_2, \dots, x_n) = \frac{\delta}{\delta iJ(x_1)} \frac{\delta}{\delta iJ(x_2)} \cdots \frac{\delta}{\delta iJ(x_n)} e^{-\frac{i}{2}JDJ} \Big|_{J=0}. \tag{8.3}$$

It is clear that $G_n^{(0)}(x_1, x_2, \dots, x_n)$ is zero for odd n and is given by a product of Feynman propagators for even n. The m-th order term in the numerator of (8.1) can be written in the form

Introduction to Quantum Field Theory with Applications to Quantum Gravity. Iosif L. Buchbinder and Ilya L. Shapiro, Oxford University Press (2021). © Iosif L. Buchbinder and Ilya Shapiro.
DOI: 10.1093/oso/9780198838319.003.0008

$$\frac{i^n}{n!}\frac{\delta}{\delta iJ(x_1)}\frac{\delta}{\delta iJ(x_2)}\cdots\frac{\delta}{\delta iJ(x_n)}\frac{i^m}{m!}\int d^4y_1 d^4y_2\dots d^4y_m\,\mathcal{L}_{int}\Big(\frac{\delta}{\delta iJ(y_1)}\Big)$$
$$\times\,\mathcal{L}_{int}\Big(\frac{\delta}{\delta iJ(y_2)}\Big)\dots\mathcal{L}_{int}\Big(\frac{\delta}{\delta iJ(y_m)}\Big)\,e^{-\frac{i}{2}JDJ}\Big|_{J=0}\,. \tag{8.4}$$

Thus, the calculation of perturbative corrections to Green function is reduced to the differentiation of the function $e^{-\frac{i}{2}JDJ}$ at the vanishing source $J(x)$. The result is given by a product of Feynman propagators and powers of coupling constant coming from the interaction Lagrangian $\mathcal{L}_{int}$, with some integrations over Minkowski space. The spacetime arguments x_i and y_j appear in the Feynman propagators $D(x_i,x_j)$, $D(x_i,y_k)$ and $D(y_k,y_l)$.

Consider the relatively simple example of the interaction Lagrangian of the form $\mathcal{L}_{int}=-V(\varphi)$, where $V(\varphi)=\lambda\varphi^k$, $k>2$. Then $\mathcal{L}_{int}\big(\frac{\delta}{\delta iJ(y)}\big)=-\lambda\big(\frac{\delta}{\delta iJ(y)}\big)^k$. Taking into account Eqs. (8.3) and (8.4), the Green function $G_n(x_1,x_2,\dots,x_n)$ can be associated with diagrams constructed according to the following rules:

1. Let us draw $n+m$ points and call n of them external, and m of them internal. The external points originate from the n functional derivatives $\frac{\delta}{\delta iJ(x_1)}\frac{\delta}{\delta iJ(x_2)}\cdots\frac{\delta}{\delta iJ(x_n)}$ or, which is the same thing, n arguments of the Green function $G_n(x_1,x_2,\dots,x_n)$. The internal points originate from the product of m integrals

$$\int d^4y_1V(\varphi(y_1))\int d^4y_2V(\varphi(y_2))\dots\int d^4y_mV(\varphi(y_m)).$$

2. All of the $n+m$ arguments $x_1,x_2,\dots,x_n$ and $y_1,y_2,\dots,y_m$ are distributed among the propagators, each of those depending only on two arguments. Each propagator corresponds to a line connecting one pair of points. Thus, to construct a diagram for a Green function, we need to connect all the points with lines. Each external point connects to the number of lines equal to the power of the field in the potential $V(\varphi)$. For example, for $V(\varphi)=\lambda\varphi^k$, an internal point is associated with an expression of the form $\dots\int d^4y\big(\frac{\delta}{\delta iJ(y)}\big)^k e^{-\frac{i}{2}JDJ}$ with k propagators containing equal or distinct arguments y. Here $\dots$ means functional derivatives with respect to other arguments. A line from an external point is called the external line, while a line connecting two internal points is called the internal line. Internal points are called vertices.

As a result, we arrive at the rules for constructing the Feynman diagrams associated with Green functions. On the other hand, if we have all the Feynman diagrams with m vertices, we can calculate the m-th order contribution to the Green function. This is done on the basis of the following Feynman rules:

1. All of the Feynman diagrams should be drawn m vertices.

2. Each vertex is associated with the factor $-i\lambda$ and an integral over the Minkowski space.

3. The line connecting points x and y is associated with the propagator $iD(x,y)$.

4. The contributions of all the lines should be multiplied to one of the vertices, to achieve the contribution of a given diagram.

5. All the possible ways to connect the points by lines (the propagators) should be summed up and multiplied by a factor of $\frac{1}{m!}$. Many of these contributions are identical; hence, the result can be obtained by using combinatorial calculus.

The Feynman rules make it possible to provide a visual representation of the perturbative contributions to Green functions and enable one to combine an analysis of the formulas with an analysis of the diagrams. Let us consider some examples of Feynman diagrams, for a few concrete models of scalar field theory.

a. *The real scalar field model with the interaction potential* $V(\varphi) = \lambda\varphi^3$.
The diagrams for two-point, three-point and four-point Green functions in the second, third and fourth orders, respectively, in λ have the following forms:

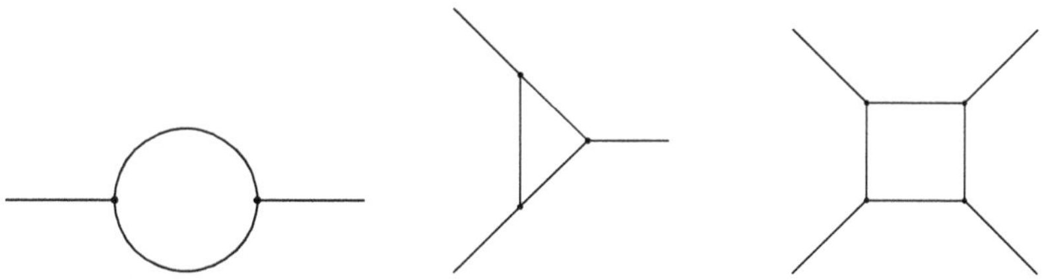

b. *Scalar field theory with the interaction potential* $V(\varphi) = \lambda\varphi^4$.
Diagrams for the two-point Green functions up to the second order in λ have the following forms:

One of the diagrams for the four-point Green function, in the sixth order in λ, is

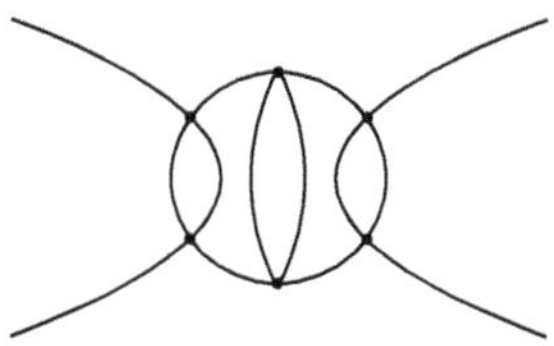

All of the examples considered above are for diagrams that are related to the numerator of the expression (8.1). The denominator of the same expression can be elaborated in the same way as for the numerator, and the result can be represented using Feynman diagrams constructed according to the rules described above. The specific feature of the diagrams associated with the denominator of (8.1) is that these diagrams do not have external points and, hence, do not have external lines. These are called vacuum diagrams.

Examples of purely vacuum diagrams in the scalar field models with interaction potentials $V(\varphi) = \lambda\varphi^3$ and $V(\varphi) = \lambda\varphi^4$ are as follows:

Let us discuss the role of vacuum diagrams in the expression (8.1) in more detail. It is important to note that not only the denominator but also the numerator of this formula contains the vacuum diagrams. The vacuum diagrams in the numerator

originate as subdiagrams, when there are no lines connecting external and internal points. In other words, vacuum diagrams appear when some of the internal points are connected by lines among themselves. One can prove that the vacuum diagrams of the numerator and the denominator cancel each other and, as a result, vacuum diagrams give no contribution to the Green functions.

To prove this important statement about the cancelation of the vacuum diagrams in Eq. (8.1), let us rewrite the relation (7.56) for Green functions in the form

$$\begin{aligned}
G_n(x_1, x_2, \dots, x_n) &= \frac{\int \mathcal{D}\varphi\, \varphi(x_1)\varphi(x_2)\dots\varphi(x_n)\, e^{iS_0[\varphi]+iS_{int}[\varphi]}}{\int \mathcal{D}\varphi\, e^{iS_0[\varphi]+iS_{int}[\varphi]}} \\
&= \frac{\int \mathcal{D}\varphi\, \varphi(x_1)\varphi(x_2)\dots\varphi(x_n) \sum\limits_{m=0}^{\infty} \big(iS_{int}[\varphi]\big)^m e^{iS_0[\varphi]}}{\int \mathcal{D}\varphi \sum\limits_{m=0}^{\infty} \big(iS_{int}[\varphi]\big)^m e^{iS_0[\varphi]}} \\
&= \frac{\sum\limits_{m=0}^{\infty} \frac{(-i)^m}{m!} \int d^4y_1 d^4y_2 \dots d^4y_m \, \langle \varphi(x_1)\dots\varphi(x_n) V(\varphi(y_1))\dots V(\varphi(y_m))\rangle}{\sum\limits_{m=0}^{\infty} \frac{(-i)^m}{m!} \int d^4y_1 d^4y_2 \dots d^4y_m \, \langle V(\varphi(y_1)) V(\varphi(y_2))\dots V(\varphi(y_m))\rangle},
\end{aligned} \tag{8.5}$$

where $\langle(\dots)\rangle = \int \mathcal{D}\varphi\, (\dots)\, e^{iS_0[\varphi]}$ is the Gaussian functional integral for the free action $S_0[\varphi]$. The numerator and denominator of the relation (8.1) include the products of the fields $\varphi(x)$. For example, if $V(\varphi) \sim \varphi^k$, the numerator contains the m-th order of the product of $n + km$ fields $\varphi(x)$. With each pair of fields $\varphi(x)$ and $\varphi(y)$, we associate the factor of $iD(x, y)$, where $D(x, y)$ is the Feynman propagator. This operation is called contraction of fields. According to the Feynman rules, contraction means that we associate a line with the two fields. In each fixed m-th order, each Feynman diagram corresponds to a concrete way of contracting all the fields in the numerator and the denominator of the relation (8.1). The full contribution in the m-th order corresponds to all possible ways of contracting the fields in (8.1).

The diagrams for the numerator and the denominator of (8.1) can be divided into two groups or types. The first group do not contain vacuum subdiagrams, while second type does. It is evident that the denominator in (8.1) contains only the diagrams from the second group, while the numerator contains diagrams of both types. Let us denote the contribution of diagrams from the first group by the label (1). Then we can write

$$\begin{aligned}
&\sum_{m=0}^{\infty} \frac{(-i)^m}{m!} \int d^4y_1 d^4y_2 \dots d^4y_m \langle \varphi(x_1)\dots\varphi(x_n) V(\varphi(y_1)) V(\varphi(y_2)\dots V(\varphi(y_m))\rangle \\
&\quad = \sum_{m=0}^{\infty} \frac{(-i)^m}{m!} \int d^4y_1 d^4y_2 \dots d^4y_m \sum_{l=0}^{m} C_m^l \, \langle \varphi(x_1)\varphi(x_2)\dots\varphi(x_n) \\
&\quad\times V(\varphi(y_1)) V(\varphi(y_2)\dots V(\varphi(y_l))\rangle^{(1)} \, \langle V(\varphi(y_{l+1})) V(\varphi(y_{l+2}))\dots V(\varphi(y_m))\rangle.
\end{aligned} \tag{8.6}$$

Here the factor $C_m^l = \frac{m!}{l!(m-l)!}$ is inserted because all the terms that differ by the rearrangements of the arguments $y_1, y_2, \dots, y_m$ have the same integrations over these arguments. The expression (8.6) can be transformed as follows:

$$\sum_{m=0}^{\infty}\sum_{l=0}^{m}\frac{(-i)^l(-i)^{m-l}}{l!(m-l)!}\int d^4y_1d^4y_2\dots d^4y_ld^4z_1d^4z_2\dots d^4z_{m-l}\,\langle\varphi(x_1)\varphi(x_2)\dots\varphi(x_n)$$
$$\times\,V(\varphi(y_1))V(\varphi(y_2)\dots V(\varphi(y_l))\rangle^{(1)}\;\langle V(\varphi(z_1))V(\varphi(z_2))\dots V(\varphi(z_{m-l}))\rangle$$
$$=\sum_{l=0}^{\infty}\frac{(-i)^l}{l!}\int d^4z_1d^4z_2\dots d^4z_l\,\langle\varphi(x_1)\dots\varphi(x_n)V(\varphi(y_1))V(\varphi(y_2)\dots V(\varphi(y_l))\rangle^{(1)}$$
$$\times\sum_{m}^{\infty}\frac{(-i)^m}{m!}\int d^4y_1d^4y_2\dots d^4y_m\langle V(\varphi(y_1))V(\varphi(y_2))\dots V(\varphi(y_m))\rangle.$$

It is easy to see that the second factor in this relation is just the denominator in the relation (8.1) and hence it cancels perfectly well with the contribution of the denominator.

Thus, we have proved that the vacuum diagrams can be omitted in the perturbative expansion of the Green functions. As a result, we arrive at the formula

$$G_n(x_1,x_2,\dots,x_n)=\sum_{m=0}^{\infty}\frac{(-i)^m}{m!}d^4y_1d^4y_2\dots d^4y_m\langle\varphi(x_1)\varphi(x_2)\dots\varphi(x_n)$$
$$\times V(\varphi(y_1))V(\varphi(y_2)\dots V(\varphi(y_m))\rangle^{(1)}.$$

In what follows, we omit the label (1), keeping in mind that all vacuum subdiagrams can be safely omitted. Therefore, the generating functional of the Green function can be written as

$$Z[J]\;=\;\int\mathcal{D}\varphi\,e^{i\left\{S[\varphi]+\int d^4x\varphi(x)J(x)\right\}},\tag{8.7}$$

where the vacuum diagrams do not need to be accounted for.

8.2 Feynman diagrams in momentum space

Usually, it is more convenient to calculate the contributions of Feynman diagrams in the momentum representation. Let us write the Fourier transform of the Green function:

$$\tilde{G}_n(p_1,p_2,\dots,p_n)=\int d^4x_1d^4x_2\dots d^4x_n\,e^{i(p_1x_1+p_2x_2\dots p_nx_n)}\,G_n(x_1,x_2\dots x_n).$$

As we pointed out, the translational invariance (6.43) requires that the Green function $\tilde{G}_n(p_1,p_2,\dots,p_n)$ be proportional to $\delta^4(p_1+p_2+\dots+p_n)$. Thus, we can write

$$\tilde{G}_n(p_1,p_2,\dots,p_n)\;=\;(2\pi)^4\delta^4(p_1+p_2+\dots+p_n)\,\bar{G}_n(p_1,p_2,\dots,p_n).\tag{8.8}$$

Here the delta function expresses the conservation of the total four-momentum. As a result, only $n-1$ momenta in the Green function $\tilde{G}_n(p_1,p_2,\dots,p_n)$ are independent.

Consider the Feynman diagram for the Green function $G_n(x_1,x_2\dots x_n)$. According to the Feynman rules, the diagram contains n external lines going from external points to the vertices, and each of these external lines is associated with the propagator. Let one of the external lines connect an external point x to a vertex y, as shown in the next figure:

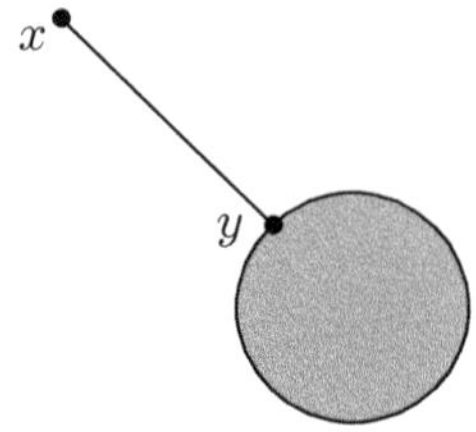

The shaded circle in this diagram means the full set of possible combinations of internal lines and vertices. The contribution of this diagram to the Green function has the form

$$\int d^4y\, i\, D(x-y) F(y, \ldots),$$

where $F(y, \ldots)$ is a contribution from this diagram without the given external line. Let us perform a Fourier transformation over the external point x:

$$\begin{aligned}
&\int d^4x\, e^{ipx} \int d^4y \int \frac{d^4q}{(2\pi)^4} \frac{i}{q^2 - m^2 + i\varepsilon} e^{-iq(x-y)} F(y, \ldots) \\
&= \int \frac{d^4q}{(2\pi)^4} \frac{i}{q^2 - m^2 + i\varepsilon} \int d^4x e^{ix(p-q)} \int d^4y\, e^{iqy} F(y, \ldots) \\
&= \frac{i}{p^2 - m^2 + i\varepsilon} \int d^4y\, e^{ipy}\, F(y, \ldots) \;=\; \frac{i}{p^2 - m^2 + i\varepsilon} \tilde{F}(p, \ldots). \qquad (8.9)
\end{aligned}$$

It is easy to see from this example that, in the momentum representation, the external line is associated with the propagator $i\tilde{D}(p)$.

Consider the vertex y in the model with interaction $\sim \varphi^4$, as shown below:

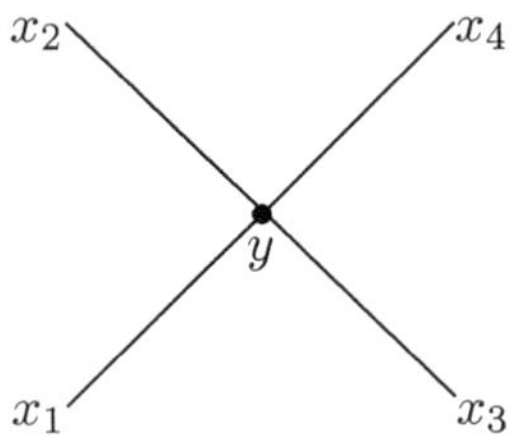

According to the Feynman rules, the contribution of this vertex to $G_4(x_1, \ldots, x_4)$ is

$$\int d^4y\; iD(x_1 - y)\, iD(x_2 - y)\, iD(x_3 - y)\, iD(x_4 - y), \qquad (8.10)$$

where $x_{1,2,3,4}$ are some of the external or internal points of the diagrams. After Fourier transformation of the propagators, this expression takes the form

$$\begin{aligned}
&\int d^4y \prod_{j=1}^{4} \int \frac{d^4k_j}{(2\pi)^4} \frac{i\, e^{-ik_j(x_j - y)}}{k_j^2 - m^2 + i\varepsilon} \\
&= (2\pi)^4\, \delta^4(k_1 + k_2 + k_3 + k_4) \prod_{j=1}^{4} \int \frac{d^4k_j}{(2\pi)^4} \frac{i\, e^{-ik_jx_j}}{k_j^2 - m^2 + i\varepsilon}. \qquad (8.11)
\end{aligned}$$

One can see that each vertex is associated with the delta function $\delta^4(k_1+k_2+k_3+k_4)$, expressing the momentum conservation in the vertex. The remaining exponentials in Eq. (8.11) can be used for Fourier transformations in other vertices.

Thus, we can formulate the following Feynman rules in the momentum representation:

a. Each external line is associated with the propagator $i\tilde{D}(p) = \frac{i}{p^2-m^2+i\varepsilon}$.

b. Each internal line is associated with the expression $\frac{d^4k}{(2\pi)^4}\frac{i}{k^2-m^2+i\varepsilon}$.

c. Each vertex includes the factor $-i\lambda(2\pi)^4\delta^4(k_1+k_2+k_3+k_4)$, where k_1, k_2, k_3 and k_4 are momenta related to this vertex.

d. Contributions of external and internal lines and vertices should be multiplied and integrated over all the momenta k.

As an example, consider the following diagram for the two-point function in the second order of perturbation theory, in the model with the interaction potential $V(\varphi) = \lambda\varphi^4$:

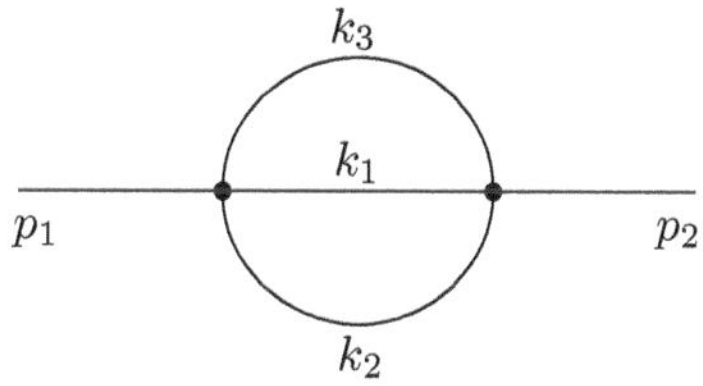

According to the Feynman rules described above, the contribution to $\tilde{G}_2(p_1,p_2)$ from this diagram has the form

$$\tilde{G}_2(p_1,p_2) = \frac{\lambda^2}{2}\frac{i}{p_1^2-m^2+i\varepsilon}\frac{i}{p_2^2-m^2+i\varepsilon}\int\frac{d^4k_1}{(2\pi)^4}\frac{d^4k_2}{(2\pi)^4}\frac{d^4k_3}{(2\pi)^4}\frac{i}{k_1^2-m^2+i\varepsilon}$$
$$\times\frac{i}{k_2^2-m^2+i\varepsilon}\frac{i}{k_3^2-m^2+i\varepsilon}(2\pi)^8\delta^4(k_1+k_2+k_3+p_1)\delta^4(k_1+k_2+k_3-p_2).$$

After integration over k_3, one gets

$$\tilde{G}_2(p_1,p_2) = (2\pi)^4\delta^4(p_1+p_2)\,\bar{G}_2(p_1), \tag{8.12}$$

where the function $\bar{G}_2(p_1)$ has the form

$$\bar{G}_2(p_1) = \frac{\lambda^2}{2}\left(\frac{i}{p_1^2-m^2+i\varepsilon}\right)^2$$
$$\times\int\frac{d^4k_1d^4k_2}{(2\pi)^8}\frac{i}{k_1^2-m^2+i\varepsilon}\frac{i}{k_2^2-m^2+i\varepsilon}\frac{i}{(p_1-k_1-k_2)^2-m^2+i\varepsilon}.$$

Thus, the contribution of this diagram is reduced to the calculation of the so-called loop integrals. In Part I of this book, we do not consider the details of such integrations, but in Part II the reader can find a few elaborated examples.

Let us make a general estimate of a diagram with I internal lines and V vertices. According to the Feynman rules, each internal line is associated with an integral over internal momenta, and each vertex contains a delta function. This means that the full diagram contains V delta functions. One of these delta functions should be factorized to express the conservation law for the momenta of external lines. As a result, for internal integrations, there are I integrals over internal four-momenta, and $(V-1)$ delta functions involving internal momenta. Owing to the presence of these delta functions, the $(V-1)$ integrals can be easily taken. In this way, we arrive at $(I-V+1)$ independent integrations over four-momenta.

The number of independent integrations over four-momenta is called the number of loops in the diagram and is denoted L. Thus, the number of loops in an arbitrary Feynman diagram is

$$L = I - V + 1. \tag{8.13}$$

This last formula is called the topological relation, since it admits a clear geometric identification. Feynman graphs without loops are called tree diagrams.

It is easy to establish the dependence of the L-loop diagram on the constant $\hbar$. The diagram technique is constructed on the basis of the functional integral for the generating functional of Green functions. This functional integral contains the expression $\exp\big\{\frac{i}{\hbar}((S[\varphi]+\int d^4x\varphi(x)J(x))\big\}$, where we provided the true dimensionless quantity in the exponential by restoring the factor of $\frac{1}{\hbar}$. The action is assumed to have a standard structure $S = S_0 + S_{int}$, where the free action S_0 is quadratic in fields and can be schematically presented as $S_0 = \frac{1}{2}\int d^4x\,\varphi K\varphi$, where K is some differential operator. The propagator D is defined as a solution to the equation $\frac{1}{\hbar}KD \sim \delta^4$. Therefore, the propagator has the form $D \sim \hbar K^{-1}\delta^4$. Thus, each propagator contains a factor of $\hbar$. The vertices are generated by the expression $\frac{1}{\hbar}S_{int}$; hence, each vertex contains a factor of $\frac{1}{\hbar}$. For a diagram with E external lines, I internal lines and V vertices, we meet the power $\hbar^{E+I-V} = \hbar^{E-1+L}$. Thus, for a given number of external lines, the power of $\hbar$ in the diagram is defined by the number of loops L.

8.3 Feynman diagrams for the $\mathcal{S}$-matrix

In this section, we discuss how the Feynman rules for the Green functions lead to the diagrams for the $\mathcal{S}$-matrix.

The matrix elements of the $\mathcal{S}$-matrix are related to the Green functions in the momentum representation by the expression (6.41). According to the Feynman rules, the diagram for Green functions in momentum representation, $\tilde{G}_{M+N}(p_1, p_2, \dots, p_M, -q_1, -q_2, \dots, -q_N)$, contains $M+N$ external lines associated with the propagators in the momentum representation. Consider one of such lines with the momentum p, corresponding to the propagator $\frac{i}{p^2-m^2+i\varepsilon}$. In the expression (6.41), the momentum p corresponds to the factor p^2-m^2, which cancels in the diagram with the inverse to $p^2-m^2+i\varepsilon$ in the propagator, such that only the factor i remains.

Using the diagram language, this means that the line with the momentum p has been removed (amputated) from the diagram. The same consideration can be carried out for any external line. A Green function without external lines is called an amputated Green function and can be denoted as $\tilde{G}_n^{(amp)}(p_1, p_2, \dots, p_n)$, as shown in the figure below:

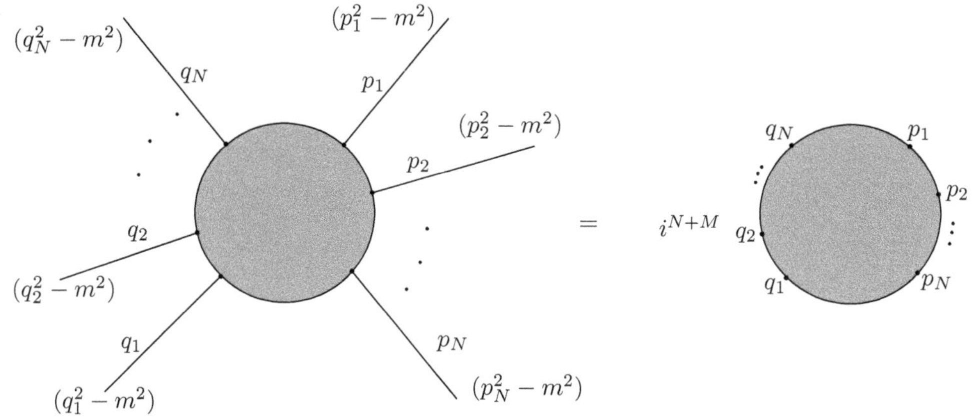

As a result, we arrive at the expression for the $\mathcal{S}$-matrix in terms of Green functions,

$$\mathcal{S}_{MN}(\mathbf{p}_1, \mathbf{p}_2, \dots, \mathbf{p}_M, \mathbf{q}_1, \mathbf{q}_2, \dots, \mathbf{q}_N)$$
$$= \frac{G^{(amp)}_{M+N}(p_1, p_2, \dots, p_M, -q_1, -q_2, \dots, -q_N)}{\big(2(2\pi)^3\big)^{M+N}\sqrt{\big(\varepsilon(\mathbf{p}_1)\varepsilon(\mathbf{p}_2)\dots\varepsilon(\mathbf{p}_M)\varepsilon(\mathbf{q}_1)\varepsilon(\mathbf{q}_2)\dots\varepsilon(\mathbf{q}_N)\big)}}. \tag{8.14}$$

All the momenta in the Green function are taken on shell, such that $p_k^2 \equiv q_l^2 \equiv m^2$. Thus, the elements of the $\mathcal{S}$-matrices can be calculated using the Feynman rules, with amputated Green functions with on-shell external momenta.

8.4 Connected Green functions

A diagram is called connected if each vertex is linked by lines to any other vertex. In other words, the diagram is connected if we can come from one vertex to another one, moving along the lines of the diagram. In the opposite case, the diagram is called disconnected. It is clear that any disconnected diagram can be divided into independent blocks which are not linked with the rest of the diagram by lines.

It is evident that the disconnected diagram consists of a number of independent connected diagrams. Since any diagram will either be connected or consist of a number of independent connected diagrams, the general properties and the structure of an arbitrary diagram is defined by the properties and structure of connected diagrams.

Consider a few examples of connected and disconnected diagrams:

a. In the model with the interaction $V(\varphi) \sim \varphi^4$, the following diagrams are connected:

and the following diagrams are disconnected:

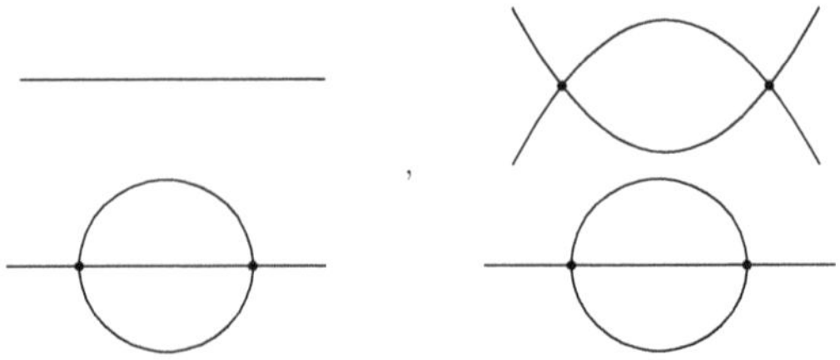

b. In the model with the potential $V(\varphi) \sim \varphi^3$, the following diagram is connected:

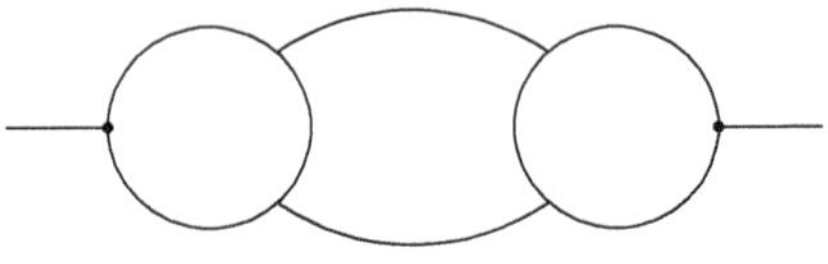

and the following diagram is disconnected:

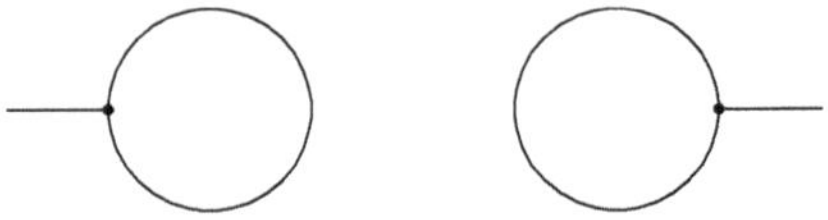

To form an arbitrary, n-point, disconnected diagram, we have to divide the given n points onto separate (disjoint) subsets with n_1, n_2, $\ldots$, n_l points, where $n_1 + n_2 + \cdots + n_l = n$. After that, using the Feynman rules, it is necessary to construct the n_1-point, n_2-point, $\ldots$, n_l-point diagrams that are not linked between themselves by lines. The number of different disconnected n-point diagrams is equal to the number of ways of dividing the n points into disjoint subsets. We can schematically write

$$G_n(x_1, x_2, \ldots, x_n) = \sum\nolimits_{\{k\}}^{(n)} \prod G_k^c(x_1, x_2, \ldots, x_k) + G_n^c(x_1, x_2, \ldots, x_n). \quad (8.15)$$

Here $\sum_{\{k\}}^{(n)}$ means a sum over all of the ways to divide the given n points onto disjoint subsets of points. Then we construct a Feynman diagram over each of such ways and multiply these contributions. The remaining term $G_n^c(x_1, x_2, \ldots, x_n)$ corresponds to the connected Feynman diagram and defines the connected Green function. In what follows, we describe the procedure for generating the connected Green functions.

Consider the generating functional of the Green functions $Z[J]$. It proves useful to introduce the Green functions depending on a source J (compare to Eq. (6.47)),

$$G_n(x_1, x_2, \ldots, x_n|J) = \frac{1}{Z[J]} \frac{\delta^n Z[J]}{\delta iJ(x_1)\delta iJ(x_2)\ldots\delta iJ(x_n)}. \quad (8.16)$$

Furthermore, the connected Green functions $G_n^c(x_1, x_2, \ldots, x_n|J)$ depend on the source J, which is defined as follows:

$$\begin{aligned}
&G_1(x|J) = G_1^c(x|J),\\
&G_2(x_1, x_2|J) = G_1^c(x_1|J)G_1^c(x_2|J) + G_2^c(x_1, x_2|J),\\
&G_3(x_1, x_2, x_3|J) = G_1^c(x_1|J)G_1^c(x_2|J)G_1^c(x_3|J) + G_1^c(x_1|J)G_2^c(x_2, x_3|J)\\
&\quad + G_1^c(x_2|J)G_2^c(x_1, x_3|J) + G_1^c(x_3|J)G_2^c(x_1, x_2|J) + G_3^c(x_1, x_2, x_3|J), \qquad (8.17)
\end{aligned}$$

etc. with the general expression

$$G_l(x_1, x_2, \ldots, x_l|J) = \sum\nolimits_{\{k\}}^{(l)} \prod G_k^c(x_1, x_2, \ldots, x_k|J) + G_l^c(x_1, x_2, \ldots, x_l|J). \quad (8.18)$$

Setting $J = 0$ in these last formulas, we obtain exactly the relations (8.15), defining the connected Green functions $G_n^c(x_1, x_2, \ldots, x_n)$.

Let us now show that the connected Green functions (8.18) can be written in terms of some functional $W[J]$ in the form

$$G_n^c(x_1, x_2, \ldots, x_n|J) = \frac{\delta^n iW[J]}{\delta iJ(x_1)\delta iJ(x_2)\ldots\delta iJ(x_n)}. \quad (8.19)$$

Switching off the source, $J = 0$, after taking variational derivatives, we can see that $W[J]$ is the generating functional of the connected Green functions, i.e.,

$$iW[J] = \sum_{n=0}^{\infty} \frac{i^n}{n!} \int d^4x_1 d^4x_2 \ldots d^4x_n G_n^c(x_1, x_2, \ldots, x_n) J(x_1)J(x_2)\ldots J(x_n). \quad (8.20)$$

To be more precise, we have to prove that each of the functions G_n^c defined by Eq. (8.19) satisfies Eq. (8.18). The statement is based on the relation between the generating functionals $Z[J]$ and $W[J]$, which is given by the following theorem:

Theorem. If $Z[J]$ is the generating functional of the Green functions and $W[J]$ is the generating functional of the connected Green functions (8.20), then

$$Z[J] = e^{iW[J]}. \quad (8.21)$$

Proof. Let us show that, by taking n variational derivatives of (8.21) to the source, on the *l.h.s.* one gets $G_n(x_1, x_2, \ldots, x_n|J)$ on the *r.h.s.*, and, on the *r.h.s.*, the relations (8.18). The proof is carried out by induction.

Consider $n = 1$. We meet
$$G_1(x|J) = \frac{\delta iW[J]}{\delta iJ(x)}. \quad (8.22)$$

The comparison with (8.18) at $n = 1$ shows that $\frac{\delta iW[J]}{\delta iJ(x)} = G_1^c(x|J)$.

Consider $n = 2$; then,

$$G_2(x_1, x_2|J) = \frac{1}{Z[J]} \frac{\delta^2 e^{iW[J]}}{\delta iJ(x_1)\delta iJ(x_2)} = G_1^c(x_1|J)G_1^c(x_2|J) + \frac{\delta^2 iW[J]}{\delta iJ(x_1)\delta iJ(x_2)}. \quad (8.23)$$

Comparison with (8.18) shows that $\frac{\delta^2 iW[J]}{\delta iJ(x_1)\delta iJ(x_2)} = G_2^c(x_1, x_2|J)$. Let us assume that Eqs. (8.18) are valid for all $l \leq n$, and prove that the expression

$$\frac{\delta^{n+1} iW[J]}{\delta iJ(x_1)\delta iJ(x_2)\ldots\delta iJ(x_{n+1})} \quad (8.24)$$

satisfies the condition (8.18) for $l = n+1$. According to this assumption, one can write

$$\frac{\delta^n Z[J]}{\delta iJ(x_1)\delta iJ(x_2)\dots\delta iJ(x_n)} = Z[J]\{\sum{}_{\{k\}}^{(n)}\prod G_k(x_1,x_2,\dots,x_k|J) + G_n^c(x_1,x_2,\dots,x_n|J)\} \tag{8.25}$$

and consider

$$G_{n+1}(x_1,x_2,\dots,x_{n+1}|J) = \frac{1}{Z[J]}\frac{\delta^{n+1}Z[J]}{\delta iJ(x_1)\delta iJ(x_2)\dots\delta iJ(x_{n+1}}$$
$$= G_1(x_{n+1}|J)\{\sum{}_{\{k\}}^{(n)}\prod G_k(x_1,x_2,\dots,x_k|J) + G_n^c(x_1,x_2,\dots,x_n|J)\}$$
$$+\sum{}_{\{k\}}^{(n)}\frac{\delta}{\delta iJ(x_{n+1})}\prod G_k^c(x_1,x_2,\dots,x_k|J) + \frac{\delta^{n+1}iW[J]}{\delta iJ(x_1)\delta iJ(x_2)\dots\delta iJ(x_{n+1})}.$$

The expression

$$\sum{}_{\{k\}}^{(n)}\frac{\delta}{\delta iJ(x_{n+1})}\prod G_k^c(x_1,x_2,\dots,x_k|J) = \tilde{\sum}{}_{\{k\}}^{(n+1)}\prod G_k^c(\dots|J), \tag{8.26}$$

where the term on the *r.h.s.* denotes the sum of all the ways of dividing the points $x_1,x_2,\dots,x_{n+1}$ onto disjoint subsets such that the point x_{n+1} is included in these subsets, together with at least one of the points $x_1,x_2,\dots,x_n$. Then

$$\tilde{\sum}{}_{\{k\}}^{(n+1)}\prod G_k^c(\dots|J) + G_1(x_{n+1})\Big\{\sum{}_{\{k\}}^{(n)}G_k(x_1,x_2,\dots,x_k|J)$$
$$+G_n^c(x_1,x_2,\dots,x_n|J)\Big\} = \sum{}_{\{k\}}^{(n+1)}\prod G_k^c(x_1,x_2,\dots,x_k|J).$$

The last sum is taken over all the ways of dividing points $x_1,x_2,\dots,x_{n+1}$ onto disjoint subsets. Thus, we get

$$G_{n+1}(x_1,x_2,\dots,x_{n+1}|J) = \sum{}_{\{k\}}^{(n+1)}\prod G_k^c(x_1,x_2,\dots,x_k|J) + \frac{\delta^{n+1}iW[J]}{\delta iJ(x_1)\delta iJ(x_2)\dots\delta iJ(x_{n+1})}. \tag{8.27}$$

Finally, comparison with (8.18) shows that

$$G_{n+1}^c(x_1,x_2,\dots,x_{n+1}|J) = \frac{\delta^{n+1}iW[J]}{\delta iJ(x_1)\,\delta iJ(x_2)\dots\delta iJ(x_{n+1})}.$$

This last relation means that the theorem is proved.

8.5 Effective action

In this section, we define the fundamental object of our interest. Taking the functional derivative of the generating functional of the connected Green functions $W[J]$, we get

$$\frac{\delta W[J]}{\delta J(x)} = \frac{1}{Z[J]}\frac{\delta Z[J]}{\delta iJ(x)} = \frac{\langle 0|T\hat{\varphi}(x)\,e^{i\int d^4x\hat{\varphi}(x)J(x)}|0\rangle}{\langle 0|T\,e^{i\int d^4x\hat{\varphi}(x)J(x)}|0\rangle} \equiv \Phi(x|J) \equiv \Phi(x). \tag{8.28}$$

The function $\Phi(x)$ (8.28) is called the source-dependent mean field, or simply the *mean field*.

We will assume that the equation

$$\Phi(x) = \frac{\delta W[J]}{\delta J(x)} \tag{8.29}$$

allows us to express the source through the mean field,

$$J = J(x|\Phi). \tag{8.30}$$

Using (8.30), we define the functional $\Gamma[\Phi]$ as

$$\Gamma[\Phi] = W[J] - \int d^4x\, \Phi(x)J(x)\,. \tag{8.31}$$

The functional $\Gamma[\Phi]$) possesses many interesting properties, as revealed below.

1. Consider

$$\frac{\delta\Gamma[\Phi]}{\delta\Phi(x)} = \int d^4y\, \frac{\delta W[J]}{\delta J(y)}\frac{\delta J(y)}{\delta\Phi(x)} - J(x) - \int d^4y\, \Phi(x)\frac{\delta J(y)}{\delta\Phi(x)} = -J(x), \tag{8.32}$$

where the relation (8.29) has been used. The result (8.32) leads to important consequences. Consider the classical theory with an extended action

$$S_J[\varphi] = S[\varphi] + \int d^4x\, \varphi(x)J(x), \tag{8.33}$$

where $S[\varphi]$ is the action of some scalar field model. The classical equation of motion for such a theory looks very much like the equation for effective action (8.32),

$$\frac{\delta S_J[\varphi]}{\delta\varphi(x)} = 0 \qquad \Longrightarrow \qquad \frac{\delta S[\varphi]}{\delta\varphi(x)} = -J(x). \tag{8.34}$$

In the next section, we will prove that the functional $\Gamma[\Phi]$ has the general structure

$$\Gamma[\Phi] = S[\Phi] + \textit{quantum corrections}. \tag{8.35}$$

Already the comparison between Eqs. (8.32) and (8.34) enables us to conclude that, in quantum theory, $\Gamma[\Phi]$ plays the same role as the functional $S[\varphi]$ plays in classical theory. For this reason, the functional $\Gamma[\Phi]$ is called the quantum effective action, or simply the *effective action*. Eq. (8.32) is the equation of motion for the mean field.

2. Taking two functional derivatives of $\Gamma[\Phi]$ gives

$$\begin{aligned} -\frac{\delta^2\Gamma[\Phi]}{\delta\Phi(x)\delta\Phi(y)} &= \frac{\delta J(x)}{\delta\Phi(y)} = \left[\frac{\delta\Phi(y)}{\delta J(x)}\right]^{-1} \\ &= \left[i\frac{\delta^2 iW[J]}{\delta iJ(x)\delta iJ(y)}\right]^{-1} = \left[iG_2^c(x,y|J)\right]^{-1}. \end{aligned} \tag{8.36}$$

In more detailed form, this relation can be written as follows:

$$\int d^4z\, \frac{\delta^2\Gamma[\Phi]}{\delta\Phi(x)\delta\Phi(z)}\, iG^c(z,y|J) = -\delta^4(x-y)\,. \tag{8.37}$$

To better understand the role of effective action, let us introduce the new functions

$$\Gamma_n(x_1, x_2, \ldots, x_n|\Phi) = \frac{\delta^n \Gamma[\Phi]}{\delta\Phi(x_1)\delta\Phi(x_2)\ldots\delta\Phi(x_n)} \tag{8.38}$$

and find their relations to the connected Green functions $G^c_n(x_1, x_2, \ldots, x_n|J)$. We recall that the source J and the mean field Φ are related to each other by Eq. (8.29).

Using Eqs (8.32) and (8.29), the functional derivative with respect to the source is transformed into a functional derivative with respect to the mean field,

$$\begin{aligned}\frac{\delta}{\delta J(x)} &= \int d^4y \frac{\delta\Phi(y)}{\delta J(x)}\frac{\delta}{\delta\Phi(y)} = \int d^4y \frac{\delta^2 W[J]}{\delta J(y)\delta J(x)}\frac{\delta}{\delta\Phi(y)} \\ &= \int d^4y\, iG^c_2(x, y|J)\,\frac{\delta}{\delta\Phi(y)}.\end{aligned} \tag{8.39}$$

This relation can be written symbolically in the form

$$\frac{\delta}{\delta J} = \text{———}\ \frac{\delta}{\delta\Phi},$$

where the bold line means the function iG^c_2. Next, the function $G^c_3(x_1, x_2, x_3|J)$ can be transformed using the identity (8.39):

$$\begin{aligned}G^c_3(x_1, x_2, x_3|J) &= \frac{\delta^3 iW[J]}{\delta iJ(x_1)\delta iJ(x_2)\delta iJ(x_3)} = \int d^4y_1 G^c_2(x_1, y_1|J)\frac{\delta}{\delta\Phi(y_1)}G^c_2(x_2, x_3|J) \\ &= -\int d^4y_1 iG^c_2(x_1, y_1|J)\frac{\delta}{\delta\Phi(y_1)}\Big[-\frac{\delta^2\Gamma[\Phi]}{\delta\Phi(x)\delta\Phi(y)}\Big]^{-1} \\ &= \int d^4y_1 d^4y_2 d^4y_3\, iG^c_2(x_1, y_1|J)\, iG^c_2(x_2, y_2|J)\, iG^c_2(x_3, y_3|J)\,\Gamma_3(y_1, y_2, y_3|\Phi),\end{aligned} \tag{8.40}$$

where we used (8.36) and (8.38). Relation (8.40) can be represented graphically as

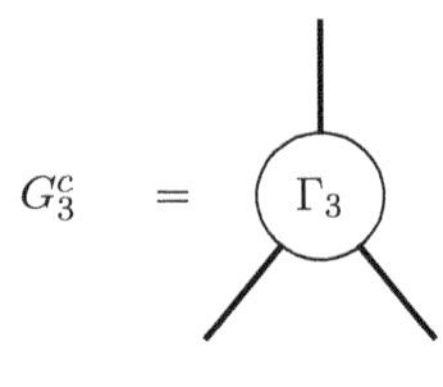

(8.41)

In particular, we have the schematic equality

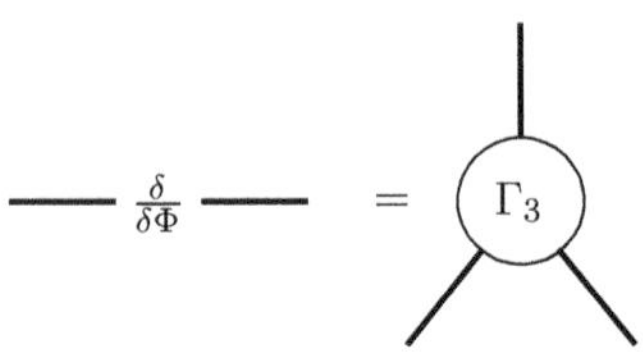

(8.42)

It is evident that the diagrams for the function Γ_3 are obtained from the diagrams for G_3^c via the amputation of external lines. Further consideration can be carried out similarly on the basis of relation (8.42). For G_4^c we have

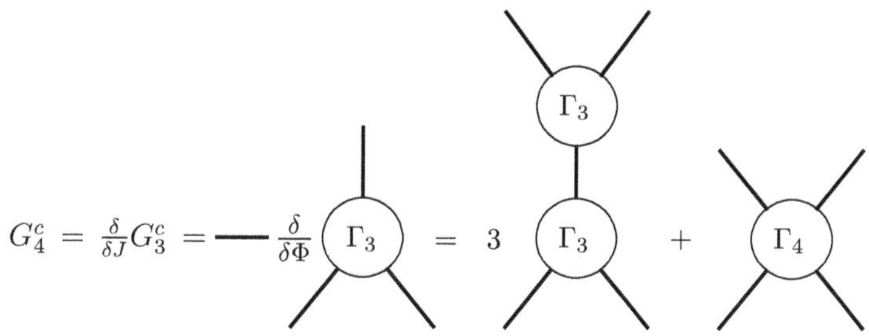

(8.43)

Here the derivative with respect to Φ can act on each of the three external lines of the diagram in the *r.h.s.* of the relation (8.41) and directly on Γ_3. It produces the three diagrams with two inserted Γ_3 (they have different arguments of external points) and the diagram for Γ_4. Furthermore, G_5^c, G_6^c, ... , G_n^c,... can be considered in the same way. Each derivative with respect to Φ acting on external or internal line produces, according to (8.42), the insertion of the Γ_3 with three lines (8.41), where one of these lines goes to another Γ_3 and another line can go to one of the Γ_3 or to another part of the diagram.

In addition, for G_n^c there will be the derivative that acts on Γ_n and produces Γ_{n+1}. In the general case, we have the diagram for G_n^c with n external lines and some number of internal lines. Picking one of the external lines, we can consider a transformation like

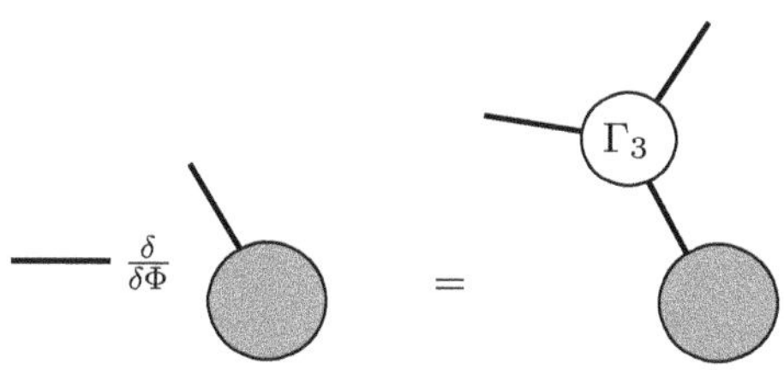

(8.44)

Here the shaded circle indicates the part of the diagram that is not involved in this operation. The operator in the *l.h.s.* of (8.44) acts only on the line that is explicitly shown.

Consider another example. Let us pick one of the internal lines and consider

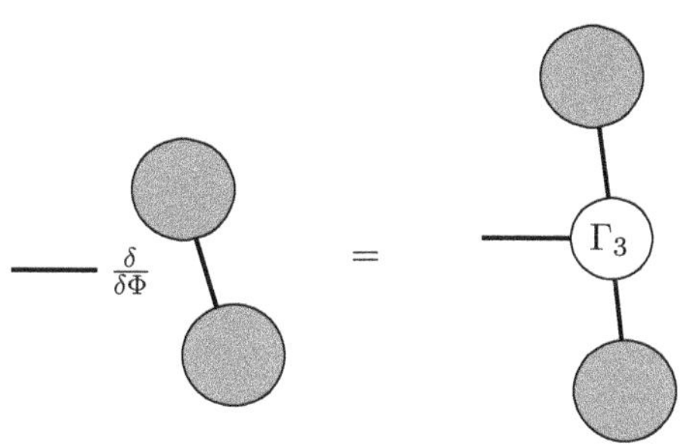

(8.45)

Here the operator in the *l.h.s.* of (8.45) acts only on this internal line.

In general, the diagrams for G_n^c include the graphs connecting different parts of the given diagram by one line and one diagram corresponding to Γ_n, with n external lines of iG_2^c. This last diagram is called one-particle reducible if it becomes disconnected after cutting one line. In the opposite case, the diagram is called one-particle irreducible. Thus, the diagrams for G_n^c include a number of one-particle reducible diagrams and a one-particle irreducible diagram corresponding to Γ_n. Finally, all one-particle reducible diagrams are generated by the equality (8.42).

The functions

$$\Gamma_n(x_1, x_2, \dots, x_n) = \Gamma_n(x_1, x_2, \dots, x_n|\Phi)\big|_{\Phi=0} \tag{8.46}$$

are called vertex Green functions, or simply vertex functions. We can conclude that the vertex functions $\Gamma_n(x_1, x_2, \dots, x_n)$ are described by n-point diagrams with the following properties: **1.** All of them are connected. **2.** None of them have external lines. **3.** All of them are one-particle irreducible.

The effective action $\Gamma[\Phi]$ can be expanded into the functional Fourier series of the form

$$\begin{aligned}\Gamma[\Phi] &= \sum_{n=1}^{\infty} \frac{1}{n!} \int d^4x_1 d^4x_2 \dots d^4x_n \frac{\delta^n \Gamma[\Phi]}{\delta\Phi(x_1)\delta\Phi(x_2)\dots\Phi(x_n)}\bigg|_{\Phi=0} \Phi(x_1)\Phi(x_2)\dots\Phi(x_n) \\ &= \sum_{n=1}^{\infty} \frac{1}{n!} \int d^4x_1 d^4x_2 \dots d^4x_n \, \Gamma_n(x_1, x_2, \dots, x_n)\, \Phi(x_1)\Phi(x_2)\dots\Phi(x_n).\end{aligned} \tag{8.47}$$

Thus, the effective action is the generating functional of the vertex functions.

An arbitrary diagram with external lines is obtained from the diagram without external lines by attaching such lines. Also, an arbitrary disconnected diagram is obtained by composing connected blocks. An arbitrary one-particle reducible diagram is obtained from one-particle irreducible diagrams by connecting them by at least one line. All in all, the diagrams for the vertex functions are the basic blocks used to compose all diagrams. Therefore, the effective action, which is the generating functional of the vertex functions, can be seen as an object of fundamental importance in quantum field theory.

8.6 Loop expansion

Let us consider loop expansion in the functional formalism.

Taking into account the relation (8.21) between the generating functional of the Green functions, $Z[J]$, and the generating functional of the connected Green functions, $W[J]$, and the functional integral representation of $Z[J]$ in (8.7), we can write

$$e^{\frac{i}{\hbar}W[J]} = \int \mathcal{D}\varphi \, e^{\frac{i}{\hbar}\{S[\varphi] + \int d^4x \varphi(x) J(x)\}}, \tag{8.48}$$

where we restored the dependence on the Plank constant $\hbar$.

The functional $W[J]$ is related to the effective action $\Gamma[\Phi]$ as

$$W[J] = \Gamma[\Phi] + \int d^4x\, \Phi(x)J(x), \tag{8.49}$$

where the source $J(x)$ is expressed through the mean field $\Phi(x)$ from the equation

$$J(x) = -\frac{\delta\Gamma[\Phi]}{\delta\Phi(x)}. \tag{8.50}$$

Substituting relation (8.49) into Eq. (8.48), one gets

$$e^{\frac{i}{\hbar}\{\Gamma[\Phi]+\int d^4x\Phi(x)J(x)\}} = \int \mathcal{D}\varphi e^{\frac{i}{\hbar}\{S[\varphi]+\int d^4x\varphi(x)J(x)\}}. \tag{8.51}$$

Now, performing the shift of the integration variable $\varphi \to \Phi + \sqrt{\hbar}\varphi$ and using (8.50) yields the equation for the effective action:

$$e^{\frac{i}{\hbar}\Gamma[\Phi]} = \int \mathcal{D}\varphi\, e^{\frac{i}{\hbar}\left\{S[\Phi+\sqrt{\hbar}\varphi]-\sqrt{\hbar}\int d^4x\varphi(x)\frac{\delta\Gamma[\Phi]}{\delta\Phi(x)}\right\}}. \tag{8.52}$$

To solve this equation, we expand the action $S[\Phi + \sqrt{\hbar}\varphi]$ in the power series in $\hbar$:

$$\begin{aligned} S[\Phi + \sqrt{\hbar}\varphi] &= S[\Phi] + \sum_{n=1}^{\infty} \frac{(\sqrt{\hbar})^n}{n!} \int d^4x_1 d^4x_2 \dots d^4x_n S_n(x_1, x_2, \dots, x_n) \\ &\times \varphi(x_1)\varphi(x_2)\dots\varphi(x_n), \end{aligned} \tag{8.53}$$

where

$$S_n(x_1, x_2, \dots, x_n) = \frac{\delta^n S[\varphi]}{\delta\varphi(x_1)\delta\varphi(x_2)\dots\delta\varphi(x_n)} \tag{8.54}$$

are the classical vertices. Of course, for a theory with a polynomial interaction term, the series (8.53) becomes a finite sum. Let introduce the compact notations

$$\begin{aligned} &\int d^4x_1 d^4x_2 \dots d^4x_n S_n(x_1, x_2, \dots, x_n)\, \varphi(x_1)\varphi(x_2)\dots\varphi(x_n) = S_{,n}[\Phi]\varphi^n \\ &\text{and} \quad \frac{\delta\Gamma[\Phi]}{\delta\Phi(x)} = \Gamma_{,1}[\Phi], \qquad \int d^4x\, \varphi(x)\frac{\delta\Gamma[\Phi]}{\delta\Phi(x)} = \varphi\Gamma_{,1}[\Phi]. \end{aligned} \tag{8.55}$$

Using these terms, one can rewrite relation (8.52) in the form

$$e^{\frac{i}{\hbar}\{\Gamma[\Phi]-S[\Phi]\}} = \int \mathcal{D}\varphi\, e^{\frac{i}{2}S_{,2}\varphi^2 + i\sum_{n=3}^{\infty}\frac{\hbar^{\frac{n}{2}-1}}{n!}S_{,n}\varphi^n - i\hbar^{-\frac{1}{2}}\varphi\left[\Gamma_{,1}[\Phi]-S_{,1}[\Phi]\right]}. \tag{8.56}$$

The effective action $\Gamma[\Phi]$ enters this equation in the combination $\Gamma[\Phi] - S[\Phi] = \bar{\Gamma}[\Phi]$. This means that the functional $\bar{\Gamma}[\Phi]$ contains the quantum correction to the classical action $S[\Phi]$. Writing this correction as a power series in $\hbar$,

$$\bar{\Gamma}[\Phi] = \sum_{k=1}^{\infty} \hbar^k \bar{\Gamma}^{(k)}[\Phi], \tag{8.57}$$

one gets

$$e^{i\sum_{k=1}^{\infty}\hbar^{k-1}\bar{\Gamma}^{(k)}[\Phi]} = \int \mathcal{D}\varphi \, \exp\Big\{\frac{i}{2}S_{,2}\varphi^2 + i\sum_{n=3}^{\infty}\frac{\hbar^{\frac{n}{2}-1}}{n!}S_{,n}\varphi^n - i\sum_{k=1}^{\infty}\hbar^{k-\frac{1}{2}}\varphi\bar{\Gamma}^{(k)}_{,1}[\Phi]\Big\}$$

$$= \int \mathcal{D}\varphi \, \exp\Big\{\frac{i}{\hbar}\Big[S[\Phi+\sqrt{\hbar}\varphi] - S[\Phi] - \sqrt{\hbar}S_{,1}[\Phi]\varphi - \sum_{k=1}^{\infty}\hbar^k\varphi\bar{\Gamma}^{(k)}_{,1}[\Phi]\Big]\Big\}. \tag{8.58}$$

Eq. (8.58) is the basis for the perturbation theory for calculating the effective action in the form of the power series in $\hbar$. The expression in the *r.h.s.* includes the integral over the field φ, depending on the classical mean field Φ, which is also called the background field. The quadratic part $\frac{1}{2}S_{,2}[\Phi]\varphi^2$ of the action defines the two-point function $G(x,y|\Phi)$ as a solution to the equation

$$\int d^4z \, \frac{\delta^2 S[\Phi]}{\delta\Phi(x)\delta\Phi(z)} \, G(z,y|\Phi) = \delta^4(x-y). \tag{8.59}$$

For instance, the boundary conditions can be chosen in such a way that $G(z,y|\Phi)$ is the Feynman propagator. The remaining part of the action with the factors of $\hbar$ corresponds to the interactions. Expanding the exponential into a power series in interaction provides the perturbation series, where the role of the expansion parameter belongs to $\hbar$. Each term of this series can be represented by Feynman diagrams, where both propagators and vertices depend on the mean field Φ. As a result, we arrive at the perturbation series for the calculation of the corrections $\bar{\Gamma}^{(k)}$.

Two observations concerning the structure of the action (8.58) are in order:

1. There are fractional powers of $\hbar$ in the exponent in the *r.h.s.*. However, as we discussed above, the non-zero contributions to the functional integral

$$\int \mathcal{D}\varphi \, e^{\frac{i}{2}S_{,2}[\Phi]\varphi^2}\varphi(x_1)\varphi(x_2)\dots\varphi(x_N) \tag{8.60}$$

appear only for even N. This feature ensures that only integer powers of $\hbar$ enter into the perturbation series.

2. Another comment concerns the term $-\sum_{k=1}^{\infty}\hbar^{n-\frac{1}{2}}\varphi\bar{\Gamma}^{(k)}_{,1}[\Phi]$ in the *r.h.s.* of the relation (8.58). After expanding the exponential into a series, this term leads only to one-particle reducible diagrams. The reason is that this term is linear in φ. In terms of the diagrams, this means there is only one line connecting different parts of the diagram. However, the *l.h.s.* of the relation (8.58) defines the effective action that is described by the one-particle irreducible diagrams only.

Thus, the term under consideration must be canceled with the one-particle reducible diagrams generated by other interaction terms in the action. In principle, such cancelation can be proved explicitly. However, we do not need to do it, since, in order to get (8.58), we performed only the identical transformations. Taking this into account, we can write

$$e^{i\sum_{k=1}^{\infty}\hbar^{l-1}\bar{\Gamma}^{(L)}[\Phi]} = \Big\{\int \mathcal{D}\varphi e^{\frac{i}{\hbar}\left[S[\Phi+\sqrt{\hbar}\varphi]-S[\Phi]-\sqrt{\hbar}S_{,1}[\Phi]\varphi\right]}\Big\}_{\text{one-particle irreducible}}$$
$$= \Big\{\int \mathcal{D}\varphi e^{i\left[\frac{1}{2}S_{,2}\varphi^2+\sum_{n=3}^{\infty}\frac{\hbar^{\frac{n}{2}-1}}{n!}S_{,n}\varphi^n\right]}\Big\}_{\text{one-particle irreducible}}. \tag{8.61}$$

Thus, only one-particle irreducible diagrams should be taken into account in the perturbation theory.

Let us clarify a sense of the parameter k in the *l.h.s.* of (8.61). We saw that an arbitrary L-loop diagram contains the factor $\hbar^{E-1+L}$, where E is a number of the external lines and L is a number of loops. Obviously, the effective action is given by the diagrams without external lines. Hence, each diagram for the effective action contains the factor $\hbar^{k-1}$. This factor is present in the *l.h.s.* of the relation (8.61). Thus, the relation (8.61) defines the effective action as a series in a number of loops in the diagrams, and hence $k \equiv L$ is exactly the number of loops, and the expansion in k is nothing else, but the loop expansion.

For this reason, the power series in $\hbar$ for the effective action is called the loop expansion. It is important to emphasize that the background field Φ in the relation (8.61) is absolutely arbitrary and does not satisfy any equations, except the equation of motion derived from the variation of the effective action.

As an example, consider the calculation of the first (one-loop) quantum correction in the effective action. From the relation (8.61), follows that

$$e^{i\bar{\Gamma}^{(1)}} = \int \mathcal{D}\varphi e^{\frac{i}{2}S_{,2}[\Phi]\varphi^2} = \big[\,\mathrm{Det}\,(S_{,2}[\Phi])\big]^{-\frac{1}{2}};$$
$$\text{therefore,}\qquad \bar{\Gamma}^{(1)}[\Phi] = \frac{i}{2}\log\,\mathrm{Det}\,(S_{,2}[\Phi]) = \frac{i}{2}\,\mathrm{Tr}\,\log(S_{,2}[\Phi]). \tag{8.62}$$

Thus, the effective action in the one-loop approximation can be written as

$$\Gamma^{(1)}[\Phi] = S[\Phi] + \frac{i}{2}\hbar\,\mathrm{Tr}\,\log(S_{,2}[\Phi]). \tag{8.63}$$

For the classical action for scalar theory,

$$S[\varphi] = \int d^4x\Big(-\frac{1}{2}\varphi\Box\varphi - \frac{1}{2}m^2\varphi^2 - V(\varphi)\Big),$$

we have

$$\Gamma^{(1)}[\Phi] = S[\Phi] + \frac{i}{2}\hbar\,\mathrm{Tr}\,\log\big(-\Box - m^2 - V''(\Phi)\big). \tag{8.64}$$

Thus, the calculation of the one-loop effective action is reduced to the calculation of the functional determinant of the differential operators. Let us note that we already discussed such calculations in section 7.7. In general, the one-loop effective action cannot be calculated exactly for an arbitrary background field $\Phi(x)$. Some examples of approximate calculations will be given in Part II.

8.7 Feynman diagrams in theories with spinor fields

Perturbation theory in terms of Feynman diagrams in models with spinor fields is constructed according to the same scheme as for scalar field theory. The free action, which is quadratic in fields, defines the propagators, and the interaction part of the action defines vertices. The generating functional of the spinor Green function was introduced in section 6.4, its functional representation constructed in section 7.4.2 and the propagator in section 7.5 [see Eq. (7.114)].

There are only three specific points to be taken into account when constructing the Feynman diagrams. First, the spinor propagator $S(x, y) = \langle 0|T\hat{\psi}(x)\hat{\bar{\psi}}(y)|0\rangle$ is not a symmetric function in x and y (6.73). Therefore, one should specify the direction in the line associated with the spinor propagator in the diagrams. Such a propagator will be denoted by a dashed line directed from the point y to the point x:

Second, since the fermionic fields $\psi(x)$ and $\bar{\psi}(x)$ are the four-component spinors, the spinor propagator is a 4×4 matrix $S_{AB}(x, y) = \langle 0|T\hat{\psi}_A(x)\hat{\bar{\psi}}_B(y)|0\rangle$ with spinor indices $A, B = 1, 2, 3, 4$. Here $\hat{\psi}_A(x)$ and $\hat{\bar{\psi}}_B(y)$ are free spinor field operators.

Third, the perturbation theory is constructed on the basis of Eqs. (6.78) and (7.108), when we act by the functional derivatives over the sources on the generating functional of free Green functions. Since the derivatives with respect to spinor sources $\eta(x)$ and $\bar{\eta}(x)$ anticommute, the contribution of the spinor loop enters with a negative sign.

In the following subsections, we consider in detail the following two models: the theory of interacting scalar and spinor fields with Yukawa coupling, and spinor electrodynamics.

8.7.1 Feynman diagrams in the theory with Yukawa coupling

The theory under consideration has the real scalar field $\varphi(x)$ and the Dirac spinor field $\psi(x)$. The Yukawa interaction term is

$$S_{int}[\varphi, \psi, \bar{\psi}] = -h \int d^4x \, \varphi \, \bar{\psi}\psi, \tag{8.65}$$

where h is the coupling constant. The perturbation theory for the Green functions

$$\begin{aligned} &G_{nkl}(x_1, x_2, \dots, x_n, y_1, y_2, \dots, y_k, z_1, z_2, \dots, z_l) \\ &= \langle 0|T\hat{\varphi}(x_1)\hat{\varphi}(x_2)\dots\hat{\varphi}(x_n)\hat{\psi}(y_1)\hat{\psi}(y_2)\dots\hat{\psi}(y_k)\hat{\bar{\psi}}(z_1)\hat{\bar{\psi}}(z_2)\dots\hat{\bar{\psi}}(z_l)|0\rangle\,, \end{aligned} \tag{8.66}$$

is constructed on the basis of the relation

$$\begin{aligned} &G_{nkl}(x_1, x_2, \dots, x_n, y_1, y_2, \dots, y_k, z_1, z_2, \dots, z_l) = \\ &\frac{\delta^n}{\delta iJ(x_1)\delta iJ(x_2)\dots\delta iJ(x_n)} \left(\frac{\overrightarrow{\delta}}{\delta i\bar{\eta}(y_k)} \cdots \frac{\overrightarrow{\delta}}{\delta i\bar{\eta}(y_1)} Z[J, \bar{\eta}, \eta] \frac{\overleftarrow{\delta}}{\delta i\eta(z_l)} \cdots \frac{\overleftarrow{\delta}}{\delta i\eta(z_1)} \right) \bigg|_{J,\eta,\bar{\eta}=0}. \end{aligned} \tag{8.67}$$

Here

$$Z[J,\bar{\eta},\eta] = e^{iS_{int}\left[\frac{\delta}{\delta iJ},\frac{\delta}{\delta i\eta},\frac{\delta}{\delta i\bar{\eta}}\right]}\, e^{-\frac{i}{2}JDJ}\, e^{-i\bar{\eta}S\eta}, \tag{8.68}$$

and S_{int} is given by the relation (8.65). Expanding $e^{iS_{int}}$ in the relation (8.68) into a power series, we arrive at the perturbation series for the Green functions (8.66). Each term of this series can be represented by Feynman diagrams with two types of propagators, namely, scalar $D(x,x')$ and spinor $S(y,z)$. The vertex is defined by the action (8.65) and has the graphic presentation

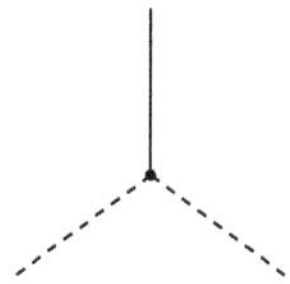

The simplest, one-loop, diagrams are as follows:

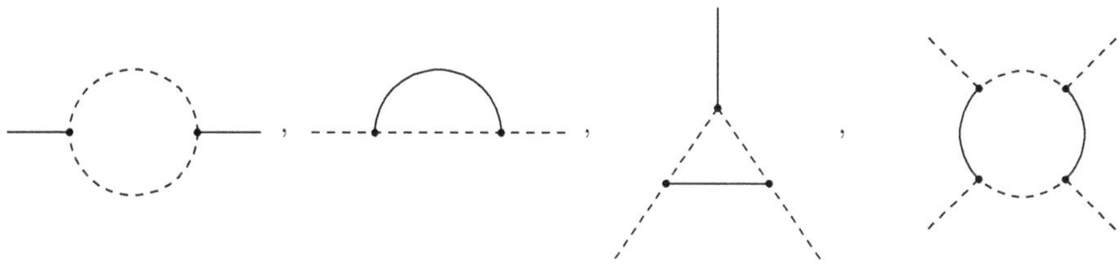

An example of a detailed presentation of the diagram is

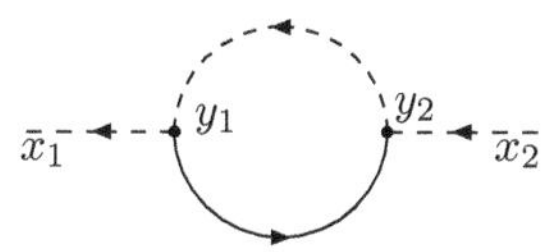

This diagram describes the one-loop correction and is called the fermionic self-energy in the second perturbative order. Its momentum representation is

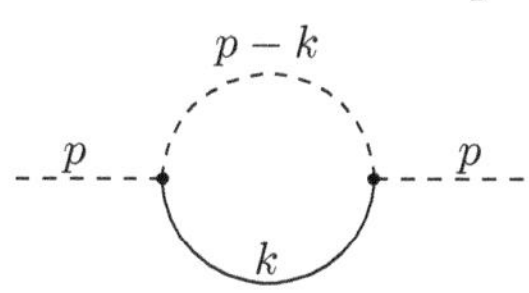

The corresponding analytic expression has the form

$$\begin{aligned} &-\frac{h^2}{2}\frac{(\gamma^\mu p_\mu + mI)_{AC}}{p^2 - m^2 + i\varepsilon}\,\frac{(\gamma^\nu p_\nu + mI)_{DB}}{p^2 - m^2 + i\varepsilon} \\ &\times \int \frac{d^4k}{(2\pi)^4}\,\frac{(\gamma^\alpha k_\alpha + mI)_{CD}}{k^2 - m^2 + i\varepsilon}\,\frac{1}{(p-k)^2 - m^2 + i\varepsilon}. \end{aligned} \tag{8.69}$$

Here I_{AB} is the unit 4×4 matrix in the spinor space; the spinor indices are written down explicitly. Certain simplifications are possible by using the properties of the gamma matrices, before calculating the integral over d^4k.

8.7.2 Feynman diagrams in spinor electrodynamics

The action of spinor electrodynamics has the form

$$S[\psi,\bar{\psi},A] = S_0[\psi,\bar{\psi},A] + S_{int}[\psi,\bar{\psi},A] = \int d^4x\{\mathcal{L}_0^{(spin)} + \mathcal{L}_0^{(em)} + \mathcal{L}_{int}\}, \tag{8.70}$$

where

$$\mathcal{L}_0^{(spin)} = \bar{\psi}i\gamma^\mu\partial_\mu\psi - m\bar{\psi}\psi \tag{8.71}$$

is the Lagrangian of a free spinor field,

$$\mathcal{L}_0^{(em)} = -\frac{1}{4}F^{\mu\nu}F_{\mu\nu} + \frac{1}{2}(\partial_\mu A^\mu)^2 = \frac{1}{2}A^\mu\Box A_\mu \tag{8.72}$$

is the Lagrangian of the free electromagnetic field in the (minimal) Lorentz gauge and

$$\mathcal{L}_{int}(A,\psi,\bar{\psi}) = e\bar{\psi}\gamma^\mu\psi A_\mu \tag{8.73}$$

is the interaction Lagrangian. The parameter e is the coupling constant.

The perturbation theory for the Green functions

$$\begin{aligned} &G_{\mu_1\mu_2\ldots\mu_n}(x_1,x_2,\ldots,x_n,y_1,y_2,\ldots,y_k,z_1,z_2,\ldots,z_l) \\ &= \langle 0|T(\hat{A}_{\mu_1}(x_1)\ldots\hat{A}_{\mu_n}(x_n)\,\hat{\psi}(y_1)\ldots\hat{\psi}(y_k)\,\hat{\bar{\psi}}(z_1)\ldots\hat{\bar{\psi}}(z_l))|0\rangle \end{aligned} \tag{8.74}$$

is constructed on the basis of the generating functional

$$Z[J,\bar{\eta},\eta] = \langle 0|Te^{i\int d^4x\left(\hat{A}_\mu(x)J^\mu(x)+\bar{\eta}(x)\hat{\psi}(x)+\hat{\bar{\psi}}(x)\eta(x)\right)}|0\rangle, \tag{8.75}$$

where $J = J^\mu(x)$ is the source for the electromagnetic field operator.

The generating functional (8.75) can be evaluated using the same methods as for the scalar and spinor field theories. The result is

$$Z[J,\bar{\eta},\eta] = e^{i\int d^4x\mathcal{L}_{int}\left(\frac{\delta}{\delta iJ^\mu(x)},\frac{\delta}{\delta i\eta(x)}\frac{\delta}{\delta i\bar{\eta}(x)}\right)}Z_0[J,\bar{\eta},\eta], \tag{8.76}$$

where

$$Z_0[J,\bar{\eta},\eta] = e^{-\frac{i}{2}\int d^4xd^4y\,J^\mu(x)\,\Delta_{\mu\nu}(x,y)\,J^\nu(y)}\,e^{-i\int d^4xd^4y\,\bar{\eta}(x)S(x,y)\eta(y)}. \tag{8.77}$$

Here $S(x,y)$ and $\Delta_{\mu\nu}(x,y)$ are the propagators of spinor and electromagnetic fields, respectively. The latter is also called the photon propagator; it satisfies the equation

$$\Box_x\Delta_{\mu\nu}(x,y) = \delta^4(x-y)\,\eta_{\mu\nu}\,. \tag{8.78}$$

In the momentum representation, this propagator has the form

$$\tilde{\Delta}_{\mu\nu}(p) = -\frac{\eta_{\mu\nu}}{p^2+i\varepsilon}. \tag{8.79}$$

The Green functions (8.74) are obtained by differentiating the generating functional (8.75) with respect to the sources in the vanishing sources limit. The perturbation theory is constructed by the expansion of the expression

$$e^{i\int d^4x\mathcal{L}_{int}\left(\frac{\delta}{\delta iJ^\mu(x)},\frac{\delta}{\delta i\eta(x)}\frac{\delta}{\delta i\bar{\eta}(x)}\right)} \tag{8.80}$$

into the power series in $\mathcal{L}_{int}$. Each term in this series can be represented by the Feynman diagrams. The theory is characterized by two propagators, $\Delta_{\mu\nu}(x,y)$ and $S(x,y)$, which are indicated by the wavy and the dotted lines, respectively. There is the only one vertex, which is defined by the interaction Lagrangian (8.73):

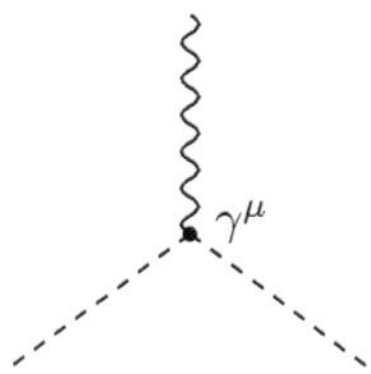

Some examples of diagrams in spinor electrodynamics are as follows:

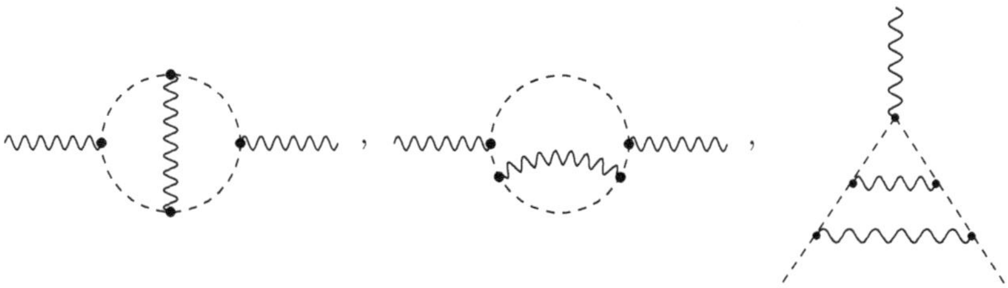

As the simplest example, we consider the diagram

which describes the one-loop correction to the photon propagator. This contribution is called the polarization operator in the second perturbative order. In the momentum representation, this diagram has the form

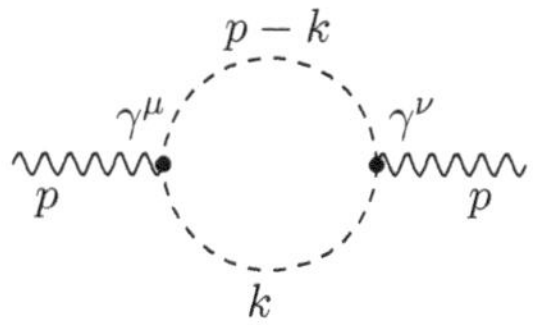

The analytic expression for the corresponding contribution is

$$-\frac{(ie)^2}{2}\frac{1}{(p^2+i\varepsilon)^2}\int\frac{d^4k}{(2\pi)^4}\,\mathrm{tr}\,\frac{\gamma_\mu(\gamma^\alpha k_\alpha+m)(\gamma^\beta[k-p]_\beta+m)\gamma_\nu}{(k^2-m^2+i\varepsilon)([p-k]^2-m^2+i\varepsilon)}. \tag{8.81}$$

Here the negative sign is stipulated by fermionic loop, since the trace is taken according to the standard "bosonic" rule.

Exercises

8.1. Using relation (8.1), calculate, in the theory with the interaction $V(\varphi)\sim\varphi^4$, the Green function $G_2(x_1,x_2)$ in the first and second orders of the perturbation theory. Show the results using Feynman diagrams, and check that the vacuum diagrams are canceled.

8.2. Using Eq. (8.1), calculate, in the theory with the interaction $V(\varphi)\sim\varphi^3$, the two-point Green function in the second order of the perturbation theory. Show the result using Feynman diagrams.

8.3. In the theory with the interaction $V(\varphi) = \lambda\varphi^3$, derive the diagram for the Green function $G_3(x_1, x_2, x_3)$ in the third order of the perturbation theory. After that, transform the result to the momentum representation, and find the contribution of this diagram in coordinate and momentum representations (without calculating the integrals).

8.4. Work out the Feynman rules for scalar electrodynamics in coordinate and momentum representations.

8.5. Develop the perturbation theory in scalar electrodynamics. Write down the one-loop effective action in terms of functional determinants.

8.6. Using the relation (8.68), calculate the Green function

$$G_3(x,y,z) = \langle 0|T(\hat{\varphi}(x)\hat{\psi}(y)\hat{\bar{\psi}}(z))|0\rangle$$

in the third order of perturbation theory. Show the result as a Feynman diagram, and find the contribution of the diagram in coordinate and momentum representations (without calculating the integrals).

8.7. Derive the relation (8.77).

8.8. Using the generating functional (8.76), calculate the Green function

$$G_{\mu\nu}(x_1, x_2) = \langle 0|T(\hat{A}_\mu(x_1)\hat{A}_\nu(x_2))|0\rangle$$

in spinor electrodynamics, in the second order of the perturbation theory. Draw the diagram and find its contribution in the coordinate and momentum representations, without calculation of the integrals.

8.9. Represent the one-loop effective action in spinor electrodynamics in terms of functional determinants.

8.10. Using the basic relation (2.93) for the Dirac gamma matrices, simplify the matrix structure in the expression (8.69).

8.11. Using the relations (4.139) for gamma matrices, calculate the matrix traces in the expression (8.81).

Comments

Constructions and calculations of the Feynman diagrams in various theories are considered in detail in many books on quantum field theory, e.g., in [57], [187], [250], [304], [274] and [345].

The effective action $\Gamma[\Phi]$ was introduced to quantum field theory in the papers by J. Goldstone, A. Salam and S. Weinberg [162], B.S. DeWitt [105] and G. Jone-Lasinio [204].

9

Renormalization

In what follows, we discuss the general theoretical aspects of renormalization in quantum field theory in flat spacetime. Some explicit examples, at the one-loop order, can be found in Part II, where we present calculations in flat and curved backgrounds. On the other hand, these calculations will explicitly or implicitly use general statements which will be discussed in the present chapter.

9.1 The general idea of renormalization

The standard procedure of evaluating the Green functions or effective action is based on the use of perturbation theory. As a result, we meet Feynman diagrams corresponding to the integrals over internal momenta. The integrands typically have some powers of the momenta in the denominator and also possibly in the numerator. One can check that, in all examples analyzed in the previous section, the powers of the momenta in the denominator are not sufficient to provide a convergence of the integrals at large momenta, which is also called the ultraviolet (UV) limit.

Typically, the contributions of relevant Feynman diagrams are UV divergent. The last means that the whole scheme of the perturbative derivation of the Green functions is mathematically ill-defined and hence its output needs an additional procedure to remove divergences and (most relevant) consistently extract the finite expressions of our interest. The purpose of this and subsequent sections is to describe the general aspects of this additional procedure called renormalization.

Let us start by considering a few examples of the divergent Feynman diagrams.

1. In the theory of a real scalar field with the interaction potential $V(\varphi) = \frac{\lambda}{4!}\varphi^4$, the first-order perturbative correction to the two-point vertex function is given by the following diagram:

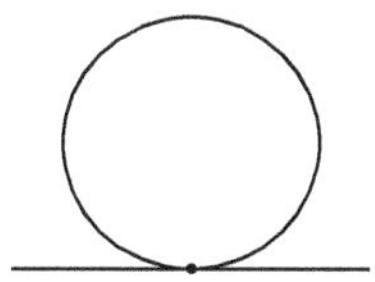

(9.1)

The corresponding contribution is proportional to $D(x-y)|_{y=x} = D(0)$, where

$$D(0) = \int \frac{d^4p}{2(\pi)^4} \frac{1}{p^2 - m^2 + i\epsilon}. \tag{9.2}$$

This integral is quadratically divergent at large momenta. In the same theory, the second-order contribution to the four-point vertex function is given by the diagram

Introduction to Quantum Field Theory with Applications to Quantum Gravity. Iosif L. Buchbinder and Ilya L. Shapiro, Oxford University Press (2021). © Iosif L. Buchbinder and Ilya Shapiro.
DOI: 10.1093/oso/9780198838319.003.0009

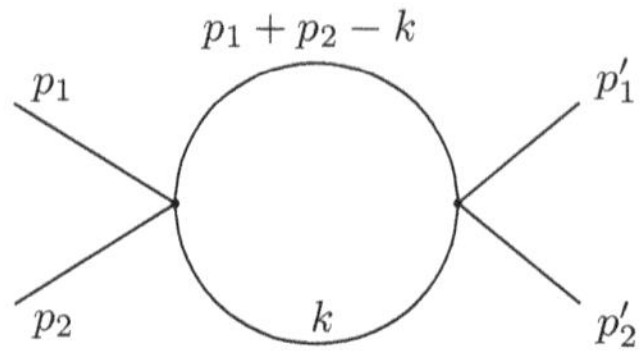

The corresponding contribution includes the integral

$$\int \frac{d^4k}{(2\pi)^4} \frac{1}{(k^2 - m^2 + i\epsilon)([p_1 + p_2 - k]^2 - m^2 + i\epsilon)},$$

which has logarithmic divergences in the UV.

2. In spinor electrodynamics, the second order perturbative correction to the photon Green function is given by the diagram presented in Eq. (8.81). It is easy to see that this integral has both quadratic and logarithmic divergences in the UV limit.

The above examples illustrate the generic situation that Feynman diagrams may be expressed through the divergent integrals over momenta. The general idea of renormalization can be understood as follows. Suppose we deal with a theory with the Lagrangian $\mathcal{L}$. The Green functions in this theory are calculated within the perturbation theory, and some of the Feynman diagrams are divergent. Then, instead of the original theory, we consider a modified version with the same set of fields but with another Lagrangian, $\mathcal{L}+\mathcal{L}_{counter}$, where the additional term $\mathcal{L}_{counter}$ has the structure

$$\mathcal{L}_{counter} = \sum_{L=1}^{\infty} \hbar^L \mathcal{L}^{(L)}_{counter}, \tag{9.3}$$

where the sum is taken over a number of loops. The Lagrangians $\mathcal{L}^{(L)}_{counter}$ are found from the condition that, in each order of the loop expansion, the divergences of all the Feynman diagrams in the theory with the Lagrangian $\mathcal{L} + \mathcal{L}_{counter}$ are canceled. The new terms $\mathcal{L}^{(L)}_{counter}$ in the relation (9.3) are called the counterterms, or the counterterm Lagrangians. The renormalization can be seen as a special way of reconstructing the theory such that all the divergences cancel in the modified version. The counterterms are calculated on the basis of the initial Lagrangian $\mathcal{L}$, utilizing of the loop expansion. The sum $\mathcal{L}+\mathcal{L}_{counter}$ is called the renormalized Lagrangian, and the theory based on this Lagrangian is also called renormalized.

9.2 Regularization of Feynman diagrams

The first step is to learn how to parameterize the divergences or, in other words, how to separate the divergent part of the diagram from the finite part. Such a separation means that we have to redefine the integral corresponding to a Feynman diagram, making it finite. This kind of redefinition is called regularization. The general approach to the regularization of integrals can be described as follows. Let I be an initial divergent Feynman integral. We replace this integral with another integral, $I(\Lambda)$, which depends on one or several parameters Λ. Until Λ does not have the special value Λ_0, the integral $I(\Lambda)$ is finite. In the limit $\Lambda \to \Lambda_0$, the integral $I(\Lambda)$ becomes the initial integral I.

Then, $I(\Lambda)$ is called the regularized Feynman integral, and the corresponding diagram is also called regularized.

It is supposed that the sum of all diagrams in the theory with the modified Lagrangian $\mathcal{L} + \mathcal{L}_{counter}$ is finite at $\Lambda \to \Lambda_0$. This limit is called the removal (or switching off) of regularization. Schematically, the renormalization consists of the two main steps. First, we introduce the regularization and calculate the counterterms. Second, calculate the observables in the modified theory. These observables depend on the regularization parameters, but, in the end, we have to remove the regularization.

As we already mentioned, regularizations can be very different. It is necessary to show that the result of the renormalization procedure satisfies some physically and mathematically sensible requirements, including preserving the relevant symmetries and the control over ambiguities in the finite expressions resulting from the whole renormalization procedure.

For example, if classical theory possesses global or local symmetries, it is natural to find the regularization that preserves these symmetries. Then there is a chance that the renormalized theory with the Lagrangian $\mathcal{L} + \mathcal{L}_{counter}$ will possess the same symmetries. In what follows, we consider some of the frequently used regularization schemes.

9.2.1 Cut-off regularization

The cut-off is the simplest and most natural regularization scheme. If the Feynman integral diverges at the upper limit, it is natural to cut-off this integral at the upper limit by a finite parameter. There are two different types of cut-off procedures for Feynman diagrams. The first of them requires the analytic continuation from Minkowski to Euclidean space. The second one is based on cutting off the integral over three-dimensional (space) momentum. We shall consider only the first of the cut-off regularizations here.

Technically the cut-off regularization is performed as follows. Consider a momentum integral $\int d^4p(\dots)$, where $p = p_\mu$ is a four-dimensional momentum in Minkowski space, with $p^2 = p_0^2 - \mathbf{p}^2$. Let us introduce the imaginary coordinate $p_{(E)0}$ by the relation $p_0 = ip_{(E)0}$. This operation is called the Wick rotation. After it, $p^2 = -p_{(E)0}^2 - \mathbf{p}^2 = -p_{(E)}^2$, where $p_{(E)}^2 = p_{(E)0}^2 + \mathbf{p}^2$ is a square of the vector in the four-dimensional Euclidean space.

After the Wick rotation, we have the integral $i \int d^4p_{(E)}(\dots)\big|_{p_0=ip_{(E)0}}$. It is convenient to make a change of variables corresponding to the four-dimensional spherical coordinates. This change leads to the integral

$$i \int_{\text{angles}} \int_0^\infty dp_{(E)} p_{(E)}^3 (\dots)\Big|_{p_0=ip_{(E)0}}, \tag{9.4}$$

where $\int_{\text{angles}}$ is an integral over the four-dimensional sphere of the unit radius. This integral is always finite (see Part II for details). Regularization by the cut-off means the modification

$$\int_0^\infty dp_{(E)} p_{(E)}^3 (\dots)\Big|_{p_0=ip_{(E)0}} \longrightarrow \int_0^\Omega dp_{(E)} p_{(E)}^3 (\dots)\Big|_{p_0=ip_{(E)0}}, \tag{9.5}$$

where Ω is a finite positive parameter with the dimension of mass, playing the role of the regularization parameter Λ. In the limit $\Omega \to \infty$, the initial integral is restored. However, until Ω is finite, the corresponding Feynman integral is finite too.

As an example, consider the calculation of the integral (9.2). The corresponding diagram is called a tadpole. After the regularization, one gets the integral

$$\int_0^\Omega \frac{dp_{(E)}p_{(E)}^3}{p_{(E)}^2+m^2} = \frac{\Omega^2}{2} - \frac{m^2}{2}\log\Big(1+\frac{\Omega^2}{m^2}\Big). \tag{9.6}$$

There are quadratic Ω^2 and logarithmic $\log\Omega^2$ divergences at the UV limit $\Omega \to \infty$. We shall consider this and some other diagrams in more detail in Part II.

9.2.2 Pauli-Villars regularization

One can think that the origin of the divergences is that the propagators tend to zero too slowly in the UV. If we could show that the propagators tend to zero sufficiently fast at large momenta, there would be no divergences. The original version of the Pauli-Villars regularization is based on speeding up the vanishing of propagators at large momenta. Let us illustrate Pauli-Villars regularization on the example of the integral (9.2). For simplicity, we avoid writing $-i\epsilon$ in the propagators in this subsection.

Consider the propagator $\tilde{D}(p) = \frac{1}{p^2-m^2}$. At large p, the propagator behaves like $\tilde{D}(p) \sim \frac{1}{p^2}$. Let us define the regularized propagator as

$$\tilde{D}_{reg}(p) = \frac{1}{p^2-m^2} + \sum_{i=1}^N \frac{c_i}{p^2-M_i^2}, \tag{9.7}$$

where c_i are numerical coefficients and M_i are auxiliary masses playing the role of the regularization parameters Λ. To explore the behaviour of $\tilde{D}_{reg}(p)$ at large p, we make the transformations

$$\begin{aligned}\tilde{D}_{reg}(p) &= \frac{1}{p^2}\bigg(\frac{1}{1-\frac{m^2}{p^2}} + \sum_{i=1}^N \frac{c_i}{1-\frac{M_i}{p^2}}\bigg) \\ &= \frac{1}{p^2}\bigg\{\Big(1+\sum_{i=1}^N c_i\Big) + \frac{1}{p^2}\Big(m^2+\sum_{i=1}^N c_iM_i^2\Big) + \frac{1}{p^4}\Big(m^4+\sum_{i=1}^N c_iM_i^2\Big) + \dots\bigg\}.\end{aligned} \tag{9.8}$$

One can choose the coefficients c_i and parameters M_i satisfying the conditions

$$1+\sum_{i=1}^N c_i = 0 \qquad \text{and} \qquad m^2+\sum_{i=1}^N c_iM_i^2 = 0. \tag{9.9}$$

In this case, $\tilde{D}_{reg}(p)$ behaves at least as $\frac{1}{p^6}$ at large p^2. Furthermore, if we add the condition $m^4+\sum_{i=1}^N c_iM_i^4 = 0$, the next-order regularized propagator $\tilde{D}_{reg}(p)$ has the asymptotic behavior of at least $\frac{1}{p^8}$ at large p^2, and so on.

In this way, Pauli-Villars regularization enables one to construct the regularized propagator $\tilde{D}_{reg}(p)$ with the desirable behavior at large momenta. When taking the limit $M_i^2 \to \infty$, the $\tilde{D}_{reg}(p)$ tends to the original propagator $\tilde{D}(p)$. Replacing $\tilde{D}(p)$ with $\tilde{D}_{reg}(p)$ provides regularized Feynman integrals. In practice, the number N can be chosen as the minimal one, providing that the relevant Feynman diagrams are finite.

For example, the integral (9.2), with the original $\tilde{D}(p) \sim \frac{1}{p^2}$, is quadratically divergent. Each step of the Pauli-Villars regularization gives an extra $\frac{1}{p^2}$ at large momenta. In the case under consideration, we need to make two steps as in Eq. (9.9). Then, the regularized propagator is

$$\tilde{D}_{reg}(p) = \frac{1}{p^2 - m^2} + \frac{c_1}{p^2 - M_1^2} + \frac{c_2}{p^2 - M_2^2},$$

where $1 + c_1 + c_2 = 0$ and $m^2 + c_1 M_1^2 + c_2 M_2^2 = 0$. Solving this system of two linear equations for the coefficients c_1 and c_2, we arrive at the regularized propagator, which behaves as $\frac{1}{p^6}$ at large momenta. Then the integral $\int d^4p\, \tilde{D}_{reg}(p)$ is UV finite.

9.2.3 Dimensional regularization

The dimensional regularization consists of two main steps. First, the field model should be formulated in the spacetime of an arbitrary dimension n. The quantization is carried out in n dimensions, and the perturbation theory is constructed for an arbitrary n. Second, the Feynman integrals in the momentum representation should be analytically continued from n-dimensional space with an integer n to the complex value of n. It turns out that the UV divergences manifest themselves in the form of simple or multiple poles, such as $\frac{1}{(n-4)^k}$. For example, for a one-loop diagram, there is always $k = 1$. The role of the regularization parameter is played by the spacetime dimension n. While n is different from an integer number, the Feynman integrals are finite. On the other hand, in the limit $n \to 4$, the regularized Feynman integrals reduce to the initial divergent integrals.

The detailed consideration of dimensional regularization for the one-loop calculations is postponed to Part II. Now we only illustrate this regularization on a simple example.

Consider the theory of a real scalar field with the action

$$S[\varphi] = \int d^4x \,\Big\{ \frac{1}{2}\partial^\alpha \varphi \partial_\alpha \varphi - \frac{1}{2} m^2 \varphi^2 - \frac{\lambda}{4!}\varphi^4 \Big\}. \tag{9.10}$$

When formulating the theory in the spacetime of an arbitrary dimension n, the dimensions of the field and parameters change. As we already know, in four dimensions the dimension of the field is $[\varphi] = 1$, the dimension of mass is $[m] = 1$ and the dimension of the coupling constant is $[\lambda] = 0$. Let us write the action of the real scalar field theory in the n-dimensional spacetime with the metric $\eta_{\alpha\beta} = diag(1, -1, -1, \dots, -1)$,

$$S^{(n)}[\varphi] = \int d^n x \,\Big\{ \frac{1}{2}\partial^\alpha \varphi_n \partial_\alpha \varphi_n - \frac{1}{2} m_n^2 \varphi_n^2 - \frac{\lambda_n}{4!}\varphi_n^4 \Big\}. \tag{9.11}$$

Taking into account that the action is dimensionless and the dimension of the coordinates is $[x^\alpha] = -1$, the dimension of the field is $[\varphi_n] = \frac{n-2}{2} = 1 + \frac{n-4}{2}$. For the mass

$[m_n] = [m] = 1$ (i.e., the dimension of mass does not depend on n), and for the coupling constant $[\lambda_n] = 4 - n$. For the sake of quantum calculations, it may be convenient to keep the dimension of the coupling constant corresponding to four-dimensional space-time. To do this, we need an arbitrary parameter μ with the dimension of mass, and redefine the coupling constant as $\lambda_n = \mu^{4-n}\lambda$, where $[\lambda] = 0$. Then the action (9.11) is rewritten in the form

$$S^{(n)}[\varphi] = \int d^n x \left\{ \frac{1}{2}\partial^\alpha \varphi_n \partial_\alpha \varphi_n - \frac{1}{2} m^2 \varphi_n^2 - \frac{\lambda}{4!}\mu^{4-n}\varphi_n^4 \right\}. \tag{9.12}$$

To illustrate the application of dimensional regularization, we consider, once again, the calculation of correction $G_2^{(1)}(x_1, x_2)$ to the two-point Green function in the first order in interaction. Such a correction is given by the diagram in (9.1), in the form

$$G_2^{(1)}(x_1, x_2) = \int d^n x \, iD(x_1, x) \, iD(x, x_2) \, \Gamma_2^{(1)}(x, x),$$

where

$$\Gamma_2^{(1)}(x, x) = -\frac{i\lambda\mu^{4-n}}{2} iD(x, x). \tag{9.13}$$

Here $D(x, x) = D(x, y)\big|_{y-x} = D(0)$. Using momentum representation, one gets

$$\Gamma_2^{(1)} = \frac{\mu^{4-n}}{2}\lambda \int \frac{d^n p}{(2\pi)^n} \frac{1}{p^2 - m^2 + i\varepsilon}. \tag{9.14}$$

Calculation of the integral (9.14) consists of the following steps:

1. Wick rotation: p_0 is replaced by ip_0,

$$p^2 = p_0^2 - p_1^2 - \ldots - p_{n-1}^2 \longrightarrow -\left(p_0 + p_1^2 + \ldots + p_{n-1}^2\right) = -p^2, \tag{9.15}$$

where now p^2 is a square of the vector in the n-dimensional Euclidean space. The integral (9.14) takes the form

$$D(0) = -i \int \frac{d^n p}{(2\pi)^n} \frac{1}{p^2 + m^2}.$$

2. Use the integral representation

$$\frac{1}{p^2 + m^2} = \int_0^\infty ds \, e^{-s(p^2 + m^2)}$$

and calculate the Gaussian integral over momenta. After that, we get

$$\Gamma_2^{(1)} = -\frac{i\mu^{4-n}\lambda}{(4\pi)^{\frac{n}{2}}} \int_0^\infty ds \, s^{-\frac{n}{2}} e^{-sm^2} = -\frac{im^2\lambda}{(4\pi)^{\frac{n}{2}}} \left(\frac{m^2}{\mu^2}\right)^{\frac{n-4}{2}} \Gamma\left(1 - \frac{n}{2}\right), \tag{9.16}$$

where $\Gamma(z)$ is a gamma function.

Relation (9.16) defines the analytic function of the complex variable n in the complex plane, except the points $n_k = 2(k+1)$, $k = 0, 1, 2, \ldots$, on the real axis. Until $n \neq n_k$, the expression $\Gamma_2^{(1)}$ is finite. Let us consider the behaviour of the expression $\Gamma_2^{(1)}$ at $n \to 4$. Using the known feature of the gamma function (see also Part II), $\Gamma(1 - \frac{n}{2}) = \frac{\Gamma(2-\frac{n}{2})}{1-\frac{n}{2}}$, and $\Gamma(z)\big|_{z\to 0} = \frac{1}{z} - C + \mathcal{O}(z)$ (where C is the Euler constant), with the definition $z^u = e^{u \log z}$, we arrive at

$$\Gamma_2^{(1)} = -\frac{im^2\lambda}{(4\pi)^2}\Big[\frac{2}{n-4} - \log\Big(\frac{m^2}{\mu^2}\Big) + C\Big]. \tag{9.17}$$

Thus, the regularized diagram in (9.1) has the pole $\frac{1}{n-4}$. This situation illustrates that the divergences can be absorbed into the poles. Let us note that the detailed calculation of this and another, more complicated diagram, will be given in Part II.

Let us make a few observations about the advantages of dimensional regularization compared with other regularization schemes.

1. The regularization parameter n is introduced into classical action, and we get the regularized functional integral for the generating functional of Green functions. Thus, we can perform formal operations with the functional integral, and the results will be finite until we take the limit $n \to 4$. For this reason, the dimensional regularization is useful for exploring the general properties of quantum field theories.

2. Dimensional regularization preserves those classical symmetries which can be formulated in the n-dimensional spacetime, especially the gauge invariance (except in a few special cases). For this reason, the dimensional regularization is well suited for the application to quantum gauge theories.

3. Dimensional regularization does not introduce massive parameters such as the cut-off in momentum space, or the auxiliary masses as in Pauli-Villars regularization. The unique dimensional parameter is μ. This makes the renormalization in dimensional regularization particularly simple.

On the other hand, it is worth pointing out that the dimensional regularization has some disadvantages.

a. In massless theories, one can expect to meet momentum-space Feynman integrals of the form

$$\int d^n p \; \frac{p_{\mu_1} p_{\mu_2} \cdots p_{\mu_N}}{(p^2)^k}. \tag{9.18}$$

Depending on the relation between N and k, this integral is divergent on the upper or lower limits, at any dimension n. The last means the dimensional regularization may need an additional prescription. For instance, one can define the integrals (9.18) as

$$\lim_{m^2 \to 0} \int d^n p \; \frac{p_{\mu_1} p_{\mu_2} \cdots p_{\mu_N}}{(p^2 + m^2)^k}, \tag{9.19}$$

where all momenta correspond to the n-dimensional Euclidean space after the Wick rotation. The mass dimension of this integral is $n + N - 2k$. Therefore, the result should be proportional to $(m^2)^{\frac{n+N-2k}{2}}$. In the case $n+N>2k$, when $m^2\to 0$, we are automatically getting zero for the integral. With this prescription, all the integrals (9.18) turn out to be equal to zero by definition.

This aspect has an interesting consequence. Consider the expression

$$\partial_{\mu_1}\partial_{\mu_2}\dots\partial_{\mu_N}\delta^n(x)\Big|_{x=0} = i^N\int\frac{d^np}{(2\pi)^n}\,e^{ipx}\,p_{\mu_1}p_{\mu_2}\dots p_{\mu_N}\Big|_{x=0}. \tag{9.20}$$

Using the above prescription, the integral in Eq. (9.20) is defined as

$$\lim_{k\to 0}\lim_{m^2\to 0}\int\frac{d^np}{(2\pi)^n}\,\frac{p_{\mu_1}p_{\mu_2}\dots p_{\mu_N}}{(p^2+m^2)^k}=0. \tag{9.21}$$

Thus, in the framework of the dimensional regularization,

$$\partial_{\mu_1}\partial_{\mu_2}\dots\partial_{\mu_N}\delta^n(x)\big|_{x=0}=0 \qquad \text{for any} \qquad N=0,1,2,\dots. \tag{9.22}$$

In particular, we have $\delta^n(0)=0$. This feature of dimensional regularization leads to significant simplifications, in many cases.

Unlike the cut-off regularization, the dimensional regularization yields only pole divergences. For example, consider the diagram (9.1) in the massless case. The result is given by the relation (9.14), where $m=0$. The corresponding momentum integral has a form similar to that of integral (9.18), and the quadratically divergent term Ω^2 is impossible. Indeed, the absence of quadratic divergences is the general feature of dimensional regularization.

b. There may be a significant problem in the application of dimensional regularization to fermionic theories. As we know, the formulation of such theories includes the gamma matrices which are defined by the Clifford algebra (2.93) with the Minkowski metric. In the framework of dimensional regularization, the basic relation for the gamma matrices should be written as $\gamma^\mu\gamma^\nu+\gamma^\nu\gamma^\mu=2\eta^{\mu\nu}I$, where I is the unit 4×4 matrix and $\eta^{\mu\nu}=diag\big(1,-1,-1,\dots,-1\big)$ is the n-dimensional Minkowski space metric. In some fermionic theories, we also need to introduce the matrix γ_5, which is defined by Eq. (4.40) using $\varepsilon_{\alpha\beta\gamma\delta}$. However, the symbol $\varepsilon_{\alpha\beta\gamma\delta}$ is well defined only in four dimensions. As a result, in order to define a matrix analogous to γ_5, we need an additional prescription. For example, one can take for arbitrary n the same definition (4.40) as for the case $n=4$. Another possibility is to define $\gamma_5=\frac{i}{n!}\varepsilon_{\alpha_1\alpha_2\dots\alpha_n}\gamma^{al_1}\gamma^{\alpha_2}\dots\gamma^{\alpha_n}$. It turns out that different versions of the matrix γ_5 may cause an ambiguity.

9.3 The subtraction procedure

In this section, we will describe a procedure allowing us to separate a regularized Feynman diagram (or, equivalently, integral) into two parts. One of them is finite in the UV limit after removal of the regularization, and the other one is divergent.

Let us consider the generic field model with the Lagrangian

$$\begin{aligned}\mathcal{L} &= \frac{1}{2}\varphi_aK_{ab}\varphi_b+\overline{\psi}_i(i\gamma^\mu\partial_\mu-m_i)\psi_i+\overline{\psi}_i(\Gamma_{ija}\psi_j)\varphi_a\\ &\quad+\lambda_{abcd}\varphi_a\varphi_b\varphi_c\varphi_d+g^\mu_{abc}\varphi_a\varphi_b\partial_\mu\varphi_c,\end{aligned} \tag{9.23}$$

where $\varphi_a \equiv (\varphi^I, A^p_\mu)$ is a set of bosonic fields (scalars and vectors) and ψ_i is a set of fermionic fields. Γ_{ija}, λ_{abcd} and $g^\mu{}_{abc}$ are coupling constants. For example, one can identify the following:

$\Gamma_{ija} \sim h$: Yakawa-type interaction of spinor-scalar coupling,

$\Gamma_{ija} \sim h\gamma_5$: Yakawa-type interaction with pseudo-scalar coupling,

$\Gamma_{ija} \sim e\gamma^\mu$: electromagnetic interaction in quantum electrodynamics,

$\lambda_{abcd}\varphi_a\varphi_b\varphi_c\varphi_d \sim \lambda\varphi^4$: scalar self-interaction, $\lambda_{abcd}\varphi_a\varphi_b\varphi_c\varphi_d \sim g^2 f^{mpq} f^{mrs} A^p{}_\mu A^q{}_\nu A^{r\mu} A^{s\nu}$: interaction in Yang-Mills theory, $g^\mu{}_{abc}\varphi_a\varphi_b\partial_\mu\varphi_c \sim ie\varphi^* A^\mu \partial_\mu \varphi$: interaction in scalar electrodynamics, $g^\mu{}_{abc}\varphi_a\varphi_b\partial_\mu\varphi_c \sim g f^{mpq}\partial_\mu A^m_\nu A^{p\mu} A^{q\nu}$: interaction in Yang-Mills theory.

The operator K_{ab} may have the following basic structures:

$K_{ab} \sim -(\Box + M^2)$: in scalar field theory,

$K_{ab} \sim (\eta_{\mu\nu}\Box)$: in massless vector field theory in the minimal Lorentz gauge.

Indeed, there may be more complicated operators, e.g., in theories with interacting bosonic and fermionic fields, or in gauge models with more complicated gauge conditions. The reader can find some examples of this kind in the last chapters of Part II, which is devoted to quantum gravity models.

The propagator of the bosonic field $D_{ab}(x, y)$ is identified by a solid line:

$$D_{ab}(p) = \frac{\kappa_{ab}}{p^2 - M^2} = \text{————} \tag{9.24}$$

where $M^2 = 0$, $\kappa_{ab} = -\eta_{\mu\nu}$ for the (Abelian) vector field, and $\kappa_{ab} = 1$ for a scalar field The propagator of the fermionic field is $S_{ij}(x, y)$ and the corresponding line is dotted:

$$S(p) = \frac{i(\gamma^\mu p_\mu + m)}{p^2 - m^2} = \text{- - - - - - - - -} \tag{9.25}$$

It is clear that, at large momenta p, these propagators behave like

$$D(p) \sim \frac{1}{p^2} \qquad \text{and} \qquad S(p) \sim \frac{1}{p}.$$

9.3.1 The substraction procedure: one-loop diagrams

Let us start by considering a one-loop correction to the fermionic propagator, which is called the fermionic mass (or self-energy) operator:

$\Sigma^{(1)}(k) =$

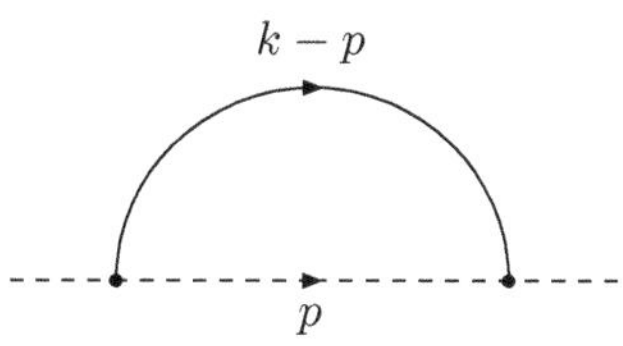

$\sim \int d^4p\, \Gamma\, S(p)\, D(k-p)\, \Gamma.$ (9.26)

Assuming that the cut-off regularization is used, the corresponding Feynman integral is finite, and one can perform some operations with it without breaking its finiteness.

Eq. (9.26) can be written as

$$\int d^4p\, Q(p,k), \qquad \text{where} \qquad Q(p,k) \sim \Gamma\, S(p) D(k-p)\, \Gamma\,.$$

In the limit $p^2 \to \infty$, we meet $Q(k,p) \sim \frac{1}{p^3}$. Thus, (9.26) becomes linearly divergent when removing the regularization. Let us rewrite the integral in the form

$$\int d^4p\, Q(p,k) \;=\; \int d^4p \Big\{ Q(p,k) - Q(p,0) - \frac{\partial Q(p,k)}{\partial k^\mu}\Big|_{k=0} k^\mu \Big\} + \int d^4p \Big\{ Q(p,0) + \frac{\partial Q(p,k)}{\partial k^\mu}\Big|_{k=0} k^\mu \Big\}. \tag{9.27}$$

Consider the behavior of each term in the integrands at large Euclidean p^2 (we assume the Wick rotation every time when use the term "large p^2" or a similar term). It is easy to see that

$$\begin{aligned} Q(p,0) \;&\sim\; \frac{\gamma p + m}{p^2 - m^2}\,\frac{1}{p^2 - M^2} \;\sim\; \frac{1}{p^3}, \\ \frac{\partial Q(p,k)}{\partial k^\mu}\Big|_{k=0} k^\mu \;&\sim\; \frac{\gamma p + m}{p^2 - m^2}\,\frac{2p_\mu k^\mu}{(p^2 - M^2)^2} \;\sim\; \frac{1}{p^4}, \\ Q(p,k) - Q(p,0) - \frac{\partial Q(p,k)}{\partial k^\mu}\Big|_{k=0} k^\mu \;&=\; \frac{\gamma p + m}{p^2 - m^2}\Big[\frac{1}{(k-p)^2 - M^2} \\ &- \frac{1}{p^2 - M^2} - \frac{2(kp)}{(p^2 - M^2)^2}\Big] \;\sim\; \frac{1}{p^5} + \dots\,. \end{aligned}$$

All in all, we have

$$\int d^4p \Big\{ Q(p,k) - Q(p,0) - \frac{\partial Q(p,k)}{\partial k^\mu}\Big|_{k=0} k^\mu \Big\} \sim \int dp\, p^3 \Big(\frac{1}{p^5} + \frac{1}{p^6} + \dots \Big). \tag{9.28}$$

The integral in the *r.h.s.* is finite in the upper limit after removal of the regularization.

Denoting

$$\Sigma_R^{(1)}(k) \;=\; \int d^4p \,\Big\{ Q(p,k) - Q(p,0) - \frac{\partial Q(p,k)}{\partial k^\mu}\Big|_{k=0} k^\mu \Big\}, \tag{9.29}$$

we see that the quantity $\Sigma_R^{(1)}(k)$ is finite. Thus, we can state that

$$\begin{aligned} &\Sigma^{(1)}(k) = \Sigma_{\text{div}}^{(1)}(k) + \Sigma_R^{(1)}(k), \\ \text{where} \qquad &\Sigma_{\text{div}}^{(1)}(k) \;=\; \int d^4p \,\Big\{ Q(p,0) + \frac{\partial Q(p,k)}{\partial k^\mu}\Big|_{k=0} k^\mu \Big\} \end{aligned} \tag{9.30}$$

is divergent after the removal of regularization. Let us introduce the quantities

$$\int d^4p Q(p,0) \;=\; i\delta m^{(1)}, \qquad \int d^4p\, k_\mu \frac{\partial Q(p,k)}{\partial k^\mu}\Big|_{k=0} \;=\; ik_\mu \delta t^{(1)\mu}, \tag{9.31}$$

which are k-independent and divergent after removal of regularization. Then

$$\Sigma^{(1)}(k) = \Sigma_R^{(1)}(k) + ik^\mu \delta t_\mu{}^{(1)} + i\delta m^{(1)}. \qquad (9.32)$$

The procedure of subtracting the first terms of expansion from the momentum-dependent Feynman integral is called the subtraction procedure. The example considered above shows that the subtraction procedure may lead to the modified Feynman integral that is finite after regularization is removed.

Consider the one-loop corrections to the two-point function for a bosonic field (also called the one-loop polarization operator, or the mass operator, depending on whether the field is massless or massive):

$\Pi^{(1)}(k) =$ 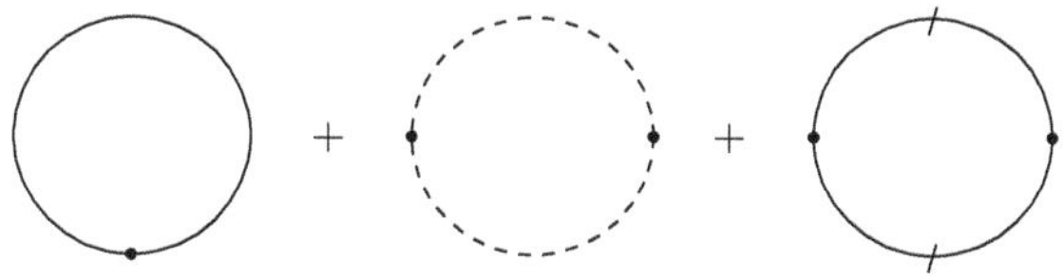 (9.33)

where 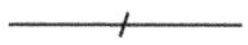 $= \partial_\mu D(x-y) = \int \frac{d^4p}{(2\pi)^4} e^{ip(x-y)} \frac{ip_\mu}{p^2 - M^2}$. (9.34)

Here a crossed line means that there is a derivative acting on the propagator. The first diagram in (9.33) originates from the $\lambda\varphi^4$-interaction, the second diagram from $\overline{\psi}\psi\varphi$-interaction and the last diagram from the $g^\mu\varphi\varphi\partial_\mu\varphi$-interaction. Let consider these diagrams in detail.

1. Consider the diagram the full version can be seen in (9.1)

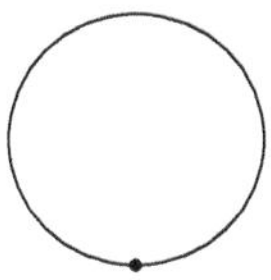 $\sim \int d^4p\, D(p)\,.$

Let us apply the subtraction procedure to this quadratically divergent expression. Since the integral under consideration does not depend on the external momenta, it should be treated as a pure divergence. From the viewpoint of the subtraction procedure, there is no finite part. Thus,

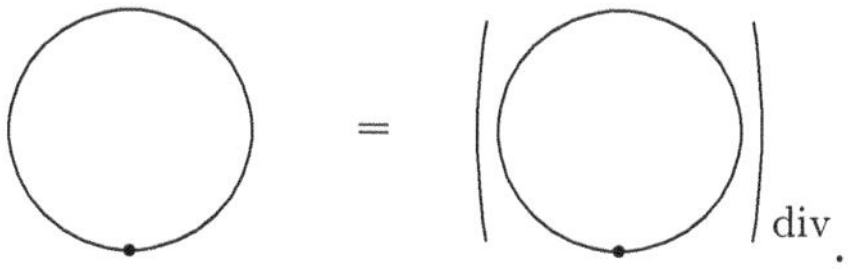

2. For the diagram,

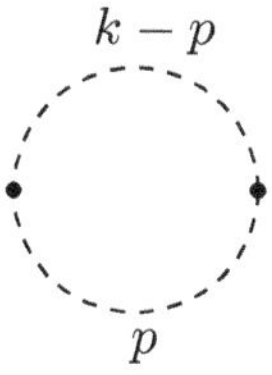

$\sim \int d^4p \Gamma S(p)\, \Gamma\, S(k-p)\,.$

At large momenta, this integral looks like $\int dp\,p$: it is quadratically divergent. After the first subtraction, we obtain

$$S(k-p)-S(-p) = \frac{(k-p)\gamma+m}{(k-p)^2-m^2} - \frac{-p\gamma+m}{p^2-m^2} \sim \frac{1}{p^2}. \tag{9.35}$$

The corresponding Feynman integral at large momentum looks like $\int dp$ and is still formally divergent. After the second subtraction, one gets

$$S(k-p)-S(-p) - \frac{\partial S(k-p)}{\partial k_\mu}\bigg|_{k=0} k_\mu \sim \frac{1}{p^3}+\frac{1}{p^4}+\dots\,. \tag{9.36}$$

The new integral $\int \frac{dp}{p}$ diverges logarithmically. After the third subtraction, we obtain

$$S(k-p)-S(-p) - \frac{\partial S(k-p)}{\partial k_\mu}\bigg|_{k=0} k_\mu - \frac{1}{2}\frac{\partial^2 S(k-p)}{\partial k_\mu \partial k_\nu)}\bigg|_{k=0} k_\mu k_\nu \sim \frac{1}{p^4}+\frac{1}{p^5}+\dots.$$

Finally, the new integral behaves in the UV as $\int \frac{dp}{p^2}$ and converges at the upper limit. Thus, to make this diagram finite, one has to perform three subtractions.

3. Consider the diagram

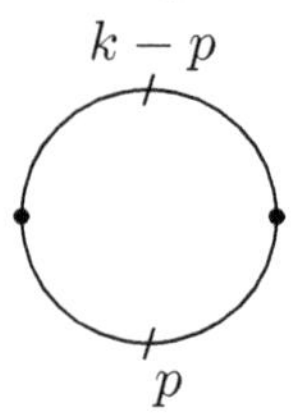

$$\sim \quad \int d^4p\, p_\mu D(p)(k-p)_\nu D(k-p)\,.$$

The corresponding integral is $\int dpp$, which diverges quadratically. We saw that each subtraction adds an extra power of momentum in the denominator of the Feynman integral. Thus, in the case under consideration, three subtractions are expected.

Assuming the three subtractions, we arrive at the result

$$\begin{aligned}
\Pi^{(1)}(k) &= \int d^4p \Big\{ \tilde{Q}(p,k) - \tilde{Q}(p,0) - \frac{\partial \tilde{Q}(p,k)}{\partial k_\mu}\bigg|_{k=0} k_\mu - \frac{1}{2}\frac{\partial^2 \tilde{Q}(p,k)}{\partial k_\mu \partial k_\nu}\bigg|_{k=0} k_\mu k_\nu \Big\} \\
&\quad + \int d^4p \left\{ \tilde{Q}(p,0) + \frac{\partial \tilde{Q}(p,k)}{\partial k_\mu}\bigg|_{k=0} k_\mu + \frac{1}{2}\frac{\partial^2 \tilde{Q}(p,k)}{\partial k_\mu \partial k_\nu}\bigg|_{k=0} k_\mu k_\nu \right\} \\
&= \Pi^{(1)}_R(k) + \Pi^{(1)}_{\text{div}}(k).
\end{aligned} \tag{9.37}$$

Here $\Pi^{(1)}_R(k)$ is finite after the removal of the regularization, and $\Pi^{(1)}_{\text{div}}(k)$ is divergent. Introducing the new notations

$$\begin{aligned}
&\int d^4p\, \tilde{Q}(p,0) = \frac{i}{2}\delta M^{(1)2}, \qquad \int d^4p \frac{\partial \tilde{Q}(p,k)}{\partial k_\mu}\bigg|_{k=0} = i\delta l_\mu{}^{(1)}, \\
&\int d^4p \frac{\partial^2 \tilde{Q}(p,k)}{\partial k_\mu \partial k_\nu}\bigg|_{k=0} = i\delta K^{(1)}_{\mu\nu},
\end{aligned} \tag{9.38}$$

it is evident that $\delta l_\mu{}^{(1)} = 0$, since, in the bosonic sector, there is no distinguishable vector from which it can be constructed. Thus,

$$\Pi^{(1)}(k) = \Pi_R^{(1)}(k) + \frac{i}{2}\delta M^{(1)2} + \frac{1}{2} i\delta K_{\mu\nu}^{(1)} k^\mu k^\nu .$$

Here $\Pi_R^{(1)}(k)$ is a finite part; the quantities $\delta M^{(1)}$ and $\delta K_{\mu\nu}^{(1)}$ are k-independent and will be divergent after the removal of the regularization.

The next step is to consider the one-loop correction to the classical vertex $\overline{\psi}\Gamma\psi\varphi$. This correction is given by the diagram

$$\Gamma^{(1)}(q,k) \quad =$$

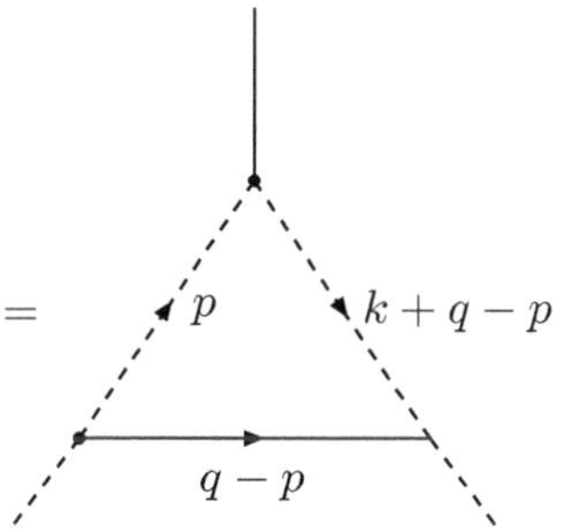

Here

$$\Gamma^{(1)}(q,k) \sim \int d^4p \Gamma\, S(p)\, \Gamma S(-p+q+k)\, \Gamma\, D(q-p). \tag{9.39}$$

At large p, one meets the logarithmically divergent integral $\int \frac{dp}{p}$. To obtain the finite integral, it is sufficient to add one extra power of momentum in the denominator; hence, we need to fulfill only one subtraction,

$$\begin{aligned}\Gamma^{(1)}(q,k) \sim & \int d^4p \left\{\Gamma\, S(p)\, \Gamma\, S(-p+k+q)\, \Gamma\, D(q-p) - \Gamma\, S(p)\, \Gamma\, S(-p)\, \Gamma\, D(-p)\right\} \\ & + \int d^4p \Gamma S(p) \Gamma S(-p) \Gamma D(-p).\end{aligned}$$

We denote

$$\begin{aligned}\Gamma_R^{(1)}(q,k) &\sim \int d^4p \left\{\Gamma S(p) \Gamma S(k-p+q) \Gamma D(k-p) - \Gamma S(p) \Gamma S(-p) \Gamma D(-p)\right\}, \\ i\delta\Gamma^{(1)} &\sim \int d^4p \Gamma S(p) \Gamma S(-p) \Gamma D(-p).\end{aligned} \tag{9.40}$$

Thus,

$$\Gamma^{(1)}(q,k) = \Gamma_R^{(1)}(q,k) + i\delta\Gamma^{(1)}.$$

The quantity $\Gamma_R^{(1)}(q,k)$ is finite and momentum dependent, while $\delta\Gamma^{(1)}$ is independent on external momenta and is divergent after the removal of the regularization.

Now we consider the one-loop correction to the classical vertex $\lambda_{abcd}\varphi_a\varphi_b\varphi_c\varphi_d$. This contribution is given by the diagram

$\tilde{\Gamma}^{(1)}(q,k) =$ 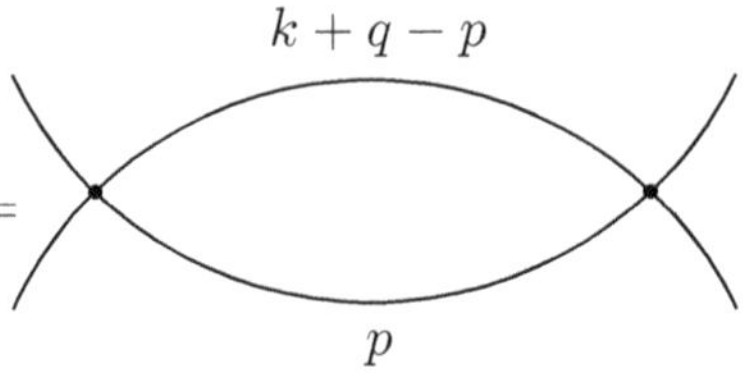

$= \int d^4p\, D(p)\, D(k+q-p)\,.$

In the UV limit, we get the integral $\int \frac{dp}{p}$, which has logarithmic divergence. To get the finite integral, we need one subtraction, which gives

$$\tilde{\Gamma}(q,k) \sim \int d^4pD(p)[D(k+q-p) - D(-p)] + \int d^4pD(p)D(-p). \qquad (9.41)$$

Hence

$$\tilde{\Gamma}^{(1)}(q,k) = \tilde{\Gamma}_R^{(1)}(q,k) + i\delta\lambda^{(1)}, \qquad (9.42)$$

where we denote

$$i\delta\lambda^{(1)} \sim \int d^4pD(p)D(-p). \qquad (9.43)$$

The quantity $\tilde{\Gamma}_R^{(1)}(q,k)$ is finite, while $\delta\lambda^{(1)}$ is independent on external momenta and divergent after the removal of the regularization.

Other one-loop diagrams can be analyzed similarly. Using the subtraction procedure, we can represent any regularized one-loop diagram as a sum of two Feynman integrals. One of them is finite after the removal of the regularization, and another is divergent. The divergent terms are always polynomial in momenta.

9.3.2 The Substraction procedure, and one-loop counterterms

We have shown that any one-loop Feynman diagram can be written as the sum of a finite Feynman integral and the divergent integral. The general prescription for obtaining a finite contribution from a Feynman diagram consists of writing the integral as the sum of the finite and the divergent parts and then throwing out the divergent part. As a result, we get a finite result for any Feynman diagram. Here we reformulate this procedure in terms of the Lagrangian.

Assume we have a theory with the classical Lagrangian

$$\mathcal{L} = \mathcal{L}(\varphi, \psi, \overline{\psi}, M, m, \Gamma, \lambda, g^\mu), \qquad (9.44)$$

where all indices are omitted. Consider another Lagrangian,

$$\begin{aligned} \mathcal{L}_R^{(1)}\ (\varphi_R, \psi_R, \overline{\psi}_R, M_R, m_R, \Gamma_R, \lambda_R, g_R^\mu) &= \mathcal{L}(\varphi_R, \psi_R, \overline{\psi}_R, M_R, m_R, \Gamma_R, \lambda_R, g_R^\mu) \\ &- \frac{1}{2}\varphi_R(\delta M^{(1)2} - \delta K^{(1)\mu\nu}\partial_\mu\partial_\nu)\varphi_R - \overline{\psi}_R(\delta m^{(1)} - t^{(1)\mu}\partial_\mu)\psi_R \\ &- \overline{\psi}_R\delta\Gamma^{(1)}\psi_R\varphi_R - \delta\lambda^{(1)}\varphi_R\varphi_R\varphi_R\varphi_R\ + \ \dots . \end{aligned} \qquad (9.45)$$

Here $\mathcal{L}(\varphi_R, \psi_R, \overline{\psi}_R, M_R, m_R, \Gamma_R, \lambda_R, g_R^\mu)$ is the Lagrangian (9.44), written in terms of the new fields φ_R, ψ_R, $\overline{\psi}_R$, and new parameters M_R, m_R, Γ_R, λ_R and g_R^μ. The dots indicate the terms which are not essential for removing divergences, and the quantities $\delta M^{(1)}$, $\delta K^{(1)\mu\nu}$, $\delta m^{(1)}$, $\delta t^{(1)\mu}$, $\delta\Gamma^{(1)}$ and $\delta\lambda^{(1)}$ were introduced in section 9.3.1.

Using the new Lagrangian (9.45), one can calculate the same one-loop diagrams as in section 9.3.1 and take into account that $\delta M^{(1)}$, $\delta k^{(1)\mu\nu}$, $\delta m^{(1)}$, $t^{(1)\mu}$, $\delta\Gamma^{(1)}$ and $\delta\lambda^{(1)}$ are taken in the one-loop approximation $\mathcal{O}(\hbar)$ and therefore should be regarded as interactions.

In what follows, we consider a few clarifying examples. Let us begin with the one-loop correction to the fermionic propagator,

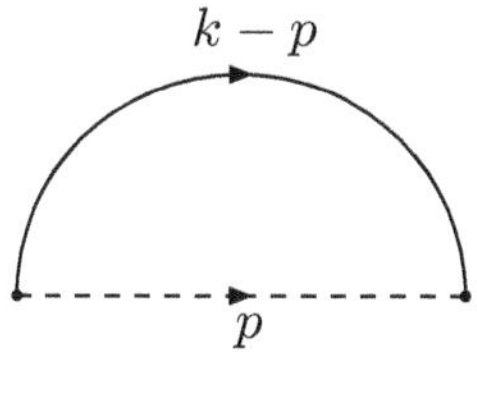

$$- i\delta m^{(1)} - it^{(1)\mu} k_\mu \,.$$

One can remember the relation which was obtained in section 9.3.1,

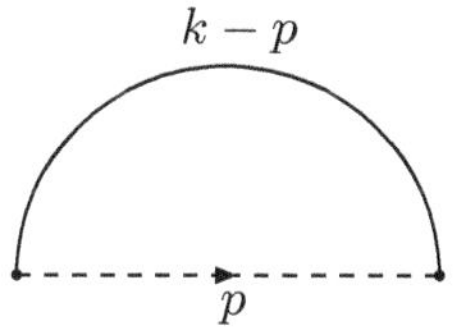

$$= \Sigma_R^{(1)}(k) + i\delta m^{(1)} + i\delta t^{(1)\mu} k_\mu \,.$$

Therefore, we arrive at the result

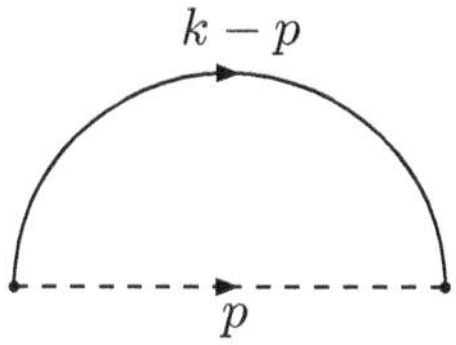

$$- i\delta m^{(1)} - it^{(1)\mu} k_\mu = \Sigma_R^{(1)}(k). \tag{9.46}$$

We see that the one-loop correction to fermionic propagator in the theory with the Lagrangian $\mathcal{L}_R^{(1)}$ (9.45) is automatically finite.

Consider the one-loop correction to the bosonic propagator in the theory with the Lagrangian (9.45),

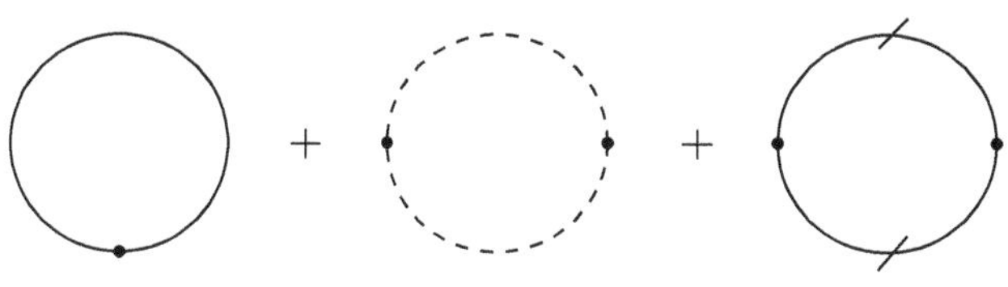

$$- i\delta M^{(1)} - ik^{(1)\mu\nu} k_\mu k_\nu.$$

Using the results from section 9.3.1, we obtain the one-loop correction

$$\Pi_R^{(1)}(k) + \delta M^{(1)} + i\delta K^{(1)\mu\nu} k_\mu k_\nu - i\delta M^{(1)} - ik^{(1)\mu\nu} k_\mu k_\nu = \Pi_R^{(1)}(k). \tag{9.47}$$

to bosonic propagator in the theory with the Lagrangian $\mathcal{L}_R^{(1)}$ (9.45), which is automatically finite.

The diagram for the one-loop correction to the classical vertex $\overline{\psi}\Gamma\psi\varphi$ in the theory (9.45) has the form

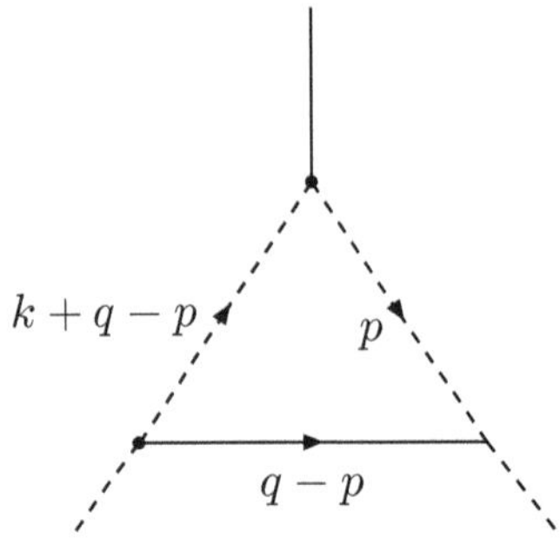

$$= \quad \Gamma_R^{(1)}(q,k) \; + \; i\delta\Gamma^{(1)} \; - \; i\delta\Gamma^{(1)} \; = \; \Gamma_R^{(1)}(q,k). \tag{9.48}$$

This contribution is automatically finite.

Consider the one-loop correction to the classical vertex $\lambda\varphi^4$ in the theory with the Lagrangian $\mathcal{L}_R^{(1)}$ (9.45). One gets

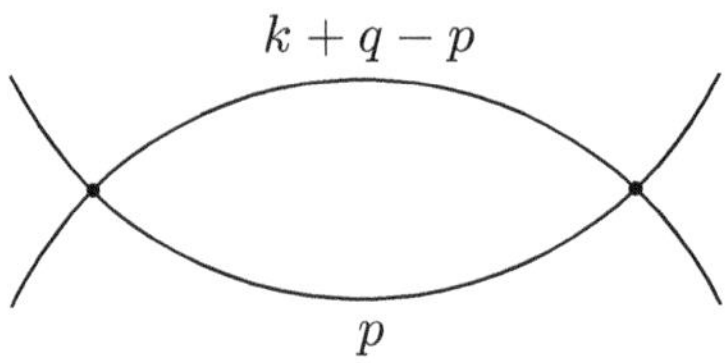

$$= \quad \tilde{\Gamma}_R^{(1)}(q,k) + i\delta\lambda^{(1)} - i\delta\lambda^{(1)} \; = \; \lambda_R^{(1)}(q,k). \tag{9.49}$$

The contribution is also automatically finite.

Usually, the following terminology is used. The Lagrangian $\mathcal{L}_R^{(1)}$ leading to finite one-loop diagrams is called the (one-loop) renormalized Lagrangian. The difference

$$\begin{aligned}\Delta\mathcal{L} &= \mathcal{L}_R^{(1)}(\varphi_R,\psi_R,\overline{\psi}_R,M_R,m_R,\Gamma_R,\lambda_R,g_R^\mu)\\ &\quad - \mathcal{L}(\varphi_R,\psi_R,\overline{\psi}_R,M_R,m_R,\Gamma_R,\lambda_R,g_R^\mu)\end{aligned} \tag{9.50}$$

is called the counterterm. By construction, counterterms depend on regularization parameters and cancel the divergent parts of diagrams, which are generated by the initial Lagrangian $\mathcal{L}$. Furthermore, φ_R, ψ_R, $\overline{\psi}_R$ are called renormalized fields, and M_R, m_R, Γ_R, λ_R, g^μ are renormalized parameters. It is evident that, to find the renormalized Lagrangian $\mathcal{L}_R$, one has to find all necessary counterterms.

Taking the theory with $K = -(\Box + M^2)$, consider $\delta K^{(1)\mu\nu}\partial_\mu\partial_\nu$. Due to the Lorentz invariance, the unique possibility is $\delta K^{(1)\mu\nu} = -\delta z_1^{(1)}\eta^{\mu\nu}$ with a constant $\delta z_1^{(1)}$. Then

$$-\frac{1}{2}\varphi_R\Box\varphi_R + \frac{1}{2}\delta z_1^{(1)}\varphi_R\eta^{\mu\nu}\partial_\mu\partial_\nu\varphi_R = -\frac{1}{2}(1+\delta z_1^{(1)})\varphi_R\Box\varphi_R = -\frac{1}{2}z_1^{(1)}\varphi_R\Box\varphi_R.$$

Thus, adding the counterterms $-\frac{1}{2}\varphi_R\delta K^{(1)\mu\nu}\partial_\mu\partial_\nu\varphi_R$ can be seen as being equivalent to replacing the initial field φ with the new field φ_R, where $\varphi = z_1^{(1)\frac{1}{2}}\varphi_R$. This relation is called the renormalization of the bosonic field. The coefficient $z_1^{(1)}$ is the one-loop renormalization constant of the bosonic field.

Next, let us consider the expression

$$-\frac{1}{2}\varphi_R M_R^2 \varphi_R - \frac{1}{2}\varphi_R \delta M^{(1)2}\varphi_R \tag{9.51}$$

in the Lagrangian $\mathcal{L}_R$ and the expression $-\frac{1}{2}\varphi M^2 \varphi$ in the initial $\mathcal{L}$. Assuming that $\delta M^{(1)2} = \delta \tilde{z}_M^{(1)} M_R^2$, we get the relation

$$-\frac{1}{2}\varphi_R M_R^2 \varphi_R - \frac{1}{2}\varphi_R \delta M^{(1)2}\varphi_R = \frac{1}{2}\varphi_R \Big[1 + \delta \tilde{z}_M^{(1)}\Big] M_R^2 \varphi_R.$$

Performing the transformation of the bosonic field $\varphi = z_1^{(1)\frac{1}{2}}\varphi_R$ and replacing M^2 by $z_M^{(1)} M_R^2$ with $z_M^{(1)} = 1 + \delta z_M^{(1)}$, we obtain

$$\frac{1}{2}\varphi M^2 \varphi = \frac{z_1^{(1)} z_M^{(1)}}{2}\varphi_R M_R^2 \varphi_R.$$

The quantity $\delta z_M^{(1)}$ is found from the equation $z_1^{(1)} z_M^{(1)} = 1 + \delta \tilde{z}_M^{(1)}$, providing $\delta z_M^{(1)} = \delta \tilde{z}_M^{(1)} + \delta z_1^{(1)}$. The expression $\frac{1}{2}\varphi_R M_R \varphi_R - \frac{1}{2}\varphi_R M^{(1)2}\varphi_R$ can be rewritten as $\frac{1}{2} z_1^{(1)} z_M^{(1)} \varphi_R M_R \varphi_R$. The last expression can be obtained from the term $\frac{1}{2}\varphi M^2 \varphi$ in the initial Lagrangian by the transformations $\varphi = z_1^{(1)\frac{1}{2}}\varphi_R$ and $M^2 = z_M^{(1)} M_R^2$. The quantity $z_M^{(1)}$ is called the one-loop renormalization constant for bosonic mass.

Analogously, the only possibility for $\delta t^{(1)\mu}$ is $\delta t^{(1)\mu} = \delta z_2^{(1)}\gamma^\mu$, where $\delta z_2^{(1)}$ is some constant. Then one can write

$$i\overline{\psi}_R \gamma^\mu \partial_\mu \psi_R - i\overline{\psi}_R \delta z_2^{(1)} \gamma^\mu \partial_\mu \psi_R = i z_2^{(1)} \overline{\psi}_R \gamma^\mu \partial_\mu \psi_R.$$

This expression is obtained from the term $i\overline{\psi}\gamma^\mu\partial_\mu\psi$ in the initial Lagrangian by using the transformation $\psi = z_2^{(1)\frac{1}{2}}\psi_R$, where $z_2^{(1)} = 1 + \delta z_2^{(1)}$.

One can write all the transformations described above in a more compact way, which we shall employ a lot in the rest of the book. It is useful to denote the set of all fields $\Phi = (\varphi, \psi, \overline{\psi})$ and the set of parameters $P = (M^2, m, \Gamma, \lambda, g^\mu)$. The relations $\Phi = z_\Phi^{\frac{1}{2}}\Phi_R$ and $P = z_P P_R$ are called the renormalization transformation, and z_ϕ, z_P comprise the constants of renormalization. Typically, z_P is a matrix, mixing different parameters.

Assuming $\mathcal{L}(\Phi, P)$ to be the initial Lagrangian, let us add the counterterms and get

$$\mathcal{L}_R(\Phi_R, P_R) = \mathcal{L}(\Phi_R, P_R) + \mathcal{L}_{counter}(\Phi_R, P_R). \tag{9.52}$$

If $\mathcal{L}_R(\Phi_R, P_R) = \mathcal{L}\big(z_\Phi^{1/2}\Phi_R, z_P P_R\big)$, the theory under consideration is called multiplicatively renormalizable. According to standard terminology, $\mathcal{L}(\Phi, P)$ is called the bare Lagrangian, and $\mathcal{L}_R(\Phi_R, P_R)$ is called the renormalized Lagrangian. The theory is multiplicatively renormalizable if the renormalized Lagrangian is obtained from the bare Lagrangian by means of the renormalization transformation with the constants of renormalization depending only on the couplings, not on the fields. It is evident that the condition for multiplicative renormalizability is that the counterterms repeat the form of the classical action of the theory.

9.3.3 The substraction procedure: two-loop diagrams

In the previous subsections, we formulated a recipe for how to divide one-loop diagrams into finite and divergent parts. Now we shall carry out similar considerations for two-loop diagrams. The extension to higher-loop orders can be done in a similar way.

We start with a diagram defining the two-loop correction to the fermionic Green function:

$$\Sigma^{(2)}(k) \quad =$$

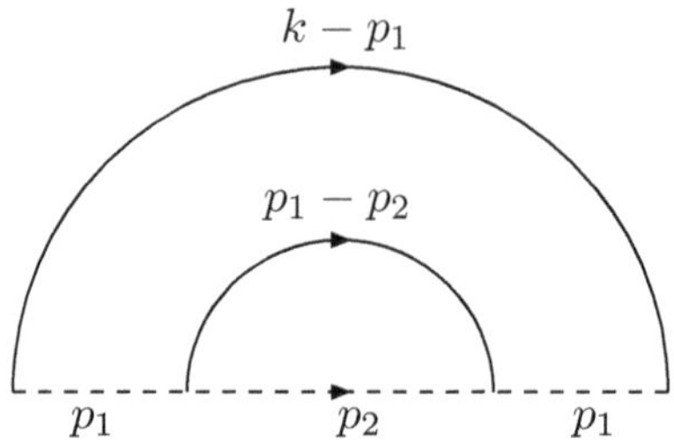

$$= \quad \int d^4p_1 d^4p_2\, Q(k, p_1, p_2). \tag{9.53}$$

In this case, $Q(k, p_1, p_2) \sim S(p_1)S(p_2)S(p_1)D(k - p_1)D(p_1 - p_2)$. At large momenta, the integral in the expression (9.53) behaves as $\int dp$ and diverges linearly. Taking into account the results of section 9.3.1, one could imagine that a finite contribution can be achieved after two subtractions in the external momentum k. However, this conclusion is not correct, as we shall see shortly.

The main new aspect in the multi-loop renormalization is that the subtraction procedure described in the previous subsections does not take into account the presence of subdiagrams. In the two-loop example under consideration, the diagram (9.53) includes the subdiagram

$$\Sigma^{(1)}(p_1) \quad =$$

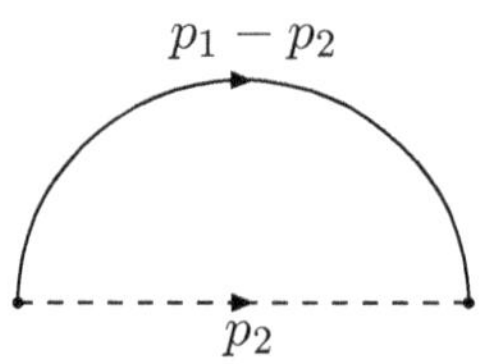

$$\tag{9.54}$$

which does not depend on the external momentum k. The diagram (9.54) was studied in section 9.3.1, and we saw that it includes divergences. Therefore, the subtraction procedure in the external momentum k for the diagram (9.53) does not lead to a finite Feynman integral, since the subdiagram (9.54) will still be divergent.

Such a situation indicates the need to use the natural procedure of renormalization, which is performed in two steps. First, we have to apply a subtraction procedure to the subdiagrams (e.g., in the given example), making all them finite. After that, we need to apply the subtraction procedure in the external momenta to the second "surface" integration. The diagram in (9.53) can be written as

$$\Sigma^{(2)}(k) = \int d^4p_1 \overline{Q}(k, p_1)\Sigma^{(1)}(p_1), \tag{9.55}$$

$\Sigma^{(1)}(p_1)$ corresponding to the subdiagram in (9.54), and $\overline{Q}(k, p_1) \sim S(p_1)S(p_1)D(k - p_1)$. Performing two subtractions from the momentum p_1, which is external for the subdiagram in (9.54), we take only its finite part. After that, we have

$$\int d^4p\,\overline{Q}(k,p_1)\,\Sigma_R^{(1)}(p_1).$$

In the second phase, we perform two subtractions from the external momentum k and obtain the finite Feynman integral

$$\Sigma_R^{(2)}(k) = \int d^4p_1 \left\{\overline{Q}(k,p_1) - Q(0,p_1) - \frac{\partial\overline{Q}(k,p_1)}{\partial k_\mu}\bigg|_{k=0} k_\mu\right\} \Sigma_R^{(1)}(p_1). \tag{9.56}$$

Let us now discuss how such a procedure is done using the counterterms. Consider the Lagrangian

$$\mathcal{L}_R^{(1)} = \mathcal{L} - \overline{\psi}_R\big[\delta_m^{(1)} - i\delta t^{(1)\mu}\partial_\mu\big]\psi_R + \dots, \tag{9.57}$$

which was discussed in section 9.3.1. Using this expression, one can compute the two-loop correction to the fermionic two-point vertex function:

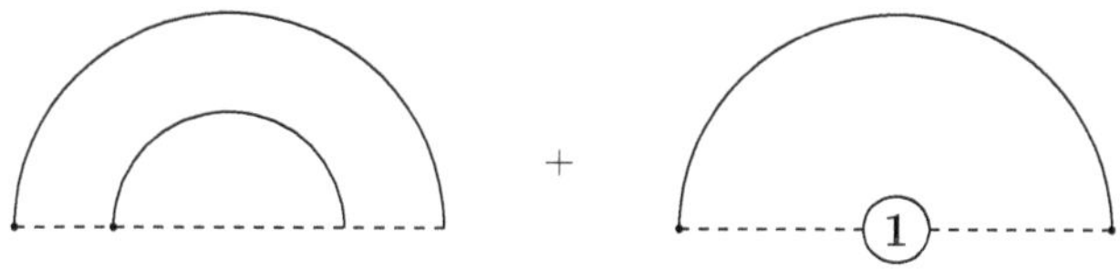

(9.58)

where

$$\textcircled{1} \;=\; -i\left[\delta_m{}^{(1)} + \delta t^{(1)\mu}\,p_{1\mu}\right].$$

The contribution corresponding to Eq. (9.58) is

$$\int d^4p_1\overline{Q}(k,p_1)\int d^4p_2\left[Q(p_2,p_1) - Q(p_2,0) - \frac{\partial Q(p_2,p_1)}{\partial p_{1\mu}}\bigg|_{p_1=0} p_{1\mu}\right]$$
$$= \int d^4p_1\overline{Q}(k,p_1)\Sigma_R^{(1)}(p_1), \tag{9.59}$$

where the function $Q(p_2,p_1)$ has been defined in section 9.3.1. Thus, taking into account the one-loop counterterm, we automatically obtain the finite subdiagram. On the other hand, the integral over d^4p_1 is still divergent, and we should perform two subtractions from the external momentum k. This is equivalent to adding the two-loop counterterm $-\overline{\psi}_R\left(\delta m^{(2)} - it^{(2)\mu}\partial_\mu\right)\psi_R$. The last means that we get the new vertex

$$\textcircled{2} \;=\; -i\left(\delta m^{(2)} + \delta t^{(2)\mu}k_\mu\right). \tag{9.60}$$

which kleads to the following renormalized diagram for the fermionic two-point function at the second-loop order:

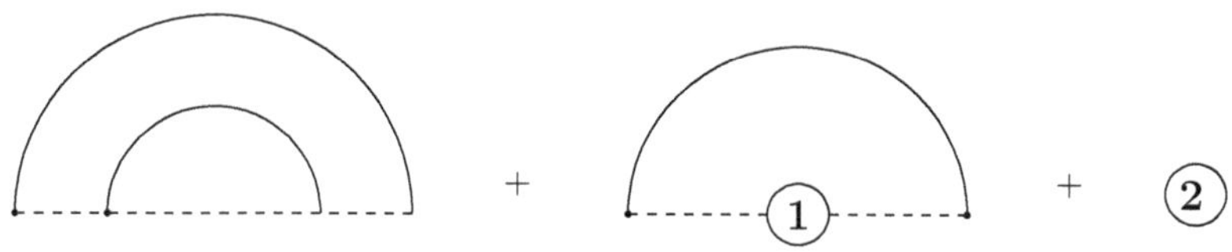

The corresponding contribution is $\Sigma_R^{(2)}(k)$, as given in Eq. (9.56).

Consider another two-loop diagram for the fermionic two-point functions,

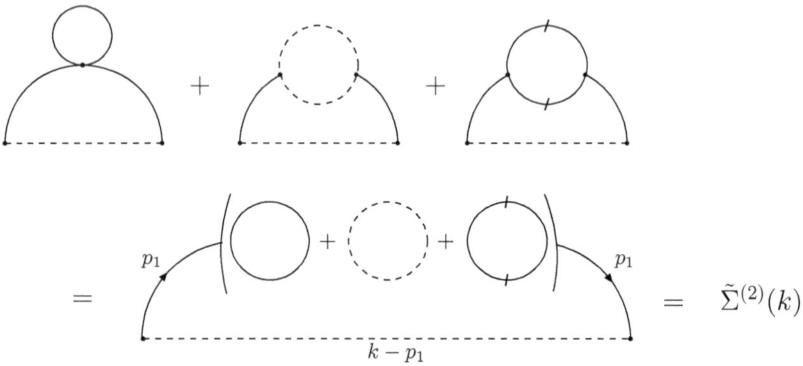

(9.61)

The corresponding contribution can be written as

$$\tilde{\Sigma}^{(2)}(k) \;=\; \int d^4p_1\, \overline{\overline{Q}}(p_1, k)\, \Pi^{(1)}(p_1). \tag{9.62}$$

Here $\Pi^{(1)}(k)$ is the one-loop correction to the bosonic two-point function considered in section 9.3.1. The function $\overline{\overline{Q}}(p_1, k)$ looks like

$$\overline{\overline{Q}}(p_1, k) \;\sim\; S(k - M)D(p_1)D(p_1). \tag{9.63}$$

The asymptotic behavior at large momentum p_1 is $\overline{\overline{Q}}(p_1, k) \sim \frac{1}{{p_1}^5}$. Since $\Pi^{(1)}(p_1)$ is quadratically divergent, $\Pi^{(1)}(p_1) \sim {p_1}^2$. Therefore, $\tilde{\Sigma}^{(2)}(k)$ behaves like $\int dp_1$, which means it is linearly divergent and also there are divergences in the subdiagrams. As usual, the subdiagrams should be made finite first, which defines $\Pi^{(1)}(p_1)$. As we already saw in section 9.3.1,

$$\Pi^{(1)}(p_1) \;=\; \Pi_R^{(1)}(p_1) \,+\, i\big[\delta M^{(1)2} + \delta K^{(1)\mu\nu} p_{1\mu} p_{1\nu}\big].$$

After that, we have to make the last subtractions, making the whole diagram finite. Taking only the finite part $\Pi_R^{(1)}(p_1)$, we get for the last integration the expression

$$\int d^4p_1\, \overline{\overline{Q}}(p_1, k)\, \Pi_R^{(1)}(p_1).$$

This integral is linearly divergent, and we should perform two subtractions,

$$\int d^4p_1 \Big\{ \overline{\overline{Q}}(p_1, k) - \overline{\overline{Q}}(p_1, 0) - \frac{\partial Q(p_1, k)}{\partial k_\mu}\bigg|_{k=0} k_\mu \Big\} \Pi_R^{(1)}(p_1) \tag{9.64}$$

$$+ \int d^4p_1 \Big\{ \overline{\overline{Q}}(p_1, 0) + \frac{\partial \overline{\overline{Q}}(p_1, k)}{\partial k_\mu}\bigg|_{k=0} k_\mu \Big\} \Pi_R^{(1)}(p_1) \;=\; \tilde{\Sigma}_R^{(2)}(k) \,+\, i\big[\delta\tilde{m}^{(2)} + \tilde{t}^{(2)\mu} k_\mu\big].$$

Let us clarify how the expression $\tilde{\Sigma}_R^{(2)}(k)$ in Eq. (9.64) can be obtained using the counterterms. Consider the one-loop renormalized Lagrangian

$$\mathcal{L}_R^{(1)} = \mathcal{L} - \frac{1}{2}\varphi_R\Big[\delta_M^{(1)2} - K^{(1)\mu\nu}\partial_\mu\partial_\nu\Big]\varphi_R + \dots \tag{9.65}$$

and calculate again the diagrams analogous to (9.61). This procedure leads to

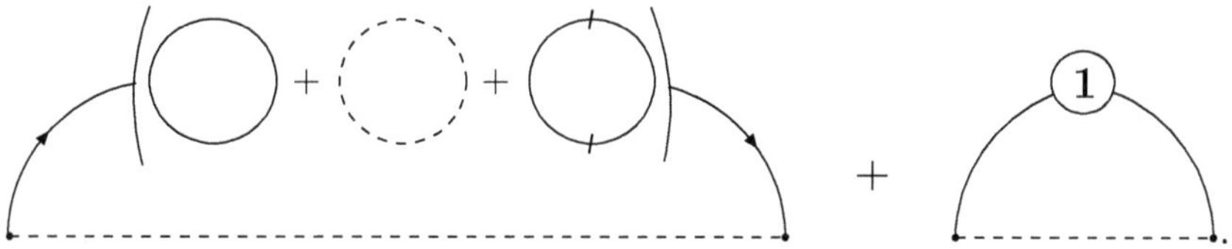

(9.66)

Here the second diagram comes from the vertex of Eq. (9.65),

$$\textcircled{1} \quad = \quad -i\Big[\delta M^{(1)2} + \delta K^{(1)\mu\nu}p_{1\mu}p_{1\nu}\Big]. \tag{9.67}$$

The contribution of the diagram in (9.66) has the form

$$\int d^4p_1\overline{\overline{Q}}(p_1,k)\Big[\Pi^{(1)}(p_1) - i\delta M^{(1)2} + i\delta K^{(1)\mu\nu}p_{1\mu}p_{1\nu}\Big] = \int d^4p_1\overline{\overline{Q}}(p_1,k)\Pi_R^{(1)}(p_1).$$

Now we should make finite the last integration over d^4p_1. To this end, one has to add to the Lagrangian (9.65) the two-loop counterterm $-\overline{\psi}_R\big[\delta m^{(2)} - it^{(2)\mu}\partial_\mu\big]\psi_R$, which corresponds to the vertex in (9.60). Finally, we arrive at the sum of the diagrams, which automatically represent the finite expression $\tilde{\Sigma}_R^{(2)}(k)$:

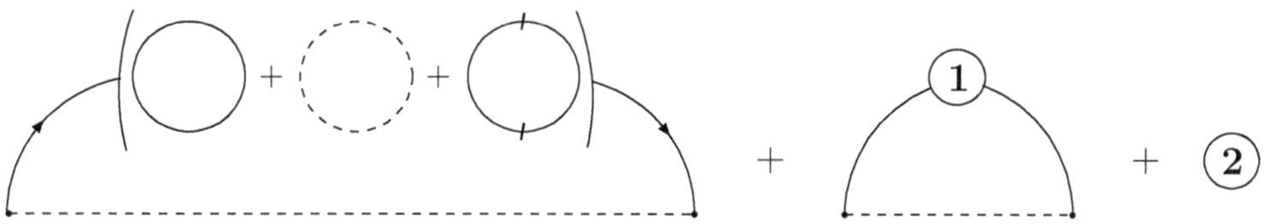

(9.68)

Consider more diagrams corresponding to the two-loop correction to the fermionic two-point function, starting from

$$\tilde{\tilde{\Sigma}}_R^{(2)}(k) \quad =$$

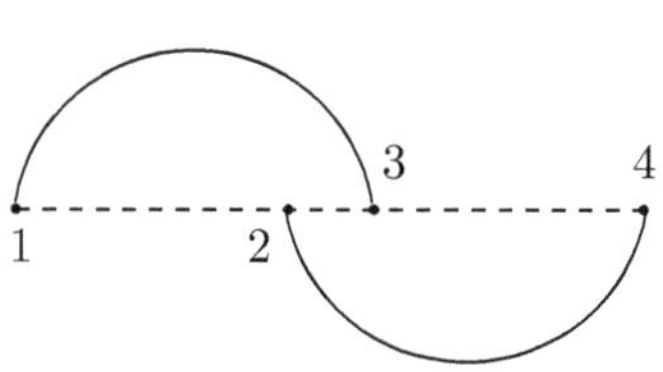

(9.69)

From the previous consideration, we know that the first step is to make the subdiagrams finite. Let us establish the subdiagrams in the case under consideration.

The diagram in (9.69) can be equivalently represented as follows:

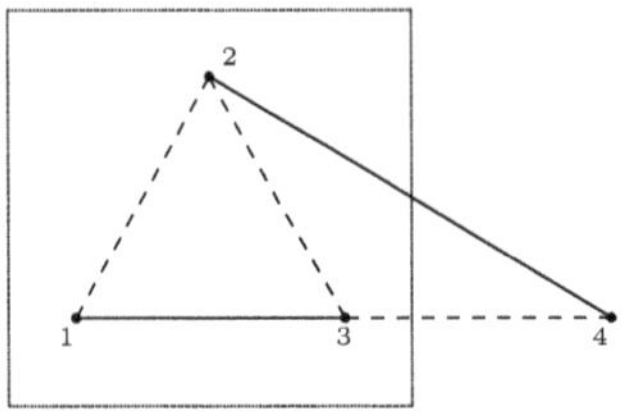

(9.70)

This diagram includes the subdiagram

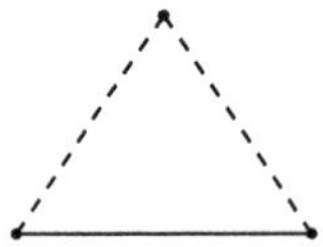

Indeed, we have already analyzed this diagram in section 9.3.1 and saw that there are logarithmic divergences. In order to make it finite, we should make one subtraction which gives a finite term plus the divergent contribution $i\delta\Gamma^{(1)}$. Using the one-loop renormalized Lagrangian,

$$\mathcal{L}_R^{(1)} = \mathcal{L} - \overline{\psi}_R \delta\Gamma^{(1)} \psi_R \varphi_R + \dots . \tag{9.71}$$

we obtain for one-loop diagram under discussion

+ ① , where ① $= -i\,\delta\Gamma^{(1)}$. (9.72)

As a result, the subdiagram will be finite. Now, the initial two-loop diagram becomes

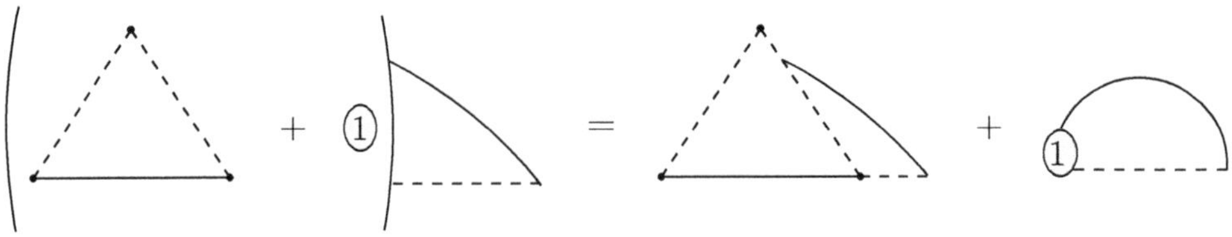

Making two subtractions in external momenta k, one may hope to arrive at the finite contribution. However, there is a complication, since the initial diagram can be represented in another form:

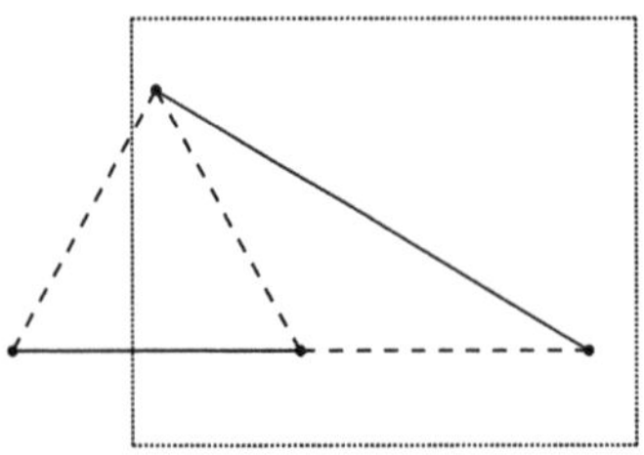

It is evident that the diagram under discussion contains another subdiagram:

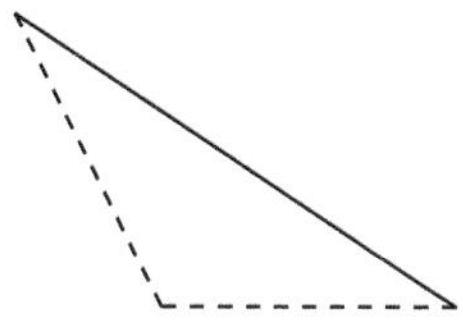

This subdiagram is analogous to the previous one. It diverges logarithmically, and one needs a single subtraction to extract a finite part. If we make only the first subdiagram finite and forget about the second one, the overall expression for the two-loop diagram will remain divergent.

Thus, our previous consideration was incomplete. This situation shows that, to make a two-loop diagram finite, we have to make finite all its subdiagrams first. In the case under consideration, this is done as follows:

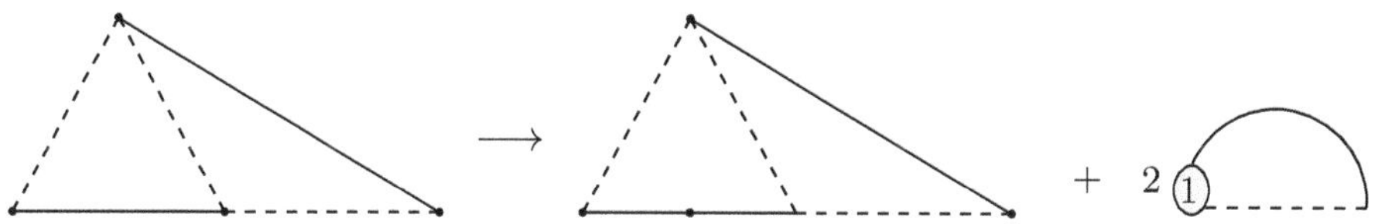

Let us clarify how the situation looks in terms of the counterterms. Consider the one-loop renormalized Lagrangian (9.71). In the two-loop Feynman diagrams starting from this Lagrangian, in the two-point fermionic sector, we meet

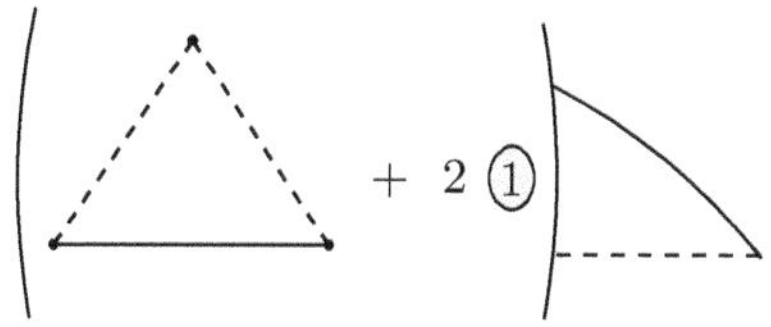

The counterterms automatically have the correct coefficient of 2.

Now, since all subdiagrams are finite, we add the two-loop counterterm,

$$-\overline{\psi}_R\big[\delta\tilde{\tilde{m}}^{(2)} - i\tilde{\tilde{t}}^{(2)\mu}\partial_\mu\big]\psi_R, \tag{9.73}$$

which is equivalent to adding the vertex

$$\textcircled{2} \quad = \quad i\Big(\delta\tilde{\tilde{m}}^{(2)} + \delta\tilde{\tilde{t}}^{(2)\mu}k_\mu\Big). \tag{9.74}$$

As a result, we get the finite contribution

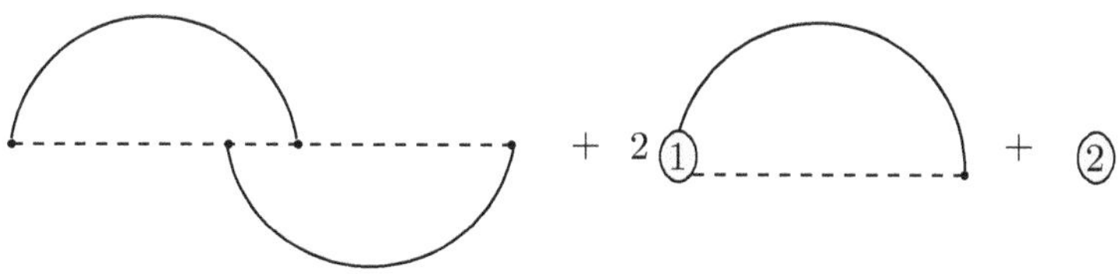

$$= \quad \tilde{\tilde{\Sigma}}_R^{(2)}(k)\,. \tag{9.75}$$

A qualitatively similar analysis can be carried out for any diagram at any loop order. In all cases, the subtraction procedure enables us to arrive at the finite results for the Feynman diagrams.

In general, the renormalization of Feynman diagrams is performed iteratively, loop by loop: first, one has to find one-loop counterterms, then one has to perform the subtraction, then one has to find the two-loop counterterms, and so on. An important aspect of the whole procedure is that the subtraction removes from the Feynman integral only the first terms of expansion in the power series in external momenta. Thus, the UV counterterms are always polynomial in the external momenta. The important consequence of this is that, in the coordinate representation, the counterterms are local functions of the fields and their derivatives. This statement constitutes Weinberg's theorem (see, e.g., [96]).

The last observation is that the divergences of Feynman diagrams are stipulated only by the loop momentum integrals. Therefore, to establish the counterterms it is sufficient to consider only the diagrams for the vertex functions.

9.4 The superficial degree of divergences

In this section, we learn how to evaluate the power of divergences in a given theory without performing explicit calculations. This task can be accomplished using the notion of the superficial degree of divergence of an arbitrary Feynman diagram. Suppose G is an arbitrary diagram for the vertex function. The superficial degree of divergence of the diagram is a number which shows how much the given diagram is divergent. This number is defined as an overall dimension of the Feynman diagram in the momenta of the *internal* lines.

We will denote the superficial degree of divergence of the diagram G as $\omega(G)$. For instance, in the case of a logarithmically divergent diagram, $\omega(G) = 0$. If there is quadratic (in the cut-off parameter) divergence, we have $\omega(G) = 2$, etc. Let us stress that, in the last case, the diagram may have logarithmic divergences too. In all cases, $\omega(G)$ is defined by the strongest divergence of the given Feynman integral.

Obviously, the degree of divergence of a given diagram $\omega(G)$ depends on the relation between the powers of integration variable (momenta) in the denominator and the numerator of the Feynman integral. Thus, the evaluation of $\omega(G)$ is conveniently based on the analysis of dimensions. Let us start by remembering the dimensions of the quantities characterizing the divergence of the diagram:

a. *Dimensions of the fields.* Let φ be a bosonic (scalar or vector) field, and ψ a fermionic field. As we already know, the dimensions of the fields can be found from the classical Lagrangian, e.g., $[\varphi] = 1$ and $[\psi] = \frac{3}{2}$. The same dimensions can be obtained from the canonical commutation relations; therefore, these dimensions are called canonical.

b. *Dimensions of the propagators.* In the momentum representation, the propagators have the form

$$\tilde{D}(p) \sim \frac{1}{p^2 - M^2}, \qquad \tilde{S}(p) \sim \frac{\gamma p + m}{p^2 - m^2}.$$

Thus, the dimensions of the propagators are $[\tilde{D}] = -2$ and $[\tilde{S}] = -1$. These two distinct dimensions can be written as a unique expression. $[\tilde{D}_\phi] = 2[\phi] - 4$ for the general field $\phi = (\varphi, \psi)$. For example, for a scalar we obtain -2 and for a spinor, -1.

c. *Dimensions of the classical vertices and coupling constants.* A vertex coming from the interaction Lagrangian has the form $g_v V$, where g_v is a coupling constant and $V \sim \partial^{l_V} \Pi\phi$. Here l_V is a number of the derivatives in the vertex, and $\Pi\phi$ is a product of some number of fields φ and ψ in the given vertex. The dimension of the vertex V is also called the degree of the vertex and is denoted $\omega(V) = [V]$. This quantity can be evaluated as $\omega(V) = l_V + \Sigma^{(V)}[\phi]$, where $\Sigma^{(V)}$ means a sum over fields ϕ in the given vertex.

The dimension of the coupling constant g_v can be found from the interaction term $g_v V$ in the classical Lagrangian. Since the dimension of the Lagrangian is $[\mathcal{L}] = 4$, we obtain $[g_v] = 4 - \omega(V)$.

d. *Calculation of* $\omega(G)$. Let G be an arbitrary diagram for a vertex function with I internal lines, E external lines (more precisely, E is a number of the external point of the vertex diagram where the external lines can be attached) and V vertices. Our aim is to calculate $\omega(G)$ in terms of I, E, V and g_V. According to the Feynman rules, each internal line with momentum p corresponds to the integration over d^4p, which has dimension 4. In addition, each internal line corresponds to the propagator $\tilde{D}_\phi$, with the dimension $2[\phi] - 4$. However, the number of independent momentum integrals is not equal to the number of internal lines, since there is a momentum conservation law in each vertex. The momentum conservation law means that each vertex contains the momentum-depending delta function. Therefore, the internal lines of G give the following contribution to the dimension of the Feynman integral:

$$\sum_I [4 + (2[\phi] - 4)] \;+\; \text{the total dimension of delta functions,}$$

where $\sum_I$ means a sum over all internal lines.

Each momentum-dependent delta function has a dimension of -4. However, we have to remember that one of the delta functions corresponds to the conservation of external momenta and does not influence the integration over momenta. Thus, the delta functions provide an overall contribution to the dimension that is equal to $-(\sum_V 4 - 4)$, where the sum is taken over all vertices in G.

Finally, all of the internal lines give the following contribution to the dimension:

$$\sum_I [4 + (2[\phi] - 4)] - \Big(\sum_V 4 - 4\Big) = 2\sum_I [\phi] + 4 - \sum_V 4. \tag{9.76}$$

Now, let us account for the vertices. Any vertex contains a certain number of fields, but the corresponding dimensions have already been taken into account in the contributions of internal lines. The coupling constants in the vertices do not depend on momenta and therefore are irrelevant for the momentum integrals. Thus, the vertices contribute with $\sum_V l_V$. As a result, the overall momentum dimension of the integrand of diagram G has the form

$$\omega(G) = 2\sum_I [\phi] + 4 - \sum_V 4 + \sum_V l_V. \tag{9.77}$$

This form of $\omega(G)$ proves very useful in quantum field theory, including in semiclassical and quantum gravity, as we will see in Part II of this book.

At the same time, Eq. (9.77) can be further transformed to a somewhat more transparent form. We denote the sum of dimensions of all fields in a given vertex V as $\sum^{(V)}[\phi]$. Then the dimension of all lines associated with all vertices of the diagram will be $\sum_V \sum^{(V)}[\phi]$. On the other hand, some of these fields will be converted into internal lines, while the remaining fields correspond to external lines. Since any internal line connects two vertices, such a line absorbs two fields from vertices. Hence, the dimension of all the fields, associated with all the vertices is $\sum_E[\phi] + 2\sum_I[\phi]$. Thus, there is an identity

$$\sum_V \sum^{(V)}[\phi] = \sum_E[\phi] + 2\sum_I[\phi]. \tag{9.78}$$

Expressing $2\sum_I[\phi]$ from this identity and substituting it into (9.77), one gets

$$\begin{aligned}\omega(G) &= \sum_V \sum^{(V)}[\phi] - \sum_E[\phi] + 4 - \sum_V 4 - \sum_V l_V \\ &= 4 - \sum_E[\phi] + \sum_V \Big[\sum^{(V)}[\phi] + l_V - 4\Big].\end{aligned} \tag{9.79}$$

On the other hand, $\sum^{(V)}[\phi] + l_V = \omega(V)$ which is the degree of a classical vertex. Then

$$\omega(G) = 4 - \sum_E[\phi] - \sum_V \big(4 - \omega(V)\big). \tag{9.80}$$

However, $4 - \omega(V) = [g_v]$, where $[g_v]$ is the dimension of the coupling constant in the given vertex V. Finally, we obtain

$$\omega(G) = 4 - \sum_E[\phi] - \sum_V[g_V]. \tag{9.81}$$

This form of the superficial degree of divergence clearly shows that theories with couplings of inverse-mass dimensions are typically badly divergent.

e) *$\omega(G)$ and the condition of finiteness for the diagram G.* The value of $\omega(G)$ can be used for the analysis of field models at the quantum level. For any Feynman integral, in the UV limit, we meet the L-loop integral in the general form $\int p^{-n} d^{4L}p$, with n depending on the number of propagators and the number of derivatives of the internal lines coming from the vertices. After the Wick rotation and using the L-dimensional spherical coordinates, we obtain the following integral:

$$\int dp\, \frac{p^{4L-1}}{p^n} \sim \int^\infty \frac{dp}{p^{1-4L+n}}. \tag{9.82}$$

The overall dimension of this integral is $4L - n$. Thus, $\omega(G) = 4L - n$.

The integral (9.82) is finite if $1 + n - 4L > 1$ or, equivalently, when $\omega(G) < 0$. If $1 + n - 4L \le 1$, the integral is divergent, which corresponds to $\omega(G) \ge 0$. All in all, a Feynman integral is finite if $\omega(G) < 0$. On the other hand, a Feynman integral corresponding to the diagram G is divergent if

$$\omega(G) \geq 0. \tag{9.83}$$

This is the condition of divergence for a diagram G, formulated in terms of $\omega(G)$, as we have expected from the very beginning.

A relevant observation is that the criterion (9.83) corresponds to the L-loop diagram on a whole, without taking into account the subdiagrams, which can be finite or divergent. The complete analysis should include the criterion (9.83) applied to all subdiagrams and to the last (surface) integration, at the same time.

The last observation is that, in some field models, the numerical coefficient in front of a particular Feynman integral may be zero, e.g., due to some symmetry, or because of the contributions of distinct diagrams cancel each other. Both of these aspects cannot be taken into account by the index $\omega(G)$ and should be carefully analyzed on the basis of the specific features of a given field theory model.

9.5 Renormalizable and non-renormalizable theories

In this section, we consider a detailed classification of field theories based on the notion of the superficial degree of divergences.

9.5.1 Analysis of the superficial degree of divergences

Consider the condition $\omega(G) \geq 0$ in more detail. From Eqs. (9.81), one gets

$$4 - \sum_E [\phi] - \sum_V [g_V] \geq 0 \tag{9.84}$$

for a divergent diagram. Let the diagram G have N_B external bosonic lines and N_F external fermionic lines. In this case, Eq. (9.84) becomes

$$N_B + \frac{3}{2} N_F + \sum_V [g_V] \leq 4. \tag{9.85}$$

If the dimensions of all coupling constants are positive, the *l.h.s.* of the last inequality is non-negative and, moreover, the values of N_B, N_F and V are bounded from above. In this case, there may be only a finite number of divergent diagrams and, in order to subtract the divergences, we need only a finite number of counterterms. This kind of theory is called superrenormalizable in power counting.

We can reduce the requirement and assume that some of the coupling constants are dimensionless. The corresponding vertices do not contribute to the inequality (9.85), which can be valid for the diagrams with any number of such vertices. In this case, there may be an infinite number of divergent diagrams. On the other hand, all required counterterms have restricted dimension. Since these counterterms are local (due to the Weinberg's theorem), we need to use only the restricted type of counterterm. This sort of theory is called renormalizable by power counting.

Let a bare Lagrangian have the form

$$\mathcal{L} = \sum_{i=1}^{n} O_i, \tag{9.86}$$

where O_i are functions of fields, their derivatives and parameters (such as masses and coupling constants). If the counterterms, which should be introduced to cancel the divergences, have the form

$$\mathcal{L}_{counter} = \sum_{i=1}^{n} \delta z_i \, O_i, \tag{9.87}$$

the theory is called multiplicatively renormalizable. The constants δz_i are calculated in the framework of loop expansion and are given by the series in $\hbar$,

$$\delta z_i = \sum_{i=1}^{\infty} \hbar^L \, \delta z_i^{(L)}, \tag{9.88}$$

where L is the order of the loop expansion, as usual.

If the dimension of at least one coupling constant is negative, there may be diagrams with $[g_V] < 0$. Then, the inequality (9.85) is fulfilled and becomes stronger with a growing number of the corresponding vertices. To cancel the divergences, one needs an infinite amount of the types of counterterms, and the dimensions of these counterterms increases with L. This kind of theory is called non-renormalizable in power counting.

Let us stress that the theory may be renormalizable or even superrenormalizable in power counting but, in the end, it may be multiplicatively non-renormalizable. We shall discuss a few examples of this in Part II, when will be dealing with different models of quantum gravity. In the next subsections, we consider a few simpler examples, just to illustrate the general notions described above.

9.5.2 Theory for a real scalar field with the interaction $\lambda\varphi^3$

This theory is described by the Lagrangian in four spacetime dimensions,

$$\mathcal{L} = \frac{1}{2}\partial^\mu\varphi\partial_\mu\varphi - \frac{1}{2}m^2\varphi^2 - \lambda\varphi^3.$$

Using the methods described in section 9.2, it is easy to see that the dimension of the coupling is $[\lambda] = 1 > 0$. The criterion (9.85) boils down to $N_\varphi \leq 4 - V$. The case $N_\varphi = 0$ corresponds to vacuum diagrams which are simply constants.[1] Assuming $V = 1$, we need to account for $N_\varphi = 1, 2, 3$ in the case of divergent diagrams. For $V = 2$, there is $N_\varphi = 1, 2$, and, for $V = 3$, only $N_\varphi = 1$.

[1] It is worth mentioning that, in curved spacetime, these kinds of diagrams are essential for renormalizing vacuum action. This aspect of quantum theory in curved spacetime will be discussed in Part II of this book.

Obviously, the theory is superrenormalizable, and there is only a finite number of divergent diagrams. Namely, only the following diagrams can be divergent:

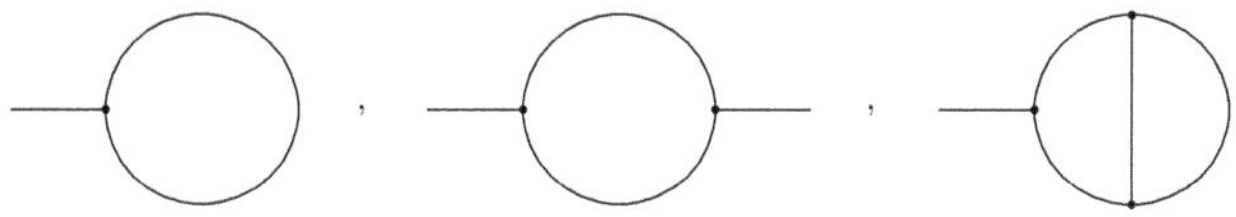

(9.89)

As usual, these diagrams can serve as subdiagrams of more complicated graphs and produce divergences, but all of them are controlled by the one-loop graphs (9.89).

It is worth noticing that the theory under consideration is not perfect, because its vacuum state is unstable. For this reason, this theory can not be associated with a physically acceptable quantum field theory.

9.5.3 A real scalar field with a $\lambda\varphi^4$ interaction

This theory is described by the Lagrangian

$$\mathcal{L} = \frac{1}{2}\partial^\mu\varphi\partial_u\varphi - \frac{1}{2}m^2\varphi^2 - \lambda\varphi^4.$$

The dimension of the coupling constant is $[\lambda] = 0$, and the inequality (9.85) has the form $N_\varphi \leq 4$, where N_φ is a number of external lines of a diagram. Since the Lagrangian is an even function of fields and their derivatives, the diagrams with an odd number of external lines vanish. Thus, we need to consider $N_\varphi = 0, 2, 4$.

Disregarding the vacuum diagrams with $N_\varphi = 0$, consider the case of $N_\varphi = 2$. The corresponding superficial degree of divergence is $\omega(G) = 2$. According to the subtraction procedure, the divergent part of these diagrams is a polynomial of the second power in the momentum, that is, $c_1p^2 + c_2m^2$, where c_1, c_2 are divergent functions of the coupling constant λ and regularization parameters. In the cut-off regularization with parameter Ω, there can be also the quadratic divergence $\sim \Omega^2$. This divergence plays an important role in the discussion of the hierarchy problem is the minimal standard model of particle physics. However, since our focus is not on the particle physics, we will mainly keep in mind the dimensional regularization, where the power divergences are absent. The corresponding counterterm has the form

$$\frac{1}{2}\delta z_1\partial_\mu\varphi_R\partial^\mu\varphi_R - \frac{1}{2}\delta z_2 m_R^2\varphi_R^2. \tag{9.90}$$

For $N_\varphi = 4$, we have divergent diagrams with four external lines. The power counting is $\omega(G) = 0$. According to the subtraction procedure, the divergent part of such diagrams is momentum independent,[2] and the corresponding counterterm has the form

$$-\,\delta z_4\,\lambda_R\,\varphi_R^4\,. \tag{9.91}$$

Taking into account (9.90) and (9.91), the renormalized Langrangian is

[2]In the explicit calculations in Part II, we shall see that each logarithmic divergence in this case is accompanied by the logarithmic dependence on external momenta in the *finite* part of the divergent diagram.

$$\mathcal{L}_R = \frac{1}{2}(1+\delta z_1)\,\partial_\mu\varphi_R\partial^\mu\varphi_R - \frac{1}{2}\,(1+\delta z_2)m_R^2\varphi_R^2 - (1+\delta z_4)\lambda_R\varphi_R^4. \quad (9.92)$$

By denoting $z_1 = 1+\delta z_1$, $z_1 z_m = 1+\delta z_2$, $z_1^2 z_\lambda = 1+\delta z_4$, one gets

$$\mathcal{L}_R = \frac{1}{2}z_1\partial^\mu\varphi_R\partial_\mu\varphi_R - \frac{1}{2}z_1 z_m m_R^2\varphi_R^2 - z_1^2 z_\lambda\lambda_R\varphi_R^4. \quad (9.93)$$

It is clear that the theory is multiplicatively renormalizable, and the renormalized Lagrangian (9.93) is related to the bare Lagrangian by the renormalization transformation

$$\varphi = z_1^{\frac{1}{2}}\varphi_R, \qquad m^2 = z_m m_R^2, \qquad \lambda = z_\lambda\lambda_R, \quad (9.94)$$

where the renormalization constants $z_{1,m,\lambda}$ depend on the coupling constant λ_R and on regularization parameters.

9.5.4 Spinor electrodynamics

This theory is described by the Lagrangian

$$\mathcal{L} = -\frac{1}{4}F_{\mu\nu}F^{\mu\nu} + i\bar{\psi}\gamma^\mu\partial_\mu\psi - m\bar{\psi}\psi + e\bar{\psi}\gamma^\mu\psi A_\mu + \frac{\alpha}{2}(\partial_\mu A^\mu)^2. \quad (9.95)$$

If we compare this equation to Eq. (4.101), we can see that the last term is responsible for the gauge fixing, where α is called the gauge-fixing parameter. In section 5.6, we used the value $\alpha = 1$.

We showed in the section 4.5 that the coupling constant is dimensionless: $[e] = 0$. The relation (9.85) has the form

$$N_A + \frac{3}{2}N_\psi \leq 4, \quad (9.96)$$

where N_A and N_ψ are the numbers of external lines of the electromagnetic and spinor fields. The divergent diagrams correspond to the following values of N_A and N_ψ:

$$N_\psi = 0, \qquad N_A = 0,1,2,3,4.$$
$$N_\psi = 2, \qquad N_A = 0,1.$$

The diagrams, with an odd number of external fermionic lines are forbidden, since the Lagrangian is an even function of fermions. Let us consider all of the relevant combinations:

a. $N_\psi = N_A = 0$. Vacuum diagrams are divergent, but the counterterms are irrelevant constants.

b. $N_\psi = 0$ and $N_A = 2$. In this case, $\omega(G) = 2$. Loop diagrams with two external photon lines are potentially divergent:

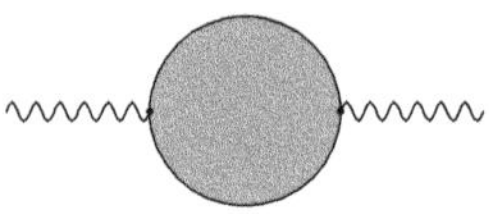

The dimension of the diagrams is 2. Since the counterterm should be quadratic in A_μ, in order to provide the dimension 2, there should be either two derivatives or the square of the mass. Therefore, the corresponding counterterms are

$$-\frac{1}{4}\delta z_A F_{R\mu\nu} F_R{}^{\mu\nu} + \frac{1}{2}\delta z_\alpha \, \alpha_R (\partial_\mu A^\mu_R)^2 + \frac{1}{2}\delta z_2 \, m^2_R \, A_{R\mu} A^\mu_R.$$

c. $N_\psi = 0$, $N_A = 3$. In this case, $\omega(G) = 1$, which means diagrams with three external photon lines are potentially divergent:

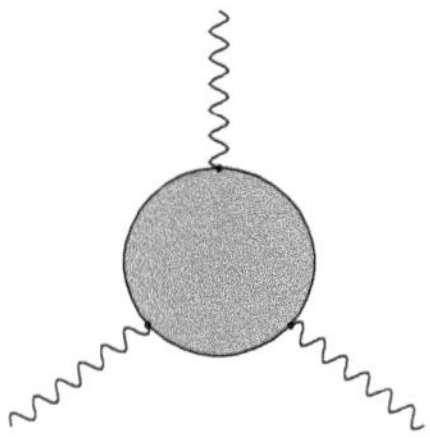

Since the dimension of this diagram is unit, the possible counterterms have the form

$$\delta z_3 \, A^\mu_R A_{R\mu} \, \partial_\nu A^\nu_R \; + \; \delta z_3' \, \partial_\mu A_{R\nu} \, A^\nu{}_R A_R{}^\mu.$$

d. $N_\psi = 0$, $N_A = 4$. In this case, $\omega(G) = 0$. Loop diagrams with four photon lines are potentially divergent. The corresponding counterterm is $\delta z_4 \, (A^\mu_R A_{R\mu})^2$.

e. $N_\psi = 2$, $N_A = 0$. In this case, $\omega(G) = 4 - 3 = 1$. This means loop diagrams with two external electron lines are potentially divergent:

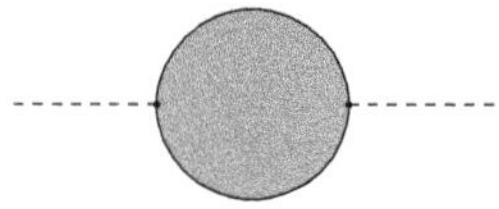

The corresponding counterterm is $\delta z_\psi \, \overline{\psi}_R \gamma^\mu \, \partial_\mu \psi_R \; - \; \delta z_m \, m_R \, \overline{\psi}_R \psi_R$.

f. $N_\psi = 2$, $N_A = 1$. In this case, $\omega(G) = 0$. Loop diagrams with two external electron lines and one external photon line are potentially divergent:

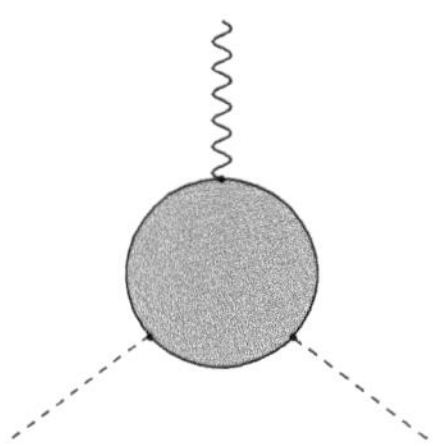

The corresponding counterterm is $\delta z_e e_R \overline{\psi}_R \gamma^\mu \psi_R A_{R\mu}$.

All in all, the total renormalized Lagrangian has the form

$$\begin{aligned}\mathcal{L}_R \; &= -\frac{1}{4}(1+\delta z_A) F_{R\mu\nu} F_R{}^{\mu\nu} + (1+\delta z_\psi)\overline{\psi}_R \gamma^\mu \partial_\mu \psi_R - (1+\delta z_m) m_R \overline{\psi}_R \psi_R \\ &\quad + (1+\delta z_e) e_R \overline{\psi}_R \gamma^\mu \psi_R A_{R\mu} + \frac{1}{2}(1+\delta z_\alpha)\alpha_R (\partial_\mu A^\mu_R)^2 + \delta z_2 m^2{}_R A_R{}^\mu A_{R\mu} \\ &\quad + \delta z_3 A^\mu_R A_{R\mu} \partial_\nu A^\nu_R + \delta z_3' \partial_\nu A_{R\mu} A^\nu_R A^\mu_R + \delta z_4 (A^\mu_R A_{R\mu})^2. \end{aligned} \tag{9.97}$$

This theory is renormalizable, since, to cancel all the divergences, only need to use restricted counterterms. However, the renormalization does not look multiplicative. The reason is that, for four of the counterterms, there are no terms in the initial Lagrangian (9.95). The list includes

$$\delta z_2 m^2{}_R A^\mu_R A_{R\mu} + \delta z_3 A^\mu_R A_{R\mu}\partial_\nu A^\nu_R + \delta z_3' \partial_\nu A^\mu_{R\mu} A_R{}^\nu A_R{}^\mu + \delta z_4 (A_R{}^\mu A_{R\mu})^2. \quad (9.98)$$

In these considerations, we did not take into account that spinor electrodynamics is a gauge theory. It is the gauge invariance of the classical theory that rules out the structures (9.98) in the classical action. In the forthcoming chapter devoted to the renormalization of gauge theories, we shall see that these terms are also forbidden at the quantum level (by the Ward identities), since the gauge invariance is preserved under quantization. In reality, we have $\delta z_2 = \delta z_3 = \delta z_3' = \delta z_4 = 0$. Because of this important point, renormalization in spinor electrodynamics is multiplicative.

Taking the gauge invariance into account, we denote

$$z_A = 1 + \delta z_A, \quad z_\psi = 1 + \delta z_\psi, \quad z_\psi z_m = 1 + \delta z_m,$$
$$z_\psi z_A^{\frac{1}{2}} z_e = 1 + \delta z_e, \quad z_A z_\alpha = 1 + \delta z_\alpha.$$

After that, the renormalized Lagrangian takes the form

$$\begin{aligned}\mathcal{L}_R &= -\frac{1}{4} z_A F_{R\mu\nu} F_R{}^{\mu\nu} + z_\psi \overline{\psi}_R i\gamma^\mu \partial_\mu \psi_R - z_\psi z_m m_R^2 \overline{\psi}_R \psi_R \\ &\quad + z_\psi z_A^{\frac{1}{2}} z_e e_R \overline{\psi}_R \gamma^\mu \psi_R A_{R\mu} + \frac{1}{2} z_\alpha z_A \alpha_R (\partial_\mu A^\mu_R)^2. \end{aligned} \quad (9.99)$$

The renormalized Lagrangian is obtained from classical Lagrangian by the following renormalization transformation:

$$\begin{aligned} &A_\mu = z_A^{\frac{1}{2}} A_{R\mu}, \quad \psi = z_\psi^{\frac{1}{2}} \psi_R, \quad \overline{\psi} = z_\psi^{\frac{1}{2}} \overline{\psi}_R, \\ &m = z_m m_R, \quad e = z_e e_R, \quad \alpha = z_\alpha \alpha_R. \end{aligned} \quad (9.100)$$

In addition, the mentioned Ward identities lead to the equalities $z_A^{\frac{1}{2}} z_e = 1 and\ z_\alpha z_A = 1$. The renormalization constants z_A, z_ψ, z_m and z_e depend on e_R and on the regularization parameters. Also, one can show that z_A and z_ψ depend on α_R.

9.5.5 The fermi (or Nambu) model

This theory is described by the Lagrangian

$$\mathcal{L} = i\overline{\psi}\gamma^\mu \partial_\mu \psi - m\overline{\psi}\psi - \lambda(\overline{\psi}\Gamma\psi)^2. \quad (9.101)$$

Here Γ is some matrix with spinor indices. For example, one can consider $\Gamma = I$, $\Gamma = \gamma^\mu$, or $\Gamma = \gamma_5\gamma^\mu$, where I is the unit matrix 4×4.

To find the dimension of the coupling constant λ, recall that the dimension of the Lagrangian is 4. Therefore, we get $4 + [\lambda] + 4[\psi]$. Thus, the coupling constant has a negative dimension $[\lambda] = -2$. According to the general approach described above, this means that loop diagrams with an even number of external lines can be divergent, and, to cancel these divergences, we need an infinite variety of counterterms. The model is non-renormalizable.

Let us note, however, that different extended versions of the model (9.101) have been successfully used as the basis of effective field theories for describing weak interactions at low energies (Fermi model), for more advanced technicolor models and for the effective approach to the interaction of quarks (the Nambu model).

9.5.6 The sigma model

This theory is described by the Lagrangian (4.19),

$$\mathcal{L} = \frac{1}{2}\, g_{ij}(\varphi)\, \eta^{\mu\nu}\partial_\mu\varphi^i\partial_\nu\varphi^j\,, \qquad \text{where} \qquad i,j = 1,2,\dots,N,$$

and $g_{ij}(\varphi)$ can be non-polynomial, in general. The scalar field has unit dimension, so functions $g_{ij}(\varphi)$ are dimensionless.

There are infinitely many vertices and, consequently, an infinite number of coupling constants. In order to see this, we expand the $g_{ij}(\varphi)$ into the power series in φ^i:

$$g_{ij}(\varphi) = \delta_{ij} + \sum_{n=1}^{\infty} \frac{\lambda^{(n)}_{ijl_1\dots l_n}}{n!}\varphi^{l_1}\dots\varphi^{l_n}, \qquad \lambda_{ijl_1\dots l_n} = \frac{\partial^n g_{ij}(\varphi)}{\partial\varphi^{l_1}\dots\partial\varphi^{l_n}}\bigg|_{\varphi=0}. \tag{9.102}$$

As a result, the Lagrangian (4.19) takes the form

$$\mathcal{L} = \mathcal{L}_0 + \mathcal{L}_{int}\,, \qquad \text{where} \qquad \mathcal{L}_0 = \frac{1}{2}\delta_{ij}\partial^\mu\varphi^i\partial_\mu\varphi^j$$

$$\text{and} \qquad \mathcal{L}_{int} = \frac{1}{2}\sum_{n=1}^{\infty}\frac{\lambda^{(n)}_{ijl_1\dots l_n}}{n!}\,\varphi^{l_1}\dots\varphi^{l_n}\,\partial^\mu\varphi^i\,\partial_\mu\varphi^j. \tag{9.103}$$

The interaction Lagrangian contains an infinite number of terms, and the parameters $\lambda^{(n)}_{ijl_1\dots l_n}$, with $(n = 1,2,\dots)$ play the role of coupling constants. Since $g_{ij}(\varphi)$ is dimensionless and the dimension of the fields $[\varphi^i] = 1$, according to (9.103), the dimensions of coupling constants $[\lambda^{(n)}] = -n$. Since the dimensions of the coupling constants are negative, the theory under consideration is non-renormalizable.

The superficial degree of divergences can be easily calculated, but we leave it as an exercise for the reader. In fact, the calculation is similar to the one for the simplest version of quantum gravity, based on Einstein's theory, which is considered in Part II. The result is $\omega(G) = 4L - 2I + 2V$, and using the topological identity $L = I - V + 1$ gives the final answer, $\omega(G) = 2 + 2L$. Thus, the L-loop counterterm with logarithmic divergences contains $2 + 2L$ derivatives.

Let us note that theory (4.19) has very different properties in two-dimensional spacetime. In this case, the fields φ^i are dimensionless, as is $g_{ij}(\varphi)$. To find the superficial degree of divergence, one takes into account that Feynman integrals have integration with d^2p. It is easy to show that $\omega(G) = 2L - 2I + 2V$ and, using the relation $I = L + V + 1$, we arrive at $\omega(G) = 2$. This means that all possible counterterms in two dimensions have two derivatives, exactly like the classical Lagrangian. Thus, the two-dimensional sigma model is a renormalizable theory.

9.6 The arbitrariness of the subtraction procedure

As we already pointed out, the subtraction procedure is designed to provide finite contributions of the Feynman diagrams. For this, we subtract from the momentum-dependent Feynman integrals, the first terms of their expansion into the Taylor series at the zero point.

Let $t^n(G)$ be the sum of the first n terms of Taylor expansion of a diagram G into the external momenta at the zero point. Schematically, the subtraction procedure can be written as

$$G - t^{\omega(G)}(G) = \big[1 - t^{\omega(G)}\big](G) \;=\; R(G). \tag{9.104}$$

Procedure (9.104) is called the R-operation and is fulfilled iteratively, loop by loop. The R-operation is formally defined as $R = 1 - t^{\omega(G)}$.

Consider the operation $R_n = 1 - t^n$. If $n < \omega(G)$, we do not cancel all divergences. If $n > \omega(G)$, all the divergences would certainly be canceled. However, in this case, we actually add more counterterms than would be necessary to make the diagram finite. In particular, this means that if the theory is multiplicative renormalizable, we may destroy this property. The reason is that adding extra terms within the subtraction procedure typically implies that one introduced higher-derivative counterterms that were not present in the initial Lagrangian and also may create problems.[3] Thus, the best option is to require that n be the minimal number of subtractions leading to a finite diagram, that is, $n = \omega(G)$.

In principle, if our purpose is just to make a diagram finite, it is possible to use a modified form of the operation,

$$\tilde{R} \;=\; 1 - t^{\omega(G)} + P^{\omega(G)} \;=\; R + P^{\omega(G)}, \tag{9.105}$$

where $P^{\omega(G)}$ is an arbitrary polynomial of the power $\omega(G)$ with constant finite coefficients. By definition, the polynomial $P^{\omega(G)}$ acts on the diagram G as follows: the diagram is shrunk into a point, and the polynomial $P^{\omega(G)}$ acts at this point. Since the polynomial $P^{\omega(G)}(G)$ is finite, the finiteness of $R(G)$ implies the same for $\tilde{R}(G)$. Thus, the subtraction procedure constructed on the base of the operation $\tilde{R}(G)$ has an arbitrariness related to the choice of the polynomial $P^{\omega(G)}$.

The subtraction operation R is based on the Taylor expansion of a diagram at zero external momenta. However, the finite result is also obtained if the subtraction procedure is fulfilled at arbitrary external momenta. Let $t_\mu^{\omega(G)}$ be the first $\omega(G)$ terms of the expansion at arbitrary momenta μ. In particular, $t^{\omega(G)} \;=\; t_0^{\omega(G)}$. We denote $R_\mu = 1 - t_\mu^{\omega(G)} + P^{\omega(G)}$ and consider

$$\tilde{R} = 1 - t_0^{\omega(G)} + P^{\omega(G)} = 1 - t_\mu^{\omega(G)} + \big[t_\mu^{\omega(G)} - t_0^{\omega(G)}\big] + P^{\omega(G)} = R_\mu + t_\mu^{\omega(G)} - t_0^{\omega(G)}.$$

Previously, we showed that $\tilde{R}(G)$ is finite. Let us prove that the difference $t_\mu^{\omega(G)} - t_0^{\omega(G)}$ is also finite. Indeed, $t_\mu^{\omega(G)}R(G)$ is a finite expression, since $R(G)$ is finite and

[3] We discuss this part in detail in Part II when consider quantum gravity models.

$t_\mu^{\omega(G)} R(G)$ is the sum of the first $\omega(G)$ terms of the Taylor expansion of the finite function. However,

$$t_\mu^\omega(G)R(G) = t_\mu^{\omega(G)}(G) - t_\mu^{\omega(G)} t_0^{\omega(G)}(G).$$

But $t_0^{\omega(G)}(G)$ is a polynomial of order $\omega(G)$. Furthermore, $t_\mu^{\omega(G)} t_0^{\omega(G)}(G)$ is the sum of the first $\omega(G)$ terms of the expansion of this polynomial. Since the Taylor expansion of a polynomial is the same polynomial, $t_\mu^{\omega(G)} t_0^{\omega(G)}(G) = t_0^{\omega(G)}(G)$. Thus,

$$t_\mu^{\omega(G)} R(G) = t_\mu^{\omega(G)}(G) - t_0^{\omega(G)}(G). \tag{9.106}$$

Here the *l.h.s.* is finite, so the *r.h.s.* is also finite.

From (9.105) we know that

$$R_\mu(G) = \tilde{R}(G) - \left[t_\mu^{\omega(G)}(G) - t_0^{\omega(G)}(G)\right]. \tag{9.107}$$

Once again, since the *r.h.s.* of this equation is finite, the *l.h.s.* is also finite. Moreover,

$$\begin{aligned}\tilde{R}(G) - \left[t_\mu^{\omega(G)}(G) - t_0^{\omega(G)}(G)\right] &= G - t_0^{\omega(G)}(G) + P^{\omega(G)}(G) \\ - \left[t_\mu^{\omega(G)}(G) - t_0^{\omega(G)}(G)\right] &= G - t_0^{\omega(G)}(G) + \tilde{P}^{\omega(G)}(G),\end{aligned}$$

where $\tilde{P}^{\omega(G)} = P^{\omega(G)} - \left[t_\mu^{\omega(G)} - t_0^{\omega(G)}\right]$ is a polynomial of the order $\omega(G)$, which differs from the polynomial $P^{\omega(G)}$ only by constant coefficients. Similarly,

$$\begin{aligned}R_{\mu_1}(G) - R_{\mu_2}(G) &= \tilde{R}(G) - \left[t_{\mu_1}^{\omega(G)}(G) - t_0^{\omega(G)}(G)\right] - \tilde{R}(G) + \left[t_{\mu_2}^{\omega(G)} - t_0^{\omega(G)}(G)\right] \\ &= t_{\mu_2}^{\omega(G)}(G) - t_{\mu_1}^{\omega(G)}(G)) \;=\; \bar{P}^{\omega(G)}(G),\end{aligned} \tag{9.108}$$

where $\bar{P}^{\omega(G)}(G)$ is a polynomial of the power $\omega(G)$ with finite coefficients.

From the considerations presented above, we can draw the following conclusions: **1.** Subtractions at any point in external momenta provide the finite Feynman integral. **2.** In general, the subtraction procedure leads to an arbitrariness parameterized by the polynomial $P^{\omega(G)}$. **3.** Two subtraction procedures with distinct subtraction points differ by a finite polynomial.

Since the subtraction procedure contains an arbitrariness, the same is true for the counterterms and for the expressions for the renormalization constants. This arbitrariness can be used to impose physically important conditions on the finite parts of the diagrams.

9.7 Renormalization conditions

The arbitrariness in the expressions for renormalization constants that we discussed in section 9.6 can lead to arbitrariness in the physical quantities calculated on the basis of the renormalized Lagrangian. In order to fix this arbitrariness, we have to impose special conditions on renormalized theory, which are called renormalization conditions.

In renormalizable theories, there is a finite number of renormalization constants. Therefore, we need a finite number of additional conditions, which is, in general, equal to the number of renormalization constants. In contrast, in non-renormalizable theories, where there is an infinite variety of counterterms, the number of renormalization conditions, formally, is also infinite.

The concrete forms of renormalization conditions are defined by physical reasons. We consider this issue for the example of scalar field theory with the Lagrangian

$$\mathcal{L} = \frac{1}{2}\,\partial_\mu\varphi\partial^\mu\varphi - \frac{1}{2}\,m^2\varphi^2 - \lambda\varphi^4.$$

The corresponding renormalized Lagrangian is

$$\mathcal{L}_R = \frac{1}{2}z_1\partial_\mu\varphi_R\partial^\mu\varphi_R - \frac{1}{2}z_1z_m m_R^2\varphi_R^2 - z_1^2 z_\lambda\lambda_R\varphi_R^4.$$

The theory contains three renormalization constants, z_1, z_m and z_λ. Therefore, we need three renormalization conditions.

Consider the exact two-point Green function in the renormalized theory, $G_R(p)$. The inverse Green function has the form

$$G_R^{-1}(p) = G_0^{-1}(p) - \Sigma_R(p)\,. \tag{9.109}$$

Here $G_0^{-1}(p) = \tilde{D}^{-1}(p) = p^2 - m_R^2$ is an inverse propagator, and the function $\Sigma_R(p)$ is affected by interactions. The $\Sigma_R(p)$ is a renormalized two-point vertex function, which is called a mass operator, contains an arbitrariness and is defined by the diagrams

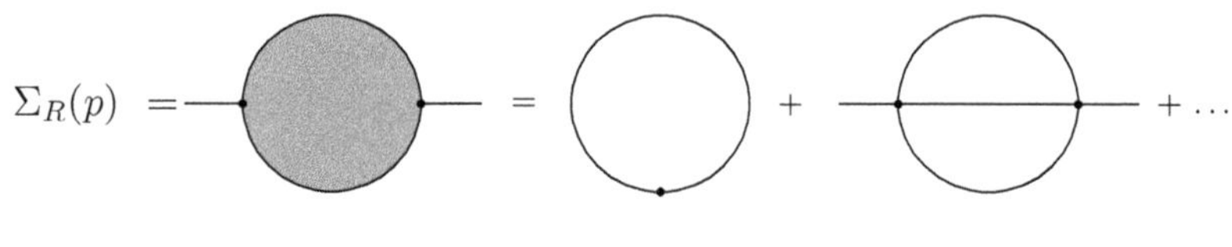

We define the observable mass of the particle as a solution to the equation

$$G_R^{-1}(p)\Big|_{p^2=m^2_{phys}} = 0. \tag{9.110}$$

Thus, the observable particle mass m^2_{phys} is defined as a pole of the exact two-point Green function. As a result, we have two mass parameters in the theory, namely, renormalized mass m_R in the Lagrangian $\mathcal{L}_R$ and the observable mass m_{phys}.

The first renormalization condition in this theory is $m_R = m_{phys}$. This means we require that the parameter m_R in the Lagrangian coincide with the observable particle mass. An equivalent form of the same renormalization condition is

$$\Sigma_R\left(p^2, m_R^2, P^{(2)}\right)\Big|_{p^2=m^2_{phys},\; m_R^2=m^2_{phys}} = 0, \tag{9.111}$$

where we took into account the fact that the diagrams for $\Sigma(p)$ have a superficial degree of divergence $\omega(G) = 2$. Thus, the result for Σ_p depends on an arbitrary second-order polynomial. The conditions (9.111) can be treated as the equation for the coefficients of the polynomial $P^2(p)$. Consider how this equation works.

Let $\Sigma(p)$ be the sum of the diagrams for the two-point function in the bare theory. Using the subtraction procedure at $p^2 = m^2_{phys}$, one gets

$$\Sigma_R = \Sigma - t^{(2)}_{m_{phys}}(\Sigma) + P^{(2)}(p). \tag{9.112}$$

Since the Lagrangian does not contain the distinguished vector, the polynomial $P^{(2)}(p)$ should be constructed from invariant quantities, so it has the general form

$$P^{(2)}(p) = c_1 + c_2\left(p^2 - m^2_{phys}\right), \tag{9.113}$$

where c_1, c_2 are arbitrary coefficients. Thus,

$$\Sigma_R\big|_{p^2=m^2_{phys}} = \left\{\Sigma - t^{(2)}_{m_{phys}}(\Sigma)\right\}\Big|_{p^2=m^2_{phys}} + c_1.$$

One can note that the expansion of the function $\Sigma - t^2_{m_{phys}}(\Sigma)$ into a Taylor series at $p^2 = {m_{phys}}^2$ has the form $b(p^2 - m^2_{phys})^2 + \mathcal{O}((p^2 - m^2_{phys})^3)$ with a constant b. Then

$$\left\{\Sigma - t^{(2)}_{m_{phys}}(\Sigma)\right\}\Big|_{p^2=m^2_{phys}} = 0, \tag{9.114}$$

and we arrive at

$$\Sigma_R\Big|_{p^2=m^2_{phys}} = c_1 = 0 \tag{9.115}$$

as a consequence of the renormalization condition (9.111), which also yields

$$\Sigma_R(p^2, m_{phys}, P^{(2)}) = (p^2 - m^2_{phys})\tilde{\Sigma}_R(p^2, m_{phys}, P^{(2)}), \tag{9.116}$$

so

$$G_R^{-1} = p^2 - m^2_{phys} - \Sigma_R = \left(p^2 - m^2_{phys}\right)\left\{1 - \tilde{\Sigma}_R(p^2, m_{phys}, P^{(2)})\right\}. \tag{9.117}$$

We can denote

$$1 - \tilde{\Sigma}_R(p^2, m, P^{(2)})\Big|_{p^2=m^2_{phys}} = z^{-1}. \tag{9.118}$$

Then the exact two-point Green function in the vicinity of the pole has the form

$$G_R(p) = \frac{z}{p^2 - m^2_{phys}}, \tag{9.119}$$

where z is a real constant. Such a constant can be eliminated by the field renormalization $\varphi = z^{\frac{1}{2}}\varphi_R$. Taking this into account, we put $z = 1$. As a result, one gets

$$\tilde{\Sigma}_R(p^2, m_{phys}, P^{(2)})\Big|_{p^2=m^2_{phys}} = 0. \tag{9.120}$$

The last formula represents the second renormalization condition.

Consider how the relation (9.120) works. Let us write again

$$\Sigma_R = \{\Sigma - t^2_{m_{phys}}(\Sigma)\} + c_2\left(p^2 - m^2_{phys}\right). \tag{9.121}$$

At the same time,

$$\Sigma - t^{(2)}_{m_{phys}}(\Sigma) = b\left(p^2 - m^2_{phys}\right) + \mathcal{O}\big((p^2 - m_{phys})^3\big), \tag{9.122}$$

where b is a constant. Then

$$\Sigma_R(p^2, m_{phys}, P^{(2)}) \;=\; (p^2 - m^2_{phys})\{b(p^2 - m^2_{phys}) + c_2 + \ldots\} \tag{9.123}$$

and hence

$$\tilde{\Sigma}_R \;=\; b\left(p^2 - m^2_{phys}\right) + \mathcal{O}\left((p^2 - m^2_{phys})^2\right) + c_2. \tag{9.124}$$

Equations (9.120), (9.124) lead to

$$\tilde{\Sigma}_R\Big|_{p^2 - m^2_{phys}} = 0 = c_2.$$

We obtain $c_2 = 0$ and, therefore, $P^{(2)}(p) = 0$. The relation $c_2 = 0$ is a consequence of the second renormalization condition.

As a result, if the subtraction procedure is fulfilled at $p^2 = m^2_{phys}$ and the renormalization conditions are $m^2_R = m^2_{phys}$ and $\tilde{\Sigma}_R\big|_{p^2 = m^2_{phys}} = 0$, then

a) An exact two-point Green function has a pole at $p^2 = m^2_{phys}$.

b) A residue of the Green function in the pole is equal to 1.

c) An arbitrary polynomial $P^{(2)}(p)$ is fixed to be zero.

There may be other renormalization conditions that can fix an arbitrary polynomial $P^{(2)}(p)$, such as, the condition

$$G^{-1}_R\left(p^2, m_{phys}, P^{(2)}\right)\Big|_{p^2=0,\ m^2_{phys}\to 0} \;\longrightarrow\; 0. \tag{9.125}$$

This condition allows us to find the coefficient c_1 at $P^{(2)}(p)$. However, the existence of such a limit of the inverse exact Green function is incompatible with the exact statement that the residue of an exact Green function in the pole $p^2 = 0$ should be equal to 1. This fact is a manifestation of the infrared singularities. To avoid this problem, the renormalization condition on $\tilde{\Sigma}_R$ should be imposed not at $p^2 = m^2_{phys}$ but at $p^2 = -\mu^2$, where μ^2 is an arbitrary massive parameter. The renormalization condition in this case is taken in the form

$$\tilde{\Sigma}_R\big|_{p^2=-\mu^2} \;=\; 0. \tag{9.126}$$

Then the limit $m^2_{phys} \to 0$ exists, but the final results may depend on an arbitrary μ^2.

Thus, we imposed two renormalization conditions. However, we have three renormalization constants and therefore need one more renormalization condition. Such a condition is imposed on the four-point vertex function.

An exact four-point vertex function is defined by the following diagrams:

$\Gamma^4_R(p_1, p_2, p_3, p_4) =$ 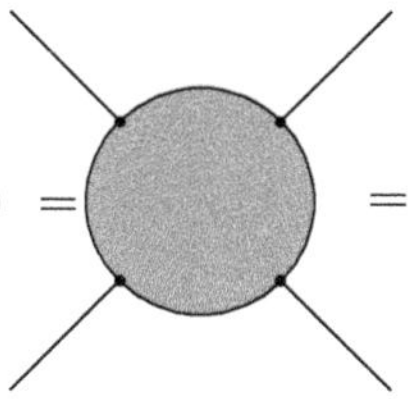 $= -\ \lambda_R +$ 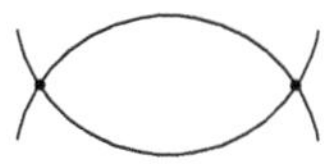 $+\ \ldots$

$+$ counterterms. (9.127)

The renormalization condition for the four-point function defines the momentum scale μ_0, where the observable coupling constant is measured. Therefore, the third renormalization condition has the form

$$\Gamma_R^{(4)}\Big|_{\lambda_R=\lambda_{phys} \text{ at } \mu=\mu_0} = -\lambda_{phys}. \tag{9.128}$$

Eq. (9.128) means that the renormalized coupling constant λ_R in Lagrangian $\mathcal{L}_R$ coincides with the observable physical coupling constant at the momenta μ_0, measured, e.g., in a scattering processes.

Usually, the following renormalization conditions are considered:

a. $\sum_{i=1}^4 p_i = 0$, $p_i^2 = m^2_{phys}$, $i = 1, 2, 3, 4$, $s = t = u$, where $s = (p_1 + p_2)^2$, $t = (p_1 - p_3)^2$, $u = (p_1 - p_4)^2$. The quantities u, s, t are called the Mandelstam parameters. However, in this case, the condition $p_i^0 > 0$ is violated for some of the momenta.

b. $p_i^2 = \mu^2$, $s = t = u$, where μ is an arbitrary massive parameter. In this case, a limit $m^2_{phys} \to 0$ exists for all Green functions, where all momenta are off shell.

To conclude, if a theory is multiplicatively renormalizable, a number of renormalization constants is finite and equal to a number of fields and parameters. To fix an arbitrariness of the subtraction procedure, we have to impose the renormalization conditions. The number of this conditions is equal to the number of renormalization constants. After that, the arbitrariness is fixed, and the renormalized theory can be used for calculating physical quantities. If the theory is non-renormalizable, formally there is an infinite arbitrariness in the counterterms, and an infinite number of the renormalization conditions is needed to fix this arbitrariness.

9.8 Renormalization with the dimensional regularization

Dimensional regularization is one of the most convenient regularization schemes in modern quantum field theory, especially for the study of the general aspects of the theory. Consider the use of dimensional regularization for the example of scalar field theory, with the action

$$S[\varphi] = \int d^4x\, \Big(\frac{1}{2}\,\partial_\alpha\varphi\partial^\alpha\varphi - \frac{1}{2}\,m^2\varphi^2 - \lambda\varphi^4\Big).$$

The theory is multiplicatively renormalizable, and the renormalized action is obtained from regularized bare action by the transformations (9.94),

Dimensional regularization assumes that the theory has been formulated in n-dimensional spacetime, where n is an arbitrary integer number. It is convenient to keep the dimensions of renormalized quantities the same as in the four-dimensional case, while the dimensions of the bare quantities correspond to n-dimensional spacetime. This simplifies the limit $n \to 4$ in the renormalized Green functions and physical quantities. In n dimensions, the field φ has dimension $[\varphi] = \frac{n-2}{2}$, the coupling constant λ has dimension $[\lambda] = 4 - n$ and the dimension of mass $[m] = 1$. The renormalization constants should be dimensionless. Let us introduce an arbitrary mass-dimensional parameter μ and redefine the renormalization transformations as follows:

$$\varphi = \mu^{\frac{n-4}{2}} z_1^{\frac{1}{2}} \varphi_R, \qquad m^2 = z_m m_R^2, \qquad \lambda = \mu^{4-n} z_\lambda \lambda_R. \tag{9.129}$$

The theory under consideration is multiplicative renormalizable, which means the renormalized action S_R can be obtained from the regularized bare action S by the renormalization transformation (9.129),

$$S[\varphi, m, \lambda, n] = S_R[\varphi_R, m_R, \lambda_R, n, \mu]. \tag{9.130}$$

Both sides here are written in n-dimensional spacetime and both actions are regularized, so all diagrams constructed on the base of these actions are finite. As a result,

$$S_R = \mu^{n-4} \int d^n x \left\{ \frac{1}{2} z_1 \partial^\alpha \varphi_R \partial_\alpha \varphi_R - \frac{1}{2} z_1 z_m m_R^2 \varphi_R^2 - z_1^2 z_\lambda \lambda_R \varphi_R^4 \right\}. \tag{9.131}$$

In dimensional regularization, the divergences of the Feynman diagrams appear in expressions of the form $\sum_{l=1}^{k} \frac{f_l}{(n-4)^l} + d_k$, where f_l and d_k are some finite constants, $k = 1, 2, \ldots$. The renormalization ambiguity means there is an arbitrariness in the choice of the constants d_k. In what follows, we use the so-called scheme of minimal subtractions, in which, in order to renormalize a theory, we simply cancel the pole terms $\frac{f_k}{(n-4)^k}$ and do not consider any finite constants d_k. Thus, in dimension regularization and the scheme of minimal subtractions, the renormalization constants are given by the Laurent series in $\frac{1}{(n-4)}$, e.g.,

$$z_1 = 1 + \sum_{k=1}^{\infty} \frac{a_k(\lambda_R)}{(n-4)^k}, \quad z_m = 1 + \sum_{k=1}^{\infty} \frac{b_k(\lambda_R)}{(n-4)^k}, \quad z_\lambda = 1 + \sum_{k=1}^{\infty} \frac{c_k(\lambda_R)}{(n-4)^k}. \tag{9.132}$$

Here all the coefficients a_k, b_k, c_k are calculated in the framework of the loop expansion and therefore have the form

$$a_k = \sum_{L=1}^{\infty} \hbar^L a_k^{(L)}, \qquad b_k = \sum_{L=1}^{\infty} \hbar^L b_k^{(L)}, \qquad c_k = \sum_{L=1}^{\infty} \hbar^L c_k^{(L)}, \tag{9.133}$$

where the index L is the order of loop expansion. One can prove that a_k, b_k, c_k do not depend on m_R and depend only on λ_R. In the one-loop approximation, $k = 1$.

The *l.h.s.* of relations (9.129) is μ-independent, hence, the *r.h.s.* is also μ-independent. This situation is possible only if the renormalized quantities φ_R, m_R

and λ_R depend on μ. In the subsequent sections, we explore how these quantities change under an infinitesimal change of μ.

Let $\mu = \mu'(1-\alpha)$, where μ' is a new parameter and α is an infinitesimal constant. Then one gets the following from the relations (9.129) for λ:

$$\begin{aligned}\lambda &= \mu^{4-n}\Big[1+\sum_{k=1}^{\infty}\frac{c_k(\lambda_R)}{(n-4)^k}\Big]\lambda_R = \mu'^{4-n}(1-\alpha)^{4-n}\Big[1+\sum_{k=1}^{\infty}\frac{c_k(\lambda_R)}{(n-4)^k}\Big]\lambda_R \\ &= \mu'^{4-n}(1+(n-4)\alpha)\Big[1+\sum_{k=1}^{\infty}\frac{c_k(\lambda_R)}{(n-4)^k}\Big]\lambda_R \qquad (9.134) \\ &= \mu'^{4-n}\Big[\lambda_R+(n-4)\alpha\lambda_R+\sum_{k=1}^{\infty}\frac{c_k(\lambda_R)\lambda_R}{(n-4)^k}+\alpha c_1(\lambda_R)\lambda_R+\alpha\sum_{k=1}^{\infty}\frac{c_{k+1}(\lambda_R)\lambda_R}{(n-4)^k}\Big].\end{aligned}$$

Let us denote

$$\tilde{\lambda}_R = \lambda_R + \alpha[(n-4)\lambda_R + c_1(\lambda_R)\lambda_R]. \qquad (9.135)$$

Substituting this relation into (9.135), after long but simple transformations, we obtain

$$\begin{aligned}\lambda &= \mu'^{4-n}\Big\{\tilde{\lambda}_R - \alpha c_1(\tilde{\lambda}_R)\lambda_R - \alpha c_1'(\tilde{\lambda}_R)\tilde{\lambda}_R^2 \qquad (9.136) \\ &+\sum_{k=1}^{\infty}\frac{c_k(\tilde{\lambda}_R)-\alpha c_{k+1}'(\tilde{\lambda}_R)\tilde{\lambda}_R-\alpha c_k(\tilde{\lambda}_R)c_1(\tilde{\lambda}_R)-\alpha c_k'(\tilde{\lambda}_R)c_1(\tilde{\lambda}_R)\tilde{\lambda}_R}{(n-4)^k}\tilde{\lambda}_R\Big\}.\end{aligned}$$

Denote

$$\overline{\lambda}_R = \lambda_R + \alpha\big[(n-4) - c_1'(\tilde{\lambda}_R)\tilde{\lambda}_R\big]\lambda_R$$

and

$$\overline{c}_k(\overline{\lambda}_R)\overline{\lambda}_R = \Big[c_k(\tilde{\lambda}_R)-\alpha c_{k+1}'(\tilde{\lambda}_R)\tilde{\lambda}_R-\alpha c_k(\tilde{\lambda}_R)c_1(\tilde{\lambda}_R)-\alpha c_k'(\tilde{\lambda}_R)c_1(\tilde{\lambda}_R)\tilde{\lambda}_R\Big]\tilde{\lambda}_R\Big|,$$

where $\Big|$ means $\tilde{\lambda}_R = \overline{\lambda}_R + \alpha c_1(\overline{\lambda}_R)\overline{\lambda}_R + \alpha c_1'(\overline{\lambda}_R)\overline{\lambda}_R^2$. After that, we get

$$\lambda = \mu'^{4-n}\Big[1+\sum_{k=1}^{\infty}\frac{\overline{c}_{k+1}(\overline{\lambda}_R)}{(n-4)^k}\Big]\overline{\lambda}_R. \qquad (9.137)$$

As a result, we obtain $\lambda = \mu'^{4-n}\overline{z}_\lambda(\overline{\lambda}_R)\overline{\lambda}_R$, which differs from $\lambda = \mu^{4-n}z_\lambda(\lambda_R)\lambda_R$ by another coupling $\overline{\lambda}_R = (1+\alpha\,[(n-4)-c_1'(\lambda_R)])\,\lambda_R$. This relation can be written as

$$\overline{\lambda}_R = \overline{\overline{z}}_\lambda(\lambda_R)\lambda_R, \qquad (9.138)$$

where $\overline{\overline{z}}_\lambda(\lambda_R)$ is finite at $n\to 4$. We see that a change in the arbitrary parameter μ means a finite renormalization of the coupling constant.

Analogously, for m_R^2 and φ_R, one gets

$$\varphi = \mu'^{\frac{n-4}{2}}\left(1+\sum_{k=1}^{\infty}\frac{\overline{a}_k(\overline{\lambda}_R)}{(n-4)^k}\right)\overline{\varphi}_R, \qquad m^2 = \left(1+\sum_{k=1}^{\infty}\frac{\overline{b}_k(\overline{\lambda}_R)}{(n-4)^k}\right)\overline{m}_R^2, \tag{9.139}$$

where

$$\overline{\varphi}_R = \overline{\overline{z}}_1^{\frac{1}{2}}(\lambda_R)\varphi_R, \qquad \overline{m}_R^2 = \overline{\overline{z}}_m(\lambda_R)m_R^2. \tag{9.140}$$

The constants $\overline{\overline{z}}_1$ and $\overline{\overline{z}}_m$ are finite at $n \to 4$.

Thus, any change of arbitrary parameter μ in (9.129) is equivalent to a finite multiplicative renormalization of the field φ_R and the parameters m_R^2, λ_R. The set of the finite renormalization transformations (9.138) and (9.140) has group properties and is called the *renormalization group.*

9.9 Renormalization group equations

Renormalization group equations are the equations for renormalized vertex functions, following from the multiplicative renormalization of the theory. We will derive the renormalization group equations within the minimal subtraction scheme of renormalization, for scalar field theory as an example, using dimensional regularization. Let us note that the reader will be able to compare the scheme considered here with the renormalization group equations in curved spacetime, and with the renormalization group equations based on the more physical momentum-subtraction scheme. The main elements of both these approaches will be discussed in Part II of the book.

9.9.1 Derivation of renormalization group equations

Consider the generating functional of the Green functions in the dimensionally regularized theory,

$$Z[J] = \int \mathcal{D}\varphi e^{i(S[\varphi,m^2,\lambda,n]+\int d^4x\varphi J)} \equiv Z[J,m^2,\lambda,n]. \tag{9.141}$$

First of all, we make a change of variable corresponding to the field renormalization in the functional integral, $\varphi = \mu^{\frac{n-4}{2}} z_1^{\frac{1}{2}}\varphi_R$. The Jacobian of this change of variable is

$$\text{Det}\left\|\frac{\delta\varphi(x)}{\delta\varphi_R(y)}\right\| = e^{\text{Tr}\left[\log\left(\mu^{\frac{n-4}{2}}z_1^{\frac{1}{2}}\right)\delta^4(x-y)\right]} = e^{\delta^{(4)}(0)\int d^4x\log\left(\mu^{\frac{n-4}{2}}z_1^{\frac{1}{2}}\right)} = 1. \tag{9.142}$$

Here we have used the property $\delta^{(4)}(0) = 0$, which is valid in dimensional regularization. After this change in variables, we get

$$Z[J,m^2,\lambda,n] = \int \mathcal{D}\varphi_R e^{i\left\{S\left[\mu^{\frac{n-4}{2}}z_1^{\frac{1}{2}}\varphi_R,m^2,\lambda,n\right]+\mu^{\frac{n-4}{2}}z_1^{\frac{1}{2}}\int d^4x\varphi_R J\right\}}.$$

Using the relations from Eq. (9.129), in the exponential of the last integrand, we obtain

$$S\left[\mu^{\frac{n-4}{2}}z_1\varphi_R,\ z_m m_R^2,\ \mu^{\frac{4-n}{4}}z_\lambda\lambda_R,\ n\right].$$

According to (9.130), this expression is equal to $S_R\left[\varphi_R, m_R, \lambda_R, \mu, n\right]$. Denoting

$$J_R = \mu^{\frac{n-4}{2}} z_1^{\frac{1}{2}} J, \tag{9.143}$$

we get

$$\begin{aligned} Z[J, m^2, \lambda, n] &= \int \mathcal{D}\varphi_R e^{i(S_R[\varphi_R, m_R^2, \lambda_R, \mu, n] + \int d^4x \varphi_R J_R)} \\ &= Z_R[J_R, m_R^2, \lambda_R, \mu, n], \end{aligned} \tag{9.144}$$

where Z_R is the generating functional of renormalized Green functions.

Let us introduce the generating functionals of connected Green functions for the bare and renormalized theories as follows:

$$Z = e^{iW}, \qquad Z_R = e^{iW_R}. \tag{9.145}$$

Then we get from (9.144) and (9.145),

$$W[J, m^2, \lambda, n] = W_R[J_R, m_R^2, \lambda_R, \mu, n]. \tag{9.146}$$

The background fields Φ and Φ_R can be defined by the relations

$$\Phi = \frac{\delta W}{\delta J}, \qquad \Phi_R = \frac{\delta W_R}{\delta J_R},$$

which gives

$$\begin{aligned} \Phi(x) = \frac{\delta W}{\delta J(x)} &= \int d^4y \frac{\delta W_R}{\delta J_R(y)} \frac{\delta J_R(y)}{\delta J_R(x)} \\ &= \int d^4y \, \mu^{\frac{n-4}{2}} z_1^{\frac{1}{2}} \Phi_R(y) \, \delta^4(x-y) \;=\; \mu^{\frac{n-4}{2}} z_1^{\frac{1}{2}} \Phi_R, \end{aligned} \tag{9.147}$$

which means that the background fields Φ and Φ_R are related by the same transformation as the fields φ and φ_R. Furthermore, using (9.143) and (9.147), for the effective action, we get

$$\begin{aligned} \Gamma[\Phi, m^2, \lambda, n] &= W[J, m^2, \lambda, n] - \int d^4x \phi J \\ &= W_R[\phi_R, m_R^2, \lambda_R, \mu, n] - \int d^4x \phi_R J_R \;=\; \Gamma_R[\Phi_R, m_R^2, \lambda_R, \mu, n]. \end{aligned}$$

Finally, we get the important relation

$$\Gamma[\Phi, m^2, \lambda, n] = \Gamma_R[\Phi_R, m_R^2, \lambda_R, \mu, n], \tag{9.148}$$

which is the starting point for deriving the renormalization group equations. Since the *l.h.s.* of the last equation is μ-independent, we have

$$\mu \frac{d}{d\mu} \Gamma_R[\Phi_R, m_R^2, \lambda_R, \mu, n] = 0, \tag{9.149}$$

which can be written in the form

$$\mu \frac{\partial \Gamma_R}{\partial \mu} + \frac{\partial \Gamma_R}{\partial \lambda_R} \mu \frac{d\lambda_R}{d\mu} + \frac{\partial \Gamma_R}{\partial m_R^2} \mu \frac{dm_R^2}{d\mu} + \int d^4x \frac{\delta \Gamma_R}{\delta \Phi_R(x)} \mu \frac{\Phi_R(x)}{d\mu} = 0. \qquad (9.150)$$

Equation (9.150) is written for renormalized effective action. Therefore, it is possible to take the limit $n \to 4$, removing the regularization.

Let us introduce the new objects

$$\beta = \beta_\lambda = \mu \frac{d\lambda_R}{d\mu}\Big|_{n=4}, \qquad \gamma_m = -\frac{1}{m_R^2} \mu \frac{dm_R^2}{d\mu}\Big|_{n=4}, \qquad \gamma = \frac{1}{\Phi_R} \mu \frac{d\Phi_R}{d\mu}\Big|_{n=4}, \qquad (9.151)$$

which are called renormalization group functions, or beta and gamma functions. $\gamma(\lambda)$ is also called the anomalous dimension. $\beta(\lambda)$ is sometimes called the Gell-Mann–Low function.

For the sake of simplicity, we redefine $\lambda_R = \lambda$, $m_R^2 = m$, $\Phi_R = \phi$. As a result, one gets Eq. (9.150) in the form

$$\left[\mu \frac{\partial}{\partial \mu} + \beta \frac{\partial}{\partial \lambda} - m^2 \gamma_m \frac{\partial}{\partial m^2} + \int d^4x \phi(x) \frac{\delta}{\delta \phi(x)}\right] \Gamma_R = 0. \qquad (9.152)$$

Here $\Gamma_R = \Gamma_R\left[\phi, m^2, \lambda, \mu, n\right]$ in the limit $n \to 4$. Eq. (9.152) is called the renormalization group equation for the (renormalized) effective action.

The renormalized vertex functions $\Gamma_R^{(2l)}(x_1, \ldots, x_{2l}; m^2, \lambda, \mu)$ are defined as follows:

$$\Gamma_R^{(2l)}(x_1, \ldots, x_{2l}; m^2, \lambda, \mu) = \frac{\delta^{2l} \Gamma_R}{\delta\phi(x_1) \ldots \delta\phi(x_{2l})}\Big|_{\phi=0}.$$

Differentiating the equation (9.152) $2l$ times and setting $\phi = 0$, one gets

$$\left(\mu \frac{\partial}{\partial \mu} + \beta \frac{\partial}{\partial \lambda} - m^2 \gamma_m \frac{\partial}{\partial m^2} + 2l\gamma\right) \Gamma_R^{(2l)}(x_1, \ldots, x_{2l}; m^2, \lambda, \mu) = 0. \qquad (9.153)$$

Equation (9.153) is called the renormalization group equation for renormalized vertex functions. After performing a Fourier transform on Eq. (9.153), one obtains

$$\left(\mu \frac{\partial}{\partial \mu} + \beta \frac{\partial}{\partial \lambda} - m^2 \gamma_m \frac{\partial}{\partial m^2} + 2l\gamma\right) \bar{\Gamma}_R^{(2l)}(p_1, \ldots, p_{2l}; m^2, \lambda, \mu) = 0, \qquad (9.154)$$

which is the renormalization group equation for vertex functions in the momentum representation.

9.9.2 Renormalization group with scale-transformed momenta

Eq. (9.154) contains the parameter μ, and it would be quite interesting to link this parameter with some physical quantity. Let us solve this problem by rewriting the mentioned equation in another form.

The vertex functions in perturbation theory are homogeneous functions of all dimensional parameters, i.e., they satisfy the conditions

$$\bar{\Gamma}_R^{(2l)}(ap_1, \dots, ap_{2l};\, a^2m^2, \lambda, a\mu) \;=\; a^{d_{\bar{\Gamma}^{(2l)}}}\bar{\Gamma}_R^{(2l)}(p_1, \dots, p_{2l};\, m^2, \lambda, \mu), \tag{9.155}$$

where $d_{\bar{\Gamma}^{(2l)}}$ is the order of homogeneity and a is an arbitrary constant. Using the Euler theorem on homogeneous functions, one gets

$$\left\{ \sum_{i=1}^{2l} p_i^\alpha \frac{\partial}{\partial p_i^\alpha} + 2m^2 \frac{\partial}{\partial m^2} + \mu \frac{\partial}{\partial \mu} - d_{\bar{\Gamma}^{(2l)}} \right\} \bar{\Gamma}_R^{(2l)}(p_1, \dots, p_{2l}; m^2, \lambda, \mu) = 0. \tag{9.156}$$

We can rewrite this equation as

$$\mu \frac{\partial \bar{\Gamma}_R^{(2l)}}{\partial \mu} \;=\; -\left[\sum_{i=1}^{2l} p_i^\alpha \frac{\partial}{\partial p_i^\alpha} + 2m^2 \frac{\partial}{\partial m^2} - d_{\bar{\Gamma}^{(2l)}} \right] \tilde{\Gamma}^{(2l)}. \tag{9.157}$$

Substituting (9.157) into (9.154), we obtain

$$\left\{ \sum_{i=1}^{2l} p_i^\alpha \frac{\partial}{\partial p_i^\alpha} - \beta \frac{\partial}{\partial \lambda} + m^2(\gamma_m + 2) \frac{\partial}{\partial m^2} \right.$$
$$\left. - \left(d_{\bar{\Gamma}^{(2l)}} + 2l\gamma \right) \right\} \bar{\Gamma}_R^{(2l)}(p_1, \dots, p_{2l}; m^2, \lambda, \mu) = 0. \tag{9.158}$$

Here the quantity $d_{\bar{\Gamma}^{(2l)}}$ is the dimension $\bar{\Gamma}^{(2l)}$, which can be found from the relation

$$\bar{\Gamma}_R^{(2l)}(p_1, \dots, p_{2l}; m^2, \lambda, \mu) = \int d^4x_1 \dots d^4x_n e^{i\sum_{j=1}^{2l} x_j^\alpha p_{j\alpha}} \left. \frac{\delta^{2l}\Gamma_R[\Phi, m^2, \lambda, \mu]}{\delta\Phi(x_1)\dots\delta\Phi(x_{2l})} \right|_{\Phi=0}.$$

Since $\tilde{\Gamma}_R$ is the effective action, it is dimensionless. Therefore, we get

$$d_{\bar{\Gamma}^{(2l)}} = -4 \cdot 2l - 2l[\varphi] = -2l([\varphi] + 4)$$

and $d_{\bar{\Gamma}^{(2l)}} + 2l\gamma = 2l(\gamma - [\varphi] - 4)$. It is useful to denote

$$\gamma_\varphi = \gamma - [\varphi] - 4 \tag{9.159}$$

and use it in Eq. (9.158) with scale-transformed momenta, $\bar{\Gamma}_R^{(2l)}(ap_1, \dots, ap_{2l}; m^2, \lambda, \mu)$. In this way, we arrive at the equation

$$a \frac{\partial}{\partial a} \bar{\Gamma}_R^{(2l)}(ap_1, \dots, ap_{2l}, \dots) = \sum_{i=1}^{2l} \frac{\partial(ap_i^\alpha)}{\partial a} a \frac{\partial}{\partial(ap_i^\alpha)} \bar{\Gamma}_R^{(2l)}(ap_1, \dots, ap_{2l}, \dots)$$
$$= \sum_{i=1}^{2l} p_i^\alpha \frac{\partial}{\partial p_i^\alpha} \bar{\Gamma}_R^{(2l)}(ap_1, \dots, ap_{2l}, \dots). \tag{9.160}$$

Combining this identity with the original Eq. (9.158), and denoting $a = e^t$, we obtain

$$\left\{\frac{\partial}{\partial t} - \beta\frac{\partial}{\partial \lambda} + m^2(\gamma_m + 2)\frac{\partial}{\partial m^2} - 2l\gamma_\phi\right\}\bar{\Gamma}_R^{(2l)}(e^t p_1, \dots, e^t p_{2l}; m^2, \lambda, \mu) = 0. \quad (9.161)$$

The physical sense of this relation is quite transparent. The limits $t \to \infty$ and $t \to -\infty$ define the behavior of the vertex function $\bar{\Gamma}_R^{(2l)}(e^t p_1, \dots, e^t p_{2l}; m^2, \lambda, \mu) = 0$ at large momenta (UV) and small momenta (infrared), respectively.

Let us note that, from the physical viewpoint, the UV limit here is better defined than the infrared one. In chapter 16 of Part II, we shall see a few examples of the situation when the effect of the masses of quantum fields changes dramatically the infrared limit, compared to the result (9.161), which is based on the minimal subtraction scheme of renormalization and does not take this aspect of the quantum theory into account.

9.9.3 The Solution to the renormalization group equation

Consider a solution to Eq. (9.161). Let us denote

$$g = \left\{\lambda, \frac{m^2}{\mu^2}\right\}, \quad \beta_g = \left\{\beta, -(\gamma_m + 2)\frac{m^2}{\mu^2}\right\} = \beta_g(g), \quad 2l\gamma_\phi = \overline{\gamma} = \overline{\gamma}(g). \quad (9.162)$$

In these notations, Eq. (9.161) takes the form

$$\left(\frac{\partial}{\partial t} - \beta_g\frac{\partial}{\partial g} - \overline{\gamma}\right)\bar{\Gamma}_R^{(2l)}(e^t p_1, \dots, e^t p_{2l}; m^2, g, \mu) = 0. \quad (9.163)$$

We will look for the solution in the form

$$\bar{\Gamma}_R^{(2l)}(e^t p_1, \dots, e^t p_{2l}; g, \mu) = f(t, g)\mathcal{F}(p_1, \dots, p_{2l}; G(t, g), \mu), \quad (9.164)$$

where $f(t, g)$, $G(t, g)$ and $\mathcal{F}(p_1, \dots, p_{2l}; G(t, g), \mu)$ are some functions. Eq. (9.163) gives

$$\frac{\partial}{\partial t}\Gamma_R^{(2l)}(e^t p_1, \dots, e^t p_{2l}; g, \mu) = \frac{\partial f}{\partial t}\mathcal{F} + f\frac{\partial \mathcal{F}}{\partial G(t, g)}\frac{\partial G(t, g)}{\partial t},$$

or

$$\beta_g\frac{\partial}{\partial g}\Gamma_R^{(2l)}(e^t p_1, \dots, e^t p_{2l}; g, \mu) = \beta_g\frac{\partial f}{\partial g}\mathcal{F} + f\frac{\partial \mathcal{F}}{\partial G(t, g)}\beta_g\frac{\partial G(t, g)}{\partial g}. \quad (9.165)$$

Substituting (9.165) into (9.163), one gets

$$\frac{\partial f}{\partial t}\mathcal{F} + f\frac{\partial \mathcal{F}}{\partial G(t, g)}\frac{\partial G(t, g)}{\partial g} - \beta_g\frac{\partial f}{\partial g}\mathcal{F} - f\frac{\partial \mathcal{F}}{\partial G(t, g)}\beta_g\frac{\partial G(t, g)}{\partial g} - \overline{\gamma}f\mathcal{F} = 0. \quad (9.166)$$

Let us assume that the function $f(t, g)$ satisfies the equation

$$\frac{\partial f}{\partial t} - \beta_g\frac{\partial f}{\partial g} - \overline{\gamma}f = 0, \quad (9.167)$$

and the function $G(t, g)$ satisfies the equation

$$\frac{\partial G(t, g)}{\partial t} = \beta_g(g)\frac{\partial G(f, g)}{\partial g}. \quad (9.168)$$

Then, Eq. (9.166) will be automatically satisfied. Our aim is to describe the solutions to Eqs. (9.167) and (9.168).

Consider (9.151). From definitions (9.151), follows that

$$\beta_g(g) = \mu \frac{dg(\mu)}{d\mu}\bigg|_{n=4}. \tag{9.169}$$

Then one can write

$$\beta_G(G(t,g)) = \mu \frac{dG(t,g(\mu))}{d\mu}\bigg|_{n=4} = \frac{\partial G(t,g)}{\partial g}\mu \frac{dg(\mu)}{d\mu}\bigg|_{n=4} = \beta_g(g)\frac{\partial G(t,g)}{\partial g}. \tag{9.170}$$

Therefore, in the *r.h.s.* of (9.168), there is $\beta_G(G(t,g))$, and Eq. (9.168) takes the form

$$\frac{\partial G(t,g)}{\partial t} = \beta_G(G(t,g)). \tag{9.171}$$

We can denote $\beta_G(G(f,g)) = \beta_g(g)|_{g=G(f,g)}$, impose the initial condition $G(t,g)|_{t=0} = g$ and redefine $G(t,g) = g(t)$. As a result, we get the differential equation

$$\frac{dg(t)}{dt} = \beta_g(g(t)) \tag{9.172}$$

with the initial condition

$$g(t)\big|_{t=0} = g. \tag{9.173}$$

The function $g(t)$ satisfying the equations (9.172) and (9.173) is called the effective charge or the running parameter. In particular, $\lambda(t)$ is called the running coupling constant.

Concerning the solution to Eq. (9.167), we impose the initial condition in the form $f(t,g)\big|_{t=0} = 1$ and get the solution in the form

$$f(t,g) = e^{\int_0^t \overline{\gamma}(g(t'))dt'}, \tag{9.174}$$

where $g(t)$ is a solution to Eq. (9.172). One can check that the function (9.174) satisfies Eq. (9.167). It is evident that $f(t,g)\big|_{t=0} = 1$. Furthermore, $\frac{\partial f}{\partial t} = \overline{\gamma}(g(t))f$ and

$$\beta_g \frac{\partial f}{\partial g} = f\int_0^t dt' \frac{\partial \overline{\gamma}(g(t'))}{\partial g(t')}\beta_g \frac{dg(t')}{\partial g} = f\int_0^t dt' \frac{\partial \overline{\gamma}(g(t'))}{\partial g(t')}\frac{dg(t')}{dt'}$$
$$= f\int_0^t dt' \frac{d\overline{\gamma}(g(t'))}{dt'} = f(\overline{\gamma}(g(t)) - f(\overline{\gamma}(g(t))\Big|_{t=0} = f(\overline{\gamma}(g(t)) - f(\overline{\gamma}(g)).$$

Therefore, Eq. (9.167) boils down to

$$\frac{\partial f}{\partial t} - \beta_g \frac{\partial f}{\partial g} - \overline{\gamma} f = \overline{\gamma}(g(t))f - \overline{\gamma}(g(t))f + f\overline{\gamma}(g) - \overline{\gamma}(g)f = 0.$$

The unique element that remains to be found is the function $\mathcal{F}(p_1,\dots,p_{2l};g(t),\mu)$ in Eq. (9.164). Fixing $t = 0$ in (9.164), one gets

$$\Gamma_R^{(2l)}(p_1,\ldots,p_{2l};g,\mu) = f(t,g)\big|_{t=0} \times \mathcal{F}(p_1,\ldots,p_{2l};g(t),\mu)\big|_{t=0},$$

leading to

$$\mathcal{F}(p_1,\ldots,p_{2l};g,\mu) = \mathcal{F}(p_1,\ldots,p_{2l};g(t),\mu) = \Gamma_R^{(2l)}(p_1,\ldots,p_{2l};g,\mu). \tag{9.175}$$

Thus, we get

$$\Gamma_R^{(2l)}(e^t p_1,\ldots,e^t p_{2l};g,\mu) = e^{\int_0^t dt'\bar{\gamma}(g(t'))}\Gamma_R^{(2l)}(p_1,\ldots,p_{2l};g(t),\mu), \tag{9.176}$$

where the function $g(t)$ satisfies Eq. (9.172) with the initial condition (9.173).

Rewriting the last equation in terms of λ and m^2, one obtains

$$\Gamma_R^{(2l)}(e^t p_1,\ldots,e^t p_{2l};\lambda,m^2,\mu) = e^{2l\int_0^t dt'\gamma_\phi(\lambda(t'))}\Gamma_R^{(2l)}(p_1,\ldots,p_{2l};\lambda(t),m^2(t),\mu), \tag{9.177}$$

where the effective coupling constant $\lambda(t)$ and the effective mass $m^2(t)$ satisfy the renormalization group equations and initial conditions

$$\frac{d\lambda(t)}{dt} = \beta(\lambda(t)), \qquad \lambda(0) = \lambda \tag{9.178}$$

$$\frac{dm^2}{dt} = -\big[\gamma_m(\lambda(t)) + 2\big]\, m^2(t), \qquad m^2(0) = m^2. \tag{9.179}$$

Now we can analyse how (9.177) provides the relation between the vertex functions with different momenta. According to this relation, the scaling of momenta in a vertex function is equivalent to the running coupling λ, the square of mass m^2 and the running of the scalar field. Taking the limit $t \to \infty$, we can explore the behavior of vertex functions at large momenta. As a result, the UV asymptotics of vertex functions are controlled by the running of the coupling constant $\lambda(t)$ and the $m^2(t)$. Formally, the limit $t \to \infty$ defines the behavior of the vertex functions in the infrared.

The form of the equations (9.178) and (9.179) indicates the order of their solution. One first has to solve the equation for $\lambda(t)$, then use this solution in the equation for $m^2(t)$ and finally use both solutions to solve the last equation for $\varphi(t)$.

9.9.4 The behavior of the running coupling constant

Our next aim is to explore the behavior of $\lambda(t)$. Eq. (9.178) can be rewritten as

$$t = \int_\lambda^{\lambda(t)} \frac{d\lambda}{\beta(\lambda)}. \tag{9.180}$$

If $\lambda(t)$ has a finite limit λ_0 when $t \to \infty$, the integral in the *r.h.s.* of this equation should be divergent, and this is possible only if $\beta(\lambda_0) = 0$. This value of the coupling is called the *fixed point*. We can assume that $\beta(\lambda)$ changes its sign when passing from $\lambda < \lambda_0$ to $\lambda > \lambda_0$. It is clear that $\lambda(t)$ decreases in the region with $\beta(\lambda) < 0$ and increases when $\beta(\lambda) > 0$.

There are two different kinds of behavior of $\beta(\lambda)$ in the vicinity of the fixed point λ_0, as illustrated in the two plots shown below.

1. In the left plot, $\beta(\lambda) > 0$ at $\lambda < \lambda_0$ and changes sign at λ_0. It is easy to see that, in this case, $\lambda(t) \to \lambda_0$, independently of the initial value of λ. In this case, λ_0 is called the UV stable fixed point. Since the integral in the *r.h.s.* of (9.180) is divergent at $t \to \infty$, its value is mainly determined by the behaviour in the vicinity of λ_0, where we can write

$$\beta(\lambda) = \beta(\lambda_0) + \beta'(\lambda_0)(\lambda - \lambda_0), \qquad \text{where} \qquad \beta'(\lambda_0) = \frac{d\lambda}{dt}\Big|_{\lambda=\lambda_0} < 0. \tag{9.181}$$

Then

$$t \sim \int_\lambda^{\lambda(t)} \frac{d\lambda}{\beta'(\lambda_0)(\lambda - \lambda_0)} \qquad \Longrightarrow \qquad \lambda(t) \sim \lambda_0 + (\lambda - \lambda_0)e^{t\beta'(\lambda_0)}. \tag{9.182}$$

When $t \to \infty$, we observe $\lambda(t) \to \lambda_0$, as we expected.

If the equation $\lambda(t) = 0$ has a single root $\lambda_0 = 0$, $\lambda(t) \to 0$ in the UV ($t \to \infty$), independently of the initial value of λ. Then the behavior of the vertex functions at extremely large momenta is defined by the theory with a vanishing coupling constant. In other words, the UV behavior corresponds to the approximately free theory. Such a situation is called *asymptotic freedom*.

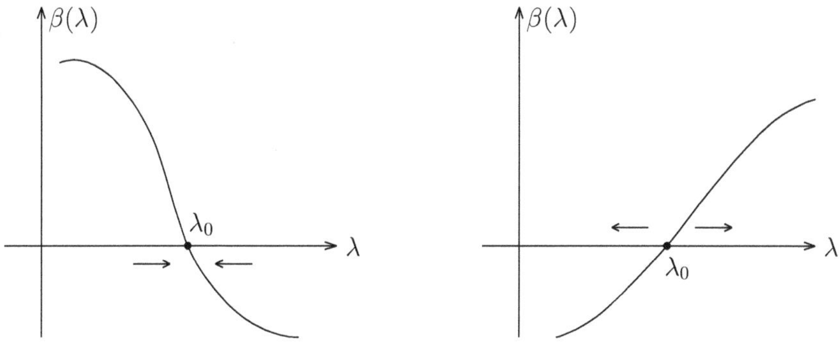

2. The right above, shows the case when $\beta(\lambda) < 0$ in the region $\lambda < \lambda_0$. As a result, $\lambda(t) \to = \pm\infty$ depends on the initial value of λ. In this case, the value λ_0 is the fixed point, which is unstable in the UV limit of large Euclidean momenta.

The analysis of the infrared behavior at $t \to -\infty$ can be formally done in a very similar way, so we leave it as an exercise for the reader. However, as we mentioned above, this analysis may have a restricted validity if at least some of the quantum fields are massive.

If $\beta(\lambda)$ is evaluated within perturbation theory, $\beta(\lambda)$ comes in the form of the series $\beta(\lambda) = a\lambda^2 + b\lambda^4 + \ldots$. In the one-loop approximation, $\beta(\lambda) = a\lambda^2$, with $a > 0$ in case **2** and with $a < 0$ in case **1**.

When $a = b^2 > 0$, the equation for the running coupling $\lambda(t)$ has the form

$$\frac{d\lambda(t)}{dt} = b^2\lambda^2(t) \qquad \lambda(0) = \lambda, \tag{9.183}$$

which can be easily solved in the form

$$\lambda(t) = \frac{\lambda}{1 - b^2\lambda t}. \tag{9.184}$$

It is easy to see that the behavior of the running coupling is pathological. When t is increasing towards the UV, at some point the denominator of the solution (9.184) vanishes and $\lambda(t) \to \infty$. Evidently, at the corresponding energy scale, the coupling becomes large, and perturbation theory badly fails. Thus, all of the approach leading to Eq. (9.183) becomes inconsistent. The only way to avoid this contradiction is to impose the initial condition $\lambda = 0$. Then the interaction is completely absent at all scales. This situation is called the zero-charge problem.

This behavior of the coupling constants is typical in quantum electrodynamics or in the scalar theory with φ^4-interaction. In both cases, the standard physical interpretation is that the theory is not fundamental, so the asymptotic behavior in the UV should change dramatically because, at high energies, the theory that leads to the conclusion that (9.183) is not valid. This is the case for quantum electrodynamics, since, at distances less than $10_{-12}cm$ the electromagnetic interaction becomes mixed with the weak interaction, and the electroweak interaction is described by the Yang-Mills field, for which the renormalization group equation changes dramatically.

Consider the situation in Yang-Mills theory. In the next chapter, it will be shown that the quantum Yang-Mills theory is multiplicatively renormalized. Therefore, we can apply renormalization group analysis to it. The explicit one-loop calculation of the beta function is considered in Part II, so let us anticipate and briefly review the main results.

In Yang-Mills theory without matter, there is a unique coupling constant g. The equation for the running coupling constant has the form

$$\frac{dg^2}{dt} = -b^2 g^2, \qquad g(0) = g. \tag{9.185}$$

It is remarkable that, in this case, the beta function has the opposite sign, compared to Eq. (9.183). The solution to Eq. (9.185) can be easily found in the form

$$g^2(t) = \frac{g^2}{1 + b^2 g^2 t}, \tag{9.186}$$

indicating the asymptotic freedom in the UV. Indeed, $g^2(t) \to 0$, independently of the initial condition of g^2. On the other hand, in the infrared limit $t \to -\infty$, we face a zero-charge problem. This indicates the non-perturbative origin of the infrared limit in Yang-Mills theory.

9.9.5 Renormalization group functions in dimensional regularization

The beta function $\beta(\lambda)$ is defined by the relation (9.151). Let us describe the calculation of this expression in dimensional regularization. Let us start from the renormalization transformation,

$$\lambda = \mu^{4-n} z_\lambda \lambda_R = \mu^{4-n}\Big[1 + \sum_{k=1}^{\infty} \frac{c_k(\lambda_k)}{(n-4)^k}\Big]\lambda_R. \tag{9.187}$$

The *l.h.s.* of this relation is μ-independent. Taking $\mu\dfrac{d}{d\mu}$ of both sides, we get

$$\mu\frac{d}{d\mu}\left[\mu^{4-n}\Big(1+\sum_{k=1}^{\infty}\frac{c_k(\lambda_k)}{(n-4)^k}\Big)\lambda_R\right] = 0. \tag{9.188}$$

After some calculations, this equation gives

$$\begin{aligned}\mu\frac{d\lambda_R}{d\mu} &= \frac{(n-4)\lambda_R + c_1(\lambda_R)\lambda_R + \sum_{k=1}^{\infty}\frac{c_{k+1}(\lambda_R)\lambda_R}{(n-4)^k}}{1+\sum_{k=1}^{\infty}\frac{c_k(\lambda_R)+c_k'(\lambda_R)\lambda_R}{(n-4)^k}}.\\ &= (n-4)\lambda_R \; - \; c_1'(\lambda_R)\lambda_R^2 \; + \; \mathcal{O}\big(\tfrac{1}{n-4}\big),\end{aligned} \tag{9.189}$$

where the prime means the derivative with respect to λ. Since λ_R is a renormalized coupling constant, the quantity $\mu\dfrac{d\lambda_R}{d\mu}$ is finite. Therefore, it is sufficient to consider a finite (at $n \to 4$) part of this expression. Following this logic, all $\mathcal{O}\big(\frac{1}{n-4}\big)$-terms vanish. Now we take $n = 4$ in the remaining expression and get

$$\beta(\lambda) \;=\; -\,c_1'(\lambda)\lambda^2. \tag{9.190}$$

We see that $\beta(\lambda)$ is defined by the term $c_1(\lambda_R)$ in Eq. (9.187).

The functions γ and γ_m are calculated in a similar way on the basis of the relations

$$\phi = \mu^{\frac{n-4}{2}}\left[1+\sum_{k=1}^{\infty}\frac{a_k(\lambda_R)}{(n-4)^k}\right]\phi_R \qquad \text{and} \qquad m^2 = \left[1+\sum_{k=1}^{\infty}\frac{b_k(\lambda_R)}{(n-4)^k}\right]m_R^2.$$

Taking the derivatives $\mu\frac{d}{d\mu}$, using the relations analogous to (9.189) and taking the limit $n \to 4$, we obtain

$$\gamma(\lambda) = -a_1'(\lambda)\lambda \qquad \text{and} \qquad \gamma_m = b_1'(\lambda)\lambda. \tag{9.191}$$

As a result, the functions $\gamma(\lambda)$ and $\gamma_m(\lambda)$ are determined by the coefficients at a_1 and b_1 in the renormalization transformations (9.129) and (9.132).

Exercises

9.1. Consider the so-called analytic regularization. For example, for the tadpole diagram, perform the Wick rotation and replace the propagator in the momentum representation $\tilde{D}(p) = \frac{1}{p^2+m^2}$ with $\tilde{D}_{(z)}(p) = \frac{1}{(p^2+m^2)^z}$. Here z is a complex number. Introduce an arbitrary parameter μ of the mass dimension in such a way that the dimension of the diagram remains the same as at $z = 1$. Calculate the corresponding Feynman integral, clarify at which z this integral is an analytic function of the complex variable z and show that the integral has a pole at $z \to 1$.

9.2. Calculate the superficial degree of divergence in D-dimensional spacetime, taking into account that the propagators in the momentum representation have the same UV behavior as in four dimensions, that is, $\tilde{D}(p) \sim \frac{1}{p^2}$ and $\tilde{S} \sim \frac{1}{p}$. Explore the conditions of convergence for the cases $D = 2,\ D = 3$ and $D > 4$.

9.3. Consider the theory of a real scalar field with interaction $\lambda\varphi^3$ in D-dimensional spacetime. Explore the structure of divergences for different D. What is the conclusion on the renormalizability of this theory in six dimensions?

9.4. Calculate the superficial degree of divergences in the theory with higher derivatives, where the propagators behave as $\tilde{D}(p) \sim \frac{1}{p^4}$ and $\tilde{S} \sim \frac{1}{p^3}$. Consider separately model A, when the interaction terms are the same as in the usual theory, and model B, with the number of derivatives in the interacting Lagrangian being the same as in the free part, defining the propagator.

9.5. Consider the theory of a real scalar field with interaction $\lambda\varphi^4$ in four dimensions. Using dimensional regularization, prove that, in the one-loop approximation, $z_1 = 1,\ z_m = 1$ and $\beta_\lambda = \frac{3\lambda^2}{16\pi^2}$.

Observation. In Part II, the reader can find several different versions of this derivation, presented with full details.

9.6. Consider the theory of a real scalar field with interaction $\lambda\varphi^3$ in six-dimensional spacetime and calculate β_λ.

Comments

Renormalization in quantum field theory, including calculations of renormalization constants, is considered in detail in many books on quantum field theory (see, e.g., [187], [250], [304], [274], [256], [345], [59]). For a mathematically rigorous consideration of renormalization, see, e.g., the book by Collins [96] and the one by Zavialov [352].

In sections 9.3 and 9.6, we followed unpublished lectures by I.V. Tyutin.

Weinberg's theorem about the locality of UV divergences was formulated and proved in [339]. Another mathematically consistent proof of this theorem can be found in the book [96].

Dimensional regularization and renormalization of quantum field theory in the framework of this regularization are discussed in detail in the review paper by Leibbrandt [213] and in the book [96]. An introduction to the modern sophisticated methods for calculating Feynman integrals can be found, e.g., in the books by Smirnov [300], where tables of the one- and two-loop integrals are given.

The asymptotical freedom of the running coupling constant in non-Abelian gauge theories was initially described in the papers by Gross and Wilczek [173] and Politzer [254]. Various approaches to the renormalization group and applications of renormalization group equations are considered from different points of view, e.g., in the books [187,250,304,274,256,345,96] and [95]. See also the review paper by Vladimirov and Shirkov [331].

10
Quantum gauge theories

The gauge theories play the central role in the description of fundamental forces. For this reason, there is a vast literature devoted to different aspects of renormalization in these theories. In what follows, we formulate and discuss the main notions used in this area.

10.1 Basic notions of Yang-Mills gauge theory

In chapter 4, we considered some examples of gauge theories, namely, electrodynamics and non-Abelian Yang-Mills theory, and also mentioned gravitation, which will be dealt with in Part II. All these theories possess the following common properties:

a. The gauge transformations of the field have spacetime-dependent parameters.

b. The action of the theory is invariant under these transformations.

Taking into account these properties, we can formulate a general field theory as a theory satisfying the following requirements:

1. The theory is described by a set of fields ϕ^i, where i is the condensed index including all the discrete indices (this means Lorenz indices, indices of internal symmetry and so on) and the spacetime argument x^μ, which is the continuous index. For example, in the case of Yang-Mills theory, the field is $\phi^i \equiv A^a_\mu(x)$, while, in the case of gravity, the field is $\phi^i \equiv g_{\mu\nu}(x)$.

2. The gauge field is subject to the infinitesimal transformations,

$$\delta\phi^i \;=\; R^i{}_\alpha[\phi]\xi^\alpha. \tag{10.1}$$

Here ξ^α are the transformation parameters. The index α is also condensed and includes discrete indices and the spacetime arguments x^μ. For example, in the case of Yang-Mills theory, this index is $\alpha = (x, a)$. The *r.h.s.* of relation (10.1) includes the generalized summation in the index α, including sums over discrete indices and integrals over continuous indices x^μ. For example, in the case of Yang-Mills theory, the expression $R^i{}_\alpha[\phi]$ has the form $\mathcal{D}^{ab}_\mu \delta^4(x-y)$, where $\mathcal{D}^{ab}_\mu = \delta^{ab}\partial_\mu + gf^{acb}A^c{}_\mu$.

3. The action $S[\phi]$ satisfies the following Noether identity for an arbitrary field ϕ^i:

$$S_{,i}[\phi]\, R^i{}_\alpha[\phi] \;=\; 0, \qquad \text{where} \qquad S_{,i}[\phi] = \frac{\delta S[\phi]}{\delta\phi^i}. \tag{10.2}$$

This identity includes generalized summation over the index i.

The theory characterized by the properties **1**, **2**, **3** is called the general gauge theory. No other properties are assumed. The field ϕ^i is called the general gauge field,

Introduction to Quantum Field Theory with Applications to Quantum Gravity. Iosif L. Buchbinder and Ilya L. Shapiro, Oxford University Press (2021).
DOI: 10.1093/oso/9780198838319.003.0010

the transformations (10.1) are called the general gauge transformations, the quantities $R^i_\alpha[\phi]$ are called the generators of the general gauge transformations, or the gauge generators, the quantities ξ^α are called gauge parameters, and the relation (10.2) is the condition of gauge invariance. The condensed notations were introduced by B.S. DeWitt. The notion of a general gauge theory is very convenient since it enables one to study various aspects of different gauge theories from a unique point of view.

It proves useful to introduce the following notations. Let $A[\phi]$ be a functional of the gauge fields $\phi = \phi^i$. Then

$$A_{,i}[\phi] = \frac{\delta}{\delta\phi^i}A[\phi], \qquad A_{,ij}[\phi] = \frac{\delta}{\delta\phi^i}\frac{\delta}{\delta\phi^j}A[\phi], \ldots . \tag{10.3}$$

In these notations, the equations of motion have the form $S_{,i}[\phi] = 0$. The fields ϕ^i can be either bosonic or fermionic. In the last case, one should distinguish the left and right derivatives. Since we intend to apply the formalism mainly to the Yang-Mills and gravitational theories, for the sake of simplicity we assume that ϕ^i are commuting (bosonic) fields.

Let us clarify some of the properties of the generators following from relations (10.1) and (10.2). Consider the gauge transformation of a field with the parameters ξ_1^α, that is, $\delta_1\phi^i = R^i_\alpha[\phi]\xi_1^\alpha$, and then the subsequent transformation with the parameters ξ_2^α, that is, $\delta_2\delta_1\phi^i = R^i_{\alpha,j}[\phi]R^j_\beta[\phi]\xi_2^\beta\xi_1^\alpha$. We can define the commutator of these two transformations in the form

$$[\delta_1, \delta_2]\phi^i = \big(R^i_{\beta,j}[\phi]R^j_\alpha[\phi] - R^i_{\alpha,j}[\phi]R^j_\beta[\phi]\big)\xi_1^\alpha\xi_2^\beta. \tag{10.4}$$

The expression

$$R^i_{\beta,j}[\phi]R^j_\alpha[\phi] - R^i_{\alpha,j}[\phi]R^j_\beta[\phi] \tag{10.5}$$

is called the commutator of the gauge generators.

Using the result (10.4), one can evaluate the expression $[\delta_1, \delta_2]S[\phi]$. On the one hand, it is equal to zero, owing to the relation (10.2). On the other hand, this quantity can be calculated independently in the form

$$[\delta_1, \delta_2]S[\phi] = S_{,i}[\delta_1, \delta_2]\phi^i = S_{,i}\left(R^i_{\beta,j}[\phi]R^j_\alpha[\phi] - R^i_{\alpha,j}[\phi]R^j_\beta[\phi]\right)\xi_1^\alpha\xi_2^\beta. \tag{10.6}$$

Since the parameters ξ_1^α and ξ_2^β are arbitrary, one gets

$$S_{,i}\left(R^i_{\beta,j}[\phi]R^j_\alpha[\phi] - R^i_{\alpha,j}[\phi]R^j_\beta[\phi]\right) = 0. \tag{10.7}$$

This relation can be treated as the condition for the commutator of gauge transformations. To solve this equation, we note that, in general gauge theory, a solution to Eq. (10.7) may depend only on the action $S[\phi]$, and on the requirements **1**, **2**, **3**, formulated above. We can write the general solution to the equation (10.7) in the form

$$R^i_{\beta,j}[\phi]R^j_\alpha[\phi] - R^i_{\alpha,j}[\phi]R^j_\beta[\phi] = f^\gamma{}_{\alpha\beta}[\phi]R^i_\gamma[\phi] + X^{ij}_{\alpha\beta}[\phi]S_{,j}[\phi], \tag{10.8}$$

where the functionals $f^{\gamma}{}_{\alpha\beta}[\phi]$ and $X^{ij}_{\alpha\beta}[\phi]$ are restricted only by the conditions

$$f^{\gamma}{}_{\alpha\beta}[\phi] = -f^{\gamma}{}_{\beta\alpha}[\phi] \tag{10.9}$$

and

$$X^{ij}_{\alpha\beta}[\phi] = -X^{ji}_{\alpha\beta}[\phi], \qquad X^{ij}_{\alpha\beta}[\phi] = -X^{ij}_{\beta\alpha}[\phi]. \tag{10.10}$$

The relation (10.8) defines a general gauge algebra. If the functional $X^{ij}_{\alpha\beta}[\phi]$ is not equal to zero, the gauge algebra is called open; in the opposite case, it is called closed. Which of these cases is realized depends on a concrete gauge model. The functionals $f^{\gamma}{}_{\alpha\beta}[\phi]$ are called the structure functions of the general gauge theory. Depending on the gauge model, the functionals $f^{\gamma}{}_{\alpha\beta}[\phi]$ may be constants or functions of the fields.

We need to introduce further specifications of gauge theory. The generators of gauge transformations are called linearly independent if the equation $R^i_{\alpha}[\phi]n^{\alpha}[\phi] = 0$ has the unique solution $n^{\alpha}[\phi] = 0$. In the opposite case, they are called linearly dependent.

In what follows, we consider only the subclass of general gauge theories with closed gauge algebra, constant structure functions and linearly independent generators. Indeed, electrodynamics, non-Abelian Yang-Mills theory and gravitation belong to such a subclass. Furthermore, the special subclass of general gauge theory with commuting gauge fields, closed gauge algebra, linearly independent generators and constant structure functions is called a Yang-Mills gauge theory.

Let us show that the constant structure functions defined by the relation (10.8) possess the standard properties of the structure constants of the Lie groups. The property $f^{\gamma}{}_{\alpha\beta} = -f^{\gamma}{}_{[\beta\alpha]}$ is fulfilled automatically, owing to Eq. (10.9). Another property is the Jacobi identity. To prove it in the case of (10.8), one introduces the operators

$$R_{\alpha}[\phi] = R^i_{\alpha}[\phi]\frac{\delta}{\delta\phi^i}. \tag{10.11}$$

Consider the commutator of the differential operators $R_{\alpha}[\phi]$ and $R_{\beta}[\phi]$. Consider the Jacobi identities in terms of double commutators,

$$\big[[R_{\alpha}[\phi],[R_{\beta}[\phi]],R_{\gamma}[\phi]\big] + \big[[R_{\beta}[\phi],[R_{\gamma}[\phi]],R_{\alpha}[\phi]\big] + \big[[R_{\gamma}[\phi],[R_{\alpha}[\phi]],R_{\beta}[\phi]\big] = 0. \tag{10.12}$$

Using the relation (10.8) with $X^{ij}_{\alpha\beta}[\phi] = 0$ and $f^{\gamma}{}_{\alpha\beta}[\phi] = \text{const}$ in the identity (10.12) we arrive at the Jacobi identity

$$f^{\lambda}{}_{\alpha\beta}f^{\sigma}{}_{\lambda\gamma} + f^{\lambda}{}_{\beta\gamma}f^{\sigma}{}_{\lambda\alpha} + f^{\lambda}{}_{\gamma\alpha}f^{\sigma}{}_{\lambda\beta} = 0 \tag{10.13}$$

for the structure constants. A direct calculation of the commutator with the help of the relation (10.8) with $X^{ij}_{\alpha\beta}[\phi] = 0$ and the constant $f^{\gamma}{}_{\alpha\beta}$ leads to

$$\big[R_{\alpha}[\phi], R_{\beta}[\phi]\big] = f^{\gamma}{}_{\alpha\beta}R_{\gamma}[\phi]. \tag{10.14}$$

The quantities $f^{\gamma}{}_{\alpha\beta}$ satisfy all of the properties of the structure constants for the Lie groups. Therefore, the operators $R_{\alpha}[\phi]$ can be treated as generators of some Lie group representation. As a result, the subclass of general gauge theory under consideration is formally related to some representation of the Lie group, locally defined by the given structure constants. We will call such a group a general gauge group.

10.2 Gauge invariance and observables

Let ϕ^i be a gauge field and let $\phi'^i = \phi^i + R^i_\alpha[\phi]\xi^\alpha$ be the transformed field. Owing to gauge invariance, $S[\phi] = S[\phi']$. Therefore, we have

$$S_{,i}[\phi] = S_{,j}[\phi']\phi'^j{}_{,i} = S_{,j}[\varphi']\left(\delta^j{}_i + R^j_{\alpha,i}[\phi]\xi^\alpha\right), \tag{10.15}$$

where $\delta^j{}_i$ is a delta symbol in discrete indices, and a delta function in continuous indices. Hence, if $S_{,i}[\phi] = 0$, then $S_{,j}[\phi'] = 0$, and vice versa. This means that both fields ϕ^i and ϕ'^i are solutions to the same equations of motion. Assume that the initial conditions for the equations of motion are defined in the time instant t_0, and let the gauge parameter ξ^α be different from zero only for times $t > t_0$. Then the functions ϕ^i and ϕ'^i satisfy the same initial conditions. Thus, in gauge theories, the solution to the equations of motion with the given initial conditions is not unique.

The condition $S[\phi] = S[\phi']$, where ϕ'^i is obtained from ϕ^i by a gauge transformation, allows us to conclude that the action in gauge theories does not depend at all on the components of a gauge field ϕ^i. For example, electrodynamics is described by the vector field $A_\mu(x)$, being subject to the gauge transformations $A'_\mu(x) = A_\mu(x)+\partial_\mu\xi(x)$. However, in fact, the action depends only on the non-covariant field $A^\perp_\mu$, which was in (5.123). In Yang-Mills theory and gravitational theory, the exact explicit construction of the gauge independent components of the field is difficult, although it can be done perturbatively starting from the linear approximation.

The natural question is, what are physical observables in the theory? Usually, the observable or a physical quantity is some real functional of the fields. However, in the gauge theory, a functional can change its value under gauge transformation, while the physical quantity has a unique value in the given state. Thus, it is natural to define an observable in gauge theory as a real gauge-invariant functional of the fields. This means that if $F[\phi]$ is an observable, then $F[\phi] = F[\phi']$, where $\phi^i = \phi^i + R^i_\alpha[\phi]\xi^\alpha$ with an arbitrary parameter ξ^α. Since the physical quantities are described by gauge-invariant functionals, they do not depend at all on the components of the gauge fields, only on gauge-independent combinations. On the other hand, gauge-dependent components can be somewhat fixed without changing the values of the physical quantities.

In the theories under consideration, gauge transformation brings an arbitrariness which is parameterized by the components of the gauge parameter ξ^α. Thus, the number of gauge-dependent components of the gauge field ϕ^i is equal to the number of components of the gauge parameter. The remaining degrees of freedom of the gauge field are gauge invariant. Let us discuss how, in principle, the gauge dependence of the field components can be fixed. The natural answer to this question is to impose some additional conditions in the number of which would be equal to the number of the the gauge parameters. One can write these conditions in the form

$$\chi^\alpha[\phi] = 0. \tag{10.16}$$

It is supposed that Eq. (10.16) allows us to express the gauge-dependent components of the gauge field ϕ^i in terms of the gauge-independent components. Eq. (10.16) is called a gauge fixing condition, or simply gauge. Since the physical quantities are gauge invariant by definition, their values are independent of the presence or absence of a gauge-fixing condition.

It proves useful to introduce additional terminology. Gauge theory is characterized by the gauge group G with the elements denoted g. The elements of a general gauge group are the general gauge transformations (10.1). Thus, each element $g \in G$ is characterized by the gauge parameter ξ^α. Let $\mathcal{M}$ be the set of the fields ϕ^i and $\phi^i_0 \in \mathcal{M}$ be a fixed element of this set. Let us call ${}^g\phi^i_0 = \phi^i_0 + R^i_\alpha[\phi_0]\xi^\alpha$. Furthermore, we denote ${}^G\phi^i_0$ the full set of the fields ${}^g\phi^i_0$, where g runs through the entire group, which is ${}^G\phi^i_0 = \{{}^g\phi^i_0,\ g \in G\}$. The set ${}^G\phi^i_0$ is called the *group orbit* of the representative ϕ^i_0 in the set $\mathcal{M}$. One can prove that the group orbit is completely defined by the representative ϕ^i_0. Next, the set of the different group orbits is called coset and denoted by $\mathcal{M}/G$. The orbit corresponding to the representative ϕ^i will be denoted as $\{\phi^i\}$, such that $\{\phi^i\} \in \mathcal{M}/G$.

By definition, the physical quantity $F[\phi]$ satisfies the condition $F[\phi] = F[{}^g\phi]$, where $\forall g \in G$. This means that the gauge-invariant quantities are functionals of the group orbits, $F[\phi] = F[\{\phi\}]$. In particular, this is true for the action, $S[\phi] = S[\{\phi\}]$. Therefore, it is natural to associate $\{\phi\}$ to the physical gauge field and to call the coset $\mathcal{M}/G$ the set of the physical gauge fields. One can prove that different orbits do not cross each other, so the set of the fields $\mathcal{M}$ consists of the uncrossing orbits.

Since the orbit can be defined by a unique representative, the condition (10.16) can be instrumental in finding such a representative for each orbit. The equation

$$\chi^\alpha[{}^g\phi] = 0 \tag{10.17}$$

enables one to find the unit element of the gauge group $g \in G$ and hence to find the representative of the orbit ${}^g\phi^i$ corresponding to the given field ϕ^i. Indeed, Eq. (10.17) can be seen as defining the surface in the set $\mathcal{M}$. This surface intersects each orbit only once and hence defines the orbit's representative.

Let us consider infinitesimal transformations in more detail. The element g of the gauge group corresponds to the infinitesimal transformation parameter ξ^α. Therefore, ${}^g\phi^i = \phi^i + R^i_\alpha[\phi]\xi^\alpha$ and Eq. (10.17) takes the form

$$\chi^\alpha[\phi^i + R^i_\beta \xi^\beta] = \chi^\alpha[\phi] + \chi^\alpha_{,i}[\phi]R^i_\beta[\phi]\xi^\beta = 0. \tag{10.18}$$

Denoting

$$M^\alpha{}_\beta[\phi] = \chi^\alpha_{,i}[\phi]R^i_\beta[\phi], \tag{10.19}$$

we arrive at the equation equivalent to (10.18),

$$M^\alpha{}_\beta[\phi]\xi^\beta = -\chi^\alpha[\phi]. \tag{10.20}$$

The matrix $M^\alpha{}_\beta[\phi]$ (10.19) is called the DeWitt-Faddeev-Popov matrix.

The gauge-fixing functions $\chi^\alpha[\phi]$ should be chosen in such a way that the matrix $M^\alpha{}_\beta[\phi]$ is invertible. In this case, Eq. (10.20) enables one to find the parameter

$$\xi^\alpha[\phi] = -(M^{-1})^\alpha{}_\beta[\phi]\,\chi^\beta[\phi], \tag{10.21}$$

defining the representative of the orbit corresponding to the field ϕ^i.

10.3 Functional integral for gauge theories

Consider a physical quantity $F[\phi]$. Since $F[\phi] = F[\{\phi\}]$, the vacuum expectation value of $F[\phi]$ is defined as

$$\langle 0|TF[\{\phi\}]|0\rangle = \int \mathcal{D}\{\phi\}e^{iS[\{\phi\}]}F[\{\phi\}] = \int \mathcal{D}\{\phi\}e^{iS[\phi]}F[\phi]. \tag{10.22}$$

This expression contains integration only over physical gauge fields.

On the other hand, consider the naive functional integral

$$\int \mathcal{D}\phi e^{iS[\phi]}F[\phi]. \tag{10.23}$$

It is evident that the integral (10.23) includes the integration over physical gauge fields (i.e., over orbits) and also integration along the orbits, that is, over the gauge group. Since $F[\phi] = F[\{\phi\}]$ and $S[\phi] = S[\{\phi\}]$, the integral over the gauge group should be factorized. As a result, we should get a common multiplier, corresponding to the volume of the gauge group, Vol(G)[1] such that

$$\int \mathcal{D}\phi e^{iS[\phi]}F[\phi] = Vol(G)\int \mathcal{D}\{\phi\}e^{iS[\phi]}F[\phi]. \tag{10.24}$$

Thus, we need to express the integral (10.22) in terms of the initial fields ϕ^i and find it's relation to (10.23). Let us analyze a possible form of the true functional integral. Introducing the gauge condition $\chi^\alpha[\phi]$, consider the equation

$$\chi^\alpha[{}^h\phi] - f^\alpha = 0, \tag{10.25}$$

where the quantity f^α is independent of the field ϕ. Assume that, for each field ϕ^i, there exists the unit element of the gauge group h, which satisfies the equation (10.25). As we pointed out above, Eq. (10.25) picks up one representative of each orbit, so it defines the set of the physical fields $\mathcal{M}/G$. As a consequence, if it is possible to write the expression (10.22) in the form of an integral over all fields, such an integral should contain the functional delta function $\delta[\chi^\alpha[\phi] - f^\alpha]$. Therefore, one can expect that the following equality occurs

$$\int \mathcal{D}\{\phi\}e^{iS[\phi]}\,F[\phi] = \int \mathcal{D}\phi e^{iS[\phi]}\,F[\phi]\,\delta(\chi^\alpha[\phi] - f^\alpha)\,\Delta[\phi], \tag{10.26}$$

with some unknown functional $\Delta[\phi]$. The next problem is to find this functional. This problem has been solved by L.D. Faddeev and V.N. Popov and independently by B.S. DeWitt. In what follows, we call it the Faddeev-Popov procedure.

[1] We postpone the definition of this mathematical notion to section 10.3.1.

10.3.1 The Faddeev-Popov method

Taking into account the relations (10.24) and (10.26), we can write

$$\int \mathcal{D}\phi\, e^{iS[\phi]} F[\phi] \;=\; Vol(G) \int \mathcal{D}\phi\, e^{iS[\phi]} F[\phi]\, \delta(\chi^\alpha[\phi] - f^\alpha)\, \Delta[\phi]. \tag{10.27}$$

The idea of further consideration is as follows. We will identically transform the *l.h.s.* of relation (10.27) to the product of the two terms, one of them being $Vol(G)$. The other one should be the *r.h.s.* of (10.26), providing the expression for $\Delta[\Phi]$.

The Faddeev-Popov procedure uses the notion of the integral over the group. Let G be a compact Lie group with the parameters ξ^1, ξ^2, ..., ξ^n, and let $f(g)$, $g \in G$ be some function defined on the group. Then $f(g) = f(\xi^1, \xi^2, \dots, \xi^n)$. The integral over the group G is written in the form

$$\int dg f(g) = \int d\xi^1 d\xi^2 \dots d\xi^n\, \mu(\xi^1, \xi^2, \dots, \xi^n)\, f(\xi^1, \xi^2, \dots, \xi^n), \tag{10.28}$$

where the function $\mu(\xi^1, \xi^2, \dots, \xi^n)$ is called a group measure which can be calculated for each concrete group. The integral over the group (10.28) possesses the two important properties

$$\begin{aligned} \text{a.} \quad & \int dg\, f(gg_0) = \int dg\, f(g_0 g) \;=\; \int dg\, f(g), \\ \text{b.} \quad & \int dg\, f(g^{-1}) = \int dg\, f(g), \end{aligned} \tag{10.29}$$

where g_0 is some fixed group element. The integral $\int dg$ is called the group volume and is denoted Vol(G).

In the case under consideration, we are interested in a gauge group G, where the parameters ξ^α are functions of the coordinates x^μ. Therefore, the integral over the group has to be understood as a functional one.

In the Faddeev-Popov procedure, the functional $\Delta[\phi]$ is introduced by the equation

$$\Delta[\phi] \int \mathcal{D}h\, \delta(\chi^\alpha[{}^h\phi] - f^\alpha) \;=\; 1, \tag{10.30}$$

where h is an element of the gauge group G, and the integration is taken over the group. One can prove that $\Delta[\phi]$ is a gauge-invariant functional,

$$\Delta[{}^g\phi]^{-1} \;=\; \int \mathcal{D}h\, \delta(\chi^\alpha[{}^{hg}\phi] - f^\alpha) \;=\; \Delta[\phi]^{-1}, \tag{10.31}$$

where the first of the properties of the integral over a group (10.29) has been used.

Consider now the naive functional integral (10.23) and multiply it by the unit factor, taken in the form (10.30). As a result, we obtain

$$\int \mathcal{D}\phi\, e^{iS[\phi]} F[\phi] \;=\; \int \mathcal{D}\phi\, \mathcal{D}h\, e^{iS[\phi]}\, \delta(\chi^\alpha[{}^h\phi] - f^\alpha)\, \Delta[\phi] F[\phi]. \tag{10.32}$$

In the *r.h.s.* of this relation, we make the functional change of variables, $\phi^i \to {}^{h^{-1}}\phi^i$,

$$\int \mathcal{D}\phi\,\mathcal{D}h\, e^{iS[\phi]}\,\delta(\chi^\alpha[\phi] - f^\alpha)\,\Delta[\phi]\,F[\phi]\,\mathrm{Det}\,\Big(\frac{\delta\;{}^{h^{-1}}\phi^i}{\delta\phi^j}\Big), \tag{10.33}$$

where we took into account the fact that the functionals $S[\phi]$, $F[\phi]$ and $\Delta[\phi]$ are gauge invariant. The functional determinant in the *r.h.s.* is the functional Jacobian of the aforementioned change of variables. This transformation is local in the theories of our interest, such as Yang-Mills, gravity, supergravity, etc. Then the expression

$$\frac{\delta^{h^{-1}}\phi^i}{\delta\phi^j} \sim \delta^4(x-y) \quad\Longrightarrow\quad \mathrm{Det}\,\Big(\frac{\delta^{h^{-1}}\phi^i}{\delta\phi^j}\Big) \;=\; \exp\Big\{\,\mathrm{Tr}\,\log\Big(\frac{\delta^{h^{-1}}\phi^i}{\delta\phi^j}\Big)\Big\} \;=\; 1,$$

since the expression in the exponential is proportional to $\delta^4(x-y)\big|_{y=x} = 0$ and we assume the use of dimensional regularization. As a result, we obtain the integral

$$\begin{aligned}&\int \mathcal{D}\phi\,\mathcal{D}h\, e^{iS[\phi]}\,\delta(\chi^\alpha[\phi] - f^\alpha)\,\Delta[\phi]F[\phi]\\ &\qquad = \mathrm{Vol}(G)\int \mathcal{D}\phi\, e^{iS[\phi]}\,\delta(\chi^\alpha[\phi] - f^\alpha)\,\Delta[\phi]F[\phi].\end{aligned} \tag{10.34}$$

Using the relations (10.24), (10.26), (10.32) and (10.34), one gets

$$\int \mathcal{D}\{\phi\}\, e^{S[\phi]}F[\phi] \;=\; \int \mathcal{D}\phi\, e^{S[\phi]}F[\phi]\,\delta(\chi^\alpha[\phi] - f^\alpha)\,\Delta[\phi]. \tag{10.35}$$

The last step of the Faddeev-Popov procedure is the calculation of $\Delta[\phi]$. This functional enters the relation (10.35) multiplied by $\delta[\chi^\alpha[\phi] - f^\alpha]$. Therefore, it is sufficient to $\Delta[\phi]$ only for the fields satisfying the equation $\chi^\alpha[\phi] - f^\alpha = 0$. Using Eq. (10.30), one gets, in this case,

$$\Delta[\phi]^{-1} \;=\; \int \mathcal{D}h\,\delta(\chi^\alpha[{}^h\phi] - f^\alpha)\Big|_{\chi^\alpha[\phi]-f^\alpha=0}. \tag{10.36}$$

This relation means that the integration over h should be taken in the vicinity of the unit element. Thus,

$$\begin{aligned}\Delta[\phi]^{-1} &= \int \mathcal{D}\xi\,\mu[\xi]\Big|_{\xi=0}\delta\big(\chi^\alpha[\phi] - f^\alpha + \chi^\alpha_{,i}R^i_\beta[\phi]\xi^\beta\big)\Big|_{\chi^\alpha[\phi]-f^\alpha=0}\\ &= \mathrm{const}\times\int \mathcal{D}\xi\delta(M^\alpha{}_\beta\xi^\beta),\end{aligned} \tag{10.37}$$

where $M^\alpha{}_\beta$ is the DeWitt-Faddeev-Popov matrix (10.19), and the constant factor is $\mu[\xi]|_{\xi=0}$. The group parameters are introduced in such a way that the group element

$g[\xi]\big|_{\xi=0} = e$. Usually, for compact groups, $\mu[\xi]\big|_{\xi=0} = 1$. In any case, this constant does not depend on the fields and can be omitted. Therefore,

$$\Delta[\phi]^{-1} = \int \mathcal{D}\xi\, \delta[\xi]\, \text{Det}^{-1} M^{\alpha}{}_{\beta}[\phi] = \text{Det}^{-1} M^{\alpha}{}_{\beta}[\phi]. \tag{10.38}$$

Thus, we meet the Faddeev-Popov-DeWitt determinant

$$\Delta[\phi] = \text{Det}\, M^{\alpha}{}_{\beta}[\phi] \tag{10.39}$$

and arrive at the following relation:

$$\langle 0|TF[\phi]|0\rangle = \int \mathcal{D}\phi\, e^{iS[\phi]} F[\phi]\, \text{Det}\,(M^{\alpha}{}_{\beta})\, \delta(\chi^{\alpha}[\phi] - f^{\alpha}), \tag{10.40}$$

which is a useful form of the functional integral for the gauge theories of our interest.

Consider the properties of the functional integral for gauge theory. It is possible to prove that the functional integral (10.40) does not depend on the choice of the gauge-fixing functions $\chi^{\alpha}[\phi] - f^{\alpha}$. Suppose $\Delta_{\chi}[\phi]$ is defined according to Eq. (10.30) on the basis of the gauge $\chi^{\alpha} - f^{\alpha}$, and let $\Delta_{\chi'}[\phi]$ be an analogous functional defined on the base of another gauge, $\chi'^{\alpha} - f^{\alpha}$. Then

$$1 = \Delta_{\chi'}[\phi] \int \mathcal{D}h \delta[\chi'^{\alpha}[{}^{h}\phi] - f'^{\alpha}]. \tag{10.41}$$

Multiplying the relation (10.40) by a unit in the form (10.41), one gets

$$\begin{aligned} &\int \mathcal{D}\phi e^{S[\phi]} F[\phi]\, \Delta_{\chi}[\phi]\, \delta[\chi^{\alpha}[\phi] - f^{\alpha}] \\ &\quad = \int \mathcal{D}\phi\, \mathcal{D}h\, e^{S[\phi]} F[\phi]\, \Delta_{\chi}[\phi]\, \delta[\chi^{\alpha}[\phi] - f^{\alpha}]\, \Delta_{\chi'}[\phi]\, \delta[\chi'^{\alpha}[{}^{h}\phi] - f'^{\alpha}]. \end{aligned}$$

Now let us make the change of variables in this integral, ${}^{h}\phi^{i} \to \phi^{i}$. As we know, the corresponding functional Jacobian is equal to unit. Taking into account that the functionals $S[\phi], F[\phi]$, $\Delta_{\chi}[\phi]$ and $\Delta_{\chi'}[\phi]$ are gauge invariant, one gets the integral

$$\int \mathcal{D}\phi e^{S[\phi]} F[\phi]\, \Delta_{\chi'}[\phi]\, \delta[\chi'^{\alpha}[\phi] - f'^{\alpha}] \int \mathcal{D}h\, \Delta_{\chi}[\phi]\, \delta[\chi^{\alpha}[{}^{h^{-1}}\phi] - f^{\alpha}]. \tag{10.42}$$

Making the change of variables $h^{-1} \to h$ and using the property of the integral over the group (10.29), one obtains

$$\int \mathcal{D}\phi\, e^{S[\phi]}\, F[\phi]\, \Delta_{\chi'}[\phi]\, \delta[\chi'^{\alpha}[\phi] - f'^{\alpha}] \int \mathcal{D}h\, \Delta_{\chi}[\phi]\, \delta[\chi^{\alpha}[{}^{h}\phi] - f^{\alpha}]. \tag{10.43}$$

The last integral over h is equal to unit, owing to the definition of $\Delta_{\chi}[\phi]$ in (10.30). Thus,

$$\int \mathcal{D}\phi e^{S[\phi]} F[\phi] \Delta_{\chi} \delta[\chi^{\alpha}[\phi] - f^{\alpha}] = \int \mathcal{D}\phi\, e^{S[\phi]} F[\phi] \Delta_{\chi'}[\phi] \delta[\chi'^{\alpha}[\phi] - f'^{\alpha}]. \tag{10.44}$$

We can see that the *r.h.s.* of relation (10.40) is gauge invariant.

An important concluding observation is in order. We have introduced the functional integral which allows us to calculate the vacuum matrix elements of gauge invariant operators. However, the fundamental objects of our interest are the Green functions, which are not gauge-invariant. The form of the functional integral (10.40) enables us to define the generating functional of Green functions in the form

$$Z[J] \;=\; \int \mathcal{D}\phi\, e^{i\left(S[\phi]+\phi^i J_i\right)}\, \mathrm{Det}\,(M^\alpha{}_\beta)\, \delta(\chi^\alpha[\phi]-f^\alpha), \tag{10.45}$$

where J_i is the source of the field ϕ^i. Since the field ϕ^i is not gauge invariant and the source is arbitrary, the functional Z[J] is not gauge invariant.

10.3.2 Faddeev-Popov ghosts

It is useful to represent the functional determinant $\mathrm{Det}\,(M^\alpha{}_\beta[\phi])$ in Eq. (10.40) by the Gaussian functional integral over anticommuting fields,

$$Det\,\big(M^\alpha{}_\beta[\phi]\big) \;=\; \int \mathcal{D}\bar{c}\,\mathcal{D}c\; e^{i\bar{c}_\alpha M^\alpha{}_\beta[\phi]c^\beta}. \tag{10.46}$$

It is important to note that the new fields $\bar{c}_\alpha$ and c^β must be anticommuting, since only in this case with the functional integral in the *r.h.s.* of relation (10.46) give the $\mathrm{Det}\,(M^\alpha{}_\beta[\phi])$ and not its inverse, $\mathrm{Det}^{-1}(M^\alpha{}_\beta[\phi])$. Substituting (10.46) in expression (10.40), one gets

$$\langle 0|TF[\phi]|0\rangle = \int \mathcal{D}\phi\mathcal{D}\bar{c}\,\mathcal{D}c\, e^{i\left(S[\phi]+\bar{c}_\alpha M^\alpha{}_\beta[\phi]c^\beta\right)} F[\phi]\,\delta(\chi^\alpha[\phi]-f^\alpha). \tag{10.47}$$

Then, using relation (10.46) in relation (10.45), we obtain the generating functional of the Green functions in gauge theories in the form

$$Z[J] \;=\; \int \mathcal{D}\phi\mathcal{D}\bar{c}\,\mathcal{D}c\, e^{i\left(S[\phi]+\bar{c}_\alpha M^\alpha{}_\beta[\phi]c^\beta+\phi^i J_i\right)}\,\delta(\chi^\alpha[\phi]-f^\alpha). \tag{10.48}$$

The anticommuting fields $\bar{c}_\alpha$ and c^β are called the Faddeev-Popov ghosts. The term "ghost" is due to the fact that all the diagrams generated by the functional (10.48) have no external lines for the fields $\bar{c}_\alpha$ and c^β. These fields occur only in internal lines. For example, to find the $\mathcal{S}$-matrix, we need to account only for the diagrams with external lines of the initial fields ϕ^i. In the expression for the $\mathcal{S}$-matrix, the external lines of the diagrams are associated with *in* and *out* particles. Since there are no external lines of $\bar{c}_\alpha$ or c^β, one cannot associate any kind of real particles with these fields.

The functional

$$S_{GH}[\phi,\bar{c},c] = \bar{c}_\alpha M^\alpha{}_\beta[\phi]c^\beta \tag{10.49}$$

is called the ghost action. Using this notation, we can write

$$Z[J] \;=\; \int \mathcal{D}\phi\,\mathcal{D}\bar{c}\,\mathcal{D}c\, e^{i\left(S[\phi]+S_{GH}[\phi,\bar{c},c]+\phi^i J_i\right)}\delta(\chi^\alpha[\phi]-f^\alpha). \tag{10.50}$$

The ghost action (10.49) with an explicit form of the Faddeev-Popov $M^\alpha{}_\beta[\phi]$ matrix can be written as follows:

$$S_{GH}[\phi,\bar{c},c] \;=\; \bar{c}_\alpha \chi^\alpha_{,i}[\phi] R^i_\beta[\phi] c^\beta. \tag{10.51}$$

In the class of theories under consideration, such as Yang-Mills theory or gravity, the gauge $\chi^\alpha[\phi]$ is usually taken in the form

$$\chi^\alpha = t^\alpha + t^\alpha_i \phi^i, \tag{10.52}$$

where the quantities t^α, t^α_i do not depend on the field ϕ^i, and the operator t^α_i is usually linear in derivatives. For example, the standard gauge in Yang-Mills theory is $\chi^a = \partial_\mu A^{a\mu}$. The generators $R^i_\alpha[\phi]$ in these theories can be taken in the form

$$R^i_\alpha[\phi] = r^i_\alpha + r^i_{\alpha j}\phi^j, \tag{10.53}$$

where the quantities r^i_α and $r^i_{\alpha j}$ are field independent and linear in derivatives. Then

$$M^\alpha{}_\beta[\phi] \;=\; t^\alpha_i\, r^i_\beta + t^\alpha_i r^i{}_{\alpha j}\phi^j. \tag{10.54}$$

Therefore, the ghost action takes the form

$$S_{GH}[\phi,\bar{c},c] \;=\; \bar{c}_\alpha \big(H^\alpha_\beta \;+\; \lambda^\alpha{}_{\beta j}\, \phi^j \big) c^\beta. \tag{10.55}$$

Here $H^\alpha_\beta \;=\; t^\alpha_i r^i_\beta$ and $\lambda^\alpha{}_{\beta j} \;=\; t^\alpha_i r^i{}_{\beta j}$. Since t^α_i and $r^i{}_\beta$ are linear in the derivatives, the operator H^α_β is quadratic in the derivatives. As a result,

$$S_{GH}[\phi,\bar{c},c] \;=\; S^{(0)}_{GH}[\bar{c},c] \;+\; S^{(int)}_{GH}[\phi,\bar{c},c], \tag{10.56}$$

where $S^{(0)}_{GH}[\bar{c},c] = \bar{c}_\alpha H^\alpha{}_\beta c^\beta$ and $S^{(int)}_{GH}[\phi,\bar{c},c] = \lambda^\alpha{}_{\beta j}\bar{c}_\alpha c^\beta \phi^j$. Thus, the ghost action has a form which is standard for quantum field theory, namely, it is the sum of the part that is quadratic in fields (free action) and the part that contains the higher powers of fields (cubic in the given case), which is an interaction. Therefore, in the case under consideration, we can apply the standard perturbation theory and Feynman diagrams.

Another reason why the fields $\bar{c}_\alpha$ and c^β are called ghosts is as follows. These fields are anticommuting, which is a characteristic feature of fermions. However, the kinetic part of the action is quadratic in the derivatives, which is the characteristic feature of bosons. Thus, the fields $\bar{c}_\alpha$ and c^β cannot be real particles since they can be neither bosons nor fermions.

10.3.3 Total action of the quantum gauge theory

The functional integral (10.40) can be transformed further into the form which is more convenient for the construction of perturbation theory. Let us rewrite this integral as

$$\int \mathcal{D}\phi\, e^{iS[\phi]} F[\phi]\, \mathrm{Det}\,(M^\alpha{}_\beta)\, \delta(\chi^\alpha[\phi] - f^\alpha)\;\, \mathrm{Det}^{\,\frac{1}{2}}(G_{\alpha\beta}[\phi])\;\mathrm{Det}^{\,-\frac{1}{2}}(G_{\alpha\beta}[\phi]), \tag{10.57}$$

where $G_{\alpha\beta}[\phi]$ is a functional satisfying the non-degeneracy condition, $\mathrm{Det}\, G_{\alpha\beta}[\phi] \neq 0$. One can represent this functional determinant by the Gaussian functional integral over new commuting fields,

$$\mathrm{Det}^{\,-\frac{1}{2}}(G_{\alpha\beta}[\phi]) = \int \mathcal{D}f\, e^{\frac{i}{2} f^\alpha G_{\alpha\beta}[\phi] f^\beta}. \tag{10.58}$$

Since the integral (10.40) is gauge independent, we can assume that the integration variable for $\mathrm{Det}^{\,-\frac{1}{2}}(G_{\alpha\beta}[\phi])$ in Eq. (10.58) coincides with f^α. As a result, one gets

$$\int \mathcal{D}\phi \mathcal{D}f\, e^{iS[\phi]} F[\phi]\, \mathrm{Det}\,(M^\alpha{}_\beta)\, e^{\frac{i}{2} f^\alpha G_{\alpha\beta}[\phi] f^\beta}\, \delta(\chi^\alpha[\phi] - f^\alpha)\, \mathrm{Det}^{\,\frac{1}{2}}(G_{\alpha\beta}[\phi]). \tag{10.59}$$

Therefore, we obtain

$$\int \mathcal{D}\phi\, e^{i\left(S[\phi] + \frac{1}{2}\chi^\alpha G_{\alpha\beta}[\phi]\chi^\beta\right)}\, F[\phi]\, \mathrm{Det}\,(M^\alpha{}_\beta)\, \mathrm{Det}^{\,\frac{1}{2}}(G_{\alpha\beta}[\phi]). \tag{10.60}$$

It proves useful to present Faddeev-Popov ghosts according to relation (10.46) and to represent the factor $\mathrm{Det}^{\,\frac{1}{2}}(G_{\alpha\beta}[\phi])$ by the Gaussian functional integral over anticommuting fields in the form

$$\mathrm{Det}^{\,\frac{1}{2}}(G_{\alpha\beta}[\phi]) = \int \mathcal{D}b e^{ib^\alpha G_{\alpha\beta}[\phi] b^\beta}. \tag{10.61}$$

Thus, we arrive at the final expression

$$\langle 0|TF[\phi]|0\rangle = \int \mathcal{D}\phi \mathcal{D}\bar{c} \mathcal{D}c \mathcal{D}b\, e^{i\left(S[\phi] + \frac{1}{2}\chi^\alpha G_{\alpha\beta}[\phi]\chi^\beta + \bar{c}_\alpha M^\alpha{}_\beta[\phi] c^\beta + b^\alpha G_{\alpha\beta}[\phi] b^\beta\right)} F[\phi]. \tag{10.62}$$

In the rest of Part I, we will consider the case when the matrix $G_{\alpha\beta}$ is a constant. Then the term $\mathrm{Det}^{\,\frac{1}{2}} G_{\alpha\beta}$ and the corresponding integral over fields b^α in the integral under consideration can be omitted. However, this matrix is essential to write the terms with gauge χ^α in the exponent of (10.62).

Finally, the generating functional of Green functions in gauge theory can be written as follows:

$$Z[J] = \int \mathcal{D}\phi\, \mathcal{D}\bar{c} \mathcal{D}c\, e^{i\left(S_{total}[\phi,\bar{c},c] + \phi^i J_i\right)}, \tag{10.63}$$

where the total action is

$$S_{total}[\phi, \bar{c}, c] = S[\phi] + S_{GF}[\phi] + S_{GH}[\phi, \bar{c}, c]. \tag{10.64}$$

and the gauge-fixing and ghost terms are, respectively,

$$S_{GF}[\phi] = \frac{1}{2}\chi^\alpha G_{\alpha\beta}\chi^\beta, \tag{10.65}$$

$$S_{GH} = \bar{c}_\alpha M^\alpha{}_\beta c^\beta. \tag{10.66}$$

Expression (10.63) has the usual form for a generating functional in quantum field theory; it can be calculated in the perturbation theory with the help of Feynman diagrams. The quadratic part of the action $S_{total}[\phi, \bar{c}, c]$ in (10.64) defines the propagators for the gauge field ϕ^i and for the Faddeev-Popov ghosts. The remaining part of the action defines vertices.

We should emphasize the following important point. Unlike the gauge-invariant action $S[\phi]$, the action $S_{total}[\phi, \bar{c}, c]$ is not gauge invariant. The variation under gauge transformation is

$$\delta\big(S[\phi] + S_{GF}[\phi]\big) = \chi^{\beta}[\phi]\, G_{\beta\gamma}\chi^{\gamma}_{,i}[\phi]\, R^{i}_{\alpha}[\phi]\xi^{\alpha} = \chi^{\beta}[\phi]G_{\beta\gamma}M^{\gamma}{}_{\alpha}[\phi]\xi^{\alpha} \neq 0. \quad (10.67)$$

This output means that the gauge-fixing term in the action $S_{total}[\phi, \bar{c}, c]$ removes the degeneracy of the initial gauge-invariant action $S[\phi]$, providing the existence of the propagator for the gauge field ϕ^i.

Sometimes, another form of the generating functional $Z[J]$, different from Eq. (10.63), can be used. As we know, integrating over ghosts in expression (10.63) gives Det $(M^{\alpha}{}_{\beta}[\phi] = e^{\mathrm{Tr}\,\log M^{\alpha}_{\beta}[\phi]}$. In this way, one gets

$$Z[J] = \int \mathcal{D}\phi\, e^{i(\tilde{S}_{total}[\phi] + \phi^i J_i)}, \quad (10.68)$$

$$\text{where} \qquad \tilde{S}_{total}[\phi] = S[\phi] + S_{GF}[\phi] - i\,\mathrm{Tr}\,\log M^{\alpha}{}_{\beta}[\phi]. \quad (10.69)$$

The advantage of this form is that the action $\tilde{S}_{total}[\phi]$ is formulated completely in terms of the initial fields ϕ^i.

In Abelian gauge theory, such as spinor and/or scalar electrodynamics, the generators of gauge transformations do not depend on the fields. If we use the Lorentz covariant gauge condition, which is linear in the fields, the Faddeev-Popov matrix is field independent. Then, the ghosts do not interact with the initial fields ϕ^i, so the integral over ghosts is factorized and can be omitted. Therefore, in these theories, the generating functional $Z[J]$ is given by an integral only over initial fields, with the action $S_{total}[\phi] = S[\phi] + S_{GF}[\phi]$.

10.4 BRST symmetry

As we have shown, the quantum theory of gauge fields can be formulated in terms of the gauge non-invariant action $S_{total}[\phi, \bar{c}, c]$ in (10.64). Nevertheless, this action is invariant under remnant global symmetry, which is called BRST symmetry, where BRST stands for Becchi, Rouet, Stora and Tyutin, who discovered this property. In what follows we formulate this symmetry and show how it can be used to explore the renormalization of gauge theories.

10.4.1 BRST transformations and the invariance of total action

Consider the global transformation of the fields ϕ^i, $\bar{c}_\alpha$, c^α of the form

$$\delta_{BRST}\phi^i = R^i_{\alpha}c^{\alpha}\mu, \quad (10.70)$$

$$\delta_{BRST}\bar{c}_{\alpha} = G_{\alpha\beta}\chi^{\beta}[\phi]\mu, \quad (10.71)$$

$$\delta_{BRST}c^{\alpha} = \frac{1}{2}f^{\alpha}{}_{\beta\gamma}c^{\beta}c^{\gamma}\mu, \quad (10.72)$$

where μ is a constant anticommuting parameter. This implies, in particular, that $\mu^2 = 0$, $c^\alpha \mu = -\mu c^\alpha$ and $\bar{c}_\alpha \mu = -\mu \bar{c}_\alpha$. Let us find out how the action $S_{total}[\phi, \bar{c}, c]$ in (10.64) changes under the transformations (10.70), (10.71) and (10.72).

The first relation (10.70) is nothing but a gauge transformation with the parameter $\xi^\alpha = c^\alpha \mu$. Owing to the gauge invariance of the initial action $S[\phi]$, we can be sure that $\delta_{BRST} S[\phi] = 0$. Therefore,

$$\begin{aligned} \delta_{BRST} S_{total}[\phi, \bar{c}, c] &= \chi^\alpha G_{\alpha\beta} \chi^\beta_{,i} R^i_\gamma c^\gamma \mu + \bar{c}_\alpha \chi^\alpha_{,ij} R^j_\gamma c^\gamma \mu R^i_\beta c^\beta \\ &\quad + \bar{c}_\alpha \chi^\alpha_{,i} R^i_{\beta,j} R^j_\gamma c^\gamma \mu c^\beta + G_{\alpha\gamma} \chi^\gamma \mu \chi^\alpha_{,i} R^i_\beta c^\beta + \frac{1}{2} \bar{c}_\alpha \chi^\alpha_{,i} R^i_\beta f^\beta{}_{\gamma\delta} c^\gamma c^\delta \mu . \end{aligned} \tag{10.73}$$

Taking into account that μ is an anticommuting parameter, the first and the third terms in the last expression cancel. The remaining terms are transformed as

$$\frac{1}{2} \bar{c}_\alpha \chi^\alpha_{,ij} \left(R^i_\beta R^j_\gamma - R^i_\gamma R^j_\beta \right) c^\beta c^\gamma \mu + \frac{1}{2} \bar{c}_\alpha \chi^\alpha_{,i} R^i_\beta f^\beta{}_{\gamma\delta} (c^\delta c^\gamma + c^\gamma c^\beta) \mu = 0.$$

The first term vanishes owing to the symmetry $\chi^\alpha_{,ij} = \chi^\alpha_{,ji}$, while the second term is zero because of the anticommutativity of the ghosts c^γ and c^δ. As a result, we have shown that the action $S_{total}[\phi, \bar{c}, c]$ in Eq. (10.64) is invariant under the transformations (10.70), (10.71) and (10.72). As we have already mentioned, these transformations are called the BRST and the corresponding symmetry is called the BRST invariance.

10.4.2 Two more BRST invariants

It is possible to show that the functionals

$$I^i = R^i_\alpha[\phi] c^\alpha \qquad \text{and} \qquad I^\alpha = \frac{1}{2} f^\alpha{}_{\beta\gamma} c^\beta c^\gamma \tag{10.74}$$

are invariants under BRST transformation. Let us prove this important statement.

Consider the transformation $\delta_{BRST} I^i$:

$$\begin{aligned} \delta_{BRST} I^i = \delta_{BRST}(R^i_\alpha c^\alpha) &= R^i_{\alpha,j} R^j_\beta c^\beta \mu c^\alpha + \frac{1}{2} f^\alpha{}_{\beta\gamma} R^i_\alpha c^\beta c^\gamma \mu \\ &= -\frac{1}{2} \left(R^i_{\alpha,j} R^j_\beta - R^i_{\beta,j} R^j_\alpha \right) c^\beta c^\alpha \mu + \frac{1}{2} f^\alpha{}_{\beta\gamma} R^i_\alpha c^\beta c^\gamma \mu = 0. \end{aligned} \tag{10.75}$$

Consider the transformation $\delta_{BRST} I^\alpha$:

$$\begin{aligned} \delta_{BRST} I^\alpha = \delta_{BRST} \left(\frac{1}{2} f^\alpha{}_{\beta\gamma} c^\beta c^\gamma \right) &= \frac{1}{2} f^\alpha{}_{\beta\gamma} f^\gamma{}_{\lambda\sigma} c^\beta c^\lambda c^\sigma \mu \\ &= \frac{1}{6} \left(f^\alpha{}_{\beta\gamma} f^\gamma{}_{\lambda\sigma} + f^\alpha{}_{\sigma\gamma} f^\gamma{}_{\beta\lambda} + f^\alpha{}_{\lambda\gamma} f^\gamma{}_{\sigma\beta} \right) c^\beta c^\lambda c^\sigma \mu = 0. \end{aligned} \tag{10.76}$$

Thus, the functionals I^α and I^i are invariant under the BRST transformations.

10.5 Ward identities

The gauge invariance of the classical action allows us to derive exact relations for the vertex functions in quantum field theory. Such relations are known as Ward identities

(Slavnov-Taylor identities, in some cases) and represent quantum analogs of classical Noether identities. Particular cases, corresponding to different symmetries, may have other specific names. Before considering how Ward identities can be obtained and used in the case of BRST symmetry in non-Abelian models, let us discuss a simpler version corresponding to Abelian gauge theory.

10.5.1 Ward identities in quantum electrodynamics

Consider the derivation of the Ward identities in spinor electrodynamics. It is assumed that quantum theory is regularized in such a way that the regularization prescription preserves the gauge invariance. For instance, this is the case for dimensional regularization, so we can simply assume it in our considerations.

As we already mentioned a few times, spinor electrodynamics is the Abelian gauge theory with the action

$$S[A,\bar{\psi},\psi] = \int d^4x\Big\{-\frac{1}{4}F_{\mu\nu}F^{\mu\nu} + i\bar{\psi}\gamma^\mu(\partial_\mu - ieA_\mu)\psi - m\bar{\psi}\psi\Big\}. \tag{10.77}$$

The gauge transformations, with the infinitesimal gauge parameter $\xi(x)$, have the form

$$A'_\mu = A_\mu + \partial_\mu\xi(x), \qquad \psi' = \psi + ie\psi\xi(x), \qquad \bar{\psi}' = \bar{\psi} - ie\bar{\psi}\xi(x). \tag{10.78}$$

The condition of gauge invariance can be written in the form

$$\partial_\mu\frac{\delta S}{\delta A_\mu} + ie\Big(\bar{\psi}\frac{\overrightarrow{\delta} S}{\delta\bar{\psi}} - \frac{S\overleftarrow{\delta}}{\delta\psi}\psi\Big) = 0. \tag{10.79}$$

The generating functional of Green functions for the theory under consideration is

$$Z[J_\mu,\bar{\eta},\eta] = \int \mathcal{D}A\mathcal{D}\bar{\psi}\mathcal{D}\psi\, e^{iS[A,\bar{\psi},\psi]+i\int d^4x\big\{\frac{1}{2}(\partial_\mu A^\mu)^2 + A_\mu J^\mu + \bar{\eta}\psi + \bar{\psi}\eta\big\}}, \tag{10.80}$$

where we used the relativistic Lorentz gauge fixing $\chi[A] = \partial_\mu A^\mu$, J_μ is a commuting source and η, $\bar{\eta}$ are anticommuting sources. Let us change the variables in the functional integral (10.80), corresponding to the transformations (10.78). Since this change of variables is local, the functional Jacobian equals unity in the dimensional regularization. Taking into account that the gauge parameter is infinitesimal, one gets

$$\begin{aligned} Z[J_\mu,\bar{\eta},\eta] &= \int \mathcal{D}A\mathcal{D}\bar{\psi}\mathcal{D}\psi\, e^{iS[A,\bar{\psi},\psi]+i\int d^4x\big\{\frac{1}{2}(\partial_\mu A^\mu)^2 + A_\mu J^\mu + \bar{\eta}\psi + \bar{\psi}\eta\big\}} \\ &\quad\times\Big\{1 + i\int d^4x\,\big(\partial_\nu A^\nu\partial^\mu\partial_\mu\xi + \partial_\mu\xi J^\mu + \bar{\eta}ie\psi\xi - ie\bar{\psi}\eta\xi\big)\Big\}. \end{aligned} \tag{10.81}$$

Since the parameter $\xi(x)$ is arbitrary, we obtain the identity for Z,

$$\frac{1}{2}\Box\partial_\mu\frac{\delta Z}{\delta J_\mu} - \partial_\mu J^\mu Z + ie\Big(\bar{\eta}\frac{\overrightarrow{\delta} Z}{\delta\bar{\eta}} - \frac{Z\overleftarrow{\delta}}{\delta\eta}\eta\Big) = 0. \tag{10.82}$$

Let us introduce the generating functional of the connected Green functions $W[J_\mu, \bar\eta, \eta]$ by the standard rule, $Z[J_\mu, \bar\eta, \eta] = e^{iW[J_\mu, \bar\eta, \eta]}$. Since (10.82) is a linear homogeneous equation for Z, the identity for W has exactly the same form,

$$\frac{1}{2}\Box\partial_\mu \frac{\delta W}{\delta J_\mu} - \partial_\mu J^\mu W + ie\Big(\bar\eta \frac{\overrightarrow{\delta} W}{\delta\bar\eta} - \frac{W\overleftarrow{\delta}}{\delta\eta}\eta\Big) = 0. \tag{10.83}$$

The mean fields are defined as

$$A_\mu = \frac{\delta W}{\delta J^\mu}, \qquad \psi = \frac{\overrightarrow{\delta} W}{\delta\bar\eta}, \qquad \bar\psi = \frac{W\overleftarrow{\delta}}{\delta\eta}. \tag{10.84}$$

The effective action $\Gamma[A_\mu, \bar\psi, \psi]$ of the mean fields is defined as

$$\Gamma[A_\mu, \bar\psi, \psi] = W[J_\mu, \bar\eta, \eta] - \int d^4x \left(A_\mu J^\mu + \bar\eta\psi + \bar\psi\eta\right), \tag{10.85}$$

which gives

$$\frac{\delta\Gamma}{\delta A^\mu} = -J_\mu, \qquad \frac{\overrightarrow{\delta}\Gamma}{\delta\bar\psi} = -\eta, \qquad \frac{\Gamma\overleftarrow{\delta}}{\delta\psi} = -\bar\eta. \tag{10.86}$$

Using the relations (10.86), (10.84) and (10.85) in (10.83), we arrive at the identity for the effective action:

$$\frac{1}{\alpha}\Box\partial_\mu A^\mu + \partial_\mu \frac{\delta\Gamma}{\delta A_\mu} + ie\Big(\bar\psi \frac{\overrightarrow{\delta}\Gamma}{\delta\bar\psi} - \frac{\Gamma\overleftarrow{\delta}}{\delta\psi}\psi\Big) = 0. \tag{10.87}$$

It proves useful to introduce the modified functional $\tilde\Gamma[A_\mu, \bar\psi, \psi]$,

$$\Gamma[A_\mu, \bar\psi, \psi] = \tilde\Gamma[A_\mu, \bar\psi, \psi] + \frac{1}{\alpha}\int d^4x(\partial_\mu A^\mu)^2. \tag{10.88}$$

Substituting (10.88) into Eq. (10.87), we get

$$\partial_\mu \frac{\delta\tilde\Gamma}{\delta A_\mu} + ie\Big(\bar\psi \frac{\overrightarrow{\delta}\tilde\Gamma}{\delta\bar\psi} - \frac{\tilde\Gamma\overleftarrow{\delta}}{\delta\psi}\psi\Big) = 0, \tag{10.89}$$

which is the final form of the Ward identity in spinor electrodynamics.

Eq. (10.89) has the same form as the condition of classical gauge invariance (10.83) and shows that the functional $\tilde\Gamma[A_\mu, \bar\psi, \psi]$ is gauge invariant. Let us write $\tilde\Gamma[A, \bar\psi, \psi] = S[A, \bar\psi, \psi] + \bar\Gamma[A, \bar\psi, \psi]$, where $S[A, \bar\psi, \psi]$ is a classical action (10.77), and the functional $\bar\Gamma[A, \bar\psi, \psi]$ contains all quantum corrections to the classical action. It is evident that the functional $\bar\Gamma[A, \bar\psi, \psi]$ satisfies the equation (10.89). Therefore, it is also gauge invariant. In the framework of dimensional regularization and the scheme of minimal subtractions,

$$\bar\Gamma[A, \bar\psi, \psi] = \bar\Gamma[A, \bar\psi, \psi]_{finite} + \bar\Gamma[A, \bar\psi, \psi]_{divergent}. \tag{10.90}$$

Here $\bar\Gamma[A, \bar\psi, \psi]_{\text{finite}}$ is a finite part of effective action, and $\bar\Gamma[A, \bar\psi, \psi]_{\text{divergent}}$ is a divergent part, containing the poles of the form $\frac{1}{(n-4)^k}$, for $k = 1, 2, \dots$. It is clear that both $\bar{\Gamma}[A,\bar{\psi},\psi]_{\text{finite}}$ and $\bar{\Gamma}[A,\bar{\psi},\psi]_{\text{divergent}}$ satisfy the identity (10.89) separately. Therefore, within the regularization scheme under consideration, the divergent part of the effective action is gauge invariant. Recall that this property was used in section 9.5.4 when we analyzed divergent structure in quantum electrodynamics.

Differentiating several times the relation (10.89) with respect to the fields A_μ, $\bar{\psi}$, ψ and setting these fields to zero, one can obtain a system of Ward identities for vertex functions. This system of identities was derived independently by E.S. Fradkin and Y. Takahashi.

10.5.2 Ward identities for non-Abelian gauge theory

The Ward identities for the non-Abelian gauge theory of the Yang-Mills type can be formulated in a way equivalent to (10.89). However, it proves more efficient to elaborate another set of Ward identities, corresponding to BRST symmetry, since these identities for effective action have a more useful form.

The effective action is defined in the standard way on the basis of the generating functional of the Green functions (10.63). Remember that the total action $S_{total}[\phi,\bar{c},c]$ in gauge theory is given by the relation (10.64). To derive the Ward identities, it is convenient to introduce the more general effective action depending on the two auxiliary fields, or sources, K_i and L_α,

$$\bar{Z}[J,\bar{\eta},\eta,K,L] = \int \mathcal{D}\phi\mathcal{D}\bar{c}\mathcal{D}c\, e^{i\left\{S_{total}[\phi,\bar{c},c]+\phi^i J_i+\eta_\alpha c^\alpha+\bar{\eta}^\alpha \bar{c}_\alpha+K_i I^i+L_\alpha I^\alpha\right\}}, \tag{10.91}$$

where $I^i = R^i_\alpha[\phi]c^\alpha$ and $I^\alpha = \frac{1}{2}f^\alpha{}_{\beta\gamma}c^\beta c^\gamma$ are the BRST invariants defined in (10.74). Furthermore, J_i are the commuting sources to the gauge fields ϕ^i, while η_α and $\bar{\eta}^\alpha$ are anticommuting sources to the ghosts c^α and $\bar{c}_\alpha$.

We denote

$$\bar{S}[\phi,\bar{c},c,K,L] \;=\; S_{total}[\phi,\bar{c},c] + K_i R^i_\alpha c^\alpha + L_\alpha \frac{1}{2} f^\alpha{}_{\beta\gamma}c^\beta c^\gamma. \tag{10.92}$$

The condition of BRST invariance of this action has the form

$$\frac{\delta\bar{S}}{\delta\phi^i}R^i_\alpha c^\alpha + \frac{\delta\bar{S}}{\delta c^\alpha}\frac{1}{2}f^\alpha{}_{\beta\gamma}c^\beta c^\gamma + \frac{\delta\bar{S}}{\delta\bar{c}_\alpha}G_{\alpha\beta}\chi^\beta \;=\; 0. \tag{10.93}$$

Taking into account that

$$R^i_\alpha c^\alpha = \frac{\delta\bar{S}}{\delta K_i} \quad \text{and} \quad \frac{1}{2}f^\alpha{}_{\beta\gamma}c^\beta c^\gamma = \frac{\delta\bar{S}}{\delta L_\alpha}, \tag{10.94}$$

we can rewrite the relation (10.93) as a closed equation for $\bar{S}$,

$$\frac{\delta\bar{S}}{\delta\phi^i}\frac{\delta\bar{S}}{\delta K_i} + \frac{\delta\bar{S}}{\delta c^\alpha}\frac{\delta\bar{S}}{\delta L_\alpha} + \frac{\delta\bar{S}}{\delta\bar{c}_\alpha}G_{\alpha\beta}\chi^\beta \;=\; 0. \tag{10.95}$$

Here the derivatives with respect the K_i, L_α and the ghosts are left. This relation is called the Zinn-Justin equation.

In what follows, we consider only the class of the linear gauges, with $\chi^\alpha = \chi_i^\alpha \phi^i$. In terms of the action $\bar{S}$ from Eq. (10.92), the relation (10.91) is rewritten as

$$\bar{Z}[J,\bar{\eta},\eta,K,L] = \int \mathcal{D}\phi\,\mathcal{D}\bar{c}\,\mathcal{D}c\,e^{i\{\bar{S}[\phi,\bar{c},c,K,L]+\phi^i J_i+\eta_\alpha c^\alpha+\bar{\eta}^\alpha \bar{c}_\alpha\}}. \tag{10.96}$$

This expression can be treated as the generating functional of the theory with the action $\bar{S}[\phi,\bar{c},c,K,L]$ depending on the set of fields ϕ^i, c^α, $\bar{c}_\alpha$. When $\eta = \bar{\eta} = K = L = 0$, we arrive at the initial generating functional (10.63).

At the quantum level, Eq. (10.95) leads to the identity

$$\int \mathcal{D}\phi\mathcal{D}\bar{c}\mathcal{D}c e^{i\left\{\bar{S}[\phi,\bar{c},c,K,L]+\phi^i J_i+\eta_\alpha c^\alpha+\bar{\eta}^\alpha \bar{c}_\alpha\right\}}$$
$$\times\left[\frac{\delta\bar{S}}{\delta\phi^i}\frac{\delta\bar{S}}{\delta K_i} + \frac{\delta\bar{S}}{\delta c^\alpha}\frac{\delta\bar{S}}{\delta L_\alpha} + \frac{\delta\bar{S}}{\delta\bar{c}_\alpha}G_{\alpha\beta}\chi^\beta\right] = 0. \tag{10.97}$$

The functional derivatives in Eq. (10.97) can be expressed through the functional $\bar{Z}$ (10.96) and its derivatives with respect to the sources. Then, (10.97) becomes

$$J_i\frac{\delta\bar{Z}}{\delta K_i} + \eta_\alpha\frac{\delta\bar{Z}}{\delta L_\alpha} + \bar{\eta}^\alpha G_{\alpha\beta}\chi_i^\beta\frac{\delta\bar{Z}}{\delta J_i} = 0. \tag{10.98}$$

All the derivatives with respect to the anticommuting sources are left. Let us introduce the generating functional of the connected Green functions $i\bar{W}[J,\bar{\eta},\eta,K,L]$ by the standard rule $\bar{Z} = e^{i\bar{W}}$ and rewrite (10.98) in the form

$$J_i\frac{\delta\bar{W}}{\delta K_i} + \eta_\alpha\frac{\delta\bar{W}}{\delta L_\alpha} + \bar{\eta}^\alpha G_{\alpha\beta}\,\chi_i^\beta\frac{\delta\bar{W}}{\delta J_i} = 0. \tag{10.99}$$

The extended effective action $\Gamma[\Phi,C,\bar{C},K,L]$ is

$$\Gamma[\Phi,C,\bar{C},K,L] = \bar{W}[J,\bar{\eta},\eta,K,L] - \Phi^i J_i - \eta_\alpha C^\alpha - \bar{\eta}^\alpha\bar{C}_\alpha, \tag{10.100}$$

where the sources J_i, η_α, $\bar{\eta}^\alpha$ are expressed through the mean fields Φ^i, C^α, $\bar{C}_\alpha$ from

$$\frac{\delta\bar{W}}{\delta J_i} = \Phi^i, \qquad \frac{\delta\bar{W}}{\delta\eta_\alpha} = C^\alpha, \qquad \frac{\delta\bar{W}}{\delta\bar{\eta}^\alpha} = \bar{C}_\alpha. \tag{10.101}$$

Here all the derivatives with respect to the sources are left. Then

$$\frac{\delta\Gamma}{\delta\Phi^i} = -J_i, \qquad \frac{\delta\Gamma}{\delta C^\alpha} = -\eta_\alpha, \qquad \frac{\delta\Gamma}{\delta\bar{C}_\alpha} = -\bar{\eta}^\alpha, \tag{10.102}$$

where all the derivatives with respect to the mean fields are right. Let us derive an auxiliary relation,

$$\frac{\delta\bar{W}}{\delta K_i} = \frac{\delta\Gamma}{\delta K_i} + \frac{\delta\Gamma}{\delta\Phi^j}\frac{\delta\Phi^j}{\delta K_i} + \frac{\overrightarrow{\delta}\Gamma}{\delta C^\alpha}\frac{\delta C^\alpha}{\delta K_i} + \frac{\overrightarrow{\delta}\Gamma}{\delta\bar{C}_\alpha}\frac{\delta\bar{C}_\alpha}{\delta K_i} + J_j\frac{\delta\Phi^j}{\delta K_i} - \eta_\alpha\frac{\delta C^\alpha}{\delta K_i} - \bar{\eta}^\alpha\frac{\delta\bar{C}_\alpha}{\delta K_i}$$
$$= \frac{\delta\Gamma}{\delta K_i} - J_j\frac{\delta\Phi^j}{\delta K_i} + \eta_\alpha\frac{\delta C^\alpha}{\delta K_i} + \bar{\eta}^\alpha\frac{\delta\bar{C}_\alpha}{\delta K_i} + J_j\frac{\delta\Phi^j}{\delta K_i} - \eta_\alpha\frac{\delta C^\alpha}{\delta K_i} - \bar{\eta}^\alpha\frac{\delta\bar{C}_\alpha}{\delta K_i} = \frac{\delta\Gamma}{\delta K_i}.$$

Here we took into account the relations

$$\frac{\overrightarrow{\delta}\Gamma}{\delta C^\alpha} = -\frac{\delta\Gamma}{\delta C^\alpha} = \eta_\alpha, \qquad \frac{\overrightarrow{\delta}\Gamma}{\delta \bar{C}_\alpha} = -\frac{\delta\Gamma}{\delta \bar{C}_\alpha} = \bar{\eta}_\alpha, \qquad \frac{\delta}{\delta K_i}(\eta_\alpha C^\alpha) = -\eta_\alpha \frac{\delta C^\alpha}{\delta K_i}$$

where, in the first two cases, there is a change from the left derivatives with respect to the ghosts to the rights ones. Analogously, we obtain

$$\frac{\delta \bar{W}}{\delta L_\alpha} = \frac{\delta \Gamma}{\delta L_\alpha}. \tag{10.103}$$

Using the relations (10.100)–(10.103), we rewrite the identity (10.99) in the form

$$\frac{\delta\Gamma}{\delta\Phi^i}\frac{\delta\Gamma}{\delta K_i} + \frac{\delta\Gamma}{\delta C^\alpha}\frac{\delta\Gamma}{\delta L_\alpha} + \frac{\delta\Gamma}{\delta \bar{C}_\alpha} G_{\alpha\beta}\, \chi^\beta = 0. \tag{10.104}$$

The equation (10.104) can be supplemented as follows. The generating functional $\bar{Z}[J,\bar{\eta},\eta,K,L]$ from Eq. (10.91) is invariant under the shift $\bar{c}_\alpha \to \bar{c}_\alpha + \delta\bar{c}_\alpha$, so

$$\bar{Z}[J,\bar{\eta},\eta,K,L] = \int \mathcal{D}\phi\mathcal{D}\bar{c}\mathcal{D}c\, e^{i\left(\bar{S}+\phi^i J_i + \eta_\alpha c^\alpha + \bar{\eta}^\alpha \bar{c}_\alpha\right)}\left(1 + \lambda_\alpha \chi_i^\alpha R^i_\beta c^\beta - \lambda_\alpha \bar{\eta}^\alpha\right).$$

This relation leads to the identity

$$i\chi_i^\alpha \frac{\delta \bar{Z}}{\delta K_i} + \bar{\eta}^\alpha = 0. \tag{10.105}$$

The last formula can be easily rewritten in terms of the functional $\bar{W}[J,\bar{\eta},\eta,K,L]$, and subsequently in terms of the effective action $\Gamma[\Phi,C,\bar{C},K,L]$, in the form

$$\chi_i^\alpha[\Phi]\frac{\delta\Gamma}{\delta K_i} + \frac{\delta\Gamma}{\delta\bar{C}_\alpha} = 0. \tag{10.106}$$

As a result, we obtain two independent identities for the effective action $\Gamma[\Phi,C,\bar{C},K,L]$, namely, (10.104) and (10.106). One can rewrite (10.104) in the form

$$\frac{\delta\Gamma}{\delta\Phi^i}\frac{\delta\Gamma}{\delta K_i} + \frac{\delta\Gamma}{\delta C^\alpha}\frac{\delta\Gamma}{\delta L_\alpha} - \chi_i^\alpha \frac{\delta\Gamma}{\delta K_i} G_{\alpha\beta}\chi^\beta = 0. \tag{10.107}$$

It is useful to introduce the functional $\tilde{\Gamma}[\Phi,C,\bar{C},K,L]$:

$$\Gamma[\Phi,C,\bar{C},K,L] = \tilde{\Gamma}[\Phi,C,\bar{C},K,L] + \frac{1}{2}\chi^\alpha[\Phi]G_{\alpha\beta}\chi^\beta[\Phi]. \tag{10.108}$$

After that, the identities (10.104) and (10.106) become

$$\frac{\delta\tilde{\Gamma}}{\delta\Phi^i}\frac{\delta\tilde{\Gamma}}{\delta K_i} + \frac{\delta\tilde{\Gamma}}{\delta C^\alpha}\frac{\delta\tilde{\Gamma}}{\delta L_\alpha} = 0, \tag{10.109}$$

$$\chi_i^\alpha \frac{\delta\tilde{\Gamma}}{\delta K_i} + \frac{\delta\tilde{\Gamma}}{\delta\bar{C}_\alpha} = 0. \tag{10.110}$$

The equation (10.110) can be solved in the form

$$\tilde{\Gamma}[\Phi,C,\bar{C},K,L] = \mathcal{T}[\Phi,C,L,K_i + \chi_i^\alpha \bar{C}_\alpha], \tag{10.111}$$

where $\mathcal{T}$ is a functional of the variables Φ^i, C^α, L_α and K_i. The form of the dependence of the $\bar{C}_\alpha$ is obtained by the simple change of the source K_i.

The validity of the relation (10.111) can be easily checked by substituting this into Eq. (10.110). Using the relation (10.111), one rewrites the identity (10.109) in the form

$$\frac{\delta\mathcal{T}}{\delta\Phi^i}\frac{\delta\mathcal{T}}{\delta K_i} + \frac{\delta\mathcal{T}}{\delta C^\alpha}\frac{\delta\mathcal{T}}{\delta L_\alpha} = 0. \tag{10.112}$$

The relation (10.112) is a final form of the identity for the effective action, called the Ward identity. If the functional $\mathcal{T}[\Phi, C, L, K_i]$ is found, the effective action $\tilde{\Gamma}[\Phi, C, \bar{C}, K, L]$ is obtained from the relation (10.111). Differentiating the effective action $\tilde{\Gamma}[\Phi, C, \bar{C}, K, L]$ with respect to the fields Φ^i, C^α, $\bar{C}_\alpha$ and sources K_i, L^α and then setting fields and sources equal to zero, one can obtain various relations for the vertex functions. All these relations can be also called the Ward identities.

Eq. (10.112) is quite general, and it is applicable to all Yang-MIlls gauge theories. The construction of the identities presented above is based on BRST symmetry. Originally, the Ward identities for Yang-Mills theory were derived independently by A.A. Slavnov and J.C. Taylor. These identities play a central role in the proof of renormalizability of Yang-Mills theory and, in general, of the gauge-invariant renormalizability of gauge theory.

10.5.3 Ward identities and the renormalizability of gauge theories

The Ward identities (Slavnov-Taylor identities in Yang-Mills) corresponding to BRST symmetry can be used to explore the gauge-invariant renormalization of gauge theories The identity (10.112) enables one to develop a general procedure of describing a renormalization structure of the gauge theories. In this subsection, we discuss the basic elements of such a procedure.

Let us start by introducing useful notions and notations. Consider two commuting functionals, $\mathcal{A}[\phi, c, K, L]$ and $\mathcal{B}[\phi, c, K, L]$. We note that the ghosts c^α, $\bar{c}^\alpha$ and the field K_i are anticommuting, while the quantities ϕ^i and L_α are commuting. One can associate with the two functionals $(\mathcal{A}, \mathcal{B})$ the following brace operation:[2]

$$(\mathcal{A}, \mathcal{B}) = \frac{1}{2}\left(\frac{\delta\mathcal{A}}{\delta\phi^i}\frac{\delta\mathcal{B}}{\delta K_i} + \frac{\delta\mathcal{A}}{\delta c^\alpha}\frac{\delta\mathcal{B}}{\delta L_\alpha} + \frac{\delta\mathcal{A}}{\delta K_i}\frac{\delta\mathcal{B}}{\delta\phi^i} + \frac{\delta\mathcal{A}}{\delta L^\alpha}\frac{\delta\mathcal{B}}{\delta c^\alpha}\right). \tag{10.113}$$

It is evident by construction that the expression (10.113) is symmetric in $\mathcal{A}$ and $\mathcal{B}$:

$$(\mathcal{A}, \mathcal{B}) = (\mathcal{B}, \mathcal{A}). \tag{10.114}$$

Consider the action $\bar{S}[\phi, \bar{c}, c, K, L]$ from Eq. (10.92) and rewrite it in the form

$$\bar{S}[\phi, \bar{c}, c, K, L] = S^{(0)}[\phi, c, K_i + \chi_i^\alpha \bar{c}_\alpha, L] + \frac{1}{2}\chi^\alpha[\phi]G_{\alpha\beta}[\phi]\chi^\beta[\phi], \tag{10.115}$$

where

$$S^{(0)}[\phi, c, K, L] = S[\phi] + K_i R^i_\alpha[\phi]c^\alpha + \frac{1}{2}L_\alpha f^\alpha{}_{\beta\gamma}c^\beta c^\gamma. \tag{10.116}$$

[2]In Batalin-Vilkovisky formalism [43], this object is called an antibracket and plays a prominent role in studying the structure of general gauge theory.

It is easy to show that the reduced action $S^{(0)}[\phi, c, K, L]$ satisfies the equation

$$\left(S^{(0)}, S^{(0)}\right) = 0. \tag{10.117}$$

The next step is to define the operator Ω, which is associated with the brace (10.113), by its action on the functionals $\mathcal{A}[\phi, c, K, L]$:

$$\Omega\mathcal{A} = (S^{(0)}, \mathcal{A}). \tag{10.118}$$

One can show that the operator Ω is nilpotent:

$$\Omega^2 = 0. \tag{10.119}$$

To prove this statement, one can rewrite the brace operation $(S^{(0)}, \mathcal{A})$ as

$$(S^{(0)}, \mathcal{A}) = \frac{1}{2}\Big(\frac{\delta S^{(0)}}{\delta b_I}\frac{\delta \mathcal{A}}{\delta f_I} + \frac{\delta S^{(0)}}{\delta f_I}\frac{\delta \mathcal{A}}{\delta b_I}\Big), \tag{10.120}$$

where $b_I = \{\phi^i, L_\alpha\}$ is a set of commuting (bosonic) quantities and $f_I = \{c^\alpha, L_\alpha\}$ is a set of anticommuting (fermionic) quantities. Then

$$\begin{aligned}
\Omega^2 &= \frac{1}{4}\Big(\frac{\delta S^{(0)}}{\delta b_I}\frac{\delta}{\delta f_I} + \frac{\delta S^{(0)}}{\delta f_I}\frac{\delta}{\delta b_I}\Big)\Big(\frac{\delta S^{(0)}}{\delta b_J}\frac{\delta}{\delta f_J} + \frac{\delta S^{(0)}}{\delta f_J}\frac{\delta}{\delta b_J}\Big) \\
&\quad + \frac{\delta S^{(0)}}{\delta b_I}\Big[\frac{\delta^2 S^{(0)}}{\delta f_I \delta b_J}\frac{\delta}{\delta f_J}\frac{\delta S^{(0)}}{\delta b_J}\frac{\delta}{\delta f_I}\frac{\delta}{\delta f_J} + \frac{\delta^2 S^{(0)}}{\delta f_I \delta b_J}\frac{\delta}{\delta b_J} - \frac{\delta S^{(0)}}{\delta f_J}\frac{\delta}{\delta f_I}\frac{\delta}{\delta b_J}\Big] \\
&\quad + \frac{\delta S^{(0)}}{\delta f_I}\Big[\frac{\delta^2 S^{(0)}}{\delta b_I \delta b_J}\frac{\delta}{\delta f_J} + \frac{\delta S^{(0)}}{\delta b_J}\frac{\delta}{\delta b_I}\frac{\delta}{\delta f_J} + \frac{\delta^2 S^{(0)}}{\delta b_I \delta f_J}\frac{\delta}{\delta b_J} + \frac{\delta S^{(0)}}{\delta f_J}\frac{\delta}{\delta b_I}\frac{\delta}{\delta b_J}\Big] \\
&= \frac{\delta^2 S^{(0)}}{\delta f_I \delta b_J}\frac{\delta S^{(0)}}{\delta b_I}\frac{\delta}{\delta f_J} + \frac{\delta^2 S^{(0)}}{\delta b_I \delta f_J}\frac{\delta S^{(0)}}{\delta f_I}\frac{\delta S^{(0)}}{\delta f_I}\frac{\delta}{\delta b_J} \\
&\quad + \frac{\delta^2 S^{(0)}}{\delta f_I \delta f_J}\frac{\delta S^{(0)}}{\delta b_J}\frac{\delta}{\delta b_I} + \frac{\delta^2 S^{(0)}}{\delta b_I \delta b_J}\frac{\delta S^{(0)}}{\delta f_I}\frac{\delta}{\delta f_J}.
\end{aligned} \tag{10.121}$$

In these transformations, we have used the facts that the derivatives with respect to b_I commute and those with respect to f_I anticommute.

To prove that expression (10.121) vanishes, one takes the derivative of the identity (10.117) with respect to b_I and with respect to f_I. This leads to the identities

$$\begin{aligned}
\frac{\delta^2 S^{(0)}}{\delta b_I \delta f_J}\frac{\delta S^{(0)}}{\delta b_J} &= -\frac{\delta^2 S^{(0)}}{\delta b_I \delta b_J}\frac{\delta S^{(0)}}{\delta f_J}, \\
\frac{\delta^2 S^{(0)}}{\delta f_I \delta b_J}\frac{\delta S^{(0)}}{\delta f_J} &= -\frac{\delta^2 S^{(0)}}{\delta f_I \delta f_J}\frac{\delta S^{(0)}}{\delta b_J}.
\end{aligned} \tag{10.122}$$

Substituting these identities into (10.121), we obtain $\Omega^2 = 0$.

The procedure for establishing the renormalization structure of Yang-Mills gauge theory is essentially based on the solutions to the equation

$$\Omega\mathcal{A} = 0. \tag{10.123}$$

It is evident that if the functional $\mathcal{A}_0$ is a solution, then the functional $\mathcal{A} = \mathcal{A}_0 + \Omega X$ is also a solution, for an arbitrary functional X. One can prove that the general solution to the linear equation (10.123) has the form

$$\mathcal{A}[\phi, c, K, L] = \mathcal{A}_0[\phi] + \Omega X[\phi, c, K, L], \tag{10.124}$$

where $\mathcal{A}_0$ is an arbitrary gauge-invariant functional, which means $\mathcal{A}_{0\,,i}[\phi]R^i_\alpha[\phi] = 0$. The functional $X[\phi, c, K, L]$ is also defined in a non-unique way, since it admits the transformation

$$X[\phi, c, K, L] \longrightarrow X[\phi, c, K, L] + \Omega\tilde{X}[\phi, c, K, L]. \tag{10.125}$$

We can proceed to the analysis of divergences on the basis of relation (10.112). Using the notation (10.113), one can rewrite (10.112) in the form

$$(\mathcal{T}, \mathcal{T}) = 0. \tag{10.126}$$

We will look for the solution to Eq. (10.126) in the framework of the loop expansion

$$\mathcal{T}[\phi, c, K, L] = \sum_{L=0}^{\infty} \hbar^L \Gamma^{(L)}[\phi, c, K, L], \tag{10.127}$$

where $\Gamma^{(0)} = S^{(0)}$, the tree-level expression $S^{(0)}$ is given by (10.116) and $\Gamma^{(L)}[\phi, c, K, L]$ is the L-loop contribution to the reduced effective action $\mathcal{T}[\phi, c, K, L]$. Substituting relation (10.127) into Eq. (10.126), one gets

$$\sum_{L'=0}^{L} \big(\Gamma^{(L')}, \Gamma^{(L-L')}\big) = 0, \qquad \text{where} \qquad L = 0, 1, \dots \tag{10.128}$$

In general, for the L-loop contribution, we have

$$\Gamma^{(L)} = \Gamma^{(L)}_{div} + \Gamma^{(L)}_{fin}, \qquad L = 0, 1, \dots. \tag{10.129}$$

We assume the use of dimensional regularization. In this context, $\Gamma^{(L)}_{div}$ is divergent after removing the regularization, while $\Gamma^{(L)}_{fin}$ remains finite.

Eq. (10.128) can be analyzed recursively, loop by loop. For $L = 0$, the equation is fulfilled identically owing to relation (10.117) for the reduced action $S^{(0)}$. Since $S^{(0)}$ is finite, in the one-loop approximation, we get

$$\big(S^{(0)}, \Gamma^{(1)}\big) = 0, \tag{10.130}$$

where $\Gamma^{(1)} = \Gamma^{(1)}(S^{(0)})$ is the one-loop effective action constructed on the basis of the action $S^{(0)}$. This equation leads to independent equations for $\Gamma^{(1)}_{div}(S^{(0)})$ and $\Gamma^{(1)}_{fin}(S^{(0)})$, in the forms

$$\begin{aligned}(S^{(0)},\Gamma^{(1)}_{div}(S^{(0)})) &= 0, &(10.131)\\ (S^{(0)},\Gamma^{(1)}_{fin}(S^{(0)})) &= 0. &(10.132)\end{aligned}$$

Suppose $\Gamma^{(1)}_{div}(S^{(0)})$ is the solution to Eq. (10.131). Then we define the action $S^{(1)}$ by the relation

$$\tilde{S}^{(1)} = S^{(0)} - \Gamma^{1}_{div}(S^{(0)}). \tag{10.133}$$

In the theory with the action $\tilde{S}^{(1)}$, all the one-loop divergences are absent by construction. Taking into account the relations (10.117) and (10.133), one gets

$$(\tilde{S}^{(1)},\tilde{S}^{(1)}) = (\Gamma^{(1)}_{div}(S^{(0)}),\Gamma^{(1)}_{div}(S^{(0)})). \tag{10.134}$$

The action $\tilde{S}^{(1)}$ satisfies the same equation (10.117) as the action $S^{(0)}$, up to the two-loop approximation. Let us define the action $S^{(1)}$ by the rule

$$S^{(1)} = \tilde{S}^{(1)} + \Delta_1 = S^{(0)} - \Gamma^{(1)}_{div}(S^{(0)}) + \Delta_1, \tag{10.135}$$

where the functional Δ_1 corresponds to the two-loop approximation and is defined by the equation

$$2(S^{(0)},\Delta_1) + (\Gamma^{(1)}_{div}(S^{(0)}),\Gamma^{(1)}_{div}(S^{(0)})) = 0. \tag{10.136}$$

Owing to the relations (10.116) and (10.131), (10.134), the action $S^{(1)}$ satisfies the equation

$$(S^{(1)},S^{(1)}) = 0, \tag{10.137}$$

up to the terms of the third-loop order. We emphasize that the fulfilment of the equation (10.137) is essentially based on the presence of the term Δ_1 in relation (10.135). The theory with the action $S^{(1)}$ is finite in the one-loop approximation.

Introduce the effective action $\Gamma(S^{(1)})$ constructed on the base of the action $S^{(1)}$,

$$\Gamma(S^{(1)}) = S^{(1)} + \Gamma^{(1)}(S^{(1)}) + \Gamma^{(2)}(S^{(1)}). \tag{10.138}$$

Here $\Gamma^{(1)}(S^{(1)})$ is the one-loop correction to the action $S^{(1)}$, which is finite by construction. The functional $\Gamma^{(2)}(S^{(1)})$ is the two-loop correction, which can be divergent:

$$\Gamma^{(2)}(S^{(1)}) = \Gamma^{(2)}_{div}(S^{(1)}) + \Gamma^{(2)}_{fin}(S^{(1)}). \tag{10.139}$$

Since the action $S^{(1)}$ (10.135) satisfies Eq. (10.137), the effective action $\Gamma(S^{(1)})$ satisfies the equation

$$(\Gamma(S^{(1)}),\Gamma(S^{(1)})) = 0. \tag{10.140}$$

The derivation of Eq. (10.138) on the basis of Eq. (10.137) is a literal repetition of the derivation of Eq. (10.126) on the basis of Eq. (10.117). Substituting the relation (10.138) into (10.140) and taking into account (10.139), one gets for the $\Gamma^{(2)}_{div}(S^{(1)})$

$$(S^{(0)}, \Gamma^{(2)}_{div}(S^{(1)})) = 0. \tag{10.141}$$

Let the functional $\Gamma^{(2)}_{div}(S^{(1)})$ be a solution to Eq. (10.141). We introduce the action

$$\tilde{S}_2 = S^{(1)} - \Gamma^{(2)}_{div}(S^{(1)}). \tag{10.142}$$

In the theory with the action $\tilde{S}^{(1)}$, all the two-loop divergences are absent by construction. Taking into account the relations (10.137) and (10.142), one gets

$$(\tilde{S}_2, \tilde{S}_2) = (\Gamma^{(2)}_{div}(S^{(1)}, \Gamma^{(2)}_{div}(S^{(1)}). \tag{10.143}$$

The action $\tilde{S}^{(2)}$ satisfies the same equation (10.137) as the action $S^{(1)}$, up to the four-loop approximation. Define the action $S^{(2)}$ by the rule

$$\begin{aligned} S^{(2)} &= \tilde{S}^{(2)} + \Delta_2 = S^{(1)} - \Gamma^{(2)}_{div}(S^{(1)}) + \Delta_2 \\ &= S^{(0)} - \Gamma^{(1)}_{div}(S^{(0)}) + \Delta_1 - \Gamma^{(2)}(S^{(1)}) + \Delta_2, \end{aligned} \tag{10.144}$$

where the functional Δ_2 corresponds to the four-loop approximation and is defined by the equation

$$2(S^{(0)}, \Delta_2) + (\Gamma^{(2)}_{div}(S^{(0)}), \Gamma^{(2)}_{div}(S^{(0)})) = 0. \tag{10.145}$$

Owing to the relations (10.135), (10.141) and (10.143), the functional $S^{(2)}$ (10.144) satisfies the equation

$$(S^{(2)}, S^{(2)}) = 0 \tag{10.146}$$

up to the higher-order terms. The theory with action $S^{(2)}$ is finite in the two-loop approximation. The next step is to introduce the effective action $\Gamma(S^{(2)})$ constructed with the action $S^{(2)}$,

$$\Gamma(S^{(2)}) = S^{(2)} + \Gamma^{(1)}(S^{(2)}) + \Gamma^{(2)}(S^{(2)}) + \Gamma^{(3)}(S^{(2)}), \tag{10.147}$$

where $\Gamma^{(1)}(S^{(2)})$ and $\Gamma^{(2)}(S^{(2)})$ are one- and two-loop corrections to the action $S^{(2)}$. These corrections are finite by construction. The functional $\Gamma^{(3)}(S^{(2)})$ is the higher-loop correction, which can be divergent, so

$$\Gamma^{(3)}(S^{(2)}) = \Gamma^{(3)}_{div}(S^{(2)}) + \Gamma^{(3)}_{fin}(S^{(2)}). \tag{10.148}$$

The relation (10.146) leads to the equation

$$(\Gamma(S^{(2)}), \Gamma(S^{(2)})) = 0. \tag{10.149}$$

In its turn, this equation, together with (10.144), yields the equation for $\Gamma^{(3)}_{div}(S^{(2)})$,

$$(S^{(0)}, \Gamma^{(3)}_{div}(S^{(2)})) = 0. \tag{10.150}$$

After that, the procedure can be continued to any desirable order of the loop expansion.

To complete the general description of renormalization, one has to consider the equation (10.128) and assume that the above procedure has been carried out for all orders $L' \leq L-1$, such that the theory is finite up to $(L-1)$-loop approximation. The equation for $\Gamma^{(L)}_{div}(S^{(L-1)})$ is

$$\big(S^{(0)}, \Gamma^{(L)}_{div}(S^{(L-1)})\big) = 0, \tag{10.151}$$

which is the same basic equation $\Omega\Gamma = 0$ for the divergent part of effective action.

The equation $\big(S^{(0)}, \Gamma\big) = 0$ does not define whether the divergences may depend on the ghost $\bar{C}_\alpha$. However, such dependence is ruled out by Eq. (10.111). By applying this equation to the described recursive procedure, we obtain, in the L-loop approximation,

$$\tilde{\Gamma}^{(L)}_{div}[..., \bar{C}, K, ...] = \mathcal{T}^{(L)}_{div}[..., K_i + \chi^\alpha_{,i}\bar{C}_\alpha, ...], \tag{10.152}$$

where the dots mean other functional arguments. Thus, to obtain the divergent part of the effective action $\tilde{\Gamma}[\phi, C, \bar{C}, K, L]$, it is sufficient to find the divergent part of the effective action $\mathcal{T}[\phi, C^\alpha, K_i + \chi^\alpha_{,i}\bar{C}_\alpha, L]$.

To conclude this section, let us make a small survey of the general scheme of renormalization of gauge theories. We begin with the action $S^{(0)}$ and calculate the one-loop correction $\Gamma^{(1)}(S^{(0)})$ with the divergent part $\Gamma^{(1)}_{div}(S^{(0)})$. Using this correction, we construct the action $S^{(1)}$, leading to the finite theory at one loop. Starting from $S^{(1)}$, we calculate the two-loop correction $\Gamma^{(2)}(S^{(1)})$, which can have two-loop divergent part $\Gamma^{(2)}_{div}(S^{(1)})$, while all the one-loop contributions to $\Gamma^{(2)}(S^{(1)})$ are finite. Using the $\Gamma^{(2)}_{div}(S^{(1)})$, we construct the $S^{(2)}$ leading to a finite theory at two loops.

As a result, the divergent terms in the effective action satisfy the universal equation

$$\Omega\Gamma^{(L)}_{div}(S^{(L-1)}) = 0, \qquad L = 1,\, 2, \dots. \tag{10.153}$$

The solution to this equation can be described in terms of a general gauge theory. The iterative procedure described above enables us to construct the action $S^{(L)}$ in such a way that the corresponding effective action $\Gamma(S^{(L)})$ is finite for all $L = 1, 2, \dots$. On the other hand, to fix the renormalization arbitrariness one should use the details of the theory, i.e., the action $S^{(0)}$.

In section 10.10, we apply Eq. (10.153) to describe the renormalization structure of Yang-Mills theory in the Lorentz gauge.

10.6 The gauge dependence of effective action

The functional integral for the generating functional of Green functions (10.63) contains the gauge-fixing function $\chi^\alpha[\phi]$. Therefore, the effective action constructed on the base of $Z[J]$ may depend on the choice of gauge, such that $\Gamma[\Phi] = \Gamma_\chi[\Phi]$. To explore this dependence, let us consider the functional $Z_\chi[J]$ in the form (10.68),

$$Z_\chi[J] = \int \mathcal{D}\phi\, e^{i\left\{S[\phi] + \frac{1}{2}\chi^\alpha[\phi]G_{\alpha\beta}\chi^\beta[\phi] - i\,\mathrm{Tr}\,\log\left(\chi^\alpha_{,i}[\phi]R^i_\beta[\phi]\right) + \phi^i J_i\right\}}. \tag{10.154}$$

The subscript χ means that the generating functional depends on the gauge-fixing procedure. Consider the generating functional $Z_{\chi+\delta\chi}[J]$, which corresponds to the modified gauge $\chi^\alpha + \delta\chi^\alpha$, and find its relation to $Z_\chi[J]$. As in the previous sections, we consider the class of the linear gauges $\chi^\alpha = \chi^\alpha_i \phi^i$, where χ^α_i is field independent.

Using (10.154) and taking into account the fact that $\delta\chi^\alpha$ are infinitesimal, we can write

$$Z_{\chi+\delta\chi}[J] = \int \mathcal{D}\phi\, e^{i\left\{S[\phi]+\frac{1}{2}\chi^\alpha[\phi]G_{\alpha\beta}\chi^\beta[\phi]-i\,\mathrm{Tr}\,\log\left(\chi^\alpha_{,i}[\phi]R^i_\beta[\phi]\right)+\phi^i J_i\right\}} \times \left\{1+\chi^\alpha G_{\alpha\beta}\delta\chi^\beta - i(M^{-1})^\alpha{}_\beta\delta\chi^\beta_{,i}R^i_\alpha\right\}.$$

Let us make the change of variables

$$\phi'^i = \phi^i + R^i_\beta\xi^\beta, \qquad \text{with} \qquad \xi^\beta = -\left(M^{-1}\right)^\beta{}_\gamma\delta\chi^\gamma. \tag{10.155}$$

The functional Jacobian of this change of variables is

$$J = \mathrm{Det}\left(\frac{\delta\phi'^i}{\delta\phi^j}\right) = \mathrm{Det}\left(\delta^i_j + \frac{\delta R^i_\alpha\xi^\alpha}{\delta\phi^j}\right) = 1 + (R^i_\alpha\xi^\alpha)_{,i} = 1 + R^i_\alpha\chi^\alpha_{,i}, \tag{10.156}$$

where we took into account the fact that $R^i_{\alpha,i}$ is proportional to $\delta^4(0)$ or $\partial_\mu\delta^4(x - x')\big|_{x'=x}$, which are equal to zero in dimensional regularization. After that, one gets

$$\begin{aligned} Z_{\chi+\delta\chi}[J] &= Z_\chi[J] + \int \mathcal{D}\phi\, e^{i\left\{S[\phi]+\frac{1}{2}\chi^\alpha[\phi]G_{\alpha\beta}\chi^\beta[\phi]-i\,\mathrm{Tr}\,\log\left(\chi^\alpha_{,i}[\phi]R^i_\beta[\phi]\right)+\phi^i J_i\right\}} \\ &\times \Big\{i\Big[\chi^\alpha G_{\alpha\beta}\delta\chi^\beta + \chi^\alpha G_{\alpha\beta}\chi^\beta_{,i}R^i_\gamma\xi^\gamma\Big] + (M^{-1})^\alpha{}_\beta\delta\chi^\beta_{,i}R^i_\alpha \\ &-\left(M^{-1}\right)^\alpha{}_\beta\chi^\beta_{,i}R^i_{\alpha,j}R^j_\gamma(M^{-1})^\gamma{}_\delta\delta\chi^\delta - \left(M^{-1}\right)^\alpha{}_\beta\delta\chi^\beta_{,i}R^i_\alpha \\ &+R^i_\alpha\left(M^{-1}\right)^\alpha{}_\beta\chi^\beta_{,j}R^j_{\gamma,i}\left(M^{-1}\right)^\gamma{}_\delta\delta\chi^\delta + iR^i_\alpha\xi^\alpha J_i\Big\}. \end{aligned} \tag{10.157}$$

The last two terms here originated from $R^i_\alpha(\xi^\alpha)_{,i}$. The first terms, in the square braces, cancel each other owing to the choice of ξ^α in (10.155). Two other terms cancel automatically. The two terms containing $\delta\chi^\delta$ are transformed as

$$\begin{aligned} &(M^{-1})^\alpha{}_\beta\chi^\beta_{,i}(R^i_{\alpha,j}R^j_\gamma - R^i_{\gamma,j}R^j_\alpha)(M^{-1})^\gamma{}_\delta\chi^\delta = (M^{-1})^\alpha{}_\beta\chi^\beta_{,i}f^\lambda{}_{\alpha\gamma}R^i_\lambda(M^{-1})^\gamma{}_\delta\chi^\delta \\ &= (M^{-1})^\alpha{}_\beta(M)^\beta{}_\lambda f^\lambda{}_{\alpha\gamma}(M^{-1})^\gamma{}_\delta\chi^\delta = f^\alpha{}_{\alpha\gamma}(M^{-1})^\gamma{}_\delta\chi^\delta = 0. \end{aligned} \tag{10.158}$$

The last expression equals zero since the trace $f^\alpha{}_{\alpha\gamma} \sim \delta^4(0)$ vanishes in Yang-Mills theory. As a result, we obtain

$$Z_{\chi+\delta\chi}[J] - Z_\chi[J] = -iJ_i\,\langle R^i_\alpha\left(M^{-1}\right)^\alpha{}_\beta\delta\chi^\beta\rangle\, Z_\chi[J]. \tag{10.159}$$

Here the following notation for the quantum average is used:

$$\langle A\rangle = \frac{1}{Z_\chi[J]}\int \mathcal{D}\phi\, A\, e^{i\left\{S[\phi]+\frac{1}{2}\chi^\alpha[\phi]G_{\alpha\beta}\chi^\beta[\phi]-i\,\mathrm{Tr}\,\log\left(\chi^\alpha_{,i}[\phi]R^i_\beta[\phi]\right)+\phi^i J_i\right\}}. \tag{10.160}$$

Introducing the generating functional of the connected Green functions and then the effective action (we leave this as an exercise for the reader), one finally gets

$$\delta_\chi \Gamma[\Phi] = \frac{\delta \Gamma[\Phi]}{\delta \Phi^i} \langle R^i_\alpha (M^{-1})^\alpha{}_\beta \delta\chi^\beta \rangle. \tag{10.161}$$

This remarkable relation shows that the variation of the effective action, stipulated by the variation of the gauge-fixing condition, is proportional to the effective equations of motion. Therefore, for solutions to effective equations of motion, the effective action independent of gauge fixing. We shall intensively use this fundamental property of gauge theories in Part II of the book, when discussing practical renormalization in the models of quantum field theory and quantum gravity.

10.7 Background field method

The background field method is a special procedure for defining the effective action in such a way that it becomes invariant under classical gauge transformations. In this case, Ward identities become trivial, just reflecting the gauge invariance, which is automatically satisfied. The main idea of the method consists of splitting fields into quantum and classical (or background) parts and imposing gauge-fixing conditions that maintain the covariance over the background field while removing the degeneracy in the quantum fields.

Consider the generating functional of Green functions in the gauge theory (10.63),

$$Z[J] = \int \mathcal{D}\phi\, \mathcal{D}\bar{c}\, \mathcal{D}c\, e^{\frac{i}{\hbar}\{S[\phi]+S_{GF}[\phi]+S_{GH}[\bar{c},c,\phi]+J_i\phi^i\}}\, \mathrm{Det}^{\,\frac{1}{2}}(G_{\alpha\beta}[\phi]), \tag{10.162}$$

$$\text{where} \qquad S_{GF} = \frac{1}{2}\chi^\alpha[\phi]G_{\alpha\beta}[\phi]\chi^\beta[\phi] \tag{10.163}$$

$$\text{and} \qquad S_{GH}[\bar{c},c,\phi] = \bar{c}_\alpha \chi^\alpha_{,i}[\phi] R^i{}_\beta[\phi] c^\beta. \tag{10.164}$$

Here $R^i_\alpha = r^i_\alpha + r^i_{\alpha j}\phi^j$ are the generators of gauge transformations. The quantities r^i_α and $r^i_{\alpha j}$ are assumed field independent.

From now on, the development is analogous to the loop expansion in section 8.6. First, introduce the generating functional of the connected Green functions $W[J]$ and the effective action $\Gamma[\Phi]$ with what we call now the background field $\Phi^i = \frac{\delta W[J]}{\delta J^i}$. Second, taking into account that the effective action depends on the gauge, we use the special background-dependent gauge-fixing condition, with $\chi^\alpha[\phi] = \chi^\alpha_i[\Phi]\phi^i$ and $G_{\alpha\beta} = G_{\alpha\beta}[\Phi]$, where the Φ is the background field. Third, we make the change of variables in the functional integral $\phi \to \Phi + \sqrt{\hbar}\phi$, $\bar{c}_\alpha \to \sqrt{\hbar}\bar{c}_\alpha$ and $c^\beta \to \sqrt{\hbar}c^\beta$. In what follows, ϕ, $\bar{c}$, c are quantum fields. In this way, we arrive at the expression

$$e^{\frac{i}{\hbar}\bar{\Gamma}[\Phi]} = \int \mathcal{D}\phi\, \mathcal{D}\bar{c}\, \mathcal{D}c\, e^{\frac{i}{\hbar}\tilde{S}_{total}[\Phi,\phi,\bar{c},c]}\, \mathrm{Det}^{\,\frac{1}{2}}(G_{\alpha\beta}[\Phi]), \tag{10.165}$$

where $\bar{\Gamma}[\Phi]$ includes all quantum corrections to the classical action and is defined by the relation $\Gamma[\Phi] = S[\Phi] + \bar{\Gamma}[\Phi]$. Furthermore,

$$\begin{aligned}\tilde{S}_{total}[\Phi,\phi,\bar{c},c] &= S[\Phi+\sqrt{\hbar}\phi]-S[\Phi]-\sqrt{\hbar}S_{,i}\phi^i \\ &= \hbar\tilde{S}_{GF}[\Phi,\phi]+\hbar\tilde{S}_{GH}[\Phi,\phi,\bar{c},c], \qquad (10.166)\end{aligned}$$

where

$$\tilde{S}_{GF}[\Phi,\phi] = \frac{1}{2}\chi_i^\alpha[\Phi]\phi^i G_{\alpha\beta}[\Phi]\chi_j^\beta[\Phi]\phi^j, \qquad (10.167)$$

$$\tilde{S}_{GH}[\Phi,\phi,\bar{c},c] = \bar{c}_\alpha M^\alpha{}_\beta[\Phi,\phi]c^\beta, \qquad (10.168)$$

$$M^\alpha{}_\beta[\Phi,\phi] = \chi_i^\alpha[\Phi]R_\beta^i[\Phi+\sqrt{\hbar}\phi]. \qquad (10.169)$$

Consider the properties of the functional integral (10.165) with $\tilde{S}_{total}$ given by the relation (10.166). It is useful to define two types of the transformations of the fields Φ, ϕ, $\bar{c}$, c in this functional integral:

a. Background gauge transformations:

$$\begin{aligned}\delta_B\Phi^i &= R_\alpha^i[\Phi]\xi^\alpha, & \delta_B\phi^i &= r^i_{\alpha j}\phi^j\xi^\alpha, \\ \delta_B\bar{c}_\alpha &= -f^\beta{}_{\alpha\gamma}\bar{c}_\beta\xi^\gamma, & \delta_B c^\alpha &= f^\alpha{}_{\beta\gamma}c^\beta\xi^\gamma. \qquad (10.170)\end{aligned}$$

For the background fields Φ, this is just the classical gauge transformation.

b. Quantum gauge transformations:

$$\delta_Q\Phi^i=0, \quad \delta_Q\phi^i = R_\alpha^i[\Phi+\sqrt{\hbar}\phi]\xi^\alpha, \quad \delta_Q\bar{c}_\alpha=0, \quad \delta_Q c^\alpha = 0. \qquad (10.171)$$

Note that the background field Φ is not affected by these transformations.

Let us prove that the effective action $\Gamma[\Phi]$ is invariant under the background gauge transformations. Since the classical action $S[\Phi]$ is invariant, we need to consider only the invariance of the quantum corrections $\bar{\Gamma}[\Phi]$. The proof consists of several steps.

The background field appears in the integral in the functionals $S[\Phi+\sqrt{\hbar}\phi]$, $S[\Phi]$, $\tilde{S}_{GF}[\Phi,\phi]$, $\tilde{S}_{GH}[\Phi,\phi,\bar{c},c]$ and $\mathrm{Det}\,(G_{\alpha\beta}[\Phi])$. We perform the background transformation of Φ together with the change of quantum fields ϕ, $\bar{c}$, c (10.170) in the functional integral. Consider the transformation of all terms in the exponential under the integral (10.165), one by one:

1. The background transformation of the expression $\Phi^i+\sqrt{\hbar}\phi^i$ is

$$\begin{aligned}\delta_B(\Phi^i+\sqrt{\hbar}\phi^i) &= R_\alpha^i[\Phi]\xi^\alpha+\sqrt{\hbar}r^i_{\alpha j}\phi^j\xi^\alpha \\ &= \big[r_\alpha^i+r^i_{\alpha j}(\Phi^j+\sqrt{\hbar}\phi^j)\big]\xi^\alpha \;=\; R_\alpha^i[\Phi+\sqrt{\hbar}\phi]\xi^\alpha.\end{aligned}$$

Owing to the gauge invariance of the classical action, this means $\delta_B S[\Phi+\sqrt{\hbar}\phi]=0$. The described change of variables is local, and the corresponding functional Jacobian is equal to unit.

2. The background field transformation of the expression $S[\Phi]_{,i}\phi^i$ consists of the classical gauge transformation of Φ, accompanied by the corresponding change of the quantum field, $\delta_B S_{,i}\phi^i = S_{,ij}R^j_\alpha\xi^\alpha\phi^i+S_{,i}r^i_{\alpha j}\phi^j\xi^\alpha$. Differentiating the gauge invariance $S_{,j}R^j_\alpha=0$, one gets $S_{,ij}R^j_\alpha = -S_{,j}R^i_{\alpha,j} = -S_{,j}r^j_{\alpha\, i}$. Therefore,

$$\delta_B S_{,i}\phi^i = (-S_{,j}r^j_{\alpha\, i}\phi^i+S_{,i}r^i_{\alpha j}\phi^j)\xi^\alpha = 0.$$

All in all, the expression $S[\Phi+\sqrt{\hbar}\phi]-S[\Phi]-\sqrt{\hbar}S_{,i}\phi^i$ in the exponential under the integral (10.165) is invariant under the background gauge transformations.

3. Consider the background transformations of the $\tilde{S}_{GF}[\Phi,\phi]$ (10.167). This expression contains two arbitrary functions, $\chi_i^\alpha[\Phi]$ and $G_{\alpha\beta}[\Phi]$. Therefore, to calculate $\delta_B\tilde{S}_{GF}[\Phi,\phi]$, we should additionally define the the background transformations of these functions. In general gauge theory, such transformations can be constructed only from the quantities which characterize the theory, i.e., from the generators $R^i_\alpha[\phi] = r^i_\alpha + r^i_{\alpha j}\phi^j$ and the structure constants $f^\alpha{}_{\beta\gamma}$. Taking into account this circumstance, we define

$$\delta_B\chi_i^\alpha[\Phi] = (f^\alpha{}_{\beta\gamma}\chi_i^\beta[\Phi] - \chi_j^\alpha[\Phi]r^j_{\gamma i})\xi^\gamma \tag{10.172}$$

and

$$\delta_B G_{\alpha\beta}[\Phi] = -(f^\gamma{}_{\alpha\delta}G_{\gamma\beta}[\Phi] + f^\gamma{}_{\beta\delta}G_{\alpha\gamma}[\Phi])\xi^\delta. \tag{10.173}$$

Using the relations (10.172), (10.173) and the change of the quantum field $\phi^i = r^i_{\alpha j}\phi^j\xi^\alpha$ in the functional integral (10.165), we obtain

$$\begin{aligned}
\delta_B\tilde{S}_{GF}[\Phi,\phi] &= \hbar\big\{\delta_B(\chi_i^\alpha[\Phi]\phi^i)G_{\alpha\beta}[\Phi]\chi_j^\beta[\Phi]\phi^j + \frac{1}{2}\chi_i^\alpha[\Phi]\phi^i\delta_B G_{\alpha\beta}[\Phi]\chi_j^\beta\phi^j\big\} \\
&= \hbar(f^\alpha{}_{\gamma\delta}\chi_i^\gamma\phi^i - \chi_j^\alpha r^j_{\delta i}\phi^i + \chi_i^\alpha r^i_{\delta j}\phi^j)\xi^\delta G_{\alpha\beta}\chi_j^\beta\phi^j \\
&\quad -\frac{\hbar}{2}\chi_i^\alpha\phi^i(f^\gamma{}_{\alpha\delta}G_{\gamma\beta} + f^\gamma{}_{\beta\delta}G_{\alpha\delta})\xi^\delta\chi_j^\beta\phi^j = 0.
\end{aligned} \tag{10.174}$$

Thus, the expression $\tilde{S}_{GF}[\Phi,\phi]$ is invariant under background transformation.

4. Consider the background transformation of $\tilde{S}_{GH}[\Phi,\phi,\bar{c},c]$, where $M^\alpha{}_\beta$ is given by (10.169). The background gauge transformation of $M^\alpha{}_\beta$ is

$$\begin{aligned}
\delta_B M^\alpha{}_\beta &= \delta_B\chi_i^\alpha[\Phi]R^i_\beta[\Phi+\sqrt{\hbar}\phi] = \big(f^\alpha{}_{\gamma\delta}\chi_i^\gamma\xi^\delta - \chi_j^\alpha r^j_{\delta i}\xi^\delta\big)R^i_\beta[\Phi+\sqrt{\hbar}\phi] \\
&= \chi_i^\alpha\, R^i_{\beta,j}[\Phi+\sqrt{\hbar}\phi]\, R^j_\delta[\Phi+\sqrt{\hbar}\phi] \\
&= f^\alpha{}_{\gamma\delta}M^\gamma{}_\beta\xi^\delta + \chi_i^\alpha\big(R^i_{\beta,j}[\Phi+\sqrt{\hbar}\phi]R^j_\delta[\Phi+\sqrt{\hbar}\phi] \\
&\quad - - R^i_{\delta\,,j}[\Phi+\sqrt{\hbar}\phi]R^j_\beta[\Phi+\sqrt{\hbar}\phi]\big)\xi^\beta \\
&= (f^\alpha{}_{\gamma\delta}M^\gamma{}_\beta - f^\gamma{}_{\beta\delta}M^\alpha{}_\gamma)\xi^\delta.
\end{aligned} \tag{10.175}$$

Taking into account this result and the background transformation of the ghosts $\bar{c}_\alpha$ and c^β in (10.170), we obtain

$$\begin{aligned}
\delta_B\tilde{S}_{GH} &= \delta_b\big(\bar{c}_\alpha M^\alpha{}_\beta c^\beta\big) = -f^\gamma{}_{\alpha\delta}\bar{c}_\gamma\xi^\delta M^\alpha{}_\beta c^\beta \\
&\quad + \bar{c}_\alpha M^\alpha{}_\beta f^\beta{}_{\gamma\delta}c^\gamma\xi^\delta + c_\alpha(\delta_B M^\alpha{}_\beta)c^\beta = 0.
\end{aligned} \tag{10.176}$$

5. Finally, the background transformation of Det $(G_{\alpha\beta}[\Phi])$ gives

$$\delta_B\,\mathrm{Det}\,(G_{\alpha\beta}[\Phi]) = -\,\mathrm{Tr}\,G^{\alpha\beta}\big(f^\gamma{}_{\alpha\delta}G_{\gamma\beta} + f^\gamma{}_{\alpha\gamma}\big)\xi^\delta = -\,2\,\mathrm{Tr}\,(f^\gamma{}_{\gamma\delta})\xi^\delta = 0. \tag{10.177}$$

In summary, all five summands of the action (10.167) are invariant under background gauge transformation. Thus, the functional integral (10.165) is invariant under classical gauge transformation. Since we assume the use of dimensional regularization, which preserves gauge symmetry, it is possible to state that the effective action, including all the loop contributions, is invariant under classical gauge transformation.

As a particular case, consider the one-loop effective action in the framework of the background field method. In this approximation, $\Gamma[\Phi] = \Gamma^{(1)}[\Phi] = S + \hbar\bar{\Gamma}^{(1)}[\Phi]$. Then the relation (10.165) yields

$$\begin{aligned} \bar{\Gamma}^{(1)}[\Phi] &= \frac{i}{2}\,\text{Tr}\,\log\Big(S_{2,ij}[\Phi] + \chi_i^\alpha[\Phi]G_{\alpha\beta}[\Phi]\chi_j^\beta[\Phi]\Big) \\ &\quad -i\,\text{Tr}\,\log\big(M^\alpha{}_\beta[\Phi]\big) - \frac{i}{2}\,\text{Tr}\,\log(G_{\alpha\beta}[\Phi]), \end{aligned} \tag{10.178}$$

where $S_{2,ij}[\Phi]$ is a background-dependent bilinear form of the classical gauge-invariant action. The last expression represents a useful base for one-loop calculations in gauge theory, including models of quantum gravity. In models with higher derivatives, this formula may be modified, as we will discuss in Part II. However, these modifications do not affect the main conclusion about the invariance of effective action, which we formulated above.

Final observations. It is important to distinguish two different notions. Although the effective action in the background gauge is invariant under classical gauge transformation, it is still dependent on gauge fixing. The background gauge contains the functions $\chi_i^\alpha[\Phi]$. Different choices of these functions, in general, lead to different effective actions. At the same time, the off-shell effective action is unique, according to Eq. (10.161). In particular, this feature provides the gauge independence of the $\mathcal{S}$-matrix, which has a fundamental physical importance. In particular, all generalizations of the standard quantum field theory approach, which do not provide the universality of the $\mathcal{S}$-matrix, should be regarded with a special caution, to say the least.

For the one-loop effective action, we note that the quantum average in Eq. (10.161) already has a factor of $\hbar$. Thus, in order to stay within the $\mathcal{O}(\hbar)$-approximation, we can trade the effective equation of motion for its classical counterpart. In this way, we arrive at the useful property of the one-loop effective action $\Gamma^{(1)}[\Phi]$, namely, that this object is gauge invariant on the classical mass shell. This feature will prove very useful and fruitful in discussing the renormalization of quantum field theory in curved spacetime and in quantum gravity, as the reader will see in Part II.

10.8 Feynman diagrams in Yang-Mills theory

In the next two sections, we apply the general constructions considered previously to pure Yang-Mills theory.

The total action (10.64) of the quantum Yang-Mills theory is

$$\begin{aligned} S_{total}[A,\bar{c},c] &= S_{YM}[A] + \frac{1}{2}\chi^\alpha[A]G_{\alpha\beta}\chi^\beta[A] + \bar{c}_\alpha M^\alpha{}_\beta[A]c^\beta \\ &= \int d^4x\,\mathcal{L}_{total} = \int d^4x\,\{\mathcal{L}_{YM} + \mathcal{L}_{GF} + \mathcal{L}_{GH}\}, \end{aligned} \tag{10.179}$$

with the Yang-Mills Lagrangian

$$\mathcal{L}_{YM} = -\frac{1}{4}G^{a\mu\nu}G^{a}_{\mu\nu} = -\frac{1}{4}F^{a\mu\nu}F^{a}_{\mu\nu} - gf^{abc}A^{b\mu}A^{c\nu}\partial_\mu A^a_\nu - \frac{1}{4}g^2 f^{abc}f^{adf}A^{b\mu}A^{c\nu}A^d_\mu A^f_\nu, \tag{10.180}$$

and $F^a_{\mu\nu} = \partial_\mu A^a_\nu - \partial_\nu A^a_\mu$. We will use the Lorentz type gauge $\chi^a = \partial_\mu A^{a\mu}$. Then

$$\mathcal{L}_{GF} = \frac{\alpha}{2}\,\partial_\mu A^{a\mu}\partial_\nu A^{a\nu}, \tag{10.181}$$

where α is a gauge-fixing parameter. The ghost action $\bar{c}_\alpha M^\alpha{}_\beta c^\beta$ has the form

$$\int d^4x\, d^4y\, d^4z\, \bar{c}^a(x)\,\frac{\delta\chi^a(x)}{\delta A^c_\mu(z)}\,\frac{\delta A^c_\mu(z)}{\delta\xi^b(y)}\,c^b(y), \tag{10.182}$$

where $\xi^b(y)$ are parameters of the gauge transformation $\delta A^c_\mu = \mathcal{D}^{cb}{}_\mu\xi^b$. Therefore,

$$\frac{\delta\chi^a(x)}{\delta A^c_\mu(z)} = \delta^{ac}\partial^\mu\delta^4(x-z), \qquad \frac{\delta A^c_\mu(z)}{\delta\xi^b(y)}c^b(y) = \mathcal{D}^{cb}{}_\mu\delta^4(x-y)c^b(y). \tag{10.183}$$

Thus, the ghost action has the form $\int d^4x\,\bar{c}^a\partial^\mu\mathcal{D}^{ab}{}_\mu c^b$. Integrating by parts, we get

$$\mathcal{L}_{GH} = -\partial^\mu\bar{c}^a\mathcal{D}^{ab}{}_\mu c^b. \tag{10.184}$$

As a result, the total Lagrangian of the quantum Yang-Mills theory can be written as

$$\mathcal{L}_{total} = \mathcal{L}_{YM} + \frac{\alpha}{2}\partial_\mu A^{a\mu}\partial_\nu A^{a\nu} - \partial^\mu\bar{c}^a\mathcal{D}^{ab}{}_\mu c^b. \tag{10.185}$$

Consider the two first terms in the last expression, with $\mathcal{L}_{YM}$ given by (10.180). After omitting the total divergences and a small algebra, we have

$$\frac{1}{2}A^{a\mu}\Box A^a_\mu + \frac{1+\alpha}{2}(\partial_\mu A^{a\mu})^2 + \dots .$$

Taking $\alpha = -1$ and using the operator $\mathcal{D}^{ab}_\mu$ from (4.116), we arrive at the total Lagrangian of the quantum Yang-Mills theory,

$$\mathcal{L}_{total} = \frac{1}{2}A^{a\mu}\Box A^a_\mu - gf^{abc}A^{b\mu}A^{c\nu}\partial_\mu A^a_\nu - \frac{1}{4}g^2 f^{abc}f^{adf}A^{b\mu}A^{c\nu}A^d_\mu A^f_\nu - \partial^\mu\bar{c}^a\partial_\mu c^a - gf^{abc}\partial^\mu\bar{c}A^b_\mu c^b. \tag{10.186}$$

The last expression has the standard form of the field theory Lagrangian,

$$\mathcal{L}_{total} = \mathcal{L}^{(0)} + \mathcal{L}_{int}, \tag{10.187}$$

where one can identify both the free and the interacting terms as follows:

$$\mathcal{L}^{(0)} = \frac{1}{2}A^{a\mu}\Box A^a_\mu - \partial^\mu\bar{c}^a\partial_\mu c^a, \tag{10.188}$$

$$\mathcal{L}_{int} = -gf^{abc}A^{b\mu}A^{c\nu}\partial_\mu A^a_\nu - \frac{1}{4}g^2 f^{abc}f^{adf}A^{b\mu}A^{c\nu}A^d_\mu A^f_\nu - gf^{acb}\partial^\mu\bar{c}^a A^c_\mu c^b.$$

Finally, we obtain a theory of two interacting fields, i.e., vectors A^a_μ and scalars $\bar{c}^a$, c^a.

The perturbation theory in the quantum Yang-Mills theory is constructed according to the standard methods, using Feynman diagrams. As usual, the free Lagrangian $\mathcal{L}^{(0)}$ produces the propagators, and the interacting Lagrangian $\mathcal{L}_{int}$ defines the vertices. Since there are two types of fields, we meet two propagators, namely, $\Delta^{ab}_{\mu\nu}(p)$ for the field A^a_μ and $D^{ab}(p)$ for the fields $\bar{c}^a$ and c^a. Taking into account the form of the Lagrangian $\mathcal{L}^{(0)}$, we can write

$$\Delta^{ab}_{\mu\nu}(p) = \frac{\delta^{ab}\eta_{\mu\nu}}{p^2 + i\varepsilon}, \qquad \text{and} \qquad D^{ab}(p) = \frac{\delta^{ab}}{p^2 + i\varepsilon}, \tag{10.189}$$

both corresponding to the massless fields.

The form of the Lagrangian $\mathcal{L}_{int}$ indicates the three types of vertices, two of which describe the self-interaction of the vector fields and the third describing the coupling between the vector fields to the Faddeev-Popov ghosts, as shown in the figures below:

$$-gf^{abc}\,\partial^\mu \bar{c}\, A^b_\mu\, c^b \;=$$

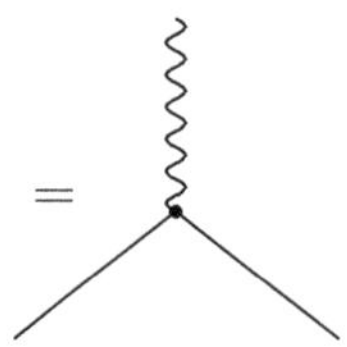

$$-gf^{abc}A^{b\mu}A^{c\nu}\partial_\mu A^a_\nu \;=$$

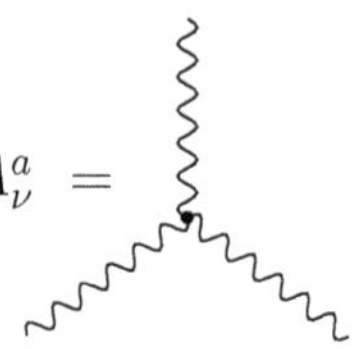

$$-\tfrac{1}{4}\,g^2 f^{abc} f^{adf} A^{b\mu} A^{c\nu} A^d_\mu A^f_\nu \;=$$

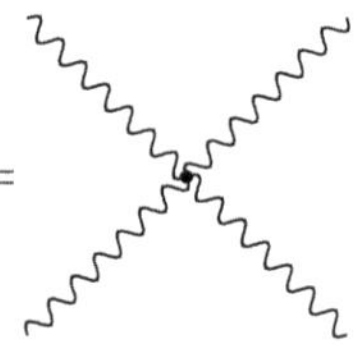

Here the wavy line corresponds to vector particles, and the solid line to ghosts. The Feynman rules are formulated as in other theories. Since the ghost fields anticommute, in the diagrams, the directions of these lines should be indicated.

10.9 The background field method for Yang-Mills theory

Let us apply the general formulation of the background field method, as discussed in section 10.7, to the quantum Yang-Mills theory. Here we discuss only the general formalism; the practical one-loop calculations using the heat-kernel method are postponed until Part II.

Following the relations (10.165)–(10.169), the loop corrections $\bar{\Gamma}[A]$ to the classical action are described by the following expression:

$$e^{\frac{i}{\hbar}\bar{\Gamma}[A]} = \int \mathcal{D}a\,\mathcal{D}\bar{c}\,\mathcal{D}c\; e^{\frac{i}{\hbar}\tilde{S}_{total}[A,a,\bar{c},c]}. \tag{10.190}$$

Here A^a_μ is the background Yang-Mills field, and a^a_μ is its quantum counterpart. The expression (10.190) includes the total action

$$\tilde{S}_{total}[A,a,\bar{c},c] = S_{YM}[A + \sqrt{\hbar}a] - S_{YM}[A] - \sqrt{\hbar}\int d^4x\, \frac{\delta S_{YM}[A]}{\delta A^a_\mu(x)}\, a^a_\mu(x)$$
$$= \hbar\tilde{S}_{GF}[A,a] + \hbar\tilde{S}_{GH}[A,a,\bar{c},c]. \tag{10.191}$$

The background-dependent gauge-fixing action has the form

$$\tilde{S}_{GF}[A,a] = -\frac{\alpha}{2}\int d^4x\, \chi^a(A,a)\chi^a(A,a), \tag{10.192}$$

where we set $\alpha = 1$, the background-dependent gauge is

$$\chi^a(A,a) = \big(\delta^{ab}\partial_\mu + gf^{acb}A^c_\mu\big)a^{b\,\mu} \equiv \mathcal{D}^{ab}_\mu(A)a^{b\,\mu}, \tag{10.193}$$

and $\mathcal{D}^{ab}_\mu$ is the background-dependent covariant derivative in the Yang-Mills theory. The weight functional $G_{\alpha\beta}[\phi]$ from the general relation (10.167) in the present case is just δ^{ab}. For this reason, there is no need to account for the term $\operatorname{Det} G_{\alpha\beta}[\Phi]$ in the integrand of (10.190). The background-dependent ghost action is

$$\tilde{S}_{GH}[A,a,\bar{c},c] = \int d^4x\; \bar{c}^a\, \mathcal{D}^{ac}_\mu(A)\mathcal{D}^{\mu cb}(A+\sqrt{\hbar}a)\, c^b. \tag{10.194}$$

Let us formulate the background and quantum gauge transformations in the theory under consideration:

a. Background gauge transformations:

$$\begin{aligned} \delta_B A^a_\mu &= \mathcal{D}^{ab}_\mu(A)\xi^a, & \delta_B a^a_\mu &= gf^{acb}a^c_\mu\xi^b, \\ \delta_B\bar{c}^a &= gf^{abc}\bar{c}^b\xi^c, & \delta_B c^a &= gf^{abc}c^b\xi^c. \end{aligned} \tag{10.195}$$

It is evident that $\delta_B(A^a_\mu + a^a_\mu) = \mathcal{D}^{ab}_\mu(A+a)\xi^b$.

b. Quantum gauge transformations:

$$\delta_Q A^a_\mu = 0, \quad \delta_Q a^a_\mu = \mathcal{D}^{ab}_\mu[A+\sqrt{\hbar}a]\xi^b, \quad \delta_Q\bar{c}^a = 0, \quad \delta_Q c^a = 0. \tag{10.196}$$

According to the general construction described in section 10.7, the effective action is invariant under the classical gauge transformation if the conditions (10.172) and (10.173) are fulfilled. These conditions are needed to ensure the invariance of the gauge-fixing term under the background transformation (10.195). Instead of checking each of these conditions separately, we will directly show that $\delta_B\tilde{S}_{GF} = 0$. Consider

$$\begin{aligned} \delta_B\chi^a(A,a) &= \delta_B\mathcal{D}^{ab}_\mu a^{b\,\mu} = \partial_\mu(\delta_B a^{a\,\mu}) + gf^{acb}\delta_B(A^c_\mu a^{b\,\mu}) + gf^{acb}\partial_\mu a^{c\,\mu}\xi^b \\ &\quad + g^2\big(f^{acd}f^{cmb} + f^{amc}f^{cdb}\big)A^m_\mu a^{d\,\mu}\xi^b = gf^{acb}\chi^c\xi^b, \end{aligned} \tag{10.197}$$

where the Jacobi identity (10.13) has been used. Taking into account (10.197), we get

$$\delta_B\tilde{S}_{GF} = \delta_B\Big\{\frac{1}{2}\chi^a(A,a)\chi^a(A,a)\Big\} = gf^{acb}\chi^a\chi^c\xi^b = 0. \tag{10.198}$$

Thus, in the framework of the background field method, the effective action in the Yang-Mills theory is invariant under classical gauge transformation.

Consider the relation (10.191) in the one-loop approximation. According to the results of section 8.6, to obtain the one-loop contribution $\bar{\Gamma}^{(1)}[A]$, it is sufficient to consider only the terms quadratic in quantum fields in the exponential of the integrand of (10.191). Then, the ghost action (10.194) boils down to

$$\tilde{S}^{(2)}_{GH}[A, a, \bar{c}, c] = \int d^4x \; \bar{c}^a \, \mathcal{D}^{ac}_{\mu}(A)\mathcal{D}^{cb\,\mu}(A)\, c^b. \tag{10.199}$$

In order to find the quadratic part of the expression

$$S_{YM}[A + \sqrt{\hbar}a] - S_{YM}[A] - \sqrt{\hbar}\int d^4x \frac{\delta S_{YM}[A]}{\delta A^a_\mu(x)} a^a_\mu(x), \tag{10.200}$$

it is sufficient to consider only $S_{YM}[A + \sqrt{\hbar}a]$ and expand it up to the second order in the quantum field a^a_μ. Thus,

$$\begin{aligned} S_{YM}[A + \sqrt{\hbar}a] &= -\frac{1}{4}\int d^4x \, G^a_{\mu\nu}(A + \sqrt{\hbar}a)G^{a\,\mu\nu}(A + \sqrt{\hbar}a) \\ = \hbar \int d^4x &\Big\{ \frac{1}{2}a^{a\mu}(\mathcal{D}^2)^{ab}a^b_\mu - \frac{1}{2}a^{a\nu}\mathcal{D}^{ac}_\mu \mathcal{D}^{cb}_\nu a^{b\mu} - \frac{g}{2} f^{abc} G^{a\,\mu\nu} a^b_\mu a^c_\nu \Big\} + \dots, \end{aligned} \tag{10.201}$$

where we kept only the quadratic in the quantum field terms and denoted $(\mathcal{D}^2)^{ab} = \mathcal{D}^{ac\,\mu}\mathcal{D}^{cb}_\mu$. Adding to the expression (10.201) the gauge-fixing term $\hbar S_{GF}$ (10.192) and using the explicit form for χ^a 10.193), one gets

$$\begin{aligned} S^{(2)}_{YM} + \hbar S_{GF} = \int d^4x \Big\{ &\frac{1}{2}\eta^{\mu\nu} a^a_\mu (\mathcal{D}^2)^{ab} a^b_\nu + \frac{g}{2} f^{abc} G^{a\,\mu\nu} a^b_\mu a^c_\nu \\ &- \frac{1}{2} a^{a\,\nu}([\mathcal{D}_\mu, \mathcal{D}_\nu])^{ab} a^{b\mu} \Big\}. \end{aligned} \tag{10.202}$$

Here $S^{(2)}_{YM}$ is the quadratic in the quantum fields part (10.201) of the action $S_{YM}[A + \sqrt{\hbar}a]$. The commutator is $([\mathcal{D}_\mu, \mathcal{D}_\nu])^{ab} = g f^{acb} G^c_{\mu\nu}$. As a result, Eq. (10.202) takes the form

$$S^{(2)}_{YM} + \hbar S_{GF} = \frac{1}{2}\int d^4x \, a^a_\mu \left\{ \eta^{\mu\nu}(\mathcal{D}^2)^{ab} + 2g f^{acb} G^{c\,\mu\nu} \right\} a^b_\nu \, . \tag{10.203}$$

The expressions (10.199) and (10.203) show that the one-loop contribution to effective action is given by the following functional integral:

$$\begin{aligned} e^{i\bar{\Gamma}^{(1)}[A]} &= \int \mathcal{D}a \, \mathcal{D}\bar{c} \, \mathcal{D}c \, e^{i\int d^4x \left\{ \frac{1}{2}a^a_\mu [\eta^{\mu\nu}(\mathcal{D}^2)^{ab} + 2g f^{acb} G^{c\,\mu\nu}] a^b_\nu + \bar{c}^a (\mathcal{D}^2)^{ab} c^b \right\}} \\ &= \left[\mathrm{Det}_{\,v}(\mathcal{H}_v)\right]^{-\frac{1}{2}} \times \left[\mathrm{Det}_{\,s}(\mathcal{H}_s)\right]. \end{aligned} \tag{10.204}$$

Here

$$\mathcal{H}_v = \mathcal{H}_v(A)^{ab}_{\mu\nu} = \eta_{\mu\nu}(\mathcal{D}^2)^{ab} + 2g f^{acb} G^c_{\mu\nu}, \qquad \mathcal{H}_s = \mathcal{H}_s(A)^{ab} = (\mathcal{D}^2)^{ab}. \tag{10.205}$$

The labels v and s mean that the corresponding functional determinants are defined for the operators acting on vector and scalar fields, respectively. Thus, we arrive at

$$\bar{\Gamma}^{(1)}[A] = \frac{i}{2}\,\mathrm{Tr}\,\log \mathcal{H}_v(A) - i\,\mathrm{Tr}\,\log \mathcal{H}_s(A), \tag{10.206}$$

which is invariant under classical gauge transformation by construction.

10.10 Renormalization of Yang-Mills theory

In this section, we explore the structure of UV counterterms in the quantum non-Abelian gauge theory. In this analysis, we shall extensively use the covariance of the counterterms, which has been proved above.

10.10.1 Power counting in the Yang-Mills theory

The total bare Lagrangian for the quantum Yang-Mills theory has the form (10.187),

$$\mathcal{L}_{total} = \mathcal{L}_{YM} + \frac{\alpha}{2}(\partial_\mu A^{a\,\mu})(\partial_\nu A^{a\,\nu}) - \partial^\mu \bar{c}^a \partial_\mu c^a - g f^{acb} \partial^\mu \bar{c}^a A^c_\mu c^b. \quad (10.207)$$

Here α is an arbitrary gauge-fixing parameter in the action $S_{GF}[A]$. In the present section, it is more convenient to keep this parameter arbitrary, instead of fixing it as we did before. The Lagrangian $\mathcal{L}_{YM}$ can be written in the form (10.180).

We start the analysis of the possible counterterms with the superficial degree of divergences $\omega(G)$, which were defined in (9.81). In Yang-Mills theory, the coupling constant g is dimensionless and the dimensions of the fields are $[A^a_\mu] = 1$, $[c^a] = 1$ and $[\bar{c}^a] = 1$. Thus, from the general expression (9.81), it follows that $\omega(G)$ has the form

$$\omega(G) = 4 - N_A - N_c, \quad (10.208)$$

where N_A is the number of external lines of the vector field and N_c is the number of external lines of the ghost fields.

The diagram G can diverge if $\omega(G) \geq 0$. This condition leaves the following possibilities for the external lines:

$$\begin{array}{lll} \textbf{a.} & N_c = 0, & N_A = 0,1,2,3,4. \\ \textbf{b.} & N_c = 2, & N_A = 0,1,2. \\ \textbf{c.} & N_c = 4, & N_A = 0. \end{array} \quad (10.209)$$

We define the ghost numbers of the fields of our interest as

$$\text{gh}(A) = 0, \qquad \text{gh}(\bar{c}) = -1, \qquad \text{gh}(c) = 1. \quad (10.210)$$

Since the ghosts $\bar{c}$ and c enter the action S_{total} in an even number, it is easy to see that $\text{gh}(S_{YM}) = \text{gh}(S_{GF}) = \text{gh}(S_{GH}) = 0$. Dimensional regularization does not violate the ghost number, so the ghost fields can enter into the counterterms only in pairs of $\bar{c}$ and c.

Consider the cases (10.209) one by one, assuming dimensional regularization.

a₁. $N_c = 0$, $N_A = 0$. The corresponding counterterm is a non-essential constant, which can be omitted.

a₂. $N_c = 0$, $N_A = 3$. In this case, $\omega(G) = 3$. The corresponding counterterm should have the form $k^a \partial^\mu \Box A^a_\mu$, where k^a is a divergent constant with the group index a. The only quantity with a single group index that can be constructed from the elements of the Lagrangian is the trace $f^{abb} = 0$. On the top of that, such a counterterm is a total divergence and can be omitted.

a₃. $N_c = 0,\ N_A = 2$. In this case, $\omega(G) = 2$. This version corresponds to the divergent two-point diagrams, defining the polarization operator of the vector field. The counterterm has the form

$$-\frac{1}{4}\Delta z_A F^a_{\mu\nu} F^{a\,\mu\nu} + \frac{1}{2}\Delta z'_A\,\alpha\,\partial_\mu A^{a\mu}\partial_\nu A^{a\nu}, \tag{10.211}$$

where $F^a_{\mu\nu} = \partial_\mu A^a_\nu - \partial_\nu A^a_\mu$.

a₄. $N_c = 0,\ N_A = 3$. In this case, $\omega(G) = 1$; it corresponds to divergent vector field three-point diagrams. The counterterm should have the form

$$2\Delta z_3\, g f^{abc}\,\partial^\mu A^{a\,\nu} A^b{}_\mu A^c{}_\nu \;=\; \Delta z_3\, g f^{abc}\, F^{a\,\mu\nu} A^b{}_\mu A^c{}_\nu. \tag{10.212}$$

a₅. $N_c = 0,\ N_A = 4$. In this case, $\omega(G) = 0$, it corresponds to divergent diagrams for the vector field four-point function. The counterterm has the form

$$\Delta z_4\, g^2 f^{abc} f^{amn} A^{b\mu} A^{c\nu} A^m{}_\mu A^n{}_\nu. \tag{10.213}$$

b₁. $N_c = 2,\ N_A = 0$. In this case, $\omega(G) = 2$. It corresponds to divergent diagrams for the ghost two-point function. The counterterm should have the form

$$-\Delta z_c\, \partial^\mu \bar{c}^a\, \partial_\mu c^a. \tag{10.214}$$

b₂. $N_c = 2,\ N_A = 2$. This corresponds to diagrams with two external vector field lines and two ghost external lines. In this case, formally, $\omega(G) = 0$. The expected counterterm should look like

$$\Delta z'\,\bar{c}^a c^b A^{a\,\mu} A^b_\mu + \Delta z''\bar{c}^a c^a A^{b\,\mu} A^b_\mu. \tag{10.215}$$

However, this case requires more careful analysis. This kind of four-point vertex can be stipulated only by the following term in the interaction Lagrangian: $-g f^{acb}\partial^\mu \bar{c}^a A^c_\mu c^b$. This means that one spacetime derivative must act either on an external ghost line or on the external vector line. When calculating $\omega(G)$, it was assumed that all the derivatives act only on the internal lines. This means that, in the case under consideration, we have overestimated the $\omega(G)$ by at least a unit. Therefore, for the given vertex functions, actually, $\omega(G) < 0$ and the diagrams converge by power counting. Thus, case **b₂** does not lead to divergence and does not require a counterterm.

b₃. $N_c = 2,\ N_A = 1$. In this case, $\omega(G) = 1$, which corresponds to divergent diagrams for the three-point function with two ghost external lines and one vector external line. The counterterm should have the form

$$-\,\Delta z_5\, g f^{acb}\partial^\mu \bar{c}^a A^c_\mu c^b \;-\; \Delta z'_5\, g f^{acb}\bar{c}^a \partial^\mu A^c_\mu c^b. \tag{10.216}$$

c₁. $N_c = 4,\ N_A = 0$. This corresponds to diagrams of the ghost four-point function. In this case, formally, $\omega(G) = 0$. However, such vertex functions can be stipulated only by the term $g f^{acb}\partial^\mu \bar{c}^a A^c_\mu c^b$ in the interaction Lagrangian. Then, at least one spacetime derivative must act on the external lines. This means that, once again, we have overestimated the $\omega(G)$ by at least a unit. In fact, for such vertices, $\omega(G) < 0$, and the corresponding diagrams converge by power counting. No counterterms are required.

Taking into account the relations (10.211), (10.212), (10.213), (10.214) and (10.216), the renormalized Lagrangian has the form

$$\begin{aligned}\mathcal{L}_R &= -\frac{1}{4}G_{R\,\mu\nu}^{\,a}G_R^{\,a\,\mu\nu} + \frac{1}{2}(\partial^\mu A_{R\,\mu}^{\,a})(\partial^\nu A_{R\,\nu}^{\,a}) - \partial^\mu \bar{c}_R^{\,a}\partial_\mu c_R^{\,a} - g_R f^{acb}\partial^\mu \bar{c}_R^{\,a} A_{R\,\mu}^{\,c} c_R^{\,b} \\ &\quad -\frac{1}{4}\Delta z_A F_{R\,\mu\nu}^{\,a} F_R^{\,a\,\mu\nu} + \frac{1}{2}\Delta z'_A \alpha_R (\partial^\mu A_{R\,\mu}^{\,a})(\partial^\nu A_{R\,\nu}^{\,a}) - \Delta z'_5 g_R f^{acb}\bar{c}_R^{\,a}\partial_\mu A_R^{\,c\,\mu} c_R^{\,b} \\ &\quad +\Delta z_3 g_R f^{abc} F_R^{\,a\,\mu\nu} A_{R\,\mu}^{\,b} A_{R\,\nu}^{\,c} + \Delta z_4 g_R^2 f^{abc} f^{amn} A_R^{\,b\,\mu} A_R^{\,c\,\nu} A_{R\,\mu}^{\,m} A_{R\,\nu}^{\,n} \\ &\quad -\Delta z_c \partial^\mu \bar{c}_R^{\,a}\partial_\mu c_R^{\,a} - \Delta z_5 g_R f^{acb}\partial^\mu \bar{c}_R^{\,a} A_{R\,\mu}^{\,c} c_R^{\,b}.\end{aligned} \tag{10.217}$$

Here the label R indicates renormalized fields and parameters. At first sight, this Lagrangian may give the impression that the theory is not multiplicatively renormalizable. However, we have to take into account that the effective action in the Yang-Mills theory satisfies the Ward identity (10.112). This identity imposes restrictions on the structure of the counterterms. We will discuss these restrictions in detail in the next sections, and here present only the final result for the renormalized Lagrangian,

$$\begin{aligned}\mathcal{L}_R &= -\frac{1}{4}z_A\left(\partial_\mu A_{R\,\nu}^{\,a} - \partial_\nu A_{R\,\mu}^{\,a} + z_1 z_A^{\frac{1}{2}} g_R f^{abc} A_{R\,\mu}^{\,b} A_{R\,\nu}^{\,c}\right)^2 + \frac{1}{2}\alpha_R(\partial_\mu A_R^{\,a\mu})^2 \\ &\quad - z_c\left(\partial_\mu \bar{c}_R^{\,a}\partial^\mu c_R^{\,a} + z_1 z_A^{\frac{1}{2}} g_R f^{acb}\partial_\mu \bar{c}_R^{\,a} A_R^{\,c\,\mu} c_R^{\,b}\right).\end{aligned} \tag{10.218}$$

Unlike (10.217), this expression includes only the gauge-invariant terms, compatible with the Ward identities. z_A, z_1, z_c, $\tilde{z}_1$ are those renormalization constants which are necessary, since the corresponding terms are compatible with the gauge symmetry. These constants depend on the renormalized coupling constant g_R.

The renormalized Lagrangian (10.218) can be obtained from the bare Lagrangian (10.207) with the help of the following renormalization transformations:

$$A_\mu^a = z_A^{\frac{1}{2}} A_{R\,\mu}^{\,a}, \quad \bar{c}^a = z_c^{\frac{1}{2}}\bar{c}_R^{\,a}, \quad c^a = z_c^{\frac{1}{2}} c_R^{\,a}, \quad \alpha = z_A^{-1}\alpha_R, \quad g = z_1 g_R. \tag{10.219}$$

10.10.2 Renormalization in the background field method

According to the general construction of the background field method in section 10.7 and its realization for Yang-Mills theory in section 10.9, the effective action has the form

$$\Gamma[A] = S_{YM}[A] + \bar{\Gamma}[A], \tag{10.220}$$

where the functional $\bar{\Gamma}[A]$ is given by the functional integral (10.190) and includes all loop corrections to the classical action. The fundamental property of the effective action in the background field method is that the functional $\bar{\Gamma}[A]$ and, therefore, the whole effective action are invariant under classical gauge transformation.

The action of the quantum theory (10.190) depends on both the Yang-Mills field A_μ^a and the ghost fields $\bar{C}^a$ and C^a. Indeed, renormalization requires that the counterterms depend not only on the background vector but also on the ghost fields. The reason is that, starting from the second loop, there are divergent subdiagrams with ghost external lines. Therefore, to study the structure of divergences in Yang-Mills theory, we need to consider the effective action depending on both types of fields.

Let us follow the procedures described in sections 8.6 and 10.7. The starting point is the generating functional

$$Z[J,\eta,\bar{\eta}] = \int \mathcal{D}\phi\,\mathcal{D}\bar{c}\,\mathcal{D}c\, e^{\frac{i}{\hbar}\{S_{total}[\phi,\bar{c},c]+J_i\phi^i+\eta_\alpha c^\alpha+\bar{\eta}^\alpha\bar{c}_\alpha\}}\ \mathrm{Det}^{\frac{1}{2}}\left(G_{\alpha\beta}[\phi]\right). \tag{10.221}$$

Introducing the generating functional of the connected Green functions $W[J,\eta,\bar{\eta}]$, we define the background (mean) fields Φ^i, C^α and $\bar{C}_\alpha$:

$$\Phi^i = \frac{\delta W[J,\eta,\bar{\eta}]}{\delta J^i}, \quad C^\alpha = \frac{\delta W[J,\eta,\bar{\eta}]}{\delta \eta_\alpha}, \quad \bar{C}_\alpha = \frac{\delta W[J,\eta,\bar{\eta}]}{\delta \bar{\eta}^\alpha}. \tag{10.222}$$

Here the derivatives with respect to the sources $\bar{\eta}$, η are left, and the ones with respect to the ghost fields are right. After that, the transformations are almost literally the same as in section 10.7, except for the shifts of the ghost fields in the functional integral,

$$\bar{c}_\alpha \longrightarrow \bar{C}_\alpha + \hbar^{\frac{1}{2}}\bar{c}_\alpha, \qquad c^\alpha \longrightarrow C^\alpha + \hbar^{\frac{1}{2}}c^\alpha. \tag{10.223}$$

After that, we arrive at the background-dependent effective action $\Gamma[\Phi,\bar{C},C]$,

$$e^{\frac{i}{\hbar}\bar{\Gamma}[\Phi,\bar{C},C]} = \int \mathcal{D}\phi\,\mathcal{D}\bar{c}\,\mathcal{D}c\, e^{\frac{i}{\hbar}\tilde{S}_{total}[\Phi\bar{C},C,\phi,\bar{c},c]}, \tag{10.224}$$

where $\bar{\Gamma}[\Phi,\bar{C},C]$ includes loop corrections to the classical action $S[\Phi]+S_{GH}[\Phi,\bar{C},C]$, with $S_{GH}[\Phi,\bar{C},C]=\bar{C}_\alpha M^\alpha{}_\beta[\Phi]C^\beta$. The action $\tilde{S}_{total}[\Phi,\bar{C},C,\phi,\bar{c},c]$ has the form

$$\begin{aligned}
\tilde{S}_{total}[\Phi,\bar{C},C,\phi,\bar{c},c] &= S[\Phi+\sqrt{\hbar}\phi] - S[\Phi] - S_{,i}[\Phi]\sqrt{\hbar}\phi^i \\
&\quad +(\bar{C}_\alpha+\sqrt{\hbar}\bar{c}_\alpha)M^\alpha{}_\beta[\Phi+\sqrt{\hbar}\phi](C^\beta+\sqrt{\hbar}c^\beta) \\
&\quad -\bar{C}_\alpha M^\alpha{}_\beta[\Phi]C^\beta - \bar{C}_\alpha M^\alpha{}_{\beta\,,i}[\Phi]\sqrt{\hbar}\phi^i C^\beta.
\end{aligned} \tag{10.225}$$

The action of the Yang-Mills theory is written as follows:

$$\begin{aligned}
\tilde{S}_{total}[A,\bar{C},C,a,\bar{c},c] &= S_{YM}[A+\sqrt{\hbar}a] - S_{YM}[A] \\
&\quad - \int d^4x \frac{\delta S_{YM}[A]}{\delta A^a_\mu(x)}\sqrt{\hbar}a^a_\mu(x) + S_{GH}[A,\bar{C},C,a,\bar{c},c],
\end{aligned} \tag{10.226}$$

where

$$\begin{aligned}
S_{GH}[A,\bar{C},C,a,\bar{c},c] &= \int d^4x\Big\{(\bar{C}^a+\sqrt{\hbar}\bar{c}^a)\mathcal{D}^{ac\,\mu}(A)\mathcal{D}^{cb}_\mu(A+\sqrt{\hbar}a)(C^b+\sqrt{\hbar}c^b) \\
&\quad -\bar{C}^a\mathcal{D}^{ac\,\mu}(A)\mathcal{D}^{cb}_\mu(A)C^b \\
&\quad -g\bar{C}^a\mathcal{D}^{ac\,\mu}(A)f^{cdb}\sqrt{\hbar}a^d_\mu C^b\Big\}.
\end{aligned} \tag{10.227}$$

Here $A^a_\mu(x)$ is the background vector field and $a^a_\mu(x)$ is the quantum vector field, while $\bar{C}^a(x)$, $C^a(x)$ are the background ghost fields and $\bar{c}^a(x)$, $c^a(x)$ are the quantum ghost fields. In the relation (10.227), we used the notation

$$\mathcal{D}^{ab\,\mu}(A+\sqrt{\hbar}a) = \delta^{ab}\partial_\mu + gf^{acb}(A^c_\mu + \sqrt{\hbar}a^c_\mu) \;=\; \mathcal{D}^{ab\,\mu}(A) + gf^{acb}\sqrt{\hbar}a^c_\mu. \quad (10.228)$$

The background transformations of the fields $A, a, \bar{c}, c$ (10.195) should be supplemented by the background transformations of the fields $\bar{C}^a$, C^a,

$$\delta_B\bar{C}^a = gf^{abc}\bar{C}^b\xi^c, \qquad \delta_B C^a = gf^{abc}C^b\xi^c. \quad (10.229)$$

The quantum transformations (10.196) remain the same.

Let us denote $\bar{\Gamma}[A,\bar{C},C]$ the quantum corrections to the classical action $S_{YM}[A] + \int d^4x\bar{C}^a\mathcal{D}^{ac\,\mu}(A)\mathcal{D}^{cb}_\mu(A)C^b$. We can demonstrate that $\bar{\Gamma}[A,\bar{C},C]$ is invariant under the background transformations (10.195) and (10.229). To do this, we have to prove the invariance of the action $\tilde{S}[A,\bar{C},C,a,\bar{c},c]$ from (10.225). Since the transformations (10.229) have the same form as the transformations $\delta_B\bar{c}$ and $\delta_B c$ in relations (10.195), we only need to consider the invariance of the following term in (10.225):

$$-\int d^4x g\bar{C}^a\mathcal{D}^{ac\,\mu}(A)f^{cdb}a^d_\mu C^b \;=\; gf^{adb}\int d^4x(\mathcal{D}^{ac\,\mu}(A)\bar{C}^c)a^d_\mu C^b, \quad (10.230)$$

where the integration by parts was done. The invariance of the expression (10.230) under background transformation can be proved in the same way as the invariance of the ghost action in section 10.9. First, one has to derive the identity

$$\delta_B\mathcal{D}^{ac}_\mu\bar{C}^c) \;=\; gf^{acb}(\mathcal{D}^{cd}_\mu(A)C^d)\xi^b \quad (10.231)$$

and then use it together with the transformations (10.195), (10.229) and the Jacobi identity (10.13) for the structure constants. As a result, the expression (10.230) is invariant. Thus, the functional $\bar{\Gamma}[A,\bar{C},C]$ is also invariant under the background transformation.

Using dimensional regularization and the minimal subtractions scheme, we can state that all the counterterms should be background invariant. Since the divergences are local and have dimension 4, there are two independent counterterms,

$$\mathcal{L}_{counter} \;=\; \Delta z_A S_{YM}(A) + \Delta z_c\bar{C}^a\mathcal{D}^{ac}_\mu(A)\mathcal{D}^{cb\,\mu}(A)C^b, \quad (10.232)$$

where the quantities Δz_A and Δz_c are given by the power series in $\frac{1}{n-4}$ with the coefficients depending on the coupling constant. Thus, the renormalized Yang-Mills Lagrangian, written in terms of renormalized fields A_R, $\bar{C}_R$, C_R and the renormalized coupling constant g_R, has the form

$$\begin{aligned}\mathcal{L}_{YM,R} \;&=\; -\frac{1}{4}z_A\left(\partial_\mu A^a_{R\,\nu} - \partial_\nu A^a_{R\,\mu} + g_R f^{abc}A^b_{R\,\mu}A^c_{R\,\nu}\right)^2 \\ &\quad - z_c(\partial_\mu\bar{C}^a_R + g_R f^{acd}A^c_{R\,\mu}\bar{C}^d_R)(\partial^\mu C^a_R + g_R f^{alb}A^{l\,\mu}_R C^b_R),\end{aligned} \quad (10.233)$$

where z_A and z_c are renormalization constants.

On the other hand, let us consider the renormalization transformation

$$A = z_A^{\frac{1}{2}} A_R, \qquad \bar{C} = z_c^{\frac{1}{2}} \bar{C}_R, \qquad C = z_c^{\frac{1}{2}} C_R, \qquad g = z_1 g_R, \tag{10.234}$$

where z_1 is the renormalization constant of the coupling. After such a transformation, the Lagrangian $\mathcal{L}_{YM} + \mathcal{L}_{GH}$ takes the form

$$\begin{aligned} \mathcal{L}_{YM,R} &= -\frac{1}{4} z_A \left(\partial_\mu A^a_{R\nu} - \partial_\nu A^a_{R\mu} + z_1 z_A^{\frac{1}{2}} g_R f^{abc} A^b_{R\mu} A^c_{R\nu} \right)^2 \\ &\quad - z_c (\partial_\mu \bar{C}^a_R + z_1 z_A^{\frac{1}{2}} g_R f^{acd} A^c_{R\mu} \bar{C}^d_R)(\partial^\mu C^a_R + z_1 z_A^{\frac{1}{2}} g_R f^{alb} A^{l\,\mu}_R C^b_R). \end{aligned} \tag{10.235}$$

A comparison of expressions (10.233) and (10.235) yields the relation

$$z_1 z_A^{\frac{1}{2}} = 1. \tag{10.236}$$

We see that, in the framework of the background field method, Yang-Mills theory is multiplicatively renormalizable and is characterized by two independent renormalization constants, namely, one for the Yang-Mills field and another one for the ghosts. The last renormalization is relevant for multi-loop diagrams, where one has to account for graphs with external lines for C and $\bar{C}$.

10.10.3 Renormalization in the Lorentz gauge

The proof of renormalizability for Yang-Mills theory is based on the Ward identity (10.112). In Yang-Mills theory, the gauge field is $A^a_\mu(x)$ and the ghost fields are $\bar{c}^a(x)$ and $c^a(x)$. The sources of the BRST invariants are $K^{\mu a}(x)$ and $L^a(x)$. In this notation the identity (10.112) has the form

$$\int d^4x \left\{ \frac{\delta \mathcal{T}}{\delta A^a_\mu(x)} \frac{\delta \mathcal{T}}{\delta K^{\mu a}(x)} + \frac{\delta \mathcal{T}}{\delta c^a(x)} \frac{\delta \mathcal{T}}{\delta L^a(x)} \right\} = 0. \tag{10.237}$$

Here $\mathcal{T} = \mathcal{T}[A, c, K, L]$. The dependence on the ghost $\bar{c}$ is defined by the relation (10.111), which in the case under consideration becomes

$$\tilde{\Gamma}[A, c, \bar{c}, K, L] = \mathcal{T}[A, c, K^a_\mu + \partial_\mu \bar{c}^a, L], \tag{10.238}$$

According to the analysis carried out in section 10.5, the renormalization structure of a gauge theory is defined by the universal equation (10.153), which, in the case under consideration, takes the form

$$\begin{aligned} \int d^4x \bigg\{ & \frac{\delta S^{(0)}}{\delta A^a_\mu(x)} \frac{\delta \Gamma_{div}}{\delta K^{\mu a}(x)} + \frac{\delta S^{(0)}}{\delta c^a(x)} \frac{\delta \Gamma_{div}}{\delta L^a(x)} \\ & + \frac{\delta S^{(0)}}{\delta K^{\mu a}(x)} \frac{\delta \Gamma_{div}}{\delta A^a_\mu(x)} + \frac{\delta S^{(0)}}{\delta L^a(x)} \frac{\delta \Gamma_{div}}{\delta c^a(x)} \bigg\} = 0, \end{aligned} \tag{10.239}$$

where the action $S^{(0)}$ is

$$S^{(0)} = S_{YM}[A] + \int d^4x \Big(K^{\mu a} \mathcal{D}^a_\mu c^a + \frac{1}{2} L^a f^{abc} c^b c^c \Big) \tag{10.240}$$

and $S_{YM}[A]$ is the action of Yang-Mills theory. The possible counterterms are defined by the solutions of Eq. (10.239).

In principle, there are at least two ways to find Γ_{div} from Eq. (10.239). First, one can use a general solution to equation $\Omega\Gamma = 0$ from (10.124) in general terms, to find a suitable functional X. Second, one can try to construct the solution to the equation $\Omega\Gamma = 0$, using the specific properties of the action S_{YM}. We shall follow the second approach.

Consider the total Lagrangian (10.185) for Yang-Mills theory. As in section 10.10.1, it is convenient to use the notion of the ghost number. First of all, we need to extend the definition of the ghost numbers (10.210) to the auxiliary fields K and L. According to (10.210), we have $\mathrm{gh}(\mathcal{L}_{total}) = 0$. The action $\bar{S}$ (10.92) can be written as

$$\bar{S}[A, \bar{c}, c, K, L] = \int d^4x\, \bar{\mathcal{L}}, \qquad \text{where}$$
$$\bar{\mathcal{L}} = \mathcal{L}_{total} + K^{\mu a}\mathcal{D}^{ab}_{\mu}c^b + \frac{1}{2}L^a f^{abc}c^b c^c. \tag{10.241}$$

Since the Lagrangian $\bar{\mathcal{L}}$ should have zero ghost number, we put $\mathrm{gh}(K^{\mu a}) = -1$ and $\mathrm{gh}(L^a) = -2$. In addition, since the Lagrangian $\bar{\mathcal{L}}$ has dimension $[\bar{\mathcal{L}}] = 4$, the dimensions of the auxiliary fields are $[K^{\mu a}] = 2$ and $[L^a] = 2$. It is evident that the Lagrangian $\bar{\mathcal{L}}$ is Lorentz invariant. Furthermore, it is easy to check that the Lagrangian $\bar{\mathcal{L}}$ is invariant under the global gauge transformations

$$\delta A^a_\mu = f^{abc}A^b_\mu \xi^c \quad \delta c^a = f^{abc}c^b\xi^c, \quad \delta \bar{c}^a = f^{abc}\bar{c}^b\xi^c, \quad \xi^c = const. \tag{10.242}$$

Assuming the Lorentz invariance, conservation of the ghost number and global gauge symmetry and taking into account the locality of divergences and the dimensions of $K^{\mu a}$ and L^a, we can write down the possible expression for the Γ_{div} in the form

$$\Gamma_{div}[A, c, K, L] = \Gamma^{(0)}_{div}[A]$$
$$+ \int d^4x\big(a_1 K^{\mu a}\partial_\mu c^a + a_2 K^{\mu a} g f^{abc}A^b_\mu c^c + a_3 L^a g f^{abc}c^b c^c\big), \tag{10.243}$$

where $\Gamma^{(0)}_{div}[A]$ is a functional, depending only on A^a_μ. Furthermore, a_1, a_2 and a_3 are dimensionless numbers, depending on the regularization parameter and diverging after removing the regularization. Due to the locality of divergences, one can write

$$\Gamma^{(0)}_{div}[A] = \int d^4x\, \mathcal{L}^{(0)}_{div}. \tag{10.244}$$

The functional $\Gamma^{(0)}_{div}[A]$ and the constants a_1, a_2, a_3 characterize the arbitrariness in the definition of $\Gamma_{div}[A, c, K, L]$. We emphasize that the expression (10.243) has been written only on the basis of Lorentz invariance, global gauge invariance, the conservation of ghost number and dimension. It is not a surprise that these conditions do not fix the functional structure of counterterms in a unique way. To fix the arbitrariness, one has to use the basic equation (10.239) for the divergent part of effective action.

Eq. (10.239) contains the derivatives of the functionals $S^{(0)}$ and Γ_{div}. Indeed, the classical action is known, and the form of the last expression is fixed by Eq. (10.243). Thus, the derivatives can be easily calculated. After these derivatives are substituted into Eq. (10.239), we arrive at the sum of the terms with one, two and three ghost fields. The terms with three ghosts cancel owing to the Jacobi identity (10.13) for the structure constants. The cancellation of the terms with two ghost fields yields $a_3 = a_2$, and the cancellation of the terms with one ghost leads to the constraint

$$\mathcal{D}_\mu^{ab} \frac{\delta \Gamma_{div}^{(0)}[A]}{\delta A_\mu^b(x)} + (a_2 - a_1) f^{abc} \frac{\delta S_{YM}[A]}{\delta A_\mu^b(x)} A_\mu^c(x) = 0. \tag{10.245}$$

The general solution to this equation is written in the form

$$\Gamma_{div}^{(0)}[A] = \tilde{\Gamma}_{div}^{(0)}[A] + (a_2 - a_1) \int d^4x \, A_\mu^a(x) \frac{\delta S_{YM}[A]}{\delta A_\mu^a(x)}, \tag{10.246}$$

where $\tilde{\Gamma}_{div}^{(0)}[A]$ is an arbitrary gauge-invariant functional, such that $\mathcal{D}_\mu^{ab} \frac{\delta \tilde{\Gamma}_{div}^{(0)}[A]}{\delta A_\mu^b(x)} = 0$. As a result, we have

$$\Gamma_{div}[A, c, K, L] = \tilde{\Gamma}_{div}^{(0)}[A] + \int d^4x \Big\{ (a_2 - a_1) A_\mu^a(x) \frac{\delta S_{YM}[A]}{\delta A_\mu^a(x)}$$
$$+ a_1 K^{\mu a} \partial_\mu c^a + a_2 K^{\mu a} g f^{abc} A_\mu^b c^c + a_2 L^a f^{abc} c^b c^c \Big\}. \tag{10.247}$$

Taking into account that the divergent part of the effective action is local and that the counterterms have dimension 4, the only possible gauge-invariant functional $\tilde{\Gamma}_{div}^{(0)}[A]$ should be proportional to the Yang-Mills action $S_{YM}[A]$, which is

$$\tilde{\Gamma}_{div}^{(0)}[A] = a_0 S_{YM}[A], \tag{10.248}$$

where a_0 is a dimensionless divergent constant. Thus, the functional $\Gamma_{div}[A, c, K, L]$ has the form

$$\Gamma_{div}[A, c, K, L] = \int d^4x \Big\{ a_0 \mathcal{L}_{YM} + (a_2 - a_1) A_\mu^a(x) \frac{\delta S_{YM}[A]}{\delta A_\mu^a(x)}$$
$$+ a_1 K^{\mu a} \partial_\mu c^a + a_2 K^{\mu a} g f^{abc} A_\mu^b c^c + a_2 L^a f^{abc} c^b c^c \Big\}. \tag{10.249}$$

We can conclude that, in general, the structure of divergences in Yang-Mills theory is defined by the three dimensionless divergent constants a_0, a_1 and a_2, depending on the regularization parameter and the coupling constant. It remains to establish whether the form of (10.249) can be defined by the renormalization of the fields A_μ^a, c^a and $\bar{c}^a$ in the Yang-Mills Lagrangian.

The term $A_\mu^a(x) \frac{\delta S_{YM}[A]}{\delta A_\mu^a(x)}$ in the integrand of Eq. (10.249) is transformed as

$$A_\mu^a(x) \frac{\delta S_{YM}[A]}{\delta A_\mu^a(x)} = A_\mu^a \mathcal{D}_\nu^{ab} G^{b\mu\nu} = -\frac{1}{2} G_{\mu\nu}^a G^{a\mu\nu} - \frac{1}{2} g f^{abc} A_\mu^b A_\nu^c G^{a\mu\nu}, \tag{10.250}$$

where a total divergence has been omitted. Substituting (10.250) in (10.249), we get

$$\Gamma_{div}[A,c,K,L] = \int d^4x \Big\{ (a_0 - 2a_1 + 2a_2)\mathcal{L}_{YM} + \frac{a_1 - a_2}{2} f^{abc} A^b_\mu A^c_\nu G^{a\mu\nu}$$
$$+ a_1 K^{\mu a} \partial_\mu c^a + a_2 K^{\mu a} g f^{abc} A^b_\mu c^c + a_2 L^a f^{abc} c^b c^c \Big\}. \qquad (10.251)$$

According to the relation (10.238), the dependence of the effective action on the ghost field $\bar{c}^a$ can be obtained by the change $K^{\mu a} \to K^{\mu a} + \partial^\mu \bar{c}^a$. After that, we can switch off the sources $K^{\mu a}$ and L^a and arrive at the divergent part of the effective action for the pure Yang-Mills field in the form

$$\Gamma_{div}[A,\bar{c},c] = \int d^4x \Big\{ -\frac{1}{4}(a_0 - 2a_1 + a_2) G^{a\mu\nu} G^a_{\mu\nu} + \frac{1}{2}(a_1 - a_2) f^{abc} A^b_\mu A^c_\nu G^{a\mu\nu}$$
$$+ a_1 \partial^\mu \bar{c}^a \partial_\mu c^a + a_2 g f^{abc} \partial^\mu \bar{c}^a A^b_\mu c^c \Big\} = -\int d^4x\, \mathcal{L}_{counter}, \qquad (10.252)$$

where $\mathcal{L}_{counter}$ includes all the counterterms. Thus, in each order of the loop expansion, the divergent part of the effective action has the form (10.252) and is defined by the three dimensionless divergent constants.

The renormalized Lagrangian is defined as $\mathcal{L}_R = \mathcal{L}_{YM} + \mathcal{L}_{counter}$, where the *r.h.s.* is expressed in terms of the renormalized fields $A^{\mu a}_R$, c^a_R, $\bar{c}^a_R$ and the renormalized constant coupling g_R. The form of the divergent part of the effective action (10.252) shows that the renormalized Lagrangian is obtained from the bare Lagrangian $\mathcal{L}_{YM}$ with the help of the following renormalization transformations:

$$A^{\mu a} = z_A^{\frac{1}{2}} A^{\mu a}_R, \qquad c^a = z_c^{\frac{1}{2}} c^a_R, \qquad \bar{c}^a = z_c^{\frac{1}{2}} \bar{c}^a, \qquad g = z_g g_R. \qquad (10.253)$$

Here z_A, z_c and z_g are the renormalization constants of the gauge field, the ghosts and the coupling constant, respectively. All of them have the form $z = 1 + \delta z$, where δz_A, δz_c and δz_g are expressed in each order of the loop expansion through the coefficients a_0, a_1 and a_2 in the divergent part of the effective action (10.252).

Concerning the gauge fixing-term $\frac{\alpha}{2}(\partial^\mu A^a_\mu)^2$, the corresponding term in the divergent part of the effective action is not renormalized. It means that if we define the renormalization transformation for the parameter α in the form $\alpha = z_\alpha \alpha_R$, then

$$\frac{1}{2}\alpha(\partial^\mu A^a_\mu)^2 = \frac{1}{2}\alpha_R(\partial_\mu A^{\mu a}_R)^2. \qquad (10.254)$$

Therefore, we always have $z_\alpha z_z = 1$.

Finally, let us note that the renormalization of Yang-Mills theory coupled to the set of matter fields (e.g., fermions and scalars) can be formulated following the same general scheme described above. However, the gauge-fixing dependence of the renormalization constants in the matter sector requires a special study, which can be based on the relation (10.161). The one-loop version of this general formula is the subject of Exercise 10.8.

Exercises

10.1. Apply formally the Faddeev-Popov method to the massive vector field theory and construct the corresponding functional integral.

Hint. In this case, the theory is not invariant under gauge transformation and therefore the integral over the orbit does not factorize. This problem has been elaborated for Yang-Mills theory in [299].

10.2. Consider massive electrodynamics in the Stüeckelberg formulation, as described in section 4.4.3, and construct the corresponding functional integral. Consider the versions with scalar and fermion fields.

10.3. Construct the action S_{total} for Yang-Mills theory in the Coulomb gauge $\chi^a = \partial_i A_i^a$, and explore the form of the ghost action.

10.4 Construct the action of Yang-Mills theory coupled to complex scalar fields in i) fundamental and ii) adjoint representations of the gauge group $SU(2)$. In both cases, construct the action S_{total} and write down the BRST transformations explicitly.

10.5 Discuss the most general form of the linear gauge-fixing condition in quantum electrodynamics, scalar electrodynamics and Yang-Mills theory coupled to scalar and fermion fields.

10.6 Consider a theory with the action

$$S'_{total}[\phi, \bar{c}, c, b] = S[\phi] + \chi^\alpha[\phi] b_\beta - \frac{1}{2} b_\alpha G^{\alpha\beta}[\phi] b_\beta + S_{GH}[\phi, \bar{c}, c], \qquad (10.255)$$

where $G^{\alpha\beta}[\phi]$ is the inverse matrix to $G_{\alpha\beta}[\phi]$, b_α is an auxiliary field (the Nakanishi-Lautrup field) and $S_{GH}[\phi\bar{c}, c]$ is given by relation (10.49). Fulfill the following program:

a) Express the b-field from the classical equations of motion, substitute it into the action (10.255) and show that such an action coincides with action $S_{total}[\phi, \bar{c}, c]$.

b) Prove that the action (10.255) is invariant under the following BRST transformations:

$$\delta_{BRST}\phi^i = R^i_{\ \alpha} c^\alpha \mu, \qquad \delta_{BRST} c^\alpha = \frac{1}{2} f^\alpha_{\ \beta\gamma} c^\beta c^\gamma \mu,$$
$$\delta_{BRST}\bar{c}_\alpha = G^{\alpha\beta} b_\beta \mu, \qquad \delta_{BRST} b_\alpha = 0. \qquad (10.256)$$

c) Using Noether's theorem in the classical Yang-Mills theory, construct the BRST charge, corresponding to the transformations (10.256), and then prove that this charge is a nilpotent operator.

10.7 Consider Yang-Mills theory coupled to fermionic fields in some representation of the gauge group G, construct the BRST transformations and derive the corresponding Ward identities.

10.8 Use the relation (10.161), the loop expansion and the locality of divergences to demonstrate that *one-loop* divergences and counterterms are independent of gauge fixing on the *classical* mass shell, which means there is $\frac{\delta S[\Phi]}{\delta \Phi^i}$ in the *r.h.s.* instead of the general $\frac{\delta \Gamma[\Phi]}{\delta \Phi^i}$.

Observations. This aspect of gauge-fixing dependence is very operational in quantum gravity, as we shall see in Part II.

10.9 Starting from the previous exercise, try to analyze the gauge-fixing dependence of two-loop divergences. Start from the superrenormalizable models, where the divergences have fewer counterterms than classical action does. Then consider renormalizable theories and try to identify which parts of the divergences are invariant.

Comments

The need for ghost fields in quantum Yang-Mills theory has been understood since the pioneering work by R. Feynman [132]. The general form of the functional integral and the Feynman rules for Yang-Mills theory were formulated independently by L.D. Faddeev and V.N. Popov [125] and B.S. DeWitt [106]. Ghost fields as we discussed them here were introduced in [125].

The Faddeev-Popov construction is considered in detail, e.g., in the books [127], [187], [250], and in the second volume of the fundamental monograph [345]. The canonical approach to the quantization of gauge theory was introduced by L.D. Faddeev [126]; see also the books [127] and [160]. The most general approach to the canonical quantization of general gauge fields of the Yang-Mills type has been formulated in the framework of the Batalin-Fradkin-Vilkovisky (BFV) construction [37, 41, 38–40]; see also the review [181] and the book [182].

BRST symmetry was introduced independently by C. Becci, A. Rouet, R. Stora [44], [45] and I.V. Tyutin [321].

The first example of the exact relations between different Green functions was found by J.C. Ward [338] for the on-shell four-point photon vertex in quantum electrodynamics. The complete system of Ward identities in quantum electrodynamics was derived independently by E.S. Fradkin [137] and Y. Takahashi [311].

The Zinn-Justin equation (10.95) was introduced in the paper [355]. The Ward identities for Yang-Mills theory were derived independently by A.A. Slavnov [297] and J.C. Taylor [312]. Our derivation of the identities for the effective action in section 10.5.2, and the analysis of these identities in section 10.5.3, are based on the works of B.L. Voronov and I.V. Tyutin [335] and [336].

Further generalization of the Ward identities for gauge theory leads to the Batalin-Vilkovisky (BV) master equation [42], [43]; see also the review [64] and the second volume of Weinberg's monograph [345].

The background field method was proposed by B.S. DeWitt [106] (see also [109]) and was used in quantum Yang-Mills theory, quantum gravity and the supersymmetric generalizations by many authors. Further developments of this formalism can be found in the review [2]; see also the books [80], [250] the second volume [345] and the references therein. The use of non-linear gauges is discussed in [157].

The consistent derivation of the relation (10.161) was given in the paper [334], although equivalent results had been known before, starting from [106]. In particular, an interesting discussion of related problems can be found in the first paper of [328].

The renormalization structure of Yang-Mills theory has been studied by many authors. For more details, see, e.g., the books [127,187,356], the second volume of [345] and the references therein.

Part II

Semiclassical and Quantum Gravity Models

11
A brief review of general relativity

Along with its more or less direct extensions, Albert Einstein's theory of general relativity (GR) forms the modern basis for all investigations in the area of semiclassical gravity (quantum field theory in curved spacetime) or perturbative quantum gravity. Therefore, it is quite natural to start the second part of the book with a brief review of GR.

The initial motivation for creating GR is the need to formulate a gravity theory which possesses local Lorentz invariance but also boils down to the usual Newton's gravitational law in the classical non-relativistic limit. Indeed, these two requirements can be met in many different ways. However, if we start listing the possible relativistic theories of gravity from the simplest options on to consider more and more complicated versions, then GR will be the first consistent theory that we meet in this way. The word "consistent" here implies both theoretical self-consistency and consistency with experimental and observational data.

This chapter is not supposed to serve as a textbook on GR. Instead, we include here only basic considerations and formalism of the theory, as an opportunity to introduce notions and formulas which will be useful in the rest of the second part of the book. The reader is referred to the existing textbooks on GR (e.g. [202, 340, 219, 110]) and differential geometry (e.g., [116, 252]) for a more complete introduction to the subject and many more details of the relativistic gravitational theory. A recent textbook on tensors and relativity [292] can be regarded as an extended version of this brief introductory chapter and is recommended for readers who need an extended elementary introduction.

11.1 Basic principles of general relativity

The main statement of GR is that the gravitational field is not field as others, but the fundamental geometrical characteristic of spacetime (see Weinberg's book [340] for a more "physical," - albeit mathematically equivalent interpretation). Spacetime is regarded as being pseudo-Riemannian, that is, a $(1+3)$-dimensional manifold M_{1+3}, endowed by the metric $g_{\mu\nu}$. In GR the gravitational field is described by metric components.

The most important physical postulate of GR is the equivalence principle, which states that, one can always choose a coordinate system on M_{1+3}, eliminating locally the gravitational field. This can be done in an arbitrary spacetime point by choosing the freely falling reference frame. Mathematically, this means that in any point on M_{1+3} one can find special spacetime coordinates X^a - in which the metric becomes the

Introduction to Quantum Field Theory with Applications to Quantum Gravity. Iosif L. Buchbinder and Ilya L. Shapiro, Oxford University Press (2021). © Iosif L. Buchbinder and Ilya Shapiro.
DOI: 10·1093/oso/9780198838319·003·0011

(flat) Minkowski one $\eta_{\alpha\beta} = \text{diag}(+,-,-,-)$. Transformation to any other coordinate system (also called reference frame) is performed as usual:

$$\eta_{\alpha\beta} = \frac{\partial x^\mu}{\partial X^\alpha}\frac{\partial x^\nu}{\partial X^\beta}\, g_{\mu\nu}, \qquad g_{\mu\nu} = \frac{\partial X^\alpha}{\partial x^\mu}\frac{\partial X^\beta}{\partial x^\nu}\, \eta_{\alpha\beta}. \tag{11.1}$$

In this formula, the Greek indices run the values $\alpha, \beta, \mu ... = 0, 1, 2, 3$. In the subsequent text, the Latin indices $i, j, k...$ mostly indicate three-dimensional space coordinates and hence $i, j, k... = 1, 2, 3$.

For the general case of the M_{1+3} manifold, one can not choose the coordinates which cover all the manifold (or even some its finite domain) and provide that the metric is flat everywhere. Thus, the possibility of having Lorentz coordinates X^α is a purely local property. In this respect, the general M_{1+3} differs from the Minkowski spacetime, in curvilinear coordinates. In the Minkowski spacetime, there are always such coordinates that the metric is flat everywhere (i.e., globally). One can prove that if the spacetime is flat in the vicinity of a point, with certain assumptions about the smoothness of the metric and its derivatives, the whole space is flat. It is easy to verify this is not true for many Riemann spaces, which are not flat.

The next cornerstone of GR is the general covariance principle. According to it, all physical laws do not depend on the coordinate system, and the transition from one reference frame to another can be done in a controllable way. Mathematically, this means that the fundamental physically relevant laws can be expressed as tensor equations. Therefore, relevant physical quantities must be tensors with respect to the general coordinate transformations

$$x^\mu \to x'^\mu = x^\mu + \xi^\mu(x)\,. \tag{11.2}$$

11.2 Covariant derivative and affine connection

Since physical equations may contain derivatives, we have to construct a kind of derivative operator which, when acting on a tensor, produces another tensor. The construction of such a derivative will be given in this section.

It is easy to verify that the partial derivative ∂_α of a scalar field is a covariant vector. On the other hand, partial derivatives of any other tensor field do not form a tensor. However, one can modify partial derivative by introducing an additional term, so that the resulting sum becomes a tensor. Such a construction is called the covariant derivative ∇_α. It is important that the covariant derivative of a tensor satisfy the Leibnitz rule. For instance, acting on the product of n tensors $A_1 \cdot A_2 \cdot A_3 \cdot ... \cdot A_n$, the covariant derivative gives

$$\nabla(A_1 \cdot A_2 \cdot A_3 \cdot ... \cdot A_n) = (\nabla A_1) \cdot A_2 \cdot A_3 \cdot ... \cdot A_n + ... + A_1 \cdot A_2 \cdot A_3 \cdot ... \cdot (\nabla A_n).$$

An observation about notations is in order. The parentheses are used, in particular, to show the limits of the action of the derivative (partial or covariant). For example, the covariant derivative in the first term of the last equation acts only on A_1, and not on the subsequent terms.

Let us consider a simple example of covariant derivative. In the case of a contravariant vector A^α, the covariant derivative looks like

$$\nabla_\beta\, A^\alpha = \partial_\beta\, A^\alpha + \Gamma^\alpha_{\ \beta\gamma}\, A^\gamma\,, \tag{11.3}$$

where the last term is a necessary addition to provide the tensor nature of the whole expression. It is an easy exercise to show that the covariant derivative (11.3) is a tensor if and only if the coefficients $\Gamma^\alpha_{\ \beta\gamma}$ transform in a special non-tensor way,

$$\Gamma'^\alpha_{\ \beta\gamma}(x') = \frac{\partial x'^\alpha}{\partial x^\lambda}\,\frac{\partial x^\mu}{\partial x'^\beta}\,\frac{\partial x^\nu}{\partial x'^\gamma}\,\Gamma^\lambda_{\ \mu\nu} - \frac{\partial x^\mu}{\partial x'^\beta}\,\frac{\partial x^\nu}{\partial x'^\gamma}\,\frac{\partial^2 x'^\alpha}{\partial x^\mu \partial x^\nu}\,. \tag{11.4}$$

The coefficients $\Gamma^\alpha_{\ \beta\lambda}$ form what is called affine connection.

The rule for constructing the covariant derivatives of other tensors immediately follows from the following facts:

i) The product of the co- and contravariant vectors A^α and B_α is a scalar. Therefore $\nabla_\beta(A^\alpha B_\alpha) = \partial_\beta(A^\alpha B_\alpha)$, and hence

$$\nabla_\beta\, B_\alpha = \partial_\beta\, B_\alpha - \Gamma^\gamma_{\ \beta\alpha}\, B_\gamma. \tag{11.5}$$

ii) Similarly, the contraction of any tensor with an appropriate set of vectors is a scalar. Using this fact, one can easily arrive at the standard expression for the covariant derivative of an arbitrary tensor,

$$\begin{aligned}\nabla_\beta\, T^{\alpha_1 \ldots \alpha_n}_{\ \ \gamma_1 \ldots \gamma_m} &= \partial_\beta\, T^{\alpha_1 \ldots}_{\ \ \gamma_1 \ldots} + \Gamma^{\alpha_1}_{\ \beta\lambda}\, T^{\lambda \ldots \alpha_n}_{\ \ \gamma_1 \ldots \gamma_m} \ldots + \Gamma^{\alpha_n}_{\ \beta\lambda}\, T^{\alpha_1 \ldots \lambda}_{\ \ \gamma_1 \ldots \gamma_m} \\ &- \Gamma^\tau_{\ \beta\gamma_1}\, T^{\alpha_1 \ldots \alpha_n}_{\ \ \tau \ldots \gamma_m} \ - \ \ldots \ - \Gamma^\tau_{\ \beta\gamma_m}\, T^{\alpha_1 \ldots \alpha_n}_{\ \ \gamma_1 \ldots \tau}.\end{aligned} \tag{11.6}$$

Indeed, (11.3) and (11.5) can be seen as particular cases of (11.6).

It is easy to verify that adding to affine connection an arbitrary tensor $C^\alpha_{\ \beta\lambda}$,

$$\Gamma^\alpha_{\ \beta\lambda} \longrightarrow \Gamma'^\alpha_{\ \beta\lambda} = \Gamma^\alpha_{\ \beta\lambda} + C^\alpha_{\ \beta\lambda}, \tag{11.7}$$

does not break down the transformation rule (11.4), and hence the new coefficients also form an affine connection. This possibility leads to gravitational theories with so-called torsion and/or non-metricity, but GR is based on the minimalist approach, to eliminate the ambiguity related to (11.7). Namely, the affine connection of GR satisfies the two additional conditions:

i) Symmetry, equivalent to the vanishing torsion, $\Gamma^\alpha_{\ \beta\gamma} = \Gamma^\alpha_{\ \gamma\beta}$, and
ii) The metricity condition of the covariant derivative,

$$\nabla_\alpha\, g_{\mu\nu} = \partial_\alpha\, g_{\mu\nu} - \Gamma^\lambda_{\ \mu\alpha} g_{\lambda\nu} - \Gamma^\lambda_{\ \nu\alpha} g_{\mu\lambda} = 0. \tag{11.8}$$

If these conditions are satisfied, (11.6) provides the unique solution for $\Gamma^\alpha_{\ \beta\gamma}$,

$$\Gamma^\alpha_{\ \beta\gamma} = \left\{ {\alpha \atop \beta\gamma} \right\} = \frac{1}{2}\, g^{\alpha\lambda}\, \left(\partial_\beta\, g_{\lambda\gamma} + \partial_\gamma\, g_{\lambda\beta} - \partial_\lambda\, g_{\beta\gamma}\right)\,. \tag{11.9}$$

The expression (11.9) is called the Christoffel symbol; it is the simplest version of affine connection in a Riemann space.

Any non-trivial choice of the tensor $C^\alpha_{\ \beta\gamma}$ means that there is some other (gravitational or not) field besides metric. As we know from Part I, all gauge interactions (e.g., electroweak and strong) can be introduced through extended connections and corresponding modified covariant derivatives. This approach requires that the fields (tensors or spinors, etc.) have specific charges. The purely geometric extensions of affine connection which we mentioned above, are also possible. This type of extension does not depend on the specific charges, and hence those are not related to electroweak, strong or other possible interactions of particles and fields. This type of modified affine connection defines interesting extensions of GR, including gravity with torsion (see, e.g., [80, 179, 284]), or non-metricity. In what follows, we are going to consider only metric-dependent gravity theories. The term metric-dependent here means that the affine connection is the Christoffel symbol, and therefore we require that the geometric part of the tensor $C^\alpha_{\ \beta\gamma}$ is zero. Quantum field theory in spaces with torsion is well developed (see, e.g., [80, 284]), but is beyond the textbook format that we adopt here.

In the purely metric gravity (mathematically, this means Riemann space), the covariant infinitesimal element of a spacetime volume is defined uniquely, as

$$d^4x\sqrt{-g}, \quad \text{where} \quad g = \det\big(g_{\mu\nu}\big). \tag{11.10}$$

The generalization to the n-dimensional case can be performed by trading $d^4x \to d^nx$. In the case of Euclidean signature, the invariant volume element is $d^nx\sqrt{g}$.

From the geometric viewpoint, the affine connection $\Gamma^\alpha_{\ \beta\gamma}$ defines the parallel transport of any vector (tensor) on the manifold M_{1+3}. For instance, consider a curve $x^\mu(t)$, where t is an appropriate parameter along the curve (see, e.g., the book [252], where the conditions on this parameter are discussed in detail). The equation for the parallel transport of the vector T^α along this curve is

$$\frac{dx^\mu}{dt}\,\nabla_\mu T^\alpha = 0, \tag{11.11}$$

and the same is true for any tensor. One can formulate the geometrical definition of the parallel displacement of a field $T^\alpha(t)$, requiring that the covariant derivative of this field in the direction of the velocity $u^\mu = dx^\mu/ds$ be zero.

If the spacetime is (pseudo)-Euclidean, then the result of the parallel displacement of a tensor from point A to point B does not depend on the curve connecting two points, nor on the coordinates. On the contrary, if the manifold is really curved, then the result of the parallel transport may be different for two different curves. This is a very important statement, and it can be expressed, using the infinitesimal closed curves, in the following way. The manifold is flat if the commutator of the two covariant derivatives $\big[\nabla_\alpha, \nabla_\beta\big]T \equiv 0$ for any tensor T, regardless of whether the metric looks non-trivial in some curvilinear coordinates. It turns out that the commutator $\big[\nabla_\alpha, \nabla_\beta\big]$ of any tensor can be expressed via single universal object called the curvature tensor $R^\alpha{}_{\mu\beta\nu}$. Then, the spacetime is flat if and only if $R^\alpha{}_{\mu\beta\nu}(x) \equiv 0$.

11.3 The curvature tensor and its properties

The explicit expression for the curvature (Riemann) tensor $R^\lambda{}_{\tau\beta\tau}$ can be easily found from direct calculation of the commutator of two covariant derivatives acting on a vector,

$$\left[\nabla_\beta, \nabla_\alpha\right] A^\lambda = -R^\lambda{}_{\tau\alpha\beta}\, A^\tau\,, \tag{11.12}$$

where

$$R^\lambda{}_{\tau\alpha\beta} = \partial_\alpha\, \Gamma^\lambda{}_{\tau\beta} - \partial_\beta\, \Gamma^\lambda{}_{\tau\alpha} + \Gamma^\lambda{}_{\gamma\alpha}\, \Gamma^\gamma{}_{\tau\beta} - \Gamma^\lambda{}_{\gamma\beta}\, \Gamma^\gamma{}_{\tau\alpha}\,. \tag{11.13}$$

Let us note that the commutator of two covariant derivatives acting on a scalar is zero (this is a direct consequence of our decision to consider only symmetric affine connections). Then, using the scalar nature of the product $A^\lambda B_\lambda$ and Eq. (11.12), one can easily arrive at the commutator of covariant derivatives acting on the covariant vector

$$\left[\nabla_\alpha, \nabla_\beta\right] B_\lambda = R^\tau{}_{\lambda\beta\alpha}\, B_\tau\,. \tag{11.14}$$

In the same manner, one can use the fact that the product

$$T^{\mu_1\ldots\mu_n}{}_{\nu_1\ldots\nu_m}(x)\, A^{\nu_1}(x)\ldots A^{\nu_m}(x) B_{\mu_1}(x)\ldots B_{\mu_n}(x)$$

is a scalar, and thus obtain the commutator for an arbitrary tensor, e.g.,

$$\left[\nabla_\alpha, \nabla_\beta\right] T^\mu{}_\nu = -R^\mu{}_{\lambda\beta\alpha}\, T^\lambda{}_\nu + R^\lambda{}_{\nu\beta\alpha}\, T^\mu{}_\lambda\,. \tag{11.15}$$

In this way, we can establish the general rule for the commutator of two covariant derivatives acting on an arbitrary tensor and, at the same time, prove the universality of the curvature tensor.

The curvature tensor has many interesting symmetries. Let us list them without proofs, which can be found in the standard textbooks on GR.

Let us define

$$R^\lambda{}_{\nu\alpha\beta} = g^{\mu\lambda} R_{\mu\nu\alpha\beta}$$

(as usual, all indexes can be raised or lowered, using the metric). Then the Riemann tensor satisfies the relations

$$\text{i)}\quad R_{\mu\nu\beta\alpha} = -R_{\nu\mu\beta\alpha} = -R_{\mu\nu\alpha\beta}\,. \tag{11.16}$$

$$\text{ii)}\quad R_{\mu\nu\beta\alpha} = R_{\beta\alpha\mu\nu}\,. \tag{11.17}$$

$$\text{iii)}\quad R_{\mu\nu\alpha\beta} + R_{\mu\beta\nu\alpha} + R_{\mu\alpha\beta\nu} = 0\,. \tag{11.18}$$

$$\text{iv)}\quad \nabla_\lambda R_{\mu\nu\alpha\beta} + \nabla_\nu R_{\lambda\mu\alpha\beta} + \nabla_\mu R_{\nu\lambda\alpha\beta} = 0\,. \tag{11.19}$$

The last property is called the first Bianchi identity.

Along with the Riemann tensor, there are its useful contractions possessing their proper names. The first one is the Ricci tensor (sometimes also called Ricci curvature)

$$R^\alpha{}_{\mu\alpha\nu} = R_{\mu\nu} \tag{11.20}$$

and its trace $R = R_{\mu\nu} g^{\mu\nu}$, which is called scalar curvature, or the Ricci scalar. Making contractions in the first Bianchi identity (11.19), we obtain two other Bianchi identities involving Ricci tensor and the scalar curvature,

$$\text{v)}\quad \nabla_\alpha R_{\lambda\beta} - \nabla_\beta R_{\lambda\alpha} = \nabla_\tau R^\tau{}_{\lambda\alpha\beta}\,. \tag{11.21}$$

$$\text{vi)}\quad \nabla_\alpha R^\alpha{}_\beta = \frac{1}{2}\nabla_\beta R\,. \tag{11.22}$$

One can evaluate the number of independent components of Riemann and Ricci tensors. We leave this calculation as an exercise for the interested reader and only quote the main results. In the dimension $n = 2$, there is only one independent component in both cases, and, for any metric, one can write

$$R_{\alpha\beta\mu\nu} = \frac{1}{2} R \left(g_{\alpha\mu} g_{\beta\nu} - g_{\alpha\nu} g_{\beta\mu} \right) \qquad \text{and} \qquad R_{\mu\nu} = \frac{1}{2} R g_{\mu\nu} \,. \tag{11.23}$$

In the case of $n = 3$, both Ricci and Riemann tensors have six independent components, but, already for $n = 4$, the Ricci tensor has ten such components, while the Riemann tensor has twenty.

One can construct a tensor that describes the components of the Riemann tensor which are not covered by the Ricci tensor, and therefore should be zero in $n \leq 3$. Consider a linear combination of all these curvatures, and require that its trace in any pair of indices is zero. This new curvature is called the Weyl tensor, and we will denote it as $C^{\alpha}{}_{\mu\beta\nu}$. The definition in the n-dimensional spacetime is

$$\begin{aligned} C_{\alpha\beta\mu\nu} = R_{\alpha\beta\mu\nu} &+ \frac{1}{n-2} \left(g_{\beta\mu} R_{\alpha\nu} - g_{\alpha\mu} R_{\beta\nu} + g_{\alpha\nu} R_{\beta\mu} - g_{\beta\nu} R_{\alpha\mu} \right) \\ &+ \frac{1}{(n-1)(n-2)} R \left(g_{\alpha\mu} g_{\beta\nu} - g_{\alpha\nu} g_{\beta\mu} \right). \end{aligned} \tag{11.24}$$

It is easy to see that, with this definition, the tracelessness $C^{\alpha}{}_{\mu\alpha\nu} = 0$ is satisfied.

The Weyl tensor has a remarkable property with respect to the local conformal transformation of the metric

$$g_{\mu\nu} = \bar{g}_{\mu\nu}\, e^{2\sigma}, \qquad \sigma = \sigma(x) \,. \tag{11.25}$$

It proves useful to write down the set of relevant formulas in an arbitrary spacetime dimension. The transformations of the inverse metric and metric determinant are

$$g^{\mu\nu} = \bar{g}^{\mu\nu}\, e^{-2\sigma}, \qquad g = \bar{g}\, e^{2n\sigma}. \tag{11.26}$$

The transformation rule for the Christoffel symbol is

$$\begin{aligned} \bar{\Gamma}^{\lambda}{}_{\alpha\beta} &= \Gamma^{\lambda}{}_{\alpha\beta} + \delta\Gamma^{\lambda}{}_{\alpha\beta}, \\ \delta\Gamma^{\lambda}{}_{\alpha\beta} &= \delta^{\lambda}_{\alpha}(\bar{\nabla}_{\beta}\sigma) + \delta^{\lambda}_{\beta}(\bar{\nabla}_{\alpha}\sigma) - \bar{g}_{\alpha\beta}(\bar{\nabla}^{\lambda}\sigma). \end{aligned} \tag{11.27}$$

In the last formula, $\bar{\nabla}_{\alpha}$ is covariant derivative constructed from the connection $\bar{\Gamma}^{\lambda}{}_{\alpha\beta}$ compatible with the fiducial metric $\bar{g}_{\alpha\beta}$.

Let us note that the two Christoffel symbols, $\Gamma^{\lambda}{}_{\alpha\beta}$ and $\Gamma'^{\lambda}{}_{\alpha\beta}$, are different because they are constructed with different metrics, $g_{\mu\nu}$ and $g'_{\mu\nu}$. In general, as we know from Eq. (11.4), the difference between two affine connections is a tensor, and the difference which can be seen in (11.27) confirms this rule. One can use this feature to obtain the general rule of variation of the Riemann tensor (11.13), depending only on the affine connection and not on the metric,

$$\delta R^{\lambda}{}_{\tau\alpha\beta} = \bar{\nabla}_{\alpha}\, \delta\Gamma^{\lambda}{}_{\tau\beta} - \bar{\nabla}_{\beta}\, \delta\Gamma^{\lambda}{}_{\tau\alpha} + \delta\Gamma^{\lambda}{}_{\gamma\alpha}\, \delta\Gamma^{\gamma}{}_{\tau\beta} - \delta\Gamma^{\lambda}{}_{\gamma\beta}\, \delta\Gamma^{\gamma}{}_{\tau\alpha} \,. \tag{11.28}$$

This formula will prove to be very useful on many occasions. Concerning the conformal transformation, one can use (11.28) to derive the transformation rules of the curvature tensors and scalar,

$$R^{\alpha}{}_{\beta\mu\nu} = \bar{R}^{\alpha}{}_{\beta\mu\nu} + \delta^{\alpha}_{\nu}\,(\bar{\nabla}_{\mu}\bar{\nabla}_{\beta}\sigma) - \delta^{\alpha}_{\mu}(\bar{\nabla}_{\nu}\bar{\nabla}_{\beta}\sigma) + \bar{g}_{\mu\beta}(\bar{\nabla}_{\nu}\bar{\nabla}^{\alpha}\sigma)$$
$$- \bar{g}_{\nu\beta}(\bar{\nabla}_{\mu}\bar{\nabla}^{\alpha}\sigma) + \delta^{\alpha}_{\nu}\bar{g}_{\mu\beta}(\bar{\nabla}\sigma)^2 - \delta^{\alpha}_{\mu}\bar{g}_{\nu\beta}(\bar{\nabla}\sigma)^2 + \delta^{\alpha}_{\mu}(\bar{\nabla}_{\nu}\sigma)(\bar{\nabla}_{\beta}\sigma)$$
$$- \delta^{\alpha}_{\nu}(\bar{\nabla}_{\mu}\sigma)(\bar{\nabla}_{\beta}\sigma) + \bar{g}_{\nu\beta}(\bar{\nabla}_{\mu}\sigma)(\bar{\nabla}^{\alpha}\sigma) - \bar{g}_{\mu\beta}(\bar{\nabla}_{\nu}\sigma)(\bar{\nabla}^{\alpha}\sigma)\,, \tag{11.29}$$

$$R_{\mu\nu} = \bar{R}_{\mu\nu} - \bar{g}_{\mu\nu}(\bar{\nabla}^2\sigma) + (n-2)\left[(\bar{\nabla}_{\mu}\sigma)(\bar{\nabla}_{\nu}\sigma) - (\bar{\nabla}_{\mu}\bar{\nabla}_{\nu}\sigma) - \bar{g}_{\mu\nu}(\bar{\nabla}\sigma)^2\right], \tag{11.30}$$

$$R = e^{-2\sigma}\Big[\bar{R} - 2(n-1)(\bar{\nabla}^2\sigma) - (n-1)(n-2)(\bar{\nabla}\sigma)^2\Big]. \tag{11.31}$$

Finally, after some algebra, one can check that the conformal transformation of the Weyl tensor (11.24) is much simpler than for the other curvature tensors,

$$C'_{\alpha\beta\mu\nu} = e^{2\sigma}\,C_{\alpha\beta\mu\nu} \qquad \text{and} \qquad C'^{\alpha}{}_{\beta\mu\nu} = C^{\alpha}{}_{\beta\mu\nu}\,. \tag{11.32}$$

The last observation is that one of the most important properties of the curvature tensor is its uniqueness (see, e.g., [252] or [340]). This means, in particular, that the Riemann tensor, the Ricci tensor and the Ricci scalar are the only possible covariant second-order in derivatives combinations of metric that have the given algebraic symmetries. Especially important for the construction of GR is the fact that R is the unique possible scalar of mass dimension 2, constructed from the metric and its derivatives.

Exercises

11.1. Verify equations (11.12) and (11.13) by directly calculating the commutator of the covariant derivatives. Derive Eq. (11.14) using (11.12), as is explained in the text, and verify it by direct calculation of the commutator.
11.2. Verify whether the Weyl tensor has the symmetry (11.19). Derive the Bianchi identities for the Weyl tensor, and explain the results for the analogs of Eqs. (11.21) and (11.22), using qualitative arguments.
11.3. Using combinatorial methods, derive the number of independent components of the Riemann, Ricci and Weyl tensors in n-dimensional spacetime.
11.4. What is the curvature tensor in $n = 1$? Explain what happens with Eq. (11.24) when $n = 2$ or $n = 1$.
11.5. Prove that an arbitrary variation of affine connection is a third-rank tensor. Prove that the variation law for the Riemann tensor (11.28) is valid for an arbitrary variation of affine connection. Derive similar expressions for the variations of the Ricci tensor, scalar curvature and the Weyl tensor. Consider separately the three following cases: (i) Only the metric is subject of variation, (ii) when only the connection varies, and this variation is metric-independent, and (iii) when the variation of the connection is caused by the variation of the metric, according to Eq. (11.9).
11.6. Verify all of the formulas for conformal transformations, starting from (11.26) and (11.27) and finally arriving at (11.32).

11.4 The covariant equation for a free particle: the classical limit

Let us consider a free particle moving in M_{1+3}. The word "free" means there are no other forces except gravity, while the last is regarded as a geometric characteristic of the spacetime. The action of a particle is supposed to be covariant and reduce to the known expression in the flat space limit. Therefore, such an action is postulated in the form

$$S = -m \int ds = -m \int d\tau \sqrt{g_{\mu\nu} \frac{dx^\mu}{d\tau} \frac{dx^\nu}{d\tau}}, \tag{11.33}$$

where τ is a smooth parameter along the curve. In the following, we identify τ with the interval s in units with $c = 1$, when it is not indicated explicitly.

Performing the variation with respect to $x^\mu(t)$, we arrive at the equation of motion in the form

$$\frac{d^2x^\mu}{d\tau^2} + \Gamma^\mu{}_{\alpha\beta} \frac{dx^\alpha}{d\tau} \frac{dx^\beta}{d\tau} = 0, \tag{11.34}$$

or, equivalently, the equation of parallel transport for $dx^\nu/d\tau$,

$$\frac{dx^\mu}{d\tau} \nabla_\mu \frac{dx^\nu}{d\tau} = 0\,. \tag{11.35}$$

The last two relations are also called the equation of the geodesic line. According to the least action principle, it defines the motion of the particle. It is important to note that the action (11.33) leads to consistent equation even for massless particle, when $d\tau = 0$, but, in this case, τ cannot be identified with the interval s. In this case, one should use another affine parameter, such as the time coordinate t, in the equation of motion (see, e.g., [202, 340])

It is remarkable that Eq. (11.34) depends on the metric only through the affine connection, or the Christoffel symbol. One can note that since the affine connection is not a tensor, "the gravitational force" $-\Gamma^\mu{}_{\alpha\beta} \frac{dx^\alpha}{d\tau} \frac{dx^\beta}{d\tau}$ is essentially dependent on the choice of the reference frame, which is the choice of spacetime coordinates. This fact is the mathematical manifestation of the equivalence principle.

The principle of equivalence states that one can eliminate gravitational force *locally* by using an accelerated frame, which produces the force of inertia. This means we can choose such coordinates, where $\Gamma^\mu{}_{\alpha\beta}$ in Eq. (11.34) vanish at the given spacetime point, which we shall call P. At the same time, this quantity does not vanish in the vicinity of the point P. It is easy to understand why this is so. If $\Gamma^\mu{}_{\alpha\beta} \equiv 0$ in the vicinity of P, then the partial derivatives of $\Gamma^\mu{}_{\alpha\beta}$ in the whole of this vicinity should be zero. But, in this case, the Riemann tensor (11.13) would also vanish. And this would mean that the space is flat. Therefore, the compensation of gravity by inertia can be achieved only locally, in one singular spacetime point, and never in the vicinity of such a point.

Eq. (11.35) is a covariant generalization of the equation for a particle in flat spacetime. The equation of motion (11.35) also agrees with the equivalence principle, because it does not depend on the particle mass. As a result, test massive particles accelerate identically in a given gravitational field, independent of the magnitude of their masses.

Let us verify that the second main requirement for the new equation is satisfied, and derive the Newtonian limit in the equation for the geodesic line (11.34). Consider the weak-field, non-relativistic and static gravitational field, by taking

$$g_{\mu\nu} = \eta_{\mu\nu} + h_{\mu\nu}, \quad \text{where} \quad |h_{\mu\nu}| \ll 1 \quad \text{and} \quad h_{\mu\nu} = h_{\mu\nu}(\mathbf{x}). \tag{11.36}$$

Furthermore, we assume a small velocity of the test particle (in units with $c = 1$)

$$\left|\frac{d\mathbf{x}}{d\tau}\right| \ll \left|\frac{dt}{d\tau}\right|, \qquad \text{where} \qquad t = x^0\,. \tag{11.37}$$

The relevant terms in (11.34) are

$$\frac{d^2x^\mu}{d\tau^2} + \Gamma^\mu{}_{00}\left(\frac{dt}{d\tau}\right) = 0\,. \tag{11.38}$$

In the first order in $h_{\mu\nu}$, one can easily find (we postpone the details of this and similar expansions to the next section)

$$\Gamma^\mu{}_{00} = -\frac{1}{2}\,\eta^{\mu\nu}\,\frac{\partial h_{00}}{\partial x^\nu} = -\frac{1}{2}\,\eta^{\mu i}\,\frac{dh_{00}}{dx^i}\,. \tag{11.39}$$

After using this result in (11.38), we arrive at the equations

$$\frac{d^2t}{d\tau^2} = 0, \qquad \frac{d^2\mathbf{x}}{d\tau^2} = -\frac{1}{2}\nabla\, h_{00}. \tag{11.40}$$

The first equation shows the linear dependence between proper time τ and physical time t. Without loss of generality, we can simply set $\tau = t$ in the non-relativistic limit. Then, choosing $h_{00} = 2\varphi$, the second equation in (11.40) becomes the second Newton's law for the classical particle moving in the gravitational field with the potential energy $U(\mathbf{r}) = m\varphi(\mathbf{r})$. For example, in the case when gravity is due to the point-like source with the mass M, the potential is one of Newton's gravitational force,

$$\varphi = -\frac{GM}{r}\,. \tag{11.41}$$

Thus, we have found that the general covariant equation for the "free" particle (11.34) is consistent with Newton's law in the static, weak-field and non-relativistic limit. As a by-product, we have found that, in the leading order, the (00)-component of the metric is expressed through the Newtonian potential as $g_{00} = 1 + 2\varphi(\mathbf{r})$.

In the next chapter, we shall see that, using the covariant generalization of the relativistic flat-space actions, one can construct the actions of different fields in curved spacetime. Let us now turn to the dynamics of proper gravity.

Exercises

11.7. Derive the equation for the particle (11.34) from the action principle (11.33).
11.8. Show that the same equation follows from the action which *looks* different,

$$S' = -\frac{1}{2}\int_a^b d\tau\; g_{\mu\nu}\,\frac{dx^\mu}{d\tau}\,\frac{dx^\nu}{d\tau}\,. \tag{11.42}$$

Solution. The variation of δx^λ with the boundary conditions $\delta x^\lambda\big|_{a,b} = 0$, gives

$$\delta S' = -\frac{1}{2}\int_a^b d\tau\Big(\frac{\partial g_{\mu\nu}}{\partial x^\lambda}\frac{dx^\mu}{d\tau}\frac{dx^\nu}{d\tau} + 2g_{\mu\nu}\frac{dx^\mu}{d\tau}\frac{d\,\delta x^\nu}{d\tau}\Big) \tag{11.43}$$
$$= g_{\mu\nu}\frac{dx^\mu}{d\tau}\delta x^\nu\Big|_a^b - \int_a^b d\tau\,\delta x^\lambda\Big[\frac{1}{2}\frac{\partial g_{\mu\nu}}{\partial x^\lambda}\frac{dx^\mu}{d\tau}\frac{dx^\nu}{d\tau} - \frac{d}{d\tau}\Big(g_{\mu\lambda}\frac{dx^\mu}{d\tau}\Big)\Big] = 0.$$

Taking into account the arbitrariness of δx^λ and the boundary conditions, after some algebra, we arrive at the equation

$$g_{\mu\lambda}\frac{d^2x^\mu}{d\tau^2} + \frac{\partial g_{\mu\lambda}}{\partial x^\nu}\frac{dx^\mu}{d\tau}\frac{dx^\nu}{d\tau} - \frac{1}{2}\frac{\partial g_{\mu\nu}}{\partial x^\lambda}\frac{dx^\mu}{d\tau}\frac{dx^\nu}{d\tau} = 0. \tag{11.44}$$

Finally, using the symmetry over $(\mu\nu)$ in the last expression, we get (11.34).
11.9. Discuss the equivalence between the actions (11.42) and (11.33). Is it possible to find other forms of the action which produce the same equations of motion?

11.5 Classical action for the gravity field

Let us formulate the dynamical equations for the metric. Together with the covariant equations for matter (particles and fields), the gravitational equations form a complete basis for the consistent relativistic description of the gravitational phenomena. The relation between these two components can be beautifully described through a quote from John Archibald Wheeler: “Spacetime tells matter how to move; matter tells spacetime how to curve.”

As we already mentioned, relativistic gravitational theories can be constructed in different ways. The ambiguity is mainly due to the possibility of choosing different actions for gravity. In order to construct the simplest version of relativistic gravity, which is GR, one can impose the following restrictions on the action and equations of motion:

i) Gravity is described only by the metric. This requirement is non-trivial. In principle, one can introduce different additional fields and show that the theory remains relativistic, with the correct classical (Newtonian) limit and even successfully passing all existing experimental and observational tests of GR. As examples of such extended theories, we can remember metric-scalar models, gravity with torsion and more complicated options, such as supergravity. Some of these models have strong theoretical motivations in the quantum theory of matter fields, quantum gravity or string theory. However, in most of the cases, the correspondence with the experimental and observational data is achieved only by requiring that the masses of additional fields are large, or by assuming very weak interaction with the observable objects. Taking this aspect into account, it is clear that GR is the simplest satisfactory theory that is a reference point for all mentioned extensions. Thus, it is worthwhile to start from GR and only after that introduce other fields or consider modified models of gravity.

ii) The dynamical equations for the metric must be tensor equations, and hence the action for gravity should be a general covariant scalar.

iii) The equations for the metric must be of the second order in derivatives, which means they must be free of higher derivatives. The violation of this point also leads to an interesting generalization of GR. In the next chapters, we will see that such a generalization is really difficult to avoid in the framework of the quantum theory of matter on a classical gravitational background, or in quantum gravity.

iv) The action of gravity must be a local functional. In the scientific literature, one can find gravity theories which do not follow this requirement, and some of those will be discussed in the subsequent chapters devoted to quantum aspects of gravity. Nevertheless, GR remains a reference point in the construction of all these theories; hence, at the first stage, we will hold on to the condition of locality.

Taking all four points together, we recall that there is only one possible local second-derivative scalar constructed from the metric and its derivatives, namely, the scalar curvature R. Thus, there is a single candidate for the role of the action of Einstein's gravity, the Einstein-Hilbert action,

$$S_{EH} = -\frac{1}{\gamma}\int d^4x\sqrt{-g}\,(R + 2\Lambda)\,, \tag{11.45}$$

where γ and Λ are some constants. The Λ-term is called the cosmological constant, and has to be included because it is not ruled out by the conditions listed above. The next step is to consider the total action, including the matter counterpart to gravity, S_m,

$$S_t \;=\; S_{EH} + S_m. \tag{11.46}$$

Both the gravitational S_{EH} and the matter S_m parts of the action depend on the metric (as otherwise, it is difficult to provide covariance), but S_m also depends on the variables Φ_i describing matter particles and fields. The equations of motion are defined from the variational principle,

$$\frac{\delta S_t}{\delta g_{\mu\nu}} = 0 \qquad \text{and} \qquad \frac{\delta S_t}{\delta \Phi_i} = 0. \tag{11.47}$$

In order to take the first variational derivative (11.47), consider an infinitesimal variation of the metric, $\delta g_{\mu\nu} = h_{\mu\nu}$,

$$g_{\mu\nu} \quad\longrightarrow\quad g'_{\mu\nu} \;=\; g_{\mu\nu} + h_{\mu\nu}, \tag{11.48}$$

and derive δS_t, in the first order in $h_{\mu\nu}$. Indeed, the action contains not only the metric $g_{\mu\nu}$, but also other objects, such as $g^{\mu\nu}$, $\sqrt{-g}$ and $R_{\mu\nu}$. Let us show how to perform their variation. Although in this section, we need only the first order in $h_{\mu\nu}$, it is useful to show how to perform the expansion to an arbitrary order. These skills will be very useful later on, especially when we discuss quantum gravity.

1. Using an identity $g'_{\mu\nu}\,g'^{\nu\lambda} = \delta^{\lambda}_{\;\mu}$, we find

$$g'^{\mu\nu} = g^{\mu\nu} - h^{\mu\nu} + h^{\mu}_{\;\lambda}h^{\nu\lambda} - h^{\mu}_{\;\lambda}h^{\lambda}_{\;\tau}h^{\nu\tau} + .. \tag{11.49}$$

2. For the determinant, we can start from the general formula

$$\delta g = g g^{\mu\nu}\cdot h_{\mu\nu}\,. \tag{11.50}$$

If one would like to go further, then

$$\delta(\delta g) = \delta g \cdot g^{\mu\nu} \cdot h_{\mu\nu} + g \cdot \delta g^{\mu\nu} \cdot h_{\mu\nu} = gh^2 - gh_{\mu\nu}h^{\mu\nu}, \tag{11.51}$$

etc. where $h = g^{\mu\nu}h_{\mu\nu}$. Thus, the series expansion gives

$$g' = g \cdot \Big(1 + h + \frac{1}{2}\,h^2 - \frac{1}{2}\,h_{\mu\nu}h^{\mu\nu} + ...\Big). \tag{11.52}$$

From Eq. (11.52), one can easily get

$$\sqrt{-g'} = \sqrt{-g} \cdot \Big(1 + \frac{1}{2}\,h + \frac{1}{8}\,h^2 - \frac{1}{4}\,h_{\mu\nu}h^{\mu\nu} + ...\Big), \tag{11.53}$$

which can be extended to any order in the perturbation $h_{\mu\nu}$. One can also obtain the infinite-order expansion by using Liouville's formula (see, e.g., [292]).

3. The expansion of the Christoffel symbol is quite simple, and can be derived, using Eqs. (11.49) and (11.8), in an arbitrary order in $h_{\mu\nu}$:

$$\Gamma^\alpha_{\ \beta\gamma} \to \Gamma'^\alpha_{\ \beta\gamma} = \Gamma^\alpha_{\ \beta\gamma} + \delta\Gamma^\alpha_{\ \beta\gamma}\,, \tag{11.54}$$

where

$$\begin{aligned}\delta\Gamma^\alpha_{\ \beta\gamma} &= \frac{1}{2}g'^{\alpha\lambda}\big(\partial_\beta h_{\gamma\lambda} + \partial_\gamma h_{\beta\lambda} - \partial_\lambda h_{\gamma\beta} + \partial_\beta g_{\gamma\lambda} + \partial_\gamma g_{\beta\lambda} - \partial_\lambda g_{\gamma\beta}\big) \\ &= \frac{1}{2}\big(g^{\alpha\lambda} - h^{\alpha\lambda} + h^\alpha_\kappa h^{\lambda\kappa} - h^\alpha_\kappa h^\kappa_\tau h^{\tau\lambda} + ...\big)\big(\nabla_\beta h_{\gamma\lambda} + \nabla_\gamma h_{\beta\lambda} - \nabla_\lambda h_{\gamma\beta}\big).\end{aligned} \tag{11.55}$$

Here the covariant derivative ∇_α is constructed using the non-perturbed connection $\Gamma^\alpha_{\ \beta\gamma}$. Once again, the variation $\delta\Gamma^\alpha_{\ \beta\gamma}$ is a tensor, as the difference between two affine connections. In the first order, we obtain

$$\delta^{(1)}\Gamma^\alpha_{\ \beta\gamma} = \frac{1}{2}\big(\nabla_\beta h^\alpha_\gamma + \nabla_\gamma h^\alpha_\beta - \nabla^\alpha h_{\gamma\beta}\big). \tag{11.56}$$

4. The expansion of the curvature tensor can start from the general relation (11.28). After a relatively simple algebra based on this result and contractions with (11.49), one can expand any curvature-dependent expression to any desirable order. At the moment, we restrict the consideration by the following expansion for scalar curvature:

$$\begin{aligned}\delta R = R' - R &= R^{(1)} + R^{(2)} + R^{(3)} + \dots, \\ R^{(1)} &= \nabla_\alpha\nabla_\beta h^{\alpha\beta} - \Box h - R^{\alpha\beta}h_{\alpha\beta}.\end{aligned} \tag{11.57}$$

Now we are in a position to perform the variation of the action. In the matter part, we use the dynamical definition of the energy-momentum tensor (stress tensor),

$$T^{\mu\nu} = -\frac{2}{\sqrt{-g}}\frac{\delta S_m}{\delta g_{\mu\nu}}. \tag{11.58}$$

Then, postulating that the variations $\delta g_{\mu\nu}$ vanish on the boundary of the manifold, we arrive at the following expression:

$$-\frac{2}{\sqrt{-g}}\,\frac{\delta S_g}{\delta g_{\mu\nu}} = \frac{1}{\gamma}\left[R^{\mu\nu} - \frac{1}{2}\,Rg^{\mu\nu} - \Lambda g^{\mu\nu}\right]. \tag{11.59}$$

All in all, the dynamical equations that follow from the total action S_t are

$$R^{\mu}_{\ \nu} - \frac{1}{2}\,R\delta^{\mu}_{\ \nu} = \frac{\gamma}{2}\,T^{\mu}_{\ \nu} - \Lambda\delta^{\mu}_{\ \nu}. \tag{11.60}$$

11.6 Einstein equations and the Newton limit

To make use of Eq. (11.60), we need to understand the physical sense of the constants γ and Λ and calibrate them. Let us start with Λ. Consider the stress-tensor for the ideal fluid in the co-moving coordinates (see, e.g., [203]),

$$T^{\mu}_{\nu} = \operatorname{diag}\left(\rho, -p, -p, -p\right), \tag{11.61}$$

where ρ is the energy density and p is the pressure of the fluid. The comparison of how ρ, p and Λ enter the equation (11.60), enable one to regard the cosmological constant contribution as a perfect fluid. Let us stress that this identification is purely formal, as the cosmological constant does not correspond to any fluid, but is an independent fundamental characteristic of the spacetime. Anyway, it is a very useful identification and is frequently used for different purposes. Looking at the (00)-component, we can see that the expression $-2\Lambda/\gamma$ can be seen as a sort of vacuum energy. Furthermore, the "equation of state" for the cosmological constant is different from that of any matter source: $\rho_\Lambda = -p_\Lambda$. In any case, there is nothing like the cosmological constant in the Newton's theory; hence, the cosmological constant is something typical for relativistic theory and has no analogs at the classical level. If we want to achieve correspondence with Newton's gravity, we better set $\Lambda = 0$, for a while.

As we have learned in the previous section, the correspondence with Newton's gravity can be achieved in the weak-field, static and non-relativistic limits. In this case,

$$g_{00} = 1 + 2\varphi(\mathbf{x}). \tag{11.62}$$

Our purpose is to find the correspondence between Eq. (11.60) with zero cosmological constant and the Poisson equation $\Delta\varphi = 4\pi G\rho$ that holds in Newton's theory. Substituting the metric (11.62) into (11.60) and taking the (00)-component with $T_{00} = \rho$, we find that the "correct" coefficient in the Poisson equation is achieved in the case $\gamma = 16\pi G$. Thus, we arrive at the standard form of the Einstein equations with the cosmological constant term,

$$G_{\mu\nu} \equiv R_{\mu\nu} - \frac{1}{2}\,Rg_{\mu\nu} = 8\pi G\,T_{\mu\nu} - \Lambda\,g_{\mu\nu}. \tag{11.63}$$

$G_{\mu\nu}$ is called the Einstein tensor, which can be considered, along with the Riemann, Ricci, Weyl tensors and R, as one more useful representation of the curvature tensor.

The Bianchi identity (11.22) results in the identity for the Einstein tensor, $\nabla_\mu G^{\mu\nu} \equiv 0$. Furthermore, the cosmological term in (11.63) satisfies the same identity as a consequence of the metricity condition (11.8). Therefore, the consistency of Eq. (11.63) can be achieved only if the stress tensor satisfies the conservation law $\nabla_\mu T^{\mu\nu} \equiv 0$. And this is, indeed, true, because of the general covariance.

It is a useful exercise to check that the transformation of the metric under general coordinate transformation (11.2) is

$$\delta g_{\mu\nu} = g'_{\mu\nu}(x) - g_{\mu\nu}(x) = -\nabla_\mu \xi_\nu - \nabla_\nu \xi_\mu \,. \tag{11.64}$$

Using the notation Φ_i for the field and particle variables of the matter action S_m, we denote the corresponding coordinate transformations as $\delta\Phi_i = R_i,^\mu \xi_\mu$. The operator $R_i,^\mu$ is called a generator of diffeomorphism transformations. Its form depends on the type of the variables Φ_i. Then the functional form of the general covariance of the total action (11.46) can be written as

$$\nabla_\mu \left(\frac{2}{\sqrt{-g}} \frac{\delta S_{EH}}{\delta g_{\mu\nu}} \right) + \nabla_\mu \left(\frac{2}{\sqrt{-g}} \frac{\delta S_m}{\delta g_{\mu\nu}} \right) + \frac{1}{\sqrt{-g}} \frac{\delta S_m}{\delta \Phi_i} R_i,^\nu = 0 \,. \tag{11.65}$$

Using the equations of motion for the matter variables $\frac{\delta S_g}{\delta \Phi_i} = 0$, the Bianchi identity (11.22) and the definition (11.58), we arrive at the conservation law $\nabla_\mu T^{\mu\nu} \equiv 0$. Let us note that the same identity in flat spacetime was derived in Part I as the Noether current of the spacetime symmetry.

Exercises

11.10. Derive the generator $R_i,^\mu$ for scalar ϕ and vector A^α fields.
11.11. Verify Eq. (11.64) and also derive its generalization for an arbitrary symmetric tensor $\varphi_{\mu\nu}$ that does not necessary satisfy the metricity condition.
11.12. Show that Eq. (11.63) can be written in the form

$$R^\mu_{\ \nu} = 8\pi G \left(T^\mu_{\ \nu} - \frac{1}{2} T \delta^\mu_{\ \nu} \right) + \Lambda \delta^\mu_{\ \nu} \,. \tag{11.66}$$

11.13. Derive the 00 component of Eq. (11.66) for the non-relativistic metric (11.62) and the energy-momentum tensor defined by (11.61). Show that, in the limit $\Lambda = 0$, the classical equation $\Delta\varphi = 4\pi G \rho$ follows. Discuss the physical significance of the result with positive and negative values of Λ. Evaluate the effect of a non-zero Λ at the solar-system scale and at the galactic scale, taking into account that the present estimate is $\Lambda \propto 10^{-82} G^{-1}$. Show that this estimate corresponds to $\rho_\Lambda \approx 10^{-120} G^{-2}$ for the energy density of the cosmological constant term, $\rho_\Lambda = \Lambda/(8\pi G)$.

11.7 Some physically relevant solutions and singularities

In this section, we consider the simplest solutions to Einstein equations, namely, those for the spherically symmetric and the homogeneous and isotropic metrics.

11.7.1 The spherically-symmetric Schwarzschild solution

This solution can be achieved if we consider spherically symmetric compact mass distribution and assume the same symmetry for the classical solution of vacuum Einstein equations in the region out of this distribution. For the sake of simplicity, we consider a source (star) that represents a point-like mass. Let us settle the origin of the coordinate system into the center of the mass distribution. One can prove that, in this case, the interval may depend on the angles φ and θ only through $d\Omega$, the differential element of the solid angle,

$$d\Omega \;=\; d\theta^2 + \sin^2\theta\, d\varphi^2\,. \tag{11.67}$$

Besides that, the metric may depend on the central distance r and time t.

The Schwarzschild solution can be written in the form

$$ds^2 = e^{\nu(r,t)}dt^2 - e^{\lambda(r,t)}dr^2 - r^2 d\Omega\,, \tag{11.68}$$

where $\nu(r,t)$ and $\lambda(r,t)$ are functions that can be found from the vacuum Einstein equations with the given symmetry in the form [202],

$$e^{\nu} = e^{-\lambda} = 1 + \frac{A}{r}, \qquad A = const. \tag{11.69}$$

To find the integration constant A, we require that Newton's law be asymptotically the leading part of (11.69) at very long distances. Due to (11.62) and (11.41), this means that, at $r \to \infty$, we have

$$g_{00} = e^{\nu(r,t)} \approx 1 + 2\varphi \;=\; 1 - \frac{2GM}{r}. \tag{11.70}$$

Comparing (11.69) with (11.70), we get $A = -2GM$, where M is the mass of the star. Now, the Schwarzschild solution can be written in the standard form,

$$ds^2 \;=\; \Big(1 - \frac{2GM}{r}\Big)\, dt^2 \;-\; \Big(1 - \frac{2GM}{r}\Big)^{-1} dr^2 \;-\; r^2 d\Omega\,. \tag{11.71}$$

Let us note that if we perform a simple expansion of the above solution into the series in the parameter $1/r$, we can get the relativistic corrections to the classical gravitational potential in the case of a point-like source,

$$\varphi(r) \;=\; -\frac{GM}{r} + \frac{G^2M^2}{2r^2} + \dots . \tag{11.72}$$

The Schwarzschild solution contains two singularities: one at the surface of the gravitational radius $r_g = 2GM$, and the other at the origin $r = 0$. It is well known that the first singularity is coordinate dependent (see, e.g., [202]), since it indicates the existence of the horizon for the observer, which is very far from the point $r = 0$. One can prove that particles and even the light signals can not propagate from the interior of the horizon to its exterior; that is why this solution has been called a black hole. Nevertheless, the horizon at $r = r_g$ looks like a singularity when observed from the long distance $r \gg r_g$. In contrast there may be no singularity for observers in other reference frames. For instance, in Kruskal coordinates (see [147] for details),

$$u = t - r - r_g \log \left| \frac{r}{r_g} - 1 \right|, \qquad v = t + r + r_g \log \left| \frac{r}{r_g} - 1 \right|, \tag{11.73}$$

the interval is

$$ds^2 = -\Big(1 - \frac{r_g}{r}\Big) du dv + r^2 d\Omega^2 \,, \tag{11.74}$$

and the metric is regular at $r = r_g$,

There is also a coordinate-independent way to check that $r = r_g$ does not have a physical singularity. To do this, one has to derive two independent curvature-dependent scalars (called the Kretschmann and Chern–Pontryagin invariants),

$$\text{a)}\ R_{\alpha\beta\mu\nu}R^{\alpha\beta\mu\nu} \qquad \text{and} \qquad \text{b)}\ P_4 \,=\, R_{\alpha\beta\mu\nu}\,\varepsilon^{\alpha\beta\rho\sigma}\,R_{\rho\sigma\,\cdot\,\cdot}^{\ \ \ \mu\nu} \tag{11.75}$$

and then verify that these expressions are finite at the $r = r_g$. Let us note that the two invariants in Eq. (11.75) are the only non-trivial ones for the Schwarzschild metric, because of corresponding space is *Ricci flat*, $R_{\mu\nu} = 0$. Then the invariant a) in (11.75) can be traded to the integrand E_4 of the Gauss-Bonnet topological term,

$$E_4 = R_{\alpha\beta\mu\nu}R^{\alpha\beta\mu\nu} - 4R_{\alpha\beta}R^{\alpha\beta} + R^2. \tag{11.76}$$

Let us mention that the expression b) in (11.75) is an integrand of another topological invariant which is called the Pontryagin term. Both E_4 and P_4 do not diverge at the horizon, which shows that the singularity at $r = r_g$ is coordinate dependent. Thus, the existence of the gravitational horizon around the black hole does not indicate the physical singularity and hence does not pose a fundamental problem. On the other hand, the $r = 0$ singularity does, since one can check that E_4 is singular at the origin.

Should we regard this fact as something physically relevant? Indeed, the Schwarzschild solution is valid only in a vacuum, and one should not expect the point-like masses to exist in reality. Moreover, the spherically symmetric solution inside matter does not have a singularity [202]. However, the situation with the $r = 0$ singularity is still not safe. The fundamental significance of the $r = 0$ singularity becomes clear when we consider the phenomena of a gravitational collapse (see, e.g., [202,147,148]). Qualitatively, the situation looks as follows. When a sufficiently massive star starts to lose its energy by emitting radiation and becomes cooler, its internal pressure decreases and its size gets smaller while the strength of the gravitational force on its surface increases. This process can proceed until the star becomes very small, becoming a white dwarf or a neutron star, depending on the initial conditions. After that, if the size of the star is large enough, the gravitational force on the surface can be greater than the limit, set by the nuclear forces (Chandrasekhar for a white dwarf or Tolman-Oppenheimer-Volkoff for a neutron star). In this case, the collapse of the star will continue and, eventually, its radius becomes smaller than its gravitational radius $r_g = 2GM$. Then, the external observer will not see the star anymore, but only the horizon, and thus the star converts into a black hole.

Then the process of collapse will continue inside the black hole horizon [147,148]). After all, if GR is valid at absolutely all length scales, we arrive at the situation when the $r = 0$ singularity becomes a real thing, with an infinitely high-density energy of matter as well as divergent curvature invariants (11.75). It is important to note that the situation is qualitatively the same for the more realistic models of the Kerr (rotating) black hole and the rotating and electrically charged Kerr-Newman solution [147,148]. Physical intuition tells us that the existence of a singularity is not an acceptable theoretical situation and that something should be modified.

11.7.2 The standard cosmological model

Let us consider another important solution to GR. According to the cosmological principle, the universe at the large scale is described by a homogeneous and isotropic metric. One can show that the most general form of such a metric is

$$ds^2 = dt^2 - a^2(t)\Big(\frac{dr^2}{1-kr^2} + r^2 d\Omega\Big)\,, \tag{11.77}$$

where $a(t)$ is the unique unknown function, and $k = (0, 1, -1)$ is a constant which is introduced to distinguish spatially flat, closed and open models. The curvature ${}^{(3)}\mathcal{K}$ of the space section of the four-dimensional spacetime manifold M_{1+3} is

$${}^{(3)}\mathcal{K} = \frac{k}{a^2}\,. \tag{11.78}$$

In the case $k = 0$, the space section is flat, in the case $k = 1$, it is a three-dimensional sphere and, for $k = -1$, the space section is a 3-dimensional hypersphere.

For the homogeneous and isotropic metric (11.77), Einstein equations boil down to the Friedmann - Lemaître equations (Friedmann equations in the $\Lambda = 0$ case),

$$H^2 + \frac{k}{a^2} = \frac{8\pi G}{3}\rho + \frac{\Lambda}{3}, \tag{11.79}$$

$$2\dot{H} + 3H^2 + \frac{k}{a^2} = -\,8\pi G p + \Lambda, \tag{11.80}$$

where $H = \frac{\dot{a}}{a}$ is the Hubble parameter. Along with (11.79) and (11.80), it is necessary to consider the conservation law for matter,

$$\dot{\rho} + 3H(\rho + p) = 0. \tag{11.81}$$

In realistic situations, ρ and p are the sums of the contributions of different fluids that represent the matter contents of the universe. However, for the sake of simplicity, we consider only three particular versions, all with $k = 0$.

A cosmological constant-dominated epoch is when we have only a positive cosmological constant, without matter or radiation. Within the standard ΛCDM cosmological model[1], this situation corresponds to the far future of our universe. As we

[1]The abbreviation ΛCDM means the standard cosmological model, which assumes the presence of the cosmological constant (i.e., Λ) term, baryonic matter, with the almost pressureless equation of state, and one more component called dark matter (DM). In the case of "cold" dark matter the equation of state is approximately pressureless.

mentioned in section 11.6, for the cosmological constant term, the "pressure" and "energy density" satisfy the relations

$$\rho_\Lambda = \frac{\Lambda}{8\pi G}, \qquad p_\Lambda = -\rho_\Lambda. \tag{11.82}$$

When used in (11.81), the last relation immediately shows that ρ_Λ does not depend on the conformal factor $a(t)$. Then, from Eq. (11.79), follows that the conformal factor behaves according to

$$a(t) = a_0\, e^{H_0 t}, \qquad \text{where} \qquad H_0 = \sqrt{\frac{\Lambda}{3}}. \tag{11.83}$$

Matter-dominated epoch is characterized by zero pressure ($p = 0$) and a zero cosmological constant term, $\Lambda = 0$. From Eq. (11.81) follows

$$\frac{d\rho}{\rho} = -\frac{3da}{a}, \qquad \text{which has the solution} \qquad \rho a^3 = \rho_0 a_0^3 = const. \tag{11.84}$$

Starting from (11.84), Eq. (11.79) gives

$$a^3 \dot{a}^2 = \frac{8\pi G}{3}\rho_0 a_0^3, \tag{11.85}$$

implying the time dependence

$$a(t) \sim t^{2/3} \quad \text{and} \quad H(t) \sim t^{-1}. \tag{11.86}$$

Radiation-dominated epoch is characterized by the dominating radiation with the relativistic relation between energy density and pressure $p = \rho/3$, such that $T^\mu_\mu = 0$. It is thought that this epoch is a good approximation for the "early" Universe. The equation of state (11.81) gives, in this case,

$$\frac{d\rho_R}{\rho_R} = -\frac{4da}{a}; \qquad \text{hence,} \qquad \rho_R\, a^4 = \rho_0\, a_0^4 = const. \tag{11.87}$$

It is an easy exercise to solve Eq. (11.79) to arrive at

$$a(t) \sim t^{1/2} \quad \text{and} \quad H(t) \sim t^{-1}. \tag{11.88}$$

The three solutions we have considered above are very simple, but they lead to an important general conclusion which holds in all known generalizations concerning matter contents of the universe and can not be changed even in the framework of anisotropic and nonhomogeneous models of the universe.

The observational data show that the energy balance of the present-day Universe is dominated by a cosmological constant at about 70%, with pressureless matter at about 30% and radiation at about 0.01%. However, looking backward in time, when the universe was many orders of magnitude smaller than now, the energy balance was quite different. For instance, for $a/a_0 \sim 100$, we get an energy balance in which the cosmological constant term is irrelevant compared to the dominating amount of pressureless matter. As we know from the solution (11.86), the universe was expanding at that time. At the epoch when $a/a_0 \sim 10^6$, the radiation was dominating over dust and the cosmological constant, and the solution was approximately (11.88). Therefore, in the framework of GR the solution (11.88) can be regarded a universal description of the primordial universe. And this solution is singular at the point $t = 0$. This singularity is qualitatively similar to the one we observed for a black hole at the point $r = 0$, at the final stage of gravitational collapse. The curvature invariants are singular, while the energy of radiation is divergent. Once again, we met an unacceptable situation when the physically relevant solution has a singularity, and hence it is natural to think about modifying GR, at least in the vicinity of singular spacetime points.

Let us note that the cosmological solution in the primordial universe can be very different from (11.88) in theories that are modified compared to GR. There are many observational indications that, before the radiation-dominated period, there was even much faster expansion of the universe, called inflation. The simplest way to achieve inflation is to assume that, in the very early universe, there was an induced cosmological constant, which later on decayed into matter and radiation. The most successful, phenomenologically, model of inflation is based on a modified action of gravity that includes the R^2-term with the coefficient at about 5×10^8. This model is called the Starobinsky model, after A. Starobinsky, who proposed it in 1980 [305] (see also its further development in [306]). In the next chapters, we shall discuss the status of R^2 term in the framework of quantum field theory. Let us mention the attempts to construct the inflationary model in a theory based on quantum corrections to GR [305, 327, 123, 178, 247, 230] (see also references therein).

11.8 The applicability of GR and Planck units

The significance of singularities is not negligible, because they emerge in the most important and simple solutions, which correspond to the main areas of application of GR. The most natural explanation of the problem of singularities is that GR is not valid at all scales. One can suppose that, at very short distances and/or when the curvature becomes very large, gravitational phenomena must be described by some other theory that is extended compared to GR.

When thinking about the possible extensions, it is good to remember the great success of GR in the explanation of a variety of gravitational phenomena (see, e.g., [348]). Therefore, a modified theory should coincide with GR at the large-distance limit. And then, the most likely origin of the modifications are quantum effects.

Usually, the expected scale of the effects of quantum gravity is associated with the Planck units of length, time and mass. The idea of the Planck units is the following. In the Gaussian system of units, there are three fundamental constants, namely the speed of light, the Planck constant and the Newton constant,

$$c = 3 \cdot 10^{10}\, cm, \qquad \hbar = 1.054 \cdot 10^{-27}\, erg \cdot sec, \qquad G = 6.67 \cdot 10^{-8}\, \frac{cm^3}{sec^2\, g}. \tag{11.89}$$

It turns out that one can use these in a unique way to construct quantities with dimensions of length l_P, time t_P and mass M_P. The result has the form

$$\begin{aligned} l_P &= G^{1/2}\,\hbar^{1/2}\,c^{-3/2} \approx 1.4\cdot 10^{-33}\,cm; \\ t_P &= G^{1/2}\,\hbar^{1/2}\,c^{-5/2} \approx 0.7\cdot 10^{-43}\,sec; \\ M_P &= G^{-1/2}\,\hbar^{1/2}\,c^{1/2} \approx 0.2\cdot 10^{-5}\,g \approx 10^{19}\,GeV\,. \end{aligned} \tag{11.90}$$

One can use the fundamental Planck units (11.90) in different ways. As we know from Part I, there are natural units with $c = \hbar = 1$, with which one can measure everything in powers of GeV. In these units, the Newton constant is $G = 1/M_P^2$, while the Planck time and length intervals are $t_P = l_P = 1/M_P$. It is easy to see that M_P is huge compared to the energy scale of the Standard Model, with the scale defined by the Fermi mass $M_F \approx 300\,GeV$. Indeed, M_P is huge, even compared to the grand unification theories (GUT) scale M_X, which varies between $10^{14}\,GeV$ and $10^{16}\,GeV$. The maximal energies available at present are those of ultra high-energy cosmic rays, about $10^{11}\,GeV$, which is about seven orders of magnitude greater than the maximal energy from the Large Hadron Collider and still huge eight orders below M_P. Taking into account these arguments, it is clear we can meet the energies of the Planck order only when close to the gravitational singularities, which we described above for the simplest examples. Thus, it is natural to assume that the Planck scale of energies will be characterized by a "new physics" that will cure the singularities. At the same time, one cannot rule out the possibility that the theory that may be critically important at the Planck scale may produce some remnants at much lower energies, e.g., at the scale typical for inflation, which is a few orders below the GUT energy scale M_X.

One may suppose that the existence of the fundamental units (11.90) implies a kind of fundamental physics. The usual logic of dimensional analysis tells that, since the construction of the units (11.90) involves, simultaneously, c, $\hbar$ and G, this should be a scale which corresponds to an unknown combination of relativistic, quantum and gravitational effects.

Unfortunately, as is usually the case with dimensional analysis, this interpretation is ambiguous. The description of "relativistic quantum gravitational effects" is too general, and, in fact, one can imagine different things that can fit this description. Indeed, there is a variety of different approaches to "quantum gravity" and those approaches do not have too much in common. In our opinion, all these approaches can be classified into three big groups, depending on the object of quantization:

1. In the semiclassical gravity approach we have to study the quantum effects of matter fields on the classical (by definition) gravitational background. Let us stress that this background is not "given" or "fixed", - since the equations of motion that define its dynamics should account for the contributions of quantum matter fields to the gravitational action.

In some sense, this is the most important approach, since it is safe. In this case, the object of the study certainly exists. There are many experimental confirmations of the validity of the quantum theory of matter fields. So, we can claim that we really know that the matter fields must be quantized. Also, we know that the concept of curved space is confirmed by a great amount of experimental and observational data [348]. Therefore, the question of whether the quantization of matter fields change the form of Einstein equations (11.63) is certainly an absolutely legitimate inquiry that one can pose. On top of that, quantum field theory in curved space is an extremely useful approach, owing to its relative simplicity and clarity.

2. The quantization of the metric itself looks to be an interesting and in some sense natural possibility. It is clear that the most fundamental option would be to quantize matter and gravity at the same time.

3. In the framework of an effective approach to quantum field theory one can think about "quantizing something else". This means we should expect that the classical laws of gravity, and classical and quantum description for the matter fields should be a low-energy manifestations of a more fundamental physics. Needless to say, the palace of knowledge called string theory belongs to this group of approaches for the description of "relativistic quantum gravitational effects."

The general situation in the fundamental theory of quantum gravity strongly depends on the choice between the three options listed above. In what follows, we start by discussing the quantum field theory of matter on the classical gravitational background. We shall see that, within this approach, basic problems such as renormalization can be successfully dealt with. One can say that there is a robust theoretical background for the applications of quantum field theory results, but there many unsolved problems and difficulties in many physically relevant calculations.

In contrast, the quantization of the gravitational field itself in the framework of the known methods of perturbative quantum field theory faces serious difficulties. Owing to the dimension of the gravitational constant, the perturbative version of GR is non-renormalizable. There were many attempts to generalize GR in such a way that the new theory would become renormalizable; some of these approaches look promising, and we shall describe them in the next chapters. At the same time, there is no model of quantum gravity which could be called completely consistent and free of problems at the formal level.

An alternative possibility is to assume that gravity is different from other forces and that it should not be quantized at all. There are certain shortcomings in this point of view, and the main issue is that it leads to some quantum mechanical inconsistencies at the Planck scale [109]. However, it might happen that, in the vicinity of the singularities, the quantum effects of matter fields modify the action of gravity in such a way that the singularities completely disappear. Then the quantum gravity effects, even if they exist, would be impossible to observe and measure, and we come to the fundamental importance of semiclassical gravity, which is supposed to tell us whether this scenario is true or not.

In the next chapters we will describe free and interacting quantum fields on a curved background, discuss the renormalization of quantum field theory in the presence of an external metric and consider related problems and applications. After that, we shall summarize some of the available results on perturbative quantum gravity.

Exercises

11.14. Construct the generalization of the Christoffel symbol (11.9) for the theory with non-zero torsion $T^{\lambda}{}_{\alpha\beta} = \Gamma^{\lambda}{}_{\alpha\beta} - \Gamma^{\lambda}{}_{\beta\alpha}$. Use the result to show that the covariant derivatives with torsion do not commute when acting on a scalar field. Prove that $T^\lambda{}_{\alpha\beta}$ is a tensor.

11.15. Prove that the Christoffel symbol can be made zero in a given point P of a Riemann space, by changing the coordinates. Why can this not be done in the vicinity of the point P? Why can this not be done for the affine connection in the presence of a non-zero torsion?

11.16. Consider the two metrics in the *r.h.s.* of Eqs. (11.25) and (11.48), with the flat fiducial (or background) metric $g_{\mu\nu} = \eta_{\mu\nu}$. Verify the identities (11.16)-(11.19), (11.21) and (11.22) in both cases.

Observation. In case of (11.48), one has to obtain the expansion for the Riemann and Ricci tensors first. It is better to make it for an arbitrary background metric $g_{\mu\nu}$.

11.17. Using the results of the previous exercise, derive the equations of motion for the actions

$$I_1 = \int d^4x\sqrt{-g}\, R^2_{\mu\nu\alpha\beta}\,, \quad I_2 = \int d^4x\sqrt{-g}\, R^2_{\mu\nu}\,, \quad I_3 = \int d^4x\sqrt{-g}\, R^2. \tag{11.91}$$

The variational derivatives which will be obtained in this way are

$$\frac{1}{\sqrt{-g}}\frac{\delta I_1(4)}{\delta g_{\mu\nu}} = \frac{1}{2}g^{\mu\nu}R^2_{\rho\sigma\alpha\beta} - 2R^{\mu\sigma\alpha\beta}R^{\nu}_{\sigma\alpha\beta} - 4R^{\mu\alpha\nu\beta}R_{\alpha\beta} + 4R^{\mu}_{\alpha}R^{\nu\alpha} \tag{11.92}$$

$$+ 2\nabla^\mu\nabla^\nu R - 4\Box R^{\mu\nu}, \tag{11.93}$$

$$\frac{1}{\sqrt{-g}}\frac{\delta I_2(4)}{\delta g_{\mu\nu}} = \frac{1}{2}g^{\mu\nu}R^2_{\rho\sigma} - 2R^{\mu\alpha\nu\beta}R_{\alpha\beta} + \nabla^\mu\nabla^\nu R - \frac{1}{2}g^{\mu\nu}\Box R - \Box R^{\mu\nu}, \tag{11.94}$$

$$\frac{1}{\sqrt{-g}}\frac{\delta I_3(4)}{\delta g_{\mu\nu}} = \frac{1}{2}g^{\mu\nu}R^2 + 2\nabla^\mu\nabla^\nu R - 2g^{\mu\nu}\Box R - 2RR^{\mu\nu}. \tag{11.95}$$

11.18. Derive the traces of the equations of motion for the previous exercise. Explain why the traces of the two combinations

$$\int d^4x\sqrt{-g}E_4 = I_1 - 4I_2 + I_3\,, \qquad \int d^4x\sqrt{-g}C^2 = I_1 - 2I_2 + \frac{1}{3}\,I_3\,. \tag{11.96}$$

do vanish. How can one obtain these two vanishing traces without explicit calculations? See also the next exercises for a better understanding of this issue.

11.19. Verify that the integrand of the second expression in (11.96) is nothing else but the square of the Weyl tensor (11.24) for $n = 4$. Calculate the expression for the square of the Weyl tensor in an arbitrary dimension n.

11.20. Consider a metric-dependent functional $A[g_{\mu\nu}]$. Use the conformal parametrization (11.25) to prove the identity

$$\frac{\delta A}{\delta\sigma(x)} = \int d^4y\,\frac{\delta g_{\mu\nu}(y)}{\delta\sigma(x)}\,\frac{\delta A}{\delta g_{\mu\nu}(y)} = 2\,g_{\mu\nu}\,\frac{\delta A}{\delta g_{\mu\nu}}\,.$$

Verify this identity for the cosmological constant and Einstein-Hilbert terms. The solution to this important exercise can be found in section 17.2.

11.21. The de Sitter (dS) metric corresponds to the maximally symmetric space, with Riemann and Ricci tensors that can be presented as (in n-dimensional space)

$$R_{\mu\nu\alpha\beta} = \frac{1}{n(n-1)}\Lambda(g_{\mu\alpha}g_{\nu\beta} - g_{\mu\beta}g_{\nu\alpha}), \quad R_{\mu\nu} = \frac{1}{n}\Lambda g_{\mu\nu}, \quad \Lambda = const. \tag{11.97}$$

Prove the relation which holds for the Gauss-Bonnet term [128]

$$\frac{1}{\sqrt{-g}}\frac{\delta I_{GB}(n)}{\delta g_{\mu\nu}}\bigg|_{dS} = \frac{(n-2)(n-3)(n-4)}{2n^2(n-1)}\Lambda^2 g^{\mu\nu}\,, \tag{11.98}$$

where $I_{GB}(n) = \int d^n x\sqrt{-g}E_4 = I_1 - 4I_2 + I_3$ and the integrals in (11.91) are defined as n-dimensional.

11.22. Derive Eq. (11.90) from Eq. (11.89) and prove that the solution is unique. Evaluate the masses of all known elementary particles in the units of M_P.

11.23. Prove that $I_{GB}(4)$ is the integral of a total covariant derivative, $E_4 = \nabla_\mu \chi^\mu$.

Hint. One has to express E_4 as the contraction of two Riemann tensors with two Levi-Civita tensors and use special coordinates with the locally flat metric and vanishing Christoffel symbol. More hints can be found in [292].

12
Classical fields in curved spacetime

In this chapter, we formulate spin-0, spin-1/2 and spin-1 classical field models in an arbitrary curved spacetime. These types of fields are most relevant for particle physics, including the Standard Model and Grand Unified Theories.

As we shall see in brief, for each of the cases, there are many ways to write down the actions in curved space, generalizing the flat space models. The conventional approach to reducing this ambiguity is to use as a guide the arguments of simplicity and also remember that our final purpose is a quantum theory. The last point implies that the theory of classical fields in curved space should be based on Lagrangians that coincide with the flat spacetime counterparts in the limit of flat metric and, on the other hand, lead to a consistent (e.g., renormalizable and unitary) quantum theory. Then the requirement for maximal possible simplicity tells us not to include terms that are not necessary for the purpose formulated above. Starting from these basic requirements, it is possible to formulate a set of general principles that enable us to arrive at curved space actions that have just a few new parameters, compared to the flat space analogs. In the next sections, we discuss these general principles and then turn to the practical construction of the models.

12.1 General considerations

As far as we intend to formulate classical fields in curved spacetime, for further use in constructing quantum theory, it makes sense to follow the standard requirements for the field models in quantum field theory, that we discussed in the first part of the book. On top of this, we have to comply with the principles of GR. Thus, to define the actions of matter fields in an external gravitational field, we impose the principles of locality and general covariance. Furthermore, in order to preserve the fundamental features of the original theory in flat spacetime, one has to require the most important symmetries of a given theory (e.g., gauge invariance) in flat spacetime to be preserved in the theory in curved spacetime.

Let us remember that the renormalizable flat space theory cannot have the parameters of the inverse mass dimensions. Thus, it looks reasonable to forbid such parameters in curved spacetime too. Later on, we will see that this approach is perfectly well justified, but, for a while, we introduce it as an ad hoc definition.

The principles listed above lead to the so-called nonminimal actions of matter fields. At the same time, one can use a different approach. Since the fundamental feature of a field model in curved spacetime is general covariance, the simplest way to construct such a model is via a procedure of direct covariantization of the flat space theory. This

Introduction to Quantum Field Theory with Applications to Quantum Gravity. Iosif L. Buchbinder and Ilya L. Shapiro, Oxford University Press (2021). © Iosif L. Buchbinder and Ilya Shapiro.
DOI: 10.1093/oso/9780198838319.003.0012

procedure is called the minimal inclusion of interaction with gravity and consists of the following steps:

1) Minkowski metric $\eta_{\mu\nu}$ is replaced by $g_{\mu\nu}$.

2) The fields which are tensors and spinors under global Lorentz transformations are replaced by the tensors and spinors related to general coordinate transformations and/or local Lorentz transformations.

3) Partial derivatives ∂_μ are replaced by covariant ones ∇_μ.

4) The volume element d^4x is traded for the covariant expression $d^4x\sqrt{-g}$.

Let us note that the described minimal scheme was applied in the previous chapter to the construction of the action for the free particle. In what follows we will see how the minimal procedure works for different fields, and in which cases it is also possible to construct a nonminimal generalization.

12.2 Scalar fields

The minimal version of the action for the real scalar field in curved background is

$$S_0 = \int d^4x\,\sqrt{-g}\Big\{\,\frac{1}{2}\,g^{\mu\nu}\,\partial_\mu\varphi\,\partial_\nu\varphi - U_{min}(\varphi)\,\Big\}, \tag{12.1}$$

where φ is a scalar field under the general coordinate transformations and

$$U_{min}(\varphi) = \frac{m^2}{2}\varphi^2 + V(\varphi) \tag{12.2}$$

is a minimal potential term. The unique possible nonminimal addition, depending on the scalar field, is

$$S_{non-min} = \frac{1}{2}\int d^4x\,\sqrt{-g}\,\xi\,\varphi^2\,R\,. \tag{12.3}$$

The action (12.3) involves a new dimensionless quantity of ξ which is called the *nonminimal parameter* or *nonminimal coupling* of the scalar field to gravity. Let us stress that this is a coupling with an external gravitational field. For this reason, the theory with $\xi \neq 0$ and without the interaction term $V(\varphi)$ can be regarded as a free theory of a scalar in curved spacetime.

It is easy to check that all other covariant terms are either nonlocal or require coupling parameters with an inverse mass dimension. For instance, this is the case for the term $\eta\int\sqrt{-g}R^{\mu\nu}\partial_\mu\varphi\partial_\nu\varphi$, where $[\eta] = \text{mass}^{-2}$. These terms are not necessary for the construction of the consistent quantum theory, and that is why we do not include them in what follows. Therefore, (12.3) is the unique nonminimal generalization of (12.1) that is compatible with the principles described in the previous section.

In what follows, we will sometimes use the condensed notations

$$\int = \int_x = \int d^4x\sqrt{-g}. \tag{12.4}$$

In the case of a multi-component scalar field theory $S[\varphi^i]$, the unique form of a nonminimal term is $\int \xi_{ij}\varphi^i\varphi^j\,R$. In particular, for the complex scalar, it is $\int \xi\varphi^*\varphi R$. Until this is not relevant, we shall write all formulas for the real single scalar case, just for the sake of simplicity. Since the nonminimal term does not contain derivatives of the scalar field, it can be most naturally included in the potential part, and thus we arrive at the new definition of the classical potential,

$$U(\varphi) = U_{min}(\varphi) - \frac{\xi}{2}\varphi^2 R = \frac{m^2}{2}\varphi^2 - \frac{\xi}{2}\varphi^2 R + \frac{f}{4!}\varphi^4. \tag{12.5}$$

In the last equality, we assume that the minimal potential corresponds to the standard quartic self-interaction term.

In addition to the nonminimal term (12.3), our principles admit several new terms that do not depend on the scalar field or fields. The corresponding actions are constructed only from the metric. These terms are, indeed, nonminimal, but they are conventionally called "vacuum action," and their general form is

$$S_{vac} = S_{EH} + S_{HD}, \tag{12.6}$$

where S_{EH} is the Einstein-Hilbert action with the cosmological constant term

$$S_{EH} = -\frac{1}{16\pi G}\int d^4x\sqrt{-g}\,(R + 2\Lambda) \tag{12.7}$$

and S_{HD} includes higher derivative terms. Owing to the restrictions coming from locality and dimension, these terms are as follows:

$$S_{HD} = \int d^4x\sqrt{-g}\,\{\alpha_1 R^2_{\mu\nu\alpha\beta} + \alpha_2 R^2_{\alpha\beta} + \alpha_3 R^2 + \alpha_4 \Box R\}, \tag{12.8}$$

where the parameters $\alpha_{1,2,3,4}$ are dimensionless. The most useful, albeit completely equivalent, representation is

$$S_{HD} = \int d^4x\sqrt{-g}\,\{a_1 C^2 + a_2 E_4 + a_3 \Box R + a_4 R^2\}\,. \tag{12.9}$$

Here $C^2 = C^2(4)$ is the square of the Weyl tensor in four spacetime dimensions. It proves useful to note the corresponding expression for an arbitrary dimension n,

$$C^2(n) = C_{\alpha\beta\mu\nu}C^{\alpha\beta\mu\nu} = R^2_{\mu\nu\alpha\beta} - \frac{4}{n-2}R^2_{\mu\nu} + \frac{2}{(n-1)(n-2)}R^2. \tag{12.10}$$

Another important combination of the curvatures E_4 is the Gauss-Bonnet topological (only in $n = 4$) invariant (11.96). The basis E_4, C^2, R^2 is, in many respects, more useful than the one that consists of $R^2_{\mu\nu\alpha\beta}$, $R^2_{\alpha\beta}$ and R^2. Thus, we shall be mainly using (12.9) in what follows.

In $n = 4$, the action (12.9) includes one conformal invariant, $\int C^2$, and two surface terms ($\int E_4$ and $\int \Box R$) that do not contribute to the classical equations of motion for the metric. However, as we shall see later on, both these terms become important at the quantum level, because they contribute to the dynamical equations for the metric through the conformal anomaly.

12.2.1 Conformal symmetry of the metric-scalar model

As we shall see later on, conformal symmetry and its violation by the quantum conformal (or trace) anomaly is an important subject. For this reason, we shall consider in detail conformal symmetry in the scalar field case. Let us consider only the four-dimensional example, leaving the generalization to n spacetime dimensions for an exercises.

The first observation is that the massless version of the action (12.1), with the potential $\sim \varphi^4$ and the special value of the nonminimal parameter $\xi = 1/6$,

$$S_{cs}(\varphi, g_{\mu\nu}) = \int d^4x \sqrt{-g}\,\Big\{\frac{1}{2}\, g^{\mu\nu}\, \partial_\mu\varphi\, \partial_\nu\varphi + \frac{1}{12}\, R\varphi^2 - \frac{f}{4!}\varphi^4\Big\}, \tag{12.11}$$

possesses symmetry under local conformal transformation of the metric (11.25) and of the scalar field,

$$g_{\mu\nu} = \bar{g}_{\mu\nu}\, e^{2\sigma}, \qquad \varphi = \bar{\varphi}\, e^{-\sigma}, \qquad \sigma = \sigma(x), \tag{12.12}$$

such that $S_{cs}(\varphi, g_{\mu\nu}) = S_{cs}(\bar{\varphi}, \bar{g}_{\mu\nu})$. The proof of this identity can be easily done by using the relations (11.31) and (11.26) in the $n = 4$ case. The generalization for an arbitrary dimension is an easy exercise too.

The Noether identity corresponding to the symmetry under the transformation (12.12) has the form

$$-\frac{2}{\sqrt{-g}}\, g_{\mu\nu}\, \frac{\delta S_m}{\delta g_{\mu\nu}} + \frac{1}{\sqrt{-g}}\, \varphi\, \frac{\delta S_m}{\delta\varphi} = 0. \tag{12.13}$$

The reader will easily check this fact after solving the first two exercises of this section.

It is worth noting that the theory (12.11) and the transformation (12.12) admit an interesting generalization [280]. Consider the sigma model

$$S = \int d^4x\sqrt{-g}\,\{A(\phi)g^{\mu\nu}\partial_\mu\phi\partial_\nu\phi + B(\phi)R + C(\phi)\}, \tag{12.14}$$

where $A(\phi)$, $B(\phi)$ and $C(\phi)$ are some functions. It is useful to start the analysis from the simple particular case of the general action (12.14), called the O'Hanlon action [234],

$$S = \int d^4x\sqrt{-g'}\,\{R'\Phi + V(\Phi)\}. \tag{12.15}$$

Here the curvature R' corresponds to the metric $g'_{\mu\nu}$ and $g' = \det\,(g'_{\mu\nu})$. Let us transform this action to the new variables $g_{\mu\nu}$ and ϕ, according to

$$g'_{\mu\nu} = g_{\mu\nu}e^{2\sigma(\phi)}, \qquad \Phi = \Phi(\phi), \tag{12.16}$$

where $\sigma(\phi)$ and $\Phi(\phi)$ are arbitrary functions of ϕ. After using (11.31) and (11.26), disregarding the total derivative terms, the action becomes

$$S = \int d^4x\sqrt{-g}\{\Phi(\phi)Re^{2\sigma(\phi)} + 6(\nabla\phi)^2e^{2\sigma(\phi)}[\Phi\sigma' + \Phi']\sigma' + V(\Phi(\phi))e^{4\sigma(\phi)}\}, \tag{12.17}$$

where $(\nabla\phi)^2 = g^{\alpha\beta}\partial_\alpha\phi\partial_\beta\phi$ and the prime indicates the derivative with respect to ϕ. Indeed, when transformed, the O'Hanlon action (12.15) becomes the sigma model type action (12.14), with the functions

$$A(\phi) = 6e^{2\sigma(\phi)}[\Phi\sigma' + \Phi']\sigma_1, \qquad B(\phi) = \Phi(\phi)e^{2\sigma}. \tag{12.18}$$

One can find such $\sigma(\phi)$ and $\Phi(\phi)$, which correspond to the given $A(\phi)$ and $B(\phi)$, by requiring

$$A = 6B'\sigma' - 6B(\sigma')^2, \qquad \Phi = Be^{-2\sigma}. \tag{12.19}$$

Placing (12.19) into (12.17), the action takes the form of (12.14) with

$$C(\phi) = \left(\frac{B}{\Phi}\right)^2 V(\Phi(\phi)), \tag{12.20}$$

where the last term is nothing but $C(\phi)$ from (12.14).

It is easy to see that the expression (12.17) enables us to make a classification of the models (12.14) into three classes, two of them being equivalent to the two distinct types of O'Hanlon actions.

i) If we have $\Phi = const$ in Eq. (12.15), this action is nothing but the Einstein-Hilbert action with a cosmological constant. After mapping to the general model (12.14), this case produces the relation between the functions $A(\phi)$, $B(\phi)$ and $C(\phi)$,

$$A = \frac{3B'^2}{2B} \quad \text{and} \quad C = \lambda B^2, \quad \text{where} \quad \lambda = const. \tag{12.21}$$

It is easy to see that the conformal scalar model (12.11) satisfies this condition. Thus, we discovered that both Einstein's gravity with the cosmological constant and the conformal scalar model are particular cases of the models of the first class. An extra scalar degree of freedom in the action (12.15) is compensated by an additional conformal symmetry (12.12). In the general version of (12.14) satisfying (12.21), the symmetry consists of the conformal transformation of the metric plus a special reparametrization of the scalar field.

An important detail is that the positiveness of the sign of the scalar mode in (12.11) requires the negative sign of the R-term in the model (12.15) with constant Φ. This does not mean that either one of the two theories is ill defined because in GR the scalar mode is *not* a physical degree of freedom. Anyway, the equivalence of the two theories should be utilized with proper caution.

ii) The theory (12.15) with a non-constant Φ is equivalent to (12.14) with the functions $A(\phi)$ and $B(\phi)$ that do not satisfy the first of the conditions (12.21). Such a theory has one more degree of freedom compared to GR and can be mapped to a sum of the action of GR and the action of a minimal scalar field.

iii) The model (12.14), with the functions $A(\phi)$ and $B(\phi)$ that satisfy the first conditions of (12.21) and the function $C(\phi)$ that does not satisfy the second condition, is somehow similar to the Proca model of a massive vector field, but there is a difference. Consider the simplest example of this sort, with the action

$$S_{ex} = \int d^4x\sqrt{-g}\Big\{\frac{1}{2}\,g^{\mu\nu}\,\partial_\mu\phi\partial_\nu\phi + \frac{1}{12}\,R\phi^2 - \Lambda\Big\}, \tag{12.22}$$

where $\Lambda \neq 0$. In this case, the conformal Noether identity has form, showing the violation of the local conformal symmetry by the cosmological constant term:

$$-\frac{1}{\sqrt{-g}}\,\phi\,\frac{\delta S_{ex}}{\delta\phi} + \frac{2}{\sqrt{-g}}\,g_{\mu\nu}\,\frac{\delta S_{ex}}{\delta g_{\mu\nu}} = 4\Lambda. \tag{12.23}$$

Since the first two terms in (12.22) possess local conformal symmetry, this identity boils down to $\Lambda = 0$. This example shows that the dynamical equations of motion for ϕ and $g_{\mu\nu}$ in the theory of the third type have no consistent solutions.

Exercises

12.1. Derive the dynamical equation of motion for the nonminimal scalar field. Use direct variation of the scalar field in the action and also Lagrange equations, and show that the results are the same.

12.2. Derive the dynamical energy-momentum tensor for a general massive scalar field with nonminimal term, according to

$$T^{\mu\nu} = -\frac{2}{\sqrt{-g}}\,\frac{\delta S_m}{\delta g_{\mu\nu}}. \tag{12.24}$$

The calculation should be done for an arbitrary spacetime dimension n.

12.3. For the same situation as in **12.1**, derive the canonical energy-momentum tensor, as was defined in Part I of the book. Show that the results of canonical and dynamical procedures coincide.

Observation. In fact, the two definitions give the same result in all known examples; regardless, the complete general proof of this is not known.

12.4. Use the expression derived in the previous two exercises and also the result of the first exercise to prove that the trace $T^\mu{}_\mu = g_{\mu\nu}T^{\mu\nu}$ vanish *on-shell*, if and only if the mass of the scalar field is zero and the nonminimal parameter is

$$\xi_n = \frac{n-2}{4(n-1)}. \tag{12.25}$$

12.5. Verify that, under the same conditions of $m = 0$ and $\xi = \xi_n$, the action of the scalar field is invariant under the simultaneous conformal transformation

$$g_{\mu\nu} = \bar{g}_{\mu\nu}\,\exp\{2\sigma\}, \quad \varphi = \bar{\varphi}\,\exp\Big\{\frac{2-n}{2}\sigma\Big\}, \quad \sigma = \sigma(x). \tag{12.26}$$

12.6. Derive the Noether identity for the symmetry (12.26) and show how this explains the vanishing trace $T^\mu{}_\mu$ *on shell* in this theory.

12.7. Find the linear relations between the dimensionless parameters of the generalizations of the fourth-derivative actions (12.8) and (12.9) for an arbitrary dimension n in Eq. (12.10) and for the particular case $n = 4$.

12.8. Consider in details the models of the type ii), based on the action (12.14) with the functions $A(\phi)$ and $B(\phi)$ not satisfying (12.21). Find the transformation of the corresponding O'Hanlon action and the sum of the Einstein-Hilbert action and the the minimal scalar field.

12.3 Spontaneous symmetry breaking in curved space and induced gravity

Let us discuss an important aspect of classical field models in curved spacetime, involving scalar fields with the nonminimal term (12.3) in the scalar potential.

Let us remember that the quantum field models of our main concern are the Standard Model of particle physics and its extensions and generalizations, such as GUTs. One of the most important elements of these theories is the scalar Higgs field, which generates mass for elementary particles, including vector bosons, due to the spontaneous symmetry breaking (see, e.g., [345]). In this section, we will discuss some remarkable consequences of spontaneous symmetry breaking in the classical scalar field theory nonminimally coupled to an external gravitational field [168].

Let us begin with the basic notions of spontaneous symmetry breaking in flat space. Consider a field model containing the scalar fields $\varphi = \varphi^I$ and also spinor and vector fields. The Lorentz-invariant field configuration with the minimal energy is called the classical vacuum. Since Lorentz invariance assumes homogeneity and isotropy of space, the classical vacuum corresponds to constant scalar fields, which can be found by minimizing the scalar potential. Assuming that the Lagrangian of the model under consideration is invariant under the group of internal symmetry, we get the following two possibilities. First, the classical vacuum is invariant under the group mentioned above. This situation means that the symmetry is unbroken. However, if the classical vacuum is not invariant under the symmetry group of the Lagrangian, then the symmetry is spontaneously broken. Our aim is to study some aspects of the this situation in curved spacetime. For the sake of simplicity, consider the complex scalar field theory with the classical potential of the form

$$U(\varphi) = -\mu_0^2\varphi^*\varphi + \lambda(\varphi^*\varphi)^2 - \xi R\varphi^*\varphi. \tag{12.27}$$

For definiteness, one can consider this potential as a part of the Abelian gauge vector A_μ model, with the action of the following form:

$$S = \int d^4x\sqrt{-g}\Big\{-\frac{1}{4}F_{\mu\nu}F^{\mu\nu} + g^{\mu\nu}(\partial_\mu - ieA_\mu)\varphi^*(\partial_\nu + ieA_\mu)\varphi - U(\varphi)\Big\}. \tag{12.28}$$

Scalar field with such a potential can be treated as a part of the $U(1)$ gauge vector model coupled to the complex scalar field in curved spacetime. The generalization for the non-Abelian theory would be straightforward, so there is no reason to consider it here.

Let $v(x)$ be a solution to the equation of motion for the scalar field in curved spacetime, then

$$-\Box v + \mu_0^2 v + \xi R v - 2\lambda v^3 = 0. \tag{12.29}$$

In flat space this equation defines the classical vacuum of the system, corresponding to the minimum of the potential. However, in curved spacetime the equation (12.29) has specific features, which will be discussed below.

If the interaction between the scalar and the metric is minimal, i.e., $\xi = 0$ Eq. (12.29) has a constant solution which coincides with the classical vacuum in flat space,

$$v_0^2 = \frac{\mu_0^2}{2\lambda}. \tag{12.30}$$

In the last expression, we introduced the special notation v_0 for the case of minimal interaction between the scalar and gravity, in order to distinguish it from the solution v of the general equation (12.29). Starting from (12.30), the conventional scheme of spontaneous symmetry breaking does not require serious modification because of the presence of an external gravitational field. The solution (12.30) corresponds to the non-trivial minimum of the potential [as shown in Fig. (12.31)] because of the "wrong" choice of the sign of the mass term in the potential (12.27). Indeed, μ_0 is not a physical mass; the last is defined from the small oscillations of the scalar near the minimum. Typical scalar potential with the spontaneous symmetry breaking in flat spacetime is shown below.

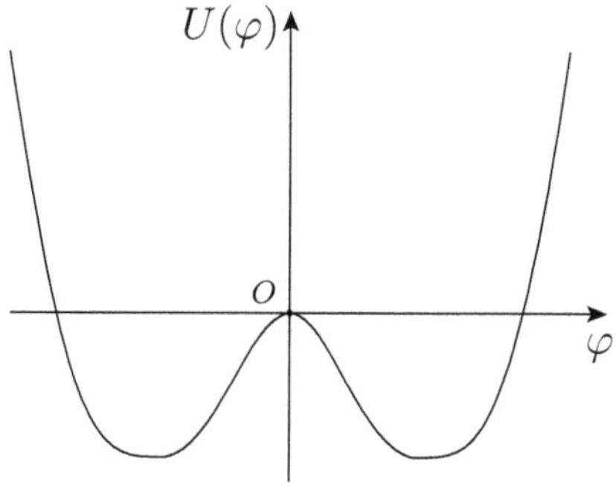

(12.31)

Now we consider how this scheme changes when we formulate the theory in curved spacetime. For the general case of a non-zero ξ and a non-constant scalar curvature R, one gets, instead of Eq. (12.30), a spacetime dependent solution $v(x) \neq const.$ In this situation, it is impossible to ignore the $\Box v$-term and therefore, the solution for the vacuum expectation value can not be obtained in a closed form. In fact, even the use of the term "classical vacuum" in curved space is not obvious. One of the manifestations of this is that the solution $v(x)$, which replaces v_0 in curved space, is not a constant. Thus, strictly speaking we can not use the notion of spontaneous symmetry breaking in the same sense as in flat space.

Thus, treating the solution of Eq. (12.29) as a classical vacuum has a clear physical sense only in flat spacetime. In curved spacetime, we can use it only for theory which is close to the flat limit, meaning that the effects of curvature are assumed to be small. Thus, we consider a slowly varying curvature and start from Eq. (12.30) as the zero-order approximation. At the next stage, we can try to find the solution of Eq. (12.29) in the form of a power series in the curvature tensor or, equivalently, in ξ,

$$v(x) = v_0 + v_1(x) + v_2(x) + \dots , \tag{12.32}$$

where v_0 is defined from (12.30).

In the first-order term $v_1(x)$, one gets the equation

$$-\Box v_1 + \mu^2 v_1 + \xi R v_0 - 6\lambda v_0^2 v_1 = 0. \tag{12.33}$$

The solution of this equation has the form

$$v_1 = \frac{\xi v_0}{\Box - \mu^2 + 6\lambda v_0^2} R = \frac{\xi v_0}{\Box + 4\lambda v_0^2} R, \tag{12.34}$$

where we used the zero-order value (12.30). In a similar way, in the next order of approximation, we find the solution

$$v_2 = \frac{\xi^2 v_0}{\Box + 4\lambda v_0^2} R \frac{1}{\Box + 4\lambda v_0^2} R - \frac{6\lambda \xi^2 v_0^3}{\Box + 4\lambda v_0^2} \Big(\frac{1}{\Box + 4\lambda v_0^2} R \Big)^2. \tag{12.35}$$

As usual, here and below, the parenthesis restrict the action of the differential or inverse differential (such as $\big[\Box + 4\lambda v_0^2\big]^{-1}$) operators. In particular, the operator inside each parenthesis in Eq. (12.35) acts only on the curvature inside the parenthesis. In contrast, the left operator $\big[\Box + 4\lambda v_0^2\big]^{-1}$ in the first term on the *r.h.s.* of (12.35) acts on all of the expression to the right of it. To be more precise, the inverse operator here is understood as the corresponding Green function. One can assume appropriate boundary conditions, which define the unique Green function for the given operator.

One can continue the expansion of v in (12.32) to any desirable order. Some observations are in order. Unlike the usual spontaneous symmetry breakingcase, here the vacuum expectation value of the scalar field is not a constant. Instead, it varies due to the variable curvature. This variation is completely negligible within particle physics owing to the extremely small value of the curvature compared to that of other dimensional quantity such as, e.g., v_0. Hence, the impact of the gravitational interaction on particle physics applications cannot be relevant, at least in the low-energy domain.

There is, however, another aspect of spontaneous symmetry breaking, where the gravitational effect is not so trivial. If one replaces the spontaneous symmetry breaking solution $v(x)$ back into the scalar sector of the action (12.28), we arrive at the following purely gravitational action, which can be called the induced low-energy action of vacuum:

$$S_{ind} = \int d^4x \sqrt{-g} \Big\{ g^{\mu\nu} \partial_\mu v \, \partial_\nu v + (\mu_0^2 + \xi R) v^2 - \lambda v^4 \Big\}, \tag{12.36}$$

where $v(x)$ is a solution to the equation (12.29). In the lowest order of the power expansion (12.32), we get the induced Einstein-Hilbert and cosmological constant terms.

What we have done is one of the simplest possible ways to derive the induced gravity. The induced values of the Newton and cosmological constants are, respectively,

$$\frac{1}{16\pi G_{ind}} = -\xi v_0^2 , \qquad \rho_\Lambda^{ind} = \frac{\Lambda_{ind}}{8\pi G_{ind}} = -\mu_0^2 v_0^2 + \lambda v_0^4 = -\lambda v_0^4, \tag{12.37}$$

in the zero order of the expansion (12.32). It is worthwhile commenting on the different relevance of these induced values. The induced action (12.36) should be summed up with the vacuum action S_{EH} in (12.7). Then the effect of the induced contribution G_{ind}^{-1} is small in the framework of either the Standard Model or GUTs. The value of v_0 is 246 GeV in the Minimal Standard Model or about $10^{16}\,GeV$ in some versions of GUTs. The squares of these values should be compared to the square of the Planck mass, $M_P \approx 10^{19}\,GeV$, to give a safe margin for claiming that the induced Einstein-Hilbert term is not critically important.

In contrast, the value of ρ_Λ, which is observed in cosmology mainly from the acceleration of the universe, is extremely small compared to the magnitude of λv_0^4, e.g., in the Standard Model of particle physics. Hence, there should be an extremely precise cancelation between the vacuum and induced cosmological constants. The precision of this cancelation is the main part of the famous and unsolved cosmological constant problem (see, e.g., [344] for a standard review of this problem). We leave the detailed discussion of this issue for the later chapters and now consider other terms in the induced gravitational action.

Starting from the complete expression (12.36), and taking the space and time dependence of the curvature into account, one arrives at an infinite series of nonlocal expressions due to the non-locality of the non-zero order terms in (12.32). In the second order, we obtain

$$S_{ind} = \int d^4x\sqrt{-g}\Big\{ - v_1\Box v_1 + \mu^2\,(v_0^2 + 2v_0v_1 + 2v_0v_2 + v_1^2) - \lambda\,(v_0^4 + 4v_0^3v_1 + 4v_0^3v_2 + 6v_0^2v_1^2) + \xi R\,(v_0^2 + 2v_0v_1)\Big\} + \mathcal{O}(R^3). \quad (12.38)$$

Now, using Eq. (12.33), after a small algebra, we arrive to an expansion in the powers of ξ for the action of induced gravity. The expression which follows at the first order in ξ is

$$S_{ind} = \int d^4x\sqrt{-g}\Big\{ \lambda v_0^4 + \xi R v_0^2 + \xi^2\, v_0^2\, R\,\frac{1}{\Box + 4\,\lambda v_0^2}\,R + \dots\Big\}. \quad (12.39)$$

The first and second terms here are exactly those we discussed above, but now there is also a qualitatively new nonlocal term. In the next orders of expansion, the induced tree-level gravitational action includes an infinite amount of nonlocal terms, owing to the non-constant vacuum expectation value of the scalar field. Although the coefficients of these terms are small compared to the vacuum Einstein-Hilbert term, the non-localities do not mix with the local terms and, in principle, can lead to some physical effects. For instance, assuming that there is an extremely light scalar (e.g., quintessence), whose mass is of the order of the Hubble parameter and has a potential admitting spontaneous symmetry breaking, the non-localities may become relevant and lead to observable consequences.

Exercises

12.9. Calculate the explicit form of the next nonlocal terms in the expansion (12.39), up to the order ξ^3. Verify that the $R(\Box + 4\lambda v_0^2)^{-2}R$ term cancels and try to explain this fact from qualitative considerations.

12.10. Consider the nonlocal terms, such as the one in (12.39) or the one mentioned in the previous exercise. Consider the limiting case of the vanishing (or extremely small) mass μ_0. Using the global scaling arguments

$$g_{\mu\nu} \longrightarrow g_{\mu\nu}\, e^{2\lambda}\,, \qquad \lambda = const, \tag{12.40}$$

explain why, when used as a basis of cosmological solutions, these two terms can be seen as additions to, or even replacements for the Einstein and cosmological constant terms, respectively. Suggest other nonlocal terms which can assume the same role.
12.11. Consider the large mass μ_0 limit in the terms obtained in **12.9**. Show that, in this limit, these terms become effectively local, and calculate how they contribute to the R^2 and R^3 structures. Discuss in detail the definition of the large mass limit in this framework. To which quantities should we compare μ_0?

12.4 Spinor fields in curved space

In flat space, the Dirac field is a four-component spinor under Lorentz transformation. The equation for this spinor uses the Dirac gamma matrices related to the flat space metric. The minimal inclusion of interaction with gravity assumes introducing the field $\psi(x)$, which is a spinor under local Lorentz transformation. The minimal procedure requires us to generalize the flat space gamma matrices on curved spacetime gamma-matrices related to curved space metric and define the appropriate spinor covariant derivative. As a result, we arrive at the action for spinor field in curved spacetime.

$$S_f = \int d^4x\sqrt{-g}\,\big(\,\bar{\psi}\, i\gamma^\mu\, \nabla_\mu\psi - m\bar{\psi}\psi\,\big)\,. \tag{12.41}$$

In Eq. (12.41), γ^μ and ∇_μ are, respectively, gamma matrices and covariant derivatives of a spinor in curved spacetime. In order to define both of these objects, we need to start from the tetrad formalism in curved spacetime.[1]

12.4.1 Tetrad formalism and covariant derivatives

As we know, GR is a theory of the metric field. However, in many cases, it proves useful to use other variables to describe the gravitational field. In particular, to define the covariant derivative of a spinor field in curved spacetime one needs an object called a tetrad. The German name *vierbein* is also frequently used for this object. The names suggest that the object is four dimensional, but the formalism described below can be easily generalized to the space of any dimension and, in addition, does not depend on the signature of the metric. For definiteness, we use four dimensional, Minkowski space notation in what follows.

Let us start by defining the tetrad in (pseudo) Riemann space. Locally, at point P, one can introduce the flat metric η_{ab}. This means that the vector basis includes four orthonormal vectors $\mathbf{e}_a$, such that $\mathbf{e}_a \cdot \mathbf{e}_b = \eta_{ab}$. Here $\mathbf{e}_a$ are four dimensional basis vectors in the tangent space to the manifold of our interest at the point P. Furthermore, let X^a be local coordinates on the manifold, which can be also seen as

[1] An alternative approach to the covariant derivative of fermions can be found, e.g., in [267, 291].

coordinates in the tangent space in close vicinity of the point P. These coordinates are defined up to local Lorentz transformations $X'^a = \Lambda^a{}_b(x)X^b$, where $\Lambda^a{}_b(x)$ is the matrix of local Lorentz transformations, i.e., $\Lambda^T(x)\eta\Lambda(x) = \eta$. The infinitesimal form of such a matrix is $\Lambda^a{}_b(x) = \delta^a{}_b + \omega^a{}_b(x)$, $\omega_{ab}(x) = -\omega_{ba}(x)$. With respect to the general coordinates x^μ, one can write $\mathbf{e}_a = e_a{}^\mu \mathbf{e}_\mu$, where $\mathbf{e}_\mu$ is a corresponding local basis and $e_a{}^\mu$ are transition coefficients from one basis to another. These coefficients form a tetrad and are related to the metric by the relations

$$e_a{}^\mu e^{a\nu} = g^{\mu\nu}\,, \qquad e^a{}_\mu e_{a\nu} = g_{\mu\nu}\,, \qquad e^a{}_\mu e_b{}^\mu = \delta^a{}_b\,, \qquad e^a{}_\mu e_a{}^\alpha = \delta^\alpha{}_\mu\,. \tag{12.42}$$

The tetrad is used to convert spacetime indices into local Lorentz indices, and vice versa, e.g.,

$$T^a_{\cdot\mu} = T^\nu_{\cdot\mu}\, e^a_\nu = T^a_{\cdot b}\, e^b_\mu\,. \tag{12.43}$$

According to (12.42), the descriptions in terms of the metric and the tetrad are equivalent. In particular,

$$\det\left(e^a{}_\mu\right) = \sqrt{|g|}\,, \qquad g = \det\left(g_{\mu\nu}\right). \tag{12.44}$$

Now we consider the construction of the covariant derivative in tetrad formalism. In the metric formalism, the covariant derivative of a vector V^μ is

$$\nabla_\nu V^\mu = \partial_\nu V^\mu + \Gamma^\mu{}_{\nu\lambda} V^\lambda, \tag{12.45}$$

where $\Gamma^\mu_{\nu\lambda}$ is the Christoffel symbol. Our purpose is to construct a consistent version of the covariant derivative of a tetrad, which has not only a general covariant but also a local Lorentz index. Such a covariant derivative is defined on the basis of the general Yang-Mills covariant derivative formulated in section 4.6 in Part I. This covariant derivative is constructed using the generators of gauge transformations. In the case under consideration, the role of gauge symmetry is played by the local Lorentz symmetry. Therefore, to construct the corresponding covariant derivative, we need to use the generators of Lorentz transformations.

Let $V^a = e^a{}_\mu V^\mu$ be a vector with respect to the local Lorentz transformations, i.e., $V'^a = \frac{\partial X'^a}{\partial X^b} V^b$. The infinitesimal form of this transformation has the form described in section 2.4, Eq. (2.46),

$$\delta V^a = \omega^a{}_b V^b = -\frac{i}{2}\omega^{cd}(S_{cd})^a{}_b V^b, \tag{12.46}$$

where

$$(S_{cd})^a{}_b = i(\delta_c{}^a \eta_{db} - \delta_d{}^a \eta_{cb}) \tag{12.47}$$

are the generators of the Lorentz transformations in contravariant vector representation. Using the general construction of the Yang-Mills covariant derivative, one gets

$$\nabla_\mu V^a = \partial_\mu V^a + \frac{i}{2}\omega_\mu{}^{cd}(S_{cd})^a{}_b V^b. \tag{12.48}$$

Let us stress that the sign in the covariant derivative is opposite to the one in the gauge transformations. It is evident that $\omega_\mu{}^{cd} = -\omega_\mu{}^{dc}$ by construction. The real field $\omega_\mu{}^{cd}$

is the gauge field for Lorentz group; usually, this field is called the *spinor connection.* Substituting the generators (12.47), we obtain

$$\nabla_\mu V^a = \partial_\mu V^a - \omega_\mu{}^a{}_b V^b. \tag{12.49}$$

This is the Lorenz covariant derivative of the vector field. The gauge transformations of the connection $\omega_\mu{}^{ab}$ can be found using the standard Yang-Mills consideration. This means

$$\delta(\nabla_\mu V^a) = \omega^a{}_b(\nabla_\mu V^b). \tag{12.50}$$

Substituting relation (12.49) into (12.50), one gets

$$\delta\omega_\mu{}^{ab} = \partial_\mu \omega^{ab} + \omega^a{}_c\omega_\mu{}^{cb} - \omega_\mu{}^a{}_c\omega^{cb}. \tag{12.51}$$

Relation defines the transformation of the spinor connection under the local Lorentz transformation.

Relation (12.49) allows us to find the covariant derivative of the tetrad. To do that, we substitute $V^a = e^a{}_\nu V^\nu$ in (12.49) and use the standard covariant derivative of the vector field. Then one gets

$$\nabla_\mu e^a{}_\nu = \partial_\mu e^a{}_\nu + \omega_\mu{}^a{}_b e^b{}_\nu - e^a{}_\lambda \Gamma^\lambda{}_{\mu\nu}. \tag{12.52}$$

We note that the spinor connection $\omega_\mu{}^{cd}$ is still an independent geometrical object. To establish the explicit form for the spinor connection, we use the metricity relation $\nabla_\mu g_{\alpha\beta} = 0$. Since $e^a{}_\mu e_{a\nu} = g_{\mu\nu}$, one can impose the constraint $\nabla_\mu e^a{}_\nu = 0$, providing

$$\omega_\mu{}^{ab} = e^{b\nu}\partial_\mu e^a{}_\nu - e^a{}_\lambda e^{b\nu}\Gamma^\lambda{}_{\mu\nu}. \tag{12.53}$$

Now the spin connection is not an independent geometrical object but is a function of the tetrad and its derivative. Relation (12.53) is analogous to the expression for the Christoffel symbol, (11.9). One can show that the spin connection given by relation (12.53) is antisymmetric in the local Lorentz indices a, b. For convenience, relation (12.53) can be antisymmetrized, and we arrive at

$$\omega_\mu{}^{ab} = \frac{1}{2}(e^{b\nu}\partial_\mu e^a{}_\nu - e^{a\nu}\partial_\mu e^b{}_\nu) - \frac{1}{2}(e^a{}_\lambda e^{b\nu} - e^b{}_\lambda e^{a\nu})\Gamma^\lambda{}_{\mu\nu}. \tag{12.54}$$

12.4.2 The covariant derivative of a Dirac spinor

The tetrad can be used to define the curved-space gamma matrices as $\gamma^\mu = e_a{}^\mu\gamma^a$. The indices of the matrices γ^μ are lowered and raised, respectively, by means of the metrics $g_{\mu\nu}$ and $g^{\mu\nu}$. The basic relation for curved-space gamma matrices has the form

$$\gamma_\mu\gamma_\nu + \gamma_\nu\gamma_\mu = 2g_{\mu\nu}I, \tag{12.55}$$

where I is a unit 4×4 matrix. Because of (12.53), we have $\nabla_\alpha\gamma^\mu(x) = 0$.

Consider the following definition of a Dirac spinor field in curved spacetime. In flat space, the four component spinor is defined by the transformation rule under the

Lorentz group, as described in section 4.3. It is natural to define the spinor field $\psi(x)$ in curved spacetime as a spinor under local Lorentz transformation, and as a scalar under general coordinate transformations. The variation of the spinor field under local Lorentz transformation is written as follows:

$$\delta\psi(x) = -\frac{i}{2}\omega^{ab}(x)\Sigma_{ab}\psi, \tag{12.56}$$

where $\omega^{ab}(x)$ are the parameters of the local Lorentz transformations and Σ_{ab}, as defined in (4.37),

$$\Sigma_{ab} = \frac{i}{4}(\gamma_a\gamma_b - \gamma_b\gamma_a), \tag{12.57}$$

are the generators of the Lorentz transformations in the four-component spinor representation. The spinor covariant derivative is defined according to the general Yang-Mills covariant derivative formulated in section 4.6 in Part I. It leads to

$$\nabla_\mu\psi = \partial_\mu\psi + \frac{i}{2}\omega_\mu{}^{ab}\Sigma_{ab}\psi, \tag{12.58}$$

where $\omega_\mu{}^{ab}$ is the spinor connection or the real gauge field corresponding to the local Lorentz group in Dirac spinor representation. Here we took into account that the sign in the covariant derivative must be opposite to one in the gauge transformation. Actually, this $\omega_\mu{}^{ab}$ is just the same spinor connection (12.53) that was derived on the basis of vector field consideration. We emphasize that gauge field is defined by the gauge group only, and the details of representation are only in the form of gauge generators. For vector representation, these generators are S_{cd} as in (12.47); for spinor representation, the generators are Σ_{ab} as in (4.37).

The Dirac conjugate spinor in curved spacetime is defined analogously to flat space case in the form $\bar{\psi}(x) = \psi^\dagger\gamma^0$, where γ^0 is the flat space gamma matrix. Performing conjugation in relation (12.58) and using properties of the gamma matrices, one gets

$$\nabla_\mu\bar{\psi} = \partial_\mu\bar{\psi} - \frac{i}{2}\,\omega_\mu{}^{ab}\,\Sigma_{ab}\,. \tag{12.59}$$

It is easy to check that $\nabla_\mu(\bar{\psi}\psi) = \partial_\mu(\bar{\psi}\psi)$, independently of the choice of $\omega_\mu{}^{ab}$.

As a result, the Dirac equation in curved spacetime is written in the form

$$(i\gamma^\mu\nabla_\mu - m)\psi = 0, \tag{12.60}$$

where the spinor covariant derivative is given by the relation (12.58).

12.4.3 The commutator of covariant derivatives

The next step is to derive the commutator of two covariant derivatives acting on a Dirac spinor. Consider

$$\begin{aligned}[\nabla_\mu, \nabla_\nu] &= \Big[\partial_\mu + \frac{i}{2}\omega_\mu{}^{ab}\Sigma_{ab},\, \partial_\nu + \frac{i}{2}\omega_\nu{}^{cd}\Sigma_{cd}\Big] \\ &= \frac{i}{2}(\partial_\mu\omega_\nu{}^{cd} - \partial_\nu\omega_\mu{}^{ab})\Sigma_{ab} - \frac{1}{4}\omega_\mu{}^{ab}\omega_\nu{}^{ac}[\Sigma_{ab}, \Sigma_{cd}].\end{aligned} \tag{12.61}$$

Let us use the commutation relation for the Lorentz group generators, which were derived in Part I,

$$[\Sigma_{ab}, \Sigma_{cd}] = i\big(\eta_{ad}\Sigma_{bc} + \eta_{bc}\Sigma_{ad} - \eta_{ac}\Sigma_{bd} - \eta_{bd}\Sigma_{ac}\big). \tag{12.62}$$

Substituting the relation (12.62) to the relation (12.61) one rewrites last relation in the form

$$[\nabla_\mu, \nabla_\nu] = \frac{i}{2} R_{\mu\nu}{}^{ab}\Sigma_{ab}, \tag{12.63}$$

where

$$R_{\mu\nu}{}^{ab} = \partial_\mu \omega_\nu{}^{ab} - \partial_\nu \omega_\mu{}^{ab} + \omega_\mu{}^{ac}\omega_{\nu\, c}{}^{b} - \omega_\nu{}^{ac}\omega_{\mu\, c}{}^{b}. \tag{12.64}$$

The expression (12.64) is usually called the strength tensor, or curvature tensor, associated with the Lorentz group. It is easy to see that this tensor is antisymmetric in the Lorentz indices a, b. Substituting the explicit form of the generators Σ_{ab} (4.37) into relation (12.63), one gets

$$[\nabla_\mu, \nabla_\nu] = \frac{1}{4} R_{\mu\nu}{}^{ab}\gamma_a\gamma_b. \tag{12.65}$$

Using relations (12.64) and (12.53) , we can relate the curvature tensor $R_{\mu\nu}{}^{ab}$ and Riemann curvature tensor $R_{\mu\nu\lambda\sigma}$ as

$$R_{\mu\nu\lambda\sigma} = R_{\mu\nu}{}^{ab} e_{a\lambda} e_{b\sigma}. \tag{12.66}$$

One of the applications of (12.66) is the possibility of a "doubling" of the general covariant Dirac equation (12.60) corresponding to the action (12.41). Consider the product

$$-\big(i\gamma^\mu\nabla_\mu - m\big)\big(i\gamma^\nu\nabla_\nu + m\big)\psi = \big(\gamma^\mu\nabla_\mu\gamma^\nu\nabla_\nu + m^2\big)\psi. \tag{12.67}$$

The first term in this relation transforms as

$$\begin{aligned}
\gamma^\mu\nabla_\mu\gamma^\nu\nabla_\nu\psi &= \frac{1}{2}\gamma^\mu\gamma^\nu\big(\nabla_\mu\nabla_\nu + \nabla_\mu\nabla_\nu\big)\psi + \frac{1}{2}\gamma^\mu\gamma^\nu\big(\nabla_\mu\nabla_\nu - \nabla_\mu\nabla_\nu\big)\psi \\
&= \frac{1}{2}\big(\gamma^\mu\gamma^\nu + \gamma^\nu\gamma^\mu\big)\nabla_\mu\nabla_\nu\psi + \frac{1}{2}\gamma^\mu\gamma^\nu\big(\nabla_\mu\nabla_\nu - \nabla_\mu\nabla_\nu\big)\psi \\
&= g^{\mu\nu}\,\nabla_\mu\nabla_\nu\psi - \frac{1}{8}R_{\mu\nu}{}^{ab}\gamma^\mu\gamma^\nu\gamma_a\gamma_b\psi \;=\; \Big(\Box - \frac{1}{4}R\Big)\psi.
\end{aligned} \tag{12.68}$$

In the last step, we used the identity

$$R_{\mu\nu}{}^{ab}\,\gamma^\mu\gamma^\nu\gamma_a\gamma_b = 2R, \tag{12.69}$$

which can be proved by using the properties of the Dirac matrices and the algebraic properties of the Riemann tensor. We leave the verification of this identity as an exercise for the reader.

12.4.4 Local conformal transformation

Consider the conformal transformation of the metric and the spinor field.

It is easy to see that the massless part of the action (12.41) is invariant under the *global* conformal transformation

$$g_{\mu\nu} \to g_{\mu\nu}\, \exp{(2\lambda)}\,, \qquad \psi \to \psi\, \exp\Big(-\frac{3}{2}\lambda\Big)\,, \qquad \lambda = const. \tag{12.70}$$

The conformal transformation for the conjugated fermion $\bar{\psi}$ is always the same as for ψ. Using the global symmetry as a hint, let us consider the *local* conformal transformation

$$g_{\mu\nu} \,=\, \bar{g}_{\mu\nu}\, e^{2\sigma(x)}\,, \qquad \psi \,=\, \psi_*\, e^{-\frac{3}{2}\,\sigma(x)}\,. \tag{12.71}$$

In what follows, we will omit the spacetime argument but always assume that $\sigma = \sigma(x)$. Furthermore, all quantities with bars are constructed using the fiducial metric $\bar{g}_{\mu\nu}$. Also, we use compact notation for the partial derivative $\sigma_{,\lambda} = \partial_\lambda \sigma$.

Our aim is to find the transformation rule for the action (12.41), and this requires transformations of all intermediate quantities. Direct replacement of (12.71) gives

$$g^{\mu\nu} = \bar{g}^{\mu\nu} e^{-2\sigma}, \quad \sqrt{-\bar{g}} = \sqrt{-g} e^{4\sigma}, \quad e_a{}^{\mu} = \bar{e}_a{}^{\mu} e^{-\sigma}, \quad e^b{}_{\mu} = \bar{e}^b{}_{\mu} e^{\sigma} \tag{12.72}$$

and, consequently,

$$\gamma^\mu \,=\, \gamma^a e_a{}^\mu \,=\, \bar{\gamma}^\mu\, e^{-\sigma}\,, \qquad \gamma_\mu \,=\, \gamma^a e_{a\mu} \,=\, \bar{\gamma}_\mu\, e^{\sigma}\,. \tag{12.73}$$

For the Christoffel symbols and spinor connection, we find

$$\begin{aligned} \Gamma^\lambda_{\alpha\beta} &= \bar{\Gamma}^\lambda_{\alpha\beta} + \delta^\lambda_\alpha \sigma_{,\beta} + \sigma_{,\alpha}\delta^\lambda_\beta - \sigma_{,\tau}\bar{g}^{\lambda\tau}\bar{g}_{\alpha\beta}, \\ \omega_\mu{}^{ab} &= \bar{\omega}_\mu{}^{ab} \;-\; (\bar{e}^a_\mu\, \bar{e}^{b\lambda} - \bar{e}^b{}_\mu\, \bar{e}^{a\lambda})\, \sigma_{,\lambda}\,. \end{aligned} \tag{12.74}$$

Furthermore,

$$\begin{aligned} \gamma^\mu \nabla_\mu \psi &= \bar{\gamma}^\mu e^{-\frac{5}{2}\sigma} \Big\{ e^{\frac{3}{2}\sigma} \partial_\mu (e^{-\frac{3}{2}\sigma} \psi_*) + \frac{i}{2}\,\bar{\omega}_\mu{}^{ab}\, \Sigma_{ab}\psi_* \\ &\quad - \frac{i}{2}\,(\bar{e}^a_\mu \bar{e}^{\lambda b} - \bar{e}^b_\mu \bar{e}^{\lambda a})\sigma_{,\lambda} \cdot \frac{i}{4}\,(\gamma_a\gamma_b - \gamma_b\gamma_a)\psi_* \Big\} \\ &= \bar{\gamma}^\mu e^{-\frac{5}{2}\sigma} \Big\{ \bar{\nabla}\psi_* - \frac{3}{2}\sigma_{,\mu}\psi_* + \frac{1}{4}(\bar{\gamma}_\mu\bar{\gamma}^\lambda - \bar{\gamma}^\lambda\bar{\gamma}_\mu)\, \sigma_{,\lambda}\psi_* \Big\} \;=\; e^{-\frac{5}{2}\sigma} \bar{\nabla}\psi_*\,. \end{aligned} \tag{12.75}$$

By substituting (12.75), (12.71) and (12.72) into the action (12.41), we arrive at the transformation law

$$\int d^4x \sqrt{-g}\,\bar{\psi}(i\gamma^\mu \nabla_\mu - m)\psi = \int d^4x \sqrt{-\bar{g}}\,\bar{\psi}_*(i\bar{\gamma}^\mu \bar{\nabla}_\mu - m\, e^{\sigma})\psi_*\,. \tag{12.76}$$

One can see that the massless action is invariant under local conformal transformation.

12.4.5 The energy-momentum tensor for the Dirac field

Let us consider the derivation of the dynamical energy-momentum tensor $T_{\mu\nu}$ for the Dirac field. The corresponding definition is

$$T^{\mu\nu} = -\frac{2}{\sqrt{-g}}\,\frac{\delta S_f}{\delta g_{\mu\nu}}\,. \tag{12.77}$$

Now we have two problems. The first problem is that, the Lagrangian corresponding to the action (12.41), in the form we wrote it, is real up to total divergence. For further analysis, it is convenient to rewrite this action in the equivalent form with the Hermitian operator in the action,

$$S_f = \frac{1}{2}\int d^4x\sqrt{-g}\big(\bar{\psi}\gamma^\mu i\nabla_\mu\psi - i\nabla_\mu\bar{\psi}\,\gamma^\mu\psi - 2m\bar{\psi}\psi\big)\,. \tag{12.78}$$

The second problem is that the variation in (12.77) should be taken with respect to the metric, but the action (12.41) is constructed in terms of the tetrad and spin connection. To avoid this problem, we consider the following variation of the metric

$$g_{\alpha\beta} \to g'_{\alpha\beta} = g_{\alpha\beta} + h_{\alpha\beta}, \tag{12.79}$$

and find the corresponding variations for all of the involved quantities. Solving this problem in the (requested here) first order in $h_{\alpha\beta}$ is quite simple, and we arrive at

$$\begin{aligned}
&\delta\sqrt{-g} = \frac{1}{2}\sqrt{-g}h\,, \quad \delta g^{\mu\nu} = -h^{\mu\nu}\,, \quad \delta e^c{}_\mu = \frac{1}{2}\,h^\nu_\mu e^c_\nu\,, \quad \delta e^\rho_b = -\frac{1}{2}\,h^\rho_\lambda e_b{}^\lambda\,,\\
&\delta\Gamma^\lambda_{\alpha\beta} = \frac{1}{2}\left(\nabla_\alpha h^\lambda_\beta + \nabla_\beta h^\lambda_\alpha - \nabla^\lambda h_{\alpha\beta}\right), \quad \delta\gamma^\mu = -\frac{1}{2}\,h^\mu_\nu\,\gamma^\nu\,.
\end{aligned} \tag{12.80}$$

Here all indices are raised and lowered, respectively, with the background metric $g^{\mu\nu}$ and $g_{\mu\nu}$. Furthermore, direct calculation using (12.80), yields

$$\delta\omega_\mu{}^{ab}_{\cdot\cdot} = -\delta\omega_\mu{}^{ba}_{\cdot\cdot} = \delta(e^b_\tau e^{a\lambda}\Gamma^\tau_{\lambda\mu} - e^{a\lambda}\partial_\mu e^b_\lambda) = \left(e^{a\tau}e^{b\lambda} - e^{b\tau}e^{a\lambda}\right)\nabla_\lambda h_{\mu\tau}. \tag{12.81}$$

Next we substitute (12.80) and (12.81) into (12.78), and expand to obtain

$$\bar{\psi}\gamma^\mu\nabla_\mu\psi - \nabla_\mu\bar{\psi}\gamma^\mu\psi = \bar{\psi}\gamma^\mu\partial_\mu\psi - \partial_\mu\bar{\psi}\gamma^\mu\psi - \frac{i}{4}\,\omega_\mu{}^{ab}_{\cdot\cdot}\,\bar{\psi}(\gamma^\mu\sigma_{ab} + \sigma_{ab}\gamma^\mu)\psi. \tag{12.82}$$

Consider the term depending on the variation $\delta\omega_\mu{}^{ab}_{\cdot\cdot}$, which was defined in (12.81),

$$\begin{aligned}
\delta_\omega S_f &= -\frac{i}{4}\int d^4x\sqrt{-g}\,\bar{\psi}(\gamma^\mu\sigma_{ab} + \sigma_{ab}\gamma^\mu)\psi\,\delta\omega_\mu{}^{ab}_{\cdot\cdot} \qquad (12.83)\\
&= -\frac{i}{8}\int d^4x\sqrt{-g}\left(e^{a\tau}e^{b\lambda} - e^{b\tau}e^{a\lambda}\right)\nabla_\lambda h_{\mu\tau}\,e^{\mu c}\,\bar{\psi}(\gamma_c\sigma_{ab} + \sigma_{ab}\gamma_c)\psi\,.
\end{aligned}$$

After partial integration, we get

$$\delta_\omega S_f = \frac{i}{8}\int d^4x\sqrt{-g}\,h_{\mu\tau}e^{c\mu}\left(e^{a\tau}e^{b\lambda} - e^{b\tau}e^{a\lambda}\right)\nabla_\lambda\big[\bar{\psi}(\gamma_c\sigma_{ab} + \sigma_{ab}\gamma_c)\psi\big].$$

The factor in the brackets is antisymmetric in ab. Hence, one can simplify the expression, using the expression for σ_{ab},

$$\delta_\omega S_f = -\frac{1}{8}\int d^4x\sqrt{-g}\,h_{\mu\nu}\,e^{c\mu}e^{a\nu}e^{b\lambda}\nabla_\lambda\big[\bar{\psi}(\gamma_c\gamma_a\gamma_b - \gamma_c\gamma_b\gamma_a + \gamma_a\gamma_b\gamma_c - \gamma_b\gamma_a\gamma_c)\psi\big]\,.$$

Replacing $\gamma_a\gamma_b + \gamma_b\gamma_a = 2\eta_{ab}$, we arrive at

$$\delta_\omega S_f = -\frac{1}{8}\int d^4x\sqrt{-g}\,h_{\mu\nu}\nabla_\lambda\big[\bar{\psi}(g^{\lambda\mu}\gamma^\nu - g^{\lambda\nu}\gamma^\mu)\psi\big] = 0\,. \tag{12.84}$$

Finally, we observe that $\delta\omega_\mu{}^{ab}{}_{\cdot\cdot}$ is irrelevant, since its contribution vanishes. Hence, we arrive at (remember $h = g^{\mu\nu}h_{\mu\nu}$)

$$\begin{aligned}\delta S_f &= \frac{i}{2}\int d^4x\sqrt{-g}\Big\{\frac{1}{2}\left(hg^{\mu\nu} - h^{\mu\nu}\right)\left(\bar{\psi}\gamma_\nu\nabla_\mu\psi - \nabla_\mu\bar{\psi}\gamma_\nu\psi\right) + ihm\bar{\psi}\psi\Big\}\\ &= \int d^4x\sqrt{-g}\,h_{\alpha\beta}\,\Big\{\frac{i}{4}g^{\alpha\beta}\left(\bar{\psi}\gamma^\lambda\nabla_\lambda\psi - \nabla_\lambda\bar{\psi}\gamma^\lambda\psi\right) - \frac{1}{2}\,g^{\alpha\beta}m\bar{\psi}\psi\\ &\quad - \frac{i}{4}\left(\bar{\psi}\gamma^\alpha\nabla^\beta\psi - \nabla^\alpha\bar{\psi}\gamma^\beta\psi\right)\Big\}.\end{aligned} \tag{12.85}$$

According to the definition (12.77), the energy-momentum tensor of the spinor field is

$$T_{\mu\nu} = \frac{i}{2}\big[\bar{\psi}\gamma_{(\mu}\nabla_{\nu)}\psi - \nabla_{(\mu}\bar{\psi}\gamma_{\nu)}\psi\big] - \frac{i}{2}g_{\mu\nu}\big[\bar{\psi}\gamma^\lambda\nabla_\lambda\psi - \nabla_\lambda\bar{\psi}\gamma^\lambda\psi\big] + m\bar{\psi}\psi g_{\mu\nu}. \tag{12.86}$$

In order to check this expression, let us take a trace,

$$T_\mu{}^\mu = T_{\mu\nu}g^{\mu\nu} = -\frac{3i}{2}\left(\bar{\psi}\gamma^\lambda\nabla_\lambda\psi - \nabla_\lambda\bar{\psi}\gamma^\lambda\psi\right) + 4m\bar{\psi}\psi\,. \tag{12.87}$$

Using Dirac equations of motion, $\gamma^\lambda\nabla_\lambda\psi = -im\psi$ and $\nabla_\lambda\bar{\psi}\gamma^\lambda = im\bar{\psi}$, the on-shell the trace is

$$T_\mu{}^\mu\Big|_{on-shell} = m\bar{\psi}\psi\,. \tag{12.88}$$

For a massless theory, this expression vanishes, meaning that the massless fermion theory is invariant under local conformal transformation.

Exercises

12.12. Consider a metric of the form $g_{\mu\nu} = g_{\mu\nu} + \kappa h_{\mu\nu}$, where κ is a small parameter. The flat metric with $\kappa = 0$ corresponds to the tetrad $\bar{e}^a_\mu$. Find the expansions of up to the order $\mathcal{O}(\kappa^2)$ of e^a_μ and e^ν_b.

12.13. i) Verify Eq. (12.58). ii) Using both (12.58) and (12.59), verify an identity $\nabla_\alpha(\bar{\psi}\psi) = \partial_\alpha(\bar{\psi}\psi)$. iii) Discuss and explain why $\nabla_\alpha\gamma^\mu = 0$.

12.14. Consider the Dirac field in n-dimensional spacetime. First make a generalization of the transformation rule (12.70). After that, repeat the considerations leading to (12.76), and prove that the massless fermion action is conformal invariant in any dimension, while the conformal weights of ψ and $\bar{\psi}$ are equal to $\frac{1-n}{2}$.

12.15. a) Consider the possibility of restoring local conformal symmetry in the general action (12.41) through replacing the mass by a scalar field, φ. Find the conformal transformation of the scalar, making the action invariant. b) Generalize this consideration for the space of an arbitrary dimension n. To do this, one has to start from the global conformal transformation and then find the form of the transformation of the fermion field [more general than the one in (12.70)], which leaves the massless fermion action invariant. Then find the transformation law for the scalar that leaves the Yukawa term invariant.

12.16. Taking the variational derivative of the action of the Dirac fermion with respect to the conformal factor $\sigma(x)$ in (12.71), derive the Noether identity for the local conformal symmetry. After that derive (12.87) using this identity and the Dirac equation. Finally, establish the general relation between local conformal symmetry and the on-shell vanishing of the trace of the energy-momentum tensor.

12.17. Prove that having nonminimal local covariant terms without inverse mass dimension parameters is impossible for fermions. Use the arguments based on the dimension of the Dirac field.

12.18. Consider the interaction term between the Dirac fermion and the vector field A_μ in the n-dimensional spacetime, $\mathcal{L}_{int} = \sqrt{-g}A_\mu\bar{\psi}\gamma^\mu\psi$. Explore the possibility to have this term be conformal invariant. Derive the necessary transformation rule for A_μ in an arbitrary-n case.

12.5 Massless vector (gauge) fields

The minimal covariant generalization of the action for a massless vector field A_μ is straightforward:

$$S_1 = -\frac{1}{4}\int d^4x\sqrt{-g}\,F_{\mu\nu}\,F^{\mu\nu}, \tag{12.89}$$

where $F_{\mu\nu} = \nabla_\mu A_\nu - \nabla_\nu A_\mu = F_{\mu\nu} = \partial_\mu A_\nu - \partial_\nu A_\mu$. In the non-Abelian case, one has to substitute

$$A_\mu \to A^a_\mu\,, \qquad F_{\mu\nu} \to G^a_{\mu\nu} = \partial_\mu A^a_\nu - \partial_\nu A^a_\mu + gf^{abc}A^b_\mu A^c_\nu\,. \tag{12.90}$$

Here a is the gauge algebra index. In both Abelian and non-Abelian cases, the minimal action is preserving the gauge symmetry. Having nonminimal covariant terms without inverse mass dimension parameters is impossible, and the vacuum terms are, of course, the same as in the scalar and fermion cases.

Exercises

12.19. Verify the local conformal symmetry of the actions of Abelian and non-Abelian vector fields, assuming the transformation rule in the $n = 4$ case,

$$g_{\mu\nu} = \bar{g}_{\mu\nu}\,e^{2\sigma}, \quad A_\mu = \bar{A}_\mu, \qquad \sigma = \sigma(x). \tag{12.91}$$

12.20. Derive the explicit form of the equations of motion for the Abelian vector field for the cosmological conformally flat metric

$$g_{\mu\nu} = a^2(\eta)\bar{g}_{\mu\nu} = e^{2\sigma(\eta)}\bar{g}_{\mu\nu}, \tag{12.92}$$

where η is the conformal time. Rewrite these equations in terms of the physical time t, where $dt = a(\eta)d\eta$.

12.6 Interactions between scalar, fermion and gauge fields

The next step is to establish the form of the interaction terms in curved spacetime. In general, gauge theory in flat spacetime has three types of interactions, namely, gauge, Yukawa and quartic scalar self-interaction. All of the corresponding coupling constants are dimensionless, and therefore only the generalizations of the minimal kind are possible in the interaction sectors. Obviously, a minimal generalization of these interactions does not meet any difficulties. It is worth emphasizing that nonminimal coupling is possible only in purely scalar sector.

And so, we have constructed the actions of scalar, spinor and vector fields in an external gravitational field. We have found that the action in curved spacetime possesses some ambiguity related to the nonminimal term of the scalar curvature interaction and to the action of the vacuum. When we set $g_{\mu\nu} \to \eta_{\mu\nu}$, the curvature becomes zero, and all these nonminimal and vacuum terms disappear. The only exception is the cosmological constant term, but, in the flat space, it becomes an irrelevant constant that does not contribute to the equations of motion and can be neglected.

In contrast, in curved spacetime, there may be a big difference between minimal and nonminimal actions. As we shall see in the next chapters, it is the nonminimal actions that make it possible to construct renormalizable quantum theory.

Exercises

12.21. Consider the gauge interactions between a massless vector and a charged scalar, and between a Dirac fermion and a massless vector. Show that the interaction terms in the actions do not violate general covariance.

12.22. Generalize the results of the previous exercise to non-Abelian theory. Formulate the action of a $SU(N)$ Yang-Mills field coupled to n copies of a charged scalar in the adjoint representation, and m copies of a Dirac fermion in the fundamental representation of the gauge group.

12.23. Consider the interacting theory with spontaneous symmetry breaking (12.28). Derive the mass of the gauge boson in the broken phase, and verify whether the separation of the Goldstone mode is affected by the fact that the spacetime is curved and by the presence of the nonminimal term.

13 Quantum fields in curved spacetime: renormalization

In this and and the next chapters, we will discuss the theory of quantum matter fields on a classical curved background. Since the metric is not quantized, this approach is also called semiclassical gravity. In what follows, one can find a practical introduction to the definition, calculational methods and the general structure of the effective action of matter fields and a vacuum, starting from the renormalization procedure in curved spacetime. Indeed, quantum field theory in the presence of external fields or, more generally, with external conditions, is a vast research area that can hardly be covered in a single book. Therefore, we refer the reader to the existing literature on the subjects which will not be discussed below or discussed very briefly, e.g., particle creation in a gravitational field (see, e.g., [56], or [225] for a more pedagogical introduction), theories with non-trivial boundary conditions (see, e.g., [19]), and the Casimir effect [63]. Also, we will not consider here different approaches to the definition of a vacuum state in curved space (see, e.g., [105, 109, 56, 317, 150]).

13.1 Effective action in curved spacetime

The construction of quantum field theory in curved spacetime includes a number of specific problems with no analogs in flat space field theory and therefore requires the development of special methods and approaches. As we saw in Part I, the most convenient (at least for many applications) formulation of quantum field theory is given in terms of functional integrals. This formulation is especially useful for exploring various quantum aspects of perturbation theory for interacting fields. Thus, in this book, we will base our consideration on the formulation of quantum field theory in terms of functional integrals.

The fundamental object of quantum field theory is the generating functional of Green functions. To define such a generating functional for the semiclassical theory in curved spacetime, we start from the more complete version of the theory. Let $S[g, \Phi]$ be the action of matter fields Φ coupled to the quantum gravitational field, where

$$S[g, \Phi] = S_g[g] + S_m[\Phi, g]. \tag{13.1}$$

Here $S_g[g]$ is the action of the pure gravitational field, and $S_m[\Phi, g]$ is the action of matter fields in curved spacetime with the metric $g = g_{\mu\nu}$. In general, the theory with the action (13.1) may have gauge symmetry, and its quantization can be realized

Introduction to Quantum Field Theory with Applications to Quantum Gravity. Iosif L. Buchbinder and Ilya L. Shapiro, Oxford University Press (2021). © Iosif L. Buchbinder and Ilya Shapiro.
DOI: 10·1093/oso/9780198838319·003·0013

by the procedure described in chapter 10 of Part I. As a result, we obtain, for the generating functional of Green functions, the expression

$$Z[J,I] = \frac{1}{\mathcal{N}} \int \mathcal{D}g\, \mathcal{D}\Phi\, e^{i(S_g[g]+S_m[\Phi,g]+Ig+\Phi J)}. \tag{13.2}$$

Here I and J are the sources for the quantum metric $g_{\mu\nu}$ and the quantum field Φ respectively. In particular,

$$\Phi J = \int d^4x\sqrt{-g}\, \Phi(x)J(x) \tag{13.3}$$

is assuming summation over all indices, and covariant integration over the continuous spacetime variables. The symbols $\mathcal{D}g$ and $\mathcal{D}\Phi$ mean integration measures including the corresponding gauges and ghost fields. The normalization constant $\mathcal{N}$ is

$$\mathcal{N} = \int \mathcal{D}g\, \mathcal{D}\Phi\, e^{i(S_g[g]+S_m[\Phi,g])}. \tag{13.4}$$

Let us rewrite the expression (13.2) in the form

$$Z[J,I] = \frac{1}{\tilde{\mathcal{N}}} \int \mathcal{D}g\; e^{igI}\; Z[J,\, g_{\mu\nu}], \tag{13.5}$$

where

$$Z[J,\, g_{\mu\nu}] = \frac{1}{\mathcal{N}_0} \int \mathcal{D}\Phi\, e^{i(S_{vac}[g]+S_m[\Phi,g]+\Phi J)}. \tag{13.6}$$

Here we renamed $S_g[g]$ as $S_{vac}[g]$ and denoted $\tilde{\mathcal{N}} = \frac{\mathcal{N}}{\mathcal{N}_0}$, where

$$\mathcal{N}_0 = \int \mathcal{D}\Phi\, e^{i\,S[\Phi,g]}\Big|_{g_{\mu\nu}=\eta_{\mu\nu}}. \tag{13.7}$$

Expression (13.6) is the definition of the generating functional of the Green functions of matter fields in curved spacetime. This functional will be the basic object of our study in the next few chapters, which are devoted to semiclassical gravity. After that, we turn our attention to full quantum gravity and recover the functional integral over the metric. For a while, the integration over the metric was needed only to justify the special form of the normalization factor (13.7) in the semiclassical approximation to the full quantum theory.

The functional integral (13.6) depends on the external sources for the matter fields J, and also on the metric $g \equiv g_{\mu\nu}$, which enters this integral as an external parameter. The action in the expression (13.6) can be presented as the sum

$$S[\Phi, g_{\mu\nu}] \;=\; S_m[\Phi, g_{\mu\nu}] \;+\; S_{vac}[g_{\mu\nu}], \tag{13.8}$$

separating the classical vacuum action S_{vac}, which depends only on the metric and not on the matter fields, and the proper matter part S_m, which is supposed to vanish when the matter fields are absent. In some cases, especially in the theories with spontaneous symmetry breaking described in section 12.3, such a separation may be a complicated issue, but let us suppose that it can be done. Obviously, the exponential with the vacuum action S_{vac} can be taken out of the functional integral in (13.6), producing

$$Z[J,\, g_{\mu\nu}] \;=\; \mathcal{N}\, e^{iS_{vac}[g_{\mu\nu}]} \int \mathcal{D}\Phi\, e^{i\,S_m[\Phi, g_{\mu\nu}] + i\Phi\, J}. \tag{13.9}$$

After this point, we follow the same scheme as for flat spacetime and define the generating functional $W[J,\, g_{\mu\nu}]$ of connected Green functions in the presence of external gravitational field,

$$e^{iW[J,\, g_{\mu\nu}]} \;=\; Z[J,\, g_{\mu\nu}]. \tag{13.10}$$

The Green functions G and the connected Green functions G^c are defined, exactly as in the flat space, as the variational derivatives

$$G(x_1, x_2,\, ...,\, x_N) \;=\; \frac{\delta}{i\,\delta J(x_1)}\,\frac{\delta}{i\,\delta J(x_2)}\;\cdots\;\frac{\delta}{i\,\delta J(x_N)}\, Z[J,\, g_{\mu\nu}]\Big|_{J=0} \tag{13.11}$$

and

$$G^c(x_1, x_2,\, ...\, x_N) \;=\; \frac{\delta}{i\,\delta J(x_1)}\,\frac{\delta}{i\,\delta J(x_2)}\;\cdots\;\frac{\delta}{i\,\delta J(x_N)}\, W[J,\, g_{\mu\nu}]\Big|_{J=0}. \tag{13.12}$$

The difference here from the usual correlation functions in flat spacetime is that the Green functions (13.11) and (13.12) depend on the external metric that enters the functional integral as an external parameter and is not an object of quantization.

The effective action of the mean field in curved spacetime is introduced in a manner similar to the way it was described in Part I. First, we define the mean field

$$\bar{\Phi}(x) \;=\; \frac{1}{\sqrt{-g(x)}}\,\frac{\delta\, W[J,\, g_{\mu\nu}]}{\delta J(x)}. \tag{13.13}$$

Now, we have to solve (13.13) and find $J = J(\bar{\Phi},\, g_{\mu\nu})$. Now we are in a position to introduce the effective action as a functional of the mean field and external metric,

$$\Gamma[\bar{\Phi},\, g_{\mu\nu}] \;=\; W[J,\, g_{\mu\nu}] - J\bar{\Phi}. \tag{13.14}$$

It is easy to see that all of the relations discussed in Part I are preserved in curved space. The main difference from the formal side is that now we need to make all of the operations covariant.

In principle, perturbation theory in curved spacetime is constructed via the same scheme as in the flat case. The quadratic part of the action defines the propagators, and other parts of the action of matter fields define vertices. It is clear that both propagators and vertices depend on an external gravitational field. Consequently, unlike the flat space case, both propagators and vertices require special definitions and need careful consideration. For instance, in flat spacetime, the propagators are defined on the basis of the T- product, or, equivalently, on the basis of Feynman's ϵ-prescription in the momentum representation. However, such a definition is not generally covariant. Moreover, it is unclear how to implement the momentum representation in an arbitrary curved spacetime. In addition, in flat spacetime, the propagator is defined as the vacuum expectation value of the product of two operators. However, since in curved spacetime, the vacuum is not unique, we face the additional problem of finding definitions for the propagators of matter fields. Thus, we can conclude that it is impossible to formulate the propagators in a closed form for an arbitrary external gravitational field. In this situation, the perturbation theory is formally incomplete and requires additional definitions.

Let us consider two examples.

1. Real scalar field theory. The propagator $D(x, y)$, corresponding to the free action of scalar field with nonminimal coupling, satisfies the equation

$$(\Box + m^2 - \xi R)D(x, y) \;=\; -\,i\delta_c(x, y), \tag{13.15}$$

where $\Box$ is a covariant scalar field d'Alembertian, acting on the argument x and

$$\delta_c(x, y) = \frac{\delta^4(x - y)}{[-g(x)]^{1/4}\,[-g(y)]^{1/4}} \tag{13.16}$$

is a covariant delta function.

2. Spinor (fermionic) field theory. The propagator $S(x, y)$, corresponding to the free action of a spinor field, satisfies the equation

$$\big\{i\gamma^\mu(x)\nabla_\mu - m\big\}\, S(x, y) \;=\; \delta^4_c(x, y)\, I, \tag{13.17}$$

where ∇_μ is a spinor field covariant derivative acting on argument x, and I is the 4×4 unit matrix in the spinor space. Both Eqs. (13.15) and (13.17) are covariant partial differential equations with non-constant coefficients. To find the unique solution of these equations, one need to use certain boundary conditions. As a consequence, the solutions of the equations (13.15), (13.17) may depends on the global structure of the spacetime. Thus, construction of the elements of the Feynman technique in curved spacetime is a highly non-trivial problem. We shall deal with this problem perturbatively in this and the next chapters.

On the other hand, the loop expansion for the effective action has no real difference when compared to the flat spacetime case. Thus, we can skip this part and just use the results from the first part of the book. The effective action can be written as a sum of the classical action and the expansion in the loop parameter, which can be associated to $\hbar$:

$$\Gamma = S + \bar{\Gamma}, \qquad \bar{\Gamma} = \sum_{n=1}^{\infty} \hbar^n \bar{\Gamma}^{(n)}\,. \tag{13.18}$$

At the one-loop level in curved spacetime, there is a standard expression,

$$\bar{\Gamma}^{(1)}[\bar{\Phi}] \;=\; \pm\,\frac{i}{2}\,\log\,\mathrm{Det}\,S_2[\bar{\Phi}] \;=\; \pm\,\frac{i}{2}\,\mathrm{Tr}\,\log\,S_2[\bar{\Phi}], \tag{13.19}$$

where $S_2(\bar{\phi})$ is the bilinear form of the classical action, which should be taken at $\phi = \bar{\phi}$, and the signs $+$ and $-$ correspond to bosonic (commuting) and fermionic (anticommuting) fields, respectively. A relevant observation is in order at this point. If the theory has both bosonic and fermionic fields, the notion of functional traces or determinants should be modified, because one has to take into account the Grassmann parities of the quantum fields. In the field configurations with mixed contents we have to use the so-called super-trace and super-determinant. A detailed description of the mathematical aspects of these notions can be found, e.g., in [49] and [81].

In Eq. (13.19), we generalized the well-known relation

$$\mathrm{tr}\,\log A = \det\,\log A \tag{13.20}$$

from linear algebra for the case of the differential operators typical for S_2. These operators can be associated to infinite-dimensional matrices and certainly cannot be regarded as subjects of linear algebra. Thus, it is not obvious that the functional trace Tr and the determinant Det will satisfy the relation Tr log = log Det ; however, we shall intensively use it in what follows. Let us give some explanations. Our immediate purpose is to investigate UV divergences, and they always correspond to the local terms in the effective action. If there is some relevant difference between Tr log and log Det , it can be related only to nonlocal terms; otherwise, it can be always reduced by an appropriate change in the renormalization conditions[1]. Thus, such a difference cannot be important for issues such as divergence, conformal anomaly, anomaly-induced effective action or effective potential - which is in all objects related to the leading logarithmic behaviour of the theory.

An important aspect of the effective action in curved space is that there is a part of this action, that does not depend on the mean matter fields Φ, but only on the external metric. The corresponding expression is called the vacuum effective action. Such an object can be constructed either as a part of (13.14), or independently, as will be described below. In some cases, these two definitions may be essentially different, typically the case for theories with spontaneous (or dynamical) symmetry breaking. Other complications may occur when we use the vacuum effective action, including an induced part, for deriving equations for the metric [15]. At the present, initial stage, we consider only relatively simple models without the induced part.

To obtain the vacuum functional, one has to switch off the source in expression (13.6), arriving at what is called the effective action of vacuum Γ_{vac},

$$Z_{vac}[g_{\mu\nu}] = e^{i\,\Gamma_{vac}[g_{\mu\nu}]} = \frac{1}{\mathcal{N}_0}\,e^{iS_{vac}[g_{\mu\nu}]}\int\mathcal{D}\Phi\,e^{i.S_m[\Phi,g_{\mu\nu}]}. \tag{13.21}$$

It is clear that if we define the normalization factor (13.7) without setting $g_{\mu\nu} = \eta_{\mu\nu}$, the vacuum effective action will be canceled. For physical applications, this means that we "forbid" quantum corrections to the classical action of gravity, which does not make much sense. Therefore, definition (13.7) is well-justified.

[1]The reader can find examples of the differences between the two expressions in the nonlocal part of effective action in Ref. [165]. This issue is too special to be dealt with in this book, so let us just say that the consideration in the mentioned paper is partially based on the fact that the relation Tr log = log Det is not violated for divergences.

According to Eq. (13.21), the effective action of vacuum is sum of the classical term and the quantum correction:

$$\Gamma_{vac} = S_{vac} + \bar{\Gamma}_{vac}, \tag{13.22}$$

where $\bar{\Gamma}_{vac}$ comes from the integral over Φ. It is clear that Γ_{vac} should be one of the main objects of our interest. The rest of quantum theory describes the effects of gravity on the interaction and dynamics of quantum matter fields. Let us remember that the dynamics of matter fields is mainly probed in laboratories, e.g., in experiments using accelerators. It is clear that gravity cannot play too important role, in these experiments. Thus, the most interesting object to explore is Γ_{vac}, which defines quantum contributions to the dynamics of proper gravity. For instance, we may expect that, taking into account the full $\Gamma_{vac}[g_{\mu\nu}]$ in the framework of consistent quantum field theory in curved space (or taking into account the quantum effects of the proper metric) resolves the problem of singularities. Another important question to answer (as we mentioned above) is whether the quantum contributions to the action of gravity, or even the conditions imposed by the consistency of quantum theory on the *classical* action of gravity, may provide a background for experimental or observational verifications of the theoretical constructions.

We expect that the choice of the classical action of vacuum S_{vac} and the quantum contributions $\bar{\Gamma}_{vac}$ will have the following properties:

1) For a correctly constructed model of semiclassical gravity from the laboratory and up to the astrophysical scale, the sum of the classical and quantum terms should be effectively close to the action of GR. In this case, the success of experimental and observational confirmations of GR is preserved within semiclassical gravity. This condition also includes the value of the cosmological constant.

2) The choice of the vacuum S_{vac} and matter S_m counterparts of the classical actions of gravity and matter fields should provide the consistency of quantum theory. For instance, the theory should be renormalizable or admit the use of effective field theory methods, if we restrict the consideration to the low-energy domain. In particular, this means that the action of vacuum must include a Einstein-Hilbert term with a cosmological constant. If other terms are added, they should be treated as small perturbations at the laboratory and astrophysical scale, except in some extreme situations, e.g., when close to a singularity.

The general requirements formulated above can be made even before providing the definition of the functional integral in curved space. Let us now discuss how this integral can be understood. The generating functional (13.6) depends on the external metric $g_{\mu\nu}$, which should be regarded as being unconstrained. Thus, one has to modify the perturbation theory in flat space, such that the new theory incorporates the external field. Indeed, we can reduce the problem to the functional integral in flat space. To this end, one has to write the metric as a sum of a flat Minkowski metric $\eta_{\mu\nu}$ and a perturbation $h_{\mu\nu}$,

$$g_{\mu\nu} = \eta_{\mu\nu} + h_{\mu\nu}\,. \tag{13.23}$$

The procedure implies that the action $S_m[\Phi, g]$ should be expanded into a power series in $h_{\mu\nu}$, such that the propagators and vertices of all the fields are the sums of the usual flat spacetime expressions and the terms which depend on $h_{\mu\nu}$. After that, one can deal with the modified Feynman diagrams in flat space. These new diagrams have extra tails of the fields $h_{\mu\nu}$, parameterizing the external metric field. This approach is useful both for performing practical calculations (especially beyond the one-loop order) and for analyzing renormalization in curved spacetime. A diagrammatic representation of the propagator in the procedure (13.23) is shown below; the propagator in flat spacetime produces an infinite set of diagrams with an unlimited number of external lines of $h_{\mu\nu}$.

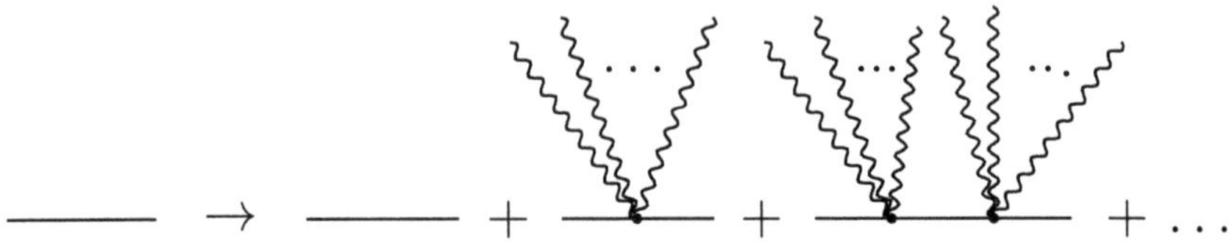

(13.24)

Exercises

13.1. Using the representation (13.23) and the expansion rules derived in chapter 12, expand the Klein-Gordon equation in curved space,

$$\hat{H}\varphi(x) = 0, \qquad \hat{H} = \hat{H}_x = \sqrt{-g(x)}\big(\Box - \xi R + m^2\big)_x, \tag{13.25}$$

up to the second order in $h_{\mu\nu}$. The representation can be written as

$$\hat{H} = \hat{H}_0 + \hat{H}_1 + \hat{H}_2 + \dots, \quad \text{where} \quad \hat{H}_k = \mathcal{O}(h^k). \tag{13.26}$$

13.2. The equation for the propagator is $\hat{H}(x)G(x,\, x') = \delta(x,\, x')$, and the solution can be found in form for the series

$$G(x,\, x') = G = G_0 + G_1 + G_2 + \dots . \tag{13.27}$$

As usual, this means $G_0 = \hat{H}_0^{-1}$, then $G_1 = -\hat{H}_0^{-1}\hat{H}_1\hat{H}_0^{-1}$, etc. i) Starting from the result of the previous exercise, find the first-order term of the propagator. ii) Derive the first few orders of the vertex expansion, which are defined by the interaction term

$$V_{int} = \frac{1}{4!}\sqrt{-g(x)}\,\lambda\varphi^4(x). \tag{13.28}$$

We recommend paying special attention to the spacetime indices and presenting both results in the momentum representation.

13.3. Using the results of chapter 12, repeat the program of the previous two exercises for Dirac fermions and massive (the action can be found in section 14.1.4) and massless vectors.

13.4. Discuss the limitations on the power of momenta of the external metric $h_{\mu\nu}$ in the expressions obtained in the previous exercises. Why are there no limitations on the order of the field $h_{\mu\nu}$, only in the number of derivatives?

13.5. Use Eq. (13.27) to explain the diagrammatic representation of the results of the propagators given in (13.24). Draw a similar diagram for the vertex. Explain why the same graphic presentation of a propagator can be used for all matter fields (scalar, spinor and vectors). Discuss how to show the derivatives of the external metric tails on these diagrams, and write the corresponding analytic expression for the fermion case.

13.2 Divergences and renormalization in curved space

In this section, we will discuss the structure of divergences in curved spacetime, show the form of the counterterms required to remove the divergences, and describe the simplest method to calculate them. It will be shown that, unlike the case for the flat spacetime theory, in order to perform the renormalization of the theory in an external gravitational field, we need not only the counterterms in the matter field sector but also the ones that depend on an external gravitational field.

Within the procedure based on (13.23), the internal lines of all the diagrams are only those of matter fields, while external lines may be of both matter and the gravitational field $h_{\mu\nu}$. As a result, any flat space diagram of the theory gives rise to an infinite set of diagrams with an increasing number of tails of the fields $h_{\mu\nu}$. The first diagrams of such a set are depicted in (13.29). Nevertheless, in spite of there being infinite amount of diagrams with $h_{\mu\nu}$ tails, the renormalization of the theory has universal structure and can be described in general terms.

One-loop self-energy diagrams for a scalar with $\lambda\varphi^3$ interaction [287] are shown below. Matter field corresponds to the continuous lines and wavy lines to the external $h_{\mu\nu}$ field. A single diagram in flat spacetime generates families of diagrams in curved spacetime.

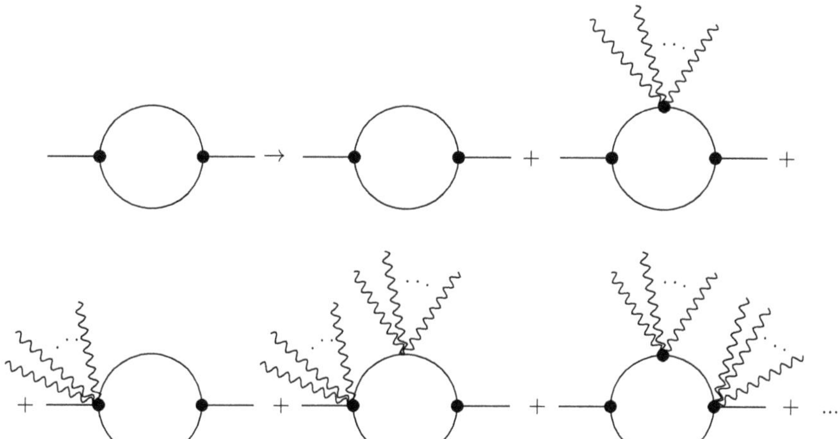

(13.29)

As was described in Part I, the construction of renormalized theory requires introducing an appropriate regularization. In general, the regularized effective action can be written as the sum of the divergent and the finite parts, $\Gamma = \Gamma_{div} + \Gamma_{fin}$, where Γ_{div} is divergent after the removal of the regularization, and Γ_{fin} is finite. The analysis of possible counterterms should correspond to Γ_{div} and can be performed using symmetries and the superficial degree of divergences, as discussed in detail in Part I. In contrast, renormalization in curved spacetime is more complicated.

The standard analysis of counterterms in flat space is based on a concrete form of the propagators in momentum space. As we discussed above, such a representation in curved spacetime is unknown, and is thought to depend on the global structure of spacetime. As we know, one of the possible ways out is to present the external gravitational metric in the form (13.23) and construct the Feynman rules in the form of perturbations in $h_{\mu\nu}$. Of course, in this case, we lose the information about the global structure of the spacetime. However, we know that the divergences originate from the singularities of the propagators, and these singularities are local. Thus, we can expect that the divergences have an universal form and do not depend on the global structure of the spacetime.

13.2.1 The general structure of renormalization in the matter sector

Suppose we have a gauge theory (it might be a version of the Standard Model or a GUT) which is renormalizable in flat spacetime. Our purpose is to construct a version of the theory that would be renormalizable in curved spacetime. The theory under discussion includes spinor, vector and scalar fields with gauge, Yukawa and four-scalar interactions, and is characterized by gauge invariance and maybe some other symmetries. It is useful to introduce, from the very beginning, nonminimal $\xi R\varphi^2$-type interactions between scalar fields and curvature. According to the discussion in the previous section, the general action can be presented in the form

$$S = \int d^4x\sqrt{-g}\Big\{ -\frac{1}{4}\,(G_{\mu\nu})^2 + \bar{\psi}\big(i\gamma^\alpha \mathcal{D}_\alpha - m - h\phi\big)\psi + \frac{1}{2}\,g^{\mu\nu}\,\mathcal{D}_\mu\phi\,\mathcal{D}_\nu\phi + \frac{1}{2}\,\big(\xi R - M^2\big)\,\phi^2 - V_{int}(\phi)\Big\} \,+\, S_{vac}. \qquad (13.30)$$

In this expression, ϕ and ψ are sets of scalar and spinor fields, respectively, in different representations of the gauge group, which also includes gauge fields A_μ with the field tensor $G_{\mu\nu}$. Furthermore, $\mathcal{D}$ denotes derivatives which are covariant with respect to both general coordinate and gauge transformations. The last term in (13.30) is the vacuum action, which can be e.g. (12.6), or have some other form. In what follows, we will see that expression (12.6) is exactly the minimal necessary version of the vacuum action that can guarantee a renormalizable quantum theory in curved space.

It is clear that the calculation of an infinite amount of diagrams like the ones presented in (13.29), is an infinitely difficult task. However, the good is that we actually do not need to calculate all these diagrams, as all of the relevant information about the divergences can be extracted from just a few diagrams. And, even better, there are functional methods which enable one to get some relevant information without deriving even a single diagram. Let us learn these step by step, starting from the analysis of the divergences.

In the minimal subtraction scheme of renormalization, which we shall follow up to a certain point, the perturbative renormalization of the theory depends on the possible counterterms. In particular, the action of a renormalizable theory has to include all the structures that can emerge as the counterterms required to cancel divergences.

Thus, we have to explore which kind of divergences one can meet in the theory in curved spacetime. We shall consider, in parallel, both general nonminimal theory (13.30) and its particular minimal version. Some remarks are in order. The general analysis of renormalization in curved spacetime, based on covariance arguments, has been given in [70, 80, 206].

The main two lessons which we can learn from the approach based on the representation (13.23) of the metric consist in the following:

i) The divergences that result from this approach possess general covariance, regardless of the non-covariant parametrization (13.23). There is an important general theorem (see, e.g., [335, 336], [334], [164], as well as the review paper [163], the book [345] and Part I of the present book) that states that the divergences of a gauge-invariant theory can be removed, at any loop order, by the gauge-invariant and local counterterms. The diffeomorphism invariance is a particular example of gauge invariance, and these general statements can be used here. In fact, the formal proof of covariance for semiclassical gravity has been given in[2] [206], generalizing the previous considerations in [70]. Thus, we are in a position to use general covariance for the analysis of the counterterms.

The covariance strongly helps to control diagrams such as those in (13.29) and many others. Here we are not going to reproduce or even discuss the general statements about renormalization of gauge theories in the presence of background fields [105, 11, 192], or the specific arguments which can be found in the papers [70, 206], or in the book [80], but instead will show in the next sections that there is an alternative, explicitly covariant perturbation technique based on the local momentum representation and Riemann normal coordinates. For a while, let us simply assume that the covariance of the divergences is guaranteed.

ii) As long as all of the diagrams that may contribute to divergences are flat space graphs with external lines of $h_{\mu\nu}$, one can rely on Weinberg's theorem [339] (see also [96]).[3] This means that the divergent contributions to effective action are given by local expressions. Taken together, covariance and locality form a solid basis for a general understanding of the possible form of divergences in curved spacetime.

Now we are in a position to start the analysis of the divergences. Let us review two relevant facts related to the notion of power counting, which was introduced in Part I.

First, consider a set of diagrams, each of them being a "descendent" of some flat space diagram with additional external lines of $h_{\mu\nu}$. As we already know, a new line of the external metric can be connected to the existing vertex or to a new vertex [see (13.24)]. In the former case, the index of divergence either does not change or decreases. On the other hand, when the number of vertices increases, the superficial degree of divergence for the given diagram only may decrease. Therefore, the insertion of new vertices of interaction with the background fields $h_{\mu\nu}$ cannot increase the superficial degree of divergence of any given diagram. In other words, by adding an external line of $h_{\mu\nu}$ to the diagram, one can either preserve the power counting, or decrease it. Such an operation can make a diagram finite or, e.g., convert the quadratically divergent diagram into a logarithmically divergent one. For example, the quadratically divergent diagram which is shown on the left of (13.31) as (a), generates the logarithmically

[2]The statements in [206] are formally valid only in the situation when there is no anomaly. In the present case, we have regularizations (say, dimensional [185,61,213], or properly used higher-derivative regularization [296, 12]) which preserve, on the quantum level, both general covariance and the gauge invariance of the model.

[3]The locality of UV divergences was considered a well-known fact even before this theorem, see, e.g., [59].

divergent ones of (13.31) shown in the plot (b). The diagrams from this one then gives rise to the logarithmically divergent $R\varphi^2$-type counterterm by the procedure presented in (13.29). We anticipated this divergence earlier using dimensional arguments. The diagrams in (13.31) will be discussed later on in exercise 13.6.

Second, since we are working with the renormalizable theory, there is a finite number of the divergent p-loop diagrams in flat spacetime. As a result, after generating the diagrams with external metric $h_{\mu\nu}$ tails, we meet a finite number of the *families* of divergent diagrams at any loop order.

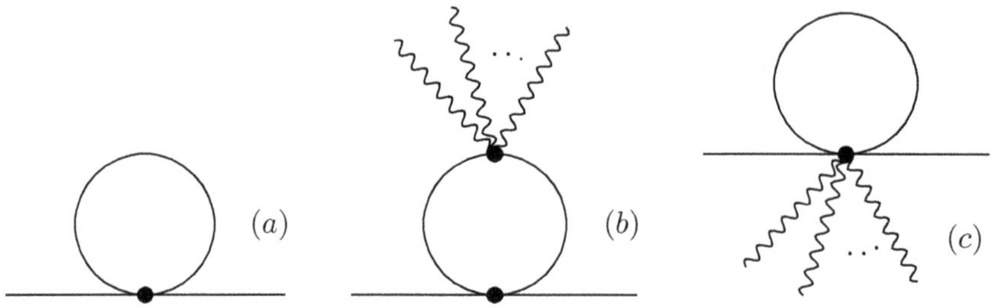

(13.31)

Taking all three points into account, we arrive at the following conclusion. The counterterms of the theory in an external gravitational background have the same dimension as the counterterms for the corresponding theory in flat spacetime. These counterterms are local and possess general covariance and gauge invariance, which are the two important symmetries of the classical action. Another important symmetry is local conformal invariance, which will be discussed in the next chapters.

At this stage, one can easily understand why the introduction of the nonminimal and vacuum terms (12.3) and (12.6) is important. The point is that the corresponding counterterms are possible since they have corresponding symmetries and proper dimensions. Let us see what will happen if we do not include these terms into the classical action and, instead, try to construct the quantum theory starting from the minimal version of the classical action, with $\xi = 0$ and also setting $S_{vac} \to 0$.

Even in the minimal case the classical action depends on the metric $g_{\mu\nu} = \eta_{\mu\nu} + h_{\mu\nu}$. Since the interaction with $h_{\mu\nu}$ will modify the diagrams, we can expect that the counterterms depending on $h_{\mu\nu}$ may appear. According to our analysis, these counterterms should be of the nonminimal and vacuum type. Then the minimal theory faces difficulty, since it is not possible to remove the nonminimal and vacuum counterterms through renormalization of the corresponding parameters in the classical action (all these parameters are zero by definition). In contrast, if we include the nonminimal and vacuum terms into the classical action, there is no problem of this kind. The corresponding divergent counterterms can be removed by renormalization transformation of ξ and the parameters of the vacuum action.

The conclusion is that a zero value of the nonminimal parameter ξ contradicts renormalizability. In the case of a multi-scalar theory with the fields φ_i, the general form of the nonminimal term is $\xi_{ij} R\varphi_i\varphi_j$. At the same time, in all known cases, the algebraic symmetries of a flat space theory can be preserved at the quantum level in curved space since the divergences that violate these symmetries do not show up.

For instance, for scalar fields with $O(N)$ symmetry it is sufficient to consider a single nonminimal parameter with a $\xi R\varphi_i\varphi_i$ nonminimal term.

The last observation is that the covariant action (13.30) with appropriate nonminimal $\xi R\varphi^2$-type terms produces quantum theory which is renormalizable in the matter sector, not only at one loop, but also at higher loops. The covariance and power counting arguments are not limited to the one-loop diagrams and have general validity.

13.2.2 The general structure of renormalization in the vacuum sector

Let us use the same approach to explore renormalization in the vacuum sector. As we discussed in the previous chapter, the principles of covariance, locality and the absence of parameters with the inverse mass dimension enable one to fix the form of the vacuum action to be (12.6), which consists of the Einstein-Hilbert action with the cosmological constant term (12.7) and the higher derivative terms (12.9). Is this sum really sufficient to provide renormalizability?

As a first example, consider the simplest one-loop vacuum diagram depicted in (13.32). For the sake of simplicity, let us forget about covariance and think only about the power counting, in the framework of cut-off regularization. It is worthwhile noting that the problem of providing covariance with cut-off is not simple to solve. We will return to this subject in the last few chapters, when we discuss the cosmological constant problem. Then we shall see explicitly that the considerations presented below are, in fact, correct and hold even in covariant formulation. For the present, however, we shall work with the non-covariant scheme, consider the Minkowski signature and use the cut-off Ω to regularize the integrals in momenta.

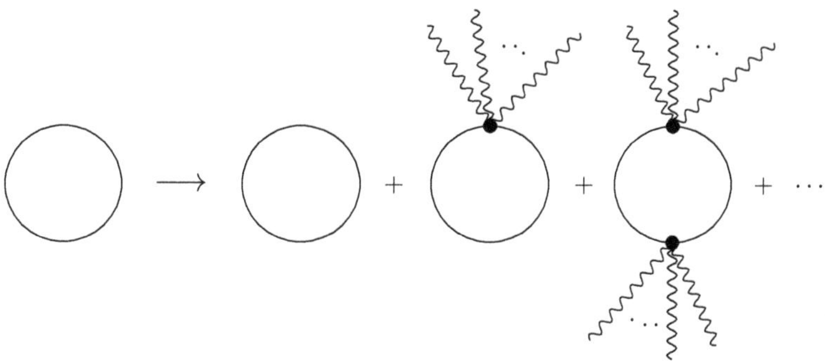

(13.32)

The original bubble at the *l.h.s.* of (13.32) gives the following contribution to the density of energy $\langle T_{00}\rangle$ of the vacuum:

$$\rho_{vac} = \frac{1}{2}\int\frac{d^3k}{(2\pi)^3}\,\sqrt{\mathbf{k}^2+m^2}\,. \tag{13.33}$$

It is clear that this integral provides a quartic divergence of $\sim \Omega^4$. According to (13.32), when we start to add new vertices with external lines of $h_{\mu\nu}$, we get extra factors of the propagator $(k^2+m^2)^{-1}$, which reduces the divergence. For example, a diagram with one vertex has quadratic rather than quartic divergence, and the diagram with two vertices has logarithmic divergence. The cut-off parameter has the dimension of mass, and, therefore, the condition of homogeneity in dimensions requires that the dimension in the diagrams with quadratic and logarithmic divergences should be compensated by derivatives of the external field $h_{\mu\nu}$. Therefore, a term with quadratic divergence has two derivatives of the metric $h_{\mu\nu}$. As we know from the first part of the book, the divergences can be always removed by adding a local counterterm to the action, so such a counterterm should be a scalar. It is well known that the unique local scalar with two derivatives of the metric is the scalar curvature R. Also, in the massive theory, there may be quartic, quadratic and logarithmic divergences with zero derivatives, since the dimension can be compensated by masses instead of metric derivatives. One can expect quadratic divergences to be removed by renormalizing the parameters of Newton and the cosmological constants in the Einstein-Hilbert action (12.7). The logarithmic divergences of the R and cosmological constant types are possible only in the massive theory.

The same logic tells us that, in a massless theory, the logarithmically divergent terms have exactly four derivatives of $h_{\mu\nu}$, while, in the theory of a field with the mass m, there may be also terms with two or zero derivatives, proportional to m^2 and m^4, correspondingly. When we add more vertices, we do not get qualitatively new divergences, because the overall dimension is restricted by four. Taking into account covariance we arrive at the conclusion that the counterterms needed to eliminate the logarithmic divergence are exactly of the forms listed in (12.9) for massless fields, and also of the forms (12.7) in the case of massive fields.

The next important observation is that the same situation also holds at higher-loop orders. The reason is that, in all the cases presented above, we used only the fact that the original flat space bubble diagram in (13.32) has dimension four. In the case for this primitive bubble, it is quartic divergence. But, even for an arbitrary multi-loop contribution, the dimension of the diagram is always 4, if the theory under consideration is renormalizable in flat space. Thus, we arrive at the conclusion that the divergences of the theory in curved space are the covariant generalizations of the ones in flat spacetime (even with the same coefficients) plus new divergences which emerge in the cases when the original flat space theory has quadratic or quartic divergences.

It is worth emphasizing that the considerations presented above are based on the notion of superficial degree of divergences, which was discussed in Part I. In curved spacetime, there can be divergent diagrams that correspond to the superficial degree of divergence $\omega = 4$ and do not dependent on the matter fields. In flat space, such divergences are irrelevant constants that can be omitted. In curved space, they depend on the external gravitational field and become relevant. The value $\omega = 4$ shows that there can be divergent local functionals constructed from the fourth power of the masses, from the squares of the masses multiplied by the scalar curvature and from the geometric invariants of the fourth power in metric derivatives. Thus, in order to arrive at a multiplicatively renormalizable theory, one has to introduce the classical action containing the cosmological constant (to absorb the divergences proportional to the fourth powers of the masses), an Einstein term with the Newton constant (to absorb the divergences proportional to the squares of masses), all possible terms quadratic in curvature tensor terms, and the d'Alembertian of scalar curvature.

All in all, in order to provide renormalizability in the vacuum sector, one needs non-zero values of the parameters of the vacuum action $a_{1,\dots,4}$, $1/G$ and Λ, including the ones for the higher-derivative terms. An important observation is that assuming zero cosmological constant Λ in the classical action of vacuum leads to a theoretical disaster, since an infinite cosmological constant would appear as a quantum correction. At the same time, one can always fix the renormalization condition such that the renormalized cosmological constant has a value which corresponds the observational data at the cosmic scale. In this way, one can provide a correspondence between renormalizable theory and the cosmological manifestations of the cosmological constant term. This possibility does not mean that the definition of the value of cosmological constant is free of problems. In fact, there is a problem there, and it is a very difficult one. We shall come back to this issue later on, when we discuss the induced gravity approach and the related cosmological constant problem.

The last observation in this section concerns higher derivatives in the vacuum action (12.6). Later on, in the chapters about quantum gravity, we shall see that since the action of vacuum has the term $\int_x C^2(x)$, the theory leads to higher derivatives in the equations for gravitational perturbations. Since the publication of the work of Ostrogradsky [238] (see [350] for recent reviews), it has been known that higher derivatives may lead to instabilities in classical solutions. On the other hand, without such terms, we get a non-renormalizable theory, which also implies instability. We will discuss all sides of this problem later on, after constructing a more solid theoretical background.

Exercises

13.6. Complement the logarithmically divergent diagrams represented by the diagram in the middle of (13.31) by adding the quadratically divergent diagrams with the same loop of external field.
Hint. These diagrams come from the expansion into power series in $h_{\mu\nu}$ of the $\sqrt{-g}$ in the vertex, which comes from the four-scalar interaction

$$-\frac{\lambda}{4!}\int d^4x\sqrt{-g}\,\varphi^4\,. \tag{13.34}$$

13.7. Consider all of the diagrams in the right plot in (13.31) using the expansion of the $\sqrt{-g}$, which is explained in section 13.1. Prove that the sum of all these graphs is proportional to $\sqrt{-g}$. Discuss the relation of this result to the covariance of divergences that was discussed in this section. Which type of counterterm is needed to remove these quadratic divergences? Calculate this counterterm.
13.8. Consider the theory of two real scalar fields, χ with the mass M and φ^k with $k = 1,\dots,N$ and the mass m. Assuming that the interaction term is $-g\int\sqrt{-g}\chi\varphi^k\varphi^k$, find the most general structure of the nonminimal terms that respect the $O(N)$ symmetry. Estimate the superficial degree of divergences in this model in flat space. Is this theory renormalizable, or superrenormalizable? Discuss whether, in this case, one can expect nonminimal divergences of the $R\chi^2$ or $R\varphi^k\varphi^k$ type. Explain the result by dimensional

arguments, and state this case is different from that for a theory with quartic scalar self-interaction.

13.9. Explain why the divergent term of a form $g^{\mu\nu}\partial_\mu\phi\partial_\nu\phi\, R_{\mu\nu}$ is forbidden in renormalizable theory.

13.3 Covariant methods: local momentum representation

The approach based on the expansion (13.23) is important because it gives us an important information about the locality of the counterterms and about the power counting. The first analysis of vacuum divergences in semiclassical gravity was done by means of this method in 1962 by Utiyama and DeWitt in the seminal paper [322]. At the same time, there are also other methods that can be used that are explicitly covariant at all intermediate stages of the calculations.

Basically, there are two available covariant techniques, both related to the background field method. One of them is called the heat-kernel method, or the Schwinger-DeWitt technique, and is the most useful method at the one-loop level. The majority of the practical calculations in this book will be performed using this method. However, the heat-kernel method is not really useful at higher loops, and we need a method for this case, e.g., to prove the general statement on the covariance of divergences. Therefore, let us consider a completely covariant method that can be used for calculating local quantities, such as counterterms, in all loop orders.

13.3.1 Riemann normal coordinates

There is a technique that, based on the local momentum representation, enables one to derive the divergencies in an explicitly covariant form. In what follows, we will introduce this method, starting from the Riemann normal coordinates. These special coordinates can be introduced on an arbitrary spacetime manifold and in an arbitrary spacetime dimension, but, for definiteness, we will write all expressions for the $(3+1)$-dimensional pseudo-Riemannian manifold, for which the metric is the unique field describing the spacetime geometry.

The construction of normal Riemann coordinates is described, e.g., in the book [252] and at a more introductory level in the recent book [292]. The application of normal coordinates and the local momentum representation for the quantum calculations in curved spacetime can be found in the papers [83, 240] (see also further references therein). Let us also mention [69], in which the one-loop divergences in quantum gravity have been derived using the local momentum representation.

Consider the manifold M_{3+1} and choose an arbitrary point P with the coordinates $x' = x'^\mu$. The normal coordinates $y^\mu = x^\mu - x'^\mu$ have to satisfy several conditions. In particular, the line of constant coordinates (except one in each line) consists of geodesics which are completely defined by the tangent vectors

$$\xi^\mu = \frac{dx^\mu}{d\tau}\bigg|_{x'}, \qquad \tau\bigg|_{x'} = 0, \tag{13.35}$$

where τ is the natural parameter along the geodesic. Furthermore, let the choice of ξ^μ be such that the metric at the point x' is the Minkowski one, $\eta_{\mu\nu}$. All of the quantities at the point x' will be denoted by primes, e.g.,

$$\Gamma'^{\lambda}_{\alpha\beta}(x) = \Gamma^{\lambda}_{\alpha\beta}(x'), \qquad \partial_\tau \Gamma'^{\lambda}_{\alpha\beta}(x) = \frac{\partial \Gamma^{\lambda}_{\alpha\beta}}{\partial x^\tau}\bigg|_{x=x'},$$

$$\partial_\tau \partial_\rho \Gamma'^{\lambda}_{\alpha\beta}(x) = \frac{\partial^2 \Gamma^{\lambda}_{\alpha\beta}}{\partial x^\tau \partial x^\rho}\bigg|_{x=x'}, \quad \text{etc}. \tag{13.36}$$

Any other coordinates z^α in the vicinity of the point x' are related to $y^\mu = \xi^\mu \tau$ by coordinate transformations, which can be presented in the form of the series

$$z^\mu = z'^\mu + A^\mu_\alpha y^\alpha + \frac{1}{2!}\, A^\mu_{\alpha\beta}\, y^\alpha y^\beta + \,\dots\, + \frac{1}{n!}\, A^\mu_{\alpha_1\alpha_2\dots\alpha_n} y^{\alpha_1} y^{\alpha_2} \dots y^{\alpha_n} + \dots, \tag{13.37}$$

where $A^\mu_\alpha, \dots, A^\mu_{\alpha_1\alpha_2\dots\alpha_n}$ are unknown constant coefficients.

It is possible to use the transformation law (11.4) to show that the coefficients $A^\mu_{\alpha_1\alpha_2\dots\alpha_n}$ can be chosen such that, in the transformed coordinates y^α[4], we have

$$\Gamma'^{\lambda}_{\mu\nu} = 0 \quad \text{and} \quad \partial_{(\mu_1\mu_2\dots\mu_n}\Gamma'^{\lambda}_{\alpha\beta)} = 0\,. \tag{13.38}$$

It is important to remember that these relations *do not* mean that *all* partial derivatives of the Christoffel symbols in the point P vanish, because these relations concern *only* the symmetrization of these partial derivatives.

For an arbitrary function $W(x)$, we can use the expansion

$$W(x) = W(x'+y) = W' + \partial_\alpha W'\, y^\alpha + \frac{1}{2}\,\partial_{\alpha\beta} W'\, y^\alpha y^\beta + \,\dots\,. \tag{13.39}$$

One can use condition (13.38) to calculate the expansion of the metric into the series in y^α. The coefficients are covariant expressions made from the Riemann tensor $R'_{\alpha\beta\mu\nu}$ and its covariant derivatives and contractions, all defined at the point P with the coordinates x'. We leave the details as exercises for the reader and present only the final result up to the sixth order in y^μ:

$$\begin{aligned}
g_{\alpha\beta}(y) &= g'_{\alpha\beta} + \frac{1}{3}\, R'_{\alpha\mu\nu\beta}\, y^\mu y^\nu + \frac{1}{3!}\, R'_{\alpha\mu\nu\beta\,;\,\sigma}\, y^\mu y^\nu y^\sigma \\
&+ \Big(\frac{1}{20} R'_{\alpha\mu\nu\beta\,;\,\rho\sigma} + \frac{2}{45}\, R'_{\alpha\mu\nu\lambda} R'^{\lambda}_{\ \cdot\,\rho\sigma\beta}\Big)\, y^\mu y^\nu y^\rho y^\sigma \\
&+ \frac{1}{6!}\Big[8R'_{\alpha\mu\nu\beta\,;\,\rho\sigma\kappa} + 16\big(R'_{\alpha\mu\nu\lambda\,;\kappa} R'^{\lambda}_{\ \cdot\,\rho\sigma\beta} + R'_{\alpha\mu\nu\lambda} R'^{\lambda}_{\ \cdot\,\rho\sigma\beta\,;\kappa}\big)\Big]\, y^\mu y^\nu y^\rho y^\sigma y^\kappa \\
&+ \frac{1}{7!}\Big[10R'_{\alpha\mu\nu\beta\,;\,\rho\sigma\kappa\omega} + 34\,\big(R'_{\alpha\mu\nu\lambda\,;\kappa\omega} R'^{\lambda}_{\ \cdot\,\rho\sigma\beta} + R'_{\alpha\mu\nu\lambda} R'^{\lambda}_{\ \cdot\,\rho\sigma\beta\,;\kappa\omega}\big) \\
&+ \frac{55}{2} R'_{\alpha\mu\nu\tau;\kappa} R'^{\tau}_{\ \cdot\,\rho\sigma\beta;\omega} + 16 R'_{\alpha\mu\nu\tau} R'^{\tau}_{\ \cdot\,\rho\sigma\lambda} R'^{\lambda}_{\ \cdot\,\kappa\omega\beta}\Big]\, y^\mu y^\nu y^\rho y^\sigma y^\kappa y^\omega + \,\dots\,.
\end{aligned} \tag{13.40}$$

Further orders of the expansion, as well as an alternative method of derivation, can be found in the papers [226]. As we have seen in the previous section, in four-dimensional spacetime, one can expect logarithmic divergences that are, at most, of the fourth order in curvature. The sixth-order terms were included in Eq. (13.40) mainly for the

[4] We use here condensed notation for multiple derivatives, namely, $\partial_{\mu_1\mu_2\dots\mu_n} A = \partial_{\mu_1}\partial_{\mu_2}\dots\partial_{\mu_n} A$.

sake of generality, while other relevant quantities will be given at lower orders. In order to obtain these expressions, one can, for instance, consider $g'_{\alpha\beta} = \eta_{\alpha\beta}$ and the whole expression (13.40) as $g_{\alpha\beta} = \eta_{\alpha\beta} + h_{\alpha\beta}$. After a certain algebra (which we leave as an exercise), one arrives at the expansions of the relevant quantities, e.g.,

$$\begin{aligned} g^{\alpha\beta}(y) &= g'^{\alpha\beta} - \frac{1}{3}R'^{\alpha\ \ \beta}_{\ \cdot\mu\nu\cdot}\, y^\mu y^\nu - \frac{1}{6}R'^{\alpha\ \ \beta}_{\ \cdot\mu\nu\cdot\,;\rho}\, y^\mu y^\nu y^\rho \\ &\quad - \frac{1}{20}R'^{\alpha\ \ \beta}_{\ \cdot\mu\nu\cdot\,;\rho\sigma}\, y^\mu y^\nu y^\rho y^\sigma + \frac{1}{15}R'^{\alpha\ \ \lambda}_{\ \cdot\mu\nu\cdot}\, R'^{\ \ \ \ \beta}_{\lambda\rho\sigma\cdot}\, y^\mu y^\nu y^\rho y^\sigma + \dots, \end{aligned} \tag{13.41}$$

$$\begin{aligned} \sqrt{-g(y)} &= 1 - \frac{1}{6}R'_{\mu\nu}\, y^\mu y^\nu - \frac{1}{12}R'_{\mu\nu;\rho}\, y^\mu y^\nu y^\rho \\ &\quad + \frac{1}{360}\Big\{5R'_{\mu\nu}R'_{\rho\sigma} - 9R'_{\mu\nu;\rho\sigma} - 2R'_{\mu\alpha\beta\nu}R'^{\ \alpha\beta}_{\rho\cdot\cdot\sigma}\Big\}\, y^\mu y^\nu y^\rho y^\sigma + \dots, \end{aligned} \tag{13.42}$$

$$\begin{aligned} R_{\mu\rho\nu\sigma}(x) &= R'_{\mu\rho\nu\sigma} + R'_{\mu\rho\nu\sigma\,;\lambda}\, y^\lambda \\ &\quad + \Big(\frac{1}{2}R'_{\mu\rho\nu\sigma\,;\lambda\tau} - \frac{1}{3}R'^{\gamma}_{\ \cdot\lambda\tau\mu}\, R'_{\rho\gamma\nu\sigma} - \frac{1}{3}R'^{\gamma}_{\ \cdot\lambda\tau\rho}\, R'_{\mu\gamma\nu\sigma}\Big)\, y^\lambda\, y^\tau + \dots. \end{aligned} \tag{13.43}$$

The fortunate feature of these power series is that all the coefficients are constructed from the curvature tensor and its covariant derivatives at the point P corresponding to $y = 0$. Therefore, all coefficients are manifestly covariant with respect to the coordinate transformations at the point x'. As we will see in what follows, normal coordinates can be used for formulating the local momentum representation. The last step is to construct the elements of the Feynman technique, using momentum representation at the unique point x'. The modified propagators and vertices depend on the curvature and its derivatives in this point. Thus, it is clear that this method can be especially useful for the calculation of local quantities such as counterterms. Let us remember that the locality of the counterterms is guaranteed by the Weinberg theorem. Then, the covariance is guaranteed by the use of the explicitly covariant technique based on normal coordinates.

Exercises

13.10. Show that the relations (13.36) can be provided by a choice of the symmetric coefficients $A^\mu_{\alpha_1\alpha_2\dots\alpha_n}$ in (13.37) . Make explicit calculations for $n = 1, 2$ and argue that the result is the same for all n.

13.11. Using (13.38) for $n = 0$ and $n = 1$, verify the relation

$$\partial_{(\mu}\Gamma'^{\lambda}_{\ \nu)\gamma} = \frac{1}{6}\left(R'^{\lambda}_{\ \cdot\mu\nu\gamma} + R'^{\lambda}_{\ \cdot\nu\mu\gamma}\right) = \frac{1}{3}R'^{\lambda}_{\ \cdot(\mu\nu)\gamma}. \tag{13.44}$$

Show that the next orders of (13.38), with $n = 3$ and $n = 4$, provide

$$\begin{aligned} \partial_{(\mu\nu}\Gamma'^{\lambda}_{\ \rho)\gamma} &= \frac{1}{2}R'^{\lambda}_{\ \cdot(\mu\nu|\gamma|\,;\rho)}, \\ \partial_{(\rho\sigma\mu}\Gamma'^{\lambda}_{\ \nu)\gamma} &= \frac{3}{5}R'^{\lambda}_{\ \cdot(\mu\nu|\gamma|;\rho\sigma)} - \frac{2}{15}R'^{\lambda}_{\ \cdot(\mu\nu|\tau|}R'^{\tau}_{\ \cdot\rho\sigma)\gamma}. \end{aligned} \tag{13.45}$$

13.12. Use the relations (13.45) to find the relation between partial and covariant derivatives of the curvature tensor up to the second order in curvature at the point P. Discuss what is supposed to change in the next orders in curvature.

13.13. Use the relations (13.45), the metricity condition (11.8) and the result of the previous exercise to verify expression (13.40) up to the fourth order in y^μ.
13.14. Verify formulas (13.41) and (13.42), at least in the first order in curvature.

13.3.2 Local momentum representation

Let us illustrate the general considerations presented at the end of the previous subsection by considering the scalar propagator. For the sake of simplicity, we shall perform calculations in the first order in curvature. Further detail and many more examples of calculations of this sort can be found in the paper by Bunch and Parker [240].

The starting action in the scalar case can be presented in the form

$$S_0 = \int d^4x \sqrt{-g}\,\Big\{\,\frac{1}{2}\,g^{\mu\nu}\partial_\mu\varphi\partial_\nu\varphi - \frac{1}{2}\,m^2\varphi^2 + \frac{1}{2}\,\xi R\,\varphi^2\,\Big\}. \tag{13.46}$$

The bilinear form (inverse propagator) is

$$\hat{H}(x,x') \;=\; -\,\frac{1}{\sqrt{-g}(x)}\,\frac{\delta^2 S_0}{\delta\varphi(x)\,\delta\varphi(x')} \;=\; \hat{H}\delta(x,x'). \tag{13.47}$$

In the scalar case, this bilinear operator is

$$\hat{H} \;=\; \big(\Box + m^2 \;-\; \xi R\big)_x\,. \tag{13.48}$$

The next step is to expand $\hat{H}$ in normal coordinates. For the d'Alembertian operator, we obtain, using Riemann normal coordinates, as described in the previous subsection,

$$\Box = \eta^{\mu\nu}\partial_\mu\partial_\nu + \frac{1}{3}\,R'^{\mu\;\;\nu}_{\;\cdot\alpha\cdot\beta}\,y^\alpha y^\beta\,\partial_\mu\partial_\nu - \frac{2}{3}\,R'^{\alpha}_{\beta}\,y^\beta\,\partial_\alpha + \dots, \tag{13.49}$$

Starting from this expression, all curvature tensor components are evaluated at the point with coordinates x'^μ; hence, we can omit primes in the expressions for the curvature. After a small algebra, we get the expression

$$\hat{H} \;=\; \eta^{\mu\nu}\partial_\mu\partial_\nu + \frac{1}{3}\,R^{\mu\;\;\nu}_{\;\cdot\alpha\cdot\beta} - \frac{2}{3}\,R^{\alpha}_{\beta}\,y^\beta\partial_\alpha \,+\, m^2 \;-\; \xi R + \;\dots\,. \tag{13.50}$$

Of course, the term $-\xi R$ must be also expanded, but, as long as we keep to only first order in curvature, this is irrelevant. Now we have the equation for the propagator,[5]

$$\hat{H}\,G(x,x') \;=\; -\delta(x,x'). \tag{13.51}$$

To be more precise, the *r.h.s* of (13.51) should contain the factor i, but we omit it for simplicity.

Our plan is to expand all of the quantities into series in the curvatures at the point x'. It is useful to make all of the calculations in the Euclidean n-dimensional case, so

[5] At this stage, we shall work with simple delta functions, as there is a significant difference between this version and the covariant version, see expression (13.67).

that we do not need to write $+i\varepsilon$ in the denominators. In this way, we look for the solution

$$G(y) = \int \frac{d^n k}{(2\pi)^n}\, e^{iky} \left[\frac{1}{k^2+m^2} + \frac{a_1 R - a_2 \xi R}{(k^2+m^2)^2} + \frac{a_3 R^{\mu\nu} k_\mu k_\nu}{(k^2+m^2)^3} + \dots , \right], \tag{13.52}$$

where a_1, a_2 and a_3 are coefficients which will be defined from Eq. (13.51).

Placing (13.52) into (13.51), we obtain the explicit form of the first and second term of this equation in the zero and first orders in curvature,

$$(\eta^{\mu\nu}\partial_\mu\partial_\nu \;-\; m^2)\, e^{iky} \;=\; -\,(k^2 \,+\, m^2)\, e^{iky}\,,$$

and

$$\begin{aligned} - \int \frac{d^n k}{(2\pi)^n}\, R^\alpha_\beta\, y^\beta \partial_\alpha\, e^{iky}\, \frac{1}{k^2+m^2} &= - \int \frac{d^n k}{(2\pi)^n}\, R^\alpha_\beta\, \frac{1}{i}\Big(\frac{\partial}{\partial k^\beta} e^{iky}\Big)\, \frac{ik_\alpha}{k^2+m^2} \\ = \int \frac{d^n k}{(2\pi)^n} R^\alpha_\beta e^{iky}\, \frac{\partial}{\partial k^\beta}\frac{ik_\alpha}{k^2+m^2} &= \int \frac{d^n k}{(2\pi)^n}\, R^\alpha_\beta e^{iky} \Big[\frac{\delta^\beta_\alpha}{k^2+m^2} - \frac{2k_\alpha k^\beta}{(k^2+m^2)^2}\Big], \end{aligned} \tag{13.53}$$

where we have integrated by parts in the momentum space at the last step.

Similar calculation for the third term gives

$$R^{\mu\ \ \nu}_{\ \ \alpha\ \ \beta} \int \frac{d^n k}{(2\pi)^n} y^\alpha y^\beta \partial_\mu\partial_\nu \frac{e^{iky}}{k^2+m^2} = \int \frac{d^n k}{(2\pi)^n} e^{iky} \left[\frac{2R_{\mu\nu}\, k^\mu k^\nu}{(k^2+m^2)^2} - \frac{R}{k^2+m^2}\right]. \tag{13.54}$$

Finally, substituting the expressions (13.53), (13.54) and (13.52) into Eq. (13.51), in the first order in curvature, we arrive at the following equation:

$$\frac{(a_2\xi - a_1)R}{k^2+m^2} - \frac{a_3 R^{\mu\nu} k_\mu k_\nu}{(k^2+m^2)^2} + \frac{(1/3)R - \xi R}{k^2+m^2} - \frac{(2/3)R^{\mu\nu}k_\mu k_\nu}{(k^2+m^2)^2} = 0, \tag{13.55}$$

which can be easily solved to define the coefficients $a_{1,2,3}$. At the end of the day, we obtain the expansion for the propagator in the first-order approximation,

$$G(k) \;=\; \frac{1}{k^2+m^2} + \frac{1}{3}\frac{(1-3\xi)R}{(k^2+m^2)^2} - \frac{2}{3}\frac{R^{\mu\nu}k_\mu k_\nu}{(k^2+m^2)^3} + \mathcal{O}(k^{-3})\,. \tag{13.56}$$

Higher-order terms would have greater negative powers of momenta, which would be compensated by higher orders of the curvature tensor and its covariant derivatives at the point P.

In principle, it is possible to continue this expansion to any order in curvature. In a similar manner, one can expand vertices to any desirable order, and construct Feynman diagrams at any loop order and in any order in curvature. Expansions of the vertices always have powers of y^μ; in this respect, they differ from expansions of the propagators. However, after being used in Feynman diagrams, the powers of y^μ can be elaborated via integration by parts in momentum space, exactly as was demonstrated above for the propagators. Thus, the local momentum representation leads to the manifestly covariant perturbation technique in curved spacetime. This technique provides covariant expression for any local quantity associated to the given point P with coordinates x'. Similar expansions can be constructed for spinors, vectors (for which many practical details can be found in [240]) and any other field of our interest.

Since the locality of the UV counterterms is guaranteed, one can draw strong conclusions from the covariant technique described above. First of all, the expansion of the propagator (13.56) and similar expansions of the vertices involve higher orders in curvature, which are compensated by higher orders in $(k^2+m^2)^{-1}$. Therefore, any flat space Feynman diagram, in curved spacetime gives rise to an infinite set of diagrams, corresponding to the expansion in the curvature and its covariant derivatives at the point P. But all these diagrams, except the zero-order one, have better UV convergence than the original "flat" diagram. If we start from the theory which is multiplicative renormalizable in flat spacetime, then total number of divergent diagrams, at any given loop order, is finite. Thus, the superficial degree of divergence for the part of the diagram which has a curvature-dependent factor is always smaller than the part that is identical to the original flat space expression.

Let us summarize the lessons we have learned from our examination of these two different techniques. Concerning the possible divergences in curved spacetime, we know that, independently on the order of the expansion in loops, the following rules hold:

i) The minimal covariant generalization of the divergences in flat spacetime has coefficients that are exactly the same as those in flat spacetime.

ii) On the top of that, there may be new divergences proportional to the components of the curvature tensor and its covariant derivatives. If the original flat space theory is renormalizable, the maximal dimension of the divergences in curved space is equal to 4. Therefore, there are counterterms of the first and second orders in the curvature tensor, while divergences of the higher orders are ruled out by power counting.

iii) The first curvature-dependent additional terms in (13.56) convert the quadratic and quartic divergent diagrams into the logarithmically divergent ones, depending on the curvature at the point x'. In this respect, the situation is very similar to the one illustrated in (13.29), (13.31) and (13.32) for the approach based on Eq. (13.23).

The last (almost trivial) observation is that the second order in curvature divergences cannot depend on matter fields, owing to the dimension arguments. In the end, we arrive at the conclusion that the list of possible counterterms includes minimal, nonminimal and vacuum ones. Therefore, the nonminimal prescription of generalizing the flat space theory leads to the renormalizable theory in curved spacetime.

Exercises

13.15. Using Eq. (13.53) as an example, verify Eq. (13.54).

13.16. Calculate the expansions for the propagators for massless and massive vector fields and the Dirac field in the local momentum representation up to the first order in curvature.

13.17. Derive the expansion of the vertices of φ^4 and Yukawa interactions, up to the first order in curvature.

13.3.3 Effective potential in curved spacetime: scalar fields

Most of our examples of explicit one-loop calculations will be done using the heat-kernel method and the Schwinger-DeWitt proper-time technique, which will described in subsequent sections. Indeed, for many one-loop calculations, this technique is making all of the work much more economical. For this reason, let us give just a very simple example (following Ref. [302]), which nicely illustrates how local momentum representation works. Consider the derivation of the effective potential of a scalar self-interacting field in the first order in curvature. This approximation is sufficient to see the main features of the method and, at the same time, is sufficiently simple.

The starting point will be the action of the real scalar field

$$S_0 = \int d^4x \sqrt{-g}\Big\{ \frac{1}{2}\, g^{\mu\nu}\, \partial_\mu\varphi\, \partial_\nu\varphi - \frac{1}{2}\,(m^2 - \xi\, R)\varphi^2 - V(\varphi) \Big\}, \qquad (13.57)$$

where the minimal potential term $V(\varphi) + \frac{1}{2}m^2\varphi^2$ is supplemented by the nonminimal term $\frac{1}{2}\xi R\varphi^2$, which is required to formulate the renormalizable theory in curved spacetime.

The effective potential V_{eff} is defined as a zero-order term in the derivative expansion of the effective action of the mean scalar field,

$$\Gamma[\varphi,\, g_{\mu\nu}] = \int d^4x \sqrt{-g}\Big\{ - V_{eff}(\varphi) + \frac{1}{2}\, Z(\varphi)\, g^{\mu\nu}\, \partial_\mu\varphi\, \partial_\nu\varphi + \dots \Big\}. \qquad (13.58)$$

Obviously, $V_{eff}(\varphi) = \frac{1}{2}m^2\varphi^2 + V(\varphi) + \bar{V}_0(\varphi)$ can be calculated for constant φ. According to the general arguments presented above, the theory with $V \sim \varphi^4$ is renormalizable in both flat and curved space; therefore, we expect to observe this feature after deriving the one-loop effective potential.

Let us briefly consider the flat space calculation first. The result for the massless scalar field is pretty well known [94]. The derivation for the massive case can be found, e.g., in the textbook [332], based on the use of Feynman diagrams. However, this method is different from our method of calculation, so we suggest the former as an exercise.

The one-loop contribution $\bar{V}^{(1)}$ represents a quantum correction to the classical expression of the potential $V(\varphi) + \frac{1}{2}m^2\varphi^2$. It proves useful to subtract the trivial vacuum term from the quantum contribution,

$$\bar{V}_0(\varphi) = \frac{1}{2}\, \mathrm{Tr}\, \log S_2(\varphi) - \frac{1}{2}\, \mathrm{Tr}\, \log S_2(\varphi = 0). \qquad (13.59)$$

Here $S_2(\varphi)$ is the bilinear form of the classical action in the background field formalism. The subtraction of the vacuum term in (13.59) can be seen as a normalization of a functional integral. This term arises naturally through the diagrammatic representation of the effective potential [332]. Here and below, we omit an infinite volume factor.

Let us use the simplest version of the cut-off regularization. By introducing the four-dimensional Euclidean momentum cut-off Ω, we arrive at the following result:[6]

[6] In all momentum integrals in this section, we assume that the Wick rotation to Euclidean space has been performed. Also, we omit the indication to the one-loop order, e.g., $\bar{V} = \bar{V}^{(1)}$.

$$\bar{V}_0(\varphi, \eta_{\mu\nu}) = \frac{1}{32\pi^2} \int_0^\Omega k^2\, dk^2 \, \log\Big(\frac{k^2 + m^2 + V''}{k^2 + m^2}\Big). \tag{13.60}$$

After taking this integral, we obtain,

$$\bar{V}_0(\varphi, \eta_{\mu\nu}) = \bar{V}_0 \,=\, \bar{V}_0^{div} + \bar{V}_0^{fin}, \tag{13.61}$$

$$\bar{V}_0^{div} \,=\, \frac{1}{32\pi^2}\Big\{\Omega^2 V'' - \frac{1}{2}\,(m^2 + V'')^2 \log\frac{\Omega^2}{m^2}\Big\}, \tag{13.62}$$

$$\bar{V}_0^{fin} \,=\, \frac{1}{32\pi^2}\Big\{\frac{1}{2}\,(m^2 + V'')^2 \,\log\big(1 + \frac{V''}{m^2}\big) - \frac{1}{4}\,(m^2 + V'')^2\Big\}. \tag{13.63}$$

In the last two expressions, we have included the φ-independent m^4-type term, which is a part of the second term in (13.59). The naive quantum contribution (13.61) must be supplemented by an appropriate local counterterm, which we choose to be in the form:[7]

$$\Delta V_0 = \frac{1}{32\pi^2}\Big\{ - \Omega^2 V'' + \frac{1}{2}\,(m^2 + V'')^2 \,\log\frac{\Omega^2}{\mu^2} + \frac{1}{4}\,(m^2 + V'')^2\Big\}. \tag{13.64}$$

Looking at the counterterm (13.64), it is easy to see that the renormalizable theory is the one with the classical potential $V(\varphi) \sim \varphi^4$. For this choice of potential, the counterterms have the same functional dependence on the scalar field as the classical potential, with an additional cosmological constant term. At the next stage, we will see that the same feature holds in curved space if the nonminimal term $\xi R\varphi^2$ is introduced. Another simple choice of potential is $V(\varphi) \sim \varphi^3$, but, in this case, the potential is not bounded from below. For any other choice of potential, the counterterms do not have the same form as the terms in the classical action.

Let us make an observation concerning the subtraction of quadratic divergences. One can replace the Ω^2-type counterterm by the more general expression $-(\Omega^2 - \mu_1^2)V''$, where μ_1 is one more dimensional parameter of renormalization, independent from μ. In this case, we arrive at the more general renormalization ambiguity with two dimensional parameters. The reason why we will not do it is that the ambiguity related to the quadratic divergence does not lead to the running parameters, unlike that for the logarithmic divergence.

Adding the counterterm (13.64), we eliminate both the quadratic and the logarithmic divergences and arrive at the simple form of the renormalized effective potential:

$$\begin{aligned} V_{eff,\,0}^{ren}(\eta_{\mu\nu}, \varphi) &= m^2\varphi^2 \,+\, V \,+\, \bar{V}_0 \,+\, \Delta V_0 \\ &= m^2\varphi^2 \,+\, V \,+\, \frac{1}{64\pi^2}\,(m^2 + V'')^2 \log\Big(\frac{m^2 + V''}{\mu^2}\Big). \end{aligned} \tag{13.65}$$

Now we are in a position to generalize this flat space result to the curved space case, by using Riemann normal coordinates and local momentum representation. The

[7]For the sake of convenience, we have included into ΔV the finite term, regardless it can be compensated by changing μ.

bilinear operator of the action (13.57) is a slight modification of (13.50), which comes from the scalar potential,

$$
\begin{aligned}
-\hat{H} &= -\frac{1}{\sqrt{-g}}\,\frac{\delta^2 S_0}{\delta\varphi(x)\,\delta\varphi(x')} = \Box + m^2 - \xi R + V'' \\
&= \eta^{\mu\nu}\partial_\mu\partial_\nu + \frac{1}{3}R^{\mu\ \ \nu}_{\ \cdot\alpha\cdot\beta}\,y^\alpha y^\beta \partial_\mu\partial_\nu - \frac{2}{3}R^\alpha_\beta\, y^\beta\partial_\alpha + m^2 - \xi R + V''.
\end{aligned}
\tag{13.66}
$$

The equation for the propagator of the scalar field has the form

$$
\hat{H}\,G(x,x') = -\,g^{-1/4}(x')\,\delta(x,x')g^{-1/4}(x) = -\,\delta_c(x,x'). \tag{13.67}
$$

Here $\delta_c(x,x')$ is the covariant delta function, which is symmetric in the arguments x and x' and also satisfies

$$
\int d^n y\,\sqrt{g(y)}\,f(y)\,\delta_c(x,y) = f(x). \tag{13.68}
$$

It proves better to work with the modified propagator (see [83]) $\bar{G}(x,x')$, where

$$
\hat{H}\,\bar{G}(x,x') = -\,\delta(x,x')\,. \tag{13.69}
$$

It is important for us that the *r.h.s.* of the last relation does not depend on the metric, because we are going to use the relation $\mathrm{Tr}\,\log\hat{H} = -\,\mathrm{Tr}\,\log G(x,x')$ to obtain the dependence on the curvature. For this reason, it is more useful for us to have the *r.h.s.* of the equation for the propagator (13.69), which does not require the expansion in normal coordinates.

The explicit form of $\bar{G}(x,x')$ has been known for a long time (see [83]) for the free $V''=0$ case. We have

$$
\begin{aligned}
g^{1/4}\Box g^{-1/4} &= \Big(1 - \frac{1}{12}R'_{\mu\nu}\,y^\mu y^\nu + \cdots\Big)\Big(\Box + \frac{1}{6}R'_{\alpha\beta}\eta^{\alpha\beta} \\
&\qquad + \frac{1}{6}R'_{\alpha\beta}\,y^\alpha\partial^2 y^\beta + \frac{1}{12}R'_{\alpha\beta}\,y^\alpha y^\beta\partial^2 + \cdots\Big) \\
&= \partial^2 + \frac{1}{6}R + \frac{1}{3}\Big(R'^{\mu\ \ \nu}_{\ \cdot\alpha\cdot\beta}\,y^\alpha y^\beta\,\partial_\mu\partial_\nu - R'^\alpha_\beta\,y^\beta\,\partial_\alpha\Big) + \cdots\,,
\end{aligned}
\tag{13.70}
$$

where the derivatives are $\partial_\alpha = \frac{\partial}{\partial y^\alpha}$, $\partial^2 = \eta^{\mu\nu}\partial_\mu\partial_\nu$ and $\cdots$ stands for the terms of higher orders in the curvature and its covariant derivatives. The calculation leading to this expression is left as an exercise.

From this point on, we shall omit the primes over the curvatures, as all of the curvature tensors are at the same point, P. As long as $V''(\varphi) = const$, we can simply replace m^2 by $\tilde{m}^2 = m^2 + V''$ and obtain, in the linear in curvature approximation,

$$
\bar{G}(y) = \int \frac{d^4k}{(2\pi)^4}\,e^{iky}\left[\frac{1}{k^2+\tilde{m}^2} - \Big(\xi - \frac{1}{6}\Big)\frac{R}{(k^2+\tilde{m}^2)^2}\right]. \tag{13.71}
$$

Now it is a simple exercise to expand $\mathrm{Tr}\,\log\hat{H} = -\,\mathrm{Tr}\,\log G(x,x')$ up to the first order in the scalar curvature. We define

$$\hat{H} = \hat{H}_0 + \hat{H}_1 R + \mathcal{O}(R^2)\,, \qquad \bar{G} = \bar{G}_0 + \bar{G}_1 R + \mathcal{O}(R^2)$$

and consider, in the first-order approximation,

$$-\frac{1}{2}\,\mathrm{Tr}\,\log \bar{G}(x,x') = \frac{1}{2}\,\mathrm{Tr}\,\log\big(\hat{H}_0 + \hat{H}_1 R\big) = \frac{1}{2}\,\mathrm{Tr}\,\log \hat{H}_0 + \frac{1}{2}\,\mathrm{Tr}\,\big(\bar{G}_0\,\hat{H}_1\,R\big). \tag{13.72}$$

The first term in the last expression has been calculated above, and the second one can be transformed as follows:

$$\begin{aligned}
\frac{1}{2}\,\mathrm{Tr}\,\big(\bar{G}_0\,\hat{H}_1\,R\big) &= -\int d^4x\, V_1\, R \;=\; \frac{1}{2}\,\mathrm{Tr}\,\big[\bar{G}_0^{-1}(x'',\,x')\,\bar{G}_1(x',\,x)\big]\,R \\
&= \frac{1}{2}\int d^4x\, d^4x'\,\big[\bar{G}_0^{-1}(x,\,x')\,\bar{G}_1(x',\,x)\big]\,R \\
&= \frac{1}{2}\int d^4x\, d^4x'\, R\int \frac{d^4k}{(2\pi)^4}e^{ik(x-x')}\int \frac{d^4p}{(2\pi)^4}e^{ip(x'-x)}\bar{G}_0^{-1}(k)\,\bar{G}_1(p) \\
&= \frac{1}{2}\int d^4x\; R\int \frac{d^4k}{(2\pi)^4}\,\bar{G}_0^{-1}(k)\,\bar{G}_1(-k).
\end{aligned} \tag{13.73}$$

The last integration is trivial, owing to the simple form of $\bar{G}_0(k)$ and to the feature $\bar{G}_1(k) = \bar{G}_1(-k)$ in (13.71). The final result reads

$$\bar{V}(\varphi,\,g_{\mu\nu}) = \bar{V}_0 + \bar{V}_1\,R\,, \tag{13.74}$$

where $\bar{V}_0$ is given by the expression (13.61) and

$$\bar{V}_1 = \bar{V}_1^{div} + \bar{V}_1^{fin}, \quad \text{with} \tag{13.75}$$

$$\bar{V}_1^{div} = \frac{1}{2(4\pi)^2}\Big(\xi - \frac{1}{6}\Big)\Big\{-\,\Omega^2 + \big(m^2 + V''\big)\,\log\frac{\Omega^2}{m^2}\Big\}, \tag{13.76}$$

$$\bar{V}_1^{fin} = -\,\frac{1}{2(4\pi)^2}\Big(\xi - \frac{1}{6}\Big)\,\big(m^2 + V''\big)\log\Big(\frac{m^2 + V''}{m^2}\Big). \tag{13.77}$$

As in the flat space case, the curvature-dependent part of the effective potential must be modified by adding the counterterm

$$\Delta V_1 = \frac{1}{2(4\pi)^2}\Big(\xi - \frac{1}{6}\Big)\Big\{\Omega^2 - \big(m^2 + V''\big)\,\log\frac{\Omega^2}{\mu^2}\Big\}. \tag{13.78}$$

As a result, we eliminate the quadratic and logarithmic divergences and arrive at the renormalized expression

$$V_{eff,\,1}^{ren}(g_{\mu\nu},\,\varphi) = -\,\xi\varphi^2 - \frac{1}{2(4\pi)^2}\Big(\xi - \frac{1}{6}\Big)\big(m^2 + V''\big)\,\log\Big(\frac{m^2 + V''}{\mu^2}\Big). \tag{13.79}$$

Obviously, the renormalizable theory has the nonminimal term in the classical expression (13.57), as, without this term, we can not deal with the corresponding counterterm (13.78). This fact is in accordance with our previous general considerations, including the analysis based on the simplest $g_{\mu\nu} = \eta_{\mu\nu} + h_{\mu\nu}$ prescription for defining curved-space quantum theory.

Generalizing the flat space result (13.65) to the covariant case and summing up with (13.79), we arrive at the complete one-loop renormalized expression

$$V_{eff}^{ren}(g_{\mu\nu}, \varphi) = \rho_\Lambda + \frac{1}{2}(m^2 - \xi R)\varphi^2 + V \\ + \frac{\hbar}{2(4\pi)^2}\Big[\frac{1}{2}\left(m^2 + V''\right)^2 - \Big(\xi - \frac{1}{6}\Big)R\left(m^2 + V''\right)\Big]\log\Big(\frac{m^2 + V''}{\mu^2}\Big). \tag{13.80}$$

Here we restored the loop expansion parameter $\hbar$ to its rightful place and included the classical density of the cosmological constant term, ρ_Λ, both for the sake of completeness.

It is instructive to show how the beta functions are obtained from the effective potential [302]. The renormalized potential is equal to the bare potential, so

$$\rho_\Lambda + (m^2 - \xi R)\varphi^2 + \frac{\lambda\varphi^4}{4!} + \Delta V_0 + R\Delta V_1 \\ = \rho_{\Lambda(0)} + \big[m_{(0)}^2 - \xi_{(0)}R\big]\varphi^2 + \frac{\lambda_{(0)}\varphi^4}{4!}\,. \tag{13.81}$$

The *l.h.s.* of this relation depends on μ explicitly, but the *r.h.s.* does not. This condition should be satisfied for all terms separately, because there are arbitrary quantities φ and R. Therefore, using (13.64) and (13.78), we arrive at the equations

$$\begin{aligned} \rho_{\Lambda(0)} &= \rho_\Lambda + \frac{m^4}{2(4\pi)^2}\log\frac{\Omega^2}{\mu^2}\,, \\ m_{(0)}^2 &= m^2 + \frac{\lambda m^2}{2(4\pi)^2}\log\frac{\Omega^2}{\mu^2}\,, \\ \xi_{(0)} &= \xi + \frac{\lambda}{2(4\pi)^2}\Big(\xi - \frac{1}{6}\Big)\log\frac{\Omega^2}{\mu^2}\,, \\ \lambda_{(0)} &= \lambda + \frac{4!\,\lambda^2}{16\,(4\pi)^2}\log\frac{\Omega^2}{\mu^2}\,. \end{aligned} \tag{13.82}$$

Taking the logarithmic derivatives $\mu\frac{d}{d\mu}$ of the bare quantities $\rho_\Lambda^{(0)}$, $m_{(0)}^2$, $\xi_{(0)}$ and $\lambda_{(0)}$, we set them to zero and arrive at the following beta functions:

$$\begin{aligned} \mu\frac{d\rho_\Lambda}{d\mu} &= \frac{m^4}{2(4\pi)^2}\,, &\quad \rho_\Lambda(\mu_0) &= \rho_{\Lambda 0}\,; \\ \mu\frac{dm^2}{d\mu} &= \frac{\lambda}{(4\pi)^2}\,m^2\,, &\quad m^2(\mu_0) &= m_0^2\,; \\ \mu\frac{d\xi}{d\mu} &= \frac{\lambda}{(4\pi)^2}\Big(\xi - \frac{1}{6}\Big)\,, &\quad \xi(\mu_0) &= \xi_0\,; \\ \mu\frac{d\lambda}{d\mu} &= \frac{3\,\lambda^2}{(4\pi)^2}\,, &\quad \lambda(\mu_0) &= \lambda_0\,, \end{aligned} \tag{13.83}$$

where the initial conditions are fixed at the scale parameter μ_0. Later on, we shall obtain the same beta functions from the heat-kernel calculations in dimensional regularization. The solution of the last equation in the leading-log approximation is

$$\lambda(\mu) \;=\; \lambda_0 + \frac{3\,\hbar\,\lambda_0^2}{(4\pi)^2}\,\log(\mu/\mu_0)\,, \qquad (13.84)$$

where we restored $\hbar$. Eqs. (13.83) can be solved in a similar way. Replacing the solutions into the renormalized effective potential (13.80), the dependence on μ completely disappears in the $\mathcal{O}(\hbar)$-terms. This definitely does not mean that the effective potential becomes trivial, because the real content of the quantum corrections is related to the dependence on φ, which did not change under the procedure described above. We conclude that μ-dependence is nothing but a useful tool for obtaining the dependence on φ or on its derivatives (or on another mean field, in other models). In general, the ambiguity related to μ can be eliminated by imposing renormalization conditions.

The relative importance of the gravitational term in (13.80) is strongly dependent on the values of the mass m and the nonminimal parameter ξ and on the magnitude of the scalar curvature R in a given physical problem. In this respect the relation between "flat" and "curved" terms in (13.80) is similar for both the classical term and the one-loop contribution. Consider a simple example of numerical evaluation. In the case when the scalar field is the Standard Model Higgs, we can assume that its mass is about $125\,GeV$. On the other hand, it is known that the magnitude of ξ needed for the Higgs inflation model [51] is about $4 \cdot 10^4$. Then it is easy to evaluate the value of curvature when the gravitational term in (13.80) become dominating. From the relation $\xi R = m^2$ we see that the critical value of curvature is $R \propto 0.3\,GeV^2$. In the cosmological setting, the corresponding value of the Hubble parameter is, therefore, $H \propto GeV$, which is much greater than the phenomenologically acceptable value for the present-day universe. This consideration shows that the aforementioned value of ξ is not unnaturally large, because the dimensional product ξR remains small, at least after the inflationary period.

13.3.4 The fermion contribution to effective potential

For the sake of completeness, consider fermionic contributions to the effective potential. The classical action of the curved-space spinor with Yukawa interaction is

$$S_Y = \int d^4x\sqrt{-g}\Big\{i\bar{\Psi}_i\,(\gamma^\mu\nabla_\mu + iM + ih\varphi)\,\delta^{ij}\,\Psi_j\Big\}, \quad i,j = 1,2,\dots,N\,. \qquad (13.85)$$

In the case of the potential, the background field φ can be treated as a constant; hence, it is useful to denote $\tilde{M} = M + h\varphi$. Taking the Grassmann parity of the quantum field into account, in the Euclidean notations we get

$$\Gamma^{(1)}_f[\varphi, g_{\mu\nu}] \;=\; -\;\mathrm{Tr}\,\log\hat{H}_f, \qquad (13.86)$$

where $\hat{H}_f = i(\gamma^\mu\nabla_\mu + i\tilde{M})\delta^{ij}$. It proves useful to consider

$$\mathrm{Tr}\,\log\hat{H}_f = \frac{1}{2}\,\mathrm{Tr}\,\log(\hat{H}_f\hat{H}^*_f), \qquad (13.87)$$

with $\hat{H}_f^* = i(\gamma^\mu \nabla_\mu - i\tilde{M})\delta_{jk}$. After some algebra this gives

$$\mathrm{Tr}\ \log \hat{H}_f = \frac{1}{2}\ \mathrm{Tr}\ \log\Big(-\Box + \frac{1}{4}R - \tilde{M}^2\Big)\delta^i_k. \tag{13.88}$$

The fermion propagator is defined from the relation

$$(\hat{H}_f \hat{H}_f^*)\,\mathcal{G}(x,x') \;=\; -\delta^c(x,x'). \tag{13.89}$$

Following the same scheme as was used in the scalar case, one can define a modified propagator

$$\mathcal{G}(x,x') \;=\; \big[-g(x)\big]^{-\frac{1}{4}}\,\bar{\mathcal{G}}(x,x'), \tag{13.90}$$

which satisfies the equation

$$(\hat{H}_f \hat{H}_f^*)\bar{\mathcal{G}}(x,x') = -\delta(x,x'). \tag{13.91}$$

One can prove that (see [83]) $\bar{\mathcal{G}}(x,x') = \bar{\mathcal{G}}(y)$ is defined as

$$\bar{\mathcal{G}}(y) = \int \frac{d^4k}{(2\pi)^4} e^{iky}\Big[1 - \frac{1}{12}R\frac{\partial}{\partial \tilde{M}^2}\Big](k^2 + \tilde{M}^2)^{-1}\hat{1} \tag{13.92}$$

in the first order in curvature. Thus,

$$-\frac{1}{2}\ \mathrm{Tr}\ \log \mathcal{G}(x,x') = \frac{1}{2}\ \mathrm{Tr}\ \log \hat{H}_f = \frac{1}{4}\ \mathrm{Tr}\ \log(\hat{H}_f\hat{H}_f^*). \tag{13.93}$$

Using the same considerations as for the scalar field, we find that

$$-\frac{1}{2}\ \mathrm{Tr}\ \log \mathcal{G}(x,x') = \frac{1}{4}\ \mathrm{Tr}\ \log(\hat{H}_f\hat{H}_f^*)_0 \;+\; \frac{1}{4}\ \mathrm{Tr}\,\bar{\mathcal{G}}_0(\hat{H}_f\hat{H}_f^*)_1 R. \tag{13.94}$$

The first term in the *r.h.s.* corresponds to the flat space case, and the second one is the contribution in the first order in contribution. For the flat space case, we get

$$\frac{1}{4}\ \mathrm{Tr}\ \log(\hat{H}_f\hat{H}_f^*)_0 \;=\; \frac{1}{4}\ \mathrm{sTr}\ \log(-\eta^{\mu\nu}\partial_\mu\partial_\nu - \tilde{M}^2).$$

In the momentum representation, this gives

$$\frac{1}{4}\ \mathrm{Tr}\ \log(\hat{H}_f\hat{H}_f^*)_0 \;=\; (-2N)\int_0^\Omega \frac{dk^2}{(4\pi)^2}k^2 \log\Big(\frac{1}{k^2 + \tilde{M}^2}\Big),$$

providing the first part of the one-loop effective potential, $V_0(fer) = V_0^{div}(fer)$,

$$V_0^{div}(fer) \;=\; -\frac{2N}{(4\pi)^2}\Big\{\frac{1}{2}\log\Big(\frac{\Omega^2}{\tilde{M}^2}\Big)\tilde{M}^4 + \frac{1}{4}\Omega^2\Big\}. \tag{13.95}$$

In order to renormalize this result, we introduce a counterterm of the form

$$\Delta V_0 = \frac{2N}{(4\pi)^2}\Big\{\frac{1}{2}\log\Big(\frac{\Omega^2}{\mu^2}\Big)\tilde{M}^4 + \frac{1}{4}\Omega^2\Big\}. \tag{13.96}$$

Thus,

$$V_0^{ren}(fer) = \frac{N}{(4\pi)^2}\log\Big(\frac{\tilde{M}^2}{\mu^2}\Big)\tilde{M}^4. \tag{13.97}$$

The contribution in the first order of curvature is

$$\frac{1}{4}\,\mathrm{Tr}\,\bar{\mathcal{G}}_0(\hat{H}_f\hat{H}_f^*)_1R = -2N\int d^4xR\int\frac{d^4k}{(2\pi)^4}\bar{\mathcal{G}}_0^{-1}(k)\bar{\mathcal{G}}_1(k),$$

which can be written in momentum space as

$$\frac{1}{4}\,\mathrm{Tr}\,\bar{\mathcal{G}}_0(\hat{H}_f\hat{H}_f^*)_1R = -\frac{N}{6(4\pi)^2}R\int_0^{\Omega}dk^2\frac{k^2}{k^2+\tilde{M}^2}.$$

Therefore,

$$V_1^{div}(fer)R = \frac{N}{6(4\pi)^2}R\Big\{\tilde{M}^2\log\frac{\Omega^2}{\tilde{M}^2} + \Omega^2\Big\}, \tag{13.98}$$

without the finite part. The divergences can be eliminated by adding the counterterm

$$\Delta V_1 = -\frac{N}{6(4\pi)^2}\Big\{\tilde{M}^2\log\frac{\Omega^2}{\mu^2} + \Omega^2\Big\}. \tag{13.99}$$

Finally, the part of the renormalized effective potential that is the contribution in the first order in curvature has the form

$$V_{ren}(fer) = -\frac{N}{(4\pi)^2}\Big\{\tilde{M}^4 - \frac{1}{6}R\tilde{M}^2\Big\}\log\Big(\frac{\tilde{M}^2}{\mu^2}\Big). \tag{13.100}$$

Summing up the scalar and fermion contributions, we arrive at the general expression for the effective potential of our model, which includes a single real sterile scalar and N copies of massive fermion fields,

$$\begin{aligned} V_{eff}^{ren}(g_{\mu\nu},\varphi) &= \rho_\Lambda + \frac{1}{2}(m^2-\xi R)\varphi^2 + V \\ &+ \frac{\hbar}{2(4\pi)^2}\Big\{\Big[\frac{1}{2}(V''+m^2)^2 - 2N(M+h\varphi)^4\log\Big[\frac{(M+h\varphi)^2}{\mu^2}\Big] \\ &- \Big(\xi-\frac{1}{6}\Big)R(V''+m^2)\Big]\log\frac{V''+m^2}{\mu^2} + \frac{N}{3}R(M+h\varphi)^2\log\frac{(M+h\varphi)^2}{\mu^2}\Big\}, \end{aligned} \tag{13.101}$$

where the interacting and odd terms of the classical potential V can be compared to those in expression (16.9).

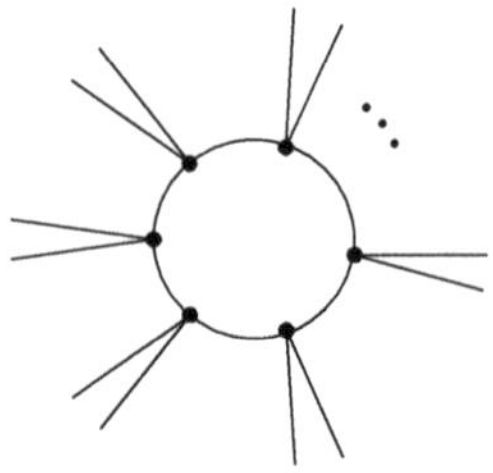

(13.102)

Exercises

13.18. Verify the result for the integral (13.60).
13.19. Repeat the calculation for the effective potential for both flat space and curved space (using local momentum representation, in the first order in curvature) using dimensional and Pauli-Villars regularizations. Also, consider another version of cut-off regularization, when the Wick rotation is not performed and the cut-off is applied to the integration over the space components of momenta. In all these cases, the minimal requirement is to reproduce the logarithmic divergences.
13.20. Consider the diagrammatic representation of the expression (13.59) in the $\lambda\varphi^4$-theory, and show that it corresponds to the one in (13.102). The external field is static, hence the momenta of all external lines is zero and hence the momenta of all internal line is the same. This makes the diagrammatic calculation of the effective potential an easy exercise.
13.21. Use both dimensional (see Exercise 13.19) and cut-off regularization to obtain the beta functions for scalar mass, nonminimal parameter ξ and the cosmological constant from the effective potential and show that they are regularization independent.
13.22. Consider the model for a quantized Dirac spinor ψ in curved space, coupled to the scalar field φ via the Yukawa interaction term $h\bar{\psi}\psi\varphi$. Using both cut-off and dimensional regularizations, derive the effective potential of the scalar field coming from
(a) Functional integration over ψ, when φ is treated as a purely classical background field, and (b) integration over both ψ and φ, when the classical background fermion is assumed to be equal to zero.

Observation. The result and details of this calculation can be found in [302].
13.23. Verify the full set of relations (13.82) and (13.83), and obtain the analogs of (13.84) for m^2 and ξ. Starting from this point, verify that the overall dependence on μ in the effective potential cancels in all sectors.
13.24. Justify the use of relation (13.87). There are several proofs of this formula, of different completeness and complexity. The interested reader can consult Ref. [242].

13.4 The heat-kernel technique, and one-loop divergences

As we know, one-loop corrections are related to the expressions of the form

$$\bar{\Gamma}^{(1)} = \frac{i}{2}\log \operatorname{Det} \hat{H} = \frac{i}{2}\operatorname{Tr}\log \hat{H}, \tag{13.103}$$

where the operator $\hat{H}$ typically depends on the metric and other background fields. This follows from the background field method which was explained in the first part of the book. The idea of the background field method is to split the field into quantum φ^i and background Φ^i parts, according to $\Phi^i \to \Phi^i + \varphi^i$. Functional integration is then performed over the quantum field, while the background field plays the role of the external parameter in the integration. The background field can be identified with the mean field, that is, the arguments of the effective action (13.14).

The one-loop contributions to the vacuum effective action are given by the expression (13.103), where $\hat{H}$ depends only on the metric. This fact follows also from the general expression (13.21) in the case of the one-loop effective action of free fields Φ,

$$e^{i\,\Gamma_{vac}[g_{\mu\nu}]} = e^{iS_{vac}[g_{\mu\nu}]} \int \mathcal{D}\Phi \, e^{i\,S_m[\Phi, g_{\mu\nu}]}, \tag{13.104}$$

$$\text{where} \qquad S_m[\Phi, g_{\mu\nu}] = \int d^4x\sqrt{-g}\, \Phi\hat{H}\Phi. \tag{13.105}$$

Here, and frequently in what follows, we use a condensed notation, where Φ is the set of matter fields of different spins. The bilinear form $\hat{H}$ is supposed to be Hermitian. As the dependence on the coupling constants always appears together with the matter background field, the one-loop contribution to the vacuum action is of the free fields, that is, the Gaussian integral (13.104), reducing to (13.103).

In the general case, when one is interested in both vacuum and matter sectors, the bilinear form of the action depends on both $g_{\mu\nu}$ and Φ. In the next sections, we learn how to reduce the contributions of fermion and vector fields to the combinations of the $\mathrm{Tr}\,\log \hat{H}$ of the operators $\hat{H}$ of standard types. As we shall see from these and other examples described below, in many cases, the practical one-loop calculation reduces to an evaluation of the functional determinant (13.19) of the operator S_2 of the form

$$S_2 = \hat{H} = \hat{1}\Box + \hat{1}m^2 + \hat{\Pi}\,. \tag{13.106}$$

Here hats indicate operators acting on the quantum fields of our interest, which can be scalars, vectors, tensors, spinors or more complicated cases including mixture of different types of fields.[8] The operator (13.106) depends on the metric and other external fields through the operator $\hat{\Pi}$. However, the following observation is in order. We can either include or do not include the mass term $\hat{1}m^2$ into $\hat{\Pi}$. Initially, we will keep the mass term separate, but at some point it will be more useful to include it into $\hat{\Pi}$.

Similar to what we did in section 7.7 in Part I, let us perform the variation with respect to external parameters (e.g., background fields) in the expression (13.103):

$$\frac{i}{2}\,\delta\;\mathrm{Tr}\;\log \hat{H} = \frac{i}{2}\;\mathrm{Tr}\;\hat{H}^{-1}\,\delta\,\hat{H}\,. \tag{13.107}$$

The propagator $G = \hat{H}^{-1}$ admits the Schwinger proper-time representation

$$\hat{H}^{-1} = \int_0^\infty ids\, e^{-is\,\hat{H}}, \tag{13.108}$$

[8]Several examples of this kind will be elaborated in the next sections and chapters; see also [80].

where we assume that the operator $\hat{H}$ has a small positive imaginary part which provides $\lim\limits_{s\to\infty} \exp\{is\hat{H}\} = 0$. Substituting representation (13.108) into the relation (13.107), the last can be written in the form

$$\frac{i}{2}\,\mathrm{Tr}\,\delta\,\hat{H}\int_0^\infty ids\,e^{-is\,\hat{H}} = -\frac{i}{2}\,\mathrm{Tr}\,\delta\int_0^\infty \frac{ds}{s}\,e^{-is\,\hat{H}}. \tag{13.109}$$

Finally, we arrive at the relation

$$\frac{i}{2}\,\mathrm{Tr}\,\log\hat{H} = const - \frac{i}{2}\,\mathrm{Tr}\int_0^\infty \frac{ds}{s}\,\hat{U}(x,x';s), \tag{13.110}$$

$$\text{where}\qquad \hat{U}(x,x';s) = e^{-is\,\hat{H}}\,\delta_c(x,x') \tag{13.111}$$

and the constant term can be, of course, safely disregarded. It is easy to note that the operator $\hat{U}(x,x';s)$ satisfies the Schrödinger equation

$$i\,\frac{\partial\hat{U}(x,x'\,;s)}{\partial s} = -\,\hat{H}_x\hat{U}(x,x';s) \tag{13.112}$$

Along with the main equation (13.112) there is the initial condition

$$\hat{U}(x,x'\,;0) = \delta_c(x,x'). \tag{13.113}$$

As in quantum mechanics, $\hat{U}(x,x';s)$ is usually called the evolution operator, even though s is not the usual time variable, but a special artificial parameter called *proper time*. Also, operator $\hat{H}$, which acts on the covariant delta function is not a Hamiltonian, but a bilinear form of the classical action with respect to quantum fields, in the background field formalism.

The effective action (13.110) requires not the whole evolution operator, but only the functional trace of its *coincidence limit*, when $x' \to x$. As we shall see in what follows, there is a well-developed technique of deriving such a trace. In the next subsections we present the details of elaborating the representation (13.110) by means of the Schwinger-DeWitt technique (also called the Fock-Schwinger-DeWitt, or simply heat-kernel technique). Many more details of this technique can be found, e.g., in the books by B. DeWitt [105, 109], in the paper [31], where one can also learn the more mighty generalized Schwinger-DeWitt technique, and also in the reviews [19, 324].

13.4.1 The world function $\sigma(x,x')$ and related quantities

In what follows, we shall, in general, follow the classical books of DeWitt [105, 109], but we will simplify the calculations when possible, and add many intermediate details. The starting point of our consideration will be the notion of bi-scalars, bi-vectors and bi-tensors.

Consider the equation of the geodesic line (11.34) in terms of the interval $s = c\tau$. In what follows, we assume the units with $c = 1$ and change from s and τ without special explanation. It proves useful to deal with the space of Euclidean signature, where s is a natural parameter along the curve $x^\lambda(s)$. Indeed, the Wick rotation in curved space may be a non-trivial issue, but, at the introductory level, it is better to assume that we are working only with the theories for which this problem can be easily solved. Thus, we have the equation

$$\frac{d^2x^\mu}{ds^2} + \Gamma^\mu_{\alpha\beta}\,\frac{dx^\alpha}{ds}\,\frac{dx^\beta}{ds} = 0. \tag{13.114}$$

Denoting the derivative with respect to s by a point, for an arbitrary tensor T, we have

$$\dot{T}^{\alpha\beta\dots\nu} = \nabla_\lambda T^{\alpha\beta\dots\nu}\,\frac{dx^\lambda}{ds}. \tag{13.115}$$

In particular, for a scalar field $\varphi(x)$, we have

$$\dot{\varphi} = \frac{D\varphi}{ds} = \nabla_\lambda\varphi\,\frac{dx^\lambda}{ds} = \varphi_{,\lambda}\,\frac{dx^\lambda}{ds}. \tag{13.116}$$

Since $ds^2 = g_{\mu\nu}dx^\mu dx^\nu$ and $\nabla_\lambda g_{\mu\nu} = 0$, one can easily verify that

$$\frac{d}{ds}\left(g_{\mu\nu}\,\dot{x}^\mu\dot{x}^\nu\right) = 0 \qquad \Longrightarrow \qquad \dot{x}_\mu\ddot{x}^\mu = 0. \tag{13.117}$$

As we know from one of the exercises in section 11.4, the geodesic equation (13.114) can be obtained from the variational principle with the action

$$\mathcal{S} = \int_a^b ds\,L, \qquad \text{where} \qquad L = \frac{1}{2}\,g_{\mu\nu}\,\frac{dx^\mu}{ds}\,\frac{dx^\nu}{ds}. \tag{13.118}$$

The canonical momentum is

$$p_\mu = \frac{\partial L}{\partial\dot{x}^\mu} = \dot{x}_\mu, \tag{13.119}$$

so conservation of the Hamilton function gives

$$H = \dot{x}^\mu p_\mu - \frac{1}{2}\,\dot{x}^\mu\dot{x}_\mu = \frac{1}{2}\,p^\mu p_\mu = \frac{1}{2}\left(\frac{ds}{d\tau}\right)^2 = const. \tag{13.120}$$

The action with the variable lower and upper limits can be written in the form

$$S(x,\tau|x',\tau') = \int_{x,\tau}^{x',\tau'} Ld\tau = \frac{\sigma(x,x')}{\tau-\tau'}, \tag{13.121}$$

where $\sigma = \sigma(x,x') = \sigma(x',x)$ is a new quantity which is called the *geodesic interval* [105] or a *world function* [310]. As an action with a variable upper or lower limit is a scalar function, and the factor of $(\tau-\tau')$ does not affect the transformation properties, σ is a bi-scalar. This means that it transforms as a scalar under the coordinate change in both points x and x'. The first partial derivative of σ can be replaced by the covariant one, $\sigma_{,\mu} = \sigma_{;\mu} = \nabla_\mu \sigma$. On the other hand, when raising the index, we have to use the symbol of the covariant derivative, $\sigma^{;\mu} = g^{\mu\nu}\sigma_{,\mu}$.

According to the Hamilton-Jacobi formalism and definition (13.121), the canonical momentum (13.119) can be written as

$$p_\mu = \frac{\partial}{\partial x^\mu} S(x,\tau|x',\tau') = \frac{\sigma_{,\mu}}{\tau - \tau'}. \tag{13.122}$$

Then, the Hamilton-Jacobi equation based on (13.120) gives

$$H + \frac{\partial S}{\partial \tau} = \frac{1}{2} p_\mu p^\mu + \frac{\partial}{\partial \tau}\Big(\frac{\sigma}{\tau - \tau'}\Big) = \frac{1}{2}\frac{\sigma_{,\mu}\sigma^{;\mu}}{(\tau-\tau')^2} - \frac{\sigma(x,x')}{(\tau-\tau')^2} = 0, \tag{13.123}$$

which means

$$\frac{1}{2}\left(\nabla\sigma\right)^2 = \frac{1}{2}\sigma_{,\mu}\sigma^{;\mu} = \sigma. \tag{13.124}$$

The last formula represents the fundamental relation that defines the properties of the world function σ. It is easy to see that, in flat space, $\sigma(x,x')$ is the half of the square of the distance between the two points, $\frac{1}{2}|x-x'|^2$, which satisfies relation (13.124) in this particular case.

The derivatives in Eq. (13.124) are taken with respect to the point x. Similar considerations can provide the relation for the point x', namely,

$$\frac{1}{2}\left(\nabla'\sigma\right)^2 = \frac{1}{2}\sigma_{,\mu'}\sigma^{;\mu'} = \sigma. \tag{13.125}$$

Let us introduce the bi-vector of parallel transport $g_{\mu\nu'} = g_{\mu\nu'}(x,x')$, defined by the relations for the parallel transport and the boundary conditions, namely,

$$\sigma^{,\lambda}\, g_{\mu\nu',\lambda} = 0\,, \qquad \lim_{x'\to x} g_{\mu\nu'}(x,x') = g_{\mu\nu}(x). \tag{13.126}$$

The first of these relations shows that the partial derivative of $g_{\mu\nu'}$ is zero along the geodesic line. Therefore, this quantity is parallel transported along this line. For an arbitrary vector A^μ, the rule of transportation from the point x' to the point x is

$$g_{\mu\nu'} A^{\nu'} = \int d^4x \sqrt{g(x')}\, g_{\mu\nu'}(x') A^{\nu'}(x') = A_\mu(x). \tag{13.127}$$

The indices of the bivector $g_{\mu\nu'}$ are raised and lowered using the metric in the corresponding points. Other properties of this bivector are as follows:

$$\begin{aligned}
&g_\mu{}^{\nu'}\,\sigma_{,\nu'} = -\sigma_{,\mu}, \qquad && g_{\mu'}{}^{\nu}\,\sigma_{,\nu} = -\sigma_{,\mu'}, \qquad && \sigma^{,\lambda'}\, g_{\mu\nu',\lambda'} = 0,\\
&g_{\mu\nu'} = g_{\nu'\mu}, && g_{\mu\lambda'}\, g_\nu{}^{\lambda'} = g_{\mu\nu}, && g_{\lambda\mu'}\, g^\lambda{}_{\nu'} = g_{\mu'\nu'}
\end{aligned} \tag{13.128}$$

and $\det\left(g_{\mu\nu'}\right) = g^{1/2}\, g'^{1/2}$.

The quantity $g_{\mu\nu'}$ can be used for the parallel transport of scalars and tensors. For spinors, we need another matrix (bispinor) $I = I(x,x')$ satisfying the relations

$$\sigma^{,\lambda} I_{,\lambda} = 0\,, \qquad \sigma^{,\lambda'} I_{,\lambda'} = 0\,, \qquad \lim_{x'\to x} I(x,x') = I_0, \tag{13.129}$$

where I_0 is an identity matrix. Exactly as in the case of tensors, $I(x,x')$ transports the curved space spinor ψ from point x to the point x'. Similar operators can be constructed for objects with internal indices, such as Yang-Mills fields.

It proves useful to consider the negative of the second derivative of the world function with respect to the two coordinates,

$$D_{\mu\nu'} = -\,\sigma_{,\mu\nu'} = -\,\frac{\partial^2\sigma}{\partial x^\mu \partial x'^\nu}, \tag{13.130}$$

and, something that is especially important for us, the Van Vleck-Morette determinant,

$$D(x,x') = \det\big(D_{\mu\nu'}\big), \tag{13.131}$$

We leave it as an exercise for the reader to show that

$$\lim_{x'\to x} D(x,x') = \det\big(g_{\mu\nu}\big) = g. \tag{13.132}$$

In what follows, we will need derivatives of $D(x,x')$ with respect to both coordinates. The starting point for the derivation of these expressions is the relation (13.124). Taking the partial derivative with respect to x'^ν, we get

$$\sigma^{;\mu}\sigma_{,\mu\nu'} = \sigma_{,\nu'}. \tag{13.133}$$

Let us note that all the derivatives in Eq. (13.133) are of the first order, because they are taken with respect to different coordinates, x^μ and x'^ν. Thus, the two derivatives can be partial, not covariant.

Taking one more partial derivative, we obtain

$$(\sigma^{;\mu})_{,\alpha}\,\sigma_{,\mu\nu'} + \sigma^{;\mu}\,\sigma_{,\mu\nu'\alpha} = \sigma_{,\nu'\alpha}. \tag{13.134}$$

Let us stress that here we use the notation

$$(\sigma^{;\mu})_{,\alpha} = \partial_\alpha(g^{\mu\nu}\partial_\mu\sigma) = \nabla_\alpha\nabla^\mu\sigma - \Gamma^\mu_{\nu\alpha}\nabla^\nu\sigma = \sigma_{;\alpha}^{\ \ \mu} - \Gamma^\mu_{\nu\alpha}\sigma^{;\nu} \tag{13.135}$$

for the *non-tensor* quantity of our interest. The relation (13.134) can be written in terms of the matrix (13.130),

$$D_{\alpha\nu'} = \sigma^{;\mu}\,D_{\alpha\nu',\mu} + (\sigma^{;\mu})_{,\alpha}\,D_{\mu\nu'}. \tag{13.136}$$

Regardless of the fact that our main interest is four-dimensional space, we will write all formulas in the dimension 2ω. This will prove useful for the future application to dimensional regularization, and also for showing that the main results (called the

Schwinger-DeWitt coefficients) are independent of the dimension of spacetime. Contracting Eq. (13.136) with the inverse matrix $(D^{-1})^{\alpha\nu'}$, we get

$$2\omega = (D^{-1})^{\alpha\nu'} D_{\alpha\nu'} = \sigma^{;\mu} (D^{-1})^{\alpha\nu'} D_{\alpha\nu',\mu} + (\sigma^{;\mu})_{,\alpha} (D^{-1})^{\alpha\nu'} D_{\mu\nu'}$$
$$= (D^{-1})^{\alpha\nu'} D_{\alpha\nu',\mu} \sigma^{;\mu} + (\sigma^{;\mu})_{,\mu}. \tag{13.137}$$

The first term in the *r.h.s.* of relation (13.137) can be linked to the partial derivative of the determinant

$$D_{,\mu} = D\,(D^{-1})^{\nu'\alpha} D_{\nu'\alpha,\,\mu}. \tag{13.138}$$

After some algebra, (13.137) becomes

$$2\omega = D^{-1}\big(D\sigma^{;\mu}\big)_{,\mu} = (\sigma^{;\mu})_{,\mu} + \sigma^{;\mu} D^{-1} D_{,\mu}\,. \tag{13.139}$$

However, the following observation is in order. As $D = D(x, x')$ is a bi-tensor density and not a tensor, formally, we cannot take its covariant derivative. For this reason, in the last equation, there are both covariant derivatives of the bi-scalar σ and partial derivatives of the quantities related to the determinant D. This situation makes things complicated, and it would be better to present the relation (13.139) in a completely covariant form by using another object instead of D. Thus, our next step is to define a new quantity $\Delta(x, x')$, such that

$$D(x, x') = g^{1/2}(x)\,\Delta(x, x')\,g^{1/2}(x'). \tag{13.140}$$

Using definitions (13.131) and (13.139), one can easily show that, at each of the points x and x', the new function $\Delta(x, x')$ transforms as a scalar under general coordinate transformation; hence, it is a bi-scalar. In what follows, we use the following condensed notations:[9]

$$\Delta = \Delta(x, x'), \qquad g = g(x) \qquad \text{and} \qquad g' = g(x'). \tag{13.141}$$

Let us now rewrite Eq. (13.139) in a covariant form, using the Van Vleck-Morette bi-scalar $\Delta(x, x')$. Consider a partial derivative of the determinant,

$$D_{,\mu} = \sqrt{g'}\,\big[\sqrt{g}\Delta_{,\mu} + \Delta\sqrt{g}_{,\mu}\big]. \tag{13.142}$$

It is easy to check that $\sqrt{g}_{,\mu} = \sqrt{g}\,\Gamma^{\lambda}_{\mu\lambda}$ and hence

$$D_{,\mu} = \sqrt{g'}\,\sqrt{g}\big[\Delta_{,\mu} + \Gamma^{\lambda}_{\mu\lambda}\Delta\big]. \tag{13.143}$$

Thus, we can use Eq. (13.135) to transform the expression in (13.139) as follows:

$$\big(D\sigma^{;\mu}\big)_{,\mu} = D_{,\mu}\,\sigma^{;\mu} + D\cdot\partial_\mu\sigma^{;\mu}$$
$$= \sqrt{g'}\,\sqrt{g}\big[\sigma^{;\mu}\Delta_{;\mu} + \sigma^{;\mu}\Gamma^{\lambda}_{\mu\lambda}\Delta + \Delta\sigma^{;\mu}{}_{;\mu} - \sigma^{;\mu}\Gamma^{\lambda}_{\mu\lambda}\Delta\big]$$
$$= \sqrt{g'}\,\sqrt{g}\big[\sigma^{;\mu}\Delta_{;\mu} + \Delta\sigma^{;\mu}{}_{;\mu}\big] = \sqrt{g'}\,\sqrt{g}\big(\sigma^{;\mu}\Delta\big)_{;\mu}. \tag{13.144}$$

[9] In order to avoid confusing this operator with the Laplace operator, we shall denote the last as $\Box$, even in Euclidean space.

All in all, (13.139) becomes the covariant relation

$$\Delta^{-1}\left(\sigma^{;\mu}\Delta\right)_{;\mu} = \Delta^{-1}\nabla_\mu(\Delta\cdot\nabla^\mu\sigma) = \Box\sigma + \Delta^{-1}\sigma^\mu\nabla_\mu\Delta = 2\omega. \qquad (13.145)$$

This formula will prove a useful starting point for the subsequent calculations.

In what follows, we will need the covariant derivatives of the world function $\sigma(x,x')$ and the Van Vleck-Morette bi-scalar $\Delta(x,x')$ in the *coincidence limit* $x'\to x$, which will be denoted by a vertical bar,

$$\Big|_{x\to x'} \equiv \Big|, \qquad (13.146)$$

until the end of this section. Directly from definitions (13.121) and (13.124), we get

$$\sigma\Big| = 0, \qquad \text{and} \qquad \sigma^\mu\Big| = 0. \qquad (13.147)$$

Concerning the coincidence limit of the Van Vleck-Morette bi-scalar, by using locally Minkowski coordinates and (13.132), it it easy to check that

$$\Delta(x,x) = \Delta\Big| = 1. \qquad (13.148)$$

In the rest of this section and in the next sections, all of the derivatives of our interest will be covariant ones. For this reason, it is useful to introduce a condensed notation which denotes covariant derivatives of σ without commas of any kind, e.g.,

$$\sigma_{\alpha\nu\mu} = \nabla_\mu\nabla_\nu\nabla_\alpha\sigma, \qquad \sigma_{\alpha\,\cdot\,\mu}^{\;\;\nu} = \nabla_\mu\nabla^\nu\nabla_\alpha\sigma, \qquad \text{etc.} \qquad (13.149)$$

Taking two covariant derivatives of (13.124), we get

$$\sigma_{\nu\alpha} = \sigma^\mu_{\cdot\,\alpha}\,\sigma_{\mu\nu} + \sigma^\mu\,\sigma_{\mu\nu\alpha} \qquad (13.150)$$

and, after multiplying it by $\left(\sigma^{-1}\right)^{\gamma\alpha}$, arrive at

$$\delta^\gamma_\nu = \delta^\gamma_\beta\, g^{\beta\mu}\sigma_{\mu\nu} + \left(\sigma^{-1}\right)^{\gamma\alpha}\sigma^\mu\sigma_{\mu\nu\alpha}\,. \qquad (13.151)$$

Using (13.147), we arrive at the relation

$$\sigma_{\mu\nu}\Big| = g_{\mu\nu}. \qquad (13.152)$$

We leave it as an exercise for the reader to show that

$$\sigma_{\mu\nu'}\Big| = -\,g_{\mu\nu}. \qquad (13.153)$$

Elaborating subsequent coincidence limits will become more and more involved, so we are going to show calculations in less and less detail, eventually presenting only the final results. Indeed, we will calculate only those limits which will prove relevant, as the reader can see in the next section.

One more set of covariant derivatives of (13.150) gives us

$$\sigma_{\nu\alpha\tau} = \sigma^{\mu}_{\cdot\,\alpha\tau}\sigma_{\mu\nu} + \sigma^{\mu}_{\cdot\,\alpha}\sigma_{\mu\nu\tau} + \sigma^{\mu}_{\cdot\,\tau}\sigma_{\mu\nu\alpha} + \sigma^{\mu}\sigma_{\mu\nu\alpha\tau}\,, \tag{13.154}$$

with the coincidence limit of the form (where all immediately vanishing terms are omitted here, and (13.152) used)

$$\begin{aligned}\sigma_{\nu\alpha\tau}\big| &= g_{\mu\nu}\,\sigma^{\mu}_{\cdot\,\alpha\tau}\big| + \delta^{\mu}_{\alpha}\,\sigma_{\mu\nu\tau}\big| + \delta^{\mu}_{\tau}\,\sigma_{\mu\nu\alpha}\big| \\ &= \sigma_{\nu\alpha\tau}\big| + \sigma_{\alpha\nu\tau}\big| + \sigma_{\tau\nu\alpha}\big| = \big(3\sigma_{\nu\alpha\tau} - \sigma_{\nu\alpha\tau} + \sigma_{\nu\tau\alpha}\big)\big| \\ &= 3\sigma_{\nu\alpha\tau}\big| + \big[\nabla_{\tau}, \nabla_{\alpha}\big]\sigma_{\nu}\big| = 3\sigma_{\nu\alpha\tau}\big| + R^{\lambda}_{\cdot\,\nu\alpha\tau}\,\sigma_{\lambda}\big| = 3\sigma_{\nu\alpha\tau}\big|\,. \end{aligned} \tag{13.155}$$

Since $\sigma_{\lambda}\big| = 0$, this result obviously implies

$$\sigma_{\nu\alpha\tau}\big| = 0. \tag{13.156}$$

Taking the next covariant derivatives of (13.154) gives

$$\begin{aligned}\sigma_{\nu\alpha\tau\rho} &= \sigma^{\mu}_{\cdot\,\alpha\tau\rho}\sigma_{\mu\nu} + \sigma^{\mu}_{\cdot\,\alpha\tau}\sigma_{\mu\nu\rho} + \sigma^{\mu}_{\cdot\,\alpha\rho}\sigma_{\mu\nu\tau} + \sigma^{\mu}_{\cdot\,\alpha}\sigma_{\mu\nu\tau\rho} \\ &\quad + \sigma^{\mu}_{\cdot\,\tau\rho}\sigma_{\mu\nu\alpha} + \sigma^{\mu}_{\tau}\sigma_{\mu\nu\alpha\rho} + \sigma^{\mu}_{\cdot\,\rho}\sigma_{\mu\nu\alpha\tau} + \sigma^{\mu}\sigma_{\mu\nu\alpha\tau\rho}. \end{aligned} \tag{13.157}$$

In the coincidence limit, one can use (13.147), (13.152) and (13.156) and, after certain amount of algebra, arrive at

$$\begin{aligned}\sigma_{\nu\alpha\tau\rho}\big| &= 3\sigma_{\nu\alpha\tau\rho}\big| + \big(\sigma_{\nu\tau\alpha\rho} - \sigma_{\nu\alpha\tau\rho}\big)\big| + \big(\sigma_{\nu\rho\alpha\tau} - \sigma_{\nu\alpha\rho\tau}\big)\big| + \sigma_{\nu\alpha\rho\tau}\big| \\ &= 3\sigma_{\nu\alpha\tau\rho}\big| - R_{\rho\nu\alpha\tau} - R_{\tau\nu\alpha\rho} + \sigma_{\nu\alpha\rho\tau}\big|, \end{aligned} \tag{13.158}$$

which is

$$\sigma_{\nu\alpha\tau\rho}\big| = \frac{1}{3}\big(R_{\rho\nu\alpha\tau} + R_{\rho\alpha\nu\tau}\big). \tag{13.159}$$

The relevant contractions of this expression are

$$\sigma_{\alpha\,\cdot\,\mu\nu}^{\;\;\alpha}\big| = \sigma_{\mu\nu\alpha}^{\;\;\;\;\;\alpha}{}_{\cdot}\big| = -\frac{2}{3}\,R_{\mu\nu}, \qquad \sigma_{\mu\alpha\,\cdot\,\nu}^{\;\;\;\;\alpha}\big| = \frac{1}{3}\,R_{\mu\nu}. \tag{13.160}$$

Taking the derivatives of higher order requires additional calculations, so we present only the final result for the relevant contraction,

$$\sigma_{\mu\,\cdot\,\nu\,\cdot\,\alpha\,\cdot}^{\;\;\mu\;\;\nu\;\;\alpha}\big| = \frac{4}{15}\,R_{\mu\nu}R^{\mu\nu} - \frac{4}{15}\,R_{\mu\nu\alpha\beta}R^{\mu\nu\alpha\beta} - \frac{8}{5}\,\Box R. \tag{13.161}$$

As we already know, in four spacetime dimensions, the one-loop divergences have at most four derivatives of the metric, so this expression is what we need, at the moment.

To derive the coincidence limits of the Van Vleck-Morette scalar, we start from relation (13.145) and recast it in the form

$$2\omega\Delta^{1/2} = \Delta^{-1/2}\,\sigma^{\mu}\,\Delta_{\mu} + \Delta^{1/2}\,\sigma^{\mu}_{\cdot\,\mu} = 2(\Delta^{1/2})_{\mu}\,\sigma^{\mu} + \Delta^{1/2}\,\sigma^{\mu}_{\cdot\,\mu}, \tag{13.162}$$

where the condensed notations for the derivatives, e.g., $(\Delta^{1/2})_{\mu\nu} = \nabla_{\mu}\nabla_{\nu}\Delta^{1/2}$ of the powers of the Van Vleck-Morette scalar are used, exactly as we did for σ.

Taking the covariant derivative of (13.162) yields

$$2\omega(\Delta^{1/2})_\nu = 2(\Delta^{1/2})_{\mu\nu}\,\sigma^\mu + 2(\Delta^{1/2})_\mu\,\sigma^\mu_{\cdot\,\nu} + (\Delta^{1/2})_\nu\,\sigma^\mu_{\cdot\,\mu} + \Delta^{1/2}\,\sigma^\mu_{\cdot\,\mu\nu}. \qquad (13.163)$$

When we omit vanishing terms, in the coincidence limit, the last relation results in

$$2\omega(\Delta^{1/2})_\nu\big| = (2+2\omega)(\Delta^{1/2})_\nu\big| \qquad \Longrightarrow \qquad (\Delta^{1/2})_\nu\big| = 0. \qquad (13.164)$$

Taking one more covariant derivative of Eq. (13.163), after a small algebra, we arrive at the coincidence limit

$$2\omega(\Delta^{1/2})_{\nu\alpha}\big| = (4+2\omega)(\Delta^{1/2})_{\nu\alpha}\big| + \Delta^{1/2}\big| \cdot \sigma_{\mu\nu}{}^{\mu}{}_{\cdot\,\alpha}\big|. \qquad (13.165)$$

Substituting into this equation the previous limit from (13.160), we get

$$(\Delta^{1/2})_{\nu\alpha}\big| = \frac{1}{6}\,R_{\nu\alpha}. \qquad (13.166)$$

As the reader will check in one of the exercises, the conjugated formula is

$$(\Delta^{-1/2})_{\mu\nu}\big| = -\frac{1}{6}\,R_{\mu\nu}. \qquad (13.167)$$

Starting from the third-order derivatives, the calculation has nothing qualitatively new, but it rapidly becomes very involved. So, here we just give the list of the coincidence limits that will be used in what follows:

$$(\Delta^{-1/2})_{\mu}{}^{\mu}{}_{\cdot\,\nu}\big| = -\frac{1}{6}\,R_\nu = -\frac{1}{6}\,\nabla_\nu R, \qquad (13.168)$$

$$(\Delta^{-1/2})_{\mu}{}^{\mu}{}_{\cdot\,\nu}{}^{\nu}\big| = \frac{1}{30}\,R_{\mu\nu\alpha\beta}R^{\mu\nu\alpha\beta} - \frac{1}{30}\,R_{\mu\nu}R^{\mu\nu} + \frac{1}{36}\,R^2 - \frac{1}{5}\Box R. \qquad (13.169)$$

Exercises

13.25. Derive the coincidence limits of the Van Vleck-Morette determinant and scalar.
Solution. We know that $g^\nu_{\cdot\,\mu'}\,\sigma_\nu = -\sigma^{\mu'}$. Taking derivative of this relation, we get

$$g^\nu_{\cdot\,\mu',\alpha}\,\sigma_\nu + g^\nu_{\cdot\,\mu'}\,\sigma_{\nu\alpha} = -\,\sigma_{,\mu'\alpha} = D_{\mu'\alpha}.$$

In the coincidence limit, this gives

$$D_{\mu'\alpha}\big| = \delta^\nu_\mu\,g_{\nu\alpha} = g_{\mu\alpha}, \qquad (13.170)$$

which directly leads to $D\big| = g$ and finally to $\Delta\big| = (g')^{-1/2}\,D\,g^{-1/2}\big| = 1$.
13.26. Verify the coincidence limits (13.159)-(13.161).
13.27. i) Using (13.161) and (13.166), derive relation (13.167). ii) Recalculate the same second derivative using the elementary method, which means deriving $\Delta_{\mu\nu}$ first and then applying the derivative of a composite function. Using this simple approach, calculate the second covariant derivative for an arbitrary power r of the Van Vleck-Morette scalar, Δ^r.

13.4.2 Coincidence limits of the first terms of the evolution operator

Now we are in a position to construct the solution of Eq. (13.112) for the evolution operator $\hat{U}(x,x'\,;s) = e^{-is\,\hat{H}}$. As we already know from Eq. (13.110), the coincidence limit of the evolution operator defines the one-loop effective action. It can be used also to define the two-point Green function (13.108).

The heat-kernel method is based on presenting the evolution operator in the form

$$\hat{U}(x,x'\,;s) = \hat{U}_0(x,x';s)\,\hat{\Omega}(x,x'), \tag{13.171}$$

where

$$\hat{\Omega}(x,x') = \sum_{n=0}^{\infty} (is)^n\, \hat{a}_n(x,x') \tag{13.172}$$

is a power series expansion with unknown coefficients $\hat{a}_n(x,x')$, and the first factor is defined in the form

$$\hat{U}_0(x,x';s) = -\frac{i}{(4\pi s)^{\omega}}\,\Delta^{1/2}(x,x')\,e^{\frac{i\sigma(x,x')}{2s}-im^2 s}. \tag{13.173}$$

In this expression, $\sigma = \sigma(x,x')$ is a world function and $\Delta(x,x')$ is a Van Vleck-Morette bi-scalar, which we discussed in the previous section. The unknown quantities $\hat{a}_n = \hat{a}_n(x,x')$ in Eq. (13.172) are called the heat-kernel, or Schwinger-DeWitt, coefficients.[10] These coefficients should be defined such that the expression (13.171) becomes a solution of the main equation, (13.112). Indeed, the form of the operator $\hat{U}_0(x,x';s)$ in Eq. (13.173) has been chosen to make the equations for $\hat{a}_n$ relatively simple and provide us the possibility of calculating their coincidence limits,

$$\hat{a}_n\big| = \lim_{x'\to x} \hat{a}_n(x,x'), \tag{13.174}$$

in a closed and universal form. In the last equation, we use a condensed notation with a vertical bar for the coincidence limit, as in the previous section.

An important observation is in order. At the initial stage, we will work with the simplest operator (13.106). As we have explained above, in the definition of $\hat{H}$, one does not have to include the mass parameter in the operator $\hat{P}$. In some cases, representation (13.173) becomes more useful if m^2 is separated. The reason is that our main interest will be related to UV divergences, and it is better to separate them from the infrared sector of the theory. Separating the massive term eliminates artificial infrared divergences, provides the convergence of the integral over s at the upper limit. In the case of a massless theory, one can introduce a small mass term artificially, just to factorize the infrared sector, which can be dealt with separately. On the other hand, the practical use of the heat-kernel technique is usually simpler if the m^2-term is included into $\hat{P}$.

[10] In the literature these expressions are also called the Hadamard-Minakshisundaram-DeWitt-Seeley, or HAMIDEW, coefficients, reflecting the contributions of the scientists who calculated them first.

Keeping all this in mind, we shall follow an economical approach and use a modified version of the representation (13.173), with the mass term included into $\hat{\Pi}$,

$$\hat{U}_0(x,x';s) = -\frac{i}{(4\pi s)^\omega}\,\Delta^{1/2}(x,x')\,e^{\frac{i\sigma(x,x')}{2s}}, \tag{13.175}$$

and the modified version of (13.106),

$$\hat{H} = \hat{1}\Box + \hat{\Pi}. \tag{13.176}$$

Let us state once again that we do not need to fix the space of the fields in which the operators $\hat{H}$ and $\hat{a}_k(x,x')$ act. The main advantage of the Schwinger-DeWitt technique is that it enables one to derive the general expressions for the divergences and sometimes even for the finite part of effective action in the universal form, independently on the type of quantum field.

From representation (13.173), it immediately follows that the proper time s has the dimension of the inverse square mass. Then, looking at representation (13.171), one can easily figure out that the Schwinger - DeWitt coefficients have the mass dimensions $\hat{a}_n(x,x') \propto [\text{mass}]^{2n}$. Therefore, according to what we already know about the limits imposed by power counting on the divergences in curved space, we can expect that the divergences will be given by the first coefficients and that the rest of the series will contribute to the finite part of the effective action.

In what follows, we shall present the main steps for deriving the traces of the first few coefficients $\hat{a}_n(x,x')$. The full details of this calculation are really cumbersome and complicated, and elaborating them should be regarded as the most difficult exercise of this book. Nonetheless, we recommend this task to the readers who intend to work in the area of semiclassical and/or quantum gravity.

Replacing the product (13.171) into the main equation (13.112) with the operator (13.176), after a certain amount of algebra we get (here, in many cases, the arguments x and x' are omitted for the sake of compactness, as we did in the previous section)

$$i\,\frac{\partial\hat{U}(x,x'\,;s)}{\partial s} = -\frac{\omega}{(4\pi)^\omega}\,\frac{\Delta^{1/2}}{s^{\omega+1}}\,e^{\frac{i\sigma}{2s}}\sum_{n=0}^{\infty}(is)^n\,\hat{a}_n$$
$$-\frac{i\,\sigma}{2(4\pi)^\omega}\,\frac{\Delta^{1/2}}{s^{\omega+2}}\,e^{\frac{i\sigma}{2s}}\sum_{n=0}^{\infty}(is)^n\,\hat{a}_n + \frac{i}{(4\pi)^\omega}\,\frac{\Delta^{1/2}}{s^{\omega}}\,e^{\frac{i\sigma}{2s}}\sum_{n=1}^{\infty}n(is)^{n-1}\,\hat{a}_n, \tag{13.177}$$

$$\Box\hat{U}(x,x'\,;s) = -\frac{i}{(4\pi s)^\omega}\,e^{\frac{i\sigma}{2s}}\sum_{n=0}^{\infty}(is)^n\,\Box(\Delta^{1/2}\,\hat{a}_n)$$
$$+\frac{1}{(4\pi)^\omega}\,\frac{1}{s^{\omega+1}}\,e^{\frac{i\sigma}{2s}}\sum_{n=0}^{\infty}(is)^n\,\sigma^\mu\,\nabla_\mu(\Delta^{1/2}\,\hat{a}_n) \tag{13.178}$$
$$+\frac{1}{(4\pi)^\omega}\,\frac{\Delta^{1/2}}{2s^{\omega+1}}e^{\frac{i\sigma}{2s}}(\Box\sigma)\sum_{n=0}^{\infty}(is)^n\,\hat{a}_n + \frac{i}{(4\pi)^\omega}\,\frac{\Delta^{1/2}}{2s^{\omega+2}}\,e^{\frac{i\sigma}{2s}}\sigma\sum_{n=0}^{\infty}(is)^n\,\hat{a}_n.$$

Substituting these two expressions together with $\hat{\Pi}\hat{U}(x,x'\,;s)$ from (13.175) into Eq. (13.112), after some cancelations we arrive at the equation

$$
\begin{aligned}
&-\frac{\omega\,\Delta^{1/2}}{(4\pi)^\omega\, s^{\omega+1}}\sum_{n=0}^{\infty}(is)^n\,\hat{a}_n \;+\;\frac{i\,\Delta^{1/2}}{(4\pi s)^\omega}\sum_{n=0}^{\infty}n(is)^{n-1}\,\hat{a}_n\\
&-\frac{i}{(4\pi s)^\omega}\sum_{n=0}^{\infty}(is)^n\,\Box(\Delta^{1/2}\,\hat{a}_n) \;+\;\frac{1}{(4\pi)^\omega\, s^{\omega+1}}\sum_{n=0}^{\infty}(is)^n\,\sigma^\mu\,\nabla_\mu(\Delta^{1/2}\,\hat{a}_n)\\
&+\frac{\Delta^{1/2}}{2\,(4\pi)^\omega\, s^{\omega+1}}\sum_{n=0}^{\infty}(is)^n\,(\Box\sigma)\hat{a}_n \;+\;\frac{i\,\Delta^{1/2}}{(4\pi s)^\omega}\,\hat{\Pi}\sum_{n=0}^{\infty}(is)^n\,\hat{a}_n \;=\;0. \qquad (13.179)
\end{aligned}
$$

This is the master equation, which is the basis for deriving the recurrent relations for the coefficients $\hat{a}_n$. In order to put this plan into practice, we have to separate terms with equal powers of the proper time s, requiring that all these terms vanish separately.

If we neglect the overall factor of $(4\pi s)^{-\omega}$, it is easy to note that the lowest power of proper time in Eq. (13.179) is s^{-1}. The corresponding terms involve only the coefficient $\hat{a}_0$. Thus, we get the first relation,

$$
\Delta^{1/2}\Big[\frac{1}{2}\,(\Box\sigma)\,\hat{a}_0 - \omega\,\hat{a}_0 + \Delta^{-1/2}\,\sigma^\mu\nabla_\mu\big(\Delta^{1/2}\hat{a}_0\big) - i\Delta^{-1/2}\Box\big(\Delta^{1/2}\hat{a}_0\big)\Big] = 0. \qquad (13.180)
$$

Using (13.145), this boils down to

$$
-\,\omega\,\hat{a}_0 \,+\, \Delta^{-1/2}\,\sigma^\mu\,\nabla_\mu\big(\Delta^{1/2}\hat{a}_0\big) \,+\, \frac{1}{2}\,\hat{a}_0\,\big(2\omega - \sigma^\mu\Delta^{-1}\Delta_\mu\big) \;=\; 0 \qquad (13.181)
$$

and, finally, to

$$
\frac{1}{2}\,\hat{a}_0\,\Delta^{-1}\sigma^\mu\Delta_\mu \,+\, \sigma^\mu\nabla_\mu\hat{a}_0 \,-\, \frac{1}{2}\,\hat{a}_0\,\Delta^{-1}\sigma^\mu\Delta_\mu \;=\; \sigma^\mu\,\nabla_\mu\hat{a}_0 \;=\; 0. \qquad (13.182)
$$

This is the first of the set of relations that will be used below for deriving the coincidence limits of the Schwinger-DeWitt coefficients.

By taking into account the vanishing mass dimension of the coefficient $\hat{a}_0(x,x')$ and the initial condition (13.113), the solution of (13.182) is

$$
\hat{a}_0(x,x') \;=\; \hat{1}\delta_c(x,x'), \qquad (13.183)
$$

where $\hat{1}$ is an identity matrix in the space of fields where the operators $\hat{H}$ and $\hat{a}_n(x,x')$ act, as in the operator (13.106). One should note that the delta function in (13.183) is covariant, which means there is some dependence on the metric. Later on, we shall see that this is an important feature of this solution.

Let us start the analysis of the terms in the master equation (13.179) with the powers of s higher than s^{-1}, always neglecting the overall factors of $(4\pi s)^{-\omega}$ and $\Delta^{1/2}\exp\{\frac{i\sigma}{2s}\}$. In this way, we arrive at the equations

$$\omega \sum_{n=0}^{\infty} (is)^n \hat{a}_n - \sum_{n=1}^{\infty} n(is)^n \hat{a}_n + \Delta^{-1/2} \sum_{n=0}^{\infty} (is)^{n+1} \Box(\Delta^{1/2}\hat{a}_n) \tag{13.184}$$
$$-\Delta^{1/2} \sum_{n=0}^{\infty} (is)^n \,\sigma^\mu \nabla_\mu (\Delta^{1/2}\hat{a}_n) - \frac{1}{2}(\Box\sigma) \sum_{n=0}^{\infty} (is)^n \hat{a}_n + \hat{\Pi} \sum_{n=0}^{\infty} (is)^{n+1}\, \hat{a}_n \;=\; 0.$$

Taking into account the linear independence of the terms with different powers of the proper time s, we get the equation for the terms of the order $n+1$,

$$\omega\hat{a}_{n+1} - (n+1)\hat{a}_{n+1} + \Delta^{-1/2}\,\Box(\Delta^{1/2}\hat{a}_n)$$
$$- \Delta^{1/2}\,\sigma^\mu \nabla_\mu(\Delta^{1/2}\hat{a}_{n+1}) - \frac{1}{2}(\Box\sigma)\hat{a}_{n+1} + \hat{\Pi}\,\hat{a}_n \;=\; 0, \tag{13.185}$$

where $n = 0, 1, 2, \ldots$. By using, once again, the relation (13.145), one can rewrite (13.185) in the final form

$$(n+1)\hat{a}_{n+1} + \sigma^\mu \nabla_\mu \hat{a}_{n+1} \;=\; \Delta^{-1/2}\,\Box(\Delta^{1/2}\hat{a}_n) + \hat{\Pi}\,\hat{a}_n. \tag{13.186}$$

The next problem is to solve the last equation by iterations in n, using the solution (13.183) as the initial condition. Indeed, for the coefficients $\hat{a}_n$ with $n \geq 1$, we need only the coincidence limits, which makes our task realistic.

For the first of equations (13.186) with $n = 0$, we have

$$\hat{a}_1 + \sigma^\mu \cdot \nabla_\mu \hat{a}_1 \;=\; \Delta^{-1/2}\,\Box(\Delta^{1/2}\hat{a}_0) + \hat{\Pi}\,\hat{a}_0. \tag{13.187}$$

In the coincidence limit (using the notations with the bar and $\sigma_\mu = \nabla_\mu \sigma$, which were introduced in the previous section), becomes

$$\hat{a}_1\big| + \sigma^\mu\big| \cdot \nabla_\mu \hat{a}_1\big| \;=\; \Delta^{-1/2}\big| \cdot \Box(\Delta^{1/2}\hat{a}_0)\big| + \hat{\Pi}\,\hat{a}_0\big|. \tag{13.188}$$

As we know, $\Delta\big| = 1$, $\sigma^\mu\big| = 0$, and, certainly, $\hat{a}_0\big| = \hat{1}$. Also, for $\Box\Delta^{1/2}\big|$, we can use (13.166), and thus we get

$$\hat{a}_1\big| \;=\; \hat{\Pi} + \frac{1}{6}\,g^{\mu\nu}\,\hat{1}R_{\mu\nu} = \hat{P} \;=\; \hat{\Pi} + \frac{\hat{1}}{6}\,R. \tag{13.189}$$

Let us say that the notation $\hat{P}$ introduced in the last formula will be used a lot in what follows. In a perfect accordance with our expectations based on the dimension of the proper time s, the coincidence limit of $\hat{a}_1$ has the dimension of $[mass]^2$. For further coefficients, one can expect growing dimension, e.g., $\hat{a}_2 \propto [mass]^4$, which means we may expect that the coincidence limit of $\hat{a}_2$ will be quadratic in curvatures.

The difficulty of deriving the next set of coincidence limits grows pretty fast with n; therefore, we present the derivation of $\hat{a}_2$ only. The corresponding particular case of Eq. (13.186) has a coincidence limit of the form

$$2\hat{a}_2 + \sigma^\mu \nabla_\mu \hat{a}_2 \;=\; \Delta^{-1/2}\,\Box(\Delta^{1/2}\,\hat{a}_1) + \hat{\Pi}\,\hat{a}_1. \tag{13.190}$$

The main problem in deriving the coincidence limit of this expression is that the first term in the *r.h.s.* requires the limit of derivatives of $\hat{a}_1$, while the expression (13.189) has only the limit of $\hat{a}_1$ itself. For this reason, one has to work hard to solve the last equation.

By substituting (13.187) into (13.190), after some algebra, we arrive at

$$2\hat{a}_2 + \sigma^\mu \nabla_\mu \hat{a}_2 = \Delta^{-1/2}\,\Box^2(\Delta^{1/2}\,\hat{a}_0) \;-\; \Delta^{-1/2}\,\Box(\Delta^{1/2}\,\hat{a}_1)$$
$$+\,\Delta^{-1/2}\,\Box(\Delta^{1/2}\hat{\Pi}) \;+\; \hat{\Pi}\,\hat{a}_1. \tag{13.191}$$

Taking derivatives in the *r.h.s.* and disregarding terms with the factors σ_μ and $\sigma_{\mu\nu\alpha}$, which vanish in the final result owing to (13.147) and (13.156), after lengthy calculation we arrive at (remember that the action of the derivative ends on the parenthesis)

$$2\hat{a}_2\Big| = \Big\{ - 2\sigma^{\mu\nu}\,\nabla_\mu\nabla_\nu\hat{a}_1 + \hat{\Pi}\,\hat{a}_1 + \Delta^{-1/2}\,(\Box^2\Delta^{1/2})\,\hat{a}_0 + 2\Delta^{-1/2}(\Box\Delta^{1/2})\,(\Box\hat{a}_0)$$
$$+\,2\Delta^{-1/2}(\nabla_\mu\Box\Delta^{1/2} + \Box\nabla_\mu\Delta^{1/2})\,\nabla^\mu\hat{a}_0 + 4\Delta^{-1/2}(\nabla^\mu\nabla^\nu\Delta^{1/2})\,(\nabla_\mu\nabla_\nu\hat{a}_0)$$
$$+\,\Box^2\hat{a}_0 + \Delta^{-1/2}(\Box^2\Delta^{1/2})\hat{\Pi}\hat{a}_0 + (\Box\hat{\Pi})\,\hat{a}_0 + (\nabla^\mu\hat{\Pi})\,\nabla_\mu\hat{a}_0 + \hat{\Pi}\Box\hat{a}_0\Big\}\Big|.$$

Let us omit the tedious but not really complicated calculations of this expression, which uses the relations (13.166)-(13.169). The result has the form

$$2\hat{a}_2\Big| = \hat{\Pi}^2 - (\Box\hat{\Pi}) - \frac{\hat{1}}{30}\,(R^2_{\mu\nu\alpha\beta} - R^2_{\mu\nu} + \Box R) + 4\Big\{(\Box\hat{a}_1) - \hat{\Pi}\Box\hat{a}_0 \tag{13.192}$$
$$-\,\Box^2\hat{a}_0 - 2(\nabla^\mu\hat{\Pi})\,(\nabla_\mu\hat{a}_0) + \frac{2}{3}(\nabla^\mu R)\,(\nabla_\mu\hat{a}_0) + \frac{2}{3}R^{\mu\nu}\,(\nabla_\mu\nabla_\nu\hat{a}_0)\Big\}\Big|.$$

Taking two covariant derivatives of Eq. (13.187) and performing a contraction, after some algebra, we obtain

$$\Box\hat{a}_1\Big| = -\frac{\hat{1}}{90}\,(R^2_{\mu\nu\alpha\beta} - R^2_{\mu\nu} - 6\Box R) + \frac{1}{3}\,\Box\hat{\Pi} + \Big\{\frac{1}{3}\,\Box^2\hat{a}_0 + \frac{1}{3}\,\hat{\Pi}\,\Box\hat{a}_0 - \frac{1}{18}\,R\,\Box\hat{a}_0$$
$$-\,\frac{2}{9}\,(\nabla^\mu R)\nabla_\mu\hat{a}_0 - \frac{2}{9}\,R^{\mu\nu}\nabla_\mu\nabla_\nu\hat{a}_0 + \frac{2}{3}\,(\nabla^\mu\hat{\Pi})\,\nabla_\mu\hat{a}_0\Big\}\Big|. \tag{13.193}$$

The last step is to learn how to take the coincidence limit of the derivatives of $\hat{a}_0$, which has a non-trivial part and is worth being shown in detail. Taking the derivatives of (13.182), we get

$$\sigma^\mu{}_\nu\,\nabla_\mu\hat{a}_0 + \sigma^\mu\,\nabla_\mu\nabla_\nu\hat{a}_0 \;=\; 0, \tag{13.194}$$

which results in

$$\nabla_\mu\hat{a}_0\Big| \;=\; 0. \tag{13.195}$$

Taking the next covariant derivative of (13.194) gives

$$\sigma^\mu{}_{\nu\alpha}\,\nabla_\mu\hat{a}_0 + \sigma^\mu{}_\nu\,\nabla_\alpha\nabla_\mu\hat{a}_0 + \sigma^\mu{}_\alpha\,\nabla_\nu\nabla_\mu\hat{a}_0 + \sigma^\mu\,\nabla_\alpha\nabla_\mu\nabla_\nu\hat{a}_0 \;=\; 0. \tag{13.196}$$

In the coincidence limit, we find

$$\big[\nabla_\nu, \nabla_\mu\big]\hat{a}_0\Big| + 2\nabla_\mu\nabla_\nu\hat{a}_0\Big| = 0. \tag{13.197}$$

It proves useful to introduce a special notation for the commutator of covariant derivatives in the space with the identity operator $\hat{1}$, e.g.,

$$\big[\nabla_\mu, \nabla_\nu\big]\hat{a}_0 = \hat{\mathcal{R}}_{\mu\nu}\,\hat{a}_0. \tag{13.198}$$

Indeed, this quantity is zero for scalar fields, but it is non-zero for vectors, spinors and tensors, as we know from the previous chapters. In general, $\hat{\mathcal{R}}_{\mu\nu} = -\hat{\mathcal{R}}_{\nu\mu}$ is a c-number operator acting in the space of quantum fields.

From Eq. (13.197), it immediately follows that

$$\nabla_\mu\nabla_\nu\hat{a}_0\Big| = \frac{1}{2}\,\hat{\mathcal{R}}_{\mu\nu} \qquad \Longrightarrow \qquad \Box\hat{a}_0\Big| = 0. \tag{13.199}$$

We leave it as an exercise for the reader to show that

$$\Box^2\hat{a}_0\Big| = \frac{1}{2}\,\hat{\mathcal{R}}_{\mu\nu}\hat{\mathcal{R}}^{\mu\nu} = \frac{1}{2}\,\hat{\mathcal{R}}^2_{\mu\nu}. \tag{13.200}$$

Finally, inserting all elements into the expression (13.192), we arrive at the result

$$\hat{a}_2\Big| = \hat{a}_2(x,x) = \frac{\hat{1}}{180}(R^2_{\mu\nu\alpha\beta} - R^2_{\alpha\beta} + \Box R) + \frac{1}{2}\hat{P}^2 + \frac{1}{6}\,(\Box\hat{P}) + \frac{1}{12}\hat{\mathcal{R}}^2_{\mu\nu}, \tag{13.201}$$

where the operator $\hat{P}$ has been defined in (13.189).

Let us make an important observation. In some theories, we have to work with a more general operator with a linear term [186, 30],

$$S_2 = \hat{H} = \hat{1}\Box + 2\hat{h}^\mu\nabla_\mu + \hat{\Pi}. \tag{13.202}$$

The good news is that this operator can be elaborated without much effort, since the linear term can be absorbed into the definition of the covariant derivative,

$$\nabla_\mu \quad \longrightarrow \quad \mathcal{D}_\mu = \nabla_\mu + \hat{h}_\mu. \tag{13.203}$$

It turns out that, in this case, all calculations will be essentially the same, with two exceptions. The commutator of the new covariant derivatives will not be $\hat{\mathcal{R}}_{\mu\nu}$ but the more complicated expression

$$\hat{S}_{\mu\nu} = \hat{\mathcal{R}}_{\mu\nu} + (\nabla_\nu\hat{h}_\mu - \nabla_\mu\hat{h}_\nu) + (\hat{h}_\nu\hat{h}_\mu - \hat{h}_\mu\hat{h}_\nu), \tag{13.204}$$

while, instead of (13.189), the operator $\hat{P}$ becomes

$$\hat{a}_1\Big| = \hat{a}_1(x,x) = \hat{P} = \hat{\Pi} + \frac{\hat{1}}{6}\,R - \nabla_\mu\hat{h}^\mu - \hat{h}_\mu\hat{h}^\mu. \tag{13.205}$$

Without the use of special calculations, we arrive at the following result for the relevant coincidence limit:

$$\hat{a}_2\Big| = \hat{a}_2(x,x) = \frac{\hat{1}}{180}\,(R^2_{\mu\nu\alpha\beta} - R^2_{\alpha\beta} + \Box R) + \frac{1}{2}\hat{P}^2 + \frac{1}{6}(\Box\hat{P}) + \frac{1}{12}\hat{S}^2_{\mu\nu}. \tag{13.206}$$

The great advantage of these expressions for $\hat{a}_1\big|$ and $\hat{a}_2\big|$ is their universality, for it enables one to analyze effective action in a wide class of the field theory models.

Further calculation of the $\hat{a}_k\big|$ coefficients is possible, although it is very difficult. The $\hat{a}_3\big|$ coefficient has been derived by Gilkey [159], and checked by Avramidy [19], who also derived the $\hat{a}_4\big|$ coefficient.

Exercises

13.28. Elaborate all of the calculations that were omitted in this section, especially the ones leading to Eq. (13.201).
13.29. Verify the relation (13.200).
13.30. Verify that the replacement of the covariant derivative by its extended version (13.203) does not change the calculations that lead to Eq. (13.201), except for the facts that the commutator of the extended covariant derivatives is now (13.204), and the new operator which stands in place of $\hat{\Pi}$ is $\hat{\Pi}' = \hat{\Pi} - \nabla_\mu \hat{h}^\mu - \hat{h}_\mu \hat{h}^\mu$, as it enters in Eq. (13.205).
13.31. Discuss whether the extended version (13.203) can be correctly called a covariant derivative. If this question creates any kind of difficulty, the reader is advised to reread the first chapter.

13.4.3 Separating the UV divergences

Let us learn how to extract the UV divergences from the integral representation (13.110) with evolution operator given by the Schwinger-DeWitt expansion (13.171), with the series defined in (13.172) and the base function (13.173).

Two observations are in order. First of all, the upper limit in the integration over the proper time s corresponds to the infrared limit. The reason is that the mass term in the exponential of the representation (13.173) makes this limit convergent, which is certainly not expected for the UV regime. Since we are interested in the UV sector, we should pay special attention to the lower limit only. In this case, the mass term in the exponential plays no role, and we can safely deal with the simpler expression (13.175) and, e.g., insert the factor $\exp\{-\epsilon s\}$, with $\epsilon > 0$, to regularize the infrared sector in both massless and massive theories.

The second point concerns the UV regularizations in the Schwinger-DeWitt technique. The usual form of dimensional regularization, which was described in the first part of the book and will be discussed again and used below, requires working with Feynman diagrams in the momentum representation, while the Schwinger-DeWitt technique deals with the proper-time integral (13.110). Let us note that the correspondence between Schwinger-DeWitt technique and Feynman diagrams has previously been studied in the literature. One can mention, for instance, the paper [214], where it was shown that the leading UV divergences are the same in the momentum cut-off for Feynman diagrams as in the cut-off in the lower limit of thye Schwinger representation, which is equivalent to (13.110) in flat spacetime (see the first part of this book). The same equivalence can be observed for the cut-off regularization in curved space, when the calculation of diagrams is formulated with the use of local momentum representation [302].

In what follows, we start from the covariant cut-off regularization (13.110), which proves useful, e.g., in quantum gravity [140]. After that, we concentrate on an alternative version of dimensional regularization, which was developed for the Schwinger-DeWitt technique by Brown and Cassidy in [65] (see also the review paper [31]).

Consider the integral representation (13.110) with the cut-off regularized UV limit,

$$\frac{i}{2}\,\mathrm{Tr}\,\log \hat{H}\Big|_{\text{cut-off}} = -\frac{i}{2}\,\mathrm{Tr}\int\limits_{1/\Omega^2}^{\infty} \frac{ds}{s}\,\frac{ie^{-\epsilon s}}{(4\pi s)^{\omega}}\Big\{\Delta^{1/2}e^{\frac{i\sigma}{2s}}\sum_{n=0}^{\infty}(is)^n\hat{a}_n(x,x')\Big\}\Big|. \tag{13.207}$$

Initially, we will replace $\omega \to 2$ since we are interested in the four dimensional Euclidean space. It is important that the coincidence limit be taken *before* the integration over the proper time s, so that $\Delta^{1/2}e^{\frac{i\sigma}{2s}} \to 1$ and Eq. (13.207) becomes

$$\frac{i}{2}\,\mathrm{Tr}\,\log \hat{H}\Big|_{\text{cut-off}} = \frac{1}{32\pi^2}\,\mathrm{Tr}\int\limits_{1/\Omega^2}^{\infty} ds\, e^{-\epsilon s}\Big\{\frac{1}{s^3}\hat{a}_0\Big| + \frac{1}{s^2}\,\hat{a}_1\Big| + \frac{1}{s}\,\hat{a}_2\Big| + \hat{a}_3\Big| + ...\Big\}$$

$$= \frac{1}{32\pi^2}\,\mathrm{Tr}\,\Big\{2\Omega^4\,\hat{a}_0\Big| + \Omega^2\,\hat{a}_1\Big| - \log\Big(\frac{\Omega^2}{\mu^2}\Big)\,\hat{a}_2\Big| - \hat{a}_3\Big| + \dots\Big\}, \tag{13.208}$$

where we introduced a fiducial fixed value of the cut-off parameter μ, for mathematical consistency. In most of the literature, this term is not indicated explicitly.

Expression (13.207) shows that the UV divergences are given by the coincidence limits of the first three coefficients, $\hat{a}_0$, $\hat{a}_1$ and $\hat{a}_2$. The logarithmic divergences correspond to the coefficient $\hat{a}_2$, which even was once called the "magic coefficient." The reason is that, as we will learn in the next chapters, most of the available physical information, and also such notions as the renormalization group, UV asymptotic behavior and conformal anomaly, are linked to $\hat{a}_2$. The logarithmic divergences are universal in the sense that the coefficient in front of a one-loop logarithmic divergence does not depend on the scheme, renormalization or regularization. This fact is known from the pioneer paper of Salam [269] and has been confirmed by many works after that.

Let us stress that the aforementioned regularization independence of the logarithmic divergences does not imply complete definiteness. For instance, as we know from Part I of this book, logarithmic divergences may depend on gauge fixing or, more generally, on the parametrization of quantum fields. We shall discuss this issue in detail, especially in the last part, devoted to quantum gravity.

The coefficient $\hat{a}_0\big|$ defines quartic divergences, and owing to the simplicity of the expression (13.183), it is clear that this divergence is given by an algebraic sum of the contributions of different fields, which enters through $\mathrm{Tr}\,\hat{1}$, e.g., a single real scalar field contributes as one, and N_s real scalars as N_s. A single massless Abelian vector contributes as 4, but one has to add the contribution of the DeWitt-Faddeev-Popov ghosts, which have odd Grassmann parity and, hence, enter with the opposite sign. In general, for a theory that includes fermions, one has to replace Tr by the *supertrace* sTr, where the fields contribute with the sign corresponding to their Grassmann parity. Thus, the overall electromagnetic field's contribution is twice that of the real scalar, the Dirac field contributes with a factor of -4, etc. It is clear that quartic divergences depend only on the particle contents of the theory. In this sense, they are completely universal.

Finally, $\hat{a}_1\big|$ defines quadratic divergences, and it is easy to see that this coefficient is less universal and much more ambiguous than $\hat{a}_0\big|$ and $\hat{a}_2\big|$. The reason is that one can trade the cut-off parameter Ω^2 for $\Omega^2 + \Omega_0^2$, where Ω_0^2 is an arbitrary quantity of the dimension $[mass]^2$. This quantity does not have to be a constant, and may depend on the background fields. Such a change of cut-off does not affect quartic and logarithmic divergences, but it can essentially modify quadratic divergence, which becomes completely ambiguous.

The next example is dimensional regularization, as formulated in [65, 31]. The starting point will be (13.207), but this time with an arbitrary complex ω, with $\omega \to 2+0$ afterwards, and with $1/\Omega^2 \to 0$ from the very beginning.[11] In the limit $\omega \to 2+0$, there will be divergent structures of the three types,

$$\int_0^\infty \frac{ds}{s^{\omega+k}}\, f(s,\omega), \qquad \text{with} \quad k = -1,\, 0,\, 1. \tag{13.209}$$

We assume that the function $f(s,\omega)$ is analytic in the whole interval $0 < s < \infty$ and provides sufficiently fast convergence of all relevant expressions in the limit $s \to \infty$, as discussed previously.

Consider in detail the simplest integral with $k = -1$, assuming the limit $\omega \to 2+0$. Integrating by parts, we obtain

$$\int_0^\infty \frac{ds}{s^{\omega-1}}\, f(s,\omega) = -\,\frac{1}{\omega-2}\,\frac{f(s,\omega)}{s^{\omega-2}}\bigg|_0^\infty + \frac{1}{\omega-2}\int_0^\infty \frac{ds}{s^{\omega-2}}\,\frac{\partial f(s,\omega)}{\partial s}. \tag{13.210}$$

Restricting our attention to the region $\operatorname{Re}\omega > 2$, the lower limit in the first term vanishes. As we have explained above, we are interested only in the UV sector and, hence, can assume that the first term in the *r.h.s.* also vanishes at $s \to \infty$. Thus, we need only the limit $\omega \to 2+0$ in the second term. This limit is performed by the following transformation:

$$s^{2-\omega} = e^{(2-\omega)\,\log s} = 1 + (2-\omega)\,\log s + \mathcal{O}\big((2-\omega)^2\big). \tag{13.211}$$

Since the last term vanishes in the limit $\omega \to 2+0$, one can disregard it, arriving at

$$\begin{aligned} \int_0^\infty \frac{ds}{s^{\omega-1}}\, f(s,\omega) &= \frac{1}{\omega-2}\int_0^\infty ds\frac{\partial f(s,\omega)}{\partial s} - \int_0^\infty ds\,\log s\,\frac{\partial f(s,\omega)}{\partial s} \\ &- \frac{1}{\omega-2}\big[f(\infty,\omega) - f(0,\omega)\big] - \int_0^\infty ds\,\log s\,\frac{\partial f(s,\omega)}{\partial s}. \end{aligned} \tag{13.212}$$

As $f(\infty,\omega) = 0$, we can expand the second term in the brackets and, finally, get

[11] Here $\omega \to 2+0$ means that the complex ω approach 2 in such a way that $\operatorname{Re}\omega > 2$.

$$\int_0^\infty \frac{ds}{s^{\omega-1}}\, f(s,\omega) = \frac{1}{2-\omega}\left[f(0,2) + (\omega-2)\,\frac{\partial f(0,\omega)}{\partial s}\right] - \int_0^\infty ds\,\log s\,\frac{\partial f(s,\omega)}{\partial s}$$
$$= \frac{f(0,2)}{\omega-2} - \frac{\partial f(0,\omega)}{\partial \omega}\Big|_{\omega=2} - \int_0^\infty ds\,\log s\,\frac{\partial f(s,\omega)}{\partial s}. \tag{13.213}$$

We leave the other two integrals as exercises and present only the final results [31],

$$\int_0^\infty \frac{ds}{s^{\omega}}\, f(s,\omega) = \Big(\frac{1}{2-\omega} + 1\Big)\,\frac{\partial f(s,2)}{\partial s}\Big|_{s=0} - \frac{\partial^2 f(s,\omega)}{\partial s\partial \omega}\Big|_{s=0,\,\omega=2}$$
$$- \int_0^\infty ds\,\log s\,\frac{\partial^2 f(s,2)}{\partial s^2}, \tag{13.214}$$

$$\int_0^\infty \frac{ds}{s^{\omega+1}}\, f(s,\omega) = \frac{1}{2}\Big(\frac{1}{2-\omega} + \frac{3}{2}\Big)\,\frac{\partial f(s,2)}{\partial s}\Big|_{s=0} - \frac{1}{2}\,\frac{\partial^3 f(s,\omega)}{\partial s^2\partial \omega}\Big|_{s=0,\,\omega=2}$$
$$- \frac{1}{2}\int_0^\infty ds\,\log s\,\frac{\partial^3 f(s,2)}{\partial s^3}. \tag{13.215}$$

It is assumed that the last integrals in all three of expressions (13.213), (13.214) and (13.215) are finite owing to the small-s behaviour of the function $f(s,\omega)$, as it is in the case of interest. Thus, dimensional regularization does not show quadratic and quartic divergences. On the other hand, the logarithmic divergences are exactly the same as in covariant cut-off regularization (13.208). The correspondence rule is

$$\frac{1}{2-\omega} \sim \log\Big(\frac{\Omega^2}{\Omega_0^2}\Big). \tag{13.216}$$

Finally, we can write the general expression for the one-loop divergences in dimensional regularization, using the main result (13.206), in the form (here we denote $n = 2\omega$ for the sake of further convenience),

$$\bar{\Gamma}^{(1)}_{div} = -\frac{\mu^{n-4}}{\varepsilon}\,\mathrm{sTr}\,\hat{a}_2(x,x')\Big| = -\frac{\mu^{n-4}}{\varepsilon}\int d^n x\sqrt{-g}\,\mathrm{str}\,\hat{a}_2(x,x), \tag{13.217}$$
$$\text{where}\qquad \varepsilon = (4\pi)^2(n-4).$$

Indeed, we can safely trade the integral here for the four-dimensional one, as the pole term would not be affected. However, the form shown in (13.217) will prove especially useful in the next chapters, when we consider the renormalization group and the trace anomaly.

Exercises

13.28. Verify the integrals (13.214) and (13.215), including the convergence of the last integrals for the functions of our interest. It is interesting note the order of taking the integrals and the coincidence limits.

13.29. Verify the correspondence rule (13.216) and the main result (13.217) for logarithmic divergences.

13.30. Try to explain the absence of power divergences in dimensional regularization using different kinds of arguments.

13.31. Evaluate quartic divergences in the Minimal Standard Model of particle physics.

14
One-loop divergences

In this chapter we consider examples of one-loop calculations and the related issues, such as practical renormalization and derivation of beta functions.

14.1 One-loop divergences in the vacuum sector

Formula (13.217), together with (13.206), represents a powerful tool for deriving divergences in various models of field theory in flat or curved spacetimes or in quantum gravity. However, in some cases, we need the more complicated generalized Schwinger-DeWitt technique developed by Barvinsky and Vilkovisky [31]. If the object of interest is a field theory models in the dimensions other than 4, other Schwinger-DeWitt coefficients should be invoked, e.g., in the two-dimensional case, one needs $\hat{a}_1(x,x)$ instead of $\hat{a}_2(x,x)$. In the six-dimensional case, we need $\hat{a}_3(x,x)$ to evaluate the logarithmic divergences, etc.

The calculation of the one-loop effective action is an essential part of the theory of quantum fields. In what follows, we consider the derivation of one-loop divergences for the free scalar, spinor, massless and massive vector fields in curved spacetime, and then discuss interacting theories.

Let us here warn the reader that, in most of this and subsequent chapters, we present results in the Minkowski signature, assuming the inverse Wick rotation from the Euclidean space, where the general formula for the trace of the $\hat{a}_2$ coefficient was obtained.

14.1.1 Scalar fields

For the multi-component scalar field, the bilinear form $S_2 = \hat{H}$ is given by (13.48),

$$\hat{H} \,=\, \delta^i_j \,\left(\Box + m_s^2 - \xi R\right)_x, \tag{14.1}$$

where $i,j \,=\, 1,2,\dots,N_s$ and m_s is the mass of the scalar. Identification with the general expression (13.202) gives $\hat{h}^\mu \,=\, 0$ and $\hat{\Pi} \,=\, \delta^i_j\,(m_s^2 - \xi R)$. Obviously, for the scalar,

$$\hat{S}_{\mu\nu} = \hat{\mathcal{R}}_{\mu\nu} = 0 \qquad \text{and} \qquad \hat{P} \,=\, \delta^i_j \left[m_s^2 - \left(\xi - \frac{1}{6}\right)R\right]. \tag{14.2}$$

Substituting these expressions into Eq. (13.217), we arrive at the result

$$\bar{\Gamma}^{(1)}_{div} = -\,\frac{\mu^{n-4}}{\epsilon}\,N_s \int d^n x\sqrt{-g}\,\Big\{\frac{1}{180}(R^2_{\mu\nu\alpha\beta} - R^2_{\alpha\beta} - \Box R) - \frac{1}{6}\Big(\xi - \frac{1}{6}\Big)\Box R$$
$$+\,\frac{1}{2}\left(\xi - \frac{1}{6}\right)^2 R^2 - m_s^2\Big(\xi - \frac{1}{6}\Big)\,R + \frac{1}{2}\,m_s^4\Big\}. \tag{14.3}$$

Introduction to Quantum Field Theory with Applications to Quantum Gravity. Iosif L. Buchbinder and Ilya L. Shapiro, Oxford University Press (2021). © Iosif L. Buchbinder and Ilya Shapiro.
DOI: 10·1093/oso/9780198838319·003·0014

For the complex N_s-component scalar, the divergent part of the effective action is twice that of expression (14.3), which is just a doubling of the overall factor N_s in Eq. (14.3). In general, as we have discussed above, free fields always give additive independent contributions to the vacuum effective action, both to the divergences and to the finite parts.

In the four-dimensional conformal case, we have $m = 0$ and $\xi = \frac{1}{6}$. Then the second line of formula (14.3) equals zero and the expression boils down to

$$\bar{\Gamma}^{(1)}_{div} = -\frac{\mu^{n-4}}{\epsilon}\frac{N_s}{180}\int d^4x\sqrt{-g}\,\big(R^2_{\mu\nu\alpha\beta} - R^2_{\alpha\beta} - \Box R\big)\,. \tag{14.4}$$

It proves useful to rewrite this expression in another basis. The formulas (12.10) in $n = 4$ and (11.76) can be easily solved to give

$$\begin{aligned} R_{\mu\nu\alpha\beta}R^{\mu\nu\alpha\beta} &= 2C^2 - E_4 + \frac{1}{3}R^2, \\ R_{\alpha\beta}R^{\alpha\beta} &= \frac{1}{2}C^2 - \frac{1}{2}E_4 + \frac{1}{3}R^2, \end{aligned} \tag{14.5}$$

where we used the compact notation $C^2 = C^2(4)$. Using these solutions, Eq. (14.4) becomes

$$\bar{\Gamma}^{(1)}_{div} = -\frac{\mu^{n-4}}{\epsilon}N_s\int d^nx\sqrt{-g}\,\Big\{\frac{1}{120}C^2 - \frac{1}{360}E_4 + \frac{1}{180}\Box R\Big\}. \tag{14.6}$$

From this form of divergence, it is evident that, in the conformal case, the pole terms are conformal invariant in $n = 4$, or surface structures. This means that the classical action of vacuum that provides the renormalizability of the vacuum sector at the one-loop level can be chosen in the form

$$S_{vac} = \int d^4x\sqrt{-g}\,\big\{a_1C^2 + a_2E_4 + a_3\Box R\big\}\,, \tag{14.7}$$

so that it satisfies the conformal Noether identity

$$-\frac{2}{\sqrt{-g}}\,g_{\mu\nu}\,\frac{\delta S_{vac}}{\delta g_{\mu\nu}} = 0. \tag{14.8}$$

More detailed discussion of the renormalization of conformal theories will be given later on, in chapter 17. Let us now give only a general explanation of why one-loop divergences are conformal invariant and why we can not expect the same beyond the one-loop level. The last statement will be discussed, initially, in dimensional regularization and for vacuum renormalization only.

Consider the regularized action of conformal theory. Such an action is formulated in an arbitrary dimension n and has no divergences of either the UV or the infrared type. The classical vacuum action can be hardly extended into $n \neq 4$ in a conformal form, but this feature is not of the utmost importance, for the following two reasons: i) the metric is *not* an object of quantization. ii) since the conformal symmetry of

S_{vac} holds in $n = 4$, the defect is proportional to the difference of $n - 4$ and therefore cannot affect the one-loop divergences that we intend to discuss.

As the matter field ϕ, we consider only scalars and spinors.[1] These matter actions, $S^{(n)}_{confm}[\phi, g_{\mu\nu}]$, satisfy the condition of local conformal invariance

$$S^{(n)}_{confm}[e^{k_\phi \sigma}\phi,\, e^{2\sigma} g_{\mu\nu}] \;=\; S^{(n)}_{confm}[\phi,\, g_{\mu\nu}], \tag{14.9}$$

where k_ϕ is the conformal weight of the field in the dimension n, see, e.g., Eq. (12.26) for the scalar field and Exercise 12.14 for the spinor field.

The vacuum effective action is given by

$$e^{i\Gamma^{(n)}[g_{\mu\nu}]} \;=\; \int \mathcal{D}\phi\, e^{iS^{(n)}_{vac}[g_{\mu\nu}]+iS^{(n)}_{confm}[\phi,g_{\mu\nu}]} \tag{14.10}$$

and has the general structure $\Gamma^{(n)}[g_{\mu\nu}] = S^{(n)}_{vac}[g_{\mu\nu}] + \bar{\Gamma}^{(n)}[g_{\mu\nu}]$. Here $\bar{\Gamma}^{(n)}[g_{\mu\nu}]$ represents quantum corrections to the classical action of the gravitational field.

One can perform the local conformal transformation of the external metric and, simultaneously, the change of variables in the functional integral (14.10),

$$g'_{\mu\nu} = g_{\mu\nu} \cdot e^{2\sigma} \quad \text{and} \quad \phi' = \phi \cdot e^{k_\phi \sigma}. \tag{14.11}$$

Taking the conformal invariance of the n-dimensional matter action into account, we arrive at the expression

$$\begin{aligned} e^{i\Gamma^{(n)}[g'_{\mu\nu}]} &= \int \mathcal{D}\phi'\; e^{iS^{(n)}_{vac}[g'_{\mu\nu}]+iS^{(n)}_{confm}[\phi',g'_{\mu\nu}]} \\ &= \int \mathcal{D}\phi\; J\; e^{iS^{(n)}_{vac}[g_{\mu\nu}]+iS^{(n)}_{confm}[\phi,g_{\mu\nu}]} \;=\; e^{i\Gamma^{(n)}[g_{\mu\nu}]}\,, \end{aligned} \tag{14.12}$$

where we took into account that the functional Jacobian $J = \text{Det}\left(\frac{\delta\phi'}{\delta\phi}\right)$ is equal to unit due to the relation $\delta^4(x)|_{x=0} = 0$ in the framework of dimensional regularization.

Thus, in dimensionally regularized theory, the quantum contributions to the vacuum effective action are conformal invariant before taking the limit $n \to 4$,

$$\bar{\Gamma}^{(n)}[e^{2\sigma} g_{\mu\nu}] \;=\; \bar{\Gamma}^{(n)}[g_{\mu\nu}]\,. \tag{14.13}$$

The next step is to take this limit. As usual, we can write

$$\bar{\Gamma}[g_{\mu\nu}] \;=\; \bar{\Gamma}_{div}[g_{\mu\nu}] + \bar{\Gamma}_{fin}[g_{\mu\nu}], \tag{14.14}$$

where $\bar{\Gamma}_{div}[g_{\mu\nu}]$ and $\bar{\Gamma}_{fin}[g_{\mu\nu}]$ are the divergent and finite parts of quantum contributions, respectively. In the one-loop approximation, the divergent part has the following structure:

$$\bar{\Gamma}_{div}[g_{\mu\nu}] \;=\; \frac{1}{n-4}\,\mathcal{F}[g_{\mu\nu}], \tag{14.15}$$

where the functional $\mathcal{F}[g_{\mu\nu}]$ corresponds to four-dimensional spacetime. Substituting (14.15) into (14.13), one can see that the functional $\mathcal{F}[g_{\mu\nu}]$ is conformal invariant,

[1] For the vector gauge field, the considerations should be slightly modified.

because its non-invariance can not be compensated by the finite part of the the one-loop effective action.

The statement about conformal invariance concerns *only* the divergent part of the one-loop effective action, while the finite part does not possess this symmetry due to the conformal anomaly. Let us conclude by saying that more formal consideration, including the vector case, can be found in the paper [70] (see also [80]).

Exercises

14.1. Verify the calculations leading to (14.3), and rewrite this expression using the basis of C^2 and E_4. What are the modifications, if we take $C^2(n)$ instead of $C^2(4)$? Verify that the difference is a finite expression.
14.2. Calculate the trace of the coincidence limit of $\hat{a}_1(x,x')$ and discuss the renormalization of vacuum in the two-dimensional case. Discuss qualitatively, which coefficient of the Schwinger-DeWitt expansion is needed to evaluate the quadratic and logarithmic divergences in the case of an arbitrary, even dimension n.

14.1.2 Spinor fields

As the multiplicity of the field enters in the divergences as an overall factor, we can take single fields of each spin, without loss of generality. For the Dirac spinor (12.41), the bilinear form of the action is the operator

$$\hat{H} = (i\gamma^\alpha \nabla_\alpha - m_f), \tag{14.16}$$

where m_f is the mass of the fermion.

The effective action is given by the expression

$$\bar{\Gamma}^{(1)} = -i\,\mathrm{Tr}\,\log \hat{H}, \tag{14.17}$$

where the negative sign is due to the odd Grassmann parity of the fermion field, while Tr is taken in the usual "bosonic" manner. To reduce the calculation to the standard situation, we have to multiply the operator $\hat{H}$ by the conjugated expression

$$\hat{H}^* = (i\gamma^\alpha \nabla_\alpha + m_f). \tag{14.18}$$

The divergent part of $\bar{\Gamma}^{(1)}$ cannot depend on the sign of the mass, in particular because of dimensional reasons. The divergent part of the vacuum effective action is local, covariant, has an overall mass dimension 4, and is constructed only from the metric and its derivatives. Terms that are odd in metric derivatives are forbidden by these restrictions; hence, the expression of our interest may have only even powers of the mass m_f. This structure of $\bar{\Gamma}^{(1)}_{\mathrm{div}}$ also follows from considerations based on Feynman diagrams, or on the local momentum representation, as we discussed earlier. Independently from the arguments given above, it makes sense to consider a general direct proof of the identity

$$\mathrm{Tr}\,\log \hat{H} = \mathrm{Tr}\,\log \hat{H}^*. \tag{14.19}$$

The main advantage of this proof is that it concerns also the finite part of $\mathrm{Tr}\,\log \hat{H}$ and includes more complicated cases when the fermion interacts not only with the metric but also with other external fields, e.g., with scalar, vector and/or axial vector fields.

Let us consider another basis for the Dirac matrices, in which the relation (14.19) becomes obvious. The new matrices (Γ^μ, Γ^4) are defined by means of chiral rotation,

$$\Gamma^\mu = i\gamma^5\gamma^\mu, \qquad \Gamma^4 = i\gamma^5, \tag{14.20}$$

where the γ^5 matrix is defined by Eq. (4.40). The new matrices Γ^μ and Γ^4 satisfy five-dimensional Clifford algebra:

$$\Gamma^\mu\Gamma^\nu + \Gamma^\nu\Gamma^\mu = 2g^{\mu\nu}, \quad \Gamma^\mu\Gamma^4 + \Gamma^4\Gamma^\mu = 0 \quad \text{and} \quad \Gamma^4\Gamma^4 = -1. \tag{14.21}$$

Consider the operator $\hat{F} = \Gamma^4\hat{H} = \Gamma^4(i\gamma^\mu D_\mu - m_f) = i\Gamma^\mu D_\mu - \Gamma^4 m_f$. It is easy to check that

$$-i \;\mathrm{Tr}\,\log(\hat{F}^2) = -i \;\mathrm{Tr}\,\log \hat{H}\hat{H}^*. \tag{14.22}$$

As $\mathrm{Tr}\,\log\left(\Gamma^4\right)$ is zero in dimensional regularization, we conclude that

$$-i \;\mathrm{Tr}\,\log(\hat{F}) = -\frac{i}{2}\;\mathrm{Tr}\,\log(\hat{F}^2) = -\frac{i}{2}\;\mathrm{Tr}\,\log(\hat{H}\hat{H}^*), \tag{14.23}$$

which is equivalent to (14.19). The practical calculations may be performed using the basis γ^μ or Γ^μ, and the result will be the same.

Now we use (14.19) and write

$$\bar{\Gamma}^{(1)} = -\frac{i}{2}\;\mathrm{Tr}\,\log \bar{H}\bar{H}^* = \mathrm{Tr}\,\log\left(\gamma^\mu\gamma^\nu\nabla_\mu\nabla_\nu + m_f^2\right), \tag{14.24}$$

where we use $\nabla_\alpha\gamma^\mu = 0$. Using the relations (12.65), (12.66) and (12.68), we arrive at the standard form of the operator (13.202) with the elements

$$\hat{h}^\mu = 0 \quad \text{and} \quad \hat{\Pi} = -\frac{1}{4}R + m_f^2,$$

so

$$\hat{P} = -\frac{1}{12}R + m_f^2 \quad \text{and} \quad \hat{S}_{\mu\nu} = -\frac{1}{4}R_{\mu\nu\lambda\tau}\gamma^\lambda\gamma^\tau. \tag{14.25}$$

Since we are dealing with fermions, the functional trace has the opposite sign, hence

$$\bar{\Gamma}^{(1)}_{div} = \frac{\mu^{n-4}}{\epsilon}\;\mathrm{tr}\,\hat{a}_2(x,x), \tag{14.26}$$

and the effect of doubling is compensated by the fact that the Dirac fermion is complex valued. After some algebra, we arrive at the expression for divergences of the Dirac fermion with multiplicity N_f,

$$\bar{\Gamma}^{(1)}_{div} = \frac{\mu^{n-4}}{\epsilon}N_f\int d^nx\sqrt{-g}\Big\{-\frac{1}{20}C^2 + \frac{11}{360}E_4 - \frac{1}{30}\Box R - \frac{1}{3}m_f^2R + 2m_f^4\Big\}, \tag{14.27}$$

where N_f is the number of the fermion fields. Notice that the two-component fermions, such as Weyl or Majorana spinors, contribute with an overall coefficient of $1/2$.

Two relevant comments can be made, if we compare the result in (14.27) with the one for the scalar case (14.6). First, in the conformal case $m_f = 0$, the divergences are conformal invariant, as one should expect. Second, contrary to naive expectations, the contributions of spinors and scalars do not have opposite signs, except for the cosmological constant term. In particular, to all three of the higher-derivative terms C^2, E_4 and $\Box R$, both scalars and spinors contribute with the same sign.

Exercises

14.3. Verify all calculations leading to (14.27).
14.4. Calculate the trace of the coincidence limit of the $\hat{a}_1(x, x')$, and discuss the renormalization of vacuum for the fermion field in the $2D$ case. What is the difference between this and the scalar case?

14.1.3 Massless vector fields

To explore the renormalization of the vacuum sector of the gauge vector fields, we do not need to distinguish the Abelian and non-Abelian cases. The point is that only the free part of the non-Abelian action is relevant for deriving the vacuum effective action. Without loss of generality, one can consider a single Abelian vector and then multiply the result by the multiplicity N_g of the gauge vector fields, be it Abelian or non-Abelian.

As we know from Part I, the action of a vector should be supplemented by the gauge fixing and ghost terms. Then, the one-loop contribution to the vacuum effective action is

$$\bar{\Gamma}^{(1)} = \frac{i}{2}\,\mathrm{Tr}\,\log \hat{H} \;-\; i\,\mathrm{Tr}\,\log \hat{H}_{gh}\,, \tag{14.28}$$

where $\hat{H}$ and $\hat{H}_{gh}$ are the bilinear forms of the gauge field action with the gauge fixing term, and of the ghost action. The coefficient -2 in front of the ghost contribution is due to the odd Grassmann parity of the two Faddeev-Popov ghosts, C and $\bar{C}$.

The action (12.89) can be written as

$$S_1 = \frac{1}{2}\int d^4x\sqrt{-g}\,(\partial_\mu A_\nu)^2 + \frac{1}{2}\int d^4x\sqrt{-g}\,A^\mu(\delta^\nu_\mu\Box - R^\nu_\mu)A_\nu, \tag{14.29}$$

Furthermore, the general form of the covariant gauge-fixing term is

$$S_{\mathrm{gf}} = -\frac{\alpha}{2}\int d^4x\sqrt{-g}\,(\nabla_\mu A^\mu)^2, \tag{14.30}$$

where α is an arbitrary gauge-fixing parameter. As we have seen in Part I, the dependence on gauge parameters vanishes on shell. In the present case, the classical equations of motion are proportional to the vector field A^μ, while the vacuum divergences do not depend on A^μ but on the external metric. Therefore, the vacuum divergences are independent of gauge-fixing and we can choose the gauge-fixing condition in such a way that the calculations are simplest.

Choosing the gauge-fixing term (14.30) with $\alpha = 1$, we arrive at the following *minimal* bilinear form:

$$\hat{H} = \delta^\nu_\mu \Box - R^\nu_\mu \,. \tag{14.31}$$

Taking into account the form of the generator for the gauge transformations $R_\alpha = \nabla_\alpha$ and the gauge fixing $\chi = \nabla_\beta A^\beta$, the bilinear form of the ghost action is

$$\hat{H}_{gh} = \frac{\delta \chi}{\delta A^\alpha} R_\alpha = \Box. \tag{14.32}$$

Therefore, the total expression (14.28) becomes

$$\bar{\Gamma}^{(1)}_{div} = \frac{i}{2} \operatorname{Tr} \log(\delta^\nu_\mu \Box - R^\nu_\mu) - i \operatorname{Tr} \log(\Box). \tag{14.33}$$

The last term is the factor -2 of the vacuum effective action for the minimal massless scalar, so we do not need special calculations. For the first term, we have

$$\hat{\Pi} = \Pi^\mu_\nu = -R^\mu_\nu, \qquad \hat{h}^\mu = 0, \tag{14.34}$$

hence

$$\hat{P} = \hat{\Pi} + \frac{\hat{1}}{6} R = P^\mu_\nu = \frac{1}{6} R\, \delta^\mu_\nu - R^\mu_\nu. \tag{14.35}$$

For the second element, we have

$$\hat{S}_{\alpha\beta} = [S_{\alpha\beta}]^\mu_\nu = \big[\mathcal{R}_{\alpha\beta}\big]^\mu_\nu, \quad \mathcal{R}_{\alpha\beta} = [\nabla_\alpha, \nabla_\beta]^\mu_\nu \quad \Longrightarrow \quad [S_{\alpha\beta}]^\mu_\nu = -R^\mu_{\ \cdot\,\nu\alpha\beta}. \tag{14.36}$$

Using these expressions in (13.206) and taking the ghost part into account, we get

$$\bar{\Gamma}^{(1)}_{div} = -\frac{\mu^{n-4}}{\epsilon} N_g \int d^n x \sqrt{-g} \left\{ \frac{1}{10} C^2 - \frac{31}{180} E_4 - \frac{1}{10} \Box R) \right\}. \tag{14.37}$$

These divergences are given by the conformal invariant and surface terms. The contributions to the higher-derivative terms C^2 and E_4 have the same sign as for fermions and scalars, while the contribution to the $\Box R$-term has the opposite sign.

Exercises

14.5. Verify expressions (14.29), (14.35), (14.36) and the final result (14.37).
14.6. Derive the nonminimal operator, corresponding to the general gauge-fixing condition (14.30). Use Eq. (5.30) of the review paper [31] to verify that the result (14.37) does not depend on the gauge parameter α.
14.7. Remember that the expression for the coincidence limit of the $\hat{a}_2(x, x')$ coefficient does not depend on the spacetime dimension. Verify whether the trace of this coefficient depends on the spacetime dimension. Consider scalar and fermion cases, at least.
14.8. Using dimensional arguments, explain why the cosmological constant and Einstein-Hilbert-type divergences do not show up in expression (14.37). Give another

explanation for the same fact by using arguments based on local conformal symmetry, as explained in section 14.1.1. Show that the last arguments also explain the absence of the R^2 term. Consider what would be needed if we were to consider the global conformal symmetry. Make a detailed comparison with the scalar case, paying special attention to the role of the nonminimal parameter ξ and the R^2 term in the divergences.

14.9. Calculate the trace of the coincidence limit of $\hat{a}_1(x,x')$, and discuss the renormalization of vacuum in the two-dimensional vector gauge model, using direct calculations and arguments based on power counting.

14.1.4 Massive vector fields (The Proca model)

In what follows, we shall describe two methods of deriving divergences for the massive vector (Proca) model, as given in [31] and [82].

The action of the theory in curved space has the form

$$S_P = \int d^4x\sqrt{-g}\,\Big\{ -\frac{1}{4}\,F_{\mu\nu}^2 + \frac{1}{2}\,M_v^2\,A_\mu^2\Big\}, \tag{14.38}$$

where $F_{\mu\nu}^2 = F_{\mu\nu}F^{\mu\nu}$ and $A_\mu^2 = A_\mu A^\mu$. The bilinear form of the action is

$$\hat{H} = H_\alpha^{\ \mu} \;=\; \delta_\alpha^{\ \mu}\Box \;-\; \nabla_\alpha\nabla^\mu \;-\; R_\alpha^{\ \mu} - \delta_\alpha^{\ \mu}M_v^2, \tag{14.39}$$

which has a degenerate derivative part although the theory does not have gauge symmetry. The gauge non-invariance forbids the use of the Faddeev-Popov method for eliminating the degeneracy. The problem can be solved in two different ways. The first one requires introducing an auxiliary operator,

$$\hat{H}^* = H^{*\ \nu}_{\ \mu} \;=\; -\nabla_\mu\nabla^\nu + \delta_\mu^{\ \nu}M_v^2\,, \tag{14.40}$$

which satisfies the following two properties [31]:

$$\begin{aligned} H_\alpha^{\ \mu}\,H^{*\ \nu}_{\ \mu} &= M_v^2\,\big(\delta_\alpha^{\ \nu}\Box - R_\alpha^{\ \nu} - \delta_\alpha^{\ \nu}M_v^2\big),\\ \mathrm{Tr}\,\log\hat{H}^* &= \mathrm{Tr}\,\log\big(\Box - M_v^2\big)\,. \end{aligned} \tag{14.41}$$

As a result we arrive at the relation

$$\frac{i}{2}\,\mathrm{Tr}\,\log\hat{H} \;=\; \frac{i}{2}\,\mathrm{Tr}\,\log\big(\delta_\alpha^{\ \mu}\Box - R_\alpha^{\ \mu} - M_v^2\delta_\alpha^{\ \mu}\big) \;-\; \frac{i}{2}\,\mathrm{Tr}\,\log\big(\Box - M_v^2\big)\,, \tag{14.42}$$

where both operators on the *r.h.s.* are not degenerate and admit a simple use of the standard Schwinger-DeWitt technique for the divergences. Formally, the difference between this and the massless case is that the second term in the *r.h.s.* of (14.42) does not have a factor of 2, as in the case of the gauge ghosts.

Let us consider another route to Eq. (14.42), the Stückelberg procedure. Introduce the new action

$$S_P' = \int d^4x\sqrt{-g}\,\Big\{ -\frac{1}{4}\,F_{\mu\nu}^2 - \frac{1}{2}\,M_v^2\,\Big(A_\mu - \frac{1}{M_v}\,\partial_\mu\varphi\Big)^2\Big\}. \tag{14.43}$$

This expression is invariant under the gauge transformations

$$A_\mu \to A'_\mu = A_\mu + \partial_\mu \xi \qquad \text{and} \qquad \varphi \to \varphi' = \varphi + \xi M_v. \tag{14.44}$$

In the special gauge fixing $\varphi = 0$, we come back to the Proca field action (14.38). And, finally, since both (14.38) and (14.43) are free field actions, the gauge-fixing dependence is irrelevant for the quantum correction, as we discussed in the previous subsection. The practical calculations can be performed in the useful special gauge $\chi = \nabla_\mu A^\mu - M\varphi$; the result is going to be the same as in the gauge $\varphi = 0$.

The action (14.43) with the gauge-fixing term $S_{gf} = -\frac{1}{2}\int d^4x\sqrt{-g}\,\chi^2$ has the factorized form

$$S' + S_{gf} = \int d^4x\sqrt{-g}\left\{A^\alpha\left(\delta^\nu_\alpha\Box - R^\mu_\alpha - \delta^\mu_\alpha M_v^2\right)A_\nu + \varphi\left(\Box - M_v^2\right)\varphi\right\}, \tag{14.45}$$

such that

$$\bar{\Gamma}^{(1)} = \frac{i}{2}\,\mathrm{Tr}\,\log\left(\delta^\nu_\alpha\Box - R^\mu_\alpha - \delta^\mu_\alpha M_v^2\right) + \frac{i}{2}\,\mathrm{Tr}\,\log\left(\Box - M_v^2\right) - i\,\mathrm{Tr}\,\log\left(\Box - M_v^2\right),$$

which is exactly Eq. (14.42). One can see that an extra scalar was indeed "hidden" in the massive term of the vector. At this point, we conclude that the Stückelberg procedure, which was described in the first part of the book, also works well in curved spacetime.

The calculation is very simple, and the result for the one-loop effective action is

$$\bar{\Gamma}^{(1)}_{div} = -\frac{\mu^{n-4}}{\epsilon}N_P\int d^nx\sqrt{-g}\,\Big\{\frac{13}{120}\,C^2 - \frac{7}{40}\,E_4 - \frac{17}{180}\,(\Box R) + \frac{1}{4}\,M^2R + \frac{3}{4}\,M^4\Big\}, \tag{14.46}$$

where N_P is the multiplicity of the Proca field. It is easy to see (see Exercise 14.5) that the massless limit of the expression (14.46) *does not* coincide with the result of the direct calculation in the massless theory. This is the well-known effect knows as *discontinuity* in the massless limit. Later on, we shall see that it takes place not only in the divergences but also in the finite part of the effective action.

Exercises

14.10. Using the expressions for vacuum divergences for the gauge vector (14.37) and the massive Proca vector, analyze the discontinuity in the massless limit (14.46). After that, try to reproduce the result *without* these expression by using only the result (14.42). What should we expect in the finite contributions? The answer can be found in [167] and [82].

14.11. Write down the general expression for one-loop vacuum divergences for an arbitrary theory that includes scalar fields with masses $m_1^{(s)}$, $m_2^{(s)}$, ..., $m_{N_s}^{(s)}$, Dirac fields with masses $m_1^{(f)}$, $m_2^{(f)}$, ..., $m_{N_f}^{(f)}$ and N_g massless vectors and massive vectors with the masses $m_1^{(P)}$, $m_2^{(P)}$, ..., $m_{N_P}^{(P)}$. Why is it that this expression does not depend on the interactions between these fields? How do the fermionic contributions change if some of the fermions are not Dirac spinors but are Weyl or Majorana ones?

14.2 Beta functions in the vacuum sector

Consider a theory involving N_s real scalars, N_f Dirac fermions and N_g massless vectors. For the sake of simplicity, we assume that the masses of the fields of each spin are equal, that ξ's for all scalars are also equal, and, in addition, do not consider massive vectors. Indeed, the generalization for the most general case is straightforward.

The divergent part of the vacuum effective action is

$$\bar{\Gamma}^{(1)}_{div} = -\frac{\mu^{n-4}}{\epsilon}\int d^nx\sqrt{-g}\{k_1C^2 + k_2E_4 + k_3\Box R + k_4R^2 + k_5R + k_6\}, \qquad (14.47)$$

where

$$k_1 = (4\pi)^2\omega = \frac{1}{120}N_s + \frac{1}{20}N_f + \frac{1}{10}N_g, \qquad (14.48)$$

$$k_2 = (4\pi)^2 b = -\frac{1}{360}N_s - \frac{11}{360}N_f - \frac{31}{180}N_g, \qquad (14.49)$$

$$k_3 = \Big[\frac{1}{180} + \frac{1}{6}\Big(\xi - \frac{1}{6}\Big)\Big]N_s + \frac{1}{30}N_f - \frac{1}{10}N_g, \qquad (14.50)$$

$$k_4 = \frac{N_s}{2}\Big(\xi - \frac{1}{6}\Big)^2, \qquad (14.51)$$

$$k_5 = -N_s m_s^2\Big(\xi - \frac{1}{6}\Big) + \frac{N_f m_f^2}{3}, \qquad (14.52)$$

$$k_6 = \frac{1}{2}N_s m_s^4 - 2N_f m_f^4, \qquad (14.53)$$

where, in Eq. (14.48) and (14.49), we introduced the definitions of the coefficients ω and b, which will be used intensively later on when we discuss the trace anomaly. Another useful notation of the same sort is

$$c = \frac{1}{(4\pi)^2}k_3\Big(\xi = \frac{1}{6}\Big) = \frac{1}{(4\pi)^2}\Big(\frac{1}{180}N_s + \frac{1}{30}N_f - \frac{1}{10}N_g\Big). \qquad (14.54)$$

Let us show how these divergences can be removed via renormalization of the parameters of the classical action of the vacuum (12.6) and how minimal subtraction beta functions can be calculated in dimensional regularization.

As a first example, consider the Weyl-squared term. The classical action and the renormalized classical action in n-dimensional space have the forms

$$S_W^{\rm cl} = \int d^nx\sqrt{-g}\,\mu^{n-4}a_1C^2 \qquad \text{and} \qquad S_W^{\rm ren} = S_W^{\rm cl} + \Delta S^{(W)}, \qquad (14.55)$$

where the counterterm has to cancel the divergence of the effective action (14.47),

$$\Delta S_W = \frac{\mu^{n-4}}{\epsilon}\int d^nx\sqrt{-g}\,k_1C^2. \qquad (14.56)$$

By definition, the renormalized action equals the bare one,

$$S_W^0 = \int d^4x\sqrt{-g}\,a_1^0C^2. \qquad (14.57)$$

Thus, we arrive at the renormalization relation for the parameter,

$$a_1^0 = \mu^{n-4}\Big(a_1 + \frac{k_1}{\epsilon}\Big). \tag{14.58}$$

In this expression, a_1^0 does not depend on μ, while a_1 does. Therefore,

$$0 = \mu\,\frac{da_1^0}{d\mu} = \mu^{n-4}\Big[(n-4)\Big(a_1 + \frac{k_1}{\epsilon}\Big) + \mu\,\frac{da_1}{d\mu}\Big]. \tag{14.59}$$

In this way, we arrive at the relation

$$\mu\,\frac{da_1}{d\mu} = -(n-4)a_1 - \frac{k_1}{(4\pi)^2}. \tag{14.60}$$

In the limit $n \to 4$, we obtain the desired beta function,

$$\beta_1 = \lim_{n\to 4}\mu\,\frac{da_1}{d\mu} = -\,\frac{k_1}{(4\pi)^2} = -\,w. \tag{14.61}$$

A little bit later, when we discuss the quantization of gravity, we shall see that the sign of a_1 should be negative (if there are no derivatives higher than 4 in the action), to provide the stability of the light modes of the gravitational perturbations. Moreover, in such a theory, the coupling constant is the inverse of a_1. It is useful to define this coupling $\lambda > 0$ by the relation $a_1 = -\frac{1}{2\lambda}$. Then, Eq. (14.61) gives

$$\mu\,\frac{d\lambda}{d\mu} = -\,\frac{k_1}{2\,(4\pi)^2}\,\lambda^2, \tag{14.62}$$

indicating the asymptotic freedom for the parameter λ in the contributions from matter fields of spin 0, $\frac{1}{2}$ and 1. In a similar way, one can derive the renormalization group equations for the parameters $a_{2,3,4,5,6}$ of the action of vacuum (12.6):

$$\begin{aligned}
\beta_\Lambda &= \mu\frac{d}{d\mu}\Big(\frac{\Lambda}{8\pi G}\Big) = \frac{1}{(4\pi)^2}\Big(\frac{N_s}{2}\,m_s^4 - 2N_f m_f^4\Big),\\
\beta_G &= \mu\,\frac{d}{d\mu}\Big(\frac{1}{16\pi G}\Big) = \frac{1}{(4\pi)^2}\Big[\frac{N_s m_s^2}{2}\Big(\xi - \frac{1}{6}\Big) + \frac{N_f m_f^2}{3}\Big]
\end{aligned} \tag{14.63}$$

for the cosmological and inverse Newton constants,

$$(4\pi)^2\,\mu\,\frac{da_4}{d\mu} = -\,k_4 = \frac{1}{(4\pi)^2}\,\frac{N_s}{2}\Big(\xi - \frac{1}{6}\Big)^2 \tag{14.64}$$

for the R^2-term coefficient and, finally,

$$\mu\,\frac{da_{2,3}}{d\mu} = -\,\frac{k_{2,3}}{(4\pi)^2}, \quad \text{or} \quad \beta_2 = -\,\frac{k_2}{(4\pi)^2} \quad \text{and} \quad \beta_3 = -\,\frac{k_3}{(4\pi)^2}, \tag{14.65}$$

for the coefficients of the terms that appear as one-loop divergences in classically conformal theories. As we shall see in the next chapters, these equations are supposed to describe the short-distance behavior of the corresponding effective charges and, in general, represent a useful starting point for further considerations.

14.3 One-loop divergences in interacting theories

The general algorithm (13.217) can be used not only in the vacuum sector but also in interacting theories. Let us consider a few such examples, in order of increasing complexity.

14.3.1 A self-interacting scalar

The action is

$$S = \int d^4x\sqrt{-g}\,\Big\{\frac{1}{2}\,(\partial\varphi)^2 - \frac{1}{2}\,m^2\varphi^2 + \frac{1}{2}\,\xi R\varphi^2 - \frac{\lambda}{4!}\,\varphi^4\Big\}, \tag{14.66}$$

with the self-interaction corresponding to the renormalizable theory.

The background field method splitting into quantum φ and classical χ scalars is

$$\varphi \longrightarrow \varphi' = \varphi + \chi. \tag{14.67}$$

Neglecting the surface terms, the part of the action that is bilinear in the quantum field is

$$S^{(2)} = \frac{1}{2}\int d^4x\sqrt{-g}\,\Big\{-\chi\Box\chi - m^2\chi^2 + \xi R\chi^2 - \frac{\lambda}{2}\,\varphi^2\chi^2\Big\}. \tag{14.68}$$

The overall negative sign can be absorbed into the operator, which produces no divergences, and therefore we can identify the elements of the operator (13.202) as

$$\hat{h}^\alpha = 0, \qquad \hat{\Pi} = m^2 - \xi R - \frac{1}{2}\,\varphi^2. \tag{14.69}$$

It is easy to see that

$$\hat{S}_{\mu\nu} = 0, \qquad \hat{P} = m^2 - \Big(\xi - \frac{1}{6}\Big)R - \frac{1}{2}\,\varphi^2. \tag{14.70}$$

Since we already know the vacuum divergences, let us present only the ones in the scalar sector,

$$\bar{\Gamma}^{(1)}_{\text{div}} = -\frac{\mu^{n-4}}{\epsilon}\int d^n x\sqrt{-g}\,\Big\{\frac{\lambda^2}{8}\,\varphi^4 + \frac{\lambda}{2}\,m^2\varphi^4 - \frac{\lambda}{2}\,\Big(\xi - \frac{1}{6}\Big)R\varphi^2\Big\}. \tag{14.71}$$

In order to remove the divergences, introduce the counterterms $\Delta S = -\bar{\Gamma}^{(1)}_{\text{div}}$. The renormalized classical action $S_R = S + \Delta S$ and the bare action S_0 should be equal:

$$\begin{aligned} S_R = \frac{\mu^{n-4}}{2}\int d^n x\sqrt{-g}\,\Big\{(\partial\varphi)^2 - m^2\Big(1 - \frac{\lambda}{\epsilon}\Big)\varphi^2 \\ + \Big[\xi - \frac{\lambda}{\epsilon}\Big(\xi - \frac{1}{6}\Big)\Big]R\varphi^2 - \frac{1}{12}\,\varphi^4\Big(\lambda - \frac{3\lambda^2}{\epsilon}\Big)\Big\} \\ = \frac{1}{2}\int d^4x\sqrt{-g}\,\Big\{(\partial\varphi_0)^2 - m_0^2\varphi_0^2 + \xi_0 R\varphi_0^2 - \frac{\lambda_0}{12}\,\varphi_0^4\Big\} = S_0. \end{aligned} \tag{14.72}$$

The last equation can be satisfied if we assume the renormalization relations

$$\varphi_0 = \mu^{\frac{n-4}{2}}\,\varphi, \tag{14.73}$$

$$m_0^2 = m^2 - \frac{\lambda}{\epsilon}\,m^2, \tag{14.74}$$

$$\xi_0 = \xi - \frac{\lambda}{\epsilon}\Big(\xi - \frac{1}{6}\Big), \tag{14.75}$$

$$\lambda_0 = \mu^{4-n}\Big(\lambda - \frac{3\lambda^2}{\epsilon}\Big). \tag{14.76}$$

Now we are in a position to derive the beta functions. In the minimal subtraction renormalization scheme, they reflect the dependence on the artificial parameter μ that can be associated to one or another physical quantity, depending on the given problem. Thus, we apply the logarithmic derivative to the both sides of relations (14.73)-(14.76). Starting from (14.76), we get

$$\mu\frac{d\lambda_0}{d\mu} = 0 = (4-n)\mu^{4-n}\Big(\lambda - \frac{3\lambda^2}{\epsilon}\Big) + \mu^{4-n}\,\mu\frac{d\lambda}{d\mu}\Big(1 - \frac{6\lambda}{\epsilon}\Big). \tag{14.77}$$

In order to solve this equation, we have to invert $\Big(1 - \frac{6\lambda}{\epsilon}\Big)^{-1} = 1 + \frac{6\lambda}{\epsilon}$. The reason for this rule is that all one-loop terms, including those with an $\frac{1}{\epsilon}$-factor, have a hidden loop expansion parameter $\hbar$. Since we are interested in the one-loop approximation, the higher-order terms can be omitted. Thus, after a small algebra, we arrive at the solution

$$\mu\frac{d\lambda}{d\mu} = (n-4)\lambda + \frac{3\lambda^2}{(4\pi)^2}. \tag{14.78}$$

The other two beta functions can be derived in a similar way, so we leave them as exercises for the reader and give just the results,

$$\mu\frac{dm^2}{d\mu} = \frac{\lambda}{(4\pi)^2}\,m^2, \qquad \mu\frac{d\xi}{d\mu} = \frac{\lambda}{(4\pi)^2}\Big(\xi - \frac{1}{6}\Big). \tag{14.79}$$

Exercises

14.12. Verify Eq. (14.78) and (14.79).
14.13. Remember that the value of $\xi = 1/6$ corresponds to a scalar theory that has local conformal symmetry in four-dimensional space. Use this fact and the arguments from the subsection 14.1.1 to prove that, for any theory with scalar, fermion and massless vector fields, the one-loop beta function for ξ is proportional to the difference $\xi - 1/6$.
14.14. Repeat the calculations for a more general model, with $\frac{\lambda}{4!}\varphi^4$ traded for an arbitrary $V(\varphi)$. Compare the results for the divergences with the ones from section 13.3.3.
14.15. Make the calculation for a complex scalar field and an arbitrary real potential. Use two different approaches: a) presenting a complex scalar as two real scalars, b) directly use the background field method for a complex field.

14.16. Consider the case of a complex scalar field that include interaction with an *external* Abelian vector field A_μ. Derive the full set of divergences, both in vacuum (depending on metric and vector) and in the scalar sector. Explain in different ways why the divergences respect gauge invariance.
14.17. Generalize the previous calculation, treating the gauge vector as quantum field. This means that the background field method starts from simultaneous splitting (14.67) and $A_\mu \to A_\mu + \xi_\mu$.
14.18. Modify the previous exercise by taking the Proca vector model instead of the massless gauge vector.
Observation. The last exercise is not very easy to solve.

14.3.2 Non-Abelian vector fields

Let us now use the heat-kernel approach to achieve the historically important result, namely, the beta function for the gauge-coupling constant in the theory with a Yang-Mills field coupled to scalars and fermions. The calculation is simpler than the one using diagrams. Let us note that part of the expansions were already done in section 10.9 in Part I. Here we repeat this small calculation, this time in a curved spacetime.

As a warm-up exercise, we derive the contribution of a charged scalar,

$$S_{cs} = \int \sqrt{-g}\Big\{g^{\mu\nu}\big(D_\mu\phi\big)^*\big(D_\nu\phi\big) - m^2\phi^*\phi + \xi R\phi^*\phi\Big\}, \tag{14.80}$$

where $g_{\mu\nu}$ and A_μ are external metric and vector fields, typically in the adjoint representation of the gauge group, e.g., $SU(N)$ or $O(N)$. Furthermore, as in (4.88),

$$D_\mu\phi = \nabla_\mu\phi - igA_\mu\phi, \qquad \big(D_\mu\phi\big)^* = \nabla_\mu\phi^* + igA_\mu\phi^*, \tag{14.81}$$

and ϕ belongs to some representation of the gauge group of our interest. Our purpose is to make the calculation as simple as possible. And the first important observation is that the calculation can be performed even for the Abelian case, the result will be the same. The reason is that the counterterm that we are interested in is $(G_{\mu\nu})^2 = G^a_{\mu\nu}G^{a\mu\nu}$. In the non-Abelian case, this term includes both a free part and vector self-interactions. However, the interaction terms are restricted by gauge invariance, and hence the overall coefficient can be defined by the free part only. Hence, without loss of generality, we can work with the action (14.80), treating A_μ as a single Abelian field and, after that, multiply the result by the number of charged scalar fields.

Since the quantum field ϕ has no self-interaction, we do not need to split the scalar field into classical and quantum parts. Therefore, neglecting the trivial contribution to $\mathrm{Tr}\log\hat{H}$, we can identify the operator of our interest as

$$\begin{aligned} \hat{H} &= \Box + 2\hat{h}^\alpha\nabla_\alpha + \hat{\Pi}, \qquad \text{where} \\ \hat{h}^\alpha &= -i\,gA^\alpha \quad \text{and} \quad \hat{\Pi} = m^2\chi^2 - \xi R - g^2A_\alpha A^\alpha - i\,g\nabla_\alpha A^\alpha. \end{aligned} \tag{14.82}$$

As was explained above, the $\mathcal{O}(A\cdot A)$ term was omitted. Then we easily arrive at

$$\hat{P} = m^2\chi^2 - \Big(\xi - \frac{1}{6}\Big)R, \qquad \hat{S}_{\alpha\beta} = ig\,G_{\alpha\beta}, \tag{14.83}$$

and, finally, to

$$\bar{\Gamma}^{(1)}_{\text{div}}(A) = -\frac{\mu^{n-4}}{\epsilon}\int d^n x\sqrt{-g}\Big\{-\frac{g^2}{12}(G_{\mu\nu})^2\Big\}, \tag{14.84}$$

where the vacuum terms were omitted. These terms are exactly twice that of Eq. (14.3), which is the result for a real scalar.

Let us derive the contribution Eq. (14.84) to the beta function for g. The calculation is similar to the one for the vacuum metric-dependent terms, as explained in section 14.2. The analogy is natural, since, in both cases, we deal with external (metric or vector) fields. On the other hand, there is an important difference, because, in the vector case, one has to include renormalization of the field, which is also called renormalization of the wave function. Later on, we offer an exercise explaining how the renormalization of A_μ can be avoided.

The equality of renormalized $S_R(A) = S(A) + \Delta S = S(A) - \bar{\Gamma}^{(1)}_{\text{div}}(A)$ and bare $S_0(A_0)$ actions can be formulated as

$$-\frac{1}{4}\mu^{n-4}\int d^n x\sqrt{-g}\Big[1+\frac{g^2}{3\epsilon}\Big](G_{\mu\nu})^2 = -\frac{1}{4}\int d^4x\sqrt{-g}\,(G^0_{\mu\nu})^2. \tag{14.85}$$

The renormalization of the field A_μ follows from (14.85),

$$A^0_\mu = \mu^{\frac{n-4}{2}}\Big(1+\frac{g^2}{6\epsilon}\Big)A_\mu \quad\Longrightarrow\quad g_0 = \mu^{\frac{4-n}{2}}\Big(1-\frac{g^2}{6\epsilon}\Big)g. \tag{14.86}$$

The product satisfies the relation

$$g_0 A^0_\mu = gA_\mu, \tag{14.87}$$

which guarantees the gauge invariance of the renormalized covariant derivative (14.81) and also the gauge-invariant form of renormalized $G^a_{\mu\nu}$. Applying the differential operator $\mu\frac{d}{d\mu}$ to both sides of (14.86), in a standard way one gets

$$\mu\frac{dg}{d\mu} = \frac{n-4}{2}\,g + \frac{N_{cs}}{6\,(4\pi)^2}\,g^3, \tag{14.88}$$

and, in the limit, $n \to 4$, we arrive at the beta function for the N_{cs}-component scalar:[2]

$$\beta_g(\text{scalar}) = \frac{N_{cs}}{6\,(4\pi)^2}\,g^3. \tag{14.89}$$

The next step is to obtain the same type of contribution to β_g, but this time from the spinor field. The action is

$$S_f = i\int\sqrt{-g}\bar{\psi}(\gamma^\mu D_\mu + im)\psi, \tag{14.90}$$

where the extended covariant derivative is defined in the same way as for scalars (14.81), of course with ∇_μ now being the fermion version of covariant derivative.

[2]Let us warn that, here, N_{cs} means the number of *complex* scalars, which is half of the N_s in the previous sections where we dealt with the vacuum contributions of a real scalar field.

After the same doubling trick which we already discussed for a purely metric background, we identify the operator of our interest as

$$\hat{H} = (\gamma^\mu D_\mu - im)(\gamma^\mu D_\mu + im) = \Box + 2\hat{h}^\alpha \nabla_\alpha + \hat{\Pi}, \qquad (14.91)$$

where, after small calculations, we obtain

$$\hat{h}^\alpha = -i\,gA^\alpha, \qquad \hat{\Pi} = m^2\chi^2 - \frac{1}{4}R - \frac{g}{2}\sigma^{\alpha\beta}G_{\alpha\beta} - g^2 A_\alpha A^\alpha - i\,g\nabla_\alpha A^\alpha. \qquad (14.92)$$

Exactly as in the scalar case, the $\mathcal{O}(A \cdot A)$ term can be omitted for the sake of practical calculations, hence we shall omit these terms. Thus, one can obtain

$$\hat{P} = m^2\chi^2 - \frac{1}{12}R - \frac{g}{2}\sigma^{\alpha\beta}G_{\alpha\beta}, \quad \hat{S}_{\alpha\beta} = -\frac{1}{4}R_{\alpha\beta\rho\sigma}\gamma^\rho\gamma^\sigma + igG_{\alpha\beta}. \qquad (14.93)$$

After some algebra we get (disregarding vacuum terms)

$$\bar{\Gamma}^{(1)}_{\rm div}(A) = -\frac{\mu^{n-4}}{\epsilon}\int d^n x\sqrt{-g}\,\Big\{-\frac{2g^2}{3}\,G^2_{\mu\nu}\Big\}, \qquad (14.94)$$

which is eight times greater in magnitude compared to scalar contribution (14.84).

The last part is the calculation for the proper Yang-Mills field. This time, the non-linear terms become important. The classical action is $S_{\rm YM} = \int d^4x\sqrt{-g}\mathcal{L}$, where the Lagrangian is defined by the expressions (4.124) and (4.123).

For the sake of simplicity we assume the semisimple Lie symmetry group, hence the structure constants of this group are absolutely antisymmetric, $f^{abc} = f^{[abc]}$, and satisfy the relation

$$f^{abc}\,f^{abd} = C_1\delta^{cd}. \qquad (14.95)$$

Here C_1 is the Casimir operator. For example, in the case of $SU(N)$ group, $C_1 = N$. Finally, the gauge transformation is

$$A^a_\mu \longrightarrow A^a_\mu + D^{ab}_\mu\zeta^b, \quad \text{where} \quad D^{ab}_\mu = \delta^{ab}\nabla_\mu - gf^{abc}A^c \qquad (14.96)$$

is the extended covariant derivative, which is exactly the same as what we should use in (14.81) if we decided to do the full calculation with the $\mathcal{O}(A^3)$ and $\mathcal{O}(A^4)$ terms.

The background field method for Yang-Mills field has been discussed in detail in Part I, and the presence of an external metric does not change things too much. The splitting into background and quantum vectors is done according to

$$A^a_\mu \longrightarrow A'^a_\mu = A^a_\mu + \xi^a_\mu, \qquad (14.97)$$

with the gauge transformation of the form

$$\delta\xi^a_\mu = D^{ab}_\mu\zeta^b \qquad (14.98)$$

and the covariant derivative (14.96) built with the background vector A^a_μ.

We choose to construct the Schwinger-DeWitt technique in a fully covariant form, based on the derivative D_μ. Then the expansion of the field tensor has the form

$$G^a_{\mu\nu} \longrightarrow G'^a_{\mu\nu} = \big(D_\mu\xi_\nu\big)^a - \big(D_\nu\xi_\mu\big)^a + gf^{abc}\,\xi^b_\mu\,\xi^c_\nu, \tag{14.99}$$

where we introduced a compact notation $\big(D_\mu\xi_\nu\big)^a = D^{ab}_\mu\xi^b_\nu$.

The bilinear in quantum field ξ^a_μ part of the action has the form

$$S^{(2)}_{\rm YM} = \frac{1}{2}\int d^4x\sqrt{-g}\,\Big\{\big(D_\mu\xi_\nu\big)^a\,\big(D^\nu\xi^\mu\big)^a - \big(D_\mu\xi_\nu\big)^a\,\big(D^\mu\xi^\nu\big)^a \\ + gf^{abc}G^{a\mu\nu}\,\xi^b_\mu\,\xi^c_\nu\Big\}. \tag{14.100}$$

The completely covariant form of the gauge-fixing term is

$$S_{\rm gf} = \frac{\alpha}{2}\int d^4x\sqrt{-g}\,\chi^a\chi^a, \tag{14.101}$$

where $\chi^a = \big(D_\mu\xi^\mu\big)^a$ and α is the gauge fixing parameter. The value $\alpha = -1$ of the gauge fixing parameter in (14.101) provides the simplest minimal operator in the vector sector; hence, we will use this choice.

The ghost action is defined by the operator

$$M^a_b = \frac{\delta\chi^a}{\delta\xi_{\mu c}}\,D^{cb}_\mu = D^{ac}_\mu\,D^{\mu cb} = \big(D^2\big)^{ab} = \hat{1}D^2, \tag{14.102}$$

where $\hat{1} = \delta^{ab}$ is the identity operator in the space of ghosts. The contribution of ghosts has the purely gravitational part that we already know, and hence we will concentrate only on the A-dependent part of the divergences. It is easy to see that

$$\hat{P}_{\rm gh} = \frac{1}{6}\,\delta^{ab}R + \mathcal{O}(A^2). \tag{14.103}$$

Furthermore,

$$\hat{S}^{\rm gh}_{\lambda\tau} = gf^{abc}\,G^c_{\lambda\tau}. \tag{14.104}$$

Therefore, the relevant contribution from the ghosts is

$$\frac{1}{12}\,{\rm tr}\,\hat{S}^{\rm gh}_{\lambda\tau}\,\hat{S}^{{\rm gh}\,\lambda\tau} = -\,\frac{1}{12}\,g^2C_1\,\big(G_{\mu\nu}\big)^2. \tag{14.105}$$

Consider now the first term of the general expression

$$\bar{\Gamma}^{(1)}_{\rm div}(A) = \frac{i}{2}\,{\rm Tr}\,\log\hat{H} - i\,{\rm Tr}\,\log\hat{H}_{\rm gh}. \tag{14.106}$$

After some integration by parts and commutations in the expression $S^{(2)}_{\rm YM} + S_{\rm gf}$, we arrive at the operator in the vector sector, which has the form

$$\hat{H} = H^{ab}_{\mu\nu} = \delta^{ab}\big(g_{\mu\nu}D^2 - R_{\mu\nu}\big) - 2gf^{abc}\,G^c_{\mu\nu}. \tag{14.107}$$

Half of the last term comes directly from the bilinear expression (14.100), and the other half comes from the commutator of derivatives in the first term of the integrand. of the same expression.

In the standard way, we get from (14.103)

$$\hat{P} = \delta^{ab}\Big(\frac{1}{6}\,Rg_{\mu\nu} - R_{\mu\nu}\Big) - 2gf^{abc}\,G^{c}_{\ \mu\nu}, \tag{14.108}$$

$$\hat{S}_{\lambda\tau} = \big[D_\tau, D_\lambda\big] = -\delta^{ab}R_{\mu\nu\lambda\tau} - gf^{abc}\,G^{c}_{\ \lambda\tau}\,g_{\mu\nu}. \tag{14.109}$$

Thus, the contributions of the two relevant parts to the first term in (14.106) are

$$\frac{1}{2}\,\mathrm{tr}\,\hat{P}^2 = 2C_1\,g^2\big(G_{\mu\nu}\big)^2 \qquad \text{and} \qquad \frac{1}{12}\,\mathrm{tr}\,\hat{S}_{\lambda\tau}\,\hat{S}^{\lambda\tau} = -\frac{1}{3}\,C_1\,g^2\big(G_{\mu\nu}\big)^2. \tag{14.110}$$

Summing up these terms with (14.105), in accordance with the formula (14.106), after a very small algebra we arrive at the final result

$$\bar{\Gamma}^{(1)}_{\mathrm{div}}(A) = -\frac{\mu^{n-4}}{\epsilon}\int d^Dx\sqrt{-g}\,\Big\{\frac{11}{6}\,C_1\,g^2\big(G_{\mu\nu}\big)^2\Big\}. \tag{14.111}$$

Summing up the contributions from scalars, fermions and the proper Yang-Mills field, the overall beta function for g has the form

$$\beta_g = \lim_{D\to 4}\,\mu\frac{dg}{d\mu} = -\frac{1}{(4\pi)^2}\Big(\frac{11}{3}\,C_1 - \frac{1}{6}\,N_{cs} - \frac{4}{3}\,N_f\Big)\,g^3. \tag{14.112}$$

According to this famous formula, in the gauge theory with not too many fermions and scalars, the sign of the beta function is negative, which signals the asymptotic freedom in the theory. The asymptotic freedom means that, at large values of μ, the running coupling constant $g(\mu)$ becomes small [see Eq. (9.186)].

Associating μ with the typical energy of the physical process, we arrive at the situation where the quantum theory becomes free of interactions in the far UV, i.e., at high energies. The importance of asymptotic freedom was recognized in the award of the 2004 Nobel Prize in Physics to Gross, Wilczek and Politzer for the works [173,254].

The main application of Eq. (14.112) is a basic explanation of confinement in the low-energy theory of strong interactions, or quantum chromodynamics (QCD). At high energies, the interaction is weak, according to asymptotic freedom. The value of $g = g_s$ at the typical QCD scale $\Lambda_{QCD} \approx 200\,MeV$, is very large, i.e., about 1. This means that the same coupling constant runs and grows very fast at the lower end of the energy scale. Since the Yang-Mills fields in QCD are gluons and they provide interaction between quarks and self-interaction, this means that, when quarks are inside a nucleon or a meson, they are almost free. But, when a quark tries to escape from the hadron, the typical distance increases, the energy scale decreases and, according to Eq. (14.112), the force between quarks becomes too strong and the the quark that is trying to escape is forced to come back. Indeed, in such an infrared regime, the one-loop equation cannot be regarded too seriously; that is why this explanation of confinement is no more than qualitative. The problem of detailed description of the physical situation in low-energy quantum chromodynamics is much more complicated. Typically, the calculations in low-energy QCD require sophisticated non-perturbative methods, such as lattice simulations and other approaches.

Exercises

14.19. Repeat the calculations for scalars and fermions by using the extended covariant derivative (14.81), as it was done in the pure Yang-Mills case.
14.20. Repeat the calculations for scalars, fermions and the pure Yang-Mills case, choosing for scalars and fermions fundamental or adjoint representations of the $SU(N)$ gauge group, and taking special care with the non-linear elements in vector field counterterms, which were omitted in our calculations for the sake of brevity.
14.21. Write an equation similar to (14.112) for the simplified version of QCD, with the $SU(3)$ gauge group, $N_{cs} = 0$ and $N_f = 6 \times 3 = 18$. The factor of 3 here comes from the color of quarks. Associating the parameter μ with energy, and assuming $g = g_s = 1$ at the QCD scale $\mu_0 = 200\,MeV$, derive the value of g_s at the energies $1\,GeV$, $10\,GeV$, and $1000\,GeV$.
14.22. Discuss whether the calculation from the previous exercise can be extended to the infrared domain. Is the one-loop result (14.112) reliable in this case? Without deriving the coefficients, write the general form of second and third loop contributions. Why do these contributions become irrelevant in UV?

14.3.3 Yukawa model

The last example we consider now, shows how to work with an interacting theory that has both bosons and fermions. The simplest theory of this kind, which is, in fact, interesting by itself,[3] is the Yukawa model with a sterile scalar.

Consider the action of N copies of Dirac fermion and a single real scalar field,

$$S = \int d^4x\sqrt{-g}\Big\{\sum_{k=1}^{N} \bar{\Psi}_k\left(i\gamma^\mu\nabla_\mu - M - h\varphi\right)\Psi_k + \frac{1}{2}(\nabla\varphi)^2 - \frac{1}{2}m^2\varphi^2 + \frac{1}{2}\xi R\varphi^2 - V(\varphi)\Big\}, \tag{14.113}$$

where $V(\varphi)$ is the potential of self-interaction of the scalar field. In order to have a renormalizable theory, one has to include in this potential all of the terms that can emerge in the divergences. The sterile real scalar is a special case in this respect, as it has a feature which does not show up for more complicated scalar models. An additional benefit of this example is that it makes the idea of constructing renormalizable theory more transparent.

There are no symmetries that protect divergences from odd-power terms in φ. On the other hand, power counting tells us that the potential $V(\varphi)$ of renormalizable theory cannot have powers greater than four for φ. Thus, the potential of a renormalizable theory has the form

$$V(\varphi) = \frac{\lambda}{4!}\varphi^4 + \frac{g}{3!}\varphi^3 + \tau\varphi + fR\varphi, \tag{14.114}$$

[3]The calculation of divergences in this kind of models was done long ago [78], but recently there was a revival of interest in a more general model involving both the usual and axial scalar fields [318], and also in exploring the special features of this model with a "sterile" scalar field, which does not belong to the gauge groups of particle physics [24]. Here we mainly follow the last reference.

where λ, g, τ and f are constant parameters. Only λ and g can be called couplings, but all these parameters are expected to get renormalized. Let us see whether this expectation will be confirmed by the explicit one-loop calculations.

Using the background fields method, we start by decomposing the fields into classical φ, $\bar{\Psi}$, Ψ and their quantum σ, $\bar{\eta}$, χ counterparts,

$$\varphi \to \varphi + i\sigma, \qquad \bar{\Psi}_k \to \bar{\Psi}_k + \bar{\eta}_k, \qquad \Psi_k \to \Psi_k - \frac{i}{2}(\nabla\!\!\!\!/ - iM)\chi_k. \qquad (14.115)$$

In the last expression, we used our previous experience with the pure scalar and Dirac fields, making the change of variables at once. Since the vacuum divergences can be easily obtained as a sum of N contributions of Dirac spinors and one of a real scalar, we can restrict our attention to the counterterms in the sector of φ and Ψ_k fields. Thus, the Jacobian of the change of variables in (14.115) has no interest for us.

The one-loop divergences are defined by the bilinear part of the action,

$$\begin{aligned} S^{(2)} &= \int_x \frac{1}{2}\sigma\big[\Box + m^2 - \xi R + V''\big]\sigma + \frac{1}{2}\bar{\eta}_k\Big[\Box - \frac{1}{4}R + M^2 + ih\varphi\nabla\!\!\!\!/ + hM\varphi\Big]\delta_{kl}\,\chi_l \\ &\quad + \frac{1}{2}\,\bar{\eta}_k\big[-2ih\Psi_k\big]\sigma + \frac{1}{2}\sigma\big[-h\bar{\Psi}_l\nabla\!\!\!\!/ + ihM\bar{\Psi}_l\big]\chi_l \\ &= \frac{1}{2}\int_x \big(\sigma H_{\sigma\sigma}\sigma + \bar{\eta}_k H_{\bar{\eta}\chi}\chi_l + \bar{\eta}_k H_{\bar{\eta}\sigma}\sigma + \sigma H_{\sigma\chi}\chi_l\big). \end{aligned} \qquad (14.116)$$

where we assume the sum is over the repeating indices $k, l = 1, \dots, N$ and use the standard notation $\nabla\!\!\!\!/ = \gamma^\alpha\nabla_\alpha$. It is useful to rewrite expression (14.116) in the matrix form as

$$S^{(2)} = \frac{1}{2}\int_x \begin{pmatrix} \sigma & \bar{\eta}_k \end{pmatrix} \begin{pmatrix} H_{\sigma\sigma} & H_{\sigma\chi} \\ H_{\bar{\eta}\sigma} & H_{\bar{\eta}\chi} \end{pmatrix} \begin{pmatrix} \sigma \\ \chi_l \end{pmatrix}, \qquad (14.117)$$

where

$$\begin{aligned} H_{\sigma\sigma} &= \Box + m^2 - \xi R + V'', \qquad H_{\sigma\chi} = -h\bar{\Psi}_l\gamma^\alpha\nabla_\alpha + ihM\bar{\Psi}_l, \\ H_{\bar{\eta}\sigma} &= -2ih\Psi_k, \qquad H_{\bar{\eta}\chi} = \delta_{kl}\Big[\Box + M^2 - \frac{1}{4}R + ih\varphi\gamma^\alpha\nabla_\alpha + h\varphi M\Big]. \end{aligned} \qquad (14.118)$$

The bilinear operator has the standard form (13.202), where

$$\begin{aligned} \hat{1} &= \begin{pmatrix} 1 & 0 \\ 0 & \delta^{kl} \end{pmatrix}, \qquad \hat{h}^\alpha = \begin{pmatrix} 0 & -\frac{1}{2}\,h\bar{\Psi}_l\gamma^\alpha \\ 0 & \frac{i}{2}\,h\varphi\gamma^\alpha\delta_{kl} \end{pmatrix}, \\ \hat{\Pi} &= \begin{pmatrix} V'' - \xi R + m^2 & ihM\bar{\Psi}_l \\ -2ih\Psi_k & \delta^{kl}\Big[M^2 - \frac{1}{4}R + hM\varphi\Big] \end{pmatrix}. \end{aligned} \qquad (14.119)$$

Consider the expressions for $\hat{P}$ and $\hat{S}_{\alpha\beta}$. In the first case, the elements are

$$\nabla_\mu\hat{h}^\mu = \begin{pmatrix} 0 & -\frac{1}{2}\,h(\nabla_\alpha\bar{\Psi}_l)\gamma^\alpha \\ 0 & \frac{i}{2}\,h\delta_{kl}\,(\nabla_\alpha\varphi)\,\gamma^\alpha \end{pmatrix}, \qquad \hat{h}_\mu\hat{h}^\mu = \begin{pmatrix} 0 & -ih^2\varphi\bar{\Psi}_l \\ 0 & -h^2\delta_{kl}\,\varphi^2 \end{pmatrix}. \qquad (14.120)$$

Then we have, according to (13.205),

$$\hat{P} = \begin{pmatrix} V'' - \xi R + m^2 + \frac{1}{6}R & ihM\bar{\Psi}_l + \frac{1}{2}h(\nabla_\alpha\bar{\Psi}_l)\gamma^\alpha + ih^2\varphi\bar{\Psi}_l \\ -2ih\Psi_k & \delta_{kl}\big[M^2 - \frac{1}{12}R + hM\varphi - \frac{i}{2}h(\nabla_\alpha\varphi)\gamma^\alpha + h^2\varphi^2\big] \end{pmatrix}.$$

In the second case, the general expression is (13.204), where the commutator of covariant derivatives $\hat{\mathcal{R}}_{\alpha\beta}$ is non-zero only in the fermion sector, according to (12.65). After a small amount of algebra we arrive at

$$\hat{S}_{\alpha\beta} = \begin{pmatrix} 0 & \frac{1}{2}h(\nabla_\alpha\bar{\Psi}_l\gamma_\beta - \nabla_\beta\bar{\Psi}_l\gamma_\alpha) + \frac{i}{2}h^2\bar{\Psi}_l\varphi\sigma_{\alpha\beta} \\ 0 & \big[\frac{i}{2}h(\nabla_\beta\varphi\gamma_\alpha - \nabla_\alpha\varphi\gamma_\beta) - \frac{i}{2}h^2\varphi^2\sigma_{\alpha\beta} - \frac{1}{4}R_{\alpha\beta\rho\sigma}\gamma^\rho\gamma^\sigma\big]\delta_{kl} \end{pmatrix},$$

where $\sigma_{\alpha\beta} = \frac{i}{2}(\gamma_\alpha\gamma_\beta - \gamma_\beta\gamma_\alpha)$.

Until now, all of the calculation was almost trivial. The qualitatively new element compared to the examples which we elaborated earlier, is that the operators of our interest, such as $\hat{1}$, $\hat{P}$ and $\hat{S}_{\alpha\beta}$, act in the space of quantum fields of mixed Grassmann parity. Thus, instead of taking a usual trace, one has to take a super-trace. This means that the scalar trace enters with the coefficient $+1$, while the fermion trace enters with the coefficient -2, just as in the derivation of vacuum divergences for the free Dirac field. In the simplest case, this procedure gives (remember $\mathrm{tr}\,\hat{1} = 4$ in the spinor sector)

$$\mathrm{str}\,\hat{P} = V'' - \Big(\xi - \frac{1}{6}\Big)R + m^2 \;-\; 2N\,\mathrm{tr}\Big\{M^2 - \frac{1}{12}R + hM\varphi - \frac{i}{2}h(\nabla_\alpha\varphi)\gamma^\alpha + h^2\varphi^2\Big\},$$

such that

$$\frac{1}{6}\,\mathrm{str}\,\Box\hat{P} = \frac{1}{6}\Box V'' + \Big[\frac{N}{9} - \frac{1}{6}\Big(\xi - \frac{1}{6}\Big)\Big]\Box R - \frac{4hNM}{3}\Box\varphi - \frac{4Nh^2}{3}\Box\varphi^2. \qquad (14.121)$$

In another part we get, using the same logic,

$$\begin{aligned} \frac{1}{2}\,\mathrm{str}\,\hat{P}^2 &= \frac{1}{2}\Big[V'' - \Big(\xi - \frac{1}{6}\Big)R + m^2\Big]^2 + \frac{1}{2}\Big(ihM\bar{\Psi}_l - \frac{h}{2}\nabla_\alpha\bar{\Psi}_l\gamma^\alpha \\ &\quad + ih^2\varphi\bar{\Psi}_l\Big)(2ih\Psi_l) - N\,\mathrm{tr}\Big[M^2 - \frac{1}{12}R + hM\varphi - \frac{i}{2}h(\nabla_\alpha\varphi)\gamma^\alpha + h^2\varphi^2\Big]^2 \\ &\quad - \mathrm{str}\,\big(-2ih\Psi_k\big)\Big[ihM\bar{\Psi}_l + \frac{h}{2}(\nabla_\alpha\bar{\Psi}_l)\gamma^\alpha + ih^2\varphi\bar{\Psi}_l\Big]. \end{aligned} \qquad (14.122)$$

The terms in the last line should be elaborated as $\mathrm{str}\,(\Psi_k\bar{\Psi}_l) = -\bar{\Psi}_k\Psi_k$, due to the anticommuting nature of fermion fields. Thus, after some algebra, we get

$$\begin{aligned} \frac{1}{2}\,\mathrm{str}\,\hat{P}^2 &= \frac{3i}{2}\,h^2\bar{\Psi}_k\not{\nabla}\Psi_k + 3h^2\bar{\Psi}_k\Psi_k(M + h\varphi) - \Big(\xi - \frac{1}{6}\Big)RV'' - m^2V'' + \frac{1}{2}\,V''^2 \\ &+ Nh^2(\partial\varphi)^2 + Nh^2\varphi^2\Big[\frac{2}{3}R - 12M^2 - 4h^2\varphi^2 - 8hM\varphi\Big] + \Big[\frac{2}{3}R - 8M^2\Big]NMh\varphi. \end{aligned}$$

The remaining super-trace, that of $\hat{S}^2_{\alpha\beta}$, is much simpler, because there are only fermion contributions. Let us give only the final expression and leave the intermediate calculations to the reader as a small exercise:

$$\frac{1}{12}\,\mathrm{str}\,\hat{S}_{\alpha\beta}\hat{S}^{\alpha\beta} = Nh^2(\partial\varphi)^2 + 2Nh^4\varphi^4 - \frac{Nh^2}{3}R\varphi^2.$$

Summing up the contributions of the three relevant terms, we arrive at

$$\begin{aligned}\Gamma^{(1)}_{div} = &-\frac{\mu^{n-4}}{\epsilon}\int d^n x\sqrt{-g}\,\Big\{\frac{3ih^2}{2}\,\bar{\Psi}_k\nabla\!\!\!\!/\,\Psi_k + 3h^3\bar{\Psi}_k\varphi\Psi_k + 3h^2\bar{\Psi}_k M\Psi_k \\ &+2Nh^2(\partial\varphi)^2 - 2Nh^4\varphi^4 - 12NM^2h^2\varphi^2 - 8NM^3h\varphi - 8Nh^3M\varphi^3 \\ &+\frac{N}{3}\left(h^2R\varphi^2 + 2hMR\varphi\right) + \frac{1}{2}V''^2 - \Big(\xi - \frac{1}{6}\Big)RV'' - m^2V'' + \frac{1}{6}\Box V'' \\ &-\frac{1}{6}\Big(\xi - \frac{1}{6}\Big)\Box R + \frac{N}{9}\Box R - \frac{4N}{3}hM\Box\varphi - \frac{4N}{3}h^2\Box\varphi^2\Big\}. \end{aligned} \tag{14.123}$$

It is easy to see that our expectations about the relevance of the odd terms are confirmed by this result. The divergences include $\mathcal{O}(\varphi)$ and $\mathcal{O}(\varphi^3)$ terms in addition to the even terms which we met earlier in the pure scalar model with φ^4-interaction.

The renormalization relations between bare and renormalizable quantities directly follow from the divergences. For the fields, we have

$$\varphi_0 = \mu^{\frac{n-4}{2}}\left(1 + \frac{2Nh^2}{\epsilon}\right)\varphi, \qquad \Psi_{k0} = \mu^{\frac{n-4}{2}}\left(1 + \frac{3}{4\epsilon}h^2\right)\Psi_k. \tag{14.124}$$

The relations for the masses have the form

$$M_0 = \left(1 - \frac{9\,h^2}{2\epsilon}\right)M, \qquad m_0^2 = m^2 - \frac{g^2 + 4Nh^2m^2 + \lambda m^2 - 24Nh^2M^2}{\epsilon}. \tag{14.125}$$

For the even couplings and nonminimal parameters, we have

$$\xi_0 = \xi - \frac{\lambda + 4Nh^2}{\epsilon}\Big(\xi - \frac{1}{6}\Big), \tag{14.126}$$

$$h_0 = \mu^{\frac{4-n}{2}}h\left(1 - \frac{4Nh^2 + 9h^2}{2\epsilon}\right), \tag{14.127}$$

$$\lambda_0 = \mu^{4-n}\left(\lambda + \frac{48Nh^4 - 8N\lambda h^2 - 3\lambda^2}{\epsilon}\right). \tag{14.128}$$

And, finally, for the odd couplings and nonminimal parameters,

$$g_0 = \mu^{\frac{4-n}{2}}\left(g + \frac{48NMh^3 - 3g\lambda - 6Nh^2g}{\epsilon}\right), \tag{14.129}$$

$$\tau_0 = \mu^{\frac{n-4}{2}}\left(\tau + \frac{8NhM^3 - 2N\tau h^2 - m^2g}{\epsilon}\right), \tag{14.130}$$

$$f_0 = \mu^{\frac{n-4}{2}}\left[f + \frac{g}{\epsilon}\Big(\xi - \frac{1}{6}\Big) - \frac{2NhM + 6Nfh^2}{3\epsilon}\right]. \tag{14.131}$$

The beta functions are defined as

$$\beta_P = \lim_{n\to 4} \mu \frac{dP}{d\mu}, \qquad \text{where} \qquad P = \{m, M, h, \lambda, \xi, g, \tau, f\} \tag{14.132}$$

are the renormalized parameters. The complete list of beta–functions is

$$\begin{aligned}
\beta_h &= \frac{(4N+9)h^3}{(4\pi)^2}, \qquad \beta_M = \frac{9h^2M}{2(4\pi)^2}, \\
\beta_\lambda &= \frac{1}{(4\pi)^2}\Big(8N\lambda h^2 + 3\lambda^2 - 48Nh^4\Big), \\
\beta_\xi &= \frac{1}{(4\pi)^2}\Big(\xi - \frac{1}{6}\Big)\Big(4Nh^2 + \lambda\Big), \\
\beta_g &= \frac{1}{(4\pi)^2}\Big(\frac{3}{2}g\lambda + 3Ngh^2 - 12NMh^3\Big), \\
\beta_{m^2} &= \frac{1}{(4\pi)^2}\Big[m^2\lambda + g^2 + (4m^2 - 24M^2)Nh^2\Big], \\
\beta_\tau &= \frac{1}{(4\pi)^2}\Big(2N\tau h^2 + gm^2 - 8NhM^3\Big), \\
\beta_f &= \frac{1}{(4\pi)^2}\Big[Nh\Big(2fh + \frac{2M}{3}\Big) - g\Big(\xi - \frac{1}{6}\Big)\Big].
\end{aligned} \tag{14.133}$$

The gamma functions are defined as usual, (9.151), where the renormalized fields are $\Phi_R = \Phi = (\varphi, \Psi_k)$. The relations (14.124) lead to

$$\gamma_\varphi = -\frac{2Nh^2}{(4\pi)^2}, \qquad \gamma_{\Psi_k} = -\frac{3h^2}{4(4\pi)^2}. \tag{14.134}$$

In the case of a conformal invariant theory, we need to set all dimensional constants m^2, M, g, τ and f to vanish and set $\xi = \frac{1}{6}$. Then, the beta functions have the corresponding conformal fixed point, as it has to be from the general perspective [70, 80].

14.3.4 Standard Model-like theories

The first calculation of one-loop divergences in the model with Yang-Mills, fermion and scalar fields in curved space was done in [71] for a relatively simple model based on the gauge group $SU(2)$. A few years after that, there were many calculations of this kind, including for different gauge groups. Most of these calculations were performed using the Schwinger-DeWitt technique (see, e.g., [80, 79] for detailed references) but in some cases, also by means of local momentum representation [240]. The main focus of these calculations was always the renormalization group behavior for the nonminimal and vacuum parameters. It is worth mentioning that the derivation of one-loop beta functions in the Minimal Standard Model has been done, in [351].

In general, these calculations are good to know but, owing to space limitations, we include only the result for a very simple example based on the $SU(2)$ gauge group, leaving all of the calculations for the reader as an exercise.

Consider the $SU(2)$ theory with the action

$$S = \int d^4x\sqrt{-g}\Big\{ -\frac{1}{4}\,G^a_{\mu\nu}G^{a\,\mu\nu} + \frac{1}{2}\,g^{\mu\nu}\big(\mathcal{D}_\mu\phi\big)^a\big(\mathcal{D}_\mu\phi\big)^a - \frac{1}{2}\,m^2\phi^a\phi^a$$
$$+ \frac{1}{2}\,\xi R\phi^a\phi^a - \frac{f}{4!}\,\big(\phi^a\phi^a\big)^2 + \sum_{k=1}^{s}\bar{\psi}^a_k\Big[i\gamma^\mu\mathcal{D}^{ab}_\mu - \delta^{ab}M - h\varepsilon^{acb}\varphi^c\Big]\psi^b_k\Big\}. \qquad (14.135)$$

Here both scalars and fermions are in the adjoint representation of the gauge group; hence, $\big(\mathcal{D}_\mu\varphi\big)^a = \nabla_\mu\varphi^a - g\varepsilon^{acb}A^c_\mu\varphi^b$ and similarly for fermions.

The one-loop divergences in this theory can be calculated with a switched-off vector field, as the last appears only in the gauge-invariant combinations that can be recovered without calculations. The result has the form

$$\Gamma^{(1)}_{div} = \frac{\mu^{n-4}}{\epsilon}\int d^nx\sqrt{-g}\Big\{4\big(h^2-g^2\big)\,\big(\nabla\varphi^a\big)^2 + \frac{1}{2}\,\phi^a\phi^a\Big[\frac{5}{3}fm^2 - 4g^2m^2 - 48h^2M^2\Big]$$
$$+ \frac{1}{2}\,R\phi^a\phi^a\Big[\frac{4}{3}\big(h^2-g^2\big) - \Big(\xi - \frac{1}{6}\Big)\Big(\frac{5}{3}f - 4g^2\Big)\Big]$$
$$+ \frac{1}{4!}\,\big(\phi^a\phi^a\big)^2\Big[\frac{11}{3}f^2 - 8g^2f + 72g^4 - 96h^4\Big]$$
$$+ \sum_{k=1}^{s}\bar{\psi}^a_k\Big[2i(h^2+2g^2)\gamma^\mu\delta^{ab}\nabla_\mu + \delta^{ab}M(h^2-4g^2) + 2h(h^2-6g^2)\varepsilon^{acb}\varphi^c\Big]\psi^b_k\Big\}$$
$$+ \text{purely metric- and } A^a_\mu\text{-dependent terms.} \qquad (14.136)$$

Full details of the calculation can be found in the book [80]. More general examples, including those based on the gauge group $SU(N)$, can be found, e.g., in the paper [79], in which complicated GUT-like models have been elaborated.

Let us note the importance of renormalization in similar models for physical applications. For example, the one- and two-loop quantum corrections in the Minimal Standard Model in curved spacetime are relevant for Higgs inflation [51]. Furthermore, the role of the loop effects on the Higgs potential was discussed in [52–54] and [27,28].

Exercises

14.23. Consider the massless version of the Yukawa model (14.113). Explain why, in this case, odd terms in the potential $V(\varphi)$ are not necessary. Verify this conclusion by the analysis of one-loop divergences (14.123).

14.24. Write down, without special calculations, the full expression for one-loop divergences in the model (14.113), including vacuum terms. What would be the change in the vacuum part of divergences if we included a complex scalar field instead of the real one? Do we need odd terms in the potential in the case of a complex scalar?

14.25. Try to reconstruct the divergences in the similar Yukawa model with a complex scalar without explicit calculations.

14.26. Consider the massless and conformal invariant limit of the Yukawa model (14.113). Verify that, in this case, the one-loop divergences (14.123) have an integrand that is also conformal invariant in the $n = 4$ dimension. Explain the result.

14.27. Starting from expression (14.123) with the extra vacuum terms restored as in the Exercise 14.14 and using the explicit form of potential (14.114), obtain the renormalization relations for all couplings and parameters, and calculate the beta functions for these parameters.
14.28. Make detailed calculation and verify Eq. (14.136).
14.29. Using gauge invariance and the results of section 14.3.2, without explicit calculations, restore all of the vector-dependent terms in Eq. (14.136).
14.30. Using the results of section 14.1, recover the purely vacuum, metric-dependent terms in Eq. (14.136).
14.31. Starting from Eq. (14.136), derive renormalization relations for all couplings, for both masses and nonminimal parameters. Using the result of Exercise 14.14 and these relations, calculate the beta functions for Yukawa, scalar self-interaction constants and for ξ.
14.32. Without explicit calculation, discuss to what extent the expressions for divergences and the beta functions depend on the choice of the gauge fixing.
Hint. Use the known theorem (10.161), from which follows that the gauge fixing dependence in the one-loop divergences vanishes in the classical equations of motion (on shell).

15 The renormalization group in curved space

The previous chapter was mainly devoted to the practical calculation of divergences, which arise in quantum field theory in curved spacetime. We also described the process of renormalization as a process of removing divergences. However, the main purpose of renormalization is not removing divergences but getting essential information about the finite part of effective action. In the present and subsequent chapters, we discuss some of the existing methods for solving this problem, which can be denoted by the common name of renormalization group. We start from the simplest version, based on the minimal subtraction procedure and global scaling. After that, we consider the renormalization group based on the momentum subtraction scheme, which enables one to take care of the quantum effects of massive fields, including at low energies. And, finally, we will discuss the local version of the renormalization group, which relies on the integration of the conformal (trace) anomaly, and is the most fruitful for many applications.

15.1 The renormalization group based on minimal subtractions

Let us start by reviewing the standard, well-known formalism of the renormalization group in curved spacetime. Further details can be found in the original papers [229, 70] or in the book [80]. In order to make things simpler, we shall use a dimensional regularization scheme, but it is possible to construct a renormalization group starting from other regularizations. We shall suppose that the theory under consideration is multiplicatively renormalizable, which can be achieved in a way described in the previous sections.

As we already know, the total action of a renormalizable theory has the form

$$S = S_{vac} + S_{min} + S_{nonmin}, \tag{15.1}$$

where S_{vac} is the vacuum action (12.6), S_{min} is the minimal action and S_{nonmin} is the integral of the scalar-curvature nonminimal term $\xi R\varphi^2$. The total action (15.1) depends on the following matter fields: scalars φ, spinors ψ and vectors A_α, plus the external metric $g_{\mu\nu}$. The list of couplings include gauge, g, Yukawa, h and four-scalar, f. There are also scalar and spinor masses, the nonminimal parameter ξ and the parameters G, Λ and $a_{1,2,3,4}$ of the vacuum action. The divergences are removed by the renormalization of the coupling constants, the masses, nonminimal parameter ξ and the vacuum parameters.

Introduction to Quantum Field Theory with Applications to Quantum Gravity. Iosif L. Buchbinder and Ilya L. Shapiro, Oxford University Press (2021). © Iosif L. Buchbinder and Ilya Shapiro.
DOI: 10.1093/oso/9780198838319.003.0015

Let us denote Φ the full set of matter fields $\Phi = (\varphi, \psi, A)$ and P the full set of parameters, including couplings, masses, ξ and vacuum parameters. Consider the bare action $S_0[\Phi_0, P_0] = S_0[\Phi_0, P_0, g_{\mu\nu}]$, which depends on the bare fields, and parameters and $S[\Phi, P] = S[\Phi, P, g_{\mu\nu}]$, the corresponding renormalized action. Multiplicative renormalizability means that

$$S_0[\Phi_0, P_0, g_{\mu\nu}] = S[\Phi, P, g_{\mu\nu}, \mu], \tag{15.2}$$

where (Φ_0, P_0) and (Φ, P) are related by the renormalization transformations. Starting from this point, the considerations can be done essentially as in flat spacetime, as explained in Part I. The unique novelty is that generating functionals, e.g.,

$$Z[J, g_{\mu\nu}] = e^{iW[J,g_{\mu\nu}]} = \int \mathcal{D}\Phi\, e^{i(S[\Phi,P,g_{\mu\nu}]+\Phi J)}, \tag{15.3}$$

and vertices depend on the metric, which plays the role of the external parameter. Finally, we arrive at the curved-space analog of Eq. (9.150),

$$\Gamma_0[g_{\alpha\beta}, \Phi_0, P_0, 4] = \Gamma[g_{\alpha\beta}, \Phi, P, n, \mu], \tag{15.4}$$

or, in the form of a differential equation,

$$\mu\frac{d}{d\mu}\Gamma[g_{\alpha\beta}, \Phi, P, n, \mu] = 0. \tag{15.5}$$

This equation shows that the *overall* dependence on μ is absent. From the physics viewpoint, this looks like something that is unavoidable, because μ is an artificial parameter, introduced as a by-product of regularization. Without identifying it with some physical quantity, μ cannot be relevant, and the last equation shows this explicitly. At the same time, Eq. (15.5) is the result of the cancelation of two μ-dependencies, one of the proper Γ and another of P and Φ. Each of these dependencies, if taken separately, can be quite informative, if we learn how to identify μ with the physical parameters of our interest.[1]

Taking into account the possible μ-dependence of P and Φ, we recast (15.5) as

$$\left\{\mu\frac{\partial}{\partial\mu} + \mu\frac{dP}{d\mu}\frac{\partial}{\partial P} + \int d^n x \mu\frac{d\Phi(x)}{d\mu}\frac{\delta}{\delta\Phi(x)}\right\}\Gamma[g_{\alpha\beta}, \Phi, P, n, \mu] = 0. \tag{15.6}$$

Now we define, exactly as in flat spacetime,

$$\beta_P(n) = \mu\frac{dP}{d\mu}, \qquad \beta_P(4) = \beta_P; \tag{15.7}$$

$$\gamma_\Phi(n) = \mu\frac{d\Phi}{d\mu}, \qquad \gamma_\Phi(4) = \gamma_\Phi\,. \tag{15.8}$$

Then, (15.6) becomes

[1] An instructive discussion of this point for the cosmological constant can be found in [288].

$$\left\{ \mu \frac{\partial}{\partial \mu} + \beta_P(n) \frac{\partial}{\partial P} + \int d^n x \gamma_\Phi(n) \frac{\delta}{\delta \Phi(x)} \right\} \Gamma[g_{\alpha\beta}, \Phi, P, n, \mu] = 0\,. \qquad (15.9)$$

Indeed, (15.9) is the general renormalization group equation, which can be used for different purposes, depending on the physical interpretation of μ. Let us follow [229,70] and consider the short-distance limit.

Let us perform a global rescaling of all quantities, including the length l and, in particular, the coordinates x^μ, according to their dimension

$$\Phi \to \Phi k^{-d_\Phi}\,, \qquad P \to P k^{-d_P}\,, \qquad \mu \to k\mu\,, \qquad l \to k^{-1} l\,. \qquad (15.10)$$

Because of its fixed dimension, the effective action Γ does not change under (15.10). On the other hand, since Γ does not depend on x^μ explicitly, one can replace $l \to lk^{-1}$ with the transformation of the metric $g_{\mu\nu} \to k^2 g_{\mu\nu}$, while x^μ does not transform. Then, in addition to (15.9), we meet another identity,

$$\Gamma[g_{\alpha\beta}, \Phi, P, n, \mu] = \Gamma[k^2 g_{\alpha\beta}, k^{-d_\Phi} \Phi, k^{-d_P} P, n, k^{-1}\mu]\,, \qquad (15.11)$$

where $\int d^n x \sqrt{-g} \to \int d^n x \sqrt{-g} k^n$ and, for the curvatures,

$$R^2_{\mu\nu\alpha\beta} \to R^2_{\mu\nu\alpha\beta} k^{-4}\,, \qquad R^2_{\alpha\beta} \to R^2_{\alpha\beta} k^{-4}\,, \qquad R \to R k^{-2}\,, \qquad \text{etc}\,. \qquad (15.12)$$

Now, if we put $k = e^{-t}$, (15.11) gives

$$\frac{d}{dt} \Gamma[e^{2t} g_{\alpha\beta}, e^{-d_\Phi t} \Phi, e^{d_P t} P, n, e^{-t}\mu]\,. \qquad (15.13)$$

For $t = 0$, this gives

$$\left\{ \int d^n x \left(2 g_{\alpha\beta} \frac{\delta}{\delta g_{\alpha\beta}} - d_\Phi \frac{\delta}{\delta \Phi(x)} \right) - d_P \frac{\partial}{\partial P} - \mu \frac{\partial}{\partial \mu} \right\} \Gamma[g_{\alpha\beta}, \Phi, P, n, \mu] = 0. \quad (15.14)$$

Taking the difference between (15.9) and (15.14), we get

$$\left\{ \int d^n x \left(2 g_{\alpha\beta} \frac{\delta}{\delta g_{\alpha\beta}} + \left[\gamma_\Phi(n) - d_\Phi \Phi \right] \frac{\delta}{\delta \Phi(x)} \right) + \left[\beta_P(n) - d_P P \right] \frac{\partial}{\partial P} \right\} \Gamma = 0, \quad (15.15)$$

where $\Gamma = \Gamma[g_{\alpha\beta}, \Phi, P, n, \mu]$. In order to obtain the relation between global scaling (15.12) and Eq. (15.15), we note that

$$-2 \int d^n x\, g_{\alpha\beta} \frac{\delta}{\delta\, g_{\alpha\beta}} \Gamma[g_{\alpha\beta} e^{-2t}, \Phi, P, n, \mu] = \frac{\partial}{\partial t} \Gamma[g_{\alpha\beta} e^{-2t}, \Phi, P, n, \mu]\,. \quad (15.16)$$

Since other terms in (15.15) are unaffected by the insertion of an extra e^{-2t}, we get

$$\left\{ \frac{\partial}{\partial t} - \left[\beta_P(n) - d_P P \right] \frac{\partial}{\partial P} - \left[\gamma_\Phi(n) - d_\Phi \right] \int d^n x\, \Phi(x) \frac{\delta}{\delta \Phi(x)} \right) \Big\} \Gamma_t = 0, \quad (15.17)$$

where $\Gamma_t = \Gamma[g_{\alpha\beta} e^{-2t}, \Phi, P, n, \mu]$. Eq. (15.17) is a special form of the renormalization group equation (15.9), designed for investigating the short-distance behaviour of the effective action.

The minimal subtraction scheme solution of Eq. (15.17) has the form

$$\Gamma[g_{\alpha\beta}e^{-2t}, \Phi, P, n, \mu] = \Gamma[g_{\alpha\beta}, \Phi(t), P(t), n, \mu]\,, \tag{15.18}$$

where $P(t)$ and $\Phi(t)$ satisfy the renormalization group equations

$$\frac{d\Phi}{dt} = (\gamma_\Phi - d_\Phi)\Phi\,, \qquad \frac{dP}{dt} = \beta_P - Pd_P. \tag{15.19}$$

The correctness of (15.19) can be checked by direct substitution into Eq. (15.17).

The limit $t \to \infty$ means $g_{\alpha\beta}e^{-2t} \to 0$, which is the limit of short distances. In view of (15.12), it is also the limit of large values of curvatures. Thus, for the version of the renormalization group we have constructed, the rescaling of the metric is equivalent to the standard rescaling of momenta in flat space quantum field theory. This equivalence shows that the procedure described above is a direct generalization of the one we were considering in section 9.9 in Part I.

One should note that the application of (15.18) and (15.19) to the particular physical situations is a non-trivial issue, requiring special attention. For example, consider the exponential expansion of the universe, which is the simplest version of inflation. The time dependence of the metric is very similar to the rescaling (here, we denote time as x^0 in order to avoid confusion with $t = \log \mu/\mu_0$),

$$g_{\alpha\beta} \to g_{\alpha\beta} \cdot e^{Hx^0}\,, \qquad \text{where} \qquad H = const\,. \tag{15.20}$$

However, this situation does not correspond to the renormalization group, because the scalar curvature does not behave as in (15.12) and remains constant, $R = -12H^2$.

15.2 The effective potential from a renormalization group

In section (13.3.3), we considered the derivation of the first curvature-dependent correction to the flat space expression for the effective potential of a real scalar field. The calculation was performed by direct integration over momenta in the local momentum representation. It is very instructive to obtain the same expression without direct calculations, using the renormalization group instead.

The general formalism of restoring effective action from the minimal subtraction renormalization group can be found in the book [80] (see also the original paper [75]) and we will not repeat this consideration. Instead, we consider a useful and sufficiently general pedagogical example, using the simplified version of a Yukawa model from section 14.3.3 (see also [24]).

The starting point is the overall μ-independence of the effective action in the form of Eq. (15.9), where we assume $\Phi = (\varphi,\, \Psi_k)$ and we set $n = 4$. The effective potential is defined as a zero-order approximation in the derivative expansion for the scalar sector (13.58). Since (15.9) is a linear homogeneous equation, we get

$$\Big\{\mu\frac{\partial}{\partial\mu} + \beta_P\frac{\partial}{\partial P} + \gamma_\varphi\,\varphi\frac{\partial}{\partial\varphi}\Big\}V_{eff}(g_{\alpha\beta}, \varphi, P, \mu) = 0. \tag{15.21}$$

We will look for the effective potential in the form $V_{eff} = V_0 + RV_1$, where V_0 is the flat space effective potential and RV_1 is the first curvature-dependent correction. It is evident that both V_0 and V_1 satisfy Eq. (15.21). We can also take into account that, in the one-loop approximation, both scalar and spinor fields give additive contributions. Therefore, we can write $V_0 = V_0^{(0)} + V_0^{(\frac{1}{2})}$ and $V_1 = V_1^{(0)} + V_1^{(\frac{1}{2})}$, where the labels (0) and $(\frac{1}{2})$ mean the contributions from quantum scalar and spinor fields, respectively.

The equation for the flat space scalar contribution, $V_0^{(0)}$, has the form

$$\left\{ \mu\frac{\partial}{\partial\mu} + \beta_P\,\frac{\partial}{\partial P} + \,\gamma_\varphi\,\varphi\,\frac{\partial}{\partial\varphi} \right\} V_0^{(0)}(\varphi, P, \mu) \,=\, 0. \tag{15.22}$$

The equation (15.22) is a complicated partial differential equation with non-constant coefficients. Before solving this equation, it proves helpful to bring qualitative considerations that simplify the solution. The parameter μ can enter into the solution for $V_0^{(0)}$ only logarithmically. Since the argument of the logarithm must be dimensionless, the dependence on μ should be through the parameter $t = \frac{1}{2}\log\frac{X}{\mu^2}$, where the quantity X has the mass dimension 2. This quantity can be constructed only from the dimensional parameters of the classical action, i.e., from m, M, φ, g, τ with arbitrary dimensionless coefficients. In principle, these coefficients should be fixed with the help of appropriate renormalization conditions for the effective potential. However, it is natural to assume that the form of effective potential should be consistent with the form of the operator $\hat{\mathcal{H}}$ (13.26) in the scalar sector. This means that the most natural choice for X is $X^{(0)} = m^2 + g\varphi + \frac{1}{2}\lambda\varphi^2$. Thus, we identify

$$t^{(0)} \;=\; \frac{1}{2}\log\frac{m^2 + \frac{1}{2}\lambda\varphi^2 + g\varphi}{\mu^2} \tag{15.23}$$

for the scalar field contribution. Since the parameters M and h do not contribute in the scalar sector, $V_0^{(0)} = V_0^{(0)}(t, m^2, g, \tau, \varphi)$ and the equation (15.22) becomes

$$\left\{\mu\frac{\partial}{\partial\mu} + \beta_{m^2}^{(0)}\frac{\partial}{\partial m^2} + \beta_\lambda^{(0)}\frac{\partial}{\partial\lambda} + \beta_g^{(0)}\frac{\partial}{\partial g} + \beta_\tau^{(0)}\frac{\partial}{\partial\tau} + \gamma_\varphi^{(0)}\varphi\,\frac{\partial}{\partial\varphi}\right\} V_0^{(0)} \;=\; 0. \tag{15.24}$$

Here $\gamma_\varphi^{(0)} = 0$ and the functions $\beta_\lambda^{(0)}$, $\beta_g^{(0)}$, β_{m^2}, $\gamma_\varphi^{(0)}$ and $\beta_\tau^{(0)}$ are taken at $M = 0$ and $h = 0$. The derivative with respect to μ can be expressed through the derivative with respect to the parameter $t^{(0)}$ defined in (15.23). As a result, Eq. (15.24) becomes

$$\left\{\frac{\partial}{\partial t^{(0)}} - \bar{\beta}_{m^2}^{(0)}\frac{\partial}{\partial m^2} - \bar{\beta}_\lambda^{(0)}\frac{\partial}{\partial\lambda} - \bar{\beta}_g^{(0)}\frac{\partial}{\partial g} - \bar{\beta}_\tau^{(0)}\frac{\partial}{\partial\tau} - \bar{\gamma}_\varphi^{(0)}\varphi\frac{\partial}{\partial\varphi}\right\} V_0^{(0)} \;=\; 0, \tag{15.25}$$

where the general expressions are

$$\big(\bar{\beta}_{m^2}^{(0)}, \bar{\beta}_\lambda^{(0)}, \bar{\beta}_g^{(0)}, \bar{\beta}_\tau^{(0)}, \bar{\gamma}_\varphi^{(0)}\big) \;=\; \frac{1}{1 - Q^{(0)}}\big(\beta_{m^2}^{(0)}, \beta_\lambda^{(0)}, \beta_g^{(0)}, \beta_\tau^{(0)}, \gamma_\varphi^{(0)}\big) \tag{15.26}$$

and
$$Q^{(0)} = 1 - \frac{\partial t^{(0)}}{\partial m^2} - \frac{\partial t^{(0)}}{\partial\lambda} - \frac{\partial t^{(0)}}{\partial\varphi} - \frac{\partial t^{(0)}}{\partial g}. \tag{15.27}$$

According to the general prescription (15.18), the solution to the equation (15.25) has the form

$$V_0^{(0)}(t^{(0)}, m^2, \lambda, g, \tau, \varphi) = V_{0\,cl}(m^2(t^{(0)}), \lambda(t^{(0)}), g(t^{(0)}), \tau(t^{(0)}), \varphi(t^{(0)})), \quad (15.28)$$

$$\text{where} \qquad V_{0\,cl} = \frac{1}{2}m^2\varphi^2 + \frac{\lambda}{4!}\varphi^4 + \frac{g}{3!}\varphi^3 + \tau\phi \quad (15.29)$$

is the classical potential and $m^2(t^{(0)})$, $\lambda(t^{(0)})$, $g(t^{(0)})$, $\tau(t^{(0)})$ and $\varphi(t^{(0)})$ are the running parameters $P(t^{(0)})$ and the scalar field satisfying the equations

$$\frac{dP(t^{(0)})}{dt^{(0)}} = \bar{\beta}_P^{(0)}(t^{(0)}), \qquad \frac{d\varphi(t^{(0)})}{dt^{(0)}} = \bar{\gamma}_\varphi(t^{(0)}) \quad (15.30)$$

with the initial conditions

$$P(t)\Big|_{t=0} = P. \quad (15.31)$$

In the present case, $P = m^2$, λ, g, τ. In the one-loop approximation, all quantum corrections are linear in $\hbar$; hence, in (15.27), we can set $Q^{(0)} = 1$. Then the solutions of the equations (15.30) can be easily found as

$$P(t^{(0)}) = P + \beta_P^{(0)} t^{(0)}, \qquad \varphi(t^{(0)}) = \varphi + \gamma_\varphi^{(0)} t^{(0)} = \varphi. \quad (15.32)$$

Here we took into account that $\gamma^{(0)} = 0$. The relations (15.32), together with the explicit forms of the functions $\beta_P^{(0)}$, represent the solution for $V_0^{(0)}$.

The analysis of $V_1^{(0)}$ can be done in a similar way, so we skip the details. The result has the form

$$V_1^{(0)} = V_{1\,cl}(P_1(t^{(0)}),\, \phi(t^{(0)})), \quad (15.33)$$

with $V_{1\,cl}R = -\frac{1}{2}\xi R\phi^2 + fR\varphi$, with $P_1(t^{(0)}) = P_1 + \beta_{P_1}^{(0)} t^{(0)}$ and $\beta_{P_1}^{(0)} = (\beta_\xi^{(0)},\, \beta_f^{(0)})$. These relations, together with (15.33), give the final solution for the first-order curvature-dependent scalar contribution to the effective potential.

Next we consider the contribution $\bar{V}_0^{(\frac{1}{2})} + R\bar{V}_1^{(\frac{1}{2})}$ to the effective potential from the quantum spinor field. Eq. (15.21) is satisfied separately for $V_0^{(\frac{1}{2})}$ and $V_1^{(\frac{1}{2})}$, and the natural choice for dimensionless parameter with the logarithm of μ comes from the operator $\hat{H}$ in the fermionic sector,

$$t^{(\frac{1}{2})} = \frac{1}{2}\log\frac{(M + h\varphi)^2}{\mu^2}. \quad (15.34)$$

The rest of the considerations are analogous to the ones for the scalar case. Thus we present only the final results for quantum corrections,

$$\bar{V}_1^{(\frac{1}{2})} = V_{0\,cl}(P^{(\frac{1}{2})}(t^{(\frac{1}{2})}),\, \varphi(t^{(\frac{1}{2})})), \quad (15.35)$$

$$\bar{V}_1^{(\frac{1}{2})} = V_{1\,cl}(P_1^{(\frac{1}{2})}(t^{(\frac{1}{2})}),\, \varphi(t^{(\frac{1}{2})})), \quad (15.36)$$

where the running parameters and the field have the forms

$$P^{(\frac{1}{2})}(t^{(\frac{1}{2})}) = \beta_P^{(\frac{1}{2})} t^{(\frac{1}{2})}, \qquad \varphi(t^{(\frac{1}{2})}) = \gamma_\varphi^{(\frac{1}{2})} t^{(\frac{1}{2})}. \quad (15.37)$$

respectively. Here we have solved the equations for the running parameters and field with zero initial conditions since the classical contribution to the effective potential was found when we calculated $V_0^{(0)}$ and $V_1^{(0)}$. The functions $\beta_P^{(\frac{1}{2})}$ and $\gamma_\varphi^{(\frac{1}{2})}$ in the relations (15.37) are the β_P and γ_φ at non-zero M and h but with zero parameters m^2, λ, g, τ, ξ.

Thus, we are in a position to write down the explicit expression for the the effective potential based on the minimal subtraction scheme

$$\begin{aligned} V_{eff} = V_{cl} + \hbar(V_1 + V_2 R) &= -\frac{1}{2}\,m^2\varphi^2 - \frac{1}{2}\,\xi R\varphi^2 + \frac{\lambda}{4!}\,\varphi^4 + \frac{g}{3!}\varphi^3 + \tau\varphi + fR\varphi \\ &- \frac{\hbar}{2(4\pi)^2}\bigg\{\Big[\frac{\lambda m^2 - g^2}{2}\,t^{(0)} + 12NM^2h^2t^{(\frac{1}{2})} + C_1\Big]\varphi^2 + \Big[\frac{\lambda}{2}\,\Big(\xi - \frac{1}{6}\Big)t^{(0)} \\ &- \frac{Nh^2}{3}\,t^{(\frac{1}{2})} + C_2\Big]R\varphi^2 - \frac{1}{3!}\,\Big[\frac{3\lambda g}{2}\,t^{(0)} - 3N\big(gh^2 - 4Mh^3\big)t^{(\frac{1}{2})} + C_3\Big]\varphi^3 \\ &- \frac{1}{4!}\,\big[3\lambda^2 t^{(0)} - 48Nh^4 t^{(\frac{1}{2})} + C_4\big]\varphi^4 + \big[gm^2 t^{(0)} + 8NhM^3 t^{(\frac{1}{2})} + C_5\big]\varphi \\ &+ \Big[\frac{2NMh}{3}\,t^{(\frac{1}{2})} - g\Big(\xi - \frac{1}{6}\Big)t^{(0)} + C_6\Big]R\varphi\bigg\}, \end{aligned} \tag{15.38}$$

where we restored the loop expansion parameter $\hbar$ and $t^{(0)}$ and $t^{(\frac{1}{2})}$ were identified in (15.23) and (15.34). The constants $C_{1\,\dots\,6}$ can be found from the renormalization conditions. For instance, two well-known values that correspond to the standard choices in the massless scalar case are $C_4 = -\frac{25}{6}$ (see [94]) and $C_2 = -3$ (see [72, 80]). We leave the calculation for the massless and massive (difficult!) theories as exercises.

It is easy to verify the perfect correspondence of the logarithmic terms in (15.38) and in the expression (13.101) derived from the local momentum representation. This fact confirms the correctness of the scale identifications (15.23) and (15.34). One may think that the renormalization group calculation is much more economical than the direct one. This is true, but there are models for which the identification of the analogs of $t^{(0)}$ and $t^{(\frac{1}{2})}$ is complicated, since the masses of the different quantum fields mix in the loops. The reader can find an elaborated example of this sort in the second paper of Ref. [24].

Exercises

15.1. Consider the the Yukawa theory without dimensional parameters. Show that the scalar potential of the theory with zero fermion mass does not need to have odd terms to provide renormalizability.
15.2. Analyze the special case of expression (15.38) when the theory has no dimensional parameters. Demonstrate that the scaling factors $t^{(0)}$ and $t^{(\frac{1}{2})}$ are identical.
15.3. In the theory without dimensional parameters, impose the renormalization conditions on the effective potential (15.38), in the form

$$\frac{d^2V_1}{d\varphi^2}\bigg|_{\varphi=0} = 0\,, \qquad \frac{d^4V_1}{d\varphi^4}\bigg|_{\varphi=\mu} = 4!f\,, \qquad \frac{d^2V_2}{d\varphi^2}\bigg|_{\varphi=\mu} = -2\xi R\,. \tag{15.39}$$

The results: $$\bar{V}_1 = \frac{1}{48}\,\varphi^4\,\Big(\beta_f + 4f\gamma\Big)\Big\{\log\Big(\frac{\varphi^2}{\mu^2}\Big) - \frac{25}{6}\Big\},$$
$$\bar{V} = -\frac{1}{4}\,\varphi^2\,\Big(\beta_\xi + 2\xi\gamma\Big)\Big\{\log\Big(\frac{\varphi^2}{\mu^2}\Big) - 3\Big\}. \qquad (15.40)$$

15.3 The global conformal (scaling) anomaly

In the next chapters, we shall consider the anomalous violation of local conformal symmetry, which is one of the most important tools for semiclassical gravity. It will be very useful to have, for the sake of comparison, the expressions for the global conformal anomaly, which can be derived on the basis of the renormalization group equation (15.9) or (15.17).

The symmetry with respect to a global scaling, or the *global* conformal invariance, takes place at the classical level in massless models. The transformations have the forms

$$g_{\mu\nu} \to g_{\mu\nu} = e^{2\lambda}\bar{g}_{\mu\nu}\,, \qquad \phi_i \to \phi_i = e^{d_i\lambda}\bar{\phi}_i\,, \qquad \lambda = const\,. \qquad (15.41)$$

Here d_i is the conformal weight of the matter fields ϕ_i, namely, -1 for scalars, $-\frac{3}{2}$ for spinors and 0 for vectors.[2] The constant scaling is essentially different from the *local* conformal transformation; in particular it does not impose restrictions on the nonminimal scalar-curvature parameter ξ and it also leaves $\int_x R^2$-term invariant.

It is easy to see that the Noether identity corresponding to the scaling (15.41) has the form (in some cases, we use condensed notation (12.4))

$$\frac{\partial}{\partial\lambda}\,S = \int_x \Big\{ \frac{2}{\sqrt{-g}}\,g_{\mu\nu}\,\frac{\delta S}{\delta g_{\mu\nu}} + \frac{d_i}{\sqrt{-g}}\,\phi^i\,\frac{\delta S}{\delta\phi^i}\Big\} = 0\,, \qquad (15.42)$$

and, in the vacuum sector, we get the identity

$$-\,\frac{2}{\sqrt{-g}}\,g_{\mu\nu}\,\frac{\delta S_{vac}}{\delta g_{\mu\nu}} = T^\mu_\mu = 0. \qquad (15.43)$$

At the one-loop quantum level, we have to consider the vacuum average

$$\int d^4x\sqrt{-g}\,\langle T^\mu_\mu\rangle = \frac{\partial\Gamma^{(1)}_R}{\partial\lambda}\Big|_{n\to 4}. \qquad (15.44)$$

Here $\Gamma^{(1)}_R$ is the renormalized one-loop vacuum effective action, which is the sum

$$\Gamma^{(1)}_R = S_{vac} + \Gamma^{(1)}_{div} + \Gamma^{(1)}_{fin} + \Delta S, \qquad (15.45)$$

where the last term is a counterterm introduced to cancel divergences. It is easy to see that the sum of the first three terms in (15.45) is scale invariant, which follows from

[2] The conformal weight and mass dimension of the vector do not coincide because scaling is performed in curved spacetime, while the vector should be defined in the flat tangent space. The connection with the covariant vector is $A_\mu = e^a_\mu A_a$, which explains the difference.

the simultaneous change of the quantum fields and the metric (15.41) in the functional integral (15.3). Thus, we obtain

$$\int d^4x\sqrt{-g}\,\langle T^\mu_\mu\rangle \;=\; \frac{\partial \Delta S}{\partial\lambda}\Big|_{n\to 4}. \tag{15.46}$$

The reason of why this term can be non-zero is that the counterterms

$$\Delta S = +\frac{\mu^{n-4}}{n-4}\int d^nx\sqrt{-g}\,\big(\beta_1 C^2 + \beta_2\, E + \beta_3 \Box R\big) \tag{15.47}$$

should be introduced before the regularization is removed, in n-dimensional spacetime. After the scaling transformation of the metric in (15.41), we get

$$\begin{aligned}
\int d^nx\sqrt{-g}C^2 \;&=\; e^{\lambda(n-4)}\int d^nx\sqrt{-\bar{g}}\bar{C}^2\,,\\
\int d^nx\sqrt{-g}E \;&=\; e^{\lambda(n-4)}\int d^nx\sqrt{-\bar{g}}\bar{E}\,,\\
\int d^nx\sqrt{-g}(\Box R) \;&=\; e^{\lambda(n-4)}\int d^nx\sqrt{-\bar{g}}(\bar{\Box}\bar{R}).
\end{aligned} \tag{15.48}$$

Substituting these relations and definition (15.47 into (15.46), we obtain

$$\int d^4x\sqrt{-g}\,\langle T^\mu_\mu\rangle \;=\; -\int d^4x\sqrt{-g}\,\big(\beta_1 C^2 + \beta_2\, E + \beta_3 \Box R\big)\,. \tag{15.49}$$

It is evident that the global (integrated) trace anomaly is completely defined by the vacuum beta functions of the theory.

Exercises

15.4. Verify that the actions of free massless scalar, spinor and vector fields are scale invariant. Consider also possible interactions between these fields, and establish the condition of scale invariance for the corresponding terms in the action.
15.5. Derive the Jacobian of the change of variables (15.41) in the path integral, and find how the sources should be transformed to keep $W[J, g_{\mu\nu}]$ invariant. Discuss the Legendre transformation, and demonstrate the invariance of the effective action.
15.6. Develop an alternative simplified approach to arrive at the main result of the previous exercise for the vacuum part of effective action only.
15.7. Discuss whether each of the terms in both sides of Eq. (15.45) are i) finite or infinite; ii) local or nonlocal; iii) scale invariant or non-invariant. Repeat the discussion for all the combinations of the terms in the *r.h.s.*.
15.8. Explore the global conformal anomaly in the interacting theory. Consider the conformal scalar with an arbitrary ξ and φ^4 potential, and calculate the anomalous violation of the Noether identity (15.42) in the interacting theory.

16
Non-local form factors in flat and curved spacetime

In the previous chapter, we saw that a renormalization group based on the minimal subtraction can be successfully formulated in curved spacetime. There are now two important questions to be answered, and we shall address them in the present chapter. The first concerns the identification of the mass-dimension renormalization parameter μ with some physically relevant quantity. We shall derive the nonlocal form factors in flat and curved space and show how μ can be associated with the modulo of the Euclidean momentum in the high-energy (UV) domain. The second question is whether the renormalization group approach can be applied to low-energy phenomena, which is how the renormalization group equations can be adapted for the in the low-energy (or infrared) domain. We shall see that, in some cases the answer to this question can be obtained from the infrared limit of the nonlocal form factors. In this chapter, we shall mainly follow the paper [313], which was prepared in relation to this book.

16.1 Non-local form factors: simple example

As the first example, we consider the derivation of a nonlocal form factor of the two-point function, using Feynman diagrams in dimensional regularization. The divergent part will be also compared to the Euclidean cut-off results. After that, we shall demonstrate how to perform calculations using the heat-kernel methods. All considerations will be restricted to the one-loop approximation.

Types of regularization. Different regularization schemes are used in the literature for different purposes, as we discussed in Part I. The examples used most are cut-off regularizations (including three-dimensional cut-off regularization in momentum space, four-dimensional Euclidean cut-off regularization, proper-time integral covariant cut-off regularization), Pauli-Villars (conventional, covariant and higher-derivative covariant), analytic regularization (different versions), ζ-regularization (which is not a regularization, properly speaking), point-splitting regularization and dimensional regularization, which has the advantages to preserve the gauge symmetry and being the most simple, in many cases.

Dimensional regularization was discussed in Part I as part of a general renormalization program, especially in gauge theories. As here we are going to use it for concrete calculations, it will be discussed in more detail, with minor repetitions.

Mathematical preliminaries. We need to introduce a few mathematical tools.

Introduction to Quantum Field Theory with Applications to Quantum Gravity. Iosif L. Buchbinder and Ilya L. Shapiro, Oxford University Press (2021). © Iosif L. Buchbinder and Ilya Shapiro.
DOI: 10·1093/oso/9780198838319·003·0016

1. Analytic continuation theorem. Consider the two regions D_1 and D_2 on a complex plane, and the two functions $F_1(z)$ and $F_2(z)$, which are defined and analytic on D_1 and D_2, correspondingly. We assume that $D_1 \cap D_2 = D$ and that $F_1(z) = F_2(z)$ on a set of points on the complex plane, that belongs to D and has at least one accumulation point. Then, $F_1(z) = F_2(z)$ on the whole of D.

Our strategy is to define the analytic continuation of the ill-defined integrals, e.g.,

$$I_4 = \int \frac{d^4p}{(p^2+m^2)[(p-k)^2+m^2]}. \tag{16.1}$$

This integral is already defined in Euclidean four-dimensional space, but we want want to extend it from dimension 4 to a complex dimension 2ω, $I_4 \to I_{2\omega}$, such that $I_{2\omega}$ is analytic on the complex plane except in some points, forming at most a countable set. Then, in the vicinity of the point $\omega = 2$, one can write

$$I_{2\omega} = \Big(\sim \frac{1}{2-\omega} \text{ pole term}\Big) + \text{a finite term} + \text{vanishing } O(2-\omega) \text{ term.}$$

Our first purpose will be to establish the divergent, term, with the pole at $\omega = 2$.

2. Gaussian integral. As we know from Eq. (7.84) in Part I, this integral reads

$$\int \frac{d^{2\omega}k}{(2\pi)^{2\omega}}\, e^{-xk^2+2kb} = \frac{1}{(2\pi)^{2\omega}} \Big(\frac{\pi}{x}\Big)^{\omega} e^{\frac{b^2}{x}}. \tag{16.2}$$

For natural values $2\omega = 1, 2, 3, 4, \dots$, this integral can be easily derived. For the complex values of ω, Eq. (16.2) should be regarded as the definition.

A typical example of applying (16.2) is related to the representation

$$\frac{1}{k^2+m^2} = \int_0^\infty d\alpha\, e^{-\alpha(k^2+m^2)}. \tag{16.3}$$

Consider the continuation of the integral (16.1) into dimension $n = 2\omega$,

$$\begin{aligned} I_{2\omega} &= \int \frac{d^{2\omega}k}{(2\pi)^{2\omega}(k^2+m^2)[(k-p)^2+m^2]} \\ &= \int \frac{d^{2\omega}k}{(2\pi)^{2\omega}} \int_0^\infty d\alpha_1 \int_0^\infty d\alpha_2\, e^{-\alpha_1(k^2+m^2)-\alpha_2[(k-p)^2+m^2]}. \end{aligned} \tag{16.4}$$

Changing the order of integration, it is easy to note that the integral over $d^{2\omega}k$ is exactly of the type (16.2); hence, we arrive at

$$\begin{aligned} I_{2\omega} &= \int_0^\infty d\alpha_1 \int_0^\infty d\alpha_2 \int \frac{d^{2\omega}k}{(2\pi)^{2\omega}}\, e^{-k^2(\alpha_1+\alpha_2)+2\alpha_2 kp-(\alpha_1+\alpha_2)m^2-\alpha_2 p^2} \\ &= \int_0^\infty d\alpha_1 \int_0^\infty d\alpha_2\, \frac{1}{(2\pi)^{2\omega}} \Big(\frac{\pi}{\alpha_1+\alpha_2}\Big)^{\omega} e^{\frac{\alpha_2^2 p^2}{\alpha_1+\alpha_2}-\alpha_2(p^2+m^2)-\alpha_1 m^2}. \end{aligned} \tag{16.5}$$

The last representation will prove useful, at some moment.

3. Some properties of the gamma function. The gamma function is defined as

$$\Gamma(z) = \int_0^\infty dt\, t^{z-1} e^{-t}. \tag{16.6}$$

The main properties (for us, at least) are as follows:

$$\begin{aligned}
&\Gamma(z+1) = z\Gamma(z) \quad \Longrightarrow \quad \Gamma(n+1) = n!\,, \\
&\Gamma\Big(\frac{1}{2}\Big) = \sqrt{\pi} \quad \Longrightarrow \quad \Gamma\Big(n+\frac{1}{2}\Big) = \frac{1\cdot 3\cdot 5\ldots(2n-1)}{2^n}\sqrt{\pi}\,, \\
&\Gamma(z) \;=\; \lim_{n\to\infty} \frac{n!\, n^z}{z(z+1)\ldots(z+n)}.
\end{aligned} \tag{16.7}$$

From the last representation it directly follows that $\Gamma(z)$ has simple poles in the points $z = 0, -1, -2, \ldots$, and nowhere else. Another representation where this fact can be seen explicitly is Weirstrass's partial fraction expansion:

$$\Gamma(z) \;=\; \Gamma_n(z) = \sum_{n=0}^{\infty} \frac{(-1)^n}{n!(n+z)} + \int_1^\infty dt\, t^{z-1} e^{-t}. \tag{16.8}$$

It is clear that $\Gamma(z)$ is analytic everywhere except $z = 0, -1, -2, \ldots$.

4. Volume of the sphere. Finally, we derive the volume of the m-dimensional sphere with the radius $R = (x_1^2 + x_2^2 + \cdots + x_m^2)^{\frac{1}{2}}$. The dimensional argument tells us that

$$V_m = C_m R^m, \tag{16.9}$$

where C_m are the numerical coefficients which we have to calculate. For this sake, let us consider the integral

$$I = \int_{-\infty}^{\infty} \cdots \int_{-\infty}^{\infty} dx_1 \ldots dx_m e^{-a(x_1^2+x_2^2+\cdots+x_m^2)} \;=\; \Big[\int_{-\infty}^{\infty} dx\, e^{-ax^2}\Big]^m \;=\; \Big(\frac{\pi}{a}\Big)^{\frac{m}{2}}. \tag{16.10}$$

On the other hand, $dV_m = mC_m R^{m-1} dR$; hence,

$$I = \int_{-\infty}^{\infty} e^{-aR^2} mC_m R^{m-1} dR.$$

Making the change of variables $z = aR^2$, we get

$$dR = \frac{1}{2a}\Big(\frac{a}{z}\Big)^{\frac{1}{2}} dz, \qquad R^{m-1} = \frac{z}{a}^{\frac{m-1}{2}}$$

and therefore

$$I \;=\; \frac{mC_m}{2a^{\frac{m}{2}}} \int_0^\infty e^{-z} z^{\frac{m}{2}-1} dz \;=\; \frac{mC_m}{2a^{\frac{m}{2}}}\, \Gamma\Big(\frac{m}{2}\Big). \tag{16.11}$$

Since (16.10) and (16.11) are the same, we get

$$C_m \;=\; \frac{\pi^{\frac{m}{2}}}{\frac{m}{2}\Gamma\big(\frac{m}{2}\big)} \;=\; \frac{\pi^{\frac{m}{2}}}{\Gamma\big(\frac{m}{2}+1\big)} \qquad \Longrightarrow \qquad V_m = \frac{\pi^{\frac{m}{2}}}{\Gamma\big(\frac{m}{2}+1\big)} R^m. \tag{16.12}$$

This last relation is valid for any natural m, but we can continue it to an arbitrary complex dimension 2ω.

The simplest loop integral. Now we are in a position to start regularizing loop integrals. The general strategy will be to continue

$$I_4 \quad \longrightarrow \quad I_{2\omega} = \int \frac{d^{2\omega}x}{(2\pi)^{2\omega}} \cdots , \tag{16.13}$$

so that $I_{2\omega}$ is defined over all of the complex plane except in some points, including $\omega = 2$. Then,

$$I_{2\omega} \;=\; (\text{pole at } \omega = 2) \,+\, \text{a regular term}.$$

Consider a scalar theory with an $\lambda\varphi^4$ interaction and the Minkowski space action

$$S = \int d^4z \Big[\frac{1}{2}(\partial\varphi)^2 - \frac{m^2}{2}\varphi^2 - \frac{\lambda}{4!}\varphi^4\Big]. \tag{16.14}$$

Let us first rewrite it as a Euclidean action, by setting $z^0 = -iz^4$. Then

$$\begin{aligned} &d^4z = dz^0 d^3z = -idz^4 d^3z = -id^4_E z \\ \text{and}\quad &(\partial\varphi)^2 \,=\, (\partial_0\varphi)^2 - (\nabla\varphi)^2 \,=\, -(\partial_4\varphi)^2 - (\nabla\varphi)^2 \,=\, -(\partial\varphi)^2_E. \end{aligned} \tag{16.15}$$

Finally,

$$S = -i\int d^4_E z\Big[-\frac{1}{2}(\partial\varphi)^2_E - \frac{m^2}{2}\varphi^2 - \frac{\lambda}{4!}\varphi^4\Big]. \tag{16.16}$$

Consider the diagram (we assume that the reader can easily reproduce this expression after reading Part I)

$$\frac{1}{2}$$ 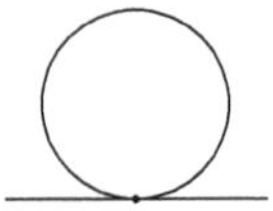 $$= -\frac{\lambda}{2}\int \frac{d^4p}{(2\pi)^4}\,\frac{1}{p^2+m^2} \;=\; -\frac{\lambda}{2}\,I_4^{tad}\,. \tag{16.17}$$

Let us start from the cut-off calculation.

$$\begin{aligned} -\frac{\lambda}{2}\,I_4^{tad} &= -\frac{\lambda}{2}\cdot\frac{1}{16\pi^2}\cdot\frac{\pi^2}{2}\int_0^{\Omega}\frac{4p^3dp}{p^2+m^2} \;=\; -\frac{\lambda}{32\pi^2}\int_0^{\Omega}\frac{p^2dp^2}{p^2+m^2} \\ &= -\frac{\lambda}{32\pi^2}\Big[\int_0^{\Omega}dp^2 - \int_0^{\Omega}\frac{m^2dp^2}{p^2+m^2}\Big] \;=\; -\frac{\lambda}{32\pi^2}\Big[\Omega^2 - m^2\log\frac{\Omega^2}{m^2}\Big], \end{aligned} \tag{16.18}$$

where the $O(\Omega^{-1})$-terms were omitted as irrelevant.

The dimensional regularization of this diagram is not as simple, but will prove very instructive for the future. We have

$$I_4^{tad} \to I_{2\omega}^{tad} = \int\frac{d^{2\omega}p}{(2\pi)^{2\omega}}\cdot\frac{1}{p^2+m^2} = \frac{2\omega\cdot\pi^{\omega}}{(2\pi)^{2\omega}\Gamma(\omega+1)}\int_0^{\infty}\frac{p^{2\omega-1}dp}{p^2+m^2}\,.$$

Remember that $\Gamma(\omega+1) = \omega\Gamma(\omega)$, so

$$I_{2\omega}^{tad} = \frac{2\pi^\omega}{(2\pi)^{2\omega}} \cdot \frac{1}{\Gamma(\omega)} \int_0^\infty \frac{p^{2\omega-1}dp}{p^2+m^2}. \tag{16.19}$$

This integral can be expressed via the beta-function

$$B(x,y) = \frac{\Gamma(x)\Gamma(y)}{\Gamma(x+y)} = \int_0^\infty dt\, t^{x-1}(1+t)^{-x-y}. \tag{16.20}$$

In (16.19), we denote $p^2 = tm^2$ and obtain

$$\begin{aligned} I_{2\omega}^{tad} &= \frac{\pi^\omega\,(m^2)^{\omega-1}}{(4\pi^2)^\omega\,\Gamma(\omega)} \int_0^\infty dt\, t^{\omega-1}(1+t)^{-1} = \frac{1}{(4\pi)^\omega}\frac{(m^2)^{\omega-1}}{\Gamma(\omega)} \cdot B(\omega, 1-\omega) \\ &= \frac{(m^2)^{\omega-1}}{(4\pi)^\omega}\frac{\Gamma(\omega)\Gamma(1-\omega)}{\Gamma(\omega)\Gamma(1)} = \frac{(m^2)^{\omega-1}}{(4\pi)^\omega}\,\Gamma(1-\omega), \end{aligned} \tag{16.21}$$

where we identified $x-1 = \omega-1$ and $-x-y = -1$ as arguments of (16.20). An explicit representation of $\Gamma(1-\omega)$ can be easily obtained from Eqs. (16.7):

$$\Gamma(2-\omega) = \lim_{n\to\infty} J_\omega, \quad \text{where} \quad J_\omega = \frac{n!\, n^{2-\omega}}{(2-\omega)(3-\omega)\dots(n+2-\omega)}. \tag{16.22}$$

The expression under the limit can be transformed as

$$J_\omega = \frac{n!\, e^{(2-\omega)\ln n}}{(2-\omega)(1+2-\omega)(2+2-\omega)\dots(n+2-\omega)}.$$

Obviously, the divergent part of J_ω is

$$J_\omega^{(div)} = \frac{n!\cdot 1}{(2-\omega)\cdot 1\cdot 2\dots n} = \frac{1}{2-\omega}.$$

The finite part can be evaluated by means of the following transformations:

$$\begin{aligned} \frac{1}{2-\omega}\, e^{(2-\omega)\log n} &= \frac{1}{2-\omega}\Big[1 + (2-\omega)\log n + \mathcal{O}(2-\omega)^2\Big] \\ &= \frac{1}{2-\omega} + \log n + \mathcal{O}\big((2-\omega)\big) \end{aligned}$$

and

$$\frac{1}{k-\omega} = \frac{1}{k-2+(2-\omega)} = \frac{1}{k-2}\cdot\frac{1}{1+\frac{2-\omega}{k-2}} = \frac{1}{k-2}\Big(1 - \frac{2-\omega}{k-2} + \dots\Big).$$

Therefore,

$$J_\omega = \frac{1}{2-\omega} + \log n - \Big(1 + \frac{1}{2} + \dots + \frac{1}{n}\Big) + \mathcal{O}(2-\omega). \tag{16.23}$$

The sum of the finite terms is

$$\gamma = \lim_{n\to\infty}\Big(1 + \frac{1}{2} + \dots + \frac{1}{n} - \ln n\Big). \tag{16.24}$$

and its value is $\gamma = 0.57721\dots$ (the Euler-Mascheroni constant, or just Euler's constant).

Thus, we can establish the relations

$$\Gamma(2-\omega) = \int_0^\infty \frac{e^{-t}\,dt}{t^{1-w}} = \frac{1}{2-\omega} - \gamma + \mathcal{O}(2-\omega)\,, \tag{16.25}$$

$$\Gamma(1-\omega) = \frac{1}{-1+(2-\omega)}\Gamma(2-\omega) = -\frac{1}{2-\omega} - 1 + \gamma + \mathcal{O}(2-\omega), \tag{16.26}$$

$$\Gamma(-\omega) = \frac{1}{2(2-\omega)} + \frac{3}{4} - \frac{\gamma}{2} + \mathcal{O}(2-\omega)\,. \tag{16.27}$$

Now we can use Eq. (16.27) to rewrite the result (16.21) as

$$\begin{aligned} I_{2\omega} &= \frac{(m^2)^{\omega-1}}{(4\pi)^\omega}\Big(-\frac{1}{2-\omega}+\gamma-1\Big) \\ &= \frac{m^2}{(4\pi)^2}(\mu^2)^{\omega-2}\Big(\frac{m^2}{4\pi\mu^2}\Big)^{\omega-2}\Big(-\frac{1}{2-\omega}+\gamma-1\Big), \end{aligned} \tag{16.28}$$

where μ is a *renormalization parameter*, with $[\mu] = [m]$. Furthermore,

$$\Big(\frac{m^2}{4\pi\mu^2}\Big)^{\omega-2} = e^{(\omega-2)\,\log\left(\frac{m^2}{4\pi\mu^2}\right)} = 1 + (2-\omega)\log\Big(\frac{4\pi\mu^2}{m^2}\Big) + \dots$$

and we finally arrive at

$$I_{2\omega} = \frac{m^2}{(4\pi)^2}(\mu^2)^{\omega-2}\Big[-\frac{1}{2-\omega}+\gamma-1-\ln\Big(\frac{4\pi\mu^2}{m^2}\Big)\Big]. \tag{16.29}$$

The last observation is that one can always redefine μ and absorb the summand $\gamma - 1$ into $\log\mu$. Of course, this is not a compulsory operation.

A comparison of (16.29) and the result in the cut-off regularization (16.18) shows that, in dimensional regularization, there is nothing like the quadratic divergences $\mathcal{O}(\Omega^2)$. A useful notation is $\varepsilon = (4\pi)^2(n-4)$, with $n-4 = -2(2-\omega)$. There is a direct relation between the leading logarithm $\log\frac{\Omega}{m}$ and the divergence ε^{-1}. The correspondence is given by the relation

$$\log\frac{\Omega^2}{m^2} \longleftrightarrow -\frac{\mu^{n-4}}{\varepsilon}, \qquad \varepsilon = (4\pi)^2(n-4), \tag{16.30}$$

which is universal and holds for all logarithmically divergent diagrams. Let us note that this is a particular manifestation of the general rule. The leading logarithms are the same in *all* regularization schemes [269]. In what follows we shall see that the pole $\frac{1}{n-4}$ corresponds to the leading logarithm in the cut-off scheme.

Derivation of the nonlocal form factor and the UV divergence. As a second example, consider the diagram

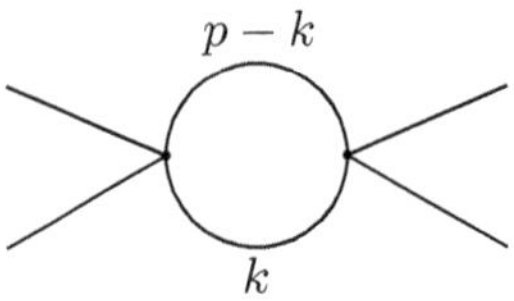

$$= \frac{\lambda^2}{2}\int\frac{d^4p}{(2\pi)^4}\cdot\frac{1}{(p^2+m^2)[(p-k)^2+m^2]} = \frac{\lambda^2 I_4}{2}. \tag{16.31}$$

As a first step, we derive the divergent part of (16.31) in the cut-off regularization. To this end, we make the following transformation:

$$\begin{aligned} I_4 &= \int \frac{d^4p}{(2\pi)^4} \frac{1}{(p^2+m^2)[(p-k)^2+m^2]} \\ &= \int \frac{d^4p}{(2\pi)^4} \frac{1}{(p^2+m^2)^2} + \int \frac{d^4p}{(2\pi)^4} \frac{1}{p^2+m^2} \Big[\frac{1}{[(p-k)^2+m^2]} - \frac{1}{p^2+m^2}\Big]. \end{aligned} \tag{16.32}$$

Remember that $d^4p = \pi^2 p^2 dp^2$ in $n=4$. Therefore, the first integral is logarithmically divergent, and the second is finite. Then

$$I_4^{div} = \int_0^{\Omega} \frac{p^2 dp^2}{(4\pi)^2(p^2+m^2)} = \frac{1}{(4\pi)^2} \log\frac{\Omega^2}{m^2} + (\text{finite terms})$$

and hence (16.31) is

$$I_4 = \frac{\lambda^2}{2(4\pi)^2} \log\frac{\Omega^2}{m^2} + (\text{finite terms}). \tag{16.33}$$

Starting with the dimensional regularization calculation, first we define

$$I_{2\omega} = \int \frac{d^{2\omega}p}{(2\pi)^{2\omega}} \cdot \frac{1}{(p^2+m^2)[(p-k)^2+m^2]}. \tag{16.34}$$

Obviously, at $\omega = 2$, the integral $I_{2\omega}$ coincides with I_4, and $I_{2\omega}$ is also analytic on a complex plane in the vicinity of $\omega = 2$, where it has a pole (as we will see in brief).

Consider the Feynman's formula (simplest version)

$$\frac{1}{ab} = \int_0^1 \frac{d\alpha}{[a\alpha + b(1-\alpha)]^2}. \tag{16.35}$$

Using (16.35), one can cast (16.34) into the form

$$I_{2\omega} = \int_0^1 d\alpha \int \frac{d^{2\omega}p}{(2\pi)^{2\omega}} \frac{1}{[(p-\alpha k)^2+a^2]^2}, \qquad a^2 = m^2 + \alpha(1-\alpha)k^2. \tag{16.36}$$

Since (16.36) is convergent on the complex plane, one can shift the integration variable, $p_\mu \to p_\mu - \alpha k_\mu$. This simple operation gives

$$I_{2\omega} = \int_0^1 d\alpha \int \frac{d^{2\omega}p}{(2\pi)^{2\omega}} \frac{1}{(p^2+a^2)^2}\,. \tag{16.37}$$

The main advantage of (16.37) is that it does not depend on angles. One can use the change of variable $p^2 = a^2 t$ to arrive at

$$\begin{aligned} I_{2\omega} &= \int_0^1 d\alpha \int \frac{2\pi^\omega}{(2\pi)^{2\omega}\Gamma(\omega)} \frac{dp\, p^{2\omega-1}}{(p^2+a^2)^{-2}} \\ &= \int_0^1 d\alpha \int_0^\infty \frac{dt}{(4\pi)^\omega \Gamma(\omega)}\, a^{2\omega-4}\, t^{\omega-1}(1+t)^{-2}. \end{aligned} \tag{16.38}$$

Comparing this to (16.20), we identify $x = \omega$ and $y = 2 - \omega$. Then,

$$I_{2\omega} = \frac{1}{(4\pi)^\omega} \int_0^1 d\alpha \, \frac{\Gamma(\omega)\Gamma(2-\omega)}{\Gamma(\omega)\Gamma(2)} \, a^{2\omega-4}.$$

Remember that $\Gamma(2) = 1$ and that a^2 is defined by (16.36). Then, using (16.25), we get

$$I_{2\omega} = \frac{1}{(4\pi)^\omega}\Big[\frac{1}{2-\omega} - \gamma\Big] \int_0^1 d\alpha [m^2 + \alpha(1-\alpha)k^2]^{\omega-2}. \tag{16.39}$$

Let us denote $\tau = \frac{k^2}{m^2}$ and transform

$$\begin{aligned} [m^2 + \alpha(1-\alpha)k^2]^{\omega-2} &= (m^2)^{\omega-2} \, e^{(\omega-2)\log[1+\alpha(1-\alpha)\tau]} \\ &= (m^2)^{\omega-2}\big[1 - (2-\omega)\log\left(1+\alpha(1-\alpha)\tau\right)\big] + \mathcal{O}\big((\omega-2)^2\big). \end{aligned} \tag{16.40}$$

Substituting this expression into (16.39), we arrive at

$$\begin{aligned} I_{2\omega} &= \frac{1}{(4\pi)^\omega}\Big(\frac{1}{2-\omega} + \gamma\Big)(m^2)^{\omega-2}\Big\{1 - (2-\omega)\int_0^1 d\alpha \log\left[1+\alpha(1-\alpha)\tau\right]\Big\} \\ &= \frac{(m^2)^{\omega-2}}{(4\pi)^\omega}\Big[\frac{1}{2-\omega} + \gamma - \int_0^1 d\alpha \log\left[1+\alpha(1-\alpha)\tau\right]\Big], \end{aligned} \tag{16.41}$$

where $\tau = \frac{k^2}{m^2}$.

In the last expression, the first term is the divergence, and the integral over α represents the nonlocal form factor, which is the desirable physical result. Also,

$$\begin{aligned} \frac{(m^2)^{\omega-2}}{(4\pi)^\omega} &= \frac{(\mu^2)^{\omega-2}}{(4\pi)^2}\Big(\frac{m^2}{4\pi\mu^2}\Big)^{\omega-2} = \frac{(\mu^2)^{\omega-2}}{(4\pi)^2}\, e^{(2-\omega)\log\left(\frac{4\pi\mu^2}{m^2}\right)} \\ &= \frac{(\mu^2)^{\omega-2}}{(4\pi)^2}\Big[1 + (2-\omega)\log\Big(\frac{4\pi\mu^2}{m^2}\Big) + \mathcal{O}\big((\omega-2)^2\big)\Big]. \end{aligned} \tag{16.42}$$

The integration can be simplified by the change of variables $z = \frac{1}{2} - \alpha$. The result is

$$\int_0^1 d\alpha \, \log\left[1+\alpha(1-\alpha)\tau\right] = -Y, \tag{16.43}$$

$$\text{where} \qquad Y = 1 - \frac{1}{a}\log\Big|\frac{2+a}{2-a}\Big| \qquad \text{and} \qquad a^2 = \frac{4\tau}{\tau+4} = \frac{4k^2}{k^2+4m^2}. \tag{16.44}$$

Substituting (16.42) and (16.43) into (16.41), we arrive at

$$\begin{aligned} I_{2\omega} &= \frac{\mu^{2\omega-4}}{(4\pi)^2}\Big[\frac{1}{2-\omega} + \gamma + \log\Big(\frac{4\pi\mu^2}{m^2}\Big) + 2Y\Big] \\ &= (\mu^2)^{\omega-2}\Big[-\frac{2}{\varepsilon} + \frac{\gamma}{(4\pi)^2} + \frac{1}{(4\pi)^2}\log\Big(\frac{4\pi\mu^2}{m^2}\Big) + \frac{2Y}{(4\pi)^2}\Big]. \end{aligned} \tag{16.45}$$

Let us conclude that the integral of our interests consists of the local divergent part and the nonlocal finite contribution. The last is called the *form factor* and, in the simplest case that we just considered, is proportional to the quantity Y.

The next thing to do is to explore the form factor Y in the two extremes, namely high- and low-energy limits:

$$\begin{aligned} &\textbf{1.} \quad \text{UV}, \quad k^2 \gg m^2, \quad \tau \gg 1, \\ &\textbf{2.} \quad \text{infrared}, \quad k^2 \ll m^2, \quad \tau \ll 1. \end{aligned} \tag{16.46}$$

1. Consider the UV regime, which means $k^2 \gg m^2$ and $\tau \gg 1$. Then,

$$a^2 = \frac{4k^2}{k^2 + 4m^2} = \frac{4}{1 + \frac{4m^2}{k^2}} = 4\Big(1 - \frac{4m^2}{k^2} + \dots\Big),$$

so $a \approx 2 - \frac{4m^2}{k^2}$. Then, $2 + a \approx 4$, and $2 - a \approx \frac{4m^2}{k^2}$, such that

$$Y \cong 1 - \frac{1}{2}\log\Big(\frac{4}{4m^2/k^2}\Big) = 1 - \frac{1}{2}\log\frac{k^2}{m^2}. \tag{16.47}$$

In this case, $I_{2\omega}$ in (16.45) includes the combination

$$\begin{aligned} I_{2\omega} &= \frac{\mu^{2\omega-4}}{(4\pi)^2}\Big[\frac{1}{2-\omega} + \log\Big(\frac{4\pi\mu^2}{m^2}\Big) + 2 + \gamma - \log\Big(\frac{k^2}{m^2}\Big)\Big] \\ &= \frac{\mu^{2\omega-4}}{(4\pi)^2}\Big[-\frac{2}{n-4} + \log\Big(\frac{\mu^2}{k^2}\Big) + \text{constant}\Big]. \end{aligned} \tag{16.48}$$

Let us stress that this is a very significant and important relation, as it shows two things at once. The first is that the large-k^2 limit means a large-μ^2 limit, and vice versa. Thus, it is sufficient to establish the limit of the large μ within the minimal subtraction scheme of renormalization, to know the physical UV limit, which is the behavior of the quantum system at high energies. Let us remember that we already know how to explore the large-μ limit throughout the usual renormalization group.

The second is that one can always restore the large-μ^2 limit from the coefficient of the divergent term with the $\frac{1}{\epsilon}$-factor. By the way, there is also a direct correspondence with the large cut-off limit in (16.33). All in all, we can say that the UV limit is pretty well controlled by the leading logarithmic divergences, which can be derived easily using the heat-kernel methods, even without the use of Feynman diagrams.

2. Consider now the infrared regime, when $k^2 \ll m^2$, or, equivalently, $\tau \ll 1$. Then $a^2 \sim \frac{k^2}{m^2} \ll 1$ and hence $a \sim \frac{k}{m}$. Consequently,

$$\log\frac{2+a}{2-a} \approx \log\frac{2+\frac{k}{m}}{2-\frac{k}{m}} \approx \log\Big(1 + \frac{k}{m}\Big) + \dots \approx \frac{k}{m},$$

and therefore

$$Y = 1 - \frac{1}{a}\log\Big|\frac{2+a}{2-a}\Big| \approx 1 - \frac{m}{k}\,\frac{k}{m} \approx 0. \tag{16.49}$$

In the zero-order approximation, we do not get a nonlocal form-factor. This means there is no $\log\Big(\frac{k^2}{m^2}\Big)$ to correspond $\frac{1}{\epsilon}$ or $\log\Omega$. The next orders of expansion read

$$Y = -\frac{1}{12}\frac{k^2}{m^2} + \frac{1}{120}\Big(\frac{k^2}{m^2}\Big)^2 + \dots. \tag{16.50}$$

The first term here is the evidence that the decoupling is quadratic in this case. The same quadratic dependence takes place in all cases when we can check it. In the infrared limit, the divergences and momentum dependence do not correlate with each other. This phenomenon is called infrared decoupling, or "the decoupling theorem". It was discovered in QED in 1975 by Appelquist and Corrazzone [10].

The last observation is that, for the tadpole diagram (16.29), there is no nonlocal form-factor. In this case, one may think that the UV divergence is "artificial," but this is not correct, because, in general, the logarithmic form factor corresponds to the *sum* of all $\mathcal{O}(\epsilon^{-1})$-contributions, including the ones from the tadpoles.

Exercises.

16.1. Verify the combinatorial coefficient and find the relations between momenta in (16.31).
16.2. Calculate the logarithmic divergence of Eq. (16.31) in Pauli-Vilars regularization, with

$$\frac{1}{p^2+m^2} \to \frac{1}{p^2+m^2} - \frac{1}{p^2+M^2},$$

and, after that, consider the limit $M \to \infty$.
16.3. Verify the integrals (16.71) and (16.43).
16.4. Verify (e.g., using Mathematica or Maple) the leading orders of expansion of Y in Eq. (16.50).
16.5. Repeat the analysis of the UV and infrared limits, and check the expressions (Y-UV), (16.48) and (16.50) using the integral representation (16.41), without taking the integral over α.
16.6. Repeat the whole procedure described in this section for the polarization operator in QED. The result for the polarization operator, given by the diagram

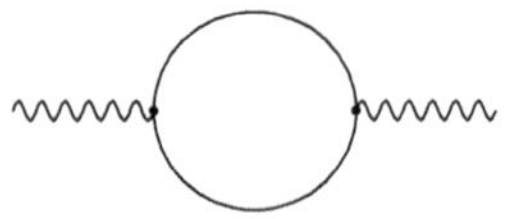

has the form

$$\bar{\Gamma}^{(1)}_{\sim F^2_{\mu\nu}} = -\frac{e^2}{2(4\pi)^2}\int d^4x\sqrt{g}\,F_{\mu\nu}\Big[\frac{2}{3\,\epsilon} + k_1^{FF}(a)\Big]F^{\mu\nu}\Big\}, \tag{16.51}$$

where the nonlocal form factor is [165]

$$k_1^{FF}(a) = Y\Big(2 - \frac{8}{3a^2}\Big) - \frac{2}{9}. \tag{16.52}$$

16.7. Starting from Eq. (16.51), consider the UV and infrared limits in QED [10].

16.2 Non-local form factors in curved spacetime

One can perform the calculations of nonlocal form factors in curved spacetime by means of Feynman diagrams in curved spacetime [167] (see also a more complete treatment including surface terms in [93]). However, there is a technically more efficient method, and, in this section, we shall demonstrate how it works in practice.

The general framework can be described as follows. As in section 13.4, we define the one-loop Euclidian effective action for a massive field as the trace of an integral of the heat kernel over the proper time s,

$$\overline{\Gamma}^{(1)} = \frac{1}{2}\,\mathrm{Tr}\,\log\Big(- \hat{1}\Box + m^2 - \hat{P} + \frac{\hat{1}}{6}\,R\Big) = \frac{1}{2}\int_0^\infty \frac{ds}{s}\,\mathrm{Tr}\,K(s)\,. \tag{16.53}$$

As usual, this formula is valid for bosonic fields in Euclidian spacetime, using the symbol $\Box = g^{\mu\nu}\nabla_\mu\nabla_\nu$ for the covariant Laplace operator in $n=4$. For fermions, one has to change the sign. The definition of $\hat{P}$ is different from what we used earlier, to fit the notations of [32], which are used below. The symbol $K(s)$ means the heat kernel of the bilinear form of the classical action of the theory. According to [32] (see also [19, 18, 93]) the solution for the trace of the heat kernel has the form

$$\begin{aligned}\mathrm{Tr}\,K(s) = \frac{(\mu^2)^{2-\omega}}{(4\pi s)^\omega}\int d^4x\sqrt{g}\,e^{-sm^2}\,\mathrm{tr}\Big\{\hat{1} + s\hat{P} + s^2\big[\hat{1}\;R_{\mu\nu}f_1(\tau)R^{\mu\nu}\\ + \hat{1}Rf_2(\tau)R + \hat{P}f_3(\tau)R + \hat{P}f_4(\tau)\hat{P} + \hat{\mathcal{R}}_{\mu\nu}f_5(\tau)\hat{\mathcal{R}}^{\mu\nu}\big]\Big\},\end{aligned} \tag{16.54}$$

where $\tau = -s\Box$ and we use notation $\hat{\mathcal{R}}_{\mu\nu} = [\nabla_\mu,\nabla_\nu]$. Expression (16.54) includes *only* the terms of the second order in "curvatures" $\hat{P}$, $\hat{\mathcal{R}}_{\mu\nu}$ (this one is zero for scalars), and $R_{\mu\nu\alpha\beta}$. The cumbersome third-order terms can be found in [33].

In Eq. (16.54), the terms between braces are matrices in the space of the fields (e.g., scalar, vector or fermion). The zero-order term is proportional to $\mathrm{tr}\,\hat{1}$ and corresponds to the quartic divergence, or to the coefficient a_0 in the Schwinger-DeWitt expansion. The term with $s\,\mathrm{tr}\,\hat{P}$ corresponds to the quadratic divergences, or to the a_1 coefficient, and the rest corresponds to the logarithmic divergences and is related to the a_2 coefficient plus finite terms, which are our main target.

The functions $f_{1...5}$ have the form (see [32])

$$f_1(\tau) = \frac{f(\tau) - 1 + \tau/6}{\tau^2},\quad f_2(\tau) = \frac{f(\tau)}{288} + \frac{f(\tau)-1}{24\tau}, \tag{16.55}$$

$$f_3(\tau) = \frac{f(\tau)}{12} + \frac{f(\tau)-1}{2\tau},\quad f_4(\tau) = \frac{f(\tau)}{2},\quad f_5(\tau) = \frac{1-f(\tau)}{2\tau}, \tag{16.56}$$

where

$$f(\tau) = \int_0^1 d\alpha\; e^{-\alpha(1-\alpha)\tau} \qquad \text{and} \qquad \tau = -s\Box\,. \tag{16.57}$$

In what follows, we shall describe the derivation of the integral in (16.53).

For the sake of brevity, we describe only the calculation for the scalar field. The cases for massive fermion and massive vector fields can be worked out similarly, and the reader can consult the papers [167, 245] and the more recent [142, 143] for the details of these calculations.

Consider a scalar field with general nonminimal coupling. The action is (13.46), and therefore

$$\hat{1} = 1, \qquad \hat{P} = -\Big(\xi - \frac{1}{6}\Big)R \qquad \text{and} \qquad \hat{\mathcal{R}}_{\mu\nu} = 0\,. \tag{16.58}$$

Let us derive the integrals (16.53), starting from the simplest one.

The zero-order term. Consider the term corresponding to the a_0 coefficient,

$$\bar{\Gamma}^{(1)}_0 = \frac{1}{2}\int_0^\infty \frac{ds}{s}\,\frac{\mu^{2(2-\omega)}}{(4\pi s)^\omega}\int d^4x\sqrt{g}\,e^{-sm^2}\,. \tag{16.59}$$

Changing the variable $s = tm^{-2}$, the integral becomes

$$\begin{aligned}\bar{\Gamma}^{(1)}_0 &= \frac{1}{2}\int d^4x\,\sqrt{g}\,\frac{\mu^{2(2-\omega)}}{(4\pi)^\omega}\,m^{2\omega}\int_0^\infty \frac{dt}{t^{1+\omega}}\,e^{-t}\\ &= \frac{1}{2(4\pi)^2}\int d^4x\sqrt{g}\,\Big[\frac{1}{2-\omega} + \log\Big(\frac{4\pi\mu^2}{m^2}\Big) + \frac{3}{2}\Big]\,\frac{m^4}{2},\end{aligned} \tag{16.60}$$

where we used relations derived in section 16.1.

It proves useful to introduce the following notation:

$$\frac{1}{\varepsilon_{\omega,\mu}} = \frac{1}{2(4\pi)^2}\Big[\frac{1}{\omega-2} - \log\Big(\frac{4\pi\mu^2}{m^2}\Big)\Big]\,. \tag{16.61}$$

Then the one-loop contribution to the cosmological constant term becomes

$$\bar{\Gamma}^{(1)}_0 = \int d^4x\sqrt{g}\,\Big[-\frac{1}{\varepsilon_{\omega,\mu}} + \frac{3}{4\,(4\pi)^2}\Big]\,\frac{m^4}{2}\,. \tag{16.62}$$

We can observe that this expression consists of the UV divergence, corresponding to the $\log\mu$ term hidden in $1/\varepsilon_{\omega,\mu}$, and the irrelevant constant term, which can be easily absorbed into $1/\varepsilon_{\omega,\mu}$ by changing μ. There is no nonlocal form factor in the expression (16.62). This is a natural result [167], since such a form factor should be constructed from $\Box$, and, when acting on m^4, this operator gives zero.

The first-order term. In a very similar way, in the next order in s, we have the result

$$\begin{aligned}\bar{\Gamma}^{(1)}_1 &= \frac{1}{2}\int_0^\infty \frac{ds}{s}\,\frac{\mu^{2(2-\omega)}}{(4\pi s)^\omega}\int d^4x\sqrt{g}\;e^{-sm^2}\;\mathrm{tr}\,(s\hat{P})\\ &= -\frac{1}{2}\,\frac{\mu^{2(2-\omega)}}{(4\pi)^\omega}\,m^{2(\omega-1)}\int d^4x\sqrt{g}\Big(\xi-\frac{1}{6}\Big)\,\Gamma(1-\omega)\,R\\ &= \Big[-\frac{1}{\varepsilon_{\omega,\mu}} + \frac{1}{2(4\pi)^2}\Big]\,\Big(\xi-\frac{1}{6}\Big)\int d^4x\sqrt{g}\,m^2 R\,.\end{aligned} \tag{16.63}$$

We can draw some initial conclusions from the two examples presented above. In both cases, the effective action is local, and the logarithmic dependence on the renormalization parameter μ is completely controlled by the pole $(2-\omega)^{-1}$. The impossibility of constructing a relevant nonlocal term in a term that is linear in R is explained by the fact that, acting by $\Box$ to R, we arrive at the surface term only.[1] Qualitatively, the situation is the same as for the cosmological constant contribution. The results (16.62) and (16.63) enable one to use minimal subtraction to construct renormalization group equations for the cosmological constant density and the Newton constant, but they do not show the nonlocal terms behind these equations.

Second-order terms. The next-order terms are more involved. We shall calculate them one by one, to find the coefficients $l^*_{1\ldots5}$ and $l_{1\ldots5}$, that define the final form factors of the $R_{\mu\nu}\cdot R^{\mu\nu}$- and $R\cdot R$-terms. The general expression in the second order in curvature is

$$\bar{\Gamma}^{(1)}_2 = \frac{1}{2}\int_0^\infty ds\, e^{-sm^2}\, s^{1-\omega}\,\frac{(\mu^2)^{2-\omega}}{(4\pi)^\omega}\int d^4x\sqrt{g}\,\Big\{R_{\mu\nu}f_1(-s\Box)R^{\mu\nu}$$
$$+R\Big[f_2(-s\Box)-\Big(\xi-\frac{1}{6}\Big)f_3(-s\Box)+\Big(\xi-\frac{1}{6}\Big)^2f_4(-s\Box)\Big]R\Big\}. \tag{16.64}$$

Using relations (16.55), (16.56) and (16.57) in (16.64) and replacing $-s\Box$ with τ, we arrive at

$$\bar{\Gamma}^{(1)}_2 = \frac{1}{2}\int_0^\infty ds\, e^{-sm^2}\, s^{1-\omega}\frac{(\mu^2)^{2-\omega}}{(4\pi)^\omega}\int d^4x\sqrt{g}\Big\{R_{\mu\nu}\Big[\frac{f(\tau)}{\tau^2}-\frac{1}{\tau^2}+\frac{1}{6\tau}\Big]R^{\mu\nu} \tag{16.65}$$
$$+R\Big[\Big(\frac{1}{288}-\frac{\tilde{\xi}}{12}+\frac{\tilde{\xi}^2}{2}\Big)f(\tau)+\Big(\frac{1}{24}-\frac{\tilde{\xi}}{2}\Big)\frac{f(\tau)}{\tau}-\frac{f(\tau)}{8\tau^2}+\Big(\frac{\tilde{\xi}}{2}-\frac{1}{16}\Big)\frac{1}{\tau}+\frac{1}{8\tau^2}\Big]R\Big\},$$

where we used the notation $\tilde{\xi}=\xi-1/6$.

Let us introduce the new coefficients

$$l_1^*=0,\quad l_2^*=0,\quad l_3^*=1,\quad l_4^*=\frac{1}{6},\quad l_5^*=-1, \tag{16.66}$$
$$l_1=\frac{1}{288}-\frac{1}{12}\tilde{\xi}+\frac{1}{2}\tilde{\xi}^2,\quad l_2=\frac{1}{24}-\frac{1}{2}\tilde{\xi},\quad l_3=-\frac{1}{8}=-l_5,\quad l_4=-\frac{1}{16}+\frac{1}{2}\tilde{\xi}.$$

The basic integrals will be denoted as (remember $\tau=-s\Box$)

$$M_1=\int_0^\infty \frac{ds}{(4\pi)^\omega}\,e^{-m^2s}\,s^{1-\omega}\,f(\tau)\;=\;\frac{\mu^{2(2-\omega)}}{(4\pi)^\omega}\int_0^\infty dt\,e^{-t}\,t^{1-w}\,f(tu)\,,$$
$$M_2=\int_0^\infty \frac{ds}{(4\pi)^\omega}\,e^{-m^2s}\,s^{-\omega}\,f(\tau)\;=\;\frac{\mu^{2(2-\omega)}}{(4\pi)^\omega}\int_0^\infty dt\,e^{-t}\,\frac{f(tu)}{u\,t^w}\,,$$
$$M_3=\int_0^\infty \frac{ds}{(4\pi)^\omega}\,e^{-m^2s}\,s^{-1-\omega}\,f(\tau)\;=\;\frac{\mu^{2(2-\omega)}}{(4\pi)^\omega}\int_0^\infty dt\,e^{-t}\,\frac{f(tu)}{u^2\,t^{1+w}}\,,$$
$$M_4=\int_0^\infty \frac{ds}{(4\pi)^\omega}\,e^{-m^2s}\,s^{-\omega}\;=\;\frac{\mu^{2(2-\omega)}}{(4\pi)^\omega}\int_0^\infty dt\,e^{-t}\,\frac{1}{u\,t^w}\,,$$
$$M_5=\int_0^\infty \frac{ds}{(4\pi)^\omega}\,e^{-m^2s}\,s^{-1-\omega}\;=\;\frac{\mu^{2(2-\omega)}}{(4\pi)^\omega}\,m^{2(w-2)}\int_0^\infty dt\,e^{-t}\,\frac{1}{u^2\,t^{1+w}}, \tag{16.67}$$

[1] These terms are elaborated in [142].

where we already made the change of the variable $s = tm^{-2}$ and denoted

$$u = -\frac{m^2}{\Box}\,. \tag{16.68}$$

A relevant observation is that all of the individual features of the given theory (like the scalar in the present case) are encoded into the coefficients (16.66), while the integrals (16.67) are universal in the sense they will be the same for any theory which provides us an operator of the form (16.53). In the new notations (16.66) and (16.67), the second-order part of the one-loop effective action can be cast into the form

$$\bar{\Gamma}^{(1)}_2 = \bar{\Gamma}^{(1)}_{R^2_{\mu\nu}} + \bar{\Gamma}^{(1)}_{R^2} = \frac{1}{2}\int d^4x\sqrt{g}\sum_{k=1}^{5}\Big\{R_{\mu\nu}\,l^*_k M_k\,R^{\mu\nu} + R\,l_k M_k\,R\Big\}. \tag{16.69}$$

Let us now calculate the integrals (16.67). Using the formulas from section 16.1, the derivation of M_4 and M_5 is a simple exercise and we get, using the notation (16.61),

$$M_4 = \Big[\frac{2}{\varepsilon_{\omega,\mu}} - \frac{1}{(4\pi)^2}\Big]\frac{1}{u}, \qquad M_5 = \Big[-\frac{1}{\varepsilon_{\omega,\mu}} + \frac{3}{4(4\pi)^2}\Big]\frac{1}{u^2}. \tag{16.70}$$

In order to calculate the remaining three integrals, we rewrite the notation (16.44) in the coordinate representation,

$$a^2 = \frac{4u}{u+4} = \frac{4\Box}{\Box - 4m^2}, \quad \text{and} \quad \frac{1}{u} = \frac{1}{a^2} - \frac{1}{4}. \tag{16.71}$$

In the momentum representation, we assume, for definiteness, that $a > 0$, so that it changes from $a = 0$ in the infrared to $a = 2$ in the UV. Furthermore, we need the integral (16.43). The remaining calculation of the first three integrals is not too complicated, and we just give the final results in terms of a and Y from (16.44),

$$M_1 = -\frac{2}{\varepsilon_{\omega,\mu}} + \frac{2Y}{(4\pi)^2}, \tag{16.72}$$

$$M_2 = \Big[-\frac{2}{\varepsilon_{\omega,\mu}} + \frac{1}{(4\pi)^2}\Big]\Big(\frac{1}{12} - \frac{1}{a^2}\Big) + \frac{1}{(4\pi)^2}\Big\{\frac{1}{18} - \frac{4Y}{3a^2}\Big\}, \tag{16.73}$$

$$M_3 = \frac{1}{(4\pi)^2}\Big\{\Big[\frac{3}{2} + \frac{1}{2-\omega} + \log\Big(\frac{4\pi\mu^2}{m^2}\Big)\Big]\Big[\frac{1}{2a^4} - \frac{1}{12a^2} + \frac{1}{160}\Big] + \frac{8Y}{15a^4} - \frac{7}{180a^2} + \frac{1}{400}\Big\}. \tag{16.74}$$

Next, we construct a useful combinations for the scalar case,

$$M_{R^2_{\mu\nu}} = l^*_3\,M_3 + l^*_4\,M_4 + l^*_5\,M_5 = M_3 + \frac{1}{6}\,M_4 - M_5 \tag{16.75}$$
$$= \frac{1}{(4\pi)^2}\Big\{\frac{1}{2-\omega}\Big(\frac{1}{60}\Big) + \log\Big(\frac{4\pi\mu^2}{m^2}\Big)\Big(\frac{1}{60}\Big) + \frac{8Y}{15a^4} + \frac{2}{45a^2} + \frac{1}{150}\Big\}$$

and

$$\begin{aligned} M_{R^2} &= l_1\, M_1 + l_2\, M_2 + l_3\, M_3 + l_4\, M_4 + l_5\, M_5 \\ &= \frac{1}{(4\pi)^2}\Big\{\Big[\frac{1}{2-\omega} + \log\Big(\frac{4\pi\mu^2}{m^2}\Big)\Big]\Big(\frac{1}{2}\,\tilde{\xi}^2 - \frac{1}{180}\Big) + Y\tilde{\xi}^2 + \frac{2Y}{3a^2}\tilde{\xi} \\ &= \frac{Y}{144} - \frac{Y}{15a^4} - \frac{Y}{6}\tilde{\xi} - \frac{Y}{18a^2} - \frac{59}{10800} - \frac{1}{180a^2} + \frac{1}{18}\tilde{\xi}\Big\}. \end{aligned} \tag{16.76}$$

Finally, we obtain

$$\bar{\Gamma}^{(1)}_2 = \frac{1}{2}\int d^4x\sqrt{g}\,\Big\{R_{\mu\nu}\, M_{R^2_{\mu\nu}}\, R^{\mu\nu} \,+\, R\, M_{R^2}\, R\Big\}. \tag{16.77}$$

Let us note that, in fact, there is a third term, which is related to the square of the Riemann tensor. However, for any integer N, one can prove, by means of the Bianchi identities and partial integrations, that

$$\int d^4x\sqrt{-g}\,\Big\{R_{\mu\nu\alpha\beta}\Box^N R^{\mu\nu\alpha\beta} - 4R_{\mu\nu}\Box^N R^{\mu\nu} + R\Box^N R\Big\} \,=\, \mathcal{O}(R^3_{\dots}). \tag{16.78}$$

This means that, in the bilinear in curvature approximation (which we use here), one can safely use the reduction formula related to the Gauss-Bonnet term,

$$R_{\mu\nu\alpha\beta}\, f(\Box)\, R^{\mu\nu\alpha\beta} = 4R_{\mu\nu}\, f(\Box)\, R^{\mu\nu} \,-\, R\, f(\Box)\, R. \tag{16.79}$$

As a result, in the curvature-squared approximation, there is no way to see the non-localities associated to the Gauss-Bonnet combination. Hence, one can use, e.g., $R^2_{\mu\nu}$ and R^2 terms, or some other equivalent basis. For various applications, the most useful basis consists from the square of the Weyl tensor instead of the square of the Ricci tensor. The transition can be done by means of the formulas

$$C^2 = E_4 + 2W\,, \;\text{ where } W = R^2_{\mu\nu} - \frac{1}{3}R^2 \quad\Longrightarrow\quad \tilde{M}_{R^2} = M_{R^2} + \frac{1}{3}\, M_{R^2_{\mu\nu}}. \tag{16.80}$$

Thus, we introduce the new version of the form factors,

$$k_W = k_{R^2_{\mu\nu}} = \frac{8Y}{15a^4} + \frac{2}{45a^2} + \frac{1}{150}, \tag{16.81}$$

$$k_R = Y\tilde{\xi}^2 + \Big(\frac{2Y}{3a^2} + \frac{1}{18} - \frac{Y}{6}\Big)\,\tilde{\xi} - \frac{Y}{18a^2} + \frac{Y}{144} + \frac{Y}{9a^4} - \frac{7}{2160} - \frac{1}{108a^2}. \tag{16.82}$$

Taking the zero-, first- and second-order terms together, one can write down the effective action up to the second order in curvatures:

$$\begin{aligned} \bar{\Gamma}^{(1)}_{scalar} = \frac{1}{2(4\pi)^2}\int d^4x\sqrt{g}\,\Big\{&\frac{m^4}{2}\Big[\frac{1}{2-\omega} + \log\Big(\frac{4\pi\mu^2}{m^2}\Big) + \frac{3}{2}\Big] \\ &+ \Big(\xi - \frac{1}{6}\Big)m^2R\Big[\frac{1}{2-\omega} + \log\Big(\frac{4\pi\mu^2}{m^2}\Big) + 1\Big] \\ &+ \frac{1}{2}\, C_{\mu\nu\alpha\beta}\Big[\frac{1}{60(2-\omega)} + \frac{1}{60}\log\Big(\frac{4\pi\mu^2}{m^2}\Big) \,+\, k_W\Big]C^{\mu\nu\alpha\beta} \\ &+ R\,\Big[\frac{1}{2}\Big(\xi - \frac{1}{6}\Big)^2\Big(\frac{1}{2-\omega} + \log\Big(\frac{4\pi\mu^2}{m^2}\Big)\Big) + k_R\Big]R\Big\}. \end{aligned} \tag{16.83}$$

Let us make some observations concerning the final result for the vacuum effective action for the scalar field (16.83), with the form factors (16.81) and (16.82). First of all, this action is essentially nonlocal in the higher-derivative sector. The expression is up to the second order in the curvature tensor but, at the same time, is exact in the derivatives of the curvature tensor.

On the other hand, the quantum corrections to the cosmological constant and the $\mathcal{O}(R)$-term do not have nonlocal parts. Indeed, these nonlocal parts would be extremely useful for applications in cosmology, since they would give the running cosmological constant in the infrared [288], but their derivation is beyond our present-day abilities to make calculations in curved spacetime. The non-localities in the cosmological constant and Einstein-Hilbert sectors cannot be obtained in the framework of expansion in the series in curvatures, because such an expansion is ultimately related to the expansion into the metric perturbations on the flat background [167]. This does not mean that there cannot be nonlocal corrections in the low-energy sectors; however, these hypothetic corrections cannot be obtained by the method described here.

In principle, instead of the form factor acting on the cosmological constant, one could have some in the nonlocal terms of the form [167]

$$R_{\mu\nu\lambda\beta}\frac{1}{(\Box+m^2)^2}R^{\mu\nu\lambda\beta}\,,\quad R_{\mu\nu}\frac{1}{(\Box+m^2)^2}R^{\mu\nu}\,,\quad R\frac{1}{(\Box+m^2)^2}R, \tag{16.84}$$

with a vanishing or very small mass parameter m. Similar replacements of the Einstein-Hilbert term may have the forms

$$R_{\mu\nu\lambda\beta}\frac{1}{\Box+m^2}R^{\mu\nu\lambda\beta}\,,\quad R_{\mu\nu}\frac{1}{\Box+m^2}R^{\mu\nu}\,,\quad R\frac{1}{\Box+m^2}R\,. \tag{16.85}$$

For the vanishing masses, the two types of expressions, (16.84) and (16.85), have a global scaling rule under (12.40) that is identical to that for the cosmological constant and Einstein-Hilbert terms. Since the cosmological evolution represents a local time-dependent conformal transformation of the metric, the difference between (16.84) and ρ_Λ is only due to the time derivatives of the conformal factor of the metric. This feature explains the successful use of (16.84) as an equivalent of the slowly varying cosmological constant (see, e.g., [46]).

As we know from Exercise 12.9, the terms (16.84) and (16.85) are not generated by spontaneous symmetry breaking in curved space. In the present section, we have seen that these terms (with or without additional logarithmic insertions in the UV) are also not generated by loop corrections. Thus, the replacement of the cosmological constant by (16.84) is not easily derived from these two curved-space approaches. Let us also note that the cosmological constant is required for the renormalizability of semiclassical gravity with massive quantum fields. Thus, it is somehow natural that the replacement is not operational.

16.3 The massless limit and leading logs vs. the infrared limit

In the massless limit, in Euclidean space, $p^2/m^2 \gg 1$. Then the form factors k_W and k_R become simpler. According to (16.71), in this limit, we have $a \to 2$. Expansion into series shows that the asymptotic behavior in the UV corresponds to Eq. (16.47). When $m^2/p^2 \to 0$, the expression Y is logarithmically divergent and this defines the UV asymptotic of the form factors. For instance, in the case of k_W, we have

$$k_W = -\frac{1}{60}\log\Big(-\frac{\Box}{m^2}\Big) + \text{constant and vanishing terms.} \tag{16.86}$$

Exactly as we discussed above, the coefficient $1/60$ here is the same as that of the pole $(2-\omega)^{-1}$ in the Weyl-squared term in (16.83) and, of course, exactly the same as the one for $\log\mu^2$. It is sufficient to know the divergent term in the effective action to arrive at the logarithmic μ-dependence and also at the leading UV logarithm of the form factor. In other words, the logarithmic UV divergence controls the renormalization group, covered by the μ-dependence, and also agrees with the physical behavior of the theory in the UV, which means the logarithmic dependence on the momenta p in the regime when $(p/m) \to \infty$.

The UV limit of the part of the one-loop effective action that is quadratic in curvatures in the extreme UV is the sum of the classical terms (12.6) and the logarithmic corrections,

$$\begin{aligned}\Gamma^{(1),\,UV}_{vac} = S_{EH} + \int d^4x\sqrt{-g}\,\Big\{ C_{\alpha\beta\rho\sigma}\Big[a_1 - \beta_1\log\Big(-\frac{\Box}{\mu^2}\Big)\Big]C^{\alpha\beta\rho\sigma} \\ + R\Big[a_4 - \beta_4\log\Big(-\frac{\Box}{\mu^2}\Big)\Big]R\Big\} + \dots .\end{aligned} \tag{16.87}$$

One can also derive logarithmic form factors in terms of the third order in curvatures [33], including those corresponding to the Gauss-Bonnet type of divergences. The nonlocal form factors were also derived for the Einstein-Hilbert term [19,142,143], but they represent nonlocal surface structures.

At another extreme of the energy scale, in the infrared limit, we assume that $p^2 \ll m^2$. Then the asymptotic behavior of Y and k_W is of the power-like form, namely,

$$Y = -\frac{1}{12}\frac{p^2}{m^2}\Big(1 - \frac{1}{10}\frac{p^2}{m^2}\Big) + \dots, \tag{16.88}$$

$$k_W = -\frac{1}{840}\frac{p^2}{m^2}\Big(1 + \frac{1}{18}\frac{p^2}{m^2}\Big) + \dots . \tag{16.89}$$

One can see that there is no logarithmic “running” in the infrared and hence there is no direct relation between the dependence on momenta and μ in the infrared region. This phenomena is called decoupling, and it can be also seen in the beta functions [167].

The last observation is that the same calculations can be done for massive spinors and vectors, and also for other external fields, such as electromagnetic[2] [165] or scalar [82] fields. The calculations in the massless theories are relatively simpler, since they reduce to the derivation of $\mathrm{sTr}\,\hat{a}_2$ and consequent writing of the expressions in a form that is similar to that of Eq. (16.87). In this case, there is a UV-infrared duality in the form factors, since the logarithmic form factor has similar divergences in the two limits.

[2]The electromagnetic example is especially interesting because it has a nonlocal multiplicative anomaly.

17
The conformal anomaly and anomaly-induced action

Generally speaking, there is no way to calculate the vacuum effective action completely. For instance, in Schwinger-DeWitt expansion, the effective action is given by sum of an infinite series in the curvature tensor and its derivatives. However, the practical calculations are restricted by the first few terms of this expansion.

On the other hand, there is a special class of four-dimensional theories, namely, the free massless fields with local conformal symmetry, for which one can obtain exactly the nonlocal terms in $\bar{\Gamma}^{(1)}$. The expression "exactly" means we can solve the problem at the one-loop level (indeed, the generalization to higher loops is extensively discussed in the literature) and the result is exact only on a special type of background, while, in many other cases, it is a useful and controllable approximation. Thus, the word "exact" is correct, as a way to emphasize a very special status of massless conformal theories.

In what follows, we will learn how to derive the conformal anomaly and obtain the anomaly-induced effective action.

The trace (conformal) anomaly was introduced in [117] (see also [102, 118]) and became a rich and prosperous area of quantum field theory in curved space. In this chapter, we do not pretend to give a comprehensive review of this area of research, only an introduction to the basic notions.

The most interesting applications of conformal anomaly are related to Hawking radiation [176] (which can be derived from anomaly [91]) and the modified Starobinsky model [305, 306] (see also the earlier work [135]). Due to the constraints on length, we did not include applications of the anomaly-induced action to the black hole physics and cosmology. The reader can consult the review paper [287] and original works, e.g., [22] and [123, 230] for further references.

17.1 Conformal transformations and invariants

The derivation and integration of the conformal anomaly become easy exercises if we learn a few formulas concerning the local conformal transformation (11.25). In the first chapter, we derived the transformations (11.29), (11.30), (11.31) and (11.32) for Riemann curvature, its contractions and the Weyl tensor. An important rule for the square of the Weyl tensor in n spacetime dimensions is

$$\sqrt{-g}\, C^2(n) \;=\; \sqrt{-\bar{g}}\, e^{(n-4)\sigma}\, \bar{C}^2(n). \tag{17.1}$$

Introduction to Quantum Field Theory with Applications to Quantum Gravity. Iosif L. Buchbinder and Ilya L. Shapiro, Oxford University Press (2021). © Iosif L. Buchbinder and Ilya Shapiro.
DOI: 10·1093/oso/9780198838319·003·0017

Another two transformations of our interest look less elegant, so we write them only in $n = 4$. The conformal transformation of the Gauss-Bonnet integrand is

$$\sqrt{-g}\,E_4 = \sqrt{-\bar{g}}\,\Big[\,\bar{E}_4 + 8\bar{R}^{\mu\nu}(\bar{\nabla}_\mu\sigma\bar{\nabla}_\nu\sigma - \bar{\nabla}_\mu\bar{\nabla}_\nu\sigma) + 8(\bar{\Box}\sigma)(\bar{\nabla}\sigma)^2 - 8(\bar{\nabla}_\mu\bar{\nabla}_\nu\sigma)^2 + 8(\bar{\nabla}^2\sigma)^2 + 16(\bar{\nabla}_\mu\sigma\bar{\nabla}_\nu\sigma)(\bar{\nabla}^\mu\bar{\nabla}^\nu\sigma) - 4R(\bar{\Box}\sigma)\Big]. \tag{17.2}$$

The next case is of the surface term, which has the following transformation rule:

$$\sqrt{-g}\,\Box R = \sqrt{-\bar{g}}\Big[\bar{\Box}\bar{R} - 2\bar{R}\bar{\Box}\sigma - 2(\bar{\nabla}^\mu\sigma)(\bar{\nabla}_\mu\bar{R}) - 6(\bar{\nabla}^4\sigma) - 6\bar{\Box}(\bar{\nabla}\sigma)^2 + 12(\bar{\nabla}^\mu\sigma)(\bar{\nabla}_\mu\bar{\Box}\sigma) + 12(\bar{\Box}\sigma)(\bar{\nabla}\sigma)^2 + 12(\bar{\nabla}^\mu\sigma)\bar{\nabla}_\mu(\bar{\nabla}\sigma)^2 + 12(\bar{\Box}\sigma)^2\Big]. \tag{17.3}$$

The full version in an arbitrary dimension can be found in Ref. [87].

It is remarkable that, in $n = 4$, the simple relation

$$\sqrt{-g}\,\Big(E_4 - \frac{2}{3}\Box R\Big) \;=\; \sqrt{-\bar{g}}\,\Big(\bar{E}_4 - \frac{2}{3}\bar{\Box}\bar{R} + 4\bar{\Delta}_4\sigma\Big) \tag{17.4}$$

takes place, where Δ_4 is the fourth-order Hermitian conformal invariant operator [139, 239] acting on a conformal invariant scalar field

$$\Delta_4 = \Box^2 + 2\,R^{\mu\nu}\nabla_\mu\nabla_\nu - \frac{2}{3}\,R\Box + \frac{1}{3}\,(\nabla^\mu R)\nabla_\mu. \tag{17.5}$$

The following observation is in order. Eq. (17.4) is obtained by a direct calculation, and looks completely non-trivial. There is no general proof that a total derivative supplement of the topological term (such as $-\frac{2}{3}\Box R$ in our case) that cancels most of the terms in the transformation (17.2) should exist. What we know is that such a supplement really exists in dimensions 2, 4 and 6 [128, 129].

In chapter 12, we learned three examples of conformal invariant scalar, fermion and vector fields. The operator (17.5) enables one to construct one more invariant action,

$$S(\bar{g}_{\mu\nu},\bar{\varphi}) = \int d^4x\sqrt{-g}\,\varphi\Delta_4\varphi \;=\; S(g_{\mu\nu},\varphi), \qquad \bar{\varphi} = \varphi. \tag{17.6}$$

The Hermiticity means that the action does not change under integrations by parts,

$$\int d^4x\sqrt{-g}\,\chi\Delta_4\varphi = \int d^4x\sqrt{-g}\,\varphi\Delta_4\chi. \tag{17.7}$$

Unlike the ordinary scalar action (12.11), in the action (17.6), the scalar field does not transform, while, in the usual scalar case, the transformation rule is (12.12). Indeed, both actions can be generalized to an arbitrary dimension n, with a corresponding change of the transformation rules [26].

Exercises

17.1. Verify relations (17.6) and (17.7).

17.2. Derive the analog of relation (17.4) in two-dimensional space, where $\Delta_2 = \Box$. Formulate the corresponding analogs of (17.6) and (17.7).

17.2 Derivation of the conformal anomaly

A quantum anomaly is a typical phenomenon in a situation, where the original theory has more than one symmetry. Before starting the formal consideration, let us note that, in the previous section, we already met the situation when the conformal symmetry is broken by quantum corrections.

17.2.1 Example of a derivation using a nonlocal form factor

In order to see this, consider the massless limit in the local form factor for the square of the Weyl term. The general expression (16.87) includes the term

$$\bar{\Gamma}^{(1)}_W = - \int d^4x\sqrt{-g}\,\beta_1\, C_{\alpha\beta\rho\sigma} \log\Big(- \frac{\Box}{\mu^2}\Big) C^{\alpha\beta\rho\sigma}\,, \tag{17.8}$$

where β_1 is the minimal subtraction beta function for the C^2-term. Under the conformal transformation (11.25), we have

$$g_{\mu\nu} = \bar{g}_{\mu\nu}\, e^{2\sigma(x)} \quad \text{and} \quad \sqrt{-g}\, C_{\alpha\beta\rho\sigma}C^{\alpha\beta\rho\sigma} = \sqrt{-\bar{g}}\, \bar{C}_{\alpha\beta\rho\sigma}\bar{C}^{\alpha\beta\rho\sigma}. \tag{17.9}$$

At the same time, the d'Alembertian operator acting on the tensor $C^{\alpha\beta\rho\sigma}$ transforms in a complicated way, making expression (17.8) non-invariant. The leading part of the transformation is

$$\Box = e^{-2\sigma}\big(\bar{\Box} + \text{ derivatives of } \sigma\big). \tag{17.10}$$

At this point, we need the useful relation

$$-\frac{2}{\sqrt{-g}}\, g_{\mu\nu} \frac{\delta\, A[g_{\mu\nu}]}{\delta\, g_{\mu\nu}} = - \frac{1}{\sqrt{-\bar{g}}}\, e^{-4\sigma} \frac{\delta\, A[\bar{g}_{\mu\nu}\, e^{2\sigma}]}{\delta\sigma}\bigg|_{\bar{g}_{\mu\nu}\to g_{\mu\nu},\sigma\to 0}. \tag{17.11}$$

This formula is valid for any functional $A[g]$. Since it will play an important role in what follows, let us prove it. According to (17.9), we take $A[g_{\mu\nu}] = A[\bar{g}_{\mu\nu}\, e^{2\sigma}]$; then,

$$\frac{\delta\, A}{\delta\sigma} = \frac{\delta g_{\mu\nu}}{\delta\sigma}\, \frac{\delta\, A}{\delta g_{\mu\nu}} = 2e^{2\sigma}\bar{g}_{\mu\nu}\, \frac{\delta\, A}{\delta g_{\mu\nu}} = 2g_{\mu\nu}\, \frac{\delta\, A}{\delta g_{\mu\nu}}, \tag{17.12}$$

which is equivalent to (17.11). Two observations are in order. First, the factor $e^{-4\sigma}$ is inserted in Eq. (17.11), but it is certainly irrelevant due to the limit $\bar{g}_{\mu\nu} \to g_{\mu\nu}$ and $\sigma \to 0$. In the rest of this section, we shall denote this limit by a vertical bar. Second, the trace of the variational derivative in the *l.h.s.* of (17.11) is nothing but the Noether identity for the functional $A[g]$. In particular, this identity is satisfied for the integral of the square of the Weyl tensor,

$$-\frac{2}{\sqrt{-g}}\, g_{\mu\nu} \frac{\delta}{\delta\, g_{\mu\nu}} \int d^4x\sqrt{-g}\, C^2_{\alpha\beta\rho\sigma} = 0. \tag{17.13}$$

On the other hand, the nonlocal form factor in (17.8) makes a dramatic change. Using (17.11) and neglecting the sub-leading terms in (17.10), we obtain

$$-\frac{2}{\sqrt{-g}}\, g_{\mu\nu} \frac{\delta\, \bar{\Gamma}^{(1)}_W}{\delta\, g_{\mu\nu}} = \beta_1\, C^2_{\alpha\beta\rho\sigma}. \tag{17.14}$$

This result shows several important things at once.

i) The conformal symmetry is violated by quantum corrections, namely the logarithmic form factor that emerges in the C^2-term at the one-loop order. In what follows, we shall see that the anomaly has other terms as well. It is important to note that, in the current consideration, the anomaly comes from the *finite* form factor.

ii) As we know, the leading logarithm is directly related to the UV divergences, and hence the anomaly may be expected to depend on divergences and regularizations.

iii) The universality of the leading logarithmic divergences (e.g., independence of the type of regularization) makes the expression for the anomaly (17.14) universal.

iv) At higher loops, there should be finite logarithmic terms of higher orders, e.g., with $\big[\log\big(-\frac{\Box}{\mu^2}\big)\big]^2$, and similar. In principle, one cannot expect that the violation of conformal symmetry at the one-loop level and beyond should have exactly the same structure. At the same time, if we pay attention only to the first logarithmic correction (which means we prefer to keep the relation with the minimal subtraction renormalization group), the structure of quantum corrections will be the same (17.8), even non-perturbatively. Therefore, the same will be true for the anomaly (17.14). The difference is that the one-loop result includes the corresponding approximation to the beta function β_1, being the sum of the contributions of free conformal fields of different spins. At higher orders, the β_1 becomes a sum (series at the non-perturbative level) of the terms with growing powers of coupling constants.

17.2.2 Derivation using dimensional regularization

There are different ways to calculate the conformal anomaly. As we saw in the previous section, the anomaly comes from the finite part of effective action, but, at the same time, this finite part is a direct mapping of the leading log UV divergences. The most traditional and standard approach is to assume that the origin of quantum anomalies (and the trace anomaly is an important example here) is the renormalization procedure. The quantum anomaly shows up when there is no regularization capable of preserving all the symmetries simultaneously. After subtracting the divergences, in the finite part of the effective action, some of the symmetries may be broken. In our case, the theory has general covariance and local conformal symmetry, and the last is broken by quantum corrections. Let us explain the same thing with formulas.

Consider the quantum theory of free, massless, conformal invariant matter fields denoted by Φ, on a classical gravitational background. As before, we assume that the set Φ includes N_s conformal scalars φ (all with $\xi = \frac{1}{6}$), N_f fermions ψ and N_v gauge vector fields A_μ. We denote k_Φ the conformal weights $k_{\varphi,\psi,A} = \big(-1, -\frac{3}{2}, 0\big)$.

At the one-loop level, it is sufficient to consider the simplified vacuum action (14.7),

$$S_{vac} = \int d^4x\sqrt{-g}\,\big\{a_1C^2 + a_2E_4 + a_3\Box R\big\}\,.$$

Let us emphasize that it is not wrong to supplement the last expression with the Einstein-Hilbert action, the cosmological constant, or the $\int_x R^2$-term. One can regard the action (14.7) as a special part of a classical action that is the subject of an infinite renormalization at the one-loop level.

The Noether identity for the local conformal symmetry has the form (14.8) and can be interpreted as $T^\mu_{\ \mu} = 0$, where $T^{\mu\nu}$ is the "energy-momentum tensor" of vacuum,

$$T^{\mu\nu} = -\frac{2}{\sqrt{-g}}\,\frac{\delta\, S_{vac}}{\delta\, g_{\mu\nu}}. \tag{17.15}$$

At the quantum level, $S_{vac}(g_{\mu\nu})$ has to be replaced by the effective action of vacuum $\Gamma_{vac}(g_{\mu\nu})$. As we already know, its divergent part is

$$\Gamma_{div} = \frac{\mu^{n-4}}{n-4}\int d^n x\sqrt{g}\,\{\beta_1 C^2 + \beta_2 E_4 + \beta_3 \Box R\}\,, \tag{17.16}$$

where n is the parameter of dimensional regularization. In the case of global conformal symmetry we already know from section 15.3 that (see also [177, 70, 80])

$$\langle T^{\mu}_{\ \mu}\rangle = \beta_1 C^2 + \beta_2 E_4 + a'\Box R, \tag{17.17}$$

where $a' = \beta_3$. In the case of local conformal invariance, there is an ambiguity in the parameter a' [56, 119, 14], as will be explained below.

One can derive the conformal anomaly in different ways, based on the different choices of regularization schemes, e.g., dimensional, adiabatic, point splitting, covariant cut-off in the proper-time integral [118, 90, 56, 14]. We shall be using dimensional regularization, in general following [118], in the slightly modified version [14,287]. Since our main interest is related to the vacuum effects at the one-loop level, it is possible to restrict consideration by the free fields case. The expression for divergences is (17.16), with the beta functions defined in (14.65). If some other fields (scalar or fermion with higher derivatives, conformal quantum gravity, etc.) are added, Eq. (14.65) gains additional terms.

The renormalized one-loop effective action has the form

$$\Gamma_R = S + \bar{\Gamma} + \Delta S\,, \tag{17.18}$$

where $\bar{\Gamma} = \bar{\Gamma}_{div} + \bar{\Gamma}_{fin}$ is the non-renormalized quantum correction to the classical action and ΔS is an infinite local counterterm. Indeed, ΔS is the unique source of the non-invariance of the effective action, since the naive contribution of the quantum matter fields $\bar{\Gamma}$ is conformal (see the discussion in section 14.1). Thus, the anomalous trace is

$$T = \langle T^{\mu}_{\ \mu}\rangle = -\frac{2}{\sqrt{-g}}\,g_{\mu\nu}\,\frac{\delta\,\Gamma_R}{\delta\, g_{\mu\nu}}\bigg|_{n=4} = -\frac{2}{\sqrt{-g}}\,g_{\mu\nu}\,\frac{\delta\,\Delta S}{\delta\, g_{\mu\nu}}\bigg|_{n=4}. \tag{17.19}$$

In order to derive this expression, it is useful to change the parametrization of the metric according to (11.25) and use the identity (17.11).

The transformation rules for the first two terms in Eq. (17.16) are given by (17.1) and (17.4). All the expressions of our interest, (17.16), have the same factor $e^{(n-4)\sigma}$ and, on the top of that, some extra terms with derivatives of $\sigma(x)$. For all terms which are not total derivatives, these terms are irrelevant, owing to the procedure (17.11).

It is easy to note that the *l.h.s.* in (17.11) gives zero when applied to the integral of the total derivative term $\int_x \Box R$. On the other hand, the value of a' in (17.17) can be modified by adding the finite term

$$S_3 = \alpha \int d^4x\sqrt{-g}\, R^2 \tag{17.20}$$

to the classical action (14.7), due to the identity

$$-\frac{2}{\sqrt{-g}}\, g_{\mu\nu}\frac{\delta}{\delta g_{\mu\nu}} \int d^4x\sqrt{-g}\, R^2 \;=\; 12\,\Box R\,. \tag{17.21}$$

The last formula can be derived either directly or through Eq. (17.11); we leave these calculations as an exercise. The same effect can be achieved by the term $\int_x R^2_{\mu\nu}$,

$$-\frac{2}{\sqrt{-g}}\, g_{\mu\nu}\frac{\delta}{\delta g_{\mu\nu}} \int d^4x\sqrt{-g}\, R_{\mu\nu}\, R^{\mu\nu} \;=\; 4\,\Box R \tag{17.22}$$

and also by the term $\int_x R^2_{\mu\nu\alpha\beta}$. For the sake of simplicity, we take only $\int_x R^2$.

Consider the term with the square of the Weyl tensor, using Eq. (17.11). First we assume that the counterterms is constructed with $C^{\alpha}{}_{\beta\rho\sigma}(n)$. Then

$$\frac{\delta}{\delta\sigma}\int \frac{d^4x\sqrt{-\bar{g}}}{n-4}\, e^{(n-4)\sigma}\, \bar{C}^2(n)\Big|_{n\to 4} \;=\; \sqrt{-g}\, C^2\,. \tag{17.23}$$

One may think that adding the classical non-conformal term (17.20) has nothing to do with the quantum corrections. However, consider in more detail how to apply the procedure (17.19) to the $\int_x C^2$-type counterterm. The point is that the Weyl tensor (11.24) depends on the dimension d; in particular, its square is

$$C^2(d) \;=\; C_{\alpha\beta\mu\nu}(d)C^{\alpha\beta\mu\nu}(d) \;=\; R^2_{\mu\nu\alpha\beta} - \frac{4}{d-2}R^2_{\mu\nu} + \frac{2}{(d-1)(d-2)}R^2. \tag{17.24}$$

Taking $d = n$, we arrive at (17.23). However, if we define the corresponding countert-erm with an arbitrary $d = n + \gamma(n-4)$,

$$\Delta S_C \;=\; \frac{\beta_1}{n-4}\int d^nx\sqrt{-g}C^2(d)\,, \tag{17.25}$$

the result is different. Indeed, for any such d, the counterterm is local, and cancels the divergent part of effective action. Thus, the renormalization is multiplicative on the basis of $R^2_{\mu\nu\alpha\beta}$, $R^2_{\mu\nu}$ and R^2. However, the anomalous $\Box R$ term depends on the choice of γ. For the particular choice $\gamma = -1$, we arrive at (see [118])

$$T(C^2) \;=\; \frac{2}{\sqrt{-g}}g_{\mu\nu}\frac{\delta}{\delta g_{\mu\nu}}\frac{\mu^{n-4}}{n-4}\int d^nx\sqrt{-g}\,\beta_1 C^2(4)\Big|_{n\to 4} = C^2 - \frac{2}{3}\Box R, \tag{17.26}$$

which is different from (17.23). For other values of γ, there is another coefficient of $\Box R$ in the anomaly. Hence, at this point, we meet an arbitrariness.

From the previous considerations, one can see that the difference between the counterterms (17.25) with different d is equivalent to the *finite* $\int_x R^2$ term. Qualitatively, there is a similar ambiguity in covariant Pauli-Villars regularization [14]. Finally, the anomaly is given by (17.17), but there is an ambiguity in the coefficient a'. In regularization schemes that correspond to a global conformal transformation, there is always $a' = \beta_3$ but, in general, fixing this coefficient requires a special renormalization condition [14] on the *finite* $\int_x R^2$ term.

Finally, the derivation of the term with E_4 in the anomaly (17.17) is not complicated, and we leave it as an important exercise for the reader.

Exercises

17.3. Verify formulas (17.21) and (17.22) in two different ways, namely, i) directly, using the variation of the metric, and ii) using Eq. (17.11) and a variation of σ.
17.4. After verifying relation (17.22), derive a similar formula for the *conformal shift* of the term $\int_x R^2_{\mu\nu\alpha\beta}$. Try to accomplish these two tasks *without* special calculations, just using Eq. (17.1) and the property of the Gauss-Bonnet invariant (11.96).
17.5. Verify result (17.26) and derive a similar expression for an arbitrary γ. Find the γ-dependent coefficient of the finite $\int_x R^2$ term that reduces the overall expression for the anomaly either to (17.26) or to (17.23). Repeat the same program for the terms $\int_x R^2_{\mu\nu}$ and $\int_x R^2_{\mu\nu\alpha\beta}$, and for an arbitrary fixed linear combination of these terms. Which of the coefficients provide a degeneracy?
17.6. Use relation (17.11) and the identity (17.4) to obtain the Gauss-Bonnet part of the anomaly (17.17).

17.3 Anomaly-induced action

Renormalized effective action (17.18) is finite, nonlocal and, in general, cannot be derived completely. However, one can use the conformal anomaly to calculate the most essential part of the one-loop contribution to the vacuum effective action, which is called the *anomaly-induced action* Γ_{ind},

$$\frac{2}{\sqrt{-g}}\, g_{\mu\nu}\frac{\delta\,\Gamma_{ind}}{\delta g_{\mu\nu}} = \omega C^2 + bE_4 + c\Box R\,. \tag{17.27}$$

The solution of this equation in $n = 4$ is straightforward; it was found for the first time in the papers by Riegert [262] and Fradkin and Tseytlin [138][1]. There are many generalizations, e.g., for the theory with background electromagnetic field, torsion [73] or a scalar field [286, 247]). In what follows, we will obtain three different (but equivalent) forms that can be useful in different situations.

The simplest possibility is to parameterize the metric as in (17.1), separating the conformal factor $\sigma(x)$ and rewriting Eq. (17.27) using (17.11). Then the solution for the effective action can be easily obtained in the form

$$\begin{aligned}\bar{\Gamma} = S_c[\bar{g}_{\mu\nu}] &+ \int d^4x\sqrt{-\bar{g}}\,\Big\{\omega\sigma\bar{C}^2 + b\sigma\Big(\bar{E}_4 - \frac{2}{3}\bar{\Box}\bar{R}\Big) + 2b\sigma\bar{\Delta}_4\sigma\Big\}\\ &- \frac{1}{12}\Big(c+\frac{2}{3}b\Big)\int d^4x\sqrt{-g}\,R^2,\end{aligned} \tag{17.28}$$

where $S_c[\bar{g}_{\mu\nu}] = S_c[g_{\mu\nu}]$ is an arbitrary conformal invariant functional of the metric, which serves as an integration constant to Eq. (17.27). The simplicity of deriving

[1] The most famous is the similar solution in the two-dimensional theory by Polyakov [255].

solution (17.28) does not mean that this result should be underestimated. The main fact about this solution is that all of the ambiguity and unknown parts of it are hidden in S_c. This means that the difference between (17.28) and any other solution of Eq. (17.27) reduces to a change in the conformal action S_c. On the other hand, the rest of the solution holds all the information about the UV limit of the theory, e.g., about the leading logarithms in the UV asymptotic. Since the theory under consideration is massless and conformal, it does not have a reference scale, and hence any energy scale can be regarded as UV.[2] Thus, one can expect that (17.28) should serve as a reliable base for some of the applications of field theory in curved spacetime.

One of the merits of the solution (17.28) is its simplicity. At the same time this solution is not covariant since it is not expressed in terms of the original metric $g_{\mu\nu}$. On the other hand, it is not difficult to to obtain the covariant solution of (17.27).

The starting point in deriving the covariant solution are Eqs. (17.4), (17.1) in $n = 4$, and the conformal invariance of the operator (17.5),

$$\sqrt{-g}\,C^2 = \sqrt{-\bar{g}}\,\bar{C}^2\,, \qquad \sqrt{-\bar{g}}\,\bar{\Delta}_4 = \sqrt{-g}\,\Delta_4\,. \tag{17.29}$$

It proves useful to introduce the Green function for this operator

$$\sqrt{-g(x)}\,\Delta_4^x\,G(x,y) \;=\; \delta(x,y). \tag{17.30}$$

By construction, from (17.29) follows that this Green function is conformal invariant, $G = \bar{G}$. Using these formulas and (17.11) one meets, for any metric-dependent conformal functional, $A(g_{\mu\nu}) = A(\bar{g}_{\mu\nu})$, the relation

$$\frac{\delta}{\delta\sigma}\int d^4x\sqrt{-g}A\big(E_4 - \tfrac{2}{3}\Box R\big)\Big|_{g_{\mu\nu}\to\bar{g}_{\mu\nu}} \;=\; 4\sqrt{-\bar{g}}\bar{\Delta}_4\,A \;=\; 4\sqrt{-g}\Delta_4 A(g_{\mu\nu}). \tag{17.31}$$

Now it is easy to find the action responsible for the $T_\omega = -\omega C^2$ term in the anomaly,

$$\Gamma_\omega = \frac{\omega}{4}\int_x\int_y C^2(x)\,G(x,y)\,\big(E_4 - \tfrac{2}{3}\Box R\big)_y, \tag{17.32}$$

where we used a compact notation (12.4) for the integral.

To verify (17.32) we note that, in this expression, only the term $E_4 - \frac{2}{3}\Box R$ depends on σ via the linear relation (17.4). Thus, the variational derivative with respect to σ depends only on $\bar{g}_{\mu\nu}$. Applying rule (17.11) gives the correct result.

Similarly, one can find the term which produces $T_b = b\,(E_4 - \frac{2}{3}\Box R)$, in the form

$$\Gamma_b = \frac{b}{8}\int_x\int_y \big(E_4 - \tfrac{2}{3}\Box R\big)_x\,G(x,y)\,\big(E_4 - \tfrac{2}{3}\Box R\big)_y\,. \tag{17.33}$$

The last constituent of the induced action follows from Eq. (17.21),

[2]This statement is correct in general, but, in practice there may be complications, because the metric background may have its own reference scale. For instance, this happens in the infrared limit for the conformal field theory in de Sitter space, where one can observe a kind of infrared decoupling, even though there are only massless quantum fields [98, 144].

$$\Gamma_c = -\frac{3c+2b}{36}\int_x R^2(x). \tag{17.34}$$

The covariant solution of Eq. (17.27) is the sum of (17.32), (17.33) and (17.34),

$$\begin{aligned}\Gamma_{ind} &= S_c(g_{\mu\nu}) + \frac{\omega}{4}\int_x\int_y C^2(x)\,G(x,y)\,\Big(E_4 - \frac{2}{3}\Box R\Big)_y \\ &+ \frac{b}{8}\int_x\int_y \Big(E_4 - \frac{2}{3}\Box R\Big)_x G(x,y)\,\Big(E_4 - \frac{2}{3}\Box R\Big)_y - \frac{3c+2b}{36}\int_x R^2(x).\end{aligned} \tag{17.35}$$

This form of induced action is covariant and depends only on the original metric $g_{\mu\nu}$. As expected, it is nonlocal, since there are Green functions of the operator (17.5) in two places. This fact is remarkable in at least two aspects. First, the non-localities are not related to Green functions of massless conformal scalars, spinors and vectors, which contributed to the beta functions ω, b and c, respectively, in the anomaly. Instead, there are only Green functions of the artificial fourth-derivative operator (17.5), which does not appear in the classical actions of matter fields. Second, the two Green functions in (17.35) are for the same operator (17.5), but this does not mean that they are identical. Remember that the Green functions depend not only on the operator, but also on the boundary conditions. And these conditions can be perfectly well chosen to be different for the two Green functions of our interest. In the section devoted to applications, we shall see that this is a significant feature of the solution (17.35).

One can reformulate the anomaly-induced action in the local covariant form. As a first step, we rewrite the nonlocal part of (17.5) in a symmetric "Gaussian" form,

$$\begin{aligned}\Gamma_{\omega,b} &= \int_x\int_y \big(E_4 - \tfrac{2}{3}\Box R\big)_x G(x,y)\,\big[\tfrac{\omega}{4}C^2 - \tfrac{b}{8}\big(E_4 - \tfrac{2}{3}\Box R\big)\big]_y \\ &= -\frac{b}{8}\int_x\int_y \Big[\Big(E_4 - \frac{2}{3}\Box R\Big) - \frac{\omega}{b}C^2\Big]_x G(x,y)\,\Big[\Big(E_4 - \frac{2}{3}\Box R\Big) - \frac{\omega}{b}C^2\Big]_y \\ &\quad - \frac{1}{2}\int_x\int_y \Big(\frac{\omega}{2\sqrt{-b}}\,C^2\Big)_x G(x,y)\,\Big(\frac{\omega}{2\sqrt{-b}}\,C^2\Big)_y .\end{aligned} \tag{17.36}$$

This expression enables one to introduce auxiliary scalar fields and arrive at the local covariant expression,

$$\begin{aligned}\bar{\Gamma} &= S_c[g_{\mu\nu}] + \int d^4x\sqrt{-g(x)}\,\Big\{\frac{1}{2}\varphi\Delta_4\varphi - \frac{1}{2}\,\psi\Delta_4\psi + l_1\,\psi\,C^2 \\ &+ \varphi\,\Big[k_1\,C^2 + k_2\,\Big(E_4 - \frac{2}{3}\,\Box R\Big)\Big]\Big\} - \frac{3c+2b}{36}\int d^4x\sqrt{-g(x)}\,R^2(x).\end{aligned} \tag{17.37}$$

Here φ and ψ are auxiliary scalars, which have kinetic terms defined by the fourth-derivative operators (17.5), while the parameters are defined as

$$k_1 = -l_1 = -\frac{\omega}{2\sqrt{-b}} \qquad \text{and} \qquad k_2 = \frac{\sqrt{-b}}{2}. \tag{17.38}$$

With these definitions, (17.37) is a covariant local solution of (17.27). As was the case with the other two equivalent solutions, (17.28) and (17.35), any other solution can differ from (17.37) only by the choice of conformal functional S_c.

Let us make a few relevant comments.

1. The local covariant form (17.37) is dynamically equivalent to the nonlocal one (17.36), or (17.35). The complete definition of the Cauchy problem in the theory with the nonlocal action requires defining the boundary conditions for the Green functions $G(x, y)$. In the local version, the same can be achieved by imposing the boundary conditions for the two auxiliary fields φ and ψ. For practical applications it is more useful to work with the scalar fields than with the Green functions.

2. The local form of the effective action with two auxiliary scalars, (17.37), was introduced in the paper [279]. Qualitatively similar manner of introducing a second scalar was suggested later on in [218]. The kinetic term for the auxiliary field φ is positive while, for ψ, the kinetic term has a negative sign. The wrong sign does not lead to a problem in this case, since both fields φ and ψ are auxiliary.

3. We introduced the term $\int C_x^2 G(x, y) C_y^2$ into the action to keep the equivalence between (17.36) and the symmetric form (17.35). In the local form (17.37), this term corresponds to the action of the auxiliary scalar ψ. It is easy to see that this term is conformal, and therefore could be included into the invariant functional $S_c[g_{\mu\nu}]$. However, from the physical point of view, this would be a mistake. When writing the non-conformal terms in the symmetric form (17.36) without $\int C_x^2 G(x, y) C_y^2$, we modify the four-point function. Therefore, by introducing this term, we preserve the fundamental structure of the anomaly and its relation to the logarithmic form factors in the UV. Correspondingly, the auxiliary scalar ψ represents an important element of the local covariant form of induced action.

Exercises

17.7. Verify solutions (17.32) and (17.33). After that, derive the equations for the auxiliary fields φ and ψ from the least action principle of (17.37). Substituting the solutions back into the action, show the equivalence with the nonlocal form (17.36).
17.8. Verify (17.28) and prove its equivalence to (17.37).
17.9. Consider the situation when there is some extra background field A, along with the metric. Assume that the one-loop divergence is a modified form of (17.16), namely,

$$\Gamma_{div} = \frac{\mu^{n-4}}{n-4} \int d^n x \sqrt{g} \left\{ \beta_1 C^2 + \beta_2 E_4 + \beta_3 \Box R + \mathcal{I}(A, g) \right\}, \tag{17.39}$$

where $\int_x \mathcal{I}(A, g)$ is a conformal expression,

$$\int d^4 x \sqrt{-g}\, \mathcal{I}(A, g_{\mu\nu}) = \int d^4 x \sqrt{-\bar{g}}\, \mathcal{I}(A, \bar{g}_{\mu\nu}). \tag{17.40}$$

Explore the modifications in the trace anomaly and in all three forms of the effective action, (17.28), (17.36) and (17.37).
17.10. Consider a more complicated version of the previous exercise, when there are several background fields A_i with non-trivial conformal weights k_i. The vacuum action of these fields can be regarded as conformal. Try to establish the modified definition of the conformal Noether identity for this new extended vacuum action, and derive the corresponding anomaly. Calculate the local non-covariant form of the effective action (17.28) in this case. An example of this sort is elaborated in the next chapter.

17.11. Verify the *global* conformal invariance of the induced actions in all three available forms, (17.28), (17.35) and (17.37). Discuss the relation of this feature with the same for the classical action of vacuum.

18
General notions of perturbative quantum gravity

Starting from this point, we begin to consider field models with a quantized gravitational field. The corresponding theory is traditionally called quantum gravity. It is worth noting that there are various and very different approaches to describing gravity at the quantum level. In this book, we will develop only the perturbative approach to quantum gravity, that is based on the traditional-style quantization of the metric. The models that we will discuss include quantum GR, fourth-derivative quantum gravity, higher-derivative polynomial quantum gravity, nonlocal quantum gravity, and, furthermore, the induced gravity approach and effective quantum gravity.

Our first purpose is to construct models with a quantized metric. From the general quantum field theory perspective, the metric should be quantized as well as the matter fields. Thus, the main contents of the next few chapters can be regarded as the completion of previous considerations where the metric was a classical background and only matter fields were quantized.

Along with the purely metric versions of quantum gravity, in the literature one can find many other versions of the theory, such as gravity with torsion, supergravity or superstring theory. All of these approaches are beyond the scope of the present book. The reason for this is that it is impossible to embrace the immensity of studies in the area of quantum gravity in a single book. On the other hand, many general methods, which are discussed here, are successfully used in other approaches to quantum gravity.

18.1 Symmetries of the classical gravitational models

As we have just mentioned, we intend to consider the quantum theory of the metric field $g_{\mu\nu}$. It is already known, that the metric is subject to the gauge transformation (11.64), also called diffeomorphism, corresponding to (11.2),

$$\delta g_{\mu\nu} = R_{\mu\nu,\,\lambda}(g)\,\xi^\lambda = -\,\nabla_\mu\xi_\nu - \nabla_\nu\xi_\mu. \tag{18.1}$$

This transformation does not depend on the choice of the classical action. On the other hand, we can consider the invariant actions, i.e., those satisfying the Noether identity,

$$\frac{\delta S}{\delta g_{\mu\nu}}\,R_{\mu\nu,\,\lambda}(g) = 0. \tag{18.2}$$

We shall see in brief that gravity belongs to the Yang-Mills type gauge theories. Therefore, in order to apply the general system of proofs of gauge-invariant renormalizability, we only need to ensure that the generators of the gauge transformations

Introduction to Quantum Field Theory with Applications to Quantum Gravity. Iosif L. Buchbinder and Ilya L. Shapiro, Oxford University Press (2021). © Iosif L. Buchbinder and Ilya Shapiro.
DOI: 10.1093/oso/9780198838319.003.0018

$$R_{\mu\nu,\,\lambda}(g) \;=\; -\,g_{\mu\lambda}\nabla_\nu \;-\; g_{\nu\lambda}\nabla_\mu \tag{18.3}$$

satisfy a closed algebra, which means (see Part I)

$$\frac{\delta R_{\mu\nu,\,\lambda}(g)}{\delta g_{\rho\sigma}}\,R_{\rho\sigma,\,\tau}(g) \;-\; \frac{\delta R_{\mu\nu,\,\tau}(g)}{\delta g_{\rho\sigma}}\,R_{\rho\sigma,\,\lambda}(g) \;=\; f^{\kappa}{}_{\lambda\tau}\,R_{\mu\nu,\,\kappa}(g). \tag{18.4}$$

Due to the special importance of Eq. (18.4), let us verify it in detail. We start by deriving (11.64). Consider the infinitesimal transformation (11.2). Taking only the terms of the first order in ξ^μ and their derivatives, we get $x^\mu \;=\; x'^\mu - \xi^\mu$ and hence

$$\delta g_{\alpha\beta}(x) = g'_{\alpha\beta}(x) \;-\; g_{\alpha\beta}(x), \qquad \text{where}$$

$$g'_{\alpha\beta}(x) = g'_{\alpha\beta}(x') - \frac{\partial g'_{\alpha\beta}(x')}{\partial x'^\lambda}\,\xi^\lambda = g'_{\alpha\beta}(x') - \partial_\lambda g_{\alpha\beta}\,\xi^\lambda.$$

On the other hand,

$$\begin{aligned} g'_{\alpha\beta}(x') &= \frac{\partial x^\rho}{\partial x'^\alpha}\,\frac{\partial x^\sigma}{\partial x'^\beta}\,g_{\rho\sigma}(x) \;=\; \big(\delta^\rho_\alpha - \partial_\alpha\xi^\rho\big)\big(\delta^\sigma_\beta - \partial_\beta\xi^\sigma\big)g_{\rho\sigma}(x) \\ &= g_{\alpha\beta} - g_{\rho\beta}\,\partial_\alpha\xi^\rho - g_{\alpha\rho}\,\partial_\beta\xi^\rho. \end{aligned} \tag{18.5}$$

Taking the two expressions together, we arrive at

$$\delta g_{\alpha\beta}(x) \;=\; -g_{\lambda\beta}(x)\,\partial_\alpha\xi^\lambda(x) - g_{\alpha\lambda}(x)\,\partial_\beta\xi^\lambda(x) - \partial_\lambda g_{\alpha\beta}(x)\,\xi^\lambda(x), \tag{18.6}$$

which is nothing but (18.3). In fact, the equivalent form (18.6) is more suitable for deriving the detailed form of the gauge (diffeomorphism) transformations, and of its generators, which can be easily obtained in the form

$$\begin{aligned} \delta g_{\mu\nu}(x) &= \int d^4y\,R_{\mu\nu,\,\lambda}(x,y;g)\,\xi^\lambda(y), \\ R_{\mu\nu,\,\sigma}(g) &= R_{\mu\nu,\,\sigma}(x,y;g) \;=\; -\,\delta(x-y)\partial_\sigma g_{\mu\nu}(x) \\ &\quad - g_{\mu\sigma}(x)\partial_\nu\delta(x-y) \;-\; g_{\sigma\nu}(x)\partial_\mu\delta(x-y). \end{aligned} \tag{18.7}$$

In this expression, the index g indicates the metric dependence of the generators. One can verify that the algebra of the generators has the following form:

$$\begin{aligned} &\int du\left\{\frac{\delta R_{\mu\nu,\,\sigma}(x,y;g)}{\delta g_{\alpha\beta}(u)}R_{\alpha\beta,\,\gamma}(u,z;g) \;-\; \frac{\delta R_{\mu\nu,\,\gamma}(x,z;g)}{\delta g_{\alpha\beta}(u)}R_{\alpha\beta,\,\sigma}(u,y;g)\right\} \\ &\quad = -\int du\,R_{\mu\nu\lambda}(x,u;g)\,f^\lambda{}_{\sigma\gamma}(u,y,z), \end{aligned} \tag{18.8}$$

where

$$f^\lambda{}_{\alpha\beta}(x,y,z) = \delta(x-y)\,\delta^\lambda_\beta\,\frac{\partial}{\partial x^\alpha}\delta(x-z) \;-\; \delta(x-z)\,\delta^\lambda_\alpha\,\frac{\partial}{\partial x^\beta}\delta(x-y). \tag{18.9}$$

It is easy to see that the structure functions of the gauge algebra do not depend on the metric tensor and are antisymmetric,

$$f^\lambda{}_{\alpha\beta}(x,y,z) \;=\; -\,f^\lambda{}_{\beta\alpha}(x,z,y)\,. \tag{18.10}$$

We conclude that, independently of the choice of action, any covariant, metric-dependent theory of gravity is a gauge theory with the closed gauge algebra (18.4). The structure functions (18.9) do not depend on the metric tensor and are antisymmetric. Thus, all covariant theories of gravity are gauge theories of the Yang-Mills type.

It proves fruitful to introduce another representation of the metric, regarding it as a perturbation (not necessarily small) over the Minkowski spacetime,

$$\begin{aligned} g_{\mu\nu} &= \eta_{\mu\nu} + h_{\mu\nu}, \\ \text{where} \qquad h_{\mu\nu} &= \bar{h}^{\perp\perp}_{\mu\nu} + \partial_\mu \epsilon^\perp_\nu + \partial_\nu \epsilon^\perp_\mu + \partial_\mu \partial_\nu \epsilon + \frac{1}{4}\, h\, \eta_{\mu\nu}. \end{aligned} \tag{18.11}$$

The tensor component (spin-2 field) is traceless and transverse, $\bar{h}^{\perp\perp}_{\mu\nu}\eta^{\mu\nu} = 0$ and $\partial^\mu \bar{h}^{\perp\perp}_{\mu\nu} = 0$, while the vector component (spin-1) satisfies $\partial_\mu \epsilon^{\perp\mu} = 0$. All indices are raised and lowered with the flat metric $\eta^{\mu\nu}$ and its inverse $\eta_{\mu\nu}$, respectively.

According to Eq. (18.11), the metric perturbations have one spin-2, one spin-1 and two scalar modes, h and ϵ. We leave the analysis of the gauge transformations of these modes as an exercise. According to the general understanding of the Lagrangian quantization of gauge theories, we can imagine that *before* the Faddeev-Popov procedure (described in the next section), only some of the components in (18.11) are dynamic. After the introduction of the gauge-fixing term, all these components become dynamic, and this is compensated by the degrees of freedom of the gauge ghosts.

Exercises

18.1. Verify relations (18.7), (18.8) and (18.9).
18.2. Derive the gauge transformation of the Christoffel symbol (11.9), find the form of the generators and show the structure constants for the corresponding gauge algebra (18.9). Discuss whether the result changes under the modification $\Gamma^\alpha_{\ \beta\gamma} \to \Gamma^\alpha_{\ \beta\gamma} + C^\alpha_{\ \beta\gamma}$, where $C^\alpha_{\ \beta\gamma}$ is a tensor function of the spacetime coordinates.
18.3. Starting from the transformation rule for the general metric (18.1), obtain the transformations for the perturbation $h_{\mu\nu}$ in Eq. (18.11) and for its independent components. Extract gauge-invariant combinations of these irreducible components.

18.2 Choice of the action and gauge fixing for quantum gravity

In this section, we describe the Lagrangian quantization of different versions of quantum gravity. In all cases, the quantization assumes the construction of the functional integral according to the DeWitt-Faddeev-Popov approach. For the sake of generality, we use the framework of the background field method.

18.2.1 Quantum general relativity (GR)

It is evident that the first candidate for the basis of quantum gravity is the Einstein-Hilbert action (11.45),

$$S_{EH} = -\frac{1}{\kappa^2}\int d^4x\sqrt{-g}\,(R + 2\Lambda). \tag{18.12}$$

In that expression, we denoted $\kappa^2 = 16\pi G$. As we shall see in brief, this is the most useful form for coupling in quantum gravity based on GR. To have a flat limit, one has to assume a vanishing cosmological constant. Then one can perform an expansion around the flat background metric,

$$g_{\mu\nu} = \eta_{\mu\nu} + \kappa h_{\mu\nu} \,. \tag{18.13}$$

This background is useful for constructing the propagator, evaluating the superficial degree of divergence of the Feynman diagrams, etc. However, as we are going to use the background field method, it is more useful to perform a general form of expansion into the background and quantum parts,

$$g_{\mu\nu} \quad \longrightarrow \quad g'_{\mu\nu} = g_{\mu\nu} + \kappa h_{\mu\nu}, \tag{18.14}$$

where $g_{\mu\nu}$ in the *r.h.s.* is an arbitrary background metric. One can note that the bilinear form of the action has the vanishing power of the parameter κ, while the higher orders in the expansion in $h_{\mu\nu}$ have positive powers of κ. Thus, κ can be seen as a loop expansion parameter, similar to the coupling constant g in Yang-Mills theory. The main difference between the two cases is that for a non-Abelian vector field, there are only $\mathcal{O}(g)$ vertices with three lines of the field, and $\mathcal{O}(g^2)$ vertices with four lines. In contrast, since gravity is a non-polynomial theory, there are vertices with an arbitrary orders of κ and the corresponding number of external lines of $h_{\mu\nu}$.

A natural question is, why should we choose the expansion of the metric $g_{\mu\nu}$, instead of, e.g., $g^{\mu\nu}$? The answer is that we are allowed to use different parameterizations of the quantum metric. In some cases, the most useful ones in the flat background case are the parametrizations

$$g^{\mu\nu} = \eta^{\mu\nu} + \kappa\phi^{\mu\nu} \qquad \text{and} \qquad \sqrt{-g}g^{\mu\nu} = \eta^{\mu\nu} + \kappa\Phi^{\mu\nu}, \tag{18.15}$$

instead of (18.13). In section 20.2, we will discuss more general parametrizations and learn how to keep the parametrization dependence under control. For the sake of simplicity, in the general considerations, we shall use (18.14) and (18.13) as the parametrization of the quantum metric.

The gauge-fixing term in the theory (18.12) can be taken in the form

$$S_{gf} = \frac{1}{\alpha}\int d^4x\sqrt{-g}\,\chi_\mu\,\chi^\mu, \quad \text{where} \quad \chi_\mu = \nabla_\nu h^\nu{}_\mu - \beta\nabla_\mu h \tag{18.16}$$

and where α and β are gauge-fixing parameters. The last expression is the most general version of linear gauge fixing in second-derivative theories such as quantum GR. An explanation of this issue is in order. Since gravity is a gauge theory, there is a degeneracy of the bilinear in the quantum fields form of the original action. The gauge-fixing term is needed to remove this degeneracy. For this reason, the number of derivatives of the metric in S_{gf} should be equal to the number of derivatives in the classical action of the theory. Choosing the gauge-fixing term as per (18.16), we provide a propagator of the metric perturbations that is homogeneous in the derivatives.

The action of ghosts (10.51) includes the operator

$$M_\alpha{}^\beta = \frac{\delta\chi_\alpha}{\delta g_{\mu\nu}}\,R_{\mu\nu}{}^\beta(g), \tag{18.17}$$

which is of the second order in the derivatives, owing to the choice of the gauge (18.16) and the form of the generator (18.3). Thus, our choice of gauge fixing (18.16) provides the inverse propagator of the metric that is homogeneous in derivatives and the propagator of Faddeev-Popov ghosts that has the same second power of momentum as the propagator of the metric perturbations.

In the next sections, we shall see how to satisfy analogous homogeneity conditions in more complicated models of quantum gravity.

18.2.2 Fourth derivative gravity

The next model of interest has the fourth-derivatives action. If the previous choice, namely, the Einstein-Hilbert action of GR is strongly motivated by the success of Einstein's classical gravitational theory, the fourth-derivative model is strongly motivated by the semiclassical theory. As we know from the previous chapters, such a theory can be renormalizable only in the presence of the fourth-derivative vacuum terms.

The classical action of the theory has the form

$$S = - \int d^4x\sqrt{-g}\,\Big\{\frac{1}{2\lambda}C^2 - \frac{1}{\rho}E + \frac{1}{\xi}R^2 + \tau\Box R + \frac{1}{\kappa^2}(R - 2\Lambda)\Big\}, \qquad (18.18)$$

where $C^2 = C_{\mu\nu\alpha\beta}C^{\mu\nu\alpha\beta}$ is the square (12.10) of the Weyl tensor (11.24) in $n = 4$ and E_4 is the integrand of the Gauss-Bonnet topological term (11.76). We changed the notation for the coefficients as compared to those for the semiclassical version (12.9). Here λ, ρ, ξ and τ are independent parameters in the fourth-derivative sector of the action. The sign of the coupling λ should be positive to provide the positively defined energy of the free tensor modes at low energy. We postpone the discussion of this issue to section 19.2.2, where it is based on the expression for the energy (19.18).

Other useful definitions of the couplings in the model (18.18) are

$$\theta = \frac{\lambda}{\rho} \qquad \text{and} \qquad \omega = -\,\frac{3\lambda}{\xi}\,. \qquad (18.19)$$

This is a natural choice, since pure λ defines an interaction by means of tensor modes and ω parameterizes the interaction by an exchange of the massive scalar mode.

Some observation concerning the classical solutions in the theory of fourth-derivative gravity (18.18) is in order. The Einstein-Hilbert term has second derivatives of the metric and is multiplied by the factor of $\frac{1}{\kappa^2} = \frac{M_P^2}{16\pi}$, where $M_P \propto 10^{19} GeV$ is the Planck mass. The fourth-derivative terms do not have dimensional parameters, which is compensated by the number of derivatives of the metric (remember the last is dimensionless). Thus, if we are interested in the classical solutions with $|\partial_\lambda g_{\alpha\beta}| \ll M_P$, the effect of the fourth-derivative terms is supposed to be small. This effect is called "Planck suppression." For instance, if $|\lambda|$ is of the order of magnitude 1, in all known laboratory experiments or astrophysical phenomena, these terms are negligible. However, this conclusion concerns only the proper solutions. As we discuss in what follows, the presence of higher derivatives may dramatically change the stability properties of these solutions, and it is unclear whether Planck suppression may be applied to the perturbations.

An alternative form for the action (18.18) is

$$S = \int d^4x\sqrt{-g}\,\Big\{xR^2_{\mu\nu\alpha\beta} + yR^2_{\mu\nu} + zR^2 + \tau\Box R - \frac{1}{\kappa^2}(R + 2\Lambda)\Big\}, \qquad (18.20)$$

where the parameters $x,\ y,\ z$ are related to $\rho,\ \lambda,\ \xi$. The two forms of the action (18.18) and (18.20) are completely equivalent, such that one can use either of the two forms, depending on convenience.

The next step is to introduce the gauge-fixing term. According to the general scheme described by Eq. (10.60), we have to use the expression

$$S_{GF} = \frac{1}{2}\int d^4x\sqrt{-g}\,\chi^\mu\,G_{\mu\nu}\,\chi^\nu\,, \tag{18.21}$$

where the gauge condition χ^μ is defined by the formula (18.16). The main change concerns the form of the weight operator $G_{\mu\nu}$. According to the arguments presented above, this choice should make the bilinear part of the action homogeneous in derivatives of the quantum metric. For this reason, in the fourth-derivative model (18.18), the action (18.21) should also be fourth derivative. Therefore, the weight function $G_{\alpha\beta}$ should be a non-degenerate operator of the second order in derivatives. In the presence of a background metric, the most general form of this operator is

$$G_{\mu\nu} = \frac{1}{\alpha}\Big(g_{\mu\nu}\Box + \gamma\nabla_\mu\nabla_\nu - \nabla_\nu\nabla_\mu + p_1 R_{\mu\nu} + p_2 R\,g_{\mu\nu}\Big)\,, \tag{18.22}$$

where the non-degeneracy requires $\gamma \neq 0$. The gauge-fixing term depends on arbitrary gauge-fixing parameters $\alpha_i = (\alpha,\ \beta,\ \gamma,\ p_1,\ p_2)$, where β comes from Eq. (18.16) and the other constants come from Eq. (18.22).

Now we face the problem of the non-homogeneity in derivatives of the bilinear form of the *total* action, including the ghost sector. Indeed, the bilinear form of the ghost action is the operator (18.17), which does not depend on the weight operator $G_{\alpha\beta}$ and therefore is only second order in derivatives. To resolve this issue, we can use the following trick [140]. Redefine the ghost action, to be

$$S_{GH} = \int d^4x\sqrt{-g}\,\bar{c}_\alpha\,\tilde{M}^\alpha{}_\beta\,c^\beta, \quad \text{where} \quad \tilde{M}^\alpha{}_\beta = M_\alpha{}^\lambda[g]\,G_\lambda{}^\beta[g]. \tag{18.23}$$

Here $M_\alpha{}^\lambda[g]$ is defined by the standard expression (18.17), and $G_\lambda{}^\beta[g]$ is the weight operator (18.22) that depends on the background metric field. With this choice, the propagators of both the quantum metric and the ghosts behave like p^{-4} in the high-energy (UV) limit. As compensation, the general equation (10.60) has to be modified to include the contribution of the extra factor of the weight operator,

$$\int \mathcal{D}h\, e^{i\left(S[h]+\frac{1}{2}\chi^\alpha G_{\alpha\beta}[g]\chi^\beta\right)}\ \text{Det}\,(\tilde{M}^\alpha{}_\beta)\ \text{Det}^{\,-\frac{1}{2}}(G_{\alpha\beta}[g]). \tag{18.24}$$

Then the third ghost b^α should be a commuting field since, instead of (10.61), we need

$$\text{Det}^{\,-\frac{1}{2}}(G_{\alpha\beta}[g]) = \int \mathcal{D}b\, e^{ib^\alpha G_{\alpha\beta}[g] b^\beta}. \tag{18.25}$$

With the choice of the ghost operator described above, both the propagator of the quantum metric and the ghosts behave as $1/p^4$ in the UV, which dramatically simplifies the analysis of possible divergences. Let us stress that the two forms (10.60) and (18.24) are equivalent. In reality, the second modified form is useful mainly for the evaluation of power counting. The practical one-loop calculations can be performed based on the standard expression (10.60).

The last relevant observation concerns the loop expansion parameter in the theory (18.18). The expansion (18.14) may not be the best choice for this model, since the highest-derivative terms in the bilinear form of the action come from the C^2 and R^2 terms in the action, not from the Einstein-Hilbert term. Thus, we can trade the previous expansion for the more general one from Eq. (11.48), without κ. After separating the bilinear in $h_{\mu\nu}$ terms, we can make an additional rescaling of the quantum metric $h_{\mu\nu}$, such that the propagator becomes free from the coupling constants. The main coupling constant in the fourth-derivative theory is $\sqrt{\lambda}$, as the reader can find out as an exercise.

18.2.3 Quantum gravity models, polynomial in derivatives

The previous two examples of classical actions that can be used to construct quantum gravity models can be regarded as minimal versions. The GR is certainly unique, since it is the only gravitational theory that has successfully passed many observational and experimental tests [348]. Indeed, all other theories can be seriously considered only if they coincide with general relativity in the low-energy domain, where all these verifications were conducted. The fourth-derivative model (18.18) is also very special, since it provides the renormalizability of semiclassical theory. One can discuss ways of interpreting the fourth-derivative terms (we postpone this to section 19.6), but, without these terms, the quantum theory of matter fields cannot be mathematically consistent. On top of that, with the coupling constant λ being the dimensionless quantity of order 1, the fourth-derivative terms do not spoil the classical tests of GR owing to Planck suppression.

A natural question is whether one can construct some kind of a "nonminimal" model of quantum gravity, one which goes beyond the fourth-derivative version. It is evident that this problem can be solved in infinitely many ways, so let us impose some additional constraints on such an extended model. First of all, let us require that the propagator of the new theory suppress the UV divergences more strongly than in the fourth-derivative theory. This means that the behavior of the Euclidean propagator in the limit $|p| \to \infty$ should be stronger than $1/p^4$, and this must be valid for all relevant components of the field $h_{\mu\nu}$ and the ghosts. The second condition is that the highest-derivative terms should be homogeneous in the derivatives, as in action (18.18). The consequences of the violation of these two conditions at the quantum level will be discussed in section 18.7.

Let us consider first the polynomial higher-derivative model with the action [13], in the modified notations of the paper [222],

$$
\begin{aligned}
S_N = \int d^4x\sqrt{-g}\,\Big\{ & \omega_{N,R} R\,\Box^N R + \omega_{N,C} C\,\Box^N C + \omega_{N,\mathrm{GB}} \mathrm{GB}_N \\
& + \omega_{N-1,R} R\,\Box^{N-1} R + \omega_{N-1,C} C\,\Box^{N-1} C + \omega_{N-1,\mathrm{GB}} \mathrm{GB}_{N-1} + \dots \\
& + \omega_{0,R} R^2 + \omega_{0,C} C^2 + \omega_{0,\mathrm{GB}} \mathrm{GB}_0 + \omega_{\mathrm{EH}} R + \omega_{\mathrm{cc}} + \mathcal{O}(R^3_{\dots}) \Big\}, \qquad (18.26)
\end{aligned}
$$

where the highest power of metric derivatives in the $\mathcal{O}(R^3_{\dots})$ terms does not exceed $4+2N$. Furthermore, $C{\cdot}C$ mean the squares of the Weyl tensor (with omitted indices), which is

$$C\,\Box^n\,C \,=\, C_{\mu\nu\alpha\beta}\,\Box^n\,C^{\mu\nu\alpha\beta} \,=\, R_{\mu\nu\alpha\beta}\Box^n R^{\mu\nu\alpha\beta} - 2R_{\mu\nu}\Box^n R^{\mu\nu} + \frac{1}{3}\,R\Box^n R. \tag{18.27}$$

The modified version of the four-dimensional Gauss-Bonnet invariant can be shown to have the property (16.78). Let us note that the reader can easily check this formula as an exercise; see also [13]. In particular, for any n,

$$\mathrm{GB}_n \,=\, R_{\mu\nu\alpha\beta}\Box^n R^{\mu\nu\alpha\beta} - 4R_{\mu\nu}\Box^n R^{\mu\nu} + R\Box^n R \,=\, \mathcal{O}(R^3_{\dots}). \tag{18.28}$$

This term is certainly not topological for $n \neq 0$, but it contributes only to the third- and higher-order terms in the curvature tensor.

It is evident that only $\mathcal{O}(R^2_{\dots})$ terms may contribute to the propagator of the gravitational perturbation on a flat background, while $\mathcal{O}(R^3_{\dots})$ terms only affect the vertices. The advantage of the form (18.26) is that $\mathcal{O}(R^2_{\dots})$ terms are written as Weyl-squared and R-squared expressions. As we shall see in the next sections, the Weyl-squared $C\dots C$ terms affect *only* the propagation of the tensor part $\bar{h}^{\perp\perp}_{\mu\nu}$ of the metric perturbations, while the R-squared $R\cdot R$ terms affect *only* the propagation of the scalar modes. Thus, when writing the higher-derivative actions in the form (18.26), we separate the propagators of the tensor and scalar modes from the very beginning.

Assuming that both $\omega_{N,R}$ and $\omega_{N,C}$ are non-zero, the propagator of the quantum metric is supposed to behave as $p^{-(4+2N)}$ in the UV. However, to achieve this result, the gauge-fixing terms should have the same highest power in derivatives, as the main action. To provide this property, the weight function in the gauge-fixing term (18.21) can be the differential operator of the form

$$G_{\mu\nu} \,=\, -\,\frac{1}{\alpha}\big(g_{\mu\nu}\Box + \gamma\nabla_\mu\nabla_\nu - \nabla_\nu\nabla_\mu\big)\Box^{N+1}. \tag{18.29}$$

One can, of course, add to this expression many terms with fewer derivatives. However, for our purposes, this expression is sufficient, since, with this choice, all components of the quantum metric (18.11) have propagators that behave homogeneously, that is, as $p^{-(4+2N)}$, in the UV.

Finally, we have to provide the same homogeneity in derivatives for the ghost sector. For this sake, one can redefine the ghost action as (18.23), but this time with the weight operator (18.29). After that, all quantum fields gain qualitatively the same UV behavior of the propagators.

18.2.4 Nonlocal models of gravity

The polynomial (in derivatives) action (18.26) can be written in an alternative form,

$$S = \int d^4x\sqrt{-g}\,\Big\{\frac{1}{2}\,C_{\mu\nu\alpha\beta}P_1(\Box)C^{\mu\nu\alpha\beta} \,+\, \frac{1}{2}\,RP_2(\Box)R \\ +\,\omega_{\mathrm{EH}}R \,+\, \omega_{\mathrm{cc}} \,+\, \mathcal{O}(R^3_{\dots})\Big\}, \tag{18.30}$$

where $P_{1,2}(x)$ are polynomials of the same order N and the $\mathcal{O}(R^3_{\dots})$ terms have no more than $4+2N$ derivatives of the metric. Owing to relation (16.78), even if starting from the Riemann-squared, Ricci-squared and R-squared terms, one can rewrite the action in terms of the $C \cdot C$ and $R \cdot R$ terms.

The propagator of metric components in this theory manifests the UV behavior $\sim p^{-(4+2N)}$. It is interesting (we discuss some of the reasons in section 19.5) to replace the polynomials in action (18.30) with functions that can be represented as infinite power series in the d'Alembert operator. Thus, we arrive at the non-polynomial action

$$S = \int d^4x\sqrt{-g}\Big\{ -\frac{1}{\kappa^2}\,(R+2\Lambda) \,+\, \frac{1}{2}\,C_{\mu\nu\alpha\beta}\,\Phi(\Box)\,C_{\mu\nu\alpha\beta} \\ + \frac{1}{2}\,R\,\Psi(\Box)\,R \,+\, \mathcal{O}(R^3_{\dots})\Big\}. \qquad (18.31)$$

Formally, the Cauchy problem in this theory requires an infinite set of initial and boundary conditions. For this reason, these models of gravity are called *nonlocal.*

To simplify the analysis, in these (and also polynomial) models, it is customary to assume a vanishing cosmological constant. Then the expansion of the metric can be performed around the flat Minkowski background, as we did in Eq. (18.11), since the last is a solution of the classical equations of motion.

The homogeneity of the propagator of the different components of metric perturbations (18.11) requires the functions $\Phi(x)$ and $\Psi(x)$ to have the same asymptotic behavior in the UV. For instance, this can be achieved if we require that

$$\lim_{x\to+\infty}\frac{\Phi(x)}{\Psi(x)} \,=\, C, \qquad (18.32)$$

where C is a finite non-zero constant. One useful choice is

$$\Phi(x) = -\frac{1}{\kappa^2}\times\frac{e^{\alpha_1 x}-1}{x} \quad \text{and} \quad \Psi(x) = -\frac{const}{\kappa^2}\times\frac{e^{\alpha_2 x}}{x}. \qquad (18.33)$$

Then condition (18.32) means that $\alpha_1 \,=\, \alpha_2 \,=\, \alpha$. In the next sections, we will see that the violation of condition (18.32) leads to the non-renormalizability of a quantum theory. For the present, we assume that this and similar (see below) conditions hold as a reflection of our interest in having Euclidian propagators that are homogeneous in the UV on a flat background. Since we are mainly interested in UV renormalization, it is possible to assume that $\Phi(x) \,=\, const \times \Psi(x)$. Then expression (18.31) can be seen as one of the most general forms of actions for higher-derivative gravity models, including (18.18) and (18.26) as particular cases.

To complete this story, we have to define an appropriate gauge-fixing term in theory (18.31). Using our previous experience with the polynomial model (18.29), let us write the Faddeev-Popov gauge fixing in the universal and standard form (18.21), but this time with the weight operator

$$G_{\mu\nu} \,=\, -\frac{1}{\alpha}\big(g_{\mu\nu}\Box + \gamma\nabla_\mu\nabla_\nu - \nabla_\nu\nabla_\mu\big)\,\Phi(\Box). \qquad (18.34)$$

It is clear that the function $\Phi(\Box)$ can be replaced by $\Psi(\Box)$ or any other function satisfying the condition (18.32). Also, we can add terms with fewer derivatives, e.g., if such terms simplify the calculations.

Finally, the homogeneity in the momentum at the UV for the Faddeev-Popov ghosts can be achieved by the standard replacement (18.23). After that, we arrive at the situation when the UV behavior of the propagators of all components of the metric perturbation and the ghost fields is qualitatively the same. Remarkably, we could achieve this homogeneity in all of the models that were formulated above, namely, in quantum GR, fourth-derivative, polynomial and nonlocal theories of gravity.

Exercises

18.4. Derive the non-linear dependence (at least up to the second order) between the three special parameterizations (18.13) and (18.15), e.g., in the forms $\phi^{\mu\nu} = \phi^{\mu\nu}(h)$, $\Phi^{\mu\nu} = \Phi^{\mu\nu}(h)$ or $\Phi^{\mu\nu} = \Phi^{\mu\nu}(\phi)$. Generalize the corresponding expressions to an arbitrary background metric, and explore whether the corresponding relations depend on the spacetime dimension.
18.5. Find the relation between the couplings and other parameters of the fourth-derivative actions (18.18) and (18.20). Assuming that only couplings define the dynamics of the theory at the classical level, classify the six parameters in each of the two cases. What happens with this classification if we change the dimension to $n \neq 4$?
18.6. Write the gauge-fixing term for quantum GR in the form (18.21), and discuss the restrictions leading to Eq. (18.16). Why it is not reasonable to introduce terms similar to the ones with p_1 and p_2 in Eq. (18.22)?
18.7. Using the full form of the generators of gauge transformations (18.7), derive the bilinear form of the ghost action (18.17). Assume that all gauge transformations of the sum (11.48) belong to the quantum metric $h_{\mu\nu}$.
Observation. A detailed discussion of this issue can be found in [208].
18.8. Verify the consistency of relations (18.24) and (18.25). Discuss the applicability of these formulas in polynomial and nonlocal models of quantum gravity.
18.9. Explain why the flat Minkowski metric is a solution of the classical equations of motion in the $\Lambda = 0$ versions of the general model (18.31), both polynomial and nonlocal. Try to extend this statement for the de Sitter space in the case of $\Lambda \neq 0$.
18.10. Explain why it makes sense to introduce the factor $\frac{1}{\alpha}$ and the parameter γ in the expression for the weight operator (18.34). Consider possible modifications of this formula, which do not violate the homogeneity of the propagator of the different modes of the metric perturbations and ghosts in the UV limit.
Observation. The reader can postpone this exercise until after sections 18.3, 20.4 and 20.5.

18.3 Bilinear forms and linear approximation

Consider the bilinear expansions of the relevant quantities depending on the curvature tensor. These expressions are required for the analysis of propagators in different models of quantum gravity. Later on, we shall need similar expansions for one-loop calculations in the background field method. Thus, it makes sense to assume the general expansion (11.48) of the metric, $g_{\mu\nu} \to g'_{\mu\nu} = g_{\mu\nu} + h_{\mu\nu}$. In what follows, we shall use the formulas subsequent to (11.48) and write further expansions without too much explanation. An elementary introduction to the technical details of these expansions can be found, e.g., in the book [292].

Since we are interested in the propagator of the metric and one-loop calculations, it is sufficient to consider the expansion up to the second order in $h_{\mu\nu}$. Using (11.49), (11.53) and (11.55), we obtain, in particular,

$$
\begin{aligned}
\Gamma'^{\lambda}_{\mu\nu} &= \Gamma^{\lambda}_{\mu\nu} + \delta\Gamma^{\lambda}_{\mu\nu}, \qquad \delta\Gamma^{\lambda}_{\mu\nu} = \Gamma^{(1)\,\lambda}_{\mu\nu} + \Gamma^{(2)\,\lambda}_{\mu\nu}, \qquad \text{where} \\
\Gamma^{(1)\,\lambda}_{\mu\nu} &= \frac{1}{2}\big(\nabla_\mu h^\lambda_\nu + \nabla_\nu h^\lambda_\mu - \nabla^\lambda h_{\mu\nu}\big) \qquad \text{and} \\
\Gamma^{(2)\,\lambda}_{\mu\nu} &= -\frac{1}{2}h^{\lambda\tau}\big(\nabla_\mu h_{\tau\nu} + \nabla_\nu h_{\tau\mu} - \nabla_\tau h_{\mu\nu}\big).
\end{aligned} \tag{18.35}
$$

Greek indices are lowered and raised with the background metric $g_{\mu\nu}$ and its inverse $g^{\mu\nu}$, respectively. Let us remember that the variations of $\Gamma^{\lambda}_{\mu\nu}$ are covariant tensor, and, therefore, can be subject to covariant differentiation. Then it is a relatively easy exercise to obtain the first- and second-order expansions of the Riemann tensor,

$$
R'^{\alpha}_{\cdot\,\beta\mu\nu} = R^{\alpha}_{\cdot\,\beta\mu\nu} + \delta R^{\alpha}_{\cdot\,\beta\mu\nu}, \quad \text{where} \quad \delta R^{\alpha}_{\cdot\,\beta\mu\nu} = R^{(1)\alpha}_{\quad\cdot\,\beta\mu\nu} + R^{(2)\alpha}_{\quad\cdot\,\beta\mu\nu}, \tag{18.36}
$$

$$
\begin{aligned}
R^{(1)\alpha}_{\quad\cdot\,\beta\mu\nu} &= \frac{1}{2}\Big(\nabla_\mu\nabla_\beta h^\alpha_\nu - \nabla_\nu\nabla_\beta h^\alpha_\mu + \nabla_\nu\nabla^\alpha h_{\mu\beta} \\
&\quad - \nabla_\mu\nabla^\alpha h_{\nu\beta} + R^\alpha_{\ \tau\mu\nu}h^\tau_\beta - R^\tau_{\ \beta\mu\nu}h_{\tau\alpha}\Big),
\end{aligned} \tag{18.37}
$$

$$
\begin{aligned}
R^{(2)\alpha}_{\quad\cdot\,\beta\mu\nu} &= \frac{1}{2}h^{\alpha\lambda}\{\nabla_\mu\nabla_\lambda h_{\nu\beta} - \nabla_\nu\nabla_\lambda h_{\mu\beta} + \nabla_\nu\nabla_\beta h_{\mu\lambda} - \nabla_\mu\nabla_\beta h_{\nu\lambda} \\
&\quad + [\nabla_\nu, \nabla_\mu]h_{\beta\lambda}\} + \frac{1}{4}h^{\alpha\lambda}\Big\{(\nabla_\mu h^{\alpha\lambda})(\nabla_\lambda h_{\nu\beta} - \nabla_\beta h_{\nu\lambda} - \nabla_\nu h_{\lambda\beta}) \\
&\quad - (\nabla_\nu h^{\alpha\lambda})(\nabla_\lambda h_{\mu\beta} - \nabla_\beta h_{\mu\lambda} - \nabla_\mu h_{\lambda\beta}) + (\nabla^\lambda h_{\nu\beta})(\nabla^\alpha h_{\mu\lambda} - \nabla_\lambda h^\alpha_\mu) \\
&\quad - (\nabla^\lambda h_{\mu\beta})(\nabla^\alpha h_{\nu\lambda} - \nabla_\lambda h^\alpha_\nu) + (\nabla_\beta h^\lambda_\nu + \nabla_\nu h^\lambda_\beta)(\nabla_\lambda h^\alpha_\mu - \nabla^\alpha h_{\lambda\mu}) \\
&\quad - (\nabla_\beta h^\lambda_\mu + \nabla_\mu h^\lambda_\beta)(\nabla_\lambda h^\alpha_\nu - \nabla^\alpha h_{\lambda\nu})\Big\}.
\end{aligned} \tag{18.38}
$$

Similar formulas for the Ricci tensor and scalar curvature are

$$
R'_{\beta\nu} = R_{\mu\nu} + \delta R_{\mu\nu}, \quad \text{where} \quad \delta R_{\mu\nu} = R^{(1)}_{\mu\nu} + R^{(2)}_{\mu\nu}, \tag{18.39}
$$

$$
R^{(1)}_{\mu\nu} = \frac{1}{2}\big(\nabla_\lambda\nabla_\mu h^\lambda_\nu + \nabla_\lambda\nabla_\nu h^\lambda_\mu - \nabla_\mu\nabla_\mu h - \Box h_{\mu\nu}\big), \tag{18.40}
$$

$$
\begin{aligned}
R^{(2)}_{\mu\nu} &= \frac{1}{2}h^{\alpha\beta}\big(\nabla_\alpha\nabla_\beta h_{\mu\nu} + \nabla_\mu\nabla_\nu h_{\alpha\beta} - \nabla_\alpha\nabla_\mu h_{\beta\nu} - \nabla_\alpha\nabla_\nu h_{\beta\mu}\big) \\
&\quad + (\nabla_\mu h_{\alpha\beta})(\nabla_\nu h^{\alpha\beta}) + \frac{1}{4}\big(2\nabla_\beta h^{\alpha\beta} - \nabla^\alpha h\big)\big(\nabla_\alpha h^{\mu\nu} - \nabla_\mu h^{\alpha\nu} - \nabla_\nu h^{\alpha\mu}\big) \\
&\quad + 2\big(\nabla_\alpha h_{\mu\beta}\big)\big(\nabla^\alpha h_\nu^{\ \beta} - \nabla^\beta h_\nu^{\ \alpha}\big)
\end{aligned} \tag{18.41}
$$

and (partially known from Eq. (11.57))

$$
R' = R + \delta R, \quad \text{where} \quad \delta R = R^{(1)} + R^{(2)}, \tag{18.42}
$$

$$
R^{(1)} = \nabla_\mu\nabla_\nu h^{\mu\nu} - \Box h - R_{\mu\nu}h^{\mu\nu}, \tag{18.43}
$$

$$R^{(2)} = h^{\alpha\beta}\nabla_\alpha\nabla_\beta h + h^{\alpha\beta}\Box h_{\alpha\beta} - h^{\alpha\beta}\nabla_\alpha\nabla_\mu h^\mu{}_\beta - h^{\alpha\beta}\nabla_\mu\nabla_\alpha h^\mu{}_\beta$$
$$- \frac{1}{4}(\nabla_\alpha h)(\nabla^\alpha h) + \frac{1}{4}(\nabla_\mu h_{\alpha\beta})\big(3\nabla^\mu h^{\alpha\beta} - 2\nabla^\alpha h^{\mu\beta}\big)$$
$$+ (\nabla_\alpha h^{\alpha\beta})(\nabla_\beta h - \nabla_\mu h^\mu{}_\beta) + R_{\mu\nu} h^\mu{}_\alpha h^{\nu\alpha}. \tag{18.44}$$

Using the expressions listed above, one can perform expansions of R^2, $R^2_{\mu\nu}$ and $R^2_{\mu\nu\alpha\beta}$. As we have explained above, these expansions are sufficient to explore propagators on the flat background, not only in the fourth-derivative, but in all polynomial and nonlocal models of quantum gravity.

Thus, let us list the bilinear forms of the actions of our interest. For the Einstein-Hilbert action, by combining Eqs (11.53), (18.43) and (18.44), after some integration by parts, disregarding surface terms and commuting covariant derivatives, we arrive at

$$\Big(\int d^4x\sqrt{-g'}R'\Big)^{(2)} = \frac{1}{4}\int d^4x\sqrt{-g}\, h^{\mu\nu}\Big[\delta_{\mu\nu,\alpha\beta}\Box - g_{\mu\nu}g_{\alpha\beta}\Box - 2g_{\mu\alpha}\nabla_\nu\nabla_\beta$$
$$+ \big(g_{\mu\nu}\nabla_\alpha\nabla_\beta - g_{\alpha\beta}\nabla_\mu\nabla_\nu\big) - \big(g_{\mu\nu}R_{\alpha\beta} - g_{\alpha\beta}R_{\mu\nu}\big)$$
$$+ 2R_{\mu\alpha\nu\beta} - (R + 2\Lambda)\Big(\delta_{\mu\nu,\alpha\beta} - \frac{1}{2}g_{\mu\nu}g_{\alpha\beta}\Big)\Big]h^{\alpha\beta}, \tag{18.45}$$

where the DeWitt notation for the unit matrix in the symmetric tensors space is

$$\delta_{\mu\nu,\alpha\beta} = \frac{1}{2}\,\big(g_{\mu\alpha}g_{\nu\beta} + g_{\nu\alpha}g_{\mu\beta}\big). \tag{18.46}$$

In what follows, we use this notation for both the covariant and the flat metric since it will never cause any problem to do so. It is important to note the following relevant technical detail. In the bilinear form in the square brackets of the expression (18.45), as well as in many similar formulas below, we break the symmetry between the indices $(\mu\nu)$ and $(\alpha\beta)$. In expression (18.45), this symmetry is automatically restored when the form is multiplied by the symmetric tensors $h^{\alpha\beta}$ and $h^{\mu\nu}$. However, when the elements of the same bilinear form are multiplied by other expressions, the symmetry should be restored. For example, this can be done by trading

$$g_{\mu\alpha}\nabla_\nu\nabla_\beta \longrightarrow \frac{1}{4}\big(g_{\mu\alpha}\nabla_\nu\nabla_\beta + g_{\nu\alpha}\nabla_\mu\nabla_\beta + g_{\mu\beta}\nabla_\nu\nabla_\alpha + g_{\nu\beta}\nabla_\mu\nabla_\alpha\big). \tag{18.47}$$

Our list of the three relevant bilinear forms starts from the simplest example,

$$\Big(\int d^4x\sqrt{-g'}R'^2\Big)^{(2)} = \int d^4x\sqrt{-g}\, h^{\mu\nu}\Big[g_{\alpha\beta}g_{\mu\nu}\Box^2 - g_{\mu\nu}\Box\nabla_\alpha\nabla_\beta - g_{\alpha\beta}\nabla_\mu\nabla_\nu\Box$$
$$+ \nabla_\mu\nabla_\nu\nabla_\alpha\nabla_\beta + \frac{1}{2}R(g_{\mu\nu}\nabla_\alpha\nabla_\beta + g_{\alpha\beta}\nabla_\mu\nabla_\nu) + \frac{1}{2}R(\delta_{\alpha\beta,\mu\nu} - g_{\alpha\beta}g_{\mu\nu})\Box$$
$$- Rg_{\mu\alpha}\nabla_\beta\nabla_\nu - R_{\mu\nu}\nabla_\alpha\nabla_\beta - R_{\alpha\beta}\nabla_\mu\nabla_\nu + (R_{\mu\nu}g_{\alpha\beta} + R_{\alpha\beta}g_{\mu\nu})\Box + 2g_{\alpha\nu}RR_{\mu\alpha}$$
$$- \frac{1}{4}R^2\Big(\delta_{\mu\nu,\alpha\beta} - \frac{1}{2}g_{\mu\nu}g_{\alpha\beta}\Big) - \frac{1}{2}R(g_{\mu\nu}R_{\alpha\beta} + g_{\alpha\beta}R_{\mu\nu}) + R_{\mu\nu}R_{\alpha\beta}\Big]h^{\alpha\beta}. \tag{18.48}$$

An important feature of this expression is the absence of the term $\delta_{\mu\nu,\alpha\beta}\Box^2$. This means that the R^2 term does not affect the propagation of the spin-2 mode $\bar{h}^{\perp\perp}_{\mu\nu}$ of the metric perturbation (18.11). In the following sections, we shall see that this term also does not produce an unphysical massive ghost. In one of the exercises, the reader can elaborate a different explanation of this property.

The next expansion is

$$
\begin{aligned}
\Big(\int d^4x\sqrt{-g'}R'^{\,2}_{\mu\nu\alpha\beta}\Big)^{(2)} &= \int d^4x\sqrt{-g}\,h^{\mu\nu}\Big[\delta_{\mu\nu,\alpha\beta}\Box^2 + \nabla_\alpha\nabla_\beta\nabla_\mu\nabla_\nu \\
& - g_{\nu\beta}\big(\nabla_\alpha\Box\nabla_\mu + \nabla^\lambda\nabla_\alpha\nabla_\mu\nabla_\lambda\big) + \delta_{\mu\nu,\alpha\beta}R^{\lambda\tau}\nabla_\lambda\nabla_\tau + 2R_{\mu\lambda\alpha\nu}\nabla_\beta\nabla^\lambda \\
& + g_{\mu\nu}R_{\alpha\;\cdot\cdot\;\beta}^{\;\;\lambda\tau}\big(\nabla_\lambda\nabla_\tau + \nabla_\tau\nabla_\lambda\big) - 4g_{\nu\beta}R_{\alpha\lambda\tau\mu}\nabla^\lambda\nabla^\tau - 4R^{\lambda}_{\;\cdot\,\alpha\beta\mu}\nabla_\nu\nabla_\lambda \\
& - \frac{1}{4}\,R^2_{\rho\sigma\lambda\tau}\Big(\delta_{\mu\nu,\alpha\beta} - \frac{1}{2}g_{\mu\nu}g_{\alpha\beta}\Big) - R_{\mu\alpha\nu\beta}\Box + \frac{7}{2}\,g_{\nu\beta}R_{\mu\lambda\rho\sigma}R_{\alpha\;\cdots}^{\;\;\lambda\rho\sigma} \\
& - \frac{1}{2}\,R_{\mu\alpha\rho\sigma}R_{\nu\beta\;\cdot\cdot}^{\;\;\;\;\rho\sigma} - g_{\mu\nu}R_{\alpha\lambda\rho\sigma}R_{\beta\;\cdots}^{\;\;\lambda\rho\sigma}\Big]h^{\alpha\beta}\,.
\end{aligned}
\tag{18.49}
$$

This last one we include for completeness:

$$
\begin{aligned}
\Big(\int d^4x\sqrt{-g'}R'^{\,2}_{\mu\nu}\Big)^{(2)} &= \frac{1}{2}\int d^4x\sqrt{-g}\,h^{\mu\nu}\Big[\frac{1}{2}\,\delta_{\mu\nu,\alpha\beta}\Box^2 + \frac{1}{2}\,g_{\mu\nu}g_{\alpha\beta}\nabla^\lambda\nabla^\tau\nabla_\tau\nabla_\lambda \\
& + g_{\nu\beta}\big(\nabla^\lambda\nabla_\mu\nabla_\alpha\nabla_\lambda - \nabla_\alpha\nabla_\mu\Box - \Box\nabla_\alpha\nabla_\mu\big) + \frac{1}{2}g_{\mu\nu}\big(\Box\nabla_\alpha\nabla_\beta - 2\nabla^\lambda\nabla_\alpha\nabla_\lambda\nabla_\beta\big) \\
& + \frac{1}{2}\,g_{\alpha\beta}\big(\Box\nabla_\mu\nabla_\nu - 2\nabla^\lambda\nabla_\mu\nabla_\lambda\nabla_\nu\big) + \nabla_\alpha\nabla_\mu\nabla_\beta\nabla_\nu - 4g_{\nu\beta}R^\lambda_\mu\nabla_\alpha\nabla_\lambda \\
& - 2R_{\mu\alpha}\nabla_\beta\nabla_\nu + 4g_{\alpha\beta}R^\lambda_\mu\nabla_\mu\nabla_\lambda + 4g_{\alpha\beta}R^\lambda_\mu\nabla_\nu\nabla_\lambda + 2g_{\nu\beta}R_{\mu\alpha}\Box \\
& + \big(\delta_{\mu\nu,\alpha\beta} - g_{\alpha\beta}g_{\mu\nu}\big)R^{\lambda\tau}\nabla_\lambda\nabla_\tau + 2R_{\mu\alpha}R_{\nu\beta} + 4g_{\beta\nu}R_{\mu\lambda}R^\lambda_\alpha \\
& - g_{\mu\nu}R^\lambda_\alpha R_{\lambda\beta} - g_{\alpha\beta}R^\lambda_\mu R_{\lambda\nu} - \frac{1}{2}R^2_{\lambda\tau}\Big(\delta_{\mu\nu,\alpha\beta} - \frac{1}{2}g_{\mu\nu}g_{\alpha\beta}\Big)\Big]h^{\alpha\beta}\,.
\end{aligned}
\tag{18.50}
$$

Two observations are in order at this point. It is easy to see that the last two expansions (18.49) and (18.50) have the $\delta_{\mu\nu,\alpha\beta}\Box^2$ term and therefore, contribute to the propagator of the transverse and traceless (spin-2) mode of the gravitational perturbation, in the flat background case. On the other hand, we know that the linear combination of the three terms (11.76) form a topological term, which is a total derivative and is supposed to be irrelevant for any dynamics. Thus, one can expect that the corresponding linear combination of (18.48), (18.50) and (18.49) should be identically zero. At the same time, an immediate inspection shows that this is something very difficult to verify.

Exercises

18.11. Consider the last statement in detail. In particular, derive the contribution to the propagator from the Gauss-Bonnet term for some linear gauge and try to find manifestations of the fact that this term is topological.

Observation. The calculation in the flat background is not difficult. The discussion of this issue can be found in [86] and subsequently in [244, 246] and in a few other works.

18.12. Repeat the calculation of the previous exercise for an arbitrary metric background. Try to find the restrictions on the background or/and perturbations (18.11) that make the non-dynamical nature of the Gauss-Bonnet term most evident.

18.13. Extend the expansions of Eq. (11.53) to an infinite order in $h_{\mu\nu}$.
Hint. This exercise is not difficult. One has to start from the solution for $g^{\mu\nu}$; then the general expansion for the Christoffel symbol follows automatically. The expansion for $\sqrt{-g}$ can be done through the Liouville formula. The solutions can be found, e.g., in the book [292].
18.14. Consider the particular version of the O'Hanlon action (12.15) with the quadratic potential $V(\Phi) = \frac{\alpha}{2}\Phi^2$. Show that this theory is dynamically equivalent to the pure gravity theory with the R^2 action. Using this result and combining the conformal transformations (12.16), show that the action

$$S = -\frac{1}{\kappa^2}\int d^4x\sqrt{-g}\left\{R + \frac{\beta}{6M^2}R^2\right\} \tag{18.51}$$

is conformally equivalent to the sum of the Einstein-Hilbert action and the action of a real scalar field. Discuss the implications of this result for the absence of the higher derivative tensor mode in the propagator of the theory (18.51). What are the physically "correct" signs of α and β?
Observation. The extended discussion of this issue can be found, e.g., in [7].

18.4 Propagators of quantum metric and Barnes-Rivers projectors

As we know from the consideration of general field theory in Part I, the propagators of quantum metric in all models are defined from the equation,

$$H^{\mu\nu,\alpha\beta}(x)\, G_{\alpha\beta,\rho\sigma}(x,y) = \delta^4(x-y)\,\delta^{\mu\nu}{}_{,\rho\sigma}\,, \tag{18.52}$$

$$\text{where} \quad H^{\mu\nu,\alpha\beta}(x)\,\delta^4(x-y) = \frac{1}{2\sqrt{-g}}\,\frac{\delta^2 S}{\delta g_{\mu\nu}(x)\,\delta g_{\alpha\beta}(y)}. \tag{18.53}$$

Here $H^{\mu\nu,\alpha\beta}(x)$ is the bilinear (in quantum fields) form of the classical action S of a model of quantum gravity. We can consider that the action in (18.53) includes the Faddeev-Popov gauge-fixing term, or it may be just the original action with the gauge symmetry. In the last case, we shall denote the degenerate bilinear form as $H^{\mu\nu,\alpha\beta}_{(0)}$.

All spacetime indices are raised and lowered with the background (general or flat) metric. Independently of the model and the choice of the linear gauge fixing and the weight operator in S_{GF}, the operator $H^{\mu\nu,\alpha\beta}(x)$ has the following tensor structure:

$$\begin{aligned} H_{\mu\nu,\alpha\beta}(x;g) &= a_1(-\Box)\delta_{\mu\nu,\alpha\beta}\Box + a_2(-\Box)g_{\mu\nu}g_{\alpha\beta}\Box + a_3(-\Box)\big(g_{\mu\nu}\nabla_\alpha\nabla_\beta + g_{\alpha\beta}\nabla_\mu\nabla_\nu\big) \\ &+ a_4(-\Box)\big(g_{\mu\alpha}\nabla_\beta\nabla_\nu + g_{\nu\alpha}\nabla_\beta\nabla_\mu + g_{\mu\beta}\nabla_\alpha\nabla_\nu + g_{\nu\beta}\nabla_\alpha\nabla_\mu\big) \\ &+ a_5(-\Box)\nabla_\alpha\nabla_\beta\nabla_\mu\nabla_\nu + \text{curvature-dependent terms}, \end{aligned} \tag{18.54}$$

where $a_{1,2,..,5}(\Box)$ are model-dependent functions of the d'Alembert operator. In the higher-derivative cases, all of these functions are proportional to the linear combinations of $\Phi(\Box)$ and $\Psi(\Box)$ in Eq. (18.31). In the case of quantum GR, one obtains the constant functions $a_{1,2,3,4}$ and $a_5 = 0$. For the fourth-derivative model (18.18), $a_{1,2,3,4}$ are linear functions of $\Box$ and $a_5 = const$. In what follows, we consider the general analysis of the propagator, which is valid for all types of models.

The expression (18.54) is written for the more complicated version with an arbitrary background metric. For the moment, let us concentrate on the flat version with $g_{\mu\nu} = \eta_{\mu\nu}$. Then the curvature-dependent terms vanish. Furthermore, one can make a Fourier transform and rewrite the bilinear form in the momentum representation

$$H_{\mu\nu,\alpha\beta}(k;\eta) = -\big[a_1(k^2)\delta_{\mu\nu,\alpha\beta}k^2 + a_2(k^2)\eta_{\mu\nu}\eta_{\alpha\beta}k^2 + a_3(k^2)\big(\eta_{\mu\nu}k_\alpha k_\beta + \eta_{\alpha\beta}k_\mu k_\nu\big) \\ + a_4(k^2)\big(\eta_{\mu\alpha}k_\beta k_\nu + \eta_{\nu\alpha}k_\beta k_\mu + \eta_{\mu\beta}k_\alpha k_\nu + \eta_{\nu\beta}k_\alpha k_\mu\big) + a_5(k^2)k_\alpha k_\beta k_\mu k_\nu\big], \quad (18.55)$$

where $k^2 = k_\mu k^\mu$ is the square of the four-dimensional momentum and $\delta_{\mu\nu,\alpha\beta}$ is similar to (18.46), but this time it is constructed from the flat metric $\eta_{\mu\nu}$.

It proves useful to rewrite (18.55) in a slightly different form,

$$\hat{H} = \frac{1}{k^2}\Big(s_1\,\hat{T}_1 + s_2\,\hat{T}_2 + s_3\,\hat{T}_3 + s_4\,\hat{T}_4 + s_5\,\hat{T}_5\Big), \quad (18.56)$$

where $\hat{T}_n = T^{(n)}_{\mu\nu,\alpha\beta}$ and

$$\hat{T}_1 = \delta_{\mu\nu,\alpha\beta}, \qquad \hat{T}_2 = \eta_{\mu\nu}\eta_{\alpha\beta}, \qquad \hat{T}_3 = \frac{1}{k^2}\big(\eta_{\mu\nu}k_\alpha k_\beta + \eta_{\alpha\beta}k_\mu k_\nu\big), \quad (18.57)$$
$$\hat{T}_4 = \frac{1}{4k^2}\big(\eta_{\mu\alpha}k_\beta k_\nu + \eta_{\nu\alpha}k_\beta k_\mu + \eta_{\mu\beta}k_\alpha k_\nu + \eta_{\nu\beta}k_\alpha k_\mu\big), \quad \hat{T}_5 = \frac{1}{k^4}\,k_\alpha k_\beta k_\mu k_\nu.$$

The coefficients depend on momentum, $s_l = s_l(k^2)$, and these dependencies may be complicated, especially in nonlocal models. On the other hand, the tensor structure of the expressions (18.56) is the same for all models of quantum gravity, e.g., it is the same for quantum GR and all higher-derivative polynomial and nonlocal models (18.31). At the same time, one can impose some symmetries or just some restrictions on the action, such that the bilinear form (18.56), even after the Faddeev-Popov procedure, remains degenerate. For example, this is the case for the theories in (18.31), both nonlocal or polynomial, without Einstein-Hilbert and cosmological terms and with $\Phi = 0$, or $\Psi = 0$. We discuss this aspect in more detail below.

To invert the operator (18.55) and also analyze its possible degeneracy, it is useful to consider the set of operators called Barnes-Rivers projectors [23, 265]. For the sake of generality, we present the n-dimensional versions of the formulas.

It proves useful to start from the projectors to the transverse and longitudinal subspaces of the vector space (5.121). The momentum representation is

$$\omega_{\mu\nu} = \frac{k_\mu k_\nu}{k^2}, \qquad \theta_{\mu\nu} = \eta_{\mu\nu} - \frac{k_\mu k_\nu}{k^2}, \quad (18.58)$$

with the standard properties (5.122).

The projectors to the spin-2, spin-1 and spin-0 states in the symmetric tensors space are defined, respectively,

$$\hat{P}^{(2)} = P^{(2)}_{\mu\nu\,,\,\alpha\beta} = \frac{1}{2}(\theta_{\mu\alpha}\theta_{\nu\beta} + \theta_{\mu\beta}\theta_{\nu\alpha}) - \frac{1}{n-1}\,\theta_{\mu\nu}\theta_{\alpha\beta}\,,$$
$$\hat{P}^{(1)} = P^{(1)}_{\mu\nu\,,\,\alpha\beta} = \frac{1}{2}\,(\theta_{\mu\alpha}\omega_{\nu\beta} + \theta_{\nu\alpha}\omega_{\mu\beta} + \theta_{\mu\beta}\omega_{\nu\alpha} + \theta_{\nu\beta}\omega_{\mu\alpha})\,,$$
$$\hat{P}^{(0-s)} = P^{(0-s)}_{\mu\nu\,,\,\alpha\beta} = \frac{1}{n-1}\,\theta_{\mu\nu}\theta_{\alpha\beta}, \quad \hat{P}^{(0-w)} = P^{(0-w)}_{\mu\nu\,,\,\alpha\beta} = \omega_{\mu\nu}\omega_{\alpha\beta}. \quad (18.59)$$

To arrive at the closed algebra of projectors in the scalar sector, we need the additional transfer operators

$$\hat{P}^{(ws)} = P^{(ws)}_{\mu\nu\,,\,\alpha\beta} = \frac{1}{\sqrt{n-1}}\,\theta_{\mu\nu}\omega_{\alpha\beta}, \quad \hat{P}^{(sw)} = P^{(sw)}_{\mu\nu\,,\,\alpha\beta} = \frac{1}{\sqrt{n-1}}\,\omega_{\mu\nu}\theta_{\alpha\beta}. \qquad (18.60)$$

The algebra starts with the simple relations involving vector and tensor projectors,

$$\hat{P}^{(2)}\,\hat{P}^{(i)} \;=\; \hat{P}^{(2)}\delta_{i2} \qquad \text{and} \qquad \hat{P}^{(1)}\,\hat{P}^{(i)} \;=\; \hat{P}^{(1)}\delta_{i1}\,, \qquad (18.61)$$

where $i = 2,\, 1,\, 0-w\,, 0-s\,, sw$ and ws. In the scalar sector, we need to construct the matrix projector and check the relation

$$\hat{P}_0 \;=\; \frac{1}{2}\begin{Vmatrix} P^{(0-s)} & P^{(sw)} \\ P^{(ws)} & P^{(0-w)} \end{Vmatrix} \qquad \Longrightarrow \qquad \hat{P}_0\hat{P}_0 \;=\; \hat{P}_0. \qquad (18.62)$$

The next operation we have to elaborate is the inversion of the expression

$$\hat{B} = b_2\hat{P}^{(2)} + b_1\hat{P}^{(1)} + b_{os}P^{(0-s)} + b_{ow}\hat{P}^{(0-w)} + b_{sw}\Big[P^{(0-s)} + \hat{P}^{(0-w)}\Big], \qquad (18.63)$$

which is finding such an operator

$$\hat{C} = c_2\hat{P}^{(2)} + c_1\hat{P}^{(1)} + c_{os}P^{(0-s)} + c_{ow}\hat{P}^{(0-w)} + c_{sw}\Big[P^{(0-s)} + \hat{P}^{(0-w)}\Big], \qquad (18.64)$$

that $\hat{B}\hat{C} = \hat{1} = \hat{P}^{(2)} + \hat{P}^{(1)} + P^{(0-s)} + \hat{P}^{(0-w)}$. Using the algebra of projectors, we can easily get the solution to the problem,

$$c_2 = \frac{1}{b_2}, \quad c_1 = \frac{1}{b_1}, \quad c_{os} = -\,\frac{b_{ow}}{\Delta}, \quad c_{ow} = -\,\frac{b_{os}}{\Delta}, \quad c_{sw} = \frac{b_{sw}}{\Delta}, \qquad (18.65)$$

where $\Delta = b_{sw}^2 - b_{os}b_{ow}$.

One can easily present the projectors (18.59) and the transfer operators (18.60) as the linear combinations of the expressions (18.57):

$$\begin{aligned}
\hat{P}^{(2)} \;&=\; \hat{T}_2 - \frac{1}{n-1}\hat{T}_1 + \frac{1}{n-1}\hat{T}_3 - 2\hat{T}_4 + \frac{n-2}{n-1}\hat{T}_5,\\
\hat{P}^{(1)} \;&=\; 2\big(\hat{T}_4 - \hat{T}_5\big), \qquad \hat{P}^{(0-s)} \;=\; \frac{1}{n-1}\big(\hat{T}_2 - \hat{T}_3 + \hat{T}_5\big),\\
\hat{P}^{(0-w)} \;&=\; \hat{T}_5, \qquad P^{(ws)} + P^{(sw)} \;=\; \frac{1}{\sqrt{n-1}}\big(\hat{T}_3 - 2\hat{T}_5\big).
\end{aligned} \qquad (18.66)$$

Finally, by inverting these relations, one can express the matrices in (18.57) in terms of the projectors,

$$\begin{aligned}
\hat{T}_1 &= \hat{P}^{(2)} + \hat{P}^{(1)} + P^{(0-s)} + \hat{P}^{(0-w)},\\
\hat{T}_2 &= (n-1)P^{(0-s)} + \sqrt{n-1}\Big[P^{(ws)} + \hat{P}^{(sw)}\Big] + \hat{P}^{(0-w)},\\
\hat{T}_3 &= \sqrt{n-1}\Big[P^{(ws)} + \hat{P}^{(sw)}\Big] + 2\hat{P}^{(0-w)},\\
\hat{T}_4 &= \frac{1}{2}\,\hat{P}^{(1)} + \hat{P}^{(0-w)}, \qquad \hat{T}_5 = \hat{P}^{(0-w)}.
\end{aligned} \qquad (18.67)$$

Thus, starting from the bilinear form of the total action (18.54) for an arbitrary model of quantum gravity, the solution to Eq. (18.52) consists of recasting the bilinear form as (18.57). Subsequently, one has to use the relations between matrices $\hat{T}_l$ and projectors (18.67), invert the resulting expression using Eqs. (18.65) and, finally, use relations (18.66). Since the transfer operators enter the matrices $\hat{T}_l$ only in the combination $\hat{P}^{(ws)} + \hat{P}^{(sw)}$, the inversion prescription (18.65) is sufficient to complete this program for any bilinear form (18.55).

The exceptions may occurs when the relations (18.65) do not work because some of the denominators in the *r.h.s.*'s of these expressions vanish. This situation is typical and natural for the action of gravity without the gauge-fixing term S_{GF}. In this case, it is just a direct consequence of the gauge (diffeomorphism) symmetry and the related degeneracy. Indeed, in some models, the degeneracy occurs even after the Faddeev-Popov procedure of the diffeomorphism symmetry is applied. For instance, this is the case in the pure R^2 or pure C^2 models. In the last case, the situation can be fixed by introducing an additional gauge fixing for the local conformal symmetry [140] (see also subsequent papers [77, 9, 244]). In all other cases, the corresponding models of quantum gravity are non-renormalizable, as we shall see in the future section 18.7.

The formulas listed above are very universal, as they apply to any model of quantum gravity (local or nonlocal) with an action of the form (18.31). The next important question is how the choice of the gauge-fixing condition can affect the bilinear (in quantum metric) form of the total action of gravity with the S_{GF} term. To address this problem, it is sufficient to consider only two cases, namely, the gauge-fixing term (18.16) for quantum GR and (18.21) with the weight operator (18.34), since it includes the one for the fourth-derivative theory and polynomial gravity as particular cases. We can consider only (18.21) with (18.34) even for quantum GR, since it is sufficient to take (for the flat background) $\gamma = 0$ and $\Phi(x) = 1/x$.

A simple calculation shows that the contribution of the general gauge-fixing term (18.21) with (18.34) to the bilinear form (18.55) has the form

$$
\begin{aligned}
\Delta_{GF} H_{\mu\nu,\alpha\beta}(k;\,\eta) \,&=\, \Phi(-k^2)\Big\{\frac{\beta\gamma}{\alpha}\,k^2\big(\eta_{\mu\nu}k_\alpha k_\beta + \eta_{\alpha\beta}k_\mu k_\nu\big) \,+\, \frac{1-\gamma}{\alpha}\,k_\alpha k_\beta k_\mu k_\nu \\
&- \frac{1}{4\alpha}\,k^2\big(\eta_{\mu\alpha}k_\beta k_\nu + \eta_{\nu\alpha}k_\beta k_\mu + \eta_{\mu\beta}k_\alpha k_\nu + \eta_{\nu\beta}k_\alpha k_\mu\big) \,-\, \frac{\beta^2\gamma}{\alpha}\,k^4\eta_{\mu\nu}\eta_{\alpha\beta}\Big\} \\
&= \frac{1}{\alpha}\,\Phi(-k^2)\,k^4\Big\{\beta^2\gamma\,\hat{T}_2 \,+\, \beta\gamma\,\hat{T}_3 \,+\, \hat{T}_4 \,+\, (\gamma-1)\,\hat{T}_5\Big\} \\
&= \Phi(-k^2)\,k^4\Big\{\frac{1}{2\alpha}\,\hat{P}^{(1)} \,+\, \frac{3\beta^2\gamma}{\alpha}\,\hat{P}^{(0-s)} \,+\, \frac{(\beta+1)^2\gamma}{\alpha}\,\hat{P}^{(0-w)} \\
&+ \frac{3\beta^2\gamma}{\alpha}\,\hat{P}^{(0-s)} \,+\, \frac{\sqrt{3}\beta\gamma(\gamma+1)}{\alpha}\Big[\hat{P}^{(ws)} \,+\, \hat{P}^{(sw)}\Big]\Big\},
\end{aligned}
\tag{18.68}
$$

where we used the four-dimensional version of Eq. (18.67) in the last transformation. This formula provides a lot of information about the role of the gauge-fixing parameters in the total bilinear form

$$
H_{\mu\nu,\alpha\beta}(k;\,\eta) \,=\, H_{(0)\,\mu\nu,\alpha\beta}(k;\,\eta) \,+\, \Delta_{GF} H_{\mu\nu,\alpha\beta}(k;\,\eta),
\tag{18.69}
$$

where $H_{(0)\,\mu\nu,\alpha\beta}(k;\,\eta)$ is the degenerate bilinear form of the original action with unbroken symmetry.

First of all, it is clear that the gauge fixing does not affect the coefficient a_1, of the unit matrix (18.46). Owing to relations (18.66), this means that the gauge fixing does not affect the propagation of the spin-2 component of the metric perturbation (18.11).

The second observation is that, for the higher- (at least fourth-) derivative model, we have three arbitrary parameters, α, β and γ, for the gauge fixing. As a consequence of this fact, in expression (18.68) one can eliminate the term with $\hat{P}^{(1)}$ and the sum of the transfer operators $\hat{P}^{(ws)} + \hat{P}^{(sw)}$, making the matrix (18.62) diagonal. Also, the remaining third parameter can eliminate one linear combination of the scalar projectors $\hat{P}^{(0-s)}$ and $\hat{P}^{(0-w)}$. Then the bilinear form will depend on the remaining scalar mode, which is another linear combination of the same two projectors.

Let us repeat that all these considerations apply to all models of the type (18.31) with $\Lambda = 0$. One can say that, in all models of quantum gravity, the vector mode is always a pure gauge degree of freedom, at least on the flat spacetime background. Furthermore, all these models have a tensor (spin-2) degree of freedom that is independent of gauge fixing and is also independent of the parametrization of the quantum metric [166].

The third observation is that, in Eq. (18.69), there are three linearly independent constructions, $\frac{1}{4\alpha}$, $\frac{\beta\gamma}{\alpha}$ and $\frac{1-\gamma}{\alpha}$. Thus, by using the choice of the three parameters α, β and γ, one can eliminate terms with the coefficients $a_{3,4,5}$ in the total bilinear form (18.69). As a result, in any higher-derivative model, one can achieve the so-called minimal form of the total bilinear operator,

$$H^{total\,,\,minimal}_{\mu\nu,\alpha\beta}(k;\,\eta) \;=\; -\,\big[a_1(k^2)\delta_{\mu\nu,\alpha\beta} \,+\, a_2'(k^2)\eta_{\mu\nu}\eta_{\alpha\beta}\big]k^2, \qquad (18.70)$$

where a_2' is typically different from a_2 in Eq. (18.55) because of the contribution of the gauge-fixing term. This property of the bilinear form holds also for an arbitrary background metric. For this reason, it is important for the calculations in quantum gravity.

The advantage of the minimal form (18.70) compared to the one with projectors is that (18.70) is directly suited for the use of the heat-kernel technique. Indeed, it is possible to work with the nonminimal operators in quantum gravity using the generalized Schwinger-DeWitt technique [31] (see, e.g., [100] for an example of a practical calculation), but it is much simpler to work with the minimal bilinear form.

In some models, the choice of parameters is more restricted. For instance, in quantum GR, there are only two gauge-fixing parameters, α and β, but there are also less nonminimal terms, because of $a_5 = 0$. Thus, one can always choose α and β to achieve the minimal form (18.70).

The minimal operator in quantum GR enables one to use the standard expression for the trace of the a_2 heat kernel coefficient, as described above. At the same time, working out the nonminimal operators is always a difficult task, which has been solved, in the general form, only for the vector operator [31] (see also an earlier diagrams-based derivation of the equivalent result in [140].

The third observation is that the coefficient functions $a_1(k^2)$ and $a_2'(k^2)$ depend on the choice of the classical action of quantum gravity, and *not* on the gauge fixing (assuming that the last is minimal). In the case of quantum GR, a_1 and a'_2 are constants, for fourth-derivative gravity, they are linear functions of k^2, etc.

18.5 Gravitational waves, quantization and gravitons

The gravitational wave is a *classical* solution of Einstein's GR, or a version of modified gravity. However, since it has a direct relation to the quantization of gravitational field and requires bilinear expansions and gauge fixing, which we have just considered above, we placed this section in the part of the book describing quantum gravity.

18.5.1 Gravitational waves in GR

As a first step, consider the gravitational wave solutions on a flat background in Einstein's general relativity. For this, we start with an action of gravity without a cosmological constant term and go directly to the bilinear expansion (18.45) with the background metric $g_{\mu\nu} = \eta_{\mu\nu}$. Also, it is useful to add the action of matter and consider it in the approximation that gives a linear equation for the perturbation $h_{\mu\nu}$. In this way, we arrive at the action of GR in the linearized regime,

$$
\begin{aligned}
S^{(linear)}_{total} &= -\frac{1}{32\pi G}\int d^4x\, h^{\mu\nu}\Big\{\frac{1}{2}\,\delta_{\mu\nu,\alpha\beta}\Box - \frac{1}{2}\,\eta_{\mu\nu}\eta_{\alpha\beta}\Box - \eta_{\mu\alpha}\partial_\nu\partial_\beta \\
&\quad + \frac{1}{2}\big(\eta_{\mu\nu}\partial_\alpha\partial_\beta - \eta_{\alpha\beta}\partial_\mu\partial_\nu\big)\Big\}h^{\alpha\beta} \;-\; \frac{1}{2}\int d^4x\, h^{\mu\nu}T_{\mu\nu},
\end{aligned}
\qquad (18.71)
$$

where $\Box = \eta^{\mu\nu}\partial_\mu\partial_\nu$ and the source $T_{\mu\nu}$ is the energy-momentum tensor of matter in flat spacetime background. Let us remember that, in GR, matter means everything that is not the gravitational field (metric), including, e.g., electromagnetic radiation. The dynamical equation for metric perturbations has the form

$$
\Big\{\delta_{\mu\nu,\alpha\beta}\Box - \eta_{\mu\nu}\eta_{\alpha\beta}\Box - 2\eta_{\mu\alpha}\partial_\nu\partial_\beta + \big(\eta_{\mu\nu}\partial_\alpha\partial_\beta - \eta_{\alpha\beta}\partial_\mu\partial_\nu\Big\}h^{\alpha\beta} \;=\; 16\pi G T_{\mu\nu}. \qquad (18.72)
$$

It proves useful to work with the equation for the modified stress tensor,

$$
S_{\mu\nu} \;=\; T_{\mu\nu} \;-\; \frac{1}{2}\,T^\lambda_{\ \lambda}\, g_{\mu\nu}. \qquad (18.73)
$$

Technically, the best way to make this transformation is by multiplying both sides of Eq. (18.72) by the matrix

$$
K^{\mu\nu,\rho\sigma} \;=\; \delta^{\mu\nu,\rho\sigma} \;-\; \frac{1}{2}\,\eta^{\mu\nu}\eta^{\rho\sigma}\,. \qquad (18.74)
$$

A remarkable feature of this matrix is that it represents its own inverse,[1]

$$
K^{\mu\nu,\rho\sigma}K_{\rho\sigma,\alpha\beta} \;=\; \delta^{\mu\nu}{}_{,\alpha\beta}. \qquad (18.75)
$$

In terms of (18.73), after a very small algebra, Eq. (18.72) becomes

$$
\partial_\lambda\partial_\nu h^\lambda_{\ \mu} + \partial_\lambda\partial_\nu h^\lambda_{\ \mu} - \Box h_{\mu\nu} - \partial_\mu\partial_\mu h \;=\; 16\pi G\, S_{\mu\nu}\ \ \big(\text{here } h = h_{\mu\nu}\eta^{\mu\nu}\big), \qquad (18.76)
$$

which is operational in describing both propagation and emission of the gravitational waves in the linear approximation.

[1] Indices are lowered and raised by using the corresponding metric, which is Minkowski in our case.

At this point, we recall the gauge transformation (18.1) and impose the gauge-fixing condition. It is easy to prove that the gauge transformation for $h_{\mu\nu}$ is

$$\delta h_{\mu\nu} = - \partial_\mu \xi_\nu - \partial_\nu \xi_\mu, \tag{18.77}$$

where $\xi_\mu(x)$ is an arbitrary vector field in the flat background.

Let us choose the *harmonic* gauge-fixing condition (also called the de Donder or the Fock-de Donder gauge)

$$\partial_\mu h^\mu_{\ \nu} - \frac{1}{2}\, \partial_\nu h = 0. \tag{18.78}$$

Starting from an arbitrary $\tilde{h}_{\mu\nu}(x)$ and making the gauge transformation (18.77) with ξ_μ, satisfying the condition $\Box \xi_\mu = \partial_\mu \tilde{h}^\mu_{\ \nu} - \frac{1}{2}\, \partial_\nu \tilde{h}$, we arrive at the special metric $h^\mu_{\ \nu}$, satisfying (18.78). The reader can easily recognize that this is a fundamental property of the orbits of gauge transformation, which is discussed in section 10.2. Now, using condition (18.78) in Eq. (18.76), the last is cast in the form

$$\Box h_{\mu\nu} = - 16\pi G\, S_{\mu\nu}. \tag{18.79}$$

To describe the propagation of the gravitational wave in vacuum, we have to set the *r.h.s.* of Eq. (18.79) to zero and also use condition (18.78) and the remnant gauge transformation (18.77) with $\Box \xi_\mu = 0$. The corresponding analysis is described in many textbooks on gravity, e.g., [340], so we will just mention the final result. For the plane wave propagating along the axis Oz, there are two linear-independent, gauge-invariant transverse polarizations, e.g., h_{xx} and h_{xy}. This output shows that the unique component of the metric perturbation (18.11) which is gauge invariant and can be called physical, is $\bar{h}^{\perp\perp}_{\mu\nu}$. Thus, the gravitational wave in GR is a propagation of the spin-2 state.

The emission of the gravitational wave in the linear regime can be described by Eq. (18.79); hence, the solution in the standard form of retarded potential is

$$h_{\alpha\beta}(\mathbf{x}, t) = 4G \int d^3x'\, \frac{S_{\alpha\beta}(\mathbf{x}', t - |\mathbf{x} - \mathbf{x}'|)}{|\mathbf{x} - \mathbf{x}'|}\,. \tag{18.80}$$

This expression shows why it is so difficult to detect the gravitational wave experimentally. The factor G in the *r.h.s.* of the last expression is the inverse square of the Planck mass M_P. Thus, the emission of the gravitational wave is initially suppressed by the square of the Planck mass. After that, the wave has to travel a very long distance and then go through a similar Planck suppression at the moment of its detection.[2]

18.5.2 Quantization and the notion of graviton

The canonical quantization of the linearized gravity can be performed in a way similar to the electromagnetic field. We shall skip this part almost completely and only say that

[2] In fact, the waves detected by the LIGO/Virgo collaboration [1], were emitted in a much more complicated way. Their description is very complicated, as it requires taking into account the non-linearities of the process involving the merger of two black holes or neutron stars.

the physical degrees of freedom that describe the quantum state of a free gravitational field on a flat background are exactly those of the flat gravitational wave described in the previous subsection. The corresponding particle has zero mass and spin-2, it is called a *graviton*.

In order to derive the spin-2 part of the propagator of the gravitational perturbation $h_{\mu\nu}$, we start from the bilinear expansion of the Einstein-Hilbert term (18.45). We note that, according to definitions (18.59), only the term with $\delta_{\mu\nu,\,\alpha\beta}$ contributes to the projector $P^{(2)}_{\mu\nu\,,\,\alpha\beta}$, while other terms are irrelevant. Thus, in the spin-2 sector the solution of Eq. (18.52) is especially simple and reduces to the simple inversion of the numerical coefficient $\frac{1}{4}$. For the sake of simplicity, we ignore this coefficient and hence obtain the spin-2 part of the propagator in the form

$$\langle h_{\mu\nu} h_{\alpha\beta}\rangle^{(2)} = G^{(2)}_{\mu\nu,\alpha\beta}(k) = \frac{P^{(2)}_{\mu\nu,\alpha\beta}(k)}{k^2 + i\varepsilon} \,. \tag{18.81}$$

This equation describes the propagation of the physical degrees of freedom associated with the spin-2 states, which is the graviton. Within the Faddeev-Popov procedure, all other components of the gravitational perturbation (18.11) propagate too, but these parts of the propagator are gauge (and parametrization) dependent, while (18.81) is invariant and universal.

This is the situation in the model of quantum gravity that is based on Einstein's GR. In other models, the propagator can be more complicated. For instance, including the fourth-derivative terms, there may be the following two changes:

i) Instead of the unique massless pole in the propagator (18.81), there may be another massive pole, as we discuss in detail in section 19.2.

ii) The scalar components of the metric perturbation (18.11) gain an invariant, massive, gauge-independent part of the propagator (see section 18.4).

Adding more derivatives, which means using polynomial or even nonlocal models, the modifications always concern the same two points. Namely, there will be (in the polynomial models) a growing number of poles in the spin-2 sector and the scalar sector. In contrast, some choices of nonlocal action may provide that there would not be any massive poles in the tree-level propagator, in both spin-2 and spin-0 sectors.

It is worth noting that the count of degrees of freedom, based on the simple analysis of the gravitational propagator, was confirmed by the canonical quantization of the gravitational theory in the cases of quantum GR and fourth-derivative quantum gravity (see, e.g., [160] and [80]).

Exercises

18.15. Consider the projectors to the traceless and trace modes of the metric in $n = 4$,

$$\bar{P}^{\mu\nu,\rho\sigma} = \delta_{\mu\nu,\alpha\beta} - \frac{1}{4}\eta_{\mu\nu}\eta_{\alpha\beta} = \delta_{\mu\nu,\alpha\beta} - \frac{1}{2}P^{tr}_{\mu\nu,\alpha\beta}. \tag{18.82}$$

i Prove that the two matrices are projectors, which means $\bar{P}^2 = \bar{P}$, $(P^{tr})^2 = P^{tr}$, $\bar{P} \cdot P^{tr} = 0$. Then, use them to check the identity (18.75). ii) Verify the same identity directly, without using the projectors (18.82). iii) Formulate the analogs of (18.82) and (18.75) in n-dimensional spacetime. Discuss why these formulas do not depend on the signature of the metric and on whether the metric is covariant or flat. iv) Explore what happens with all irreducible components of metric perturbation (18.11) under the action of projectors (18.82). v) Explain the qualitative difference between the two functions in Eq. (18.33).

Hint. In the last point, it is necessary to pay attention to the difference in the propagators of scalar and tensor modes in quantum GR.

18.16. Derive (18.77) from (18.1). Discuss the generalization of this relation for an arbitrary background metric.

Hint. This is a non-trivial problem, especially in the context of quantum gravity and background field method. We recommend the paper [208], where it is discussed as part of the systematic construction of the background field method in quantum gravity, based on the Blatalin-Vilkovisky formalism.

18.17. Discuss in detail the relation between the representation (18.11) of the metric perturbations on the flat background, and the projectors (18.59). In particular, identify which of the projectors or their combinations does not change each of the irreducible components of (18.11).

18.18. Verify the algebra of projectors (18.61) and (18.62). Consider in details the special dimensions $n = 2$ and $n = 3$. Explain why these two cases should be expected to possess special properties.

Hint. Start with the properties of the curvature tensors in these dimensions, and check out how this reflects on the general action (18.31). What can be the maximal number of independent coefficients s_l in the bilinear form (18.56) in these cases? Use the expression (18.48).

18.19. Verify the expression (18.68).

18.20. Try to express each of the operators in (18.55) as linear combinations of the projectors (18.59). Explain the result, in particular why it does not affect the possibility to express any *possible* structure in the $H^{total}_{\mu\nu,\alpha\beta}(k;\,\eta)$ in Eq. (18.69) as a linear combination of the projectors.

18.21. Find the combination of the projectors of the scalar modes in $n = 4$, which is independent of the gauge-fixing choice.

18.6 Gauge-invariant renormalization in quantum gravity

In section 10.10 in Part I, we showed that renormalization preserves the gauge invariance of classical field models if the algebra of generators is closed. In the case of the background field method, the structure of renormalization is especially simple. The gauge-invariant renormalization, in this case, means that the counterterms are covariant local expressions constructed from the background gauge field, as explained in section 10.10.2.

In the case of a quantum metric, we can apply all the formalism described in Part I, because the algebra of the generators is closed, according to (18.8). The renormalization of quantum gravity has been considered in the classical papers [106] and [307]. In the last reference, one can find comprehensive proof for the fourth-order gravity based on the BRST invariance and the gravitational version of the Slavnov-Taylor identities (see also [337]). In principle, the analysis of Stelle [307] can be generalized for the gravitational actions of the general form, but it is more practical to use the more sophisticated and much more economical Batalin-Vilkovisky formalism. In this way, the detailed proof of the gauge-invariant renormalization in quantum gravity of a general form, in the background field formalism, has been given in the paper [208]. This proof is technically simpler and more general than the one given in [307]. However, it requires the previous formulation of the Batalin-Vilkovisky formalism and therefore lies beyond the scope of the present textbook. Thus, we refer the reader to the aforementioned papers for a detailed treatment of the problem. On the other hand, in the rest of this book, we rely on the general consideration of the gauge theories in Part I. Accordingly, the counterterms in quantum gravity are always covariant, independently of whether the model under discussion is renormalizable or not. Let us mention that the gauge-invariant renormalization in the multi-loop diagrams imposes certain conditions on subdiagrams with external lines of the ghost fields and the antifields, which are required within the Batalin-Vilkovisky formalism [208].

The general statement about the covariance of the counterterms concerns only the last integrations in the multi-loop diagrams. These diagrams produce the counterterms corresponding to the power counting arguments, as will be described in the next chapters. Another important aspect of the general analysis of renormalization performed in [208] is that, in quantum gravity, the counterterms are local, even if the starting theory is nonlocal. Let us mention that the covariance of the one-loop counterterms has been verified for quantum GR, including that coupled to matter, in [186,101] (and in many subsequent papers that we cannot mention here). Another calculational confirmation at the one-loop level was given in fourth-derivative theory in [190,140,17] for pure gravity and for quantum gravity coupled to matter in [140, 77, 66] (see also [80] and more recent [273, 188] for alternative calculations with qualitatively similar results). The calculations in the pure gravity cases will be reproduced in subsequent chapters.

In what follows, we assume the correctness of the statement about invariant renormalizability, and use it as a basis for exploring renormalization in quantum gravity.

18.7 Power counting, and classification of quantum gravity models

The evaluation of power counting in quantum gravity is, indeed, simpler than in the models of scalars, fermions, and vectors, which were discussed in Part I. The reason is that the metric is a dimensionless field. Thus, the dimension of the counterterm that appears in the diagram of G with L loops is defined only by the number of derivatives of the metric. In what follows, we assume expansion on the flat background (13.23), without κ, since we want to deal not only with quantum GR but also with higher-derivative models, where κ is not the loop expansion parameter.

The power counting of a diagram with an arbitrary number of external lines of $h_{\mu\nu}$ is defined by the superficial degree of divergences $\omega(G)$ (also called the index of divergence) and the number $d(G)$ of the partial derivatives of the external lines of the field $h_{\mu\nu}$. According to considerations from chapter 9, the general expression is

$$\omega(G) + d(G) = \sum_{l_{int}}(4 - r_I) - 4V + 4 + \sum_{V} K_V, \tag{18.83}$$

where the first sum is over all I internal lines of the diagram, r_I is the inverse power of momentum in the propagator of the given internal line and V is the number of vertices. The second sum is taken over all the vertices, and K_V is the number of derivatives (or power of momenta, in the momentum representation) of the lines (both internal and external) coming to the given vertex.

As usual (see chapter 9 in Part I), formula (18.83) is insufficient to evaluate the renormalization feature of the given theory. However, in addition to this formula, there is the topological relation (8.13) of the form $L = I - V + 1$. The two relations together are sufficient to conduct the analysis we need. Before starting to consider concrete examples, let us make an important observation. The diagrams in quantum gravity, which we intend to analyze, have external lines of the field $h_{\mu\nu}$ only, but there are internal lines of both $h_{\mu\nu}$ and the Faddeev-Popov ghosts. However, with the modified definitions of the ghost actions, which we constructed in the section 18.2, the values of r_I are the same for the quantum metric and ghosts. For instance, in quantum GR, both fields have $r_I = 2$, in fourth-derivative gravity in both cases $r_I = 4$, in the polynomial models $r_I = 2N + 4$. In the nonlocal models of the form (18.31), both r_I and K_V are infinite, but we shall see how to deal with this special case at the end of this section. Thus, in what follows, we assume that, in all models of interest, r_I are identical for the quantum metric and ghosts.

18.7.1 Power counting in quantum gravity based on GR

As the first example, consider power counting in quantum GR, with $r_I = 2$ for all internal lines. The vertices coming from the Einstein-Hilbert term have $K_{EH} = 2$. If we include the cosmological constant term, there are also vertices $K_\Lambda = 0$. However, looking only for strongest divergences, for the moment consider the diagrams with $K_V = 2$ vertices only. Then (18.83), together with the topological relation (8.13), yields

$$\omega(G) + d(G) = 2I - 4V + 4 + 2V = 2I - 2V + 4 = 2 + 2L. \tag{18.84}$$

The last result clearly shows that the quantum gravity based on GR is not renormalizable. At the one-loop level ($L = 1$), the logarithmic divergences with $\omega(G) = 0$ have $d(G) = 4$, which means the possible covariant structures listed in the fourth-derivative action (18.20). Indeed, at the one-loop order, there are counterterms of the Einstein-Hilbert form $\sim \int \sqrt{-g}R$ with $d(G) = 2$ but with quadratic divergences only, since $\omega(G) = 2$. Logarithmic divergences of this type are also possible, but only if we introduce a vertex with $K_V = K_\Lambda = 0$, which comes from the cosmological constant term. With two such vertices, there is a logarithmic divergence without derivatives, $d(G) = 0$. In section 20.2, we shall confirm all these conclusions by direct calculations based on the heat kernel method and also analyze the gauge-fixing and parametrization dependence of the one-loop counterterms.

In the two-loop order, according to Eq. (18.84), logarithmic divergences without $K_\Lambda = 0$ vertices, have dimension 6. A detailed list of the corresponding terms has been elaborated in the publications on the conformal anomaly in six spacetime dimensions (see, e.g., [62, 130]). The full list of dynamical terms includes

$$
\begin{array}{lll}
\Sigma_1 = R_{\mu\nu}R^{\mu\alpha}R^{\nu}_{\alpha} & \Sigma_2 = (\nabla_\lambda R_{\mu\nu\alpha\beta})^2 & \Sigma_3 = R_{\mu\alpha\nu\beta}\nabla^\mu\nabla^\nu R^{\alpha\beta} \\
\Sigma_4 = R_{\mu\nu}R^{\mu\lambda\alpha\beta}R^{\nu}{}_{\lambda\alpha\beta} & \Sigma_5 = R^{\mu\nu}{}_{\alpha\beta}R^{\alpha\beta}{}_{\lambda\tau}R^{\lambda\tau}{}_{\mu\nu} & \Sigma_6 = R^{\mu}{}_{\alpha}{}^{\nu}{}_{\beta}R^{\alpha}{}_{\lambda}{}^{\beta}{}_{\tau}R^{\lambda}{}_{\mu}{}^{\tau}{}_{\nu} \\
\Sigma_7 = (\nabla_\lambda R_{\mu\nu})^2 & \Sigma_8 = R_{\mu\nu}\Box R^{\mu\nu} & \Sigma_9 = (\nabla_\mu R)^2 \\
\Sigma_{10} = R\Box R & \Sigma_{11} = (\nabla_\alpha R_{\mu\nu})\nabla^\mu R^{\nu\alpha} & \Sigma_{12} = R^{\mu\nu}\nabla_\mu\nabla_\nu R \\
\Sigma_{13} = R_{\mu\nu}R_{\alpha\beta}R^{\mu\alpha\nu\beta} & &
\end{array}
\tag{18.85}
$$

as well as the set of surface terms,

$$
\begin{array}{llll}
\Xi_1 = \Box^2 R & \Xi_2 = \Box R^2_{\mu\nu\alpha\beta} & \Xi_3 = \Box R^2_{\mu\nu} & \Xi_4 = \Box R^2 \\
\multicolumn{2}{l}{\Xi_5 = \nabla_\mu\nabla_\nu\big(R^{\mu}{}_{\lambda\alpha\beta}R^{\nu\lambda\alpha\beta}\big)} & \multicolumn{2}{l}{\Xi_6 = \nabla_\mu\nabla_\nu\big(R_{\alpha\beta}R^{\mu\alpha\nu\beta}\big)} \\
\multicolumn{2}{l}{\Xi_7 = \nabla_\mu\nabla_\nu\big(R^{\mu}_{\alpha}R^{\nu\alpha}\big)} & \multicolumn{2}{l}{\Xi_8 = \nabla_\mu\nabla_\nu\big(RR^{\mu\nu}\big),}
\end{array}
\tag{18.86}
$$

satisfying the identity [130]

$$
\Xi_2 - 4\Xi_3 + \Xi_4 - 4\Xi_5 + 8\Xi_6 + 8\Xi_7 - 4\Xi_8 = 0. \tag{18.87}
$$

All these terms can show up in the two-loop divergences, but only two of these terms are critically important. In order to understand this point, let us come back to the one-loop divergences, and write the list of possible one-loop counterterms,

$$
C^2 = E_4 + 2W \ \ \big(\text{where} \quad W = R^2_{\mu\nu} - \tfrac{1}{3}R^2\big), \quad E_4, \quad R^2, \quad \Box R. \tag{18.88}
$$

Here E_4 and $\Box R$ are surface terms, which do not affect the dynamics of the theory, and the other two terms vanish on the classical equations of motion, when $R_{\mu\nu} = 0$. Thus, the one-loop $\mathcal{S}$-matrix in the quantum GR without matter contents is finite.

Many of the two-loop terms also vanish on shell. However, there are two terms, namely, Σ_5 and Σ_6, that survive in the Ricci-flat space. Thus, the confirmation of the non-renormalizability of quantum GR requires a two-loop calculation. This calculation has been done in [170, 326] and confirmed the non-zero coefficient of Σ_5.

Thus, we can conclude that the power counting formula (18.84) really means that the theory is non-renormalizable. The last point to mention is that, for matter-gravity systems, the structure of one-loop divergences is such that these divergences do not vanish on shell [186, 101]. This result has been achieved in the aforementioned papers by direct calculation for gravity-scalar,[3] gravity-fermion and gravity-vector cases. We leave the analysis of the power counting in these cases as an exercise.

18.7.2 Power counting in fourth-derivative gravity models

Consider power counting in the fourth-derivative quantum gravity (18.18). For the sake of simplicity, we assume the Faddeev-Popov procedure with the second-order weight operator (18.22) and the modified ghost action (18.23). Then, for all modes of the gravitational perturbation $h_{\mu\nu}$ and ghosts, we have $r_l = 4$. On top of that, the vertices are $K_V = (K_{4d},\, K_{EH},\, K_\Lambda)$.

[3]In the metric-scalar case, one-loop calculations may be technically more complicated because of the nonminimal coupling of the scalar field with the scalar curvature, see, e.g., [35, 280, 195].

Let us call n_{4d} the number of vertices with four power of momenta, n_{EH} that with two and n_Λ that with zero, such that

$$n_{4d} + n_{EH} + n_\Lambda = V \quad \text{and} \quad n_{4d}K_{4d} + n_{EH}K_{EH} + n_\Lambda K_\Lambda = \sum_V K_V. \qquad (18.89)$$

Then, the general expression (18.83), together with the topological relation (8.13), provide the following result:

$$\omega(G) + d(G) \;=\; 4 - 2n_{EH} \;-\; 4n_\Lambda\,. \qquad (18.90)$$

As a starting point, consider the diagrams with the strongest divergences, where all of the vertices are of the K_{4d} type, such that $V = n_{4d}$. In this case, (18.90) means that the logarithmic divergences have $d(G) = 4$. Taking into account locality and covariance arguments, the possible counterterms are of the C^2, R^2, E_4 and $\Box R$ types. The divergences have the same form as the fourth-derivative terms in the classical action (18.18). Assuming that $V - 1 = n_{4d}$ and $n_{EH} = 1$, we obtain $d(G) = 2$, corresponding to the counterterm linear in R. Finally, for $n_{EH} = 2$ and $V - 2 = n_{4d}$, or for $n_\Lambda = 1$ and $V - 1 = n_{4d}$, there is a counterterm with $d(G) = 0$, which is the cosmological constant. Thus, the theory under consideration is multiplicatively renormalizable. As we shall see in the next section, this does not imply that the theory is consistent, since there is a high price to pay for renormalizability.

Let us now consider the particular case of the general model (18.18), without dimensional parameters. The classical action has a global conformal symmetry under the transformation $g_{\mu\nu} \to g_{\mu\nu}k^2$, with $k = const$. The power counting (18.90) can be perfectly well applied in this case, yielding $\omega(G) + d(G) = 4$. This means that the theory is multiplicatively renormalizable. The disadvantage of this version of gravity is that there is no Einstein limit in the low-energy domain. Let us remember that such a limit is one of the main conditions of consistency of any model which generalizes or modifies GR, so this situation should be seen as a problem of the model.

Another particular case, which is instructive to consider, is the $R + R^2$-gravity, which is model (18.18) without the C^2-term. As we can see from the considerations in sections 18.4 and 18.3, in this model, the traceless component of the metric $\bar{h}_{\mu\nu}$ has $1/k^2$ propagator, while the scalar mode has a propagator, that behaves as $1/k^4$ in UV. Furthermore, there are vertices K_{4d}. We leave it as an exercise for the reader to verify that the power counting in this model is dramatically different from that in the general fourth-derivative model. Let us just say that the theory is non-renormalizable and the power counting is much worse than in quantum GR.

The last example is the model (18.18) without the R^2 term. This particular model is especially interesting since the fourth-derivative part of the action possesses local conformal symmetry. This symmetry is softly broken by the Einstein-Hilbert and cosmological terms. The expression "soft breaking" means that the symmetry does not hold in the terms with dimensional parameters. Can it be that the softly broken conformal symmetry "saves" the power counting in this model? The answer is certainly negative. The propagator of the traceless mode of the metric, $\bar{h}_{\mu\nu}$, in this case, has the UV behavior $1/k^4$, and that of the scalar mode the UV behavior $1/k^2$, due to the presence of R term. At the same time, there are K_{4d} vertices that link all of the modes, and hence the power counting is qualitatively the same as in the previous $R+R^2$ case. The theory is not renormalizable.

18.7.3 Power counting in the polynomial theory

Consider power counting in the polynomial model (18.26), assuming the Faddeev-Popov quantization with the weight operator (18.29) and the modified ghost term (18.23). We will follow the original work [13], but give more details.

In the general case, which we consider first, both coefficients of the highest derivative terms $\omega_{N,R}$ and $\omega_{N,C}$ are non-zero. In this case, the propagators of both metric perturbations $h_{\mu\nu}$ and ghosts have the UV behavior of the k^{-4-2N} type, which means $r_l \equiv 4+2N$. Concerning the vertices, the generalization of Eq. (18.89) is

$$\sum_V K_V = n_{4+2N}K_{4+2N} + n_{2+2N}K_{2+2N} + n_{2N}K_{2N} + \dots + n_{EH}K_{EH} + n_\Lambda K_\Lambda$$
$$\text{and} \qquad V = n_{4+2N} + n_{2+2N} + n_{2N} + \dots + n_{EH} + n_\Lambda, \tag{18.91}$$

where $K_{4+2N} = 4+2N$, $K_{2+2N} = 2+2N$, ... $K_{EH} = 2$, $K_\Lambda = 0$ and n_{4+2N}, n_{2+2N}, n_{2N}, ..., n_{EH} and n_Λ are the numbers of the corresponding vertices.

Since there are many different types of vertices, we start by analyzing the diagrams with the strongest divergences. This means $V = n_{4+2N}$, such that the other types of vertices are absent. Then the power counting becomes very simple, because of $\sum_V K_V = V(4+2N)$. The expression (18.83) becomes

$$\begin{aligned}\omega(G) + d(G) &= (4-4-2N)I \; - \; 4V \; + \; 4 \; + \; V(4+2N) \\ &= 4 \; + \; 2N(V-I) \; = \; 4 \; + \; 2N(1-L),\end{aligned} \tag{18.92}$$

where we used the topological relation (8.13) in the form $V - I = 1 - L$. Indeed, the power counting in the four-derivative model (18.90) is a particular case of the last relation, corresponding to $N = 0$. Until the end of this section, we assume that $N \geq 1$.

According to the result (18.92), the sum $\omega(G) + d(G)$ decreases with a growing number of loops L in the diagram G. The strongest divergences occur for $L = 1$, when the aforementioned sum equals 4. Thus, the logarithmic divergences correspond to the one-loop counterterms with, at most, four derivatives. Taking the covariance and locality into account, this means that the one-loop counterterms are of the C^2, R^2, E_4 and $\Box R$ types only. In other words, all the terms with six and more derivatives do not need to be renormalized, even at the one-loop order.

If there is just a single vertex with one less derivative, $n_{2+2N} = 1$ and $n_{4+2N} = V - 1$. At the one-loop level, there is only the Einstein-Hilbert type of divergence. Finally, in the case $n_{2+2N} = 2$ or $n_{2N} = 1$, the unique divergence is that of the cosmological constant.

The features of the $L = 1$ approximation that we listed above, do not depend on N. Starting from $L \geq 2$, the structure of divergences starts to depend on the value of N. In particular, for $N \geq 3$, according to (18.92), the second- and higher-loop diagrams are all finite. This creates a situation when, for example, the one-loop beta functions are, indeed, the exact ones. In the case of $N = 2$, there are two-loop divergences, but only of the cosmological constant type. Finally, for $N = 1$, there are two-loop divergences of the Einstein-Hilbert and the cosmological constant type and also three-loop divergences, but only of the cosmological constant type.

All in all, a theory with both $\omega_{N,R} \neq 0$ and $\omega_{N,C} \neq 0$, is superrenormalizable. In contrast, in the degenerate case, when only one of these coefficients is zero, the theory is non-renormalizable. We leave the corresponding analysis as an exercise.

18.7.4 Power counting in the nonlocal models of quantum gravity

Consider the models of quantum gravity with the nonlocal action (18.31). It is evident that the power counting in these models strongly depends on the choice of the functions $\Phi(x)$ and $\Psi(x)$. Usually, these functions are exponentials of entire functions. This choice may mean that the propagator of gravitational perturbations has a very strong (stronger than in the polynomial case) UV asymptotic and, on the other hand, that the propagator does not have unphysical poles corresponding to massive ghosts.

The problem of ghosts will be discussed in detail in the next chapter, and, for the moment, we simply assume that the two functions $\Phi(x)$ and $\Psi(x)$ behave, qualitatively, as $e^{\alpha x}$. Initially, we assume that condition (18.32) is satisfied, and, after that, we will discuss the models where $\Phi(x) \sim e^{\alpha x}$ and $\Psi(x) \sim e^{\beta x}$, with $\alpha \neq \beta$.

Our intuition tells us that power counting in models with $\Phi(x) \sim \Psi(x) \sim e^{\alpha x}$ should be analogous to that in polynomial models with a very large N. However, at the first sight, the power counting formula (18.83) cannot be applied to nonlocal models. As we have $r_l = \infty$ and $K_V = \infty$, the sum $\omega(G) + d(G)$ boils down to an indefinite expression of the type $\infty - \infty$, which cannot be used in a reasonable way. The solution which we describe below is from [290].

In such a degenerate situation, when $r_l = \infty$ and $K_V = \infty$, the expression (18.83) is useless, but we still have the topological relation (8.13). Writing this formula as

$$V - I = 1 - L, \tag{18.93}$$

we note that each internal line enters the diagram with the factor $e^{-\alpha p^2}$, where, to simplify the analysis, we assume that $\alpha > 0$ and consider the Euclidean signature.[4] Thus, the contribution of internal lines to the integral is $e^{-\alpha I p^2}$. On the other hand, the contribution of the vertices is bounded from above by the factor $e^{\alpha V p^2}$, which corresponds to the situation of all derivatives in the vertices acting on internal lines. Taking into account the topological relation (18.93), we conclude that the overall exponential factor in the last integration of the L-loop diagram is $e^{\alpha(V-I)p^2} = e^{-\alpha(L-1)p^2}$. For $L \geq 2$, the UV limit in the integral over Euclidean momentum p is finite, owing to this exponential factor. This output means that all diagrams beyond the one-loop order are automatically finite. This fact was first noticed in the unpublished preprint [316] (see also the earlier publications [200, 201] and the detailed analysis in [220]).

Thus, if condition (18.32) is satisfied, the divergences in the theory (18.31) can be met only at the one-loop level. On the other hand, for the power counting of the one-loop divergences, one can safely ignore the exponential factors, since they

[4]The Wick rotation in this kind of model is not a problem, because there is a unique pole corresponding to graviton: the massless mode. The situation is completely different in the polynomial gravity models, but this complicated issue lies beyond the scope of this introductory book.

cancel out. Without these factors, the theory under consideration has the same power counting as the fourth-derivative gravity without Einstein-Hilbert and cosmological terms. The one-loop divergences are covariant local expressions (covariance has been proved in [208], where one can also find strong arguments in favor of locality) with four derivatives, which means the C^2, R^2, E_4 and $\Box R$ counterterms are needed. Does this mean that the theory is multiplicatively renormalizable? The answer is not automatic. The point is that the four-derivative terms, such as C^2 and R^2, are certainly part of the action (18.31), but their coefficients are fine-tuned to avoid massive ghosts. This means that, even though C^2 and R^2 terms are present in the action, the coefficients of these terms cannot be renormalized as independent parameters.

The rescue of renormalizability is possible, as it can come from the terms $\mathcal{O}(R^3_{\dots})$ in the action (18.31). Initially, we ignored these terms, since they contribute only to the vertices and not to the propagator of $h_{\mu\nu}$. Now it is the right moment to pay due attention to these vertex terms. First of all, it is clear from the previous analysis, that these terms should have exponential factors satisfying the conditions analogous to (18.32). Assuming that this is the case, these terms can be fine-tuned such that the one-loop C^2 and R^2 counterterms cancel. Then the theory becomes finite, up to the surface divergences. The corresponding vertex terms $\mathcal{O}(R^3_{\dots})$ were called "killers" in [220].

Thus, the main conclusions about the power counting in the nonlocal models (18.31) are as follows:

i) Assuming that the condition (18.32) is satisfied and complemented by similar relations (which may be reduced to inequalities) for the form factors in the vertices $\mathcal{O}(R^3_{\dots})$, the divergences exist only at the one-loop level.

ii) To be multiplicatively renormalizable, such a model should be finite. And the finiteness *probably* requires that the relations similar to (18.32) are satisfied for at least two vertex-generating terms $\mathcal{O}(R^3_{\dots})$. These two terms need to have precisely fine-tuned coefficients to cancel the possible C^2 and R^2 one-loop counterterms. Let us mention that the word "probably" here means that this conclusion is not supported by direct calculations at the time that this book is written. In principle, such a cancelation can occur even without the $\mathcal{O}(R^3_{\dots})$ terms, but this can be checked only by calculations. At the same time, since there are many such terms [see, e.g., Eq. (18.85)] there is no real doubt that the required cancelation can be achieved.

iii) If the relation (18.32) is *not* satisfied, the model under discussion will be non-renormalizable by power counting. The violation of this relation makes the propagators of different modes of the gravitational perturbations (18.11) non-homogeneous in the derivatives. In this situation, there will be all-L diagrams where the powers of momenta in the vertices exceed the power of momenta in the propagators.

Exercises

18.22. Assuming covariant cut-off regularization (e.g., cut-off in momenta in the local momentum representation), perform a detailed analysis of the power counting in a gravity-matter system. As a starting point, consider quantum GR coupled to a free massive scalar field with minimal interaction. Derive the generalization of Eq. (18.84), and use it to find all possible types of covariant and local one-loop counterterms for logarithmic- and power-type divergences.

18.23. Generalize the results of the previous exercise to the case of a nonminimal massless scalar. Is it possible to obtain a logarithmic counterterm that is linear in R, with a constant coefficient in a theory with $\Lambda = 0$?

18.24. Try to solve the previous exercise without making any kind of complicated considerations. One has to start by making a change of variables in the scalar sector, such that the new scalar becomes dimensionless.

18.25. Perform the same analysis in quantum GR with $\Lambda = 0$, coupled to the Dirac fermion and to the electromagnetic field (separately). Which kind of counterterms can one expect in this theory at the one-loop order?

Hint. In the fermionic case, one can start by extending expansion (12.80) from section 12.4 to the second order in $h_{\mu\nu}$ and then expanding, to the same order, the spinor connection and the whole Dirac action [80]. This calculation shows how to construct quantum gravity coupled with fermions by using the quantum metric. Let us stress that this calculation is not necessary for power counting, but it may be useful for improving the understanding of how the quantum theory is constructed.

18.26. Compare the power counting in quantum GR with the one in the nonlinear sigma model (4.19) from Part I.

18.27. Verify relation (18.90), and use it to analyze the form of quadratic divergences in the fourth-derivative theory (18.18).

18.28. Derive the power counting in the model (18.18) without the C^2 term, and compare it with the expression (18.84) for quantum GR. Confirm the result with a detailed analysis of the maximal possible divergences of the two-point and three-point functions.

18.29. Discuss power counting in pure Weyl-squared gravity with the covariant Lagrangian $\mathcal{L} = -\frac{1}{2\lambda}C^2$. The main difference with respect to the non-degenerate case discussed above is that, in Weyl-squared gravity the local conformal symmetry is unbroken at the tree level. Consider the following two types of analysis: i) the analysis that does not take the conformal anomaly into account, and ii) a full consideration taking into account the anomaly.

Warning. The last case may be complicated. It was discussed in the paper [244], where the reader can find some useful hints.

18.30. Explore power counting in the theory (18.26) with only one of the coefficients of the leading terms $\omega_{N,R}$, or $\omega_{N,C}$ being equal to zero. Compare the result with the corresponding analysis of the degenerate case in fourth-derivative gravity.

19

Massive ghosts in higher-derivative models

In the previous chapter, we saw that quantum GR, based on the Einstein-Hilbert action is a non-renormalizable theory. On the other hand, by adding the general fourth-derivative terms, we arrive at the model (18.18), which provides multiplicative renormalizability, both in the pure gravity case and when gravity is coupled to the matter fields. A very strong argument for including fourth-derivative terms is that they are required for the renormalizability of the semiclassical theory, when gravity is an external field.

Unfortunately, even if theory (18.18) is, formally, multiplicatively renormalizability, it does not make it consistent at either the quantum or even classical levels. As we shall see in brief, the spectrum of this model includes states that have negative kinetic energy. These states, or particles, are called massive unphysical ghosts. The presence of ghosts violates the unitarity of the theory at the quantum level, even though there may be various alternative formulations, including those that make the $\mathcal{S}$-matrix unitary. Worse than that, in the presence of massive ghosts or, more generally, ghost-like states, classical solutions of the theory can not be stable with respect to the general metric perturbations. Qualitatively, the same situation takes place not only in the fourth-derivative theory but in all polynomial models of quantum gravity.

The problem of ghosts is certainly the main obstacle for building a consistent theory of quantum gravity. For this reason, we consider this problem in detail in the present chapter. At the same time, some simple, albeit lengthy, calculations in this and the next chapters, are left as exercises for the interested reader.

19.1 How to meet a massive ghost

Let us start from the propagator of the transverse and traceless part $\bar{h}^{\perp\perp}_{\mu\nu}$ of the metric (18.11) in the fourth-derivative model (18.18). The dynamics of this mode does not depend on the gauge fixing and can be directly obtained from the expansions given in section 18.3. As usual, in the analysis of the propagator, we set the cosmological constant to be zero, such that the spacetime is a solution of the classical equations of motion. Thus, (18.18) boils down to

$$S = - \int d^4x\sqrt{-g}\,\Big\{ \frac{1}{\lambda}\, R^2_{\mu\nu} + \frac{\omega - 1}{3\lambda}\, R^2 + \frac{1}{\kappa^2}\, R \Big\}. \tag{19.1}$$

As we already know from the discussion of bilinear forms in section 18.3, the R^2 term does not affect tensor mode propagation. Thus, using the expansions (18.48) and

Introduction to Quantum Field Theory with Applications to Quantum Gravity. Iosif L. Buchbinder and Ilya L. Shapiro, Oxford University Press (2021). © Iosif L. Buchbinder and Ilya Shapiro.
DOI: 10·1093/oso/9780198838319·003·0019

(18.45), the action of the tensor mode becomes

$$S^{(2)}_{tensor} = \int d^4x \Big\{ -\frac{1}{4\lambda}\,(\Box h)^2 - \frac{1}{4\kappa^2}\,h\Box h \Big\} = -\frac{1}{4\lambda}\int d^4x h(\Box + m_2^2)\Box h, \qquad (19.2)$$

where $h = \bar{h}^{\perp\perp}_{\mu\nu}$ and $m_2^2 = \frac{\lambda}{\kappa^2}$ is the mass of the mode that is called a tensor ghost, a massive tensor ghost or higher-derivative ghost. The reason for this exotic name is that the Euclidean propagator of the spin-2 mode in this theory has the form

$$G_2(k) \propto \frac{P^{(2)}}{m_2^2}\Big(\frac{1}{k^2} - \frac{1}{k^2 + m_2^2}\Big). \qquad (19.3)$$

The negative sign of the second term indicates that this is an unusual particle. In fact, we have not one but the types of degrees of freedom of the tensor field $h = \bar{h}^{\perp\perp}_{\mu\nu}$. One of these degrees of freedom has positive kinetic energy and zero mass, and its contribution corresponds to the first term in Eq. (19.3). The second degree of freedom has mass m_2 and corresponds to the second term in (19.3). As we shall see in what follows, its kinetic energy is negative, and, for this reason, it is called a ghost.

The separation of the two degrees of freedom can be most simply observed by introducing an auxiliary field Φ (see, e.g., [97], a more detailed discussion in [7] and more general formulations in [266, 217]). Consider the Lagrangian

$$\mathcal{L}' = -\frac{m_2^2}{4\lambda}\,h\Box h + \lambda\Phi^2 - \Phi\Box h. \qquad (19.4)$$

The equation for Φ can be solved as $\Phi = \frac{1}{2\lambda}\Box h$. Substituting this expression back into the action (19.4), we arrive at Eq. (19.2), which shows that the Lagrangians in (19.2) and (19.4) are dynamically equivalent. The main disadvantage of (19.4) is that the two fields Ψ and h are not factorized. To improve on this issue and present both kinetic terms in the standard form, consider the following change to the new variables θ and ψ,

$$h = \frac{\sqrt{2\lambda}}{m_2}(a_1\theta + a_2\psi), \qquad \Phi = \frac{\sqrt{2\lambda}}{m_2}a_3\psi, \qquad (19.5)$$

where the unknown coefficients $a_{1,2,3}$ should provide the separation of the modes and also the coefficient $\frac{1}{2}$ or $-\frac{1}{2}$ in the kinetic terms. A small algebra shows that the condition $a_2 + a_3 = 0$ is necessary to separate the variables. Furthermore, $a_1 = a_2 = 1$ is required to provide standard normalization of the kinetic terms. Taking these conditions into account, in the new variables, the Lagrangian (19.4) becomes

$$\mathcal{L}' = \frac{1}{2}\,\eta^{\mu\nu}\partial_\mu\theta\partial_\nu\theta - \frac{1}{2}\,\Big(\eta^{\mu\nu}\partial_\mu\psi\partial_\nu\psi - m_2^2\psi^2\Big). \qquad (19.6)$$

Let us stress that the signs of the two kinetic terms cannot be modified by another choice of the coefficients $a_{1,2,3}$. Thus, we conclude that the theory (18.18) has tensor massive degrees of freedom with negative kinetic energy: the unphysical ghost. In the next section, we learn that the field ψ is classified as a non-tachyonic ghost, and, in the subsequent sections, we shall discuss the problems created by the presence of ghost(s) in the particle spectrum of quantum gravity models.

19.2 The dangers of having a ghost or a tachyon

Following [98], let us present a basic classification of ghosts and tachyons, where tachyon is another type of pathological degree of freedom. For this, we consider first the simplest second-order free actions of the field h.

19.2.1 Toy models with second derivatives

For the second-order theory, the general action of a free field $h(x) = h(t, \mathbf{r})$ is

$$S(h) = \frac{s_1}{2} \int d^4x \left\{ \eta^{\mu\nu} \partial_\mu h \partial_\nu h - s_2 m^2 h^2 \right\}$$
$$= \frac{s_1}{2} \int d^4x \left\{ \dot{h}^2 - (\nabla h)^2 - s_2 m^2 h^2 \right\}. \qquad (19.7)$$

Both s_1 and s_2 are sign factors taking values ± 1 for different types of fields. In what follows, we consider all four combinations of these signs.

It is useful to perform the Fourier transform in the space variables,

$$h(t, \mathbf{r}) = \frac{1}{(2\pi)^3} \int d^3k \, e^{i\mathbf{k}\cdot\mathbf{r}} \, h(t, \mathbf{k}), \qquad (19.8)$$

and consider the dynamics of each component $h \equiv h(t, \mathbf{k})$ separately. It is easy to see that this dynamics is defined by the action

$$S_{\mathbf{k}}(h) = \frac{s_1}{2} \int dt \left\{ \dot{h}^2 - \mathbf{k}^2 h^2 - s_2 m^2 h^2 \right\} = \frac{s_1}{2} \int dt \left\{ \dot{h}^2 - m_k^2 h^2 \right\}, \qquad (19.9)$$

where

$$\mathbf{k}^2 = \mathbf{k} \cdot \mathbf{k} \qquad \text{and} \qquad m_k^2 = s_2 m^2 + \mathbf{k}^2 . \qquad (19.10)$$

The properties of the field are defined by the signs of s_1 and s_2. The possible options can be classified as follows:

i) *Normal healthy field* corresponds to $s_1 = s_2 = 1$. The kinetic energy of the field is positive and the equation of motion has the oscillatory form,

$$\ddot{h} + m_k^2 h = 0 , \qquad (19.11)$$

with the usual periodic solutions.

ii) *A tachyon* has $s_1 = 1$ and $s_2 = -1$. The classical dynamics of tachyons is described in the literature, e.g., in [55, 314], but, for our purposes, it is sufficient to give only a basic survey.

For a relatively small momentum $k = |\mathbf{k}|$, there is $m_k^2 < 0$ in Eq. (19.10), and the equation of motion is

$$\ddot{h} - \omega^2 h = 0 , \qquad \omega^2 = \left| m_k^2 \right| , \qquad (19.12)$$

with exponential solutions

$$h = h_1 e^{\omega t} + h_2 e^{-\omega t} . \qquad (19.13)$$

However, if such a particle moves faster than light, the solution is of the normal oscillatory kind, indicating that such a motion is "natural" for this kind of particle [55].

iii) *A massive ghost* has $s_1 = -1$ and $s_2 = 1$. It is not a tachyon, because $m_k^2 \geq 0$. In this case, the kinetic energy of the field is negative, but one can postulate zero variation of the action (instead of demanding the usual least action principle, requiring the minimal action) and arrive at the normal oscillatory equation (19.11). A particle with negative kinetic energy tends to achieve a maximal speed, but a free particle can not accelerate, as this would violate energy conservation. Hence, a free ghost does not produce any harm to the environment, being isolated from it.

In the case when we admit an interaction of a ghost with healthy fields, the argument about energy conservation in a closed system does not work. A systematic study of this situation at the quantum level has been given by Veltman in [325]. Let us give just the main result. Since any physical system tends to the state with minimal action, a ghost tends to accelerate, transmitting the extra positive energy to the healthy fields interacting with it, including in the form of the quantum or classical emission of the corresponding particles.

Since gravitational theories have non-polynomial interactions, a massive ghost always couples to an infinite amount of healthy massless gravitons. This fact leads to dramatic consequences. First of all, the energy conservation does not forbid a spontaneous creation of a massive ghost from the vacuum, even in the flat Minkowski space. It is important to understand that such a spontaneous creation of a ghost also implies that the corresponding amount of positive energy should be released with the creation of massless gravitons. We shall present some additional discussion of this important feature of ghost creation later on, in section 19.7.

Second, if there were a real ghost particle, it would start to accelerate emitting and scattering gravitons. The absolute value of the energy of the ghost would increase, and hence the energy of the created and scattered gravitons would increase too, without any kind of upper bound for the emitted gravitational energy. After a short while, the ghost would acquire an infinite amount of negative energy and start to emit an infinite amount of positive energy. It is clear that if some object of this sort were around, we would certainly know about it or, rather, we would feel it. Thus, the problem is explaining why this scenario does not work in nature. Let us say at once that, at the moment, there is no solution to this problem. So, let us simply assume, for the moment, that some kind of solution exists and that it is related to the large mass of the massive ghost in higher-derivative gravity.

iv) *A tachyonic ghost* has $s_1 = s_2 = -1$. For relatively small $\mathbf{k}^2$, we have $m_k^2 < 0$. The kinetic energy is negative and the derivation of the equations of motion requires an additional definition (the principle of extremal action) similar to that for the non-tachyonic ghost case. After that, one can note that the equation of motion is of the anti-oscillatory type, Eq. (19.12), and the solutions are exponential, as in (19.13).

19.2.2 Fourth-order model

In the fourth-order gravity model (19.1), we can start from the analysis of the free tensor modes (19.2). In this expression, we use

$$\Box h \;=\; \ddot{h} - \Delta h \;\longrightarrow\; \ddot{h} + \mathbf{k}^2 h\,, \qquad (19.14)$$

where $\mathbf{k}$ is the wave vector of an individual mode of the perturbation. It is important to remember that, owing to the presence of a massive mode, the standard massless dispersion relation between the frequency and the wave vector does not hold in this case. The Lagrange function of the wave with fixed $\mathbf{k}$ can be obtained from (19.2):

$$L = -\frac{1}{4\lambda}\left(\ddot{h}+\mathbf{k}^2 h\right)^2 - \frac{1}{4\kappa^2}\,h\left(\ddot{h}+\mathbf{k}^2 h\right). \tag{19.15}$$

The Lagrange equation for $L = L(q,\,\dot{q},\,\ddot{q})$ has the form (see, e.g, [161])

$$\frac{\partial L}{\partial q} - \frac{d}{dt}\frac{\partial L}{\partial \dot{q}} + \frac{d^2}{dt^2}\frac{\partial L}{\partial \ddot{q}} = 0\,. \tag{19.16}$$

The energy (the first integral of the Lagrange equation) can be derived in the form

$$E = \dot{q}\left(\frac{\partial L}{\partial \dot{q}} - \frac{d}{dt}\frac{\partial L}{\partial \ddot{q}}\right) + \ddot{q}\,\frac{\partial L}{\partial \ddot{q}} - L\,. \tag{19.17}$$

In the case of the Lagrange function (19.15), this formula gives the energy of the individual wave with momentum $\mathbf{k}$,

$$E = \frac{1}{4\lambda}\left(2h^{(\mathrm{III})}\dot{h} - \ddot{h}^2\right) + \left(\frac{1}{4\kappa^2} + \frac{\mathbf{k}^2}{2\lambda}\right)\dot{h}^2 + \left(\frac{\mathbf{k}^2}{4\kappa^2} + \frac{\mathbf{k}^4}{4\lambda}\right)h^2\,. \tag{19.18}$$

This formula provides several important pieces of information about the fourth-derivative theory. Let us separate them into the following several points:

i) In the limit $\lambda \to \infty$, the remaining expression for the energy is positively defined, as it should be for Einstein gravity.

ii) The fourth time derivative terms are given by the first summand in (19.18). It is easy to see that this term is not positively defined. This sign indefiniteness should be expected, as a direct consequence of the presence of the massive unphysical ghost.

iii) In the model under discussion, the low-energy limit (infrared) means

$$\ddot{h}^2 \ll \mathbf{k}^2\dot{h}^2 \qquad \text{and} \qquad |hh^{(\mathrm{III})}| \ll \mathbf{k}^2\dot{h}^2. \tag{19.19}$$

In this limit, the first indefinite term in (19.18) is small, and the sign of the energy is defined by the second term, providing a relevant constraint on the action (18.18). The positivity of the theory in this limit does not depend on higher time derivatives. However, it is easy to see that the kinetic energy can be still unbounded from below for the negative sign of the coupling λ. Let us remember that, owing to the violated dispersion relation between the wave vector $\mathbf{k}$ and the time derivatives, it is possible to have a large $\mathbf{k}^2$ with the conditions (19.19) satisfied. For this reason, the sign of the coupling λ in the action (18.18) should be positive, as it was assumed, e.g., in [307,140].

The equation for tensor perturbations can be derived from (19.16),[1]

$$h^{(\mathrm{IV})} + 2\mathbf{k}^2\ddot{h} + \mathbf{k}^4 h + \frac{\lambda}{16\pi\kappa^2}\left(\ddot{h} + \mathbf{k}^2 h\right) = 0. \tag{19.20}$$

[1]The more general equation describing the dynamics of tensor perturbations on the cosmological background, will be discussed below; see Eq. (19.45).

It proves useful to introduce a new notation,

$$\frac{\lambda}{16\pi\kappa^2} = s_2 m^2, \tag{19.21}$$

where $s_2 = \operatorname{sign}\lambda$ and $m^2 > 0$. Then one can recast Eq. (19.20) into the form

$$\Big(\frac{\partial^2}{\partial t^2} + \mathbf{k}^2\Big)\Big(\frac{\partial^2}{\partial t^2} + m_k^2\Big)h \,=\, 0\,, \tag{19.22}$$

where $m_k^2 \,=\, \mathbf{k}^2 + s_2 m^2$. The solutions of the last equation can be different, depending on the sign of λ and, hence, that of s_2. The general formula for the frequencies i

$$\omega_{1,2} \approx \pm i\big(\mathbf{k}^2\big)^{1/2} \qquad \text{and} \qquad \omega_{3,4} \approx \pm\big(-\,m_k^2\big)^{-1/2}. \tag{19.23}$$

For a positive λ, there are only imaginary frequencies and, hence, oscillator-type solutions. In contrast, for $\lambda < 0$, we have $s_2 = -1$ and the roots $\omega_{3,4}$ are real, since, in this case, $-m_k^2 > 0$ for sufficiently small $\mathbf{k}^2$. Indeed, the first couple of roots corresponds to the massless graviton, and the second couple to the massive particle. According to our classification, this particle is a ghost for $\lambda > 0$ and, in contrast, is a tachyonic ghost for $\lambda < 0$.

The main difference between ghosts and tachyons is that a ghost may cause instabilities only when it couples to some healthy fields or to the background, while, with a tachyonic ghost, there is no such protection. This is an additional confirmation of the importance of the positive sign of λ, as we discussed above based on the formula for energy, (19.18).

19.3 Massive ghosts in polynomial models

As was discussed above, in both polynomial models (18.26) and in non-polynomial, nonlocal models (18.31), the tensor structure of the propagator and the role of gauge fixing on leading higher-derivative terms are the same as in the fourth-derivative theory (18.18). In particular, in all three cases, the spin-1 mode can be completely absorbed into the gauge-fixing term (18.21) with the weight operator (18.34), corresponding to the model under consideration. For polynomial models, we consider in detail the spin-2 sector and then give a few comments on the scalar sector. Also, since models in which condition (18.32) is violated are non-renormalizable, we consider only models where this important condition holds.

The analysis of the poles of the propagator in the polynomial models in (18.26) was performed in [13] for the case of real poles and in [221, 3] for the complex poles, but only in the six-derivative models. In the rest of this section, we shall closely follow [13] and prove an important statement about the structure of the propagator in the general polynomial theory with real poles. Some discussions of theories with complex conjugate poles will be presented in the next sections.

The structure of the Euclidean propagator in the spin-2 and spin-0 sectors is

$$G(k) \,=\, \big(l_{2N+4}\,k^{2N+4} + l_{2N+2}\,k^{2N+2} + l_{2N}\,k^{2N} + \,\dots\, + l_2\,k^2\big)^{-1}\,, \tag{19.24}$$

where l_i are real numbers depending on the coefficients in the action (18.26). Expression (19.24) can be decomposed in terms of simple propagators as

$$G(k) = \frac{A_0}{k^2} + \frac{A_1}{k^2+m_1^2} + \frac{A_2}{k^2+m_2^2} + \,\dots\, + \frac{A_{N+1}}{k^2+m_{N+1}^2}, \tag{19.25}$$

where the masses m_j^2 can be real or complex, depending on the values of the coefficients l_i (19.24). The complex masses are always grouped in conjugate pairs. We assume that all the poles m_j^2 are real and positive and that there is the following hierarchy of the masses:

$$0 < m_1^2 < m_2^2 < m_3^2 < \,\dots\, < m_{N+1}^2. \tag{19.26}$$

It is tempting to imagine the situation when all the coefficients A_r, except the last one, are positive and only A_{N+1} is negative. This would mean that we could choose the coefficients in action (18.26) in such a way that the unique ghost particle would have an infinite mass. Then, according to what we know from section 19.2 (and will also discuss below), the creation of such a ghost from vacuum would require an infinite energy density of gravitons, that would solve the problem of ghosts, up to some extent. Unfortunately, this scheme does not work, because one can prove that the signs of the coefficients A_j alternate, i.e., sign $[A_j] = -$ sign $[A_{j+1}]$.

Let us prove this intercalation property of normal particles and ghosts in the mass spectrum of the theory with real poles. The Euclidean propagator $G(k^2)$ has the form

$$G(k^2) = \frac{l_{2N+4}^{-1}}{\prod_{i=0}^{N+1}(k^2+m_i^2)}, \tag{19.27}$$

where the masses m_j, with $j = 0, \dots, N+1$ form a growing sequence according to Eq. (19.25). Therefore, $G(k)$ has simple poles in the points $k^2 = -m_j^2$.

Thus, $G(z)$ is a meromorphic function in the complex plane $\mathbb{C}$, with $N+2$ simple poles. Let Γ_j be a closed path around the pole $-m_j^2$, not encircling any other pole. Then, from (19.24), we have

$$\oint_{\Gamma_j} G(z) = 2\pi i \,\text{Res}\left\{\frac{l_{2N+4}^{-1}}{\prod_{i=0}^{N+1}(z+m_i^2)}, -m_j^2\right\} = \frac{2\pi i}{l_{2N+4}} \prod_{i=0,i\neq j}^{N+1} \frac{1}{m_i^2-m_j^2}. \tag{19.28}$$

On the other hand, from (19.25) it follows that

$$\oint_{\Gamma_j} G(z) = 2\pi i \,\text{Res}\left\{\sum_{i=0}^{N+1} \frac{A_i}{(z+m_i^2)}, -m_j^2\right\} = 2\pi i A_j. \tag{19.29}$$

Taking into account (19.28) and (19.29), we get

$$A_j = l_{2N+4}^{-1} \prod_{i=0,i\neq j}^{N+1} \frac{1}{m_i^2-m_j^2}. \tag{19.30}$$

Since the squares of the masses m_j^2 are real and ordered (19.26), we have $m_i^2 < m_{i+1}^2$, $\forall i = 0, \ldots, N$. Then, (19.30) tells us that the signs of the residua alternate, i.e.,

$$\text{sign}\,[A_j] \;=\; (-1)^j\,\text{sign}\,[l_{2N+4}]\,. \tag{19.31}$$

The mathematical exercise described above has several physical consequences. In the tensor sector, the lightest particle is the healthy massless graviton. After that, the lightest massive mode is a massive ghost. The next massive mode is healthy, then we have one more ghost, etc. Let us stress that this situation concerns only the models with a purely real mass spectrum. The case of complex poles is more complicated and, in general, not sufficiently well explored.

Another interesting output of the alternation rule (19.31) concerns the scalar sector of the theory. In Einstein's theory, there is no scalar mode, which appears only in the fourth-derivative theory (18.18). From the mathematical statement (19.31), follows that, with an appropriate (negative) sign of the R^2 term (remember that the C^2 term does not contribute to the propagation of the scalar mode), this lightest scalar mode is healthy. Thus, the first theory where one can meet scalar ghost is the one with six derivatives, i.e., the theory (18.26) with $N = 1$. This feature of the $R + R^2 + R\Box R$ theory makes it a useful toy model for higher-derivative gravity in general. For instance, in cosmology, this toy model was used in [88] to explore the practical consequences of the effective approach to ghosts, which we will describe below in section 19.6.

Finally, let us mention the result of the paper [4], about the possibility of having a seesaw mechanism in the spectrum of massive modes, including ghosts and healthy fields. Consider the case when all dimensional parameters in action (18.26) are related to the unique massive scale, e.g., with M_P. One can imagine the situation when these parameters combine in such a way that at least one of the masses of the massive modes in (19.26) is small. In the aforementioned paper, it was shown that this situation never takes place because the mass of the lightest mode is defined by the lowest order of the expansion of the bilinear form of the action. This may be an important feature, since it shows that even the lightest ghost has a mass of the Planck order of magnitude. If the emergence of a ghost from the vacuum has some protection, owing to its large mass, we are not left without Planck mass protection.

19.4 Complex poles and the unitarity of the $\mathcal{S}$-matrix

All in all, the problem of ghosts has two main aspects, namely, unitarity and stability. The first part is the unitarity of the quantum theory. Since this is a technically complicated issue, we shall give only a brief survey of the state of the art in this area, a brief qualitative discussion and very few relevant references. In the next section, we discuss, in much more technical detail, the issue of stability.

The general conditions that are called physical unitarity imply two different requirements, namely the absence of unphysical degrees of freedom in the spectrum of the theory and the unitarity of the $\mathcal{S}$-matrix.

The problem of unitarity in the fourth-derivative gravity model (18.18) always attracted a great deal of attention. Let us just mention the most well-known publications from the seventies and the eighties [315, 270, 8]. The main idea of these works is that the dressed propagator of the gravitational field has pairs of complex conjugate poles, instead of the real massive pole at the tree level. Assuming some special features of the complex poles, one can prove the unitarity of the theory. Thus, the proposal of these works implies the resummation of the perturbative series. One has to sum up the diagrams contributing to the propagator first, and, only after this, deal with the diagrams contributing to the vertices. Unfortunately, a detailed, unbiased analysis of this approach shows that a definite solution of this sort requires full information about the complex or real poles at the full non-perturbative level [189].

Another possibility is to start from the six- or higher-derivative polynomial model (18.26) and choose the coefficients $\omega_{k,R}$ and $\omega_{k,C}$ in such a way that the massive poles in the tree-level propagator emerge only in the complex conjugate pairs. In this case, it is possible to prove that the optical theorem is satisfied [221] within the Lee-Wick quantization approach [210]. With this choice of action, the $\mathcal{S}$-matrix is unitary without resummation of the perturbative series. One can show that the complex nature of the poles persists after the quantum corrections are taken into account.

Thus, in the six-derivative polynomial model with complex poles, we can achieve superrenormalizable models with the unitary $\mathcal{S}$-matrix. A natural question is whether these models resolve the conflict between renormalizability and unitarity. Unfortunately, the general answer to this question is no. In order to understand this, let us invoke the qualitative picture of how the unitarity is provided within the Lee-Wick approach.

Consider a *closed* system of particles that have complex conjugate poles in the propagator. These particles are free in the *in* and *out* states and are scattered owing to the interactions between them in the intermediate area between these asymptotic states. The key point is that, in the closed system, the total energy of the particles is conserved. If one of these particles starts to accelerate infinitely, this leads to the violation of energy conservation. In this way, one can successfully apply Lee-Wick quantization to the various quantum field theory models. However, when dealing with gravity, there are two important points to note: i) there are massless modes, i.e., gravitons. ii) the interactions are non-polynomial. As a result, the quantum gravitational system cannot be exactly closed. Massive particles with complex poles can accelerate, just because they can emit an infinite amount of gravitons. In some sense, the situation is close to the one with tachyonic ghosts, as we described in section 19.2.

It is easy to note that the difficulty described above has fundamental importance. In its original formulation, the $\mathcal{S}$-matrix approach is intended to describe the scattering of particles forming a *closed* system. In the case of gravity, especially in the presence of massive modes (in fact, including matter fields or particles), it is difficult to imagine how the system can be considered closed. Therefore, the applicability of the $\mathcal{S}$-matrix approach to the problem of ghosts is not evident.

19.5 Ghosts in nonlocal models

Consider the situation with ghosts in the nonlocal models (18.31). As usual, we can deal with the $\Lambda = 0$ version of the theory,

$$S = \int d^4x\sqrt{-g}\Big\{ -\frac{1}{\kappa^2}R + \frac{1}{2}\,C_{\mu\nu\alpha\beta}\,\Phi(\Box)\,C_{\mu\nu\alpha\beta} + \frac{1}{2}\,R\,\Psi(\Box)\,R + \mathcal{O}(R^3_{\dots})\Big\}. \quad (19.32)$$

As we already know from section 18.2.4, the theory (18.31) is power counting renormalizable only if condition (18.32) is satisfied. In this case, the UV asymptotic behavior of the spin-2 and spin-0 sectors of the propagator is equivalent. Also, similar UV asymptotic behavior should be required from the $\mathcal{O}(R^3_{\dots})$ terms. In the polynomial models in (18.26), we have to require that the $\mathcal{O}(R^3_{\dots})$ terms do not have more derivatives than the polynomial terms with $\Phi(\Box)$ and $\Psi(\Box)$, quadratic in curvatures. In the nonlocal case, one can formulate a qualitatively similar condition for each of the terms with n-th power of curvature. We leave this formulation to the reader as an exercise.

Assuming that $\Psi(x) = \Phi(x) \times const$, the structure of the propagator and, consequently, the spectrum of massive ghost-like states, are the same for the spin-2 and spin-0 sectors. Therefore, we can restrict our attention to the tensor part only. The choice of the function $\Phi(x)$ is supposed to satisfy the following set of conditions:

i) The propagator of the tensor perturbation should be free from massive poles.

ii) The function should grow at $|x| \to \infty$ at least as a third-order polynomial. This condition is necessary to make the theory superrenormalizable (see section 18.7).

iii) The dressed propagator in the theory should satisfy the first condition, as otherwise the number of degrees of freedom can dramatically increase after the one-loop corrections are taken into account.

iv) The theory should have a physical spectrum that is positively defined. In quantum field theory, this requirement can be formulated using the Osterwalder-Schrader reflection positivity property [237] for the Euclidean two-point function. On top of that, the Fourier transform of the two-point Schwinger function should admit a Källén-Lehmann representation [191, 211],

$$\widehat{S}_2(p) = \int_0^\infty d\mu \, \frac{\rho(\mu)}{p^2 + \mu^2}\,, \tag{19.33}$$

with a non-negative spectral density $\rho(\mu) \geqslant 0$. Those are important conditions, since their violation shows that the absence of the massive poles in the propagator may be insufficient for the physical unitarity of the theory.

Let us see whether these four conditions can be satisfied in the nonlocal theories in (19.32). The first one can be formulated as the requirement that equation

$$p^2 \left[1 + \kappa^2 p^2 \Phi(-p^2)\right] = 0 \tag{19.34}$$

should have no other solutions except $p^2 = 0$.

It is evident from Eq. (19.34) that there is always a massless pole corresponding to gravitons. For a constant Φ, there is also a massive pole corresponding to a spin-2 ghost, which may be a tachyonic ghost too. For a polynomial function $\Phi(-p^2)$, there are always ghost-like poles, real or complex. However, one can choose a non-polynomial function $\Phi(-p^2)$ in such a way that there will not be spin-2 poles except graviton $p^2 = 0$. The simplest example of this sort is (see [319])

$$1 + \kappa^2 p^2 \Phi(-p^2) = e^{\alpha p^2}\,, \tag{19.35}$$

where α is a constant of the dimension $mass^{-2}$. One can find other entire functions which have the same features [316, 220], but we consider only (19.35), since all of the other examples share the same properties concerning the four conditions listed above.

It is evident that the choice of the function (19.35) provides the first two conditions, i and ii. Consider condition iii, following [290]. The exponential function has two remarkable properties. First of all, the equation $\exp z = 0$ has no real solutions and only one very peculiar solution,

$$z = -\infty + i \times 0, \tag{19.36}$$

on the extended complex plane. On the other hand, taking any constant value of A, the modified equation $\exp z = A \neq 0$ has infinitely many complex solutions. The same is true for the modified equation of the form

$$e^z = Az^2 \log z\,. \tag{19.37}$$

Next, according to the analysis presented in section 18.7, theory (19.32) with an exponential function is superrenormalizable, and divergences are possible only at the one-loop order. These divergences have the fourth powers of the derivatives of the metric and can be removed by adding the usual C^2 and R^2 counterterms. Certainly, we can fine-tune these counterterms to give zero finite local remnants in the effective action. However, in this case, there will be the nonlocal form factors (16.87), which always come together with divergences and cannot be removed by local counterterms. If such form factors are present, the equation for the pole in the dressed propagator is exactly (19.37), and this means that the dressed propagator cannot be free of massive states if there are one-loop divergences.

And so, the theory under consideration can satisfy condition iii only if it is finite. If there are one-loop divergences, (19.37) tells us that the dressed propagator has infinitely many massive poles and at least most of them emerge in complex conjugate pairs. The finiteness can be achieved by choosing an appropriate $\mathcal{O}(R^3_{\dots})$ terms in the action (19.37). Since the purpose is to cancel only two coefficients in the divergences, those of C^2- and R^2 types, these terms should satisfy two constraints to provide the one-loop finiteness. Unfortunately, the explicit calculations in the nonlocal models of quantum gravity are not available; hence, we cannot demonstrate these constraints here.

Finally, the critical condition is iv. It was shown in [16] that this condition is not satisfied for the exponential version of the nonlocal model. The proof requires some special considerations that lie beyond the scope of the present book. Thus, we will not go into full detail and restrict ourselves to citing the result.

All in all, the nonlocal models look interesting in many respects, but they do not solve the fundamental conflict between unitarity and renormalizability in quantum gravity. Exactly like the polynomial quantum gravity with complex poles (see section 19.4), the nonlocal models may be useful in constructing the unitary $\mathcal{S}$-matrix, but they still do not provide the comprehensive consistency of the quantum theory.

19.6 Effective approach to the problem of ghosts

Owing to space restrictions, we will not describe the effective approach to quantum gravity in this book, even though this is one of the mainstream approaches for the quantum theory of gravity and its applications. Let us just explain what is conventionally understood as an effective solution to the problem of massive ghosts. The scheme described below (with some changes of presentation) was originally introduced by Simon in [294] and was elaborated further in [295] and [241]. For the last few decades, this solution to the problem of ghosts has been regarded as the standard one (see, e.g., [350] and [84]), and hence it is important to discuss it here.

The proposal in [294] is to consider the Einstein equations as the basis, regarding all higher-derivative terms in the gravitational action and equations for the metric as small perturbations. According to this treatment, the gravitational theory is described by the two physical degrees of freedom of GR, independent of what the action of the theory is and the form of the quantum corrections to this action. The propagator of the quantum metric $h_{\mu\nu}$ is derived from Einstein's gravity, and no corrections of any kind can produce additional poles in this propagator, by definition. By construction, there cannot be any kind of massive ghosts, and hence there are no problems with unitarity and instabilities at the classical or quantum levels.

This solution certainly looks appealing and simple. It has a strong motivation by the analogy with usual quantum field theories, starting from quantum electrodynamics (QED). In QED the loop corrections also produce higher-derivative contributions to the Lagrangian $-(1/4)F_{\mu\nu}F^{\mu\nu}$. Making a truncation at the fourth-order level, there is an $F_{\mu\nu}\Box F^{\mu\nu}$ term. Then we meet an unphysical massive ghost and corresponding instabilities, which are called "run-away solutions." And the resolution of the problem is essentially as follows: one has to treat the mentioned loop contributions as small corrections to the classical Lagrangian. Another interpretation contradicts the perturbation theory, which is the framework in which the loop corrections are obtained.

Moreover, at the moderate energies, these loop contributions are numerically much smaller than those for the classical part. And, at higher energies, when the fourth-derivative loop contributions become large in magnitude, we are already far beyond the electroweak scale. Then QED is not applicable, since one has to change the framework to the full Standard Model of elementary particle physics. Then, in the Standard Model, there may be, in principle, the same problem with higher-derivatives. However, in this case, the required energies are beyond the supposed scale of further unification of all elementary forces and, even though we do not know which GUT model is "correct," it is quite natural to suppose that it resolves the problem of higher-derivatives and, hence, the run-away solutions, in the same way as it is done in QED.

Thus, there is a strong temptation to believe that the same scheme should be operational for quantum or semiclassical gravity. However, there are a few serious weak points:

i) Unlike QED and Yang-Mills theories, in quantum gravity, the fourth-derivative terms in the classical action of gravity are required for multiplicative renormalizability. With a consistent treatment, one has to introduce these terms, perform their renormalization with renormalization conditions and renormalization group running, etc., and, after that, start to treat these terms as small perturbations. In semiclassical gravity this is a purely ad hoc procedure. In quantum gravity, since terms with higher-derivatives are included, one has to take into account the ghost degrees of freedom in the internal lines of the diagrams and drop them in the external lines. Even if logically consistent itself, with this interpretation, the approach used in [294] does not fit the standard procedure in field theory.

ii) The approach regarding higher-derivative terms as perturbations looks very much ad hoc. In the next section, we shall see how the equivalent result can be achieved with a less rigid set of definitions. On the other hand, the approach used in [294] produces a lot of ambiguity. For instance, we know that the R^2 term in the action does not contribute to ghosts, only to a healthy scalar mode. Should we regard this term as a small perturbation or not? In one of the first papers [295], the proposal was to use the first option. In this case, we should "forbid" Starobinsky inflation [305], which is essentially based on the $R+R^2$-type action, and the solutions are *qualitatively* different from those of GR. Since the inflationary model with a coefficient of R^2 about 5×10^8 (see [306] and the recent [230]) is the most successful candidate in the description of the very early Universe, as well as the fact that the R^2 term is "innocent," this looks like an unnecessary and badly justified sacrifice.

One can modify the approach and include an R^2 term into the main part of the action, but do not do the same with the $R\Box R$ term which produces a scalar ghost. Interestingly, the cosmological consequences of this approach [88] are essentially the same as choosing the initial conditions without ghosts [99], i.e., treating the $R\Box R$ term as a small perturbation. In section 19.7, we shall see that this output should be seen as natural, from the general point of view.

The same problem of ambiguity concerns also other structures, such as $\mathcal{O}(R^3_{\dots})$ and higher order in curvature. All these terms do not contribute to ghosts on the flat background but may be dangerous on other backgrounds. Should we treat each of them as a perturbation, independent of the magnitude of the corresponding coefficients?

iii) Higher-derivative terms can be regarded as small and be treated as perturbations only if these terms are numerically small compared to the Einstein-Hilbert term. This hierarchy holds for the small magnitudes of the metric derivatives. At the very least, this means the inequality $R_{\mu\nu\alpha\beta}R^{\mu\nu\alpha\beta} \ll M_P^4$. Thus, the procedure of avoiding ghosts is efficient almost everywhere. For instance, during the inflation epoch in cosmology, the Hubble parameter is $H \sim 10^{11} - 10^{13}\, GeV$, and, therefore, the hierarchy is, at least, $R^2_{..} \propto 10^{-24} M_P^4$. A very important exception is the region of spacetime that is close to the singularity. In this case, the energy density of the gravitational field can be above the Planck order of magnitude M_P^4.

Thus, any solution to the problem of ghosts which employs an effective approach from [294] needs both a better understanding of the universally of its implementation and some new insight into how this scheme can efficiently work at the Planck scale. Note that the Physics of extremely high energies is the main general motivation for quantum gravity.

19.7 Stable solutions in the presence of massive ghosts

In this section, we present the qualitative analysis of some physical manifestations of higher-derivative terms and their interpretation in terms of massive ghosts.

The most important issue related to massive ghosts is whether their presence can be compatible with the stability of the classical solutions of GR. In section 18.2 we discussed the Planck suppression, which causes any classical solution of GR to be a high-quality approximation of the given solution in the presence of the fourth-derivative terms. The same logic can be successfully extended to the polynomial theories (18.26), if we assume that all massive parameters are of the Planck order of magnitude. The point is that excellence in the approximation of the solution of GR does not necessarily mean the stability of this solution. In general, the problem of the stability of a gravitational solution under arbitrary small perturbations that do not have the symmetry of these solutions is not simple to solve even for a particular gravitational background. Let us mention, for instance, the study of the stability of the Schwarzschild solution in GR [259, 354, 329]. However, in the presence of C^2 term one can expect that the same solution will not be stable. In fact, regardless of existing contradictions in the literature, in general, this expectation is confirmed [347, 227]. Owing to the high level of technical difficulty, associated with this case, we cannot discuss it here. Let us just note that the instability of the black hole solutions in fourth-derivative gravity has the same origin as that for branes [171] and massive gravity [21].

In the rest of this section, we consider the stability of classical cosmological solutions in the presence of fourth-order terms. The advantage of this particular type of solutions is that it is relatively simple and therefore appropriate for the textbook presentation; in addition, the physical interpretation of the results is very explicit. In this consideration, we shall mainly follow the original works [271, 251]. In these papers (and in the first two of the related works [124, 98, 260]) one can find analysis of the higher-derivative action (18.18) with anomaly-induced semiclassical corrections (17.37). However, since the results for these two cases are very similar, we consider only the simpler version without the anomaly-induced quantum contributions to the effective action.

Let us start by describing the setting. The stability we intend to explore in the fourth-derivative gravity theory is related to the presence of massive spin-2 ghosts, which means the transverse and traceless modes of the metric perturbation on the homogeneous and isotropic, cosmological background. Since we have to deal with the cosmological background, we can use the standard formalism for the cosmological perturbations, as described, e.g., in the books [112, 224, 267].

According to the theory of cosmological perturbations, one can impose the synchronous coordinate condition $h_{\mu 0} = 0$. Remember that according to section 18.5, this is sufficient for describing the gravitational wave. The background metric is

$$ds^2 = a^2(\eta) \left[d\eta^2 - (\gamma_{ij} + h_{ij})\, dx^i dx^j \right], \tag{19.38}$$

where η is the conformal time and $a(\eta)$ corresponds to a background cosmological solution. For the sake of simplicity, we take the value $k = 0$ (hence, $\gamma_{ij} = \eta_{ij}$) and also set the cosmological constant to vanish, $\Lambda = 0$.

Furthermore, since we are interested in the gravitational wave dynamics, it is sufficient to retain only the traceless and transverse parts of h_{ij}, which are the purely tensor modes, by imposing

$$\partial_i h^{ij} = 0, \qquad h_{kk} = 0. \tag{19.39}$$

Starting from this point, there is no need to write indices, and we can use the compact notation $h = h^{ij}$. Following [123], we can rewrite the Lagrangian of the action (18.20) as

$$L = \sum_{s=0}^{3} f_s \, L_s = \sqrt{-g}\Big\{ f_0 R + f_1 R^{\alpha\beta\gamma\delta} R_{\alpha\beta\gamma\delta} + f_2 R^{\alpha\beta} R_{\alpha\beta} + f_3 R^2 \Big\} \tag{19.40}$$

and expand these expressions to the second order in the perturbations. We leave this calculation as an exercise (the reader can use the expansions from section 18.3) and present only the result, in terms of the useful physical time variable t,

$$L_0 = a^3 f_0 \left[3H^2 h^2 + h\ddot{h} + 4Hh\dot{h} + \frac{3}{4}\dot{h}^2 - \frac{h}{4}\frac{\nabla^2 h}{a^2} \right] + \mathcal{O}(h^3), \tag{19.41}$$

$$L_1 = a^3 f_1 \bigg[2H^2 \dot{h}^2 - 4H^2 h\ddot{h} - 6H^4 h^2 - 16H^3 h\dot{h} + \ddot{h}^2 + 4H\dot{h}\ddot{h}$$
$$+ \frac{1}{a^4}\nabla^2 h \nabla^2 h + 2\dot{h}\frac{\nabla^2 \dot{h}}{a^2} + (H^2 h - 2H\dot{h})\frac{\nabla^2 h}{a^2} \bigg] + \mathcal{O}(h^3), \tag{19.42}$$

$$L_2 = a^3 f_2 \bigg[- 9H^4 h^2 - 24H^3 h\dot{h} - 6H^2 h\ddot{h} - \frac{9}{4}H^2 \dot{h}^2 + \frac{3}{2}H\dot{h}\ddot{h} + \frac{\ddot{h}^2}{4}$$
$$+ \frac{1}{4a^4}\nabla^2 h \nabla^2 h - \frac{1}{2}\Big(\ddot{h} + 3H\dot{h} - 3H^2 h \Big)\frac{\nabla^2 h}{a^2} \bigg] + \mathcal{O}(h^3), \tag{19.43}$$

$$L_3 = -12H^2 a^3 f_3 \left[3H^2 h^2 + 2h\ddot{h} + 8Hh\dot{h} + \frac{3}{2}\dot{h}^2 - \frac{h}{2}\frac{\nabla^2 h}{a^2} \right] + \mathcal{O}(h^3), \tag{19.44}$$

where $H = \frac{\dot{a}}{a}$ is the Hubble parameter and $a = a(t)$. The Lagrange equation is

$$\frac{1}{3}h^{(\mathrm{IV})} + 2Hh^{(\mathrm{III})} + \Big(H^2 - \frac{\lambda M_{\mathrm{P}}^2}{16\pi} \Big)\ddot{h} + \frac{2}{3}\Big(\frac{1}{4}\frac{\nabla^4 h}{a^4} - \frac{\nabla^2 \ddot{h}}{a^2} - H\frac{\nabla^2 \dot{h}}{a^2} \Big)$$
$$- \Big(H\dot{H} + \ddot{H} + 6H^3 + \frac{3\lambda M_{\mathrm{P}}^2 H}{16\pi} \Big)\dot{h} + \Big[\frac{\lambda M_{\mathrm{P}}^2}{16\pi} + \frac{4}{3}\Big(\dot{H} + 2H^2 \Big) \Big]\frac{\nabla^2 h}{a^2}$$
$$- \Big[24\dot{H}H^2 + 12\dot{H}^2 + 16H\ddot{H} + \frac{8}{3}H^{(\mathrm{III})} + \frac{\lambda M_{\mathrm{P}}^2}{8\pi}\Big(2\dot{H} + 3H^2 \Big) \Big] h = 0 \,. \tag{19.45}$$

Let us note that the contribution of all three fourth-derivative terms in (18.20) or (19.40) is proportional to the unique parameter λ from the action (18.18). The first reason is that the Gauss-Bonnet combinations do not affect the equations of motion. Another invariant is R^2, which contributes to the equation for the conformal factor $a(t)$, but does not affect the propagation of the tensor mode in the expansion (18.11). As we can see, it also does not contribute to the dynamics of tensor perturbations (19.39) on a cosmological background. This is the main reason why the form of action (18.18) is very useful; the same is true for the superrenormalizable models (18.26).

The analysis of the dynamics of the gravitational perturbation in Eq. (19.45) can be performed in a way similar to the one we used for a flat background in section 19.2:

i) Make the Fourier transformation for the space coordinates

$$h_{\mu\nu}(\mathbf{r},t) = \int \frac{d^3k}{(2\pi)^3}\, h_{\mu\nu}(\mathbf{k},t)\, e^{i\mathbf{k}\cdot\mathbf{r}}. \tag{19.46}$$

Indeed, this can be done *only* in the conformally flat background, is what we are dealing with. The transformation (19.46) is performed in the flat spacetime, not in the physical curved space. It is easy to see that the "physical" momentum is $q = \frac{k}{a(t)}$.

As a useful approximation, we can treat the wave vector $\mathbf{k}$ as a constant and hence will be interested only in the time evolution of the perturbation $h_{\mu\nu}(\mathbf{k}, t)$. The validity of such a treatment is restricted to the linear perturbations, but this is what we need now. In this way, the complicated partial differential equation (19.45) are reduced to the much simpler ordinary differential equation for each of the individual modes.

ii) The initial conditions for the perturbations will be chosen according to the quantum fluctuations of free fields. The spectrum is identical to that of a scalar quantum field in Minkowski space, as explained in Part I (see also, e.g., [56] for a more detailed discussion of the spectrum of free fields in curved space),

$$h(x,\eta) = h(\eta)\, e^{\pm i\mathbf{k}.\mathbf{r}}\,, \qquad h(\eta) \propto \frac{e^{\pm ik\eta}}{\sqrt{2k}}\,. \tag{19.47}$$

Here $k = |\mathbf{k}|$ is the frequency of the massless field, and we employed the conformal time η. The transition to the physical time t is performed by using $a(\eta)d\eta = dt$. The normalization constant is irrelevant for the linear perturbations.

Using the notation $h = h(t, \mathbf{k}) = h(t, k)$, the equation has the form

$$\begin{aligned} & h^{(\mathrm{IV})} + 6Hh^{(\mathrm{III})} + \left(3H^2 - \frac{3\lambda M_{\mathrm{P}}^2}{16\pi}\right)\ddot{h} + \left(\frac{1}{2}\frac{k^4}{a^4}h + \frac{2k^2}{a^2}\ddot{h} + \frac{2k^2}{a^2}\,H\dot{h}\right) \\ & - 3\left(H\dot{H} + \ddot{H} + 6H^3 + \frac{3\lambda M_{\mathrm{P}}^2 H}{16\pi}\right)\dot{h} - \left[\frac{3\lambda M_{\mathrm{P}}^2}{16\pi} + 4\left(\dot{H} + 2H^2\right)\right]\frac{k^2}{a^2}h \\ & - \left[72\dot{H}H^2 + 36\dot{H}^2 + 48H\ddot{H} + 8H^{(\mathrm{III})} + \frac{3\lambda M_{\mathrm{P}}^2}{8\pi}\left(2\dot{H} + 3H^2\right)\right]h = 0\,. \end{aligned} \tag{19.48}$$

iii) The role of ghosts in the possible instability of the classical gravitational solutions can be explored using Eq. (19.48). It is sufficient to substitute into this equation the $H(t)$ corresponding to the particular cosmological solution, specify the value for the square of the frequency k^2 and solve this equation numerically, since, in this case, an analytical solution does not look possible.

A few important observations are as follows:

i) In the theory under discussion, the unique energy scale parameter is the Planck mass. Thus, all values of k that are much below the Planck scale, should be attributed to the low-energy region. Correspondingly, the long-wavelength limit is just a length which is much longer than the Planck length, $l_P \approx 10^{-43}cm$.

ii) The next issue concerns the choice of the background solution $a(t)$. Let us start with a few preliminary observations, using the semiclassical notations (12.9). As we mentioned earlier, the term with a_1C^2 in the action (12.9) does not affect the dynamics of the conformal factor $a(t)$. When choosing $a(t)$ and $H(t)$, we do not need to take the Weyl-squared term into account. Furthermore, let us assume that the main targets of the exploration of the stability in the presence of massive ghosts are the *low-energy* cosmological solutions. Thus, we can ignore the contribution of the a_4R^2 term as it is Planck suppressed. This is a good approximation until we consider the solutions for which $|a_4R^2| \ll |M_P^2 R|$. Let us stress that this hierarchy can be assumed *only* for the background $a(t)$. For the perturbations, e.g., for gravitational waves, we intend to explore all frequencies, even beyond the Planck scale.

Thus, for our purpose, it is sufficient to consider $a(t)$ only for solutions that are dominated by classical radiation, dust and cosmological constant, as they were described in section 11.7.2. Of course, we can take more complicated versions of the background cosmological solutions. However, as we shall see in brief, these three cases are sufficient for understanding the general situation.

iii) According to the known mathematical theorems about the stability of the fixed points of differential equations (see, e.g., [89]), linear stability guarantees non-linear (at least perturbative) stability for sufficiently small perturbations. This is a very subtle point, indeed. According to the Ostrogradsky theorem [238] (see also [350]), instability in classical mechanics with higher-derivatives can be observed only at the non-linear level. The reason for this is that, in the linear approximation, exponentially growing modes are inconsistent with the energy conservation that holds in the closed system.

However, in the consideration based on Eq. (19.48), the perturbations are not free, because there is a dynamic background represented by $a(t)$. Thus, the situation is not one of classical mechanics but rather one that corresponds to the aforementioned stability theorems; e.g., the perturbations are not isolated and can grow, without creating a conflict with energy conservation. So, there is no contradiction between the mentioned theorems on stability and the Ostrogradsky theorem. It is worth mentioning that the sufficiency of the linear approximation has been recently checked explicitly in [272, 260] using the Bianchi-I model as a simplified version of the gravitational perturbation.

After all these important preliminary observations, let us present the survey of the main numerical results from [271,251]. In brief, linear stability in the fourth-derivative model (18.18) is possible, but the two conditions should be fulfilled.

First of all, the signs of the parameters λ and κ^2 should be positive. For definiteness, we can fix $a_1 = -1$, which corresponds to $\lambda = 2$ in the action (18.18). For a different magnitude of $|a_1|$, the Planck mass should be rescaled. Second, the frequency k should be essentially smaller than the Planck mass. The threshold value for k slightly depends on the type of the cosmological solution (dominated by radiation, dust or cosmological constant), but, in all cases for $k < 0.1\, M_P$, there is no growth of $h(t,k)$, while such a growth is evident starting from $k \approx 0.6\, M_P$. An illustration ofthis can be seen in (19.49), where we show examples of plots for the radiation-dominated universe [271]. There are no growing modes until the frequency achieves the value k 0.5 in Planck units. Starting from this value, one can view instability as the effect of a massive ghost. For dust and the cosmological constant cases, there are no qualitative changes.

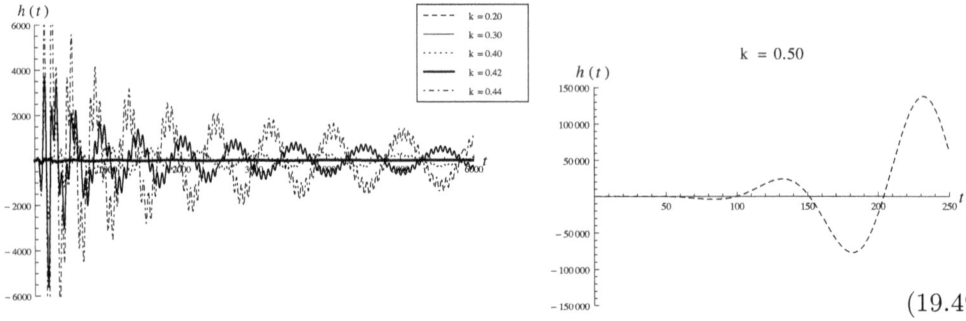

(19.49)

The explanation of this result is that, for a small frequency k, the ghost remains as a virtual mode but cannot be created as a real particle. Let us remember that the creation of a ghost from the vacuum requires that positive-energy gravitons be created with the Planck energy density (Planck energy in the space volume of a cube of the Planck-scale Compton wavelength). If the frequency of the gravitational wave is insufficient, the ghost is not created, and there is no instability.

Additional confirmations of this interpretation are as follows. In the case of $\lambda < 0$ and $\kappa^2 < 0$, the perturbations demonstrate an explosive growth for all frequencies k. This is because, in this case, the negative energy mode is massless, and the massive mode is a healthy tensor particle. Certainly, in this case, there is no energy threshold, and one meets the exponential growth of the perturbations with all frequencies. In the situation when $\lambda > 0$ and $\kappa^2 < 0$, the massive ghost is tachyonic and, according to section 19.2, there is an approximately exponential growth at all frequencies. Finally, when $\lambda < 0$ and $\kappa^2 > 0$, there are two consequences: one due to the absence of an energy gap and, at the same time, with the exponentially growing tachyonic modes. All of this can be clearly observed in numerical analysis (see [271] for detail).

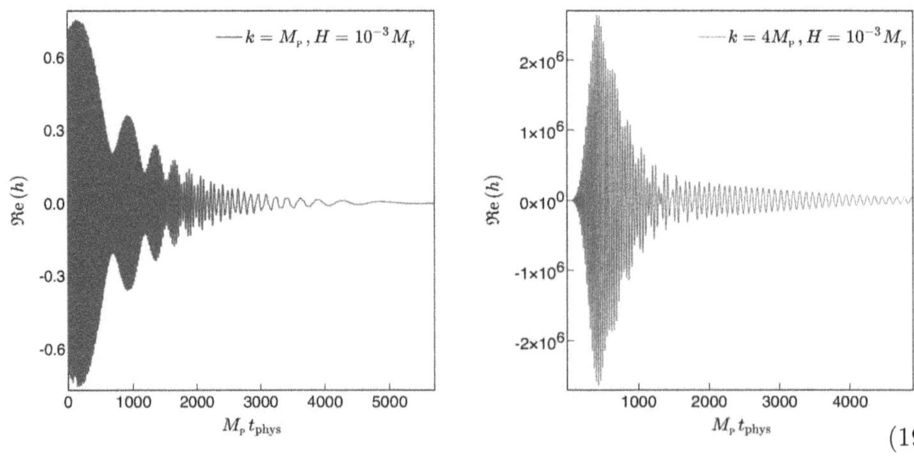

(19.50)

Another interesting thing happens when both λ and κ^2 are positive and the frequency k is greater than the energy threshold but the external gravitational background is very intensive. As we already know, in this situation, there is a very fast growth of the perturbation $h(t, k)$. It turns out that, in some cases, such growth does not last for a long time. To understand why this happens, let us take a look at the main equation, (19.48). It is easy to see that the frequency k enters this equation only in the combination $q = \frac{k}{a(t)}$. Taking this into account, for a sufficiently fast expansion of the Universe, the explosive growth of the perturbation should last only until the magnitude of this ratio becomes smaller than the energy threshold. In (19.50 one can see two examples of the plots [251] for the radiation-dominated Universe $a \propto t^{1/2}$, with a longer period of time than in Fig. (19.49). The growth stops relatively fast due to the fast expansion of the Universe.

At this point, we can make a general conclusion about how the problem of ghosts and its possible solution look from the perspective of the results described above. To avoid the problem with ghosts, we need some new physical principle forbidding the accumulation of gravitons with Planck energy density. If there where a mechanism of protection against such accumulation, it would operate as a physical cut-off at the Planck scale. In this case, the massive ghost could not be created from the vacuum, and, hence, the stability of the classical solutions of GR would not be jeopardized. Unfortunately, the origin of such a mechanism is not known, even though there are some interesting discussions of its implications and consequences [120].

It is evident that the restrictions on the frequency of the initial seeds of the metric perturbations, such as $k \ll M_P$, are qualitatively similar to the effective approach of Simon [294], which was described in section 19.6. After all, if the metric perturbations do not increase quickly, the background solution in fourth-derivative gravity maintains its symmetry and, even at the inflationary epoch, is extremely close to the classical solutions of GR, e.g., with an additional R^2 term. In this situation, there is nothing wrong in treating all higher-derivative terms as perturbations.

From this perspective, the resolution of the problem of ghosts requires a new physical principle providing protection against the accumulation of gravitons with Planck-scale energy density. On the other hand, such protection may be unnecessary for extremely intensive gravitational backgrounds, e.g., for the rapidly expanding Universe in the early cosmology.

Exercises

19.1. Explain the positive sign of λ in the action (18.18) in a more simple way that we did above, namely, by using Eq. (19.5) and the analysis from the subsequent sections. Try to generate different types of arguments.

19.2. Using the results of section 18.3, verify the expansions (19.45) on the cosmological background. In the limit $a(t) \equiv 1$, verify Eq. (19.20).
Hint. Note that the derivatives in (19.45) and (19.39) are taken for the conformally transformed flat metric.

19.3. Derive the generalized version of Eq. (19.45) for an arbitrary $k = 0, \pm 1$ and positive Λ. Discuss whether the analog of (19.48) can be constructed for the non-flat space section of a solution dominated by the cosmological constant.

19.4. Repeat the numerical analysis of Eq. (19.48) for the case of the ρ_Λ-dominated Universe. Consider both signs of λ and the sign flip of the Newton constant.

20
One-loop renormalization in quantum gravity

In this chapter, we demonstrate the basic methods of one-loop calculations in quantum gravity. Before that, we discuss the general approach for exploring the ambiguities of the results. Without this part, the practical calculations make limited sense; therefore, these two aspects of quantum gravity are equally important.

20.1 Preliminary considerations

Quantum corrections to the classical action of gravity are supposed to play a double role. First, the consistency of the theory at the quantum level is an important criterion, indicating relevant extensions of the classical GR. On the other hand, it is important to evaluate quantum corrections and explore their physical applications. In this respect, we need to know the extent to which loop corrections may be subject to ambiguities related to the choice of parametrization of quantum field and on the choice of the gauge-fixing condition. In the first part of this chapter, we explore these two ambiguities, using the general formalism described in Part I. Namely, we shall use the fact, expressed by Eq. (10.161), that the gauge dependence vanishes on shell. In particular, for one-loop divergences and, consequently, for one-loop beta functions, it is sufficient to take into account the classical equations of motion. For the parametrization dependence, the same can be easily achieved by the following consideration. Let us note that there is a version of the gauge-fixing and parametrization independent effective action [328] (see also related publications, e.g., more recent [158]). In general, this is a relatively complicated approach; hence, we will not discuss it here.

Let us start from the parametrization dependence. The one-loop expression for the non-gauge theory with the classical action $S[\Phi_i]$ has the form

$$\bar{\Gamma}^{(1)} = \frac{i}{2}\,\text{Log Det}\, S''_{ij}, \qquad \text{where} \qquad S''_{ij} = \frac{\delta^2 S}{\delta\Phi_i\,\delta\Phi_j}\,. \tag{20.1}$$

Changing the variables $\Phi_i = \Phi_i(\Phi'_k)$, we assume that this is a point (local) transformation that does not produce a relevant Jacobian. In the new variables, we have

$$\bar{\Gamma}^{(1)} = \text{Log Det}\left(\frac{\delta^2 S}{\delta\Phi'_l\,\delta\Phi'_k}\right) = \text{Log Det}\left(S''_{ij}\cdot\frac{\delta\Phi_i}{\delta\Phi'_k}\,\frac{\delta\Phi_j}{\delta\Phi'_l} + \frac{\delta S}{\delta\Phi_i}\,\frac{\delta^2\Phi_i}{\delta\Phi'_l\,\delta\Phi'_k}\right). \tag{20.2}$$

Looking at the last formula, one can easily note that the two one-loop results (20.1) and (20.2) coincide on the classical equation of motion (on shell), when $\varepsilon^i = \frac{\delta S}{\delta\Phi_i} = 0$.

Introduction to Quantum Field Theory with Applications to Quantum Gravity. Iosif L. Buchbinder and Ilya L. Shapiro, Oxford University Press (2021). © Iosif L. Buchbinder and Ilya Shapiro.
DOI: 10·1093/oso/9780198838319·003·0020

The situation is the same as in the case of gauge-fixing dependence, Eq. (10.161). Thus, the general analysis of the two cases can be done in a unified way.

20.2 Gauge-fixing dependence in quantum GR

Consider the gauge-fixing and parametrization dependence in quantum gravity based on GR with the cosmological constant. The study of gauge-fixing dependence in quantum gravity was pioneered in the paper [193] and explored further in subsequent calculations, e.g., in [194, 243]. We shall follow the analysis from [166] and later on confirm the results by direct calculations, in the way it was done in the same paper.

The general analysis may be performed in dimensional regularization and, therefore, start from n-dimensional spacetime, even though our main attention will be on $n = 4$. The metric is supposed to have the Minkowski signature $(+, -, -, \dots)$. Indeed, the practical calculations, e.g., derivation of Feynman diagrams or use of the heat kernel method, require an analytic continuation from Minkowski to Euclidean space. We assume this operation without special explanation.

The equations of motion for the classical Einstein-Hilbert action,

$$S = -\frac{1}{\kappa^2}\int d^n x\sqrt{-g}\,(R + 2\Lambda) \tag{20.3}$$

are

$$\varepsilon^{\mu\nu} = \frac{1}{\sqrt{-g}}\,\frac{\delta S}{\delta g_{\mu\nu}} = R^{\mu\nu} - \frac{1}{2}\,(R + 2\Lambda)g^{\mu\nu}\,. \tag{20.4}$$

Let us use the general statement about gauge-fixing and parametrization on-shell independence for the local part of the effective action. As we know, the power counting tells us that the divergent part of the one-loop effective action is

$$\Gamma^{(1)}_{div} = \frac{1}{\epsilon}\int d^4x\sqrt{-g}\{c_1\, R^2_{\mu\nu\alpha\beta} + c_2 R^2_{\alpha\beta} + c_3 R^2 + c_4\Box R + c_5 R + c_6\}, \tag{20.5}$$

where $\epsilon = (4\pi)^2(n-4)$ is the regularization parameter and $c_{1,2,\dots,6}$ are some coefficients. Our purpose is to explore how these coefficients depend on the gauge fixing and parametrization.

The gauge-fixing and parametrization ambiguity in $\Gamma^{(1)}_{div}$ cannot violate the locality of this expression. Taking this fact into account, and also using the dimensional arguments, the on-shell universality tells us that the ambiguity may have the form

$$\begin{aligned}\delta\Gamma^{(1)}_{div} &= \Gamma^{(1)}_{div}(\alpha_i) - \Gamma^{(1)}_{div}(\alpha_i^0) \\ &= \frac{1}{\epsilon}\int d^4x\sqrt{-g}\,(b_1 R_{\mu\nu} + b_2 R g_{\mu\nu} + b_3\Lambda g_{\mu\nu} + b_4 g_{\mu\nu}\Box + b_5\nabla_\mu\nabla_\nu)\,\varepsilon^{\mu\nu},\end{aligned} \tag{20.6}$$

where α_i represent the full set of arbitrary parameters characterizing the gauge fixing and parametrization of the quantum metric. The special values α_i^0 of these parameters correspond to those in the original paper by 'tHooft and Veltman [186] (see section 20.4), or to another fixed choice.

The parameters $b_{1,2,..,5}$ in (20.6) depend on the choice of α_i, and the explicit form of the dependence can be known only after the explicit calculations. However, one can draw relevant conclusions without this information. In the simplest case, with $\Lambda = 0$, Eq. (20.6) tells us that only the Gauss-Bonnet counterterm cannot be set to zero by choosing α_i. This result was discovered by direct calculation in [193]. The $\mathcal{S}$-matrix corresponds to the on-shell limit of the effective action and, hence, is finite.

In the general theory, with $\Lambda \neq 0$, the situation is a little bit more complicated. It is evident that the parameter b_5 has no effect on divergences, owing to the third Bianchi identity $\nabla_\mu G^\mu_{\ \nu} = 0$ and $\nabla_\mu \Lambda = 0$. Therefore, there is a four-parameter $b_{1,2,3,4}$ ambiguity for the six existing coefficients $c_{1,2,...6}$. As a result, only two combinations of these six coefficients can be expected to be gauge-fixing and parametrization independent.

Let us elaborate a little bit more on this point. A simple calculation using (20.6) shows that the parameters of the expression (20.5) vary according to

$$\begin{aligned} &c_1 \longrightarrow c_1\,, \qquad c_2 \longrightarrow c_2 + b_1\,, \qquad c_3 \longrightarrow c_3 - \big(b_2 + \tfrac{1}{2}\,b_1\big)\,, \\ &c_4 \longrightarrow c_4 - b_4\,, \quad c_5 \longrightarrow c_5 - \big(b_1 + 4b_2 + b_3\big)\Lambda\,, \qquad c_6 \longrightarrow c_6 - 4b_3\Lambda^2\,. \end{aligned} \tag{20.7}$$

After this, it is an easy exercise to show that the two gauge-fixing and parametrization invariant quantities are

$$c_1 \qquad \text{and} \qquad c_{\text{inv}} \;=\; c_6 - 4\Lambda c_5 + 4\Lambda^2 c_2 + 16\Lambda^2 c_3\,, \tag{20.8}$$

since these two quantities do not change under the modification of the parameters b_i. The considerations presented above were confirmed by a one-loop calculation in a very general (albeit, not the most general) gauge fixing and parametrization of the quantum metric in [166]. We do not include this part here, owing to space limitations.

The last observation is that the on-shell expressions for the classical action and divergences have the forms

$$\begin{aligned} S\Big|_{\text{on shell}} &= \frac{2\Lambda}{\kappa^2}\int d^4x\sqrt{-g}\,, \\ \Gamma^{(1)}_{div}\Big|_{\text{on shell}} &= \frac{1}{\epsilon}\int d^4x\sqrt{-g}\,\{c_1 R^2_{\mu\nu\alpha\beta} + c_{\text{inv}}\}\,, \end{aligned} \tag{20.9}$$

consisting only of the invariant quantities. This feature forms the basis of the so-called on-shell renormalization group equation, to be discussed below.

20.3 Gauge-fixing dependence in higher-derivative models

The general rule is that, with more derivatives in the classical action, the gauge-fixing and parametrization dependence in the divergent part of the effective action gets reduced. For purely metric, higher-derivative gravities, the unique non-trivial example is the fourth-derivative model (18.18), which was elaborated in [140] and later on, in more detail, in [279]. Let us follow the last reference and consider this model.

In the four-derivative theory, the formula analogous to (20.6) has the form

$$\Gamma^{(1)}_{div}(\alpha_i) \;-\; \Gamma^{(1)}_{div}(\alpha_i^0) \;=\; \frac{1}{\epsilon}\int d^4x\sqrt{-g}\,f_{\mu\nu}\,\varepsilon^{\mu\nu}_{(4)}, \tag{20.10}$$

where $f_{\mu\nu} = f_{\mu\nu}(\alpha_i)$ is some unknown tensor function and

$$\varepsilon^{\mu\nu}_{(4)} = \frac{1}{\sqrt{-g}}\frac{\delta S}{\delta g_{\mu\nu}} \tag{20.11}$$

is the equation of motion for the fourth-derivative gravity.

To find $f_{\mu\nu}$, let us remember that the fourth-derivative quantum gravity is a renormalizable theory. Therefore, all three of the expressions $\Gamma^{(1)}_{div}(\alpha^0_i)$, $\Gamma(\alpha_i)$ and $\varepsilon^{\mu\nu}$ have dimension 4, as the classical action. Since the divergencies are local, $f_{\mu\nu}$ is a dimensionless tensor function, hence

$$f_{\mu\nu}(\alpha_i) = g_{\mu\nu} f(\alpha_i), \tag{20.12}$$

where $f(\alpha_i)$ is an undefined (before explicit calculations) dimensionless function of the set of dimensionless parameters of gauge fixing (and change of parametrization too). Thus, the gauge/parametrization dependence of the divergent part of effective action is completely controlled by the "conformal shift" of the classical action

$$\Gamma^{(1)}_{div}(\alpha_i) - \Gamma^{(1)}_{div}(\alpha^0_i) = f(\alpha_i)\int d^4x\, g_{\mu\nu}\frac{\delta S}{\delta g_{\mu\nu}}. \tag{20.13}$$

In the case of the conformal model, the *r.h.s.* of this equation simply vanishes.

Starting from this point, the analysis becomes pretty simple for the most part. First of all, in pure (conformal) Weyl-squared gravity theory, the divergences of the effective action do not depend on the gauge-fixing parameters, because the classical action satisfies the Noether identity for the conformal invariance. For the general model (18.18), the C^2, E_4 and $\Box R$ terms in the action do not contribute to the *r.h.s.* of (20.13). Then, the gauge- and parametrization dependencies are defined by the Einstein-Hilbert, cosmological and R^2 terms. Using Einstein equations and Eq. (17.21), we get

$$\Gamma^{(1)}_{div}(\alpha_i) - \Gamma^{(1)}_{div}(\alpha^0_i) = f(\alpha_i)\int d^4x\sqrt{-g}\,\Big\{\frac{2\omega}{\lambda}\Box R - \frac{1}{\kappa^2}(R+4\Lambda)\Big\}. \tag{20.14}$$

Using this relation, we can conclude that the divergent coefficient of the $\Box R$ term depends on the gauge fixing, and the same is true for the coefficients of the Einstein-Hilbert and cosmological terms. At the same time, there are two gauge-invariant combinations of these coefficients. We leave it to the reader as an exercise to find the possible form of these combinations and the form of the classical parameters, for which they serve as beta functions.

Finally, consider the gauge and parametrization dependence of the counterterms in superrenormalizable models with more than four derivatives. It is easy to show, even without special effort, that there is no such dependence. To understand this, let us note that formula (20.10) is valid for all such models, both polynomial and nonlocal, with the replacement of $\varepsilon^{\mu\nu}_{(4)}$ for the variational derivative of the corresponding action. Furthermore, as we have seen in section 18.7, in all these models, the divergences are given by local expressions with, at most four, derivatives of the metric. At the same time, the equations of motion $\varepsilon^{\mu\nu}$ in these models have more than four derivatives of the metric. Thus, the non-zero *r.h.s.* of (20.10) is incompatible with the locality of the function $f_{\mu\nu}$, which proves the aforementioned statement about the universality of renormalization [13, 220, 290, 222].

20.4 One-loop divergences in quantum general relativity

As a warm up exercise, we start from the simplest one-loop calculation in pure four-dimensional quantum gravity. We shall repeat the original derivation of divergences that was given in the classical paper [186]. The example with a general parametrization will be dealt with in the next section.

The bilinear expansion of the action (18.12) can be easily obtained from Eq. (18.45), and the gauge-fixing term with the two gauge fixing parameters α and β is given by (18.16). The ghost action can be easily obtained from (18.17), but we postpone this until we fix the values of α and β. For this, we rewrite (18.16) as

$$S_{gf} = \frac{1}{\alpha}\int d^4x\sqrt{-g}\,h^{\mu\nu}\Big[g_{\mu\alpha}\nabla_\nu\nabla_\beta - \beta\big(g_{\mu\nu}\nabla_\alpha\nabla_\beta - g_{\alpha\beta}\nabla_\mu\nabla_\nu\big) + \beta^2 g_{\mu\nu}g_{\alpha\beta}\Box\Big]h^{\alpha\beta}. \tag{20.15}$$

Adding this expression to (18.45), we require that the sum includes the minimal operator $H_{\mu\nu,\alpha\beta}$,

$$S^{(2)}_{EH} + S_{gf} = \frac{1}{2}\int d^4x\sqrt{-g}\;h^{\mu\nu}H_{\mu\nu,\alpha\beta}\,h^{\alpha\beta}\,,$$
$$\text{with}\qquad H_{\mu\nu,\alpha\beta} = K_{\mu\nu,\alpha\beta}\Box + M_{\mu\nu,\alpha\beta}, \tag{20.16}$$

where $K_{\mu\nu,\alpha\beta}$ and $M_{\mu\nu,\alpha\beta}$ are c-number operators. This is achieved for $\alpha = 2$ and $\beta = 1/2$. After that, we arrive at the operator (20.16) with

$$K_{\mu\nu,\alpha\beta} = \frac{1}{2}\Big(\delta_{\mu\nu,\alpha\beta} - \frac{1}{2}\,g_{\mu\nu}g_{\alpha\beta}\Big), \tag{20.17}$$
$$M_{\mu\nu,\alpha\beta} = R_{\mu\alpha\nu\beta} + g_{\nu\beta}R_{\mu\alpha} - \frac{1}{2}\big(g_{\mu\nu}R_{\alpha\beta} + g_{\alpha\beta}R_{\mu\nu}\big) - \frac{1}{2}\,R\big(\delta_{\mu\nu,\alpha\beta} - \frac{1}{2}g_{\alpha\beta}g_{\mu\nu}\big)\,.$$

According to (18.75), the matrix $2K_{\rho\sigma,\alpha\beta}$ is equal to its own inverse. On the other hand, $\mathrm{Tr}\,\log\big(2K_{\rho\sigma,\alpha\beta}\big)$ does not contribute to the divergence, since this operator has no derivatives. Thus, at least regarding the divergences,

$$\mathrm{Tr}\,\log\big(H_{\rho\sigma,\alpha\beta}\big) = \mathrm{Tr}\,\log\big(2K_{\mu\nu,}{}^{\rho\sigma}\,H_{\rho\sigma,\alpha\beta}\big) = \mathrm{Tr}\,\log\big(H'_{\rho\sigma,\alpha\beta}\big)$$
$$= \mathrm{Tr}\,\log\big(\delta_{\mu\nu,\alpha\beta}\Box + \Pi_{\rho\sigma,\alpha\beta}\big). \tag{20.18}$$

A small calculation gives

$$\Pi_{\rho\sigma,\alpha\beta} = 2K_{\mu\nu,}{}^{\rho\sigma}\,M_{\rho\sigma,\alpha\beta} = M_{\mu\nu,\alpha\beta}. \tag{20.19}$$

It is evident that Eq. (20.18) enables one to use the standard Schwinger-DeWitt formula (13.206) for divergences. For this, we need to construct the operators

$$\hat{P} = \hat{\Pi} + \frac{\hat{1}}{6}\,R, \qquad \text{and} \qquad \hat{S}_{\mu\nu} = \big[\nabla_\nu, \nabla_\mu\big]. \tag{20.20}$$

A simple calculation gives

$$\begin{aligned}
&\hat{P} = P_{\mu\nu,\alpha\beta} = \hat{K}_1 + \hat{K}_2 - \frac{1}{2}\hat{K}_3 - \frac{5}{12}\hat{K}_4 + \frac{1}{4}\hat{K}_5,\\
&\hat{K}_1 = R_{\mu\alpha\nu\beta}, \qquad \hat{K}_2 = g_{\nu\beta}R_{\mu\alpha}, \qquad \hat{K}_3 = g_{\mu\nu}R_{\alpha\beta} + g_{\alpha\beta}R_{\mu\nu},\\
&\hat{K}_4 = \delta_{\mu\nu,\alpha\beta}R, \qquad \hat{K}_5 = Rg_{\alpha\beta}g_{\mu\nu},\\
&\hat{S}_{\lambda\tau} = \hat{\mathcal{R}}_{\lambda\tau} = \big[S_{\lambda\tau}\big]_{\mu\nu\alpha\beta} = -2R_{\mu\alpha\lambda\tau}\,g_{\nu\beta}.
\end{aligned} \tag{20.21}$$

It is worthwhile remembering that, in all of these expressions, we assume automatic symmetrization over the couples of indices $\mu\nu$ and $\alpha\beta$.

It is useful to construct the following multiplication table:

$$\begin{aligned}
&\operatorname{tr}\hat{K}_1\cdot\hat{K}_1 = \frac{1}{2}\,R^2_{\mu\nu\alpha\beta}, \quad \operatorname{tr}\hat{K}_2\cdot\hat{K}_2 = \frac{3}{2}\,R^2_{\mu\nu} + \frac{1}{4}\,R^2, \quad \operatorname{tr}\hat{K}_3\cdot\hat{K}_3 = 8R^2_{\mu\nu} + 2R^2,\\
&\operatorname{tr}\hat{K}_4\cdot\hat{K}_4 = 10R^2, \quad \operatorname{tr}\hat{K}_5\cdot\hat{K}_5 = 16R^2, \quad \operatorname{tr}\hat{K}_1\cdot\hat{K}_2 = -\frac{1}{2}\,R^2_{\mu\nu},\\
&\operatorname{tr}\hat{K}_1\cdot\hat{K}_3 = 2R^2_{\mu\nu}, \quad \operatorname{tr}\hat{K}_1\cdot\hat{K}_4 = -\frac{1}{2}R^2, \quad \operatorname{tr}\hat{K}_1\cdot\hat{K}_5 = R^2, \quad \operatorname{tr}\hat{K}_2\cdot\hat{K}_3 = 2R^2_{\mu\nu},\\
&\operatorname{tr}\hat{K}_2\cdot\hat{K}_4 = \frac{5}{2}\,R^2, \quad \operatorname{tr}\hat{K}_2\cdot\hat{K}_5 = R^2, \quad \operatorname{tr}\hat{K}_3\cdot\hat{K}_4 = 2R^2, \quad \operatorname{tr}\hat{K}_3\cdot\hat{K}_5 = 8R^2,\\
&\operatorname{tr}\hat{K}_4\cdot\hat{K}_5 = 4R^2, \quad \text{and} \quad \hat{S}_{\lambda\tau}\cdot\hat{S}_{\lambda\tau} = -6\,R^2_{\mu\nu\alpha\beta}.
\end{aligned} \tag{20.22}$$

Substituting these values into Eq. (13.206), we get

$$\operatorname{tr}\Big(\frac{1}{2}\,\hat{P}\cdot\hat{P} + \frac{1}{12}\,\hat{S}_{\lambda\tau}\cdot\hat{S}^{\lambda\tau}\Big) = R^2_{\mu\nu\alpha\beta} - 3\,R^2_{\mu\nu\alpha\beta} + \frac{59}{36}\,R^2. \tag{20.23}$$

In this and subsequent expressions, we ignore the surface $\Box R$ term, since it will be accounted for in the more general calculation in the next section.

For the ghost action, we obtain

$$\frac{\delta\chi^\mu}{\delta h_{\rho\sigma}} = \delta^{\mu\lambda,\rho\sigma}\nabla_\lambda - \frac{1}{2}\,g^{\rho\sigma}\nabla^\mu, \qquad R_{\rho\sigma,}{}^{\nu} = -\delta^\nu_\rho\nabla_\sigma - \delta^\nu_\sigma\nabla_\rho.$$

Then

$$M_\mu{}^\nu = \frac{\delta\chi^\mu}{\delta h_{\rho\sigma}}\,R_{\rho\sigma,}{}^{\nu} = -\big(\delta^\nu_\mu\Box + R^\nu_\mu\big). \tag{20.24}$$

Once again, we are lucky enough to meet a minimal vector operator, that enables one to use the standard Schwinger-DeWitt formula (13.206). Furthermore, for the commutator, we get, as for any vector field,

$$A_\alpha\big[\nabla_\nu, \nabla_\mu\big]A^\alpha = -\,A_\alpha\,R^\alpha_{\cdot\,\beta\lambda\tau}\,A^\beta \implies \hat{S}_{\lambda\tau} = \big[\hat{S}_{\lambda\tau}\big]_{\alpha,\beta} = -\,R_{\alpha\beta\lambda\tau}. \tag{20.25}$$

Thus,

$$\mathrm{tr}\left(\frac{1}{2}\,\hat{P}\cdot\hat{P}+\frac{1}{12}\,\hat{S}_{\lambda\tau}\cdot\hat{S}^{\lambda\tau}\right)_{ghost} = -\frac{1}{12}\,R^2_{\mu\nu\alpha\beta}+\frac{3}{2}\,R^2_{\mu\nu\alpha\beta}+\frac{2}{9}\,R^2. \qquad (20.26)$$

Finally, replacing all the expressions above into (13.206) and taking into account that, in the $h_{\mu\nu}$ sector, $\mathrm{tr}\,\hat{1} = \delta_{\mu\nu\alpha\beta}\delta^{\mu\nu\alpha\beta} = 10$ and, in the ghost sector $\mathrm{tr}\,\hat{1}_{ghost} = 4$, we arrive at the famous result of [186],

$$\begin{aligned}\Gamma^{1,\,total}_{div} &= \frac{i}{2}\,\mathrm{Tr}\,\mathrm{Log}\,\hat{H} - i\,\mathrm{Tr}\,\mathrm{Log}\,\hat{H}_{ghost}\\ &= -\frac{1}{\epsilon}\int d^nx\mu^{n-4}\sqrt{-g}\,\Big\{\frac{53}{45}\,R^2_{\mu\nu\alpha\beta} - \frac{361}{90}\,R^2_{\alpha\beta} + \frac{43}{36}\,R^2\Big\}\\ &= -\frac{2}{\epsilon}\int d^nx\mu^{n-4}\sqrt{-g}\,\Big\{\frac{53}{90}\,E_4 + \frac{7}{20}\,R^2_{\alpha\beta} + \frac{1}{120}\,R^2\Big\},\end{aligned} \qquad (20.27)$$

where, as usual, $\epsilon = (4\pi)^2(n-4)$, and μ is the renormalization parameter.

As we already know from section 20.2, only the coefficient $\frac{53}{90}$ of the Gauss-Bonnet term is invariant, while the other two coefficients can be modified by changing the gauge-fixing conditions or the parametrization of the quantum metric. In the next section, we shall check this statement by direct calculation in a general parametrization.

Exercises

20.1. Repeat the whole calculation, but this time including the cosmological constant and taking care about the $\Box R$-type divergence. The result looks like

$$\begin{aligned}\Gamma^{(1)}_{div} = -\frac{2}{\epsilon}\int d^4x\sqrt{-g}\,\Big\{&\frac{53}{90}\,E_4 + \frac{7}{20}\,R^2_{\mu\nu} + \frac{1}{120}\,R^2\\ &- \frac{19}{30}\,\Box R + \frac{13}{3}\,\Lambda R + 10\Lambda^2\Big\},\end{aligned} \qquad (20.28)$$

which is the generalization of the previous result, (20.27).

20.2. Repeat the calculation with the slightly different variables, $h_{\mu\nu} = \bar{h}_{\mu\nu} + \frac{1}{4}hg_{\mu\nu}$, where $\bar{h}_{\mu\nu}$ is traceless. Verify that the ghost action does not change, and find the equivalent of the table in (20.22). Using the arguments from section 20.2, explain why the *off-shell* result is the same as that for the original variables, Eq. (20.27).

Hints. Both these exercises are relatively simple and do not take much time. In the last part of the second exercise, it is recommended to pay attention to Eq. (20.2).

20.3. Use the intermediate formulas from this section to calculate the trace of the coincidence limit for the coefficient $\hat{a}_1$ of the Schwinger-DeWitt expansion. Explore its parametrization and/or gauge-fixing dependence off and on shell.

Hint. The solution can be found in the paper [166].

20.4. Consider the quantum GR theory coupled to the *minimal* scalar field. i) Without making explicit calculations, obtain the one-loop divergences in the purely metric sector. ii) Using the simplest parametrization and gauge fixing, calculate the full set of one-loop counterterms (one of the results of the first calculation in quantum gravity [186]). *iii)* Consider quantum GR coupled to the *nonminimal* scalar field. Without making an explicit calculation, demonstrate that the coefficient of the Gauss-Bonnet term is exactly the same as in i.

Hint. The calculations in this model can be found in the papers [35, 280, 195].
20.5. Generalize the proof in the part iii of the previous exercise, for a more general, sigma model type of scalar field. After that, discuss the difference between the Gauss-Bonnet and the Weyl-squared terms in all of the cases mentioned. Can we say that the Weyl-squared term, instead of the Gauss-Bonnet term, is parametrization/gauge-fixing invariant? Try to justify the response (negative) in several distinct ways.

20.5 One-loop divergences in a fourth-derivative model

The one-loop calculation in the fourth-derivative gravity was first performed in the pioneer paper [190]. This was the very first practical calculation in higher-derivative gravity, and the contribution $\big(\operatorname{Det} G_{\alpha\beta}\big)^{1/2}$ of the weight operator (10.60) was missed. This issue was corrected in the seminal work [140] (we believe this paper should be studied by everybody who intends to work in this area), but, owing to the large volume of calculations, there were mistakes in the final results for both the general and conformal versions of the theory. The correct results in these two versions of the theory were obtained in [17] and [9], respectively. Finally, the confirmation of the correctness of these calculations was given in Refs. [246] and [244], where another interesting aspect of the theory has been addressed.

The aforementioned interesting detail concerns the role of the Gauss-Bonnet term in quantum gravity. Originally, this issue was introduced in [86]. In $n = 4$ dimensions this term is topological, which means, locally, its integrand is a total derivative. However, when using dimensional regularization, this term becomes dynamical. Therefore, it can make a non-trivial contribution to the effective action, even starting from the one-loop order. The explicit calculations performed in [244, 246], have shown that i) the one-loop divergences do not depend on the parameter ρ in Eq. (18.18); ii) the beta functions for all parameters *before* taking the limit $n \to 4$ depend on ρ and iii) the $n \to 4$ divergences are exactly those of [17] and [9] for the general and conformal cases, respectively.

Consider the derivation of divergences in the quantum gravity model based on the action (18.18). This calculation is rather cumbersome, and full technical detail would make this section too long. Thus, we give only the main set of intermediate expressions in this calculation, so that the interested reader can reproduce all calculations, even though it may require significant effort. On the other hand, the reader can find all necessary details in the arXiv versions of the papers [246] (the full version of model (18.18)) and [244] (the purely conformal version of the model).

Combining the bilinear expansions from section 18.3 with the explicit form of the gauge-fixing term (18.21) with (18.16) and (18.22), we arrive at the following condition, which provides the minimal bilinear form of the action $S + S_{GF}$:

$$\beta = \frac{y+4z}{4(x-z)}, \qquad \alpha = \frac{1}{y+4x}, \qquad \gamma = \frac{2x-2z}{y+4x}, \qquad \delta = 1\,, \qquad p_1 = p_2 = 0. \tag{20.29}$$

With this choice of the gauge parameters, the bilinear operator $\hat{\mathcal{H}}$ takes the form

$$\hat{\mathcal{H}} = \hat{K}\Box^2 + \hat{D}^{\rho\lambda}\nabla_\rho\nabla_\lambda + \hat{N}^\lambda\nabla_\lambda - (\nabla_\lambda \hat{Z}^\lambda) + \hat{W}\,, \tag{20.30}$$

where $\hat{K}$, $\hat{D}^{\rho\lambda}$, $\hat{N}^{\lambda}$, $\hat{Z}^{\lambda}$ and $\hat{W}$ are local c-number operators acting in the $h^{\mu\nu}$-space. The matrix coefficient $\hat{K}$ of the fourth-derivative term has the following form:

$$\hat{K} = (K)_{\mu\nu,\alpha\beta} = \frac{y+4x}{4}\Big[\delta_{\mu\nu,\alpha\beta} + \frac{y+4z}{4(x-z)}\,g_{\mu\nu}\,g_{\alpha\beta}\Big]. \tag{20.31}$$

The expressions for $\hat{D}^{\rho\lambda}$ and $\hat{W}$ have the form

$$\begin{aligned}(\hat{D}^{\rho\lambda})_{\mu\nu,\alpha\beta} &= 4x\,\delta^{\rho}_{\nu}R^{\lambda}{}_{\alpha\mu\beta} - 2xg_{\nu\beta}R_{\alpha}{}^{\rho\lambda}{}_{\mu} + (3x+y)\,g^{\rho\lambda}R_{\mu\alpha\nu\beta} + 2xg_{\mu\nu}R_{\alpha}{}^{\rho\lambda}{}_{\beta}\\ &\quad - 2xg_{\nu\beta}R_{\mu\alpha}g^{\rho\lambda} + yg_{\mu\nu}\delta^{\rho}_{\alpha}R^{\lambda}{}_{\beta} - 2z\delta_{\alpha\beta}{}^{\rho\lambda}R_{\mu\nu} + 2zg_{\mu\nu}g^{\rho\lambda}R_{\alpha\beta}\\ &\quad + \Big(\frac{z}{2}R - \frac{1}{4\kappa^2}\Big)\big(\delta_{\mu\nu,\alpha\beta}\,g^{\rho\lambda} - g_{\mu\nu}g_{\alpha\beta}\,g^{\rho\lambda} - 2g_{\nu\beta}\delta_{\mu\alpha,}{}^{\rho\lambda} + 2g_{\mu\nu}\delta_{\alpha\beta,}{}^{\rho\lambda}\big)\\ &\quad + \frac{y+2x}{2}\delta_{\mu\nu,\alpha\beta}R^{\rho\lambda} + 2x\delta_{\nu\beta,}{}^{\rho\lambda}R_{\mu\alpha} - \frac{y}{4}g_{\mu\nu}g_{\alpha\beta}R^{\rho\lambda}\\ &\quad - (4x+2y)\delta^{\rho}_{\alpha}g_{\nu\beta}\,R^{\lambda}{}_{\mu}\end{aligned} \tag{20.32}$$

and

$$\begin{aligned}(\hat{W})_{\mu\nu,\alpha\beta} &= \frac{3x}{2}g_{\nu\beta}R_{\mu\sigma\tau\theta}R_{\alpha}{}^{\sigma\tau\theta} + \frac{x-y}{2}R^{\sigma}{}_{\alpha\mu}{}^{\tau}R_{\nu\beta\sigma\tau} + \frac{5x+y}{2}R^{\sigma}{}_{\alpha\mu}{}^{\tau}R_{\sigma\nu\beta\tau}\\ &\quad + \frac{y+2x}{2}R_{\mu\alpha}R_{\nu\beta} + \Big(\frac{z}{2}R - \frac{1}{4\kappa^2}\Big)\Big(R_{\mu\alpha\nu\beta} - \frac{1}{2}\,g_{\mu\nu}R_{\alpha\beta} + 3g_{\nu\beta}R_{\alpha\mu}\Big)\\ &\quad - \frac{1}{4}\Big[xR^2_{\rho\lambda\sigma\tau} + yR^2_{\rho\lambda} + zR^2 - \frac{1}{\kappa^2}(R+2\Lambda)\Big]\Big(\delta_{\mu\nu\,,\,\alpha\beta} - \frac{1}{2}\,g_{\mu\nu}g_{\alpha\beta}\Big)\\ &\quad - xg_{\mu\nu}R_{\alpha\theta\sigma\tau}R_{\beta}{}^{\theta\sigma\tau} - yg_{\mu\nu}R_{\alpha\tau}R^{\tau}{}_{\beta} + \frac{y-5x}{2}\,R_{\mu\tau}R^{\tau}{}_{\alpha\nu\beta}\\ &\quad + \frac{3y}{2}g_{\nu\beta}R_{\mu\tau}R^{\tau}{}_{\alpha} + \frac{3x+y}{2}\,R^{\sigma}{}_{\mu}{}^{\tau}{}_{\nu}R_{\sigma\alpha\tau\beta} + zR_{\mu\nu}R_{\alpha\beta}\,.\end{aligned}$$

The solution for the inverse matrix $\hat{K}^{-1}$ has the form

$$\hat{K}^{-1} = (\hat{K}^{-1})_{\mu\nu,}{}^{\theta\omega} = \frac{4}{4x+y}\,\big(\delta_{\mu\nu,}{}^{\theta\omega} - \Omega\;g_{\mu\nu}\,g^{\theta\omega}\big)\,,$$

$$\text{where}\qquad \Omega = \frac{y+4z}{4x-4z+n(y+4z)}\,.$$

In dimensional regularization, the divergences of the expressions $\log \mathrm{Det}\,\hat{\mathcal{H}}$ and $\log \mathrm{Det}\,(K^{-1}\,\hat{\mathcal{H}})$ are the same, because an extra factor $\hat{K}^{-1}$ is a c-number operator.

The minimal operator can be finally cast into the standard form,

$$\hat{H} = \hat{K}^{-1}\,\hat{\mathcal{H}} = \hat{1}\Box^2 + \hat{V}^{\rho\lambda}\nabla_\rho\nabla_\lambda + \hat{N}^{\rho}\nabla_\rho + \hat{U}\,. \tag{20.33}$$

Since we do not intend to derive the surface divergences, the explicit form of $\hat{N}^{\rho}$ will be omitted. For the matrix $\hat{V}^{\rho\lambda} = (V^{\rho\lambda})_{\mu\nu\,,\,\alpha\beta}$, the expression is

$$\hat{V}^{\rho\lambda} = \frac{4}{y+4x}\sum_{i=1}^{20} b_i\,\mathbf{k}_i\,, \tag{20.34}$$

where the following condensed notations have been used:

$$
\begin{aligned}
&\mathbf{k}_1 = g_{\nu\beta}\, g^{\rho\lambda}\, R_{\mu\alpha}\,, && \mathbf{k}_2 = \delta_{\mu\nu\,,\,\alpha\beta}\, g^{\rho\lambda}\,, && \mathbf{k}_3 = g^{\rho\lambda}\, R_{\mu\alpha\nu\beta}\,, \\
&\mathbf{k}_4 = \delta_{\nu\beta,}{}^{\rho\lambda}\, R_{\mu\alpha}\,, && \mathbf{k}_5 = \delta_{\nu\beta,}{}^{\rho\lambda}\, g_{\mu\alpha}\,, && \mathbf{k}_6 = \delta_{\mu\nu,\,\alpha\beta}\, R^{\rho\lambda}\,, \\
&\mathbf{k}_7 = \frac{1}{2}\big(\, \delta_{\nu}^{(\rho}\, R^{\lambda)}{}_{\alpha\beta\mu} + \delta_{\beta}^{(\rho}\, R^{\lambda)}{}_{\mu\nu\alpha}\,\big)\,, && && \mathbf{k}_8 = g_{\nu\beta}\,\delta^{(\rho}_{(\mu}\, R^{\lambda)}{}_{\alpha)}\,, \\
&\mathbf{k}_9 = g_{\nu\beta}\, R_{(\alpha}{}^{\rho\lambda}{}_{\mu)}\,, && \mathbf{k}_{10} = \frac{1}{2}\big(\, \delta_{\alpha\beta,}{}^{\rho\lambda}\, R_{\mu\nu} + \delta_{\mu\nu,}{}^{\rho\lambda}\, R_{\alpha\beta}\,\big)\,, \\
&\mathbf{k}_{11} = g_{\mu\nu} R_{\alpha}{}^{\rho\lambda}{}_{\beta}\,, && \mathbf{k}_{12} = g_{\alpha\beta} R_{\mu}{}^{\rho\lambda}{}_{\nu}\,, && \mathbf{k}_{13} = g_{\mu\nu}\, g^{\rho\lambda} R_{\alpha\beta}\,, \\
&\mathbf{k}_{14} = g_{\alpha\beta}\, g^{\rho\lambda} R_{\mu\nu}\,, && \mathbf{k}_{15} = g_{\mu\nu}\, \delta^{\lambda}_{\alpha}\, R^{\rho}_{\beta}\,, && \mathbf{k}_{16} = g_{\alpha\beta}\, \delta^{\lambda}_{\mu}\, R^{\rho}_{\nu}\,, \\
&\mathbf{k}_{17} = g_{\mu\nu}\, \delta_{\alpha\beta,}{}^{\rho\lambda}\,, && \mathbf{k}_{18} = g_{\alpha\beta}\, \delta_{\mu\nu,}{}^{\rho\lambda}\,, && \mathbf{k}_{19} = g_{\mu\nu}\, g_{\alpha\beta}\, g^{\rho\lambda}\,, \\
&\mathbf{k}_{20} = g_{\mu\nu}\, g_{\alpha\beta}\, R^{\rho\lambda}\,,
\end{aligned}
\tag{20.35}
$$

and

$$
\begin{aligned}
&b_1 = -2x\,, \qquad b_2 = \frac{zR}{2} - \frac{1}{4\kappa^2}\,, \qquad b_3 = 3x + y\,, \qquad b_4 = 2x \\
&b_5 = \frac{1}{2\kappa^2} - zR\,, \qquad b_6 = x + \frac{y}{2}\,, \qquad b_7 = -4x\,, \qquad b_8 = -4x - 2y\,, \\
&b_9 = -2x\,, \qquad b_{10} = -2z\,, \qquad b_{11} = 4x\,\Omega_3\,, \qquad b_{12} = x\,, \\
&b_{13} = -y\,\Omega_3, \qquad b_{14} = z, \qquad b_{15} = 2y\,\Omega_3, \qquad b_{16} = \frac{y}{2}, \\
&b_{17} = 2zR\,\Omega_3 - \frac{\Omega_1 - 2\,\Omega}{2\,\kappa^2}, \qquad b_{18} = \frac{zR}{2} - \frac{1}{4\kappa^2}, \qquad b_{19} = -b_{17}, \\
&b_{20} = -y\,\Omega_3.
\end{aligned}
\tag{20.36}
$$

The last term in the operator (20.33) is

$$
\begin{aligned}
(\hat{U})_{\mu\nu,\,\alpha\beta} = \frac{4}{y+4x}\bigg\{ & \frac{3x}{2} g_{\nu\beta} R_{\mu\rho\lambda\sigma} R_{\alpha}{}^{\rho\lambda\sigma} + \frac{5x+y}{2} R^{\lambda}{}_{\alpha\mu}{}^{\rho} R_{\lambda\nu\beta\rho} + zR_{\mu\nu}R_{\alpha\beta} \\
& + \frac{y-5x}{4}\Big(R_{\mu\sigma}R^{\sigma}{}_{\alpha\nu\beta} + R_{\alpha\sigma}R^{\sigma}{}_{\mu\beta\nu}\Big) + \Big(\frac{3zR}{2} - \frac{3}{4\kappa^2}\Big) g_{\nu\beta}R_{\alpha\mu} \\
& - \frac{1}{4}\mathcal{L}\big(\delta_{\mu\nu\,,\,\alpha\beta} - \Omega_1\, g_{\mu\nu}g_{\alpha\beta}\big) + \frac{3y}{2} g_{\nu\beta}R^{\lambda}{}_{\mu}R_{\alpha\lambda} - \frac{x}{2} g_{\alpha\beta}R_{\mu\theta\sigma\tau}R_{\nu}{}^{\theta\sigma\tau} \\
& + \frac{x+3y}{4} g_{\mu\nu}\, R^{\sigma\tau} R_{\alpha\sigma\beta\tau} + \Big(\frac{zR}{2} - \frac{1}{4\kappa^2}\Big) R_{\mu\alpha\nu\beta} + \frac{y+2x}{2} R_{\mu\alpha}R_{\nu\beta} \\
& - x\Omega_1 g_{\mu\nu} R_{\alpha\theta\sigma\tau}R_{\beta}{}^{\theta\sigma\tau} - \Omega_2 g_{\mu\nu} R_{\alpha}{}^{\tau} R_{\beta\tau} + \frac{3x+y}{2} R^{\sigma}{}_{\mu}{}^{\tau}{}_{\nu} R_{\tau\alpha\sigma\beta} \\
& + \Big(\frac{1}{4\kappa^2} - \frac{zR}{2}\Big) g_{\alpha\beta}R_{\mu\nu} - \frac{y}{2} g_{\alpha\beta}R_{\mu}{}^{\tau}R_{\nu\tau} + \frac{x-y}{2} R^{\rho}{}_{\alpha\mu}{}^{\lambda} R_{\nu\beta\rho\lambda} \\
& + \Big(\frac{1}{\kappa^2}\,\Omega_3 - zR\Omega_1\Big) g_{\mu\nu}R_{\alpha\beta} - \frac{1}{4\kappa^2}\Omega\,(4\Lambda - R)\, g_{\mu\nu}g_{\alpha\beta} \bigg\},
\end{aligned}
\tag{20.37}
$$

where $\mathcal{L} = xR^2_{\rho\lambda\sigma\tau} + yR^2_{\rho\sigma} + zR^2 - \kappa^{-2}\,(R + 2\Lambda)$, and the coefficients $\Omega_{1,2,3}$ are

$$\Omega_1 = \frac{2\,x + 3\,y + 10\,z}{4x - 4z + ny + 4z}, \qquad \Omega_2 = \frac{7\,x\,y + 9\,y^2 - 4\,x\,z + 28\,y\,z}{4(4x - 4z + ny + 4z)},$$
$$\Omega_3 = \frac{x + y + 3z}{4x - 4z + ny + 4z}. \tag{20.38}$$

The ghost action can be derived according to the general formula (18.17). For the practical calculations, there is no need to use the modified expression (18.23). Thus,

$$S_{gh} = \mu^{n-4} \int d^n x \sqrt{-g}\, \bar{C}^\mu \, (\mathcal{H}_{gh})_\mu^{\ \nu} \, C_\nu, \tag{20.39}$$

where

$$\hat{\mathcal{H}}_{gh} = (\mathcal{H}_{gh})_\mu^{\ \nu} = -\delta_\mu^\nu \Box - \nabla^\nu \nabla_\mu - 2\beta \nabla_\mu \nabla^\nu \,. \tag{20.40}$$

The one-loop effective action $\bar{\Gamma}^{(1)}$ is given by

$$\bar{\Gamma}^{(1)}[g_{\mu\nu}] = \frac{i}{2} \log\, \mathrm{Det}\, \hat{\mathcal{H}} \; - \; \frac{i}{2} \log\, \mathrm{Det}\, G_{\alpha\beta} \; - \; i \log\, \mathrm{Det}\, \hat{\mathcal{H}}_{gh}, \tag{20.41}$$

where $\hat{\mathcal{H}}$ is the bilinear form of the action (18.18) with the gauge-fixing term, and $\hat{\mathcal{H}} = M_\alpha^{\ \lambda}[g]$ is the bilinear form of the ghost action.

Thus, we meet a qualitatively new problem of deriving logarithmic divergences of the three terms in (20.41). The last two are the nonminimal vector operators (20.40) and (18.22), with the coefficients defined in Eq. (20.29). The first term in (20.41) is the functional determinant of the minimal fourth-derivative operator.

The systematic derivation of both results can be found in the review paper [31]. This derivation has a higher level of complexity, since it is based on the generalized Schwinger-DeWitt technique and the table of *universal traces* developed in the same paper.[1] Thus, owing to space limitations, we will not reproduce here the calculations and present only the final results.

The divergent contribution of the nonminimal Abelian vector operator

$$\hat{\mathcal{H}}^{vec}_{nm} = -\,\delta_\mu^\nu \Box + \sigma\, \nabla_\mu \nabla^\nu - P_\mu^\nu \tag{20.42}$$

has the form [140, 31]

$$\frac{i}{2} \log\, \mathrm{Det}\, \hat{\mathcal{H}}^{vec}_{nm}\Big|_{\mathrm{div}} = -\frac{1}{\epsilon} \int d^4 x \sqrt{-g}\, \Big\{ -\frac{11}{180} R^2_{\mu\nu\alpha\beta} + \Big(\frac{1}{24}\psi^2 + \frac{1}{12}\psi - \frac{1}{45}\Big) R^2_{\mu\nu}$$
$$+ \frac{1}{48}\psi^2 P^2 + \Big(\frac{1}{48}\psi^2 + \frac{1}{12}\psi + \frac{1}{18}\Big) R^2 + \Big(\frac{1}{12}\psi^2 + \frac{1}{3}\psi\Big) R_{\mu\nu} P^{\mu\nu}$$
$$+ \Big(\frac{1}{24}\psi^2 + \frac{1}{4}\psi + \frac{1}{2}\Big) P^2_{\mu\nu} + \Big(\frac{1}{24}\psi^2 + \frac{1}{12}\psi + \frac{1}{6}\Big) RP \Big\}, \tag{20.43}$$

where $P = P^\alpha_\alpha$ and $\psi = \frac{\sigma}{1-\sigma}$. Furthermore, the algorithm for the one-loop divergences of a minimal fourth order operator (20.33) is the following [140, 31, 174]:

[1] Additional examples of the use of this table in fourth-derivative gravity coupled to matter fields can be found in chapter 9 of [80].

$$\frac{1}{2}\log \operatorname{Det} \hat{H}\Big|_{\rm div} = -\frac{1}{\epsilon}\int d^4x\sqrt{-g}\,\Big\{\frac{\hat{1}}{90}\big(R^2_{\rho\sigma\lambda\tau} - R^2_{\rho\sigma}\big) + \frac{\hat{1}}{36}\,R^2 + \frac{1}{6}\hat{\mathcal{R}}^2_{\rho\lambda}$$
$$- \frac{1}{6}R_{\rho\lambda}\hat{V}^{\rho\lambda} + \frac{1}{12}R\hat{V}^{\rho}{}_{\rho} + \frac{1}{48}\hat{V}^{\rho}{}_{\rho}\hat{V}^{\lambda}{}_{\lambda} + \frac{1}{24}\hat{V}_{\rho\lambda}\hat{V}^{\rho\lambda} - \hat{U}$$
$$+ \frac{\hat{1}}{15}\Box R + \frac{1}{9}\Box\hat{V}^{\rho}{}_{\rho} - \frac{5}{18}\nabla_\rho\nabla_\lambda\hat{V}^{\rho\lambda} + \frac{1}{2}\nabla_\rho\hat{N}^{\rho}\Big\}, \tag{20.44}$$

where $\hat{\mathcal{R}}_{\rho\lambda}$ is the commutator (20.21) of the covariant derivatives in tensor space. For the sake of generality, we included in this formula the surface terms [174].

Substituting the operators of our interest in the last two formulas and then in the general expression (20.41), after a certain amount of algebra, we arrive at the one-loop divergences in the theory (18.18). The final expression is [140, 17, 246]

$$\Gamma^{(1)}_{\rm div} = -\frac{1}{n-4}\int d^4x\sqrt{-g}\,\Big\{\beta_1 E + \beta_2 C^2 + \beta_3 R^2 + \frac{\beta_4}{\kappa^2}\,R + \frac{\beta_5\Lambda}{\kappa^2} + \frac{\beta_6}{\kappa^4}\Big\}$$
$$= -\frac{1}{\epsilon}\int d^4x\sqrt{-g}\,\Big\{\frac{133}{20}C^2 - \frac{196}{45}E_4 + \Big(\frac{10\,\lambda^2}{\xi^2} - \frac{5\,\lambda}{\xi} + \frac{5}{36}\Big)R^2$$
$$+ \Big(\frac{\xi}{12\,\lambda} - \frac{13}{6} - \frac{10\,\lambda}{\xi}\Big)\frac{\lambda}{\kappa^2}R + \Big(\frac{2\xi}{9} - \frac{56\lambda}{3}\Big)\frac{\Lambda}{\kappa^2} + \Big(\frac{\xi^2}{72} + \frac{5\lambda^2}{2}\Big)\frac{1}{\kappa^4}\Big\}, \tag{20.45}$$

where we introduced new notations for the beta functions. It is evident that the one-loop divergences (20.45) perfectly correspond to the expectations based on the power counting, as discussed in section 18.7. At this point, it is good to remember the ambiguity of this expression, which is parameterized by the relation (20.14).

Finally, in order to complete the story, let us comment on and give the result for conformal gravity, which is the theory based on the classical action

$$S_c = \int d^4x\sqrt{-g}\,\Big\{ - \frac{1}{2\lambda}\,C^2 + \frac{1}{\rho}\,E_4 - \tau\,\Box R\Big\}. \tag{20.46}$$

This theory satisfies the Noether identities for both diffeomorphism invariance and local conformal invariance (12.13). In a perfect correspondence with what we discussed in chapter 17, the one-loop divergences in this theory should be also invariant, satisfying the same Noether identity (12.13). And the final result for these divergences confirms this statement [140, 9, 244], the divergences in this case, look like

$$\Gamma^{(1)}_{\rm div} = \frac{\mu^{n-4}}{(4\pi)^2(n-4)}\int d^nx\sqrt{-g}\,\Big\{\frac{87}{20}E_4 - \frac{199}{30}C^2\Big\}. \tag{20.47}$$

There is no $\Box R$ term in this formula because it was never calculated, and this part of the divergences is unknown as of the time when this book is written.

Even though the renormalizability of the conformal gravity (20.46) is confirmed by the result (20.47), the general renormalizability of the same theory beyond the one-loop order is a complicated problem, which was never discussed in detail. Let us just mention that the situation here is similar to the semiclassical conformal theory with scalar fields, and cite the papers [244, 14, 287] where this issue has been considered. The reader can address some aspects of this problem in the exercises suggested below.

Exercises

20.7. Perform the one-loop calculation in the conformal gravity (18.18). The points to complete in this case are as follows: i) Derive the expansion of the $\sqrt{-g}C^2$ action, using the formulas from section 18.3 (as usual, it makes sense to check these formulas first).
ii) Along with using the Faddeev-Popov procedure for fixing the diffeomorphism invariance, it is necessary to introduce an additional gauge-fixing condition for the local conformal invariance. Explain why the condition $h = 0$ can be used in the present case. Why do the gauge ghosts not appear in this case?
iii) After that, the calculations are pretty much standard and can be done using the formulas (20.44) and (20.43).
Hint. All intermediate expressions for this exercise can be found in the paper [244].
20.8. Explain why both of the coefficients in the result (20.47) are gauge and parametrization independent. Explain why this is the case for the $\Box R$ term too. Explain why, in this part, there is such a dramatic difference with the non-conformal version of higher-derivative gravity?
20.9. Derive the one-loop conformal anomaly in the case of the conformal quantum gravity (20.46). Write down the general expression for the anomaly-induced action. Discuss whether there is an ambiguity in the local part of this action.
Hint. It would be useful to consult the paper [14], which is particularly devoted to the same ambiguity in the semiclassical gravity framework.

20.6 One-loop divergences in superrenormalizable models

The calculation of divergences in nonlocal models, including those in gravity, requires new techniques, that have not yet been developed at the time of publication. For this reason, these divergences have not been calculated yet. Thus, we can present only the results for the polynomial models [13], which were introduced in section 18.2.3.[2]

As we know from the same section, the non-degenerate polynomial theory is superrenormalizable, and the divergences are composed of covariant local terms with zero-derivative (cosmological constant), two-derivative (linear in R) and four-derivative terms. Furthermore, as we know from section 20.2, all of these divergences are independent of the choice of gauge fixing and parametrization.

The practical calculations in this case, up to the present time, were published only for the two lowest-derivative sectors of the effective action. For the cosmological constant divergences, the calculation is quite simple; it was performed in the very first paper [13], where the models of this type were introduced. In contrast, derivation of an R type counterterm requires greater effort. It was done in [222], where one can also find full details and intermediate formulas. We shall give only the result for the theories (18.26), taken from this paper, in the case without "killer" $\mathcal{O}(R^3_{\dots})$ terms and for $N \geq 2$,

[2]On the other hand, a general discussion of the structure of renormalization and beta functions in nonlocal models can be found in [290, 222].

$$
\begin{aligned}
\Gamma^{(1)}_{div} &= \frac{1}{n-4}\int d^4x\sqrt{-g}\,(\beta_{\Lambda_{cc}} + \beta_G R) \\
&= -\frac{1}{\epsilon}\int d^4x\sqrt{-g}\,\bigg\{\frac{5\omega^2_{N-1,C}}{2\omega^2_{N,C}} + \frac{\omega^2_{N-1,R}}{2\omega^2_{N,R}} - \frac{5\omega_{N-2,C}}{\omega_{N,C}} - \frac{\omega_{N-2,R}}{\omega_{N,R}} \\
&\qquad + \Big(\frac{5\omega_{N-1,C}}{6\omega_{N,C}} + \frac{\omega_{N-1,R}}{6\omega_{N,R}}\Big)R\bigg\}.
\end{aligned}
\tag{20.48}
$$

The notations $\beta_{\Lambda_{cc}}$ and β_G will prove useful in the next chapter, when we discuss the renormalization group equations in this model.

Since the theory is superrenormalizable, the beta functions quoted in the last expression, for $N \geq 3$, are *exact*, even though they were derived at the one-loop level. And, on top of this, these beta functions are gauge-fixing and parametrization independent.

On the other hand, the physical interpretation of these beta functions is unclear because of i) unsolved problem of massive ghosts and ii) the problem with the application of the beta functions based on the minimal subtraction scheme, at low energies. Indeed, these kinds of problems of interpretation are typical also for renormalizable, four-derivative, gravity theory.

Exercises

20.10. Using power-counting arguments, discuss how the calculation of the cosmological constant sector in (20.48) can be performed with the same two algorithms (for minimal fourth-derivative and nonminimal massless vector operators) that we presented in section 20.5.

20.11. Try to choose the weight operator in the Faddeev-Popov procedure in such a way that the derivation of the cosmological constant divergence in (20.48) becomes technically simpler. Using the power-counting arguments from section 18.3, explain why this change in the weight operator does not affect the counterterms under discussion. Perform the practical calculations with at least two different weight operators, and show that the results are the same.

Observation. Even though these calculations are simple, they require the basic knowledge of the generalized Schwinger-deWitt technique of [31].

20.12. Derive the beta functions for the lowest-order running parameters of the action (18.26) and identify the parameters for which these beta functions are $\beta_{\Lambda_{cc}}$ and β_G from Eq. (20.48).

20.13. Using dimensional arguments, write down the general form of the *exact* (all loop) C^2 and R^2 counterterms in the polynomial theory, such that only the numerical coefficients remain undefined. Consider separately the cases with $N = 3, 2, 1$. Repeat the same consideration for the R and Λ terms. Derive the beta functions for the corresponding parameters of the theory.

21
The renormalization group in perturbative quantum gravity

In this chapter, we briefly summarize what is known about the renormalization group in perturbative quantum gravity in four spacetime dimensions.

It is worth mentioning that there are some other approaches to the renormalization group in quantum gravity that are supposed to be non-perturbative. In this respect, one can remember the $2+\epsilon$ renormalization group in quantum GR [153,341,92,196,197] and also the hypothesis of asymptotic safety in quantum gravity [341]. The practical realization of the last proposal in $n = 4$ is usually associated with the functional renormalization group [231,248]. Unfortunately, neither one of these approached solves the fundamental problems of quantum gravity.

The first approach can be successfully formulated in the vicinity of the two spacetime dimensions $n = 2+\epsilon$, but it is unclear how it can help in solving the fundamental contradiction between renormalizability and unitarity in $n = 4$. Technically, the calculations in $n = 4 + \epsilon$ are possible [244, 246], but the output cannot be regarded as non-perturbative.

The functional renormalization group is a beautiful way to implement our general understanding of how the exact (dressed) Green functions may look like [253]. This understanding has an elegant formulation as the Wetterich equation for the effective average action [346, 223] (see also the review and introduction in [50, 349]). The average effective action is supposed to coincide with the usual effective action of the quantum theory in a fixed point. As the renormalization group flow within this approach is supposed to model the non-perturbative renormalization group with full quantum corrections, it is sufficient to derive such a flow. After that, the flow indicates the non-Gaussian fixed point. In the case of gravity, this means that we should make an expansion around a certain non-flat metric that is a solution of an exact theory. Such a "correct" expansion should be around the fixed-point value $\kappa_0 \neq 0$, instead of the usual perturbative expansion (18.13) around $\kappa = 0$. This is a central idea of asymptotic safety. The coupling does not need to vanish in the UV (this would correspond to the asymptotic freedom) but tends to a special non-zero value κ_0, for which perturbative (non)renormalizability of the theory has no importance. The renormalization group flow should be non-perturbative and exact; hence, the use of the functional renormalization group is quite relevant.

Unfortunately, the application of this hypothesis to gauge theories, and, in particular, to gravity, meets serious difficulty. In the next sections we will demonstrate how on-shell universality can be used to separate the essential effective charges from the

Introduction to Quantum Field Theory with Applications to Quantum Gravity. Iosif L. Buchbinder and Ilya L. Shapiro, Oxford University Press (2021). © Iosif L. Buchbinder and Ilya Shapiro.
DOI: 10·1093/oso/9780198838319·003·0021

other running parameters, for which the running is ambiguous and therefore physically irrelevant. In a functional renormalization group this does not work this way at all. It was shown in Refs. [207, 209, 25] that, in gauge theories and especially in quantum gravity, the effective average action depends on the gauge fixing even *on shell*. In reality, this means that physical observables can be calculated in a controllable way at the fixed point only. At the same time, the renormalization group flow itself can not be consistently explored within this approach, since the result essentially depends on the choice of the gauge fixing [205]. This means that one can choose the gauge-fixing condition in such a way that the renormalization group flow gives us to the fixed point we like, while, in the real theory, such a fixed point may not exist at all. In this situation, reliable evidence for a non-perturbative fixed point can be hardly obtained. Let us note that, in [207], an alternative formulation of the functional renormalization group was proposed, that would solve this problem. However, this approach is technically more complicated, and, at present, there are no records of using it for practical calculations.

Given the dubious situation with the non-perturbative methods, in what follows, we describe only the well-established perturbative results concerning the renormalization group, applying them to different models of quantum gravity.

21.1 On-shell renormalization group in quantum GR

As we already know, quantum GR is not renormalizable; hence, there is no way to develop a traditional, completely consistent, renormalization group framework. One of the alternatives is to use the functional renormalization group, as described above. However, along with the gauge-fixing dependence problem, in this case, there is another serious difficulty. According to the power counting arguments, in quantum GR without a cosmological constant term, one never obtains logarithmic divergences that repeat the form of the classical action. The standard way of resolving this problem is to extract the renormalization group flow from the quadratic divergences. However, this procedure always brings along another ambiguity, which we discussed in section 13.4.3 (a more detailed discussion of this issue can be found in [208]). All in all, we better look for another framework.

A much simpler scheme was suggested by Fradkin and Tseytlin in the paper [140] devoted mainly to fourth-derivative gravity. Later on, it was extended to the theory with torsion [76], including with the Holst term [289]. Let us restrict our attention to the simplest original version and construct the on shell version of the renormalization group, which has the advantage of being free from all mentioned ambiguities.

The critical observation here is that in quantum GR with a cosmological constant term, there is a unique gauge-fixing and parametrization-independent combination of the divergences (20.8), which is constructed from the parameters of the classical action. And this is exactly the combination that "survives" on shell.

Our starting point will be the action (18.12) with the equations of motion (20.4), which can be rewritten as $R_{\mu\nu} = g_{\mu\nu}\Lambda$, to give the on shell form of the action (20.9). Neglecting the irrelevant topological term, we observe that the two on shell expressions have the same structure; hence, we can write the on shell renormalization transformation for the dimensionless parameter $\gamma = \Lambda\kappa^2$,

$$\gamma_0 = \mu^{n-4}\left(\gamma - \frac{58}{5}\frac{\gamma^2}{\epsilon}\right), \tag{21.1}$$

with the consequent renormalization group equation

$$(4\pi)^2\,\mu\frac{d\gamma}{d\mu} = -\,b^2\gamma^2, \qquad \text{with} \qquad b^2 = \frac{29}{5} \tag{21.2}$$

and the initial condition $\gamma_0 = \gamma(\mu_0)$. Solving the last equation, we get

$$\gamma(t) = \frac{\gamma_0}{1+\gamma_0 b^2 t}, \qquad \text{with} \qquad dt = \frac{1}{(4\pi)^2}\,\frac{d\mu}{\mu}, \tag{21.3}$$

formally indicating a UV asymptotic freedom in quantum GR in the simplest on shell framework. Let us stress that the word "formally" here is relevant, because γ is not the coupling constant in this theory. The coupling constant is $\kappa = \sqrt{16\pi G}$, and the on shell renormalization group does not describe the running of this parameter.

The last observation is that independent and consistent renormalization group equations in quantum GR with a cosmological constant can be constructed using the unique effective action of Vilkovisky [328] and effective approach to quantum gravity. Effectively, the running described by these equations is not restricted by the one-loop approximation and is valid in the range of typical energies between the low-energy cosmological scale and, at least, the Planck scale. The consideration of this application is beyond the scope of the book, but the reader can find it in the paper [158].

Exercises

21.1. Repeat the calculations leading to Eqs. (21.1) and (21.2).
21.2. Verify that the parameter λ has gauge-fixing invariant renormalization in (21.1). Use independently the arguments from section 20.2 or the results of the explicit calculation of [166].
21.3. Write the parameter λ in terms of ρ_Λ and the Planck mass M_P. Evaluate numerically the effect of the running in (21.3) at the present cosmological epoch and at the epoch of inflation, where the Hubble parameter is $H \approx \sqrt{\frac{\Lambda}{3}} \propto 10^{11} - 10^{14}\,GeV$.
Observation. The output of this exercise is a good illustration of the difficulties in applying quantum gravity to physical problems, except in the analysis of singularities.

21.2 The renormalization group in fourth-derivative gravity

The renormalization group equations for the parameters of the action (18.18) can be easily obtained from the expression for one-loop divergences (20.45), in the form

$$\frac{d\lambda}{dt} = -\,b^2\,\lambda^2, \qquad b^2 = \frac{133}{10}, \qquad \text{and} \tag{21.4}$$

$$\frac{d\xi}{dt} = -\,10\,\lambda^2 + 5\,\lambda\,\xi - \frac{5}{36}\,\xi^2, \qquad \frac{d\rho}{dt} = -\,\frac{196}{45}\,\rho^2, \tag{21.5}$$

where t has been defined in (21.3). As we know from the considerations in section 19, the initial value $\lambda_0 = \lambda(\mu_0)$ must be positive, as, otherwise, the theory is unstable even at low energies. Then the solution for the effective coupling $\lambda(t)$ is

$$\lambda(t) = \frac{\lambda_0}{1 + \lambda_0 b^2 t}, \tag{21.6}$$

indicating that the theory is asymptotically free in the UV limit $\mu \to \infty$. Similar behavior takes place for $\rho(t)$, but it can be hardly called coupling, since, in $n = 4$, no interactions correspond to this term. Let us note that if the renormalization group equations are considered in $n = 4 + \epsilon$, then the quantum effects of the Gauss-Bonnet term become relevant, and the structure of equations is more complicated. The corresponding analysis can be found in [246] for general fourth-derivative gravity and in [244] for the conformal version.

The analysis of the remaining equation for the coefficient of the R^2 term can be performed in terms of ω, which is defined in Eq. (18.19). Since this consideration has been described in papers [17,246] (in the last case, for a more complex case of $n = 4+\epsilon$) and in the book [80], we leave the corresponding analysis to the reader as an exercise.

The last two equations describe the running of the Newton constant and the cosmological constant. From Eq. (20.45) and the considerations of section 20.3, it is easy to arrive at the equations

$$\frac{d\kappa^2}{dt} = \kappa^2 \beta_4 + \kappa^2 \big[f(\alpha_i) - f(\alpha_i^{(0)})\big], \tag{21.7}$$

$$\frac{d\Lambda}{dt} = \Lambda\beta_4 + \frac{1}{2}\Lambda\beta_5 + \frac{\beta_6}{2\kappa^2} - \Lambda\,\big[f(\alpha_i) - f(\alpha_i^{(0)})\big]. \tag{21.8}$$

Here the difference $f(\alpha_i) - f(\alpha_i^{(0)})$ is an arbitrary function of the gauge-fixing parameters, as explained in section 20.2. Owing to the presence of this difference, no consistent analysis of these two equations is possible if they are considered separately.

By direct inspection, one can find the unique, dimensionless, combination

$$\gamma = \kappa^2 \Lambda \tag{21.9}$$

of κ and Λ, which has the running independent of the parameter $f(\alpha_i)$. Just as in the on shell quantum GR, the well-defined renormalization group equation is the one for γ and not for κ and Λ separately.

Using (21.7) and (21.8) we arrive at the equation

$$\frac{d\gamma}{dt} = \frac{\gamma}{2}\left(6\beta_4 + \beta_5 + \frac{\beta_6}{\gamma}\right). \tag{21.10}$$

In order to avoid lengthy consideration of trivial details, we skip the analysis of this equation, which can be found in [17, 80] and [246] (for $n = 4 + \epsilon$). Let us just say that the UV asymptotic behavior of γ may strongly depend on the initial data for the coupling ξ (or ω).

The last observation is that the renormalization group equations constructed above can be applied, in the original form, only at very high energies. The reason for this is that the $\overline{\text{MS}}$-based renormalization group applies only at the energy scale when all degrees of freedom are active and do not decouple. In the present case, the masses of the tensor ghost and the healthy massive scalar mode are of the Planck order of magnitude; hence, Eq. (21.10) is valid only above the Planck scale. For the physical process below this energy scale, one can expect that these two massive modes decouple. This decoupling has been never explored in the fourth- and higher-derivative gravity models; hence, we cannot present the corresponding discussion here. Some qualitative arguments concerning effective quantum gravity can be found in [261].

Exercises

21.4. Construct and analyze the renormalization group equation for the parameter τ in the general fourth-derivative gravity model (18.18).
21.5. Rewrite Eq. (21.5) in terms of ω from (18.19). Show that in this variable, the equation has the general structure

$$\frac{d\omega}{d\tau} = a_0(\omega - \omega_1)(\omega - \omega_2), \qquad \text{where} \qquad \tau = \frac{1}{b^2}\log\big(1 + \lambda_0 b^2 t\big). \tag{21.11}$$

Find the values of a_0, ω_1 and ω_2 and show, by assuming that the initial value of ω is positive, that i) both fixed points are real and different, $\omega_1 < \omega_2$, ii) the limit $\tau \to \infty$ corresponds to the UV limit $\mu \to \infty$, iii) show that ω_2 and ω_1 are UV stable and UV unstable fixed points, respectively, in the limit $\tau \to \infty$. and iv) discuss the possibility to explore infrared limit in this framework, from different viewpoints.
21.6. Analyze the renormalization group equation (21.10) for the case when ω is in the UV stable fixed point ω_2. Discuss physical applications of this result for the UV running, taking into account the "slow" scaling of the parameter τ defined in (21.11).
21.7. Using the result (20.47), construct and analyze the renormalization group equations in the conformal version of fourth-derivative gravity.

21.3 The renormalization group in superrenormalizable models

The beta functions for $N \geqslant 2$ in the polynomial model (18.26) follow from the expression for the divergences (20.48) and are as follows [13, 222]:

$$\beta_G = \mu\frac{d}{d\mu}\Big(-\frac{1}{16\pi G}\Big) = -\frac{1}{6(4\pi)^2}\Big(\frac{5\omega_{N-1,C}}{\omega_{N,C}} + \frac{\omega_{N-1,R}}{\omega_{N,R}}\Big), \tag{21.12}$$

$$\beta_{cc} = \mu\frac{d}{d\mu}\Big(-\frac{\Lambda_{cc}}{8\pi G}\Big) = \frac{1}{(4\pi)^2}\Big(\frac{5\omega_{N-2,C}}{\omega_{N,C}} + \frac{\omega_{N-2,R}}{\omega_{N,R}} - \frac{5\omega^2_{N-1,C}}{2\omega^2_{N,C}} - \frac{\omega^2_{N-1,R}}{2\omega^2_{N,R}}\Big). \tag{21.13}$$

As we know from power counting, for $N \geqslant 3$, these two beta functions are exact, since higher loops do not provide further contributions.

In perfect agreement with what we have learned from the power counting in section 18.7, the beta functions (21.13) do not depend on the parameters for fourth-derivative couplings but only on the highest-derivative couplings of the theory. On the other hand, as we know from section 20.3, these beta functions are universal in the sense that they do not depend on the gauge fixing or parametrization of the quantum metric. The beta functions depend only on the parameters $\omega_{k,C}$ and $\omega_{k,R}$ with $k = N-2,\, N-1,\, N$.

The next question is whether these beta functions may have a direct physical sense - in other words, whether we are able to invent a situation where these equations describe the running with some physical scale, that is, associated with μ? As usual, the answer to this question is complicated and consists of two parts.

First of all, as we have discussed in the semiclassical part of the book, the running of ρ_Λ or G^{-1} does not correspond to the nonlocal $\Box$-dependent form factors in the loop corrections to the classical action [167, 36]. However, there is no safe way to rule out the physical running of either one of the two parameters in the infrared [288]. This subject is not well understood and cannot be addressed at the textbook level, so we skip its further discussion.

On the other hand, according to our calculations in section 16.3, one should expect an infrared decoupling of, at least, some of the heavy degrees of freedom of superrenormalizable gravity (18.26). Thus, even if the low-energy running of ρ_Λ and G^{-1} takes place, it is expected that it will be different from (21.12) and (21.13). At the same time, as we have already mentioned above, this conclusion is not supported by the direct calculation of the infrared decoupling in any of the fourth- or higher-derivative models. Thus, we are not in a position to discuss solid results in this area.

All these considerations concern not only the superrenormalizable models of quantum gravity, but also the fourth-derivative model discussed above. The same decoupling is, in principle, supposed to occur also in nonlocal models such as (18.31), but in this case, the situation is more obscure, since massive poles in these theories emerge in infinite number, only in the dressed propagator and are expected to be mostly (or even completely) complex [290]. The calculation of decoupling in these models is challenging and, up to this point, has not been done.

22
The induced gravity approach

The idea of induced gravity was formulated in the paper [268] by A.D. Sakharov. For this reason, it is sometimes called Sakharov's induced gravity (see, e.g., [145, 330]). One year before the mentioned pioneer paper, Ya.B. Zeldovich published an important work that started the discussion of the cosmological constant problem [353] and can be regarded as a preliminary version of induced gravity. The main idea of the work [268] was that gravity is not a fundamental interaction, in the sense that it has no classical action. The Einstein-Hilbert action of gravity comes to life because of the quantum effects of matter fields. It is easy to see that this idea may be useful in solving the fundamental contradiction between the renormalizability and the unitarity of quantum gravity. The logic is that if gravity has no action, there is no need to quantize it and no need to insist on the unitarity of the corresponding $\mathcal{S}$-matrix. As we already know, semiclassical gravity has no problem with renormalizability, which is its main advantage.

In what follows, we shall describe the basic notions of induced gravity, keeping in focus the main problems related to massive ghosts, unitarity and stability. After that, we present a few concluding remarks on the cosmological constant problem. The interested reader is advised to look at the standard review of the subject, [5]. The general idea of induced gravity is simple, while its realization may be rather non-trivial, depending on the theory. In any case, the induced gravity concept is necessary if we consider the interaction of gravity with matter, and quantum field theory concepts.

The strong version of induced gravity assumes that, initially, there is no action for the metric and hence no predetermined equations of motion. These equations are the result of integrating out the matter degrees of freedom. Suppose we have a theory of quantum matter fields $\Phi = (\varphi,\, \psi,\, A_\mu)$ that interacts with the classical metric $g_{\mu\nu}$. The action for matter fields is $S_m[\Phi,\, g_{\mu\nu}]$. But after we integrate over the matter fields, we meet the *induced* gravitational action $S_{ind}[g_{\mu\nu}$, which is defined through

$$e^{iS_{ind}[g_{\mu\nu}]} = \int \mathcal{D}\Phi\, e^{iS_m[\Phi,\, g_{\mu\nu}]}\,. \tag{22.1}$$

After that, we gain the dynamics of the gravitational field, which corresponds to the least action principle for the total action

$$S_t = S_m[\Phi,\, g_{\mu\nu}] + S_{ind}[g_{\mu\nu}]\,. \tag{22.2}$$

Since gravity has no classical action and proper dynamics, there are no problems with its quantization. Therefore, the general scheme of induced gravity looks quite promising. The question is how it can be put into practice.

Introduction to Quantum Field Theory with Applications to Quantum Gravity. Iosif L. Buchbinder and Ilya L. Shapiro, Oxford University Press (2021).
DOI: 10.1093/oso/9780198838319.003.0022

Making the derivatives expansion in $S_{ind}(g_{\mu\nu})$, in the lowest orders we obtain

$$S_{ind}[g_{\mu\nu}] = \int d^4x\sqrt{-g}\Big(- \rho_\Lambda^{ind} - \frac{1}{16\pi\, G_{ind}}\, R + \dots \Big)\,, \tag{22.3}$$

where ... indicates higher-derivative and nonlocal terms, which are supposed to be irrelevant at low energies. The natural questions are as follows:

i) How can we evaluate the induced constants G_{ind} and ρ_Λ^{ind}?

ii) What are the ambiguities in this evaluation?

iii) Is there some certainty that the higher-derivative and nonlocal terms in the induced action will not be important and will not contradict the existing tests of GR?

iv) Is it possible to avoid massive ghosts and instabilities in induced gravity?

Indeed, there are several qualitatively different schemes for deriving the induced action; there are ambiguities in all of them, and the problems mentioned above can be solved or avoided in different ways.

22.1 Gravity induced from the cut-off

In the original paper [353] the derivation of ρ_Λ^{ind} was performed in flat space, using integration over momentum, with Λ_{QCD} regarded as a natural cut-off scale. As we shall see in what follows, in such kind of "purely induced" gravity, one can hardly go far in solving the problems listed above. So, it is wise to keep the magnitude of the cut-off arbitrary and define it later on.

Just to provide a complete discussion, let us ask another question: how one can implement a covariant cut-off regularization in curved spacetime? For instance, using cut-off regularization for the energy-momentum tensor of vacuum, we immediately break down covariance [107] (see also the more recent work [6], where the extensive discussion of this issue has been restarted). Namely, in flat spacetime, we get the following expressions for the energy density (13.33) and the pressure:

$$\rho_{vac} = \frac{1}{2}\int \frac{d^3k}{(2\pi)^3}\,\sqrt{\mathbf{k}^2+m^2}, \qquad p_{vac} = \frac{1}{6}\int \frac{d^3k}{(2\pi)^3}\,\frac{\mathbf{k}^2}{\sqrt{\mathbf{k}^2+m^2}}\,. \tag{22.4}$$

The direct calculation of the first integral gives

$$\rho = \frac{1}{16\pi^2}\Big[\Omega^4 + m^2\Omega^2 + \frac{1}{8}m^4 - \frac{1}{2}m^4\log\frac{2\Omega}{m} + \mathcal{O}\Big(\frac{m}{\Omega}\Big)\Big]\,. \tag{22.5}$$

For the Planck-scale cut-off, $\Omega = M_P$, this expression gives the famous "120 orders of magnitude discrepancy between theory and experiment." However, the second integral, i.e., the "pressure of vacuum," indicates that the situation is not that simple:

$$p = \frac{1}{48\pi^2}\Big[\Omega^4 - m^2\Omega^2 - \frac{7}{8}m^4 + \frac{3}{2}m^4\log\frac{2}{\Omega}m + \mathcal{O}\Big(\frac{m}{\Omega}\Big)\Big]\,. \tag{22.6}$$

At first sight, the result we obtained looks strange. One should expect that the flat space vacuum average $\langle T_{\mu\nu}\rangle \propto \eta_{\mu\nu}$, which would mean the following equation of state for the cosmological constant: $p_{vac} = -\rho_{vac}$. Indeed, this should be the unique result compatible with the Lorentz invariance of the flat space vacuum. But inspection of expressions (22.5) and (22.6) demonstrates that, in the leading Ω^4 terms, there is a radiation-like "equation of state" of the vacuum, instead of the one for ρ_Λ.

Indeed, this is a natural output, and this can be seen even without taking the integrals explicitly. For each mode $\mathbf{k}$, there is a plane wave, with the dispersion relation corresponding to the particle with the mass m. Taking only the leading Ω^4-type terms, we are dealing with the UV limit, when the mass of the relevant modes become negligible. In the massless limit, for each mode, there is the equation of state of radiation. Naturally, after integration with a three-dimensional space momentum cut-off Ω, we get the equation of state for the radiation in the Ω^4-type terms.

As was explained by DeWitt in [107], the fundamental reason for this result is the use of the non-covariant cut-off regularization. At the same time, this shows the difficulty of regarding the cut-off parameter as a physical quantity. A consistent treatment requires the full regularization and renormalization procedure, and, in this way, one can achieve the "correct" Lorentz-invariant equation of state for the vacuum, as was shown in [216]. In this work, the local Lorentz invariance of the *finite* expressions on the cosmological background was achieved by imposing the covariance (the conservation law on the cosmological background) step by step, by adding specially adjusted non-covariant counterterms. Let us stress again, that the physical quantity is always the finite part of effective action, and not the divergent part, which should be subtracted within a consistent treatment. Anyway, for the moment, let us continue treating the cut-off parameter as a physical quantity, just to conclude this part of describing the induced gravity approach.

It is instructive that the same "incorrect" radiation-like equation of state of the vacuum that we got from (22.5) and (22.6) can be reproduced within the completely covariant point-splitting regularization in curved space [90]. This regularization is a powerful tool, and we do not discuss it only because of its utmost technical complexity. The point-splitting provides the result which depends on the world function $\sigma(x, x')$, which we introduced in section 13.4 to construct the basis of the Schwinger-DeWitt technique. In this regularization, there are no problems with either general covariance or local Lorentz invariance. There is no dimensional cut-off parameter, and, hence, it is not possible to generate dimensional quantities such as ρ_Λ^{ind} and G_{ind}^{-1} in a direct way, just from the regularization. However, by imposing a special time-like direction of the vector between the points x and x', one can break down the local Lorentz invariance and reproduce the radiation equation of state for the vacuum [15].

Indeed, the problems with covariance and cut-off can be perfectly well solved if we use the covariant cut-off, either using normal coordinates (see section 13.3.3 for an explicit example), or within the Schwinger-DeWitt approach with a covariant cut-off.[1]

Consider the Schwinger-DeWitt approach, which is simpler. The one-loop contribution is presented as

$$\bar{\Gamma}^{(1)} = \frac{i}{2}\,\mathrm{Tr}\,\mathrm{Log}\,\hat{H}\,, \qquad \text{where, e.g.,} \qquad \hat{H} = \hat{1}\Box + \hat{P} \tag{22.7}$$

[1]At the one-loop level, the two methods are equivalent [302], but the first one can also be successfully applied at higher loops.

and operator $\hat{P}$ depends on the choice of the matter field. Thus,

$$\bar{\Gamma}^{(1)}_{\Omega} = \frac{1}{2}\int\limits_{1/\Omega^2}^{\infty} \frac{ds}{s}\,\frac{1}{(4\pi s)^2}\,\mathrm{Tr}\Big\{\hat{1} + s\hat{a}_1 + s^2\hat{a}_2 + \dots\Big\}. \tag{22.8}$$

and $$\langle T_{\mu\nu}(x)\rangle = -\frac{2}{\sqrt{-g(x)}}\,g_{\mu\alpha}(x)\,g_{\nu\beta}(x)\,\frac{\delta\bar{\Gamma}^{(1)}_{\Omega}}{\delta g_{\alpha\beta}(x)}\,. \tag{22.9}$$

Then, in the cosmological constant sector, we get

$$\langle T_{\mu\nu}(x)\rangle \sim g_{\mu\nu}\,\Omega^4 \qquad \Longrightarrow \qquad p_\Lambda = -\,\rho_\Lambda\,, \tag{22.10}$$

in perfect agreement with the covariance and, in the flat limit, the local Lorentz invariance of the vacuum.

Even after we arrive at the covariant result, the induced gravity approach in the cut-off based formulation is not free of problems. From Eq. (22.8), follows that

$$\rho_\Lambda^{ind} \propto \Omega^4 \qquad \text{and} \qquad \frac{1}{16\pi\,G_{ind}} \propto \Omega^2\,. \tag{22.11}$$

Since all of the action of gravity is presumably induced, we are forced to identify

$$G_{ind} = \frac{1}{M_P^2} \qquad \Longrightarrow \qquad \Omega \propto M_P, \quad \text{and then} \quad \rho_\Lambda^{ind} \propto \Omega^4\,. \tag{22.12}$$

In this case, it is true that the "calculated" value of the cosmological constant density is $M_P^4 \approx 10^{76}\,GeV^4$. And this is certainly way too much, compared to the observed value $\rho_\Lambda^{obs} \propto 10^{-48}\,GeV^4$ in the present-day Universe. It is remarkable that only in this way can one arrive at the famous "120 orders of magnitude difference between the theory and experiment."

Let us stress that this output cannot be seen as the "prediction of quantum field theory." When thinking about the "120 orders" problem, it is important to remember that the approach that provides this glorious prediction is very special, as it is based on the following two assumptions:

i) The regularization should be of the cut-off type, without the conventional procedure of renormalization. The cut-off result is regarded as "physical," without adding the counterterms and fixing the renormalization condition. E.g., in dimensional regularization, both of the *r.h.s.*'s of (22.11) vanish, as we know from Part I. Indeed, one of the general purposes of quantum field theory is to arrive at regularization-independent results, and this can be achieved only within a consistent renormalization procedure.

ii) All of the gravitational action is induced, which means there is no vacuum counterpart to the induced gravitational expression. In particular, this condition requires the identification of Ω with M_P.

As we have seen, this set of axioms leads to the problem of 120 orders of magnitude. The way out is to use renormalization theory instead of taking the cut-off value as a physical result. However, as we know from Part I and the semiclassical chapters of Part II of the book, this approach requires the presence of the vacuum terms with ρ_Λ^{vac} and

$(16\pi\, G_{vac})^{-1}$. Thus, the huge discrepancy with the observed value of ρ_Λ^{ind} shows that the only way out is to introduce independent vacuum parameters, renormalize them in the standard way and, finally, sum up with the induced quantities

$$\rho_\Lambda^{obs} = \rho_\Lambda^{vac} + \rho_\Lambda^{ind}, \qquad \frac{1}{16\pi\, G_{obs}} = \frac{1}{16\pi\, G_{vac}} + \frac{1}{16\pi\, G_{ind}}. \tag{22.13}$$

The *l.h.s.*'s of these two expressions represent the observed quantities. From the formal viewpoint, this relations are fine, but we are forced to abandon the original approach of the purely induced gravity.

A few more observations are in order.

i) At the quantum level, both of the induced quantities G_{ind} and Λ_{ind} of the relations (22.11) gain loop corrections, but the dependence on the cut-off will remain qualitatively the same. The only difference between one-loop and higher loop (or even non-perturbative) results will be that, in higher loops, instead of the numerical coefficient of Ω^2 and Ω^4 in Eq. (22.11), there will be series in coupling constants. This modification does not provide conceptual changes in the general situation. Thus, given the more than significant discordance between theory and observations, we have a strong reason to discard the cut-off based version of purely induced gravity.

ii) The correctness of equations (22.13) is not restricted to the cut-off method but applies to all kinds of induced gravity actions. This includes, in the first place, the *finite* contributions discussed in section 12.3, and the values induced from the spontaneous symmetry breaking (12.37). In section 22.3 we will discuss the cosmological constant problem related to Eqs. (22.13).

iii) If we continue the derivative expansion in Eq. (22.8), in the next order after the induced Einstein-Hilbert term, there will be the usual set of $R^2_{\mu\nu\alpha\beta}$, $R^2_{\mu\nu}$ and R^2 terms. This happens independently of whether we use one or the other form of regularization, since these terms correspond to logarithmic divergences and are pretty much universal. Let us remember that one of the main problems with ghosts is the instability of classical solutions generated by fourth- and higher-derivative terms. Thus, even if we are not forced to quantize gravity, the fundamental problem of ghosts and instabilities may remain unsolved.

22.2 Gravity induced from phase transitions

There are other schemes that enable one to induce the action of gravity, including cosmological and Einstein-Hilbert terms. The simplest possibility is to consider the theory with single or multiple scalar fields ϕ. As we know from chapter 13, the consistent version of the Lagrangian of such fields includes nonminimal terms of the $\xi R\phi^2$ type. Because of symmetry breaking, the field ϕ gains a non-trivial stable point v, and this automatically generates cosmological and Einstein-Hilbert terms. The simplest version of such an induced gravity has already been discussed in section 12.3, as an illustration of the importance of the nonminimal parameter ξ. The values of the induced Newton constant and the cosmological constant, as a result of spontaneous symmetry breaking, can be found in (12.37). This result can be generalized by using more complicated scalar models, including interaction with vector and spinor fields and taking the effective potential of the scalar field(s), instead of the classical potential.

As far as we know, systematic analysis of all these possibilities has never been performed, albeit there are many intersting results in the literature. For instance, in [15] it was noted that, in theories with a scalar field near the minimum of the potential, the derivation of effective equations for the metric, even after the effective action of the theory is found, requires a lot of care and non-trivial calculations.. Technically, this and other works in this area are relatively complicated and would not fit in an introductory textbook, so we suggest them as possible material for additional reading.

All of the schemes for inducing gravity from scalar fields provide outputs which are qualitatively similar to what we obtained in section 12.3. In this simplest case, the induced values of ρ_Λ^{ind} and G_{ind} depend on the vacuum expectation value v_0 for the free theory in flat space. When interacting with gravity or with other matter fields, the scalar is not exactly in the minimal position, and the symmetry that is "spontaneously broken" is not exact. However, assuming that gravity is not too strong and that the oscillations of matter fields (including the scalar itself) are also relatively weak, the value of v_0 plays the role of a reference scale. Both of the induced gravitational parameters depend on v_0. In other words, one cannot derive both ρ_Λ^{ind} and G_{ind} from "first principles" and additional input is required. In the case considered in section 12.3, this input is the value of v_0.

The situation described above reflects an important general feature of induced gravity. There is always a one-parameter ambiguity in both of the induced quantities ρ_Λ^{ind} and G_{ind}, such that only a dimensionless combination may be well defined [5].

Yet another possibility is to start with some initially massless theory with a scalar field or fields. In fact, this scalar or scalars can even be not fundamental, but consist of composite fields, e.g., constructed from fermions, as in the Nambu-Jona-Lasinio model.[2] Then all the masses of the fields are the result of the dimensional transmutation (e.g., by Coleman-Weinberg mechanism [94], which is also called dynamical mass generation). Then, in the minimum of the effective potential, there will be always induced ρ_Λ^{ind} and G_{ind}. And, as always, there will be a one-parameter ambiguity in the values of these two induced quantities [5].

Coming back to the fundamental contradiction between renormalizability and unitarity in quantum gravity, we can immediately make the following observation. The simplified scheme presented in section 12.3 is certainly oversimplified in one important aspect. The tree-level result shows that the induced gravitational action is constructed only from the scalar curvature R. However, this is a direct consequence of the fact that we dealt with the tree-level theory only. Including the loop corrections, the effective potential of the scalar field will include $R^2_{\mu\nu\alpha\beta}$, $R^2_{\mu\nu}$ and R^2 terms, which are multiplied by some functions of a scalar. At low energies, when gravity is sufficiently weak, the scalar field configuration is approximately static near the constant value v_0. At this point, we arrive at the $R^2_{\mu\nu\alpha\beta}$, $R^2_{\mu\nu}$, R^2 and further higher-derivative terms with constant coefficients. As a consequence, there will be massive ghosts and the possibility of obtaining the same instabilities that were discussed in chapter 19.

Now we can answer the main question of whether induced gravity concept can definitely solve the problems of quantum gravity, i.e., the problems of higher-derivatives,

[2]This theory is one of the simplest models with composite scalars. We do not discuss it here, but the reader can easily learn about it elsewhere, e.g., in [183].

ghosts and related instabilities? Unfortunately, the answer to this question is no, because there is no way to restrict the higher-derivative terms in the induced action.

Exercises

23.1. Discuss the ambiguity in the renormalization group equations for Λ and G^{-1}, which is described in section 21. Can we regard it as a manifestation of the common ambiguity in the induced cosmological constant and Newton constant?

23.2. Consider the one-loop corrected theory with the potential (13.80), and derive the induced values ρ_Λ^{ind} and $(16\pi\, G_{ind})^{-1}$. Assume that the loop corrections are small, that the reference scale of the effective potential is equal to the tree-level v_0 and that the quantum corrections provide small contributions to the induced values are the result of dynamic symmetry breaking.

23.3. Derive the induced values ρ_Λ^{ind} and $(16\pi\, G_{ind})^{-1}$ in the massless version of the effective potential (13.80). Note that, in this version, it is impossible to have an effect at the tree-level and the induced comes as a result of dynamical symmetry breaking. *Hint.* In both cases it is important to implement renormalization conditions (see Exercise 15.3 and [80]). It is useful to consider the flat space version first.

22.3 Once again on the cosmological constant problem

In the previous chapters, we discussed how the main theoretical problem of quantum gravity is the conflict between renormalizability and unitarity. The induced gravity paradigm does not resolve this issue. In contrast, induced gravity brings us to another field theory problem, which is equally difficult.

As we know, purely induced gravitational action leads to the kind of theoretical failure that has an "error" of the scale of 120 orders of magnitude. All of the existing attempts, which are actually repeated every few years, to consider the cosmological constant problem as a problem of vacuum fluctuations derived with the Planck scale cut-off are implicitly using the logic of purely induced gravity.

From the perspective of effective action, the general situation looks simpler. All types of vacuum fluctuations contribute to the cosmological constant term, the Einstein-Hilbert term and the fourth-derivative terms with divergent and finite contributions. It is important that the divergences should be renormalized. After the counterterms are added, one has to implement the renormalization conditions on the independent vacuum parameters ρ_Λ^{vac} and G_{vac} in the form (22.13). At this point, we can better understand the cosmological constant problem, making the following observations:

i) In general, both ρ_Λ^{vac} and G_{vac}^{-1} are, indeed, necessary in order to have a renormalizable semiclassical theory. The corresponding divergences can be consistently dealt with only if the cosmological and Einstein-Hilbert terms are introduced in the classical action.

ii) To establish the "natural" bounds of the two vacuum parameters, consider the renormalization group equations, which have the general form

$$\mu\frac{d}{\mu}\Big(\frac{1}{16\pi G_{vac}}\Big) = \sum_{i,j} a_{ij}\, m_i m_j + \frac{a_{qg}}{16\pi G_{vac}}, \tag{22.14}$$

$$\mu\frac{d\,\rho_\Lambda^{vac}}{\mu} = \sum_{i,j,k,l} b_{ijkl}\, m_i m_j m_k m_l + b_{qg}\rho_\Lambda^{vac} + \frac{b'_{qg}}{G_{vac}^2}\,. \tag{22.15}$$

Here b'_{qg}, b_{qg} and a_{qg} are the contributions of quantum gravity, being power series in the coupling constants of the corresponding model. The examples of the one-loop expressions in different models of quantum gravity can be found in section 21. Furthermore, a_{ij} and b_{ijkl} are power series in the coupling constants in the matter fields sector, and m_i represents the mass spectrum of the corresponding model.

It is important to understand at which scale Eqs. (22.14) and (22.15) may be applied. We discussed in chapter 16 that the typical situation in the infrared is the decoupling of massive degrees of freedom. However, this important feature was not proved for ρ_Λ^{vac} and G_{vac}^{-1} (see also [167, 288, 287]). While the quantum field theory arguments are not really helpful, the phenomenology tells us that the decoupling takes place, in the case of ρ_Λ^{vac}. Let us imagine that this is not true. Then the first term in Eq. (22.15) indicates that all massive fermions and bosons contribute to the running at all energy scales, including in the late Universe. The list of heavy fermions starts from the electron $m_e \approx 0.5\,MeV$ and goes up to the top quark with $m_t \approx 175\,GeV$. Then, under the assumption of non-decoupling, the running of ρ_Λ would be extremely fast. With any reasonable physical identification of μ, the Universe would look completely different from what we observe under this assumption, to say the least.

We included only the known particles into the arguments presented above. The presence of heavier fermions or bosons (e.g., those of GUT models) would make the theoretical disaster under the assumption of non-decoupling even more dramatic. On top of that, there is the last term in Eq. (22.15) with the quantum gravity contribution of the order M_P^4. Let us take a maximally cautious position and assume that gravity is purely classical and that the mass spectrum of matter is restricted by the Minimal Standard Model. Then, regardless of the necessary decoupling in the infrared, in the very early Universe, the typical energy may be greater than m_t. Then Eq. (22.15) applies in its original form, and the range of change of ρ_Λ^{vac} is roughly $\frac{1}{16\pi^2}M_F^4 \propto 10^8\,GeV^4$. This magnitude is a minimal bound for the magnitude of ρ_Λ.

iii) As we know from section 12.3, symmetry breaking leads to an induced cosmological constant. In the Standard Model, spontaneous symmetry breaking takes place at the electroweak scale $M_F \approx 300\,GeV$, contributing with the value defined in (12.37), plus the loop corrections

$$\frac{1}{16\pi\, G_{ind}} = -\,\xi\, v_0^2 + \mathcal{O}(\hbar)\,, \qquad \rho_\Lambda^{ind} = \frac{\Lambda_{ind}}{8\pi\, G_{ind}} = -\,\lambda\, v_0^4 + \mathcal{O}(\hbar). \tag{22.16}$$

It is evident that ρ_Λ^{ind} has the same order of magnitude as the vacuum term, $\rho_\Lambda^{ind} \propto 10^8\,GeV^4$. At this point, everything looks perfect, since the two terms in the sum (22.13), $\rho_\Lambda^{vac} + \rho_\Lambda^{ind}$, have the same order of magnitude.

iv) As stated above, the two terms ρ_Λ^{vac} and ρ_Λ^{ind} in the sum (22.13) are of the same order of magnitude, about $10^8\,GeV^4$. The problem is that the sum is an observed quantity and the rate of acceleration of the Universe shows that its value is $\rho_\Lambda^{obs} \propto 10^{-48}\, GeV^4$. This means that the renormalization condition on the independent parameter ρ_Λ^{vac} should be chosen with a precision of 56 orders of magnitude.

It is important that in the early Universe, the potential of the scalar field had (most likely, at least) unbroken symmetry, owing to the high-temperature contributions [58]. Thus, the 56-order or more cancelation between the vacuum cosmological constant and the induced contribution was somehow "prepared" before the change of the potential between the early and the late stages of the Universe [344]. The origin of this enormous cancelation is not clear.

The solution to the cosmological constant problem requires explaining the extremely precise fine-tuning in the renormalization condition for ρ_Λ^{vac}. In contrast, the induced Newton constant is many orders of magnitude smaller than the vacuum value, and hence there is no need for fine-tuning for G_{vac}.

To understand the size of the problem, let us recall the result for the anthropic bound on the cosmological constant density by S. Weinberg [343]. It was shown that the time of formation of the galaxies, defining the production of heavy chemical elements, depends on the value of ρ_Λ^{obs}. For example, zero or negative values of ρ_Λ^{obs} can be ruled out by the fact of our proper existence, since our civilization could not exist in a world consisting of only light elements. However, the upper theoretical bound on ρ_Λ is just two orders above the aforementioned value of ρ_Λ^{obs}, being about 70% of the present-day critical density.[3] Thus, of the 56-digits of the necessary fine-tuning of ρ_Λ^{vac}, the first 54 digits cannot be different, since this would contradict our own existence.

What is the origin of the problem? Remember that we were evaluating the magnitudes of both ρ_Λ^{vac} and ρ_Λ^{ind} at the Fermi scale, which is typical for particle physics. At the same time, the "measurement" of ρ_Λ^{vac} occurs at the cosmic scale, which can be associated with the Hubble parameter, with a modern value $H_0 \approx 10^{-42}\, GeV$. The difference between the two energy scales is about 44 orders of magnitude.

We can conclude that the cosmological constant problem is a kind of hierarchy problem between the two scales M_F and H_0. If we do not try to link the density of the cosmological constant with the notions of particle physics, and simply accept the measured value of ρ_Λ^{obs} as observational data, the problem disappears.

The last observation concerns the similarities and differences between the problem of massive ghosts in quantum gravity and the cosmological constant problem. Both problems are fundamental and are related to the presence of two different scales. Also, the comprehensive solution of both problems is not looking like a real perspective. Another similarity is that the problems of ghosts can be successfully ignored within an effective approach, as explained in section 19.6, while the cosmological constant problem can also be successfully ignored by observational cosmologists. In both cases, such an ignorance is possible because the problems are very fundamental and do not (or almost do not) affect the theoretical descriptions of observations or experiments.

[3]This purely theoretical prediction looks remarkable, especially because the non-zero value of ρ_Λ^{obs} was confirmed by observations [249, 263].

23
Final remarks on Part II

Let us present a brief general summary of Part II.

We started our discussion by stating that the singularities in GR may provide a window to see the those areas of fundamental physics that are currently unknown, e.g., some version of quantum gravity. Even in the framework of a usual perturbative quantum field theory, there are several approaches leading to theoretically satisfactory models of quantum gravitational effects, starting from the quantum field theory in curved spacetime. Here the expression "satisfactory" doesn't mean perfectness, as there is no theoretically perfect model of quantum gravity.

To be applicable at all energy scales, quantum gravity theory has to be renormalizable. Then the higher-derivative models are the main candidates. The nice feature of these models is that they do not violate the low-energy classical solutions of GR and, therefore, do not contradict existing experimental or observational tests of classical gravity. On the other hand, these theories have ghosts and lead to problems with unitarity at the quantum level. Even worse, such theories lead to instabilities in classical solutions. The last problem can be resolved within the effective approach, but this means we give up the ambitious aim of constructing a universal theory of quantum gravity for any energy scale.

Both the semiclassical and effective approaches to the quantization of the metric look rather satisfactory, if we do not require them to be applicable in *all* physical situations. As in any other branch of physics, these models of quantum gravity are consistent within the limits of their applicability. The question is whether these limits are compatible with the intended applications. And here we meet the main problem, namely, the real difficulty of experimentally or observationally testing the quantum gravity models. Already dimensional analysis shows that, outside the vicinity of spacetime singularities, the quantum effects of gravity are too weak to be observed.

Finally, the main problem of quantum gravity is not purely theoretical, but also experimental. More precisely, the real problem is that we have no experiments capable of testing quantum gravity models. And, in fact, there is no clear prospect of having such experiments in the visible future. Indeed, the absence of a clear prospect does not mean that the development of a consistent theoretical foundation for such models is senseless. Perhaps the correct viewpoint is exactly the opposite. New prospects may appear only on the basis of sound theoretical developments. Besides, the models of semiclassical and quantum gravity are rich and interesting to study, even from a purely theoretical viewpoint. We hope that the basic aspects of these models described in this book, will be useful for readers who intend to start working in this fascinating area.

Introduction to Quantum Field Theory with Applications to Quantum Gravity. Iosif L. Buchbinder and Ilya L. Shapiro, Oxford University Press (2021). © Iosif L. Buchbinder and Ilya Shapiro.
DOI: 10·1093/oso/9780198838319·003·0023

References

[1] B.P. Abbott *et al.* [LIGO Scientific and Virgo Collaborations], *Observation of Gravitational Waves from a Binary Black Hole Merger*, Phys. Rev. Lett.**116** (2016) 061102, arXiv:1602.03837.

[2] L.F. Abbott, *Introduction to the background field method*, Acta Phys. Polonica **B13** (1982) 33.

[3] A. Accioly, B.L. Giacchini and I.L. Shapiro, *Low-energy effects in a higher-derivative gravity model with real and complex massive poles*, Phys. Rev. D **96**, 104004 (2017), arXiv:1610.05260.

[4] A. Accioly, B.L. Giacchini and I.L. Shapiro, *On the gravitational seesaw in higher-derivative gravity*, Eur. Phys. J. **C77** (2017) 540, gr-qc/1604.07348.

[5] S.L. Adler, *Einstein gravity as a symmetry-breaking effect in quantum field theory*, Rev. Mod. Phys. **54** (1982) 729, Erratum: **55** (1983) 837.

[6] E.K. Akhmedov, *Vacuum energy and relativistic invariance*, hep-th/0204048.

[7] E. Alvarez, J. Anero, S. Gonzalez-Martin and R. Santos-Garcia, *Physical content of quadratic gravity*, Eur. Phys. J. **C78** (2018) 794, arXiv:1802.05922.

[8] I. Antoniadis and E.T. Tomboulis, *Gauge invariance and unitarity in higher derivative quantum gravity*, Phys. Rev. **D33** (1986) 2756.

[9] I. Antoniadis, P.O. Mazur and E. Mottola, *Conformal symmetry and central charges in four-dimensions,* Nucl. Phys. **B388** (1992) 627, hep-th/9205015.

[10] T. Appelquist and J. Carazzone, *Infrared Singularities and Massive Fields,* Phys. Rev. **D11** (1975) 2856.

[11] I.Ya. Arefeva, A.A. Slavnov and L.D. Faddeev, *Generating functional for the S-matrix in gauge-invariant theories*, Theor. Math. Phys.**21** (1975) 1165 [Version in Russian: Teor. Mat. Fiz. **21** (1974) 311].

[12] M. Asorey and F. Falceto, *Consistency of the regularization of gauge theories by high covariant derivatives*, Phys.Rev. **D54** (1996) 5290, hep-th/9502025.

[13] M. Asorey, J.L. López and I.L. Shapiro, *Some remarks on high derivative quantum gravity,* Int. Journ. Mod. Phys. **A12** (1997) 5711.

[14] M. Asorey, E.V. Gorbar and I.L. Shapiro, *Universality and ambiguities of the conformal anomaly,* Class. Quant. Grav. **21** (2004) 163.

[15] M. Asorey, P.M. Lavrov, B.J. Ribeiro and I.L. Shapiro, *Vacuum stress-tensor in SSB theories,* Phys. Rev. **D85** (2012) 104001, arXiv:1202.4235.

[16] M. Asorey, L. Rachwał and I.L. Shapiro, *Unitary issues in some higher derivative field theories,* Galaxies **6** (2018) 23, arXiv:1802.01036.

[17] I.G. Avramidi and A.O. Barvinsky, *Asymptotic freedom In higher derivative quantum gravity,* Phys. Lett. **B159** (1985) 269.

[18] I.G. Avramidi, *Covariant studies of nonlocal structure of effective action,* Sov. J. Nucl. Phys. **49**, 735 (1989) [Yad. Fiz. **49**, 1185 (1989), in Russian].

[19] I.G. Avramidi, *Covariant methods for the calculation of the effective action in quantum field theory and investigation of higher-derivative quantum gravity*, (PhD thesis, Moscow University, 1986). hep-th/9510140; *Heat kernel and quantum gravity*, (Springer-Verlag, 2000).

[20] I.G. Avramidi, *Heat Kernel Method and its Applications*, (Birkhäuser, 2015).

[21] E. Babichev and A. Fabbri, *Instability of black holes in massive gravity*, Class. Quantum Grav. **30** (2013) 152001, arXiv:1304.5992.

[22] R. Balbinot, A. Fabbri and I.L. Shapiro, *Anomaly induced effective actions and Hawking radiation*, Phys. Rev. Lett. **83** (1999) 1494, hep-th/9904074; *Vacuum polarization in Schwarzschild space-time by anomaly induced effective actions*, Nucl. Phys. **B559** (1999) 301, hep-th/9904162.

[23] K. J. Barnes, *Lagrangian theory for the second-rank tensor field*, J. Math. Phys. (N.Y.) **6** (1965) 788; Unpublished (Ph. D. Thesis at University of London, 1963).

[24] V.F. Barra, I.L. Buchbinder, J.G. Joaquim, A.R. Rodrigues and I.L. Shapiro, *Renormalization of Yukawa model with sterile scalar in curved space-time*, Eur. Phys. J. **C79** (2019) 458; arXiv:1903.11546;
I.L. Buchbinder, A.R. Rodrigues, E.A. dos Reis and I.L. Shapiro, *Quantum aspects of Yukawa model with scalar and axial scalar fields in curved spacetime*, Eur. Phys. J. **C79** (2019) 1002, arXiv:1910.01731.

[25] V.F. Barra, P.M. Lavrov, E.A. dos Reis, T. de Paula Netto and I.L. Shapiro, *Functional renormalization group approach and gauge dependence in gravity theories*, Phys. Rev. **D101** (2020) 065001. arXiv:1910.06068.

[26] J.A.de Barros and I.L.Shapiro, *Renormalization group study of the higher derivative conformal scalar model*, Phys. Lett. **B412** (1997) 242. hep-th/9706123.

[27] A.O. Barvinsky, A.Y. Kamenshchik and A.A. Starobinsky, *Inflation scenario via the Standard Model Higgs boson and LHC*, JCAP **0811** (2008) 021, arXiv:0809.2104.

[28] A.O. Barvinsky, A.Y. Kamenshchik, C. Kiefer, A.A. Starobinsky and C. Steinwachs, *Asymptotic freedom in inflationary cosmology with a non-minimally coupled Higgs field*, JCAP **0912** (2009) 003, arXiv:0904.1698.

[29] A.O. Barut, R. Raczka, *Theory of Group Representations and Applications*, (World Scientific, 1986).

[30] A.O. Barvinsky and G.A. Vilkovisky, *Divergences and anomalies for coupled gravitational and Majorana spin 1/2 fields*, Nucl. Phys. **B191** (1981) 237.

[31] A.O. Barvinsky and G.A. Vilkovisky, *The generalized Schwinger-DeWitt technique in gauge theories and quantum gravity*, Phys. Repts. **119** (1985) 1.

[32] A.O. Barvinsky and G.A. Vilkovisky, *Covariant perturbation theory. 2: Second order in the curvature. General algorithms*, Nucl. Phys. **333B** (1990) 471.

[33] A.O. Barvinsky and G.A. Vilkovisky, *Covariant perturbation theory. 3: Spectral representations of the third order form-factors*, Nucl. Phys. **B333** (1990) 512.

[34] A.O. Barvinsky, Yu.V. Gusev, G.A. Vilkovisky and V.V. Zhytnikov, *The one-loop effective action and trace anomaly in four-dimensions*, Nucl.Phys. **B439** (1995) 561, hep-th/9404187.

[35] A.O. Barvinski, A. Kamenschik and B. Karmazin, *The renormalization group for nonrenormalizable theories: Einstein gravity with a scalar field*, Phys. Rev. **D48** (1993) 3677, gr-qc/9302007.
[36] A.O. Barvinsky, *Nonlocal action for long distance modifications of gravity theory*, Phys. Lett. **B572** (2003) 109, hep-th/0304229.
[37] E.S. Fradkin and G.A. Vilkovisky, *Quantization of relativistic systems with constraints*, Phys. Lett. **B55** (1975) 224.
[38] I.A. Batalin, E.S. Fradkin and G.A. Vilkovisky, *Operator quantization of relativistic dynamical system subject to first class constraints*, Phys. Lett., **B128** (1983) 303.
[39] I.A. Batalin and E.S. Fradkin, *Operator quantization method and abelization of dynamical systems subject to first class constraints*, Riv. Nuovo Cim. **9** (1986) 1.
[40] I.A. Batalin and E.S. Fradkin, *Operatorial quantization of dynamical systems subject to constraints. A further study of the construction*, Annales de l'Institut Henri Poincaré, Theoretical Physics, **A49** (1988) 145.
[41] I.A. Batalin and G.A. Vilkovisky, *Relativistic S-matrix of dynamical systems with boson and fermion constraints*, Phys. Lett. **B69** (1977) 309.
[42] I.A. Batalin and G.A. Vilkovisky, *Gauge algebra and quantization*, Phys. Lett. **B102** (1981) 27.
[43] I.A. Batalin and G.A. Vilkovisky, *Quantization of gauge theories with linearly dependent generators*, Phys. Rev. **D28** (1983) 2567; Erratum: **D30** (1984) 508.
[44] C. Becci, A. Rouet and R. Stora, *The abelian Higgs-Kibble model. Unitarity of the S operator*, Phys. Lett., **B52** (1974) 344.
[45] C. Becci, A. Rouet, R. Stora, *Renormalization of the abelian Higgs-Kibble model*, Comm. Math. Phys., **42** (1975) 127.
[46] E. Belgacem, Y. Dirian, S. Foffa and M. Maggiore, *Nonlocal gravity. Conceptual aspects and cosmological predictions,* JCAP **1803** (2018) 002, arXiv:1712.07066.
[47] V.B. Berestetsky, E.M. Lifshitz and L.P. Pitaevskii, *Relativistic Quantum Theory. Part 1*, (Pergamon Press, 1971).
[48] F.A. Berezin, *Method of Secondary Quantization*, (Academic Press, 1966).
[49] F.A. Berezin, *Introduction to Superanalysis*, (Springer, 1987).
[50] J. Berges, N. Tetradis and C. Wetterich, *Non-perturbative renormalization flow in quantum field theory and statistical physics*, Phys. Rept. **363** (2002) 223.
[51] F.L. Bezrukov and M. Shaposhnikov, *The Standard Model Higgs boson as the inflaton.* Phys. Lett. **B659** (2008) 703.
[52] F. L. Bezrukov, A. Magnin, M. Shaposhnikov, *Standard Model Higgs boson mass from inflation*, Phys. Lett. **B675** (2009) 88, arXiv:0812.4950.
[53] F. Bezrukov, A. Magnin, M. Shaposhnikov, S. Sibiryakov, *Higgs inflation: consistency and generalisations*, JHEP **1101** (2011) 016 , arXiv:1008.5157.
[54] F. Bezrukov, M. Shaposhnikov, *Standard Model Higgs boson mass from inflation: Two loop analysis*, JHEP **0907** (2009) 089, arXiv:0904.1537.
[55] O.-M.P. Bilaniuk, V.K. Deshpande and E.C.G. Sudarshan, *"Meta" relativity*,

Americal Journal of Physics **30** (1962) 718; O.-M. Bilaniuk and E.C.G. Sudarshan, *Particles beyons the light barrier.* Physics Today **22** (1969) 43.

[56] N.D. Birell and P.C.W. Davies, *Quantum fields in curved space*, (Cambridge Univ. Press, Cambridge, 1982).

[57] J.M. Bjorken and S.D. Drell, *I. Relativistic Quantum Mechanics; II. Relativistic Quantum Fields*, (McGraw-Hill, NY, 1964; 1965).

[58] S.A. Bludman and M.A. Ruderman, *Induced cosmological constant expected above the phase transition restoring the broken symmetry*, Phys. Rev. Lett. **38** (1977) 255.

[59] N.N. Bogoliubov and D.V. Shirkov, *Introduction to the theory of quantized fields*, John Wiley & Sons, 1980.

[60] N.N. Bogoliubov, A.A. Logunov, A.I. Oksak and I. Todorov, *General principles of quantum field theory*, (Springer, 1990).

[61] C.G. Bollini and J.J. Giambiagi, *Dimensional renormalization: the number of dimension as a regularizing parameter*, Nuovo Cim. **B12** (1972) 20.

[62] L. Bonora, P. Pasti, M. Bregola, *Weyl cocycles*, Class. Quant. Grav. **3** (1986) 635.

[63] M. Bordag, G.L. Klimchitskaya, U. Mohideen and V.M. Mostepanenko, *Advances in the Casimir effect*, (Oxford University Press, 2009).

[64] F. Brandt, G. Barnich and M. Henneaux, *Local BRST cohomology in gauge theories*, Phys. Repts. **338** (2000) 439.

[65] L.S. Brown and J.P. Cassidy, *Stress tensor trace anomaly in a gravitational metric: general theory, Maxwell field*, Phys. Rev. **D15** (1977) 2810.

[66] I.L. Buchbinder, O.K. Kalashnikov, I.L. Shapiro, V.B. Vologodsky, and J.J. Wolfengaut, *The Stability of asymptotic freedom in Grand Unified models coupled to R^2 gravity*, Phys. Lett. **B216** (1989) 127.

[67] I.L. Buchbinder and S.D. Odintsov, *Asymptotical freedom and asymptotical conformal invariance in curved space-time*, Izv. VUZov. Fiz. (Sov. Phys. J.) **26** No12, (1983) 108.

[68] I.L. Buchbinder and I.L. Shapiro, *One-loop calculation of graviton self energy in the first order gravity formalism*, Yad.Fiz.- Sov.J.Nucl.Phys. **37** (1983) 248.

[69] I.L. Buchbinder, S.D. Odintsov and I.L. Shapiro, *Representation of the graviton propagator and one-loop counterterms in quantum gravity*, Izvestia VUZov, Fisica (Soviet Physics Journal), **27**,n4 (1984) 50.

[70] I.L. Buchbinder, *On Renormalization group equations in curved space-time*, Theor. Math. Phys. **61** (1984) 393.

[71] I.L. Buchbinder and S.D. Odintsov, *Asymptotical properties of nonabelian gauge theories in external gravitational fields*, Sov. J. Nucl. Phys. **40** (1984) 848.

[72] I.L. Buchbinder and S.D. Odintsov, *Effective potential in a curved space-time*, Sov. Phys. J. **27** (1984) 554; Class. Quant. Grav. *2* (1985) 721.

[73] I.L. Buchbinder, S.D. Odintsov and I.L. Shapiro, *Nonsingular cosmological model with torsion induced by vacuum quantum effects*, Phys. Lett. **B162** (1985) 92.

[74] I.L. Buchbinder and S.D. Odintsov, *Asymptotical behavior of the effective potential in an external gravitational field,* Lett. Nuovo Cim. **44** (1985) 601.

[75] I.L. Buchbinder and J.J. Wolfengaut, *Renormalization Group Equations and Effective Action in Curved Space-time,* Class. Quant. Grav. **5** (1988) 1127.

[76] I.L. Buchbinder and I.L. Shapiro, *On the asymptotical freedom in the Einstein - Cartan theory,* Sov. J. Phys. **31** (1988) 40.

[77] I. L. Buchbinder, I. L. Shapiro, *Gravitational interaction effect on behavior of the Yukawa and scalar effective coupling constants,* Sov. J. Nucl. Phys. **44**(1986)1033.

[78] I.L. Buchbinder and I.L. Shapiro, *On the renormalization group equations in curved spacetime with torsion,* Class. Quant. Grav. **7** (1990) 1197.

[79] I.L. Buchbinder, I.L. Shapiro and E.G. Yagunov, *The asymptotically free and asymptotically conformal invariant Grand Unification theories in curved spacetime,* Mod. Phys. Lett. **A5** (1990) 1599.

[80] I.L. Buchbinder, S.D. Odintsov and I.L. Shapiro, *Effective action in quantum gravity,* (IOP Publishing, Bristol, 1992).

[81] I.L. Buchbinder and S.M. Kuzenko, *Ideas and methods of supersymmetry and supergravity,* IOP Publishing, 1998).

[82] I.L. Buchbinder, G. de Berredo-Peixoto, and I.L. Shapiro, *Quantum effects in softly broken gauge theories in curved spacetimes.* Phys. Lett. **B649** (2007) 454, hep-th/0703189.

[83] T.S. Bunch and L. Parker, *Feynman propagator in curved space-time: a momentum space representation.* Phys. Rev. **D20** (1979) 2499.

[84] C.P. Burgess, *Quantum gravity in everyday life: General relativity as an effective field theory,* Living Rev. Rel. **7** (2004) 5, gr-qc/0311082.

[85] C. Callan, D. Friedan, E. Martinec and M. Perry, *Strings in background fields,* Nucl. Phys. **B272** (1985) 593.

[86] D.M. Capper and D. Kimber, *An ambiguity in one loop quantum gravity,* Journ. Phys. **A13** (1980) 3671.

[87] D.F. Carneiro, E.A. Freitas, B. Gonçalves, A.G. de Lima and I.L. Shapiro, *On useful conformal tranformations in General Relativity,* Grav. and Cosm. **40** (2004) 305; gr-qc/0412113.

[88] A.R.R. Castellanos, F. Sobreira, I.L. Shapiro and A.A. Starobinsky, *On higher derivative corrections to the $R + R^2$ inflationary model,* JCAP **1812** (2018) 007, arXiv:1810.07787.

[89] L. Cesari, *Asymptotic behavior and stability problems in ordinary differential equations,* (Springer; 3rd Edition, 2012).

[90] S.M. Christensen, *Vacuum expectation value of the stress tensor in an arbitrary curved background: the covariant point separation method,* Phys. Rev. **D14** (1976) 2490; *Regularization, renormalization, and covariant geodesic point separation,* Phys. Rev. **D17** (1978) 946.

[91] S.M. Christensen and S.A. Fulling, *Trace anomalies and the Hawking effect,* Phys. Rev. **D15** (1977) 2088.

[92] S.M. Christensen and M.J. Duff, *Quantum gravity in two+ϵ dimensions,* Phys. Lett. **B79** (1978) 213.

[93] A. Codello and O. Zanusso, *On the non-local heat kernel expansion,* J. Math. Phys. **54** (2013) 013513, arXiv:1203.2034.

[94] S.R. Coleman and E.J. Weinberg, *Radiative corrections as the origin of spontaneous symmetry breaking,* Phys. Rev. **D7** (1973) 1888.

[95] S. Coleman, *Aspects of symmetry,* (Cambridge University Press, 1985).

[96] J.C. Collins, *Renormalization.* (Cambridge University Press, 1984).

[97] P. Creminelli, A. Nicolis, M. Papucci and E. Trincherini, *Ghosts in massive gravity,* JHEP **0509** (2005) 003, hep-th/0505147.

[98] G. Cusin, F. de O. Salles and I.L. Shapiro, *Tensor instabilities at the end of the ΛCDM universe,* Phys. Rev. **D93** (2016) 044039, arXiv:1503.08059.

[99] R.R. Cuzinatto, L.G. Medeiros and P.J. Pompeia, *Higher-order modified Starobinsky inflation,* JCAP **1902** (2019) 055, arXiv:1810.08911.

[100] D.A.R. Dalvit and F.D. Mazzitelli, *Geodesics, gravitons and the gauge fixing problem,* Phys. Rev. **D56** (1997) 7779, hep-th/9708102.

[101] S. Deser and P. van Nieuwenhuisen, *One-loop divergences of quantized Einstein-Maxwell fields,* Phys. Rev. **D10** (1974) 401.

[102] S. Deser, M.J. Duff and C. Isham, *Nonlocal conformal anomalies,* Nucl. Phys. **B111** (1976) 45.

[103] S. Deser and B. Zumino, *Consistent supergravity,* Phys. Lett. **B62** (1976) 335.

[104] S. Deser and A.N. Redlich, *String induced gravity and ghost freedom,* Phys. Lett. **B176** (1986) 350.

[105] B.S. DeWitt, *Dynamical theory of groups and fields,* (Gordon and Breach, 1965).

[106] B.S. DeWitt, *Quantum theory of gravity. II. Manifestly covariant theory,* Phys. Rev. **162** (1967), 1195.

[107] B.S. DeWitt, *Quantum field theory in curved spacetime,* Phys. Repts.**C19** (1975) 297.

[108] B.S. DeWitt, *Supermanifolds,* (Cambridge University Press, 1992).

[109] B.S. DeWitt, *The global approach to quantum field theory,* (Clarendon Press, Oxford. Vol.1 and Vol. 2 - 2003).

[110] R. D'Inverno, *Introducing Einstein's relativity.* (Oxford University Press, 1998).

[111] P.A.M. Dirac, *Lectures in Quantum Mechanics,* (Belfer Graduate School of Science, Yeshiva University, 1964).

[112] S. Dodelson, *Modern cosmology.* (Academic Press, 2003).

[113] J.F. Donoghue, E. Golowich and B.R. Holstein, *Dynamics of the Standard Model,* (Cambridge University Press, 1992).

[114] J.F. Donoghue, *Leading quantum correction to the Newtonian potential,* Phys. Rev. Lett. *72* (1994) 2996, gr-qc/9310024; *General relativity as an effective field theory: The leading quantum corrections,* Phys. Rev. **D50** (1994) 3874. gr-qc/9405057.

[115] J.S. Dowker and R. Critchley, *Effective Lagrangian and energy-momentum tensor in de Sitter space,* Phys. Rev. **D 13** (1976) 3224.

[116] B.A. Dubrovin, A.T. Fomenko and S.P. Novikov, *Modern geometry-methods and applications,* (Springer-Verlag, 1984).

[117] D. M. Capper, M. J. Duff and L. Halpern, *Photon corrections to the graviton propagator,* Phys. Rev. **D10** (1974) 461; D. M. Capper and M. J. Duff, *Neutrino corrections to the graviton propagator,* Nucl. Phys. **B82** (1974) 147.

[118] M.J. Duff, *Observations On Conformal Anomalies,* Nucl.Phys. **B125** (1977) 334.

[119] M.J. Duff, *Twenty years of the Weyl anomaly,* Class. Quant. Grav. **11** (1994) 1387, hep-th/9308075.

[120] G. Dvali, S. Folkerts and C. Germani, *Physics of trans-Planckian gravity,* Phys. Rev. **D84** (2011) 024039, arXiv:1006.0984; G. Dvali and C. Gomez, *Black holes quantum N-portrait*, Fortschr. Phys. **61** (2013) 742, arXiv:1112.3359.

[121] E. Elizalde, S.D. Odintsov, A, Romeo and A.A. Bytsentko, *Zeta regularization techniques with applications*, (World Scientific, 1994).

[122] E. Elizalde, *Ten physical applications of spectral zeta functions,* (Springer, 1995).

[123] J.C. Fabris, A.M. Pelinson, I.L. Shapiro, *Anomaly induced effective action for gravity and inflation*, Grav. Cosmol. **6** (2000) 59, gr-qc/9810032; *On the gravitational waves on the background of anomaly-induced inflation,* Nucl. Phys. **B597** (2001) 539, hep-ph/0208184.

[124] J.C. Fabris, A.M. Pelinson, F. de O. Salles and I.L. Shapiro, *Gravitational waves and stability of cosmological solutions in the theory with anomaly-induced corrections,* JCAP **02** (2012) 019, arXiv: 1112.5202.

[125] L.D. Faddeev and V.N. Popov, *Feynman diagrams for Yang-Mills field*, Phys. Lett., **B 25** (1967) 29.

[126] L.D. Faddeev, *The Feynman integral for singular Lagrangians*, Theoretical and Mathematical Physics, **1** (1969) 1.

[127] L.D. Faddeev, A.A. Slavnov, *Gauge fields: An introduction to the theory of gauge fields*, (Benjamin/Commings, 1980).

[128] F.M. Ferreira, I.L. Shapiro and P.M. Teixeira, *On the conformal properties of topological terms in even dimensions* Eur. Phys. J. Plus **131** (2016) 164, arXiv:1507.03620.

[129] F.M. Ferreira and I.L. Shapiro, *Integration of trace anomaly in* $6D$, Phys. Lett. **B772** (2017) 174, arXiv:1702.06892.

[130] F.M. Ferreira and I. Shapiro, *Basis of surface curvature-dependent terms in* $6D$, Phys. Rev. **D99** (2019) 064032, arXiv:1812.01140.

[131] R. P. Feynman, *Space-time approach to non-relativistic Quantum Mechanics,* Rev. Mod. Phys. **20** (1948) 367.

[132] R.P. Feynman, *Quantum theory of gravitation*, Acta Phys. Pol. **24** (1963) 697.

[133] R.P. Feynman and A.R. Hibbs, *Quantum mechanics and path integrals,* (McGraw-Hill, 1965).

[134] M. Fierz and W. Pauli, *On relativistic wave equations for particles of arbitrary spin in an electromagnetic field*, Proc. Roy. Soc. Lond. A **173** (1939) 211.

[135] M.V. Fischetti, J.B. Hartle and B.L. Hu, *Quantum effects in the early universe. I. Influence of trace anomalies on homogeneous, isotropic, classical geometries,* Phys. Rev. **D20** (1979) 1757.

[136] V.A. Fock, *Selected Works: Quantum Mechanics and Quantum Field Theory*, (CRC Press, 2004).

[137] E.S. Fradkin, *Concerning some general relations of quantum electrodynamics*, Soviet Physics ZhETF, **2** (1956) 361.

[138] E.S. Fradkin and A.A. Tseytlin, *Conformal anomaly in Weyl theory and anomaly free superconformal theories*, Phys. Lett. **B134** (1984) 187.

[139] E.S. Fradkin, and A.A. Tseytlin, *Asymptotic freedom on extended conformal supergravities*, Phys. Lett. **B110** (1982) 117; *One-loop beta function in conformal supergravities*, Nucl. Phys. **B203** (1982) 157.

[140] E.S. Fradkin and A.A. Tseytlin, *Renormalizable asymptotically free quantum theory of gravity*, Nucl. Phys. **B201** (1982) 469.

[141] E.S. Fradkin, and A.A. Tseytlin, *Quantum string theory effective action*, Nucl. Phys. **B261** (1985) 1.

[142] S.A. Franchino-Viñas, T. de Paula Netto, I.L. Shapiro and O. Zanusso, *Form factors and decoupling of matter fields in four-dimensional gravity*, Phys. Lett. **B790** (2019) 229, arXiv:1812.00460.

[143] S. A. Franchino-Viñas, T. de Paula Netto, O. Zanusso, *Vacuum effective actions and mass-dependent renormalization in curved space*, Universe **5** (2019) 67, arXiv:1902.03167.

[144] M.B. Fröb, A. Roura and E. Verdaguer, *Riemann correlator in de Sitter including loop corrections from conformal fields*, JCAP **1407** (2014) 048, arXiv:1403.3335.

[145] V.P. Frolov and D.V. Fursaev, *Mechanism of generation of black hole entropy in Sakharov's induced gravity*, Phys. Rev. **D56** (1997) 2212, hep-th/9703178.

[146] V.P. Frolov and G.A. Vilkovisky, *Spherically symmetric collapse in quantum gravity*, Phys. Lett. **B106** (1981) 307.

[147] V.P. Frolov and I.D. Novikov, *Black hole Physics - basic concepts and new developments*, (Kluwer Academic Publishers, 1989).

[148] V.P. Frolov and A. Zelnikov, *Introduction to black hole Physics*. (Oxford University Press, 2011).

[149] K. Fujikawa, *Path integral measure for gauge invariant fermion theories*, Phys. Rev. Lett. **42** (1979) 1195.

[150] S.A. Fulling, *Aspects of Quantum Field Theory in Curved Spacetime*, (Cambridge Univ. Press, Cambridge, 1989).

[151] D. Fursaev, D. Vassilevich, *Operators, geometry and quanta: methods of spectral geometry in quantum Field Theory*, (Springer, 2011).

[152] S. Gasiorowicz, *Elementary Particle Physics*, (John Wisley & Sons, 1966).

[153] R. Gastmans, R. Kallosh and C. Truffin, *Quantum gravity near two-dimensions*, Nucl. Phys. **B133** (1978) 417.

[154] S.J. Jr. Gates, M.T. Grisaru, M. Rocek and W. Siegel, *Superspace*, (Benjamin-Cummings, New York, 1983).

[155] L. Alvarez-Gaume and M.A. Vazquez-Mozo, *Invitation to quantum field theory*, (Springer-Verlag, Berlin, 2012).

[156] I.M. Gelfand, R.A. Minlos and Z.Ya. Shapiro, *Representations of the rotation and Lorentz groups*, (Pergamon Press, 1963.)

[157] B.L. Giacchini, P.M. Lavrov and I.L. Shapiro, *Background field method and nonlinear gauges*, Phys. Lett. **B797** (2019) 134882, arXiv:1906.04767; P.M. Lavrov, *Gauge (in)dependence and background field formalism,* Phys. Lett. **B791** (2019) 293, arXiv:1805.02149.

[158] B.L. Giacchini, T. de Paula Netto, and I.L. Shapiro, *On the Vilkovisky-DeWitt approach and renormalization group in effective quantum gravity,* JHEP **10** (2020) 011, arXiv:2009.04122.

[159] P. B. Gilkey, *The spectral geometry of Riemannian manifold,* Journ. Diff. Geom., **10** (1975) 601.

[160] D.M. Gitman and I.V. Tyutin, *Quantization of fields with constraints,* (Springer, 1990).

[161] H. Goldstein, *Classical Mechanics* (Addison Wesley; 2-nd edittion, 1980).

[162] J. Goldstone, A. Salam and S. Weinberg, *Broken Symmetries*, Phys. Rev. **127** (1962) 965.

[163] J. Gomis, J. Paris and S. Samuel, *Antibracket, antifields and gauge theory quantization*, Phys. Rept. **259** (1995) 1.

[164] J. Gomis, S. Weinberg, *Are nonrenormalizable gauge theories renormalizable?*, Nucl. Phys. **B469** (1996) 473, hep-th/9510087.

[165] B. Gonçalves, G. de Berredo-Peixoto and I.L. Shapiro, *One-loop corrections to the photon propagator in the curved-space QED*, Phys. Rev. **D80** (2009) 104013, arXiv:0906.3837.

[166] J.D. Gonçalves, T. de Paula Netto and I.L. Shapiro, *On the gauge and parametrization ambiguity in quantum gravity,* Phys. Rev. **D97** (2018) 026015, arXiv:1712.03338.

[167] E.V. Gorbar and I.L. Shapiro, *Renormalization Group and Decoupling in Curved Space*, JHEP **02** (2003) 021, hep-ph/0210388; *Renormalization Group and Decoupling in Curved Space: II. The Standard Model and Beyond.* JHEP **06** (2003) 004, hep-ph/0303124.

[168] E.V. Gorbar and I.L. Shapiro, *Renormalization Group and Decoupling in Curved Space: III. The Case of Spontaneous Symmetry Breaking*, JHEP **02** (2004) 060, hep-ph/0311190.

[169] D.S. Gorbunov, and V.A. Rubakov, *Introduction to the Theory of the Early Universe: Vol. I. Hot Big Bang Theory, Vol. II. Cosmological Perturbations and Inflationary Theory.* (World Scientific Publishing, 2011).

[170] M.H. Goroff and A. Sagnotti, *The ultraviolet behavior of Einstein gravity*, Nucl. Phys. **B266** (1986) 709.

[171] R. Gregory and R. Laflamme, *Black strings and p-branes are unstable,* Phys. Rev. Lett. **70** (1993) 2837, hep-th/9301052.

[172] A.A. Grib, S.G. Mamaev and V.M. Mostepanenko, *Quantum effects in intensive external fields*, (Moscow, Atomizdat. In Russian, 1980).

[173] D.J. Gross and F. Wilczek, *Ultraviolet behavior of non-abelian gauge theories*, Phys. Rev. Lett. **30** (1973) 1343.

[174] V.P. Gusynin, *Seeley-Gilkey coefficients for the fourth order operators on a Riemannian manifold,* Nucl. Phys. **B333** (1990) 296.

[175] J.B. Hartle, *Gravity: An introduction to Einsten's general relativity*, (Pearson, 2003).
[176] S.W. Hawking, *Particle Creation by Black Holes*, Nature **248** (1974) 30; Commun. Math. Phys. **43** (1975) 199, Erratum-ibid. **46** (1976) 206.
[177] S.W. Hawking, *Zeta function regularization of path integrals in curved spacetime*, Commun. Math. Phys. **55** (1977) 133.
[178] S.W. Hawking, T. Hertog and H.S. Real, *Trace anomaly driven inflation*, Phys. Rev. **D63** (2001) 083504, hep-th/0010232.
[179] F.W. Hehl, P. Heide, G.D. Kerlick and J.M. Nester, *General relativity with spin and torsion: foundations and prospects*, Rev. Mod. Phys. **48** (1976) 393.
[180] F. W. Hehl, J.D. McCrea, E.W. Mielke and Y. Ne'eman, *Metric affine gauge theory of gravity: field equations, Noether identities, world spinors, and breaking of dilation invariance*, Phys. Rept. **258** (1995) 1, gr-qc/9402012.
[181] M. Henneaux, *Hamiltinian form of the path integral for the theories with a gauge freedom*, Phys. Repts. **126** (1985) 1.
[182] M. Henneaux and C. Teitelboim, *Quantization of gauge systems*, (Princeton University Press, 1992).
[183] C.T. Hill and D.S. Salopek, *Calculable nonminimal coupling of composite scalar bosons to gravity*, Ann. Phys. **213** (1992) 21.
[184] J. Honerkamp, *Chiral multi-loop*, Nucl. Phys. **B36** (1972) 130.
[185] G. t'Hooft and M. Veltman, *Regularization and renormalization of gauge fields*, Nucl. Phys. **B44** (1972) 189.
[186] G. t'Hooft and M. Veltman, *One-loop divergences in the theory of gravitation*, Ann. Inst. H. Poincare **A20** (1974) 69.
[187] C. Itzykson and J.-B. Zuber, *Quantum field theory*, (McGill - Hill, NY, 1980).
[188] I. Jack, *One-loop beta-functions for renormalisable gravity*, arXiv:2002.12661.
[189] D.A. Johnston, *Sedentary ghost poles in higher derivative gravity*, Nucl. Phys. **B297** (1988) 721.
[190] J. Julve and M. Tonin, *Quantum gravity with higher derivative terms*, Nuovo Cim. **B46** (1978) 137.
[191] G. Källen, *On the definition of the renormalization constants in Quantum Electrodynamics*, Helv. Phys. Acta **25**(1952) 417.
[192] R. Kallosh, *The renormalization in nonabelian gauge theories*, Nucl. Phys. **B78** (1974) 293.
[193] R.E. Kallosh, O.V. Tarasov and I.V. Tyutin, *One-loop finiteness of quantum gravity off mass shell*, Nucl. Phys. **B137** (1978) 145.
[194] M.Yu. Kalmykov, *Gauge and parametrization dependencies of the one-loop counterterms in Einstein gravity*, Class. Quant. Grav. **12** (1995) 1401, hep-th/9502152;
M.Yu. Kalmykov, K.A. Kazakov, P.I. Pronin and K.V. Stepanyantz, *Detailed analysis of the dependence of the one loop counterterms on the gauge and parametrization in the Einstein gravity with the cosmological constant*, Class. Quant. Grav. **15** (1998) 3777, arXiv: 9809169.
[195] A.Yu. Kamenshchik and C.F. Steinwachs, *Question of quantum equivalence between Jordan frame and Einstein frame*, Phys. Rev. **D91** (2015) 084033,

arXiv:1408.5769.
[196] H. Kawai and M. Ninomiya, *Renormalization group and quantum gravity*, Nucl. Phys. **B336** (1990) 115.
[197] H. Kawai, Y. Kitazawa and M. Ninomiya, *Scaling exponents in quantum gravity near two-dimensions*, Nucl. Phys. **B393** (1993) 280.
[198] T.W.B. Kibble, *Lorentz invariance and the gravitational field*, J. Math. Phys. **2** (1961) 212.
[199] C. Kiefer, *Quantum gravity*, (Oxford University Press, 3-rd Edition, 2012).
[200] N.V. Krasnikov, *Nonlocal gauge theories*, Theor. Math. Phys. **73** (1987) 1184.
[201] Y.V. Kuz'min, *The convergent nonlocal gravitation*, Sov. J. Nucl. Phys. **50** (1989) 1011.
[202] L.D. Landau and E.M. Lifshitz, *Classical theory of fields*, (Pergamon, 1971).
[203] L.D. Landau and E.M. Lifshitz, *Fluid mechanics.* (Pergamon Press, 1987).
[204] G. Jona-Lasinio, *Relativistic field theories with symmetry breaking solutions*, Nuovo Cim. **34** (1964) 1790.
[205] P.M. Lavrov, *BRST, Ward identities, gauge dependence and FRG*, arXiv: 2002.05997.
[206] P.M. Lavrov and I.L. Shapiro, *On the renormalization of gauge theories in curved spacetime*, Phys. Rev. **D81** (2010) 044026, arXiv:0911.4579.
[207] P.M. Lavrov and I.L. Shapiro, *On the functional renormalization group approach for Yang-Mills fields*, JHEP **1306** (2013) 086, arXiv:1212.2577.
[208] P.M. Lavrov and I.L. Shapiro, *Gauge invariant renormalizability of quantum gravity*, Phys. Rev. **D100** (2019) 026018, arXiv:1902.04687.
[209] P.M. Lavrov, E. A. dos Reis, T. de Paula Netto and I.L. Shapiro, *Gauge invariance of the background average effective action*, Euro. Phys. Journ. **C79** (2019) 661, arXiv:1905.08296.
[210] T.D. Lee and G.C. Wick, *Finite theory of Quantum Electrodynamics*, Phys. Rev. **D2** (1970) 1033; *Negative metric and the unitarity of the S Matrix*, Nucl. Phys. **B9** (1969) 209.
[211] H. Lehmann, *Über Eigenschaften von Ausbreitungsfunktionen und Renormierungskonstanten quantisierter Felder*, Nuovo Cim. **11** (1954) 342.
[212] H. Lehmann, K. Symanzik and W. Zimmermann, *Zur Formulierung Quantisie ter Feldtheorien*, Nuovo Cim. **1** (1955) 205.
[213] G. Leibbrandt, *Introduction to the technique of dimensional regularization*, Rev. Mod. Phys. **47** (1975) 849.
[214] S.-B. Liao, *Connection between momentum cutoff and operator cutoff regularizations*, Phys. Rev. **D53** (1996) 2020.
[215] M. Maggiore, *A modern introduction to quantum field theory*, (Oxford University Press, 2005).
[216] M. Maggiore, L. Hollenstein, M. Jaccard and E. Mitsou, *Early dark energy from zero-point quantum fluctuations.* Phys. Lett. **B704** (2011) 102, arXiv:1104.3797.
[217] S. Mauro, R. Balbinot, A. Fabbri and I.L. Shapiro, *Fourth derivative gravity in the auxiliary fields representation and application to the black hole stability,*

Eur. Phys. J. Plus **130** (2015) 135, arXiv:1504.06756.

[218] P. O. Mazur and E. Mottola, *Weyl cohomology and the effective action for conformal anomalies*, Phys. Rev. **D64** (2001) 104022.

[219] C.W. Misner, K.S. Thorne and J.A. Wheeler, *Gravitation*, (Freeman, 1973).

[220] L. Modesto, *Super-renormalizable quantum gravity*, Phys. Rev. **D86** (2012) 044005, arXiv:1107.2403; L. Modesto and L. Rachwał, *Super-renormalizable and finite gravitational theories*, Nucl. Phys. **B889** (2014) 228, arXiv:1407.8036; *Nonlocal quantum gravity: A review*, Int. J. Mod. Phys. **D26** (2017) 1730020.

[221] L. Modesto and I.L. Shapiro, *Superrenormalizable quantum gravity with complex ghosts*, Phys. Lett. **B755** (2016) 279, arXiv:1512.07600.

[222] L. Modesto, L. Rachwał and I.L. Shapiro, *Renormalization group in super-renormalizable quantum gravity*, Eur. Phys. J. **C78** (2018) 555, arXiv:1704.03988.

[223] T.R. Morris, *The exact renormalization group and approximate solutions.* Int. J. Mod. Phys. **A9** (1994) 2411, hep-ph/9308265.

[224] V. Mukhanov, *Physical foundations of cosmology.* (Cambridge Univ.Press, 2005).

[225] V. Mukhanov and S. Winitzki, *Introduction to quantum effects in gravity*, (Cambridge University Press, 2007).

[226] U. Müller, C. Schubert, and A.E.M. van de Ven, *A Closed formula for the Riemann normal coordinate expansion, Gen. Rel. and Grav.* **31** (1999) 1759.

[227] Y.S. Myung, *Unstable Schwarzschild-Tangherlini black holes in fourth-order gravity*, Phys. Rev. **D88** (2013) 084006, arXiv:1308.3907.

[228] G.L. Naber, *The geometry of Minkowski space-time*, (Springer-Verlag, 1992).

[229] B.L. Nelson and P. Panangaden, *Scaling behavior of interacting quantum fields in curved space-time*, Phys.Rev. **D25** (1982) 1019.

[230] T.d.P. Netto, A.M. Pelinson, I.L. Shapiro and A.A. Starobinsky, *From stable to unstable anomaly-induced inflation*, Eur. Phys. J. **C76** (2016) 544, arXiv:1509.08882.

[231] M. Niedermaier and M. Reuter, *The asymptotic safety scenario in quantum gravity*, Living Rev. Rel. **9** (2006) 5.

[232] S.D. Odintsov and I.L. Shapiro, *Curvature induced phase transition in quantum R^2 - gravity and the induced Einstein gravity*, Theor. Math. Phys. **90** (1992) 148.

[233] V.I. Ogievetsky and I.V. Polubarinov, *The notoph and its possible interactions*, Sov. J. Nucl. Phys. **4** (1967) 156, reprinted in *Supersymmetries and Quantum Symmetries*, (Editors J. Wess and E.A. Ivanov, Springer, 1999, p. 391–396.).

[234] J. O'Hanlon, *Intermediate-range gravity: a generally covariant model*, Phys. Rev. Lett. **29** (1972) 137.

[235] H. Osborn, *Advanced quantum field theory*, http://www.damtp.cam.ac.uk/user/ho/AQFTNotes.pdf

[236] H. Osborn, *Symmetries and Groups*, http://www.damtp.cam.ac.uk/user/ho/GNotes.pdf

[237] K. Osterwalder and R. Schrader, *Axioms for Euclidean Green's functions*, Commun. Math. Phys. **31** (1973) 83; *Axioms for Euclidean Green's functions, Part*

2., Commun. Math. Phys. **42** (1975) 281.
[238] M.V. Ostrogradsky, *Mémoires sur les équations différentielles, relatives au problème des isopérimètres*, Mem. Acad. St. Petersbourg, **VI, 4** (1850) 385.
[239] S. Paneitz, *A Quartic conformally covariant differential operator for arbitrary pseudo-Riemannian manifolds,* MIT preprint, 1983; SIGMA **4** (2008) 036, arXiv:0803.4331.
[240] L. Parker and D.J. Toms, *Renormalization group analysis of grand unified theories in curved space-time*, Phys.Rev. **29D** (1984) 1584.
[241] L. Parker and J.Z. Simon, *Einstein equation with quantum corrections reduced to second order*, Phys. Rev. **D47** (1993) 1339, gr-qc/9211002.
[242] G.B. Peixoto, *A note on the heat kernel method applied to fermions,* Mod. Phys. Lett. **A16** (2001) 2463, hep-th/0108223.
[243] G.B. Peixoto, A. Penna-Firme and I.L. Shapiro, *One loop divergences of quantum gravity using conformal parametrization,* Mod. Phys. Lett. **A15** (2000) 2335, arXiv: 0103043.
[244] G.B. Peixoto and I.L. Shapiro, *Conformal quantum gravity with the Gauss-Bonnet term*, Phys. Rev. **D70** (2004) 044024, hep-th/0307030.
[245] G.B. Peixoto, E.V. Gorbar and I.L. Shapiro, *On the renormalization group for the interacting massive scalar field theory in curved space,* Class. Quant. Grav. **21** (2004) 2281, hep-th/0311229.
[246] G. de Berredo-Peixoto and I.L. Shapiro, *Higher derivative quantum gravity with Gauss-Bonnet term,* Phys. Rev. **D71** (2005) 064005, hep-th/0412249.
[247] A.M. Pelinson, I.L. Shapiro and F.I. Takakura, *On the stability of the anomaly-induced inflation*, Nucl. Phys. **B648** (2003) 417, hep-ph/020818.
[248] R. Percacci, *Asymptotic safety,* (In D. Oriti - Editor: Approaches to quantum gravity, 111, Cambridge University Press, 2007), arXiv:0709.3851.
[249] S. Perlmutter et al. *Supernova cosmology project*, Nature **391** (1998) 51;
[250] M.E. Peskin and D.V. Schroeder, *An introduction to quantum field theory,* (Addison-Wisley, NY, 1996).
[251] P. Peter, F. de O. Salles and I.L. Shapiro, *On the ghost-induced instability on de Sitter background,* Phys. Rev. **D97** (2018) 064044, arXiv:1801.00063.
[252] A.Z. Petrov, *Einstein Spaces,* (Pergamon Press, Oxford, 1969).
[253] J. Polchinski, *Renormalization and effective lagrangians,* Nucl. Phys. **B231** (1984) 269.
[254] H.D. Politzer, *Reliable perturbative results for strong interactions*, Phys. Rev. Lett. **30** (1973) 1346.
[255] A.M. Polyakov, *Quantum geometry of bosonic strings*, Phys. Lett. **B207** (1981) 211.
[256] P. Ramond, *Field Theory: A Modern Primer,* (Westview Press; 2 edition, 2001).
[257] P. Ramond, *Group theory: a Physicist's survey*, (Cambridge Univ. Press, 2010).
[258] W. Rarita and J. Schwinger, *On a theory of particles with half-integer spin,* Phys. Rev. **60** (1941) 61.
[259] T. Regge and J. Wheeler, *Stability of a Schwarzschild singularity,* Phys. Rev. **108** (1957) 1903.
[260] S. Castardelli dos Reis, G. Chapiro and I.L. Shapiro, *Beyond the linear analysis*

of stability in higher derivative gravity with the Bianchi-I metric, Phys. Rev. **D100** (2019) 066004, arXiv:1903.01044.
[261] T.G. Ribeiro and I.L. Shapiro, *Scalar model of effective field theory in curved space,* JHEP **1910** (2019) 163, arXiv:1908.01937.
[262] R.J.Riegert, *A non-local action for the trace anomaly*, Ph.Lett. **B134**(1984)56.
[263] A. Riess et al, *High-Z Supernova Search*, Astron. J. *116* (1998) 1009.
[264] W. Rindler, *Introduction to Special Relativity*, (Oxford University Press, 1991).
[265] R. J. Rivers, *Lagrangian theory for neutral massive spin-2 fields,* Nuovo Cim. **34** (1964) 386.
[266] D.C. Rodrigues, F. de O. Salles, I.L. Shapiro and A.A. Starobinsky, *Auxiliary fields representation for modified gravity models,* Phys. Rev. **D83** (2011) 084028, arXive:1101.5028.
[267] V. Rubakov, *Classical theory of gauge fields*, (Princeton University, Press, 2002).
[268] A.D. Sakharov, *Vacuum quantum fluctuations in curved space and the theory of gravitation,* Sov. Phys. Dokl. **12** (1968) 1040; Gen. Rel. Grav. **32** (2000) 365.
[269] A. Salam, *Divergent integrals in renormalizable field theories,* Phys. Rev. **84** (1951) 426.
[270] A. Salam and J. Strathdee, *Remarks on high-energy stability and renormalizability of gravity theory*, Phys. Rev. **D18** (1978) 4480.
[271] F. de O. Salles and I.L. Shapiro, *Do we have unitary and (super)renormalizable quantum gravity below the Planck scale?.* Phys. Rev. **D89** 084054 (2014); **90**, 129903 (2014) [Erratum], arXiv:1401.4583.
[272] A. Salvio, *Metastability in quadratic gravity,* Phys. Rev. **D99** (2019) 103507, arXiv:1902.09557.
[273] A. Salvio and A. Strumia, *Agravity,* JHEP **06** (2014) 080, arXiv:1403.4226.
[274] M.D. Schwartz, *Quantum field theory and the Standard Model,* (Cambridge University Press, Cambridge, 2013).
[275] P.M. Schwarz and J.H. Schwarz, *Special relativity: from Einstein to strings,* (Cambridge University Press, 2004).
[276] S.S. Schweber, *Introduction to relativistic quantum field theory,* (Dover Books on Physics, 2-nd edition, 2005).
[277] J. Schwinger, *On gauge invariance and vacuum polarization*, Phys. Rev. **82** (1951) 664.
[278] I.L. Shapiro and E.G. Yagunov, *New set of finite and asymptotically finite GUTs,* Int. J. Mod. Phys. **A8** (1993) 1787.
[279] I.L. Shapiro and A.G. Jacksenaev, *Gauge dependence in higher derivative quantum gravity and the conformal anomaly problem*, Phys. Lett. **B324** (1994) 286.
[280] I.L. Shapiro and H. Takata, *One loop renormalization of the four-dimensional theory for quantum dilaton gravity*, Phys. Rev. **D52** (1995) 2162, hep-th/9502111.
[281] I.L. Shapiro and H. Takata, *Conformal transformation in gravity*, Phys. Lett. **B361** (1996) 31, hep-th/9504162.

[282] I.L. Shapiro, *On the conformal transformation and duality in gravity*, Class. Quantum Grav. **14** (1997) 391, hep-th/9610129.
[283] G. Cognola and I.L. Shapiro, *Back reaction of vacuum and the renormalization group flow from the conformal fixed point*, Class. Quant. Grav. **15** (1998) 3411, hep-th/9804119.
[284] I.L. Shapiro, *Physical aspects of the space-time torsion*, Phys. Rep. **357** (2002) 113, hep-th/0103093.
[285] I.L. Shapiro, J. Solà, *Scaling behavior of the cosmological constant: Interface between quantum field theory and cosmology*, JHEP **02** (2002) 006, hep-th/0012227.
[286] I.L. Shapiro, *The graceful exit from the anomaly-induced inflation: Supersymmetry as a key*, Int. Journ. Mod. Phys. **D11** (2002) 1159, hep-ph/0103128; I.L. Shapiro, J. Solà, *Massive fields temper anomaly-induced inflation: the clue to graceful exit?*, Phys. Lett. **B530** (2002) 10, hep-ph/0104182.
[287] I.L. Shapiro, *Effective action of vacuum: semiclassical approach*, Class. Quant. Grav. **25** (2008) 103001, arXiv:0801.0216.
[288] I.L. Shapiro, J. Solà, *On the possible running of the cosmological 'constant'*, Phys. Lett. **B682** (2009) 105, hep-th/0910.4925.
[289] I.L. Shapiro and P.M. Teixeira, *Quantum Einstein-Cartan theory with the Holst term*, Class. Quant. Grav. **31** (2014) 185002, arXiv:1402.4854.
[290] I.L. Shapiro, *Counting ghosts in the "ghost-free" non-local gravity*. Phys. Lett. **B744** (2015) 67, arXive:1502.00106.
[291] I.L. Shapiro, *Covariant derivative of fermions and all that,* arXiv:1611.02263.
[292] I.L. Shapiro, *Primer in tensor analysis and relativity,* (Springer, NY, 2019).
[293] W. Siegel, *Fields,* hep-th/99122005.
[294] J.Z. Simon, *Higher-derivative Lagrangians, nonlocality, problems, and solutions*, Phys. Rev. **D41** (1990) 3720.
[295] J.Z. Simon, *No Starobinsky inflation from selfconsistent semiclassical gravity,* Phys. Rev. **D45** (1992) 1953.
[296] A.A. Slavnov, *Invariant regularization of non-linear chiral theories*, Nucl. Phys. **B31** (1971) 301.
[297] A.A. Slavnov, *Ward Identities in gauge theoories*, Th.Math.Phys. **10** (1972) 99.
[298] A.A. Slavnov, *Functional integral in perturbation theory*, Theor. Math. Phys., **22** (1975) 123.
[299] A.A. Slavnov and L.D. Faddeev, *Massless and massive Yang-Mills field,* Theor. Math. Phys. **3** (1970) 312.
[300] V.A. Smirnov, *Evaluating Feynman Integrals*, (Springer, 2004); *Feynman integral calculus*, (Springer, 2006); *Analytic tools for Feynman integrals*, (Springer, 2012).
[301] V.A. Smirnov, *Analytic tools for Feynman integrals*, (Springer, 2012).
[302] F. Sobreira, B.J. Ribeiro and I.L. Shapiro, *Effective potential in curved space and cut-off regularizations*, Phys. Lett. **B705** (2011) 273, arXive: 1107.2262.
[303] R.F. Streater and A.S. Wightman, *PCT, spin and statistics, and all that,* (Princeton University Press, 2000).
[304] M. Srednicki, *Quantum field theory*, (Cambridge University Press, 2007).

[305] A.A. Starobinski, *A new type of isotropic cosmological models without singularity*, Phys. Lett. **B91** (1980) 99.
[306] A.A. Starobinsky, *The perturbation spectrum evolving from a nonsingular initially de-Sitter cosmology and the microwave background anisotropy*, Sov. Astron. Lett. **9** (1983) 302.
[307] K.S. Stelle, *Renormalization of higher derivative quantum gravity*, Phys. Rev. **D16** (1977) 953.
[308] K.S. Stelle, *Classical gravity with higher derivatives*, Gen.Rel.Grav. **9**(1978) 353.
[309] R.F. Streater and A.S. Wightman, *PCT, Spin and statistics, and all that,* (Princeton University Press, 2000).
[310] J. L. Synge, *Relativity: The general Theory.* (North-Holland Publishing, 1960).
[311] Y. Takahashi, *On the generalized Ward identity*, Nuovo Cim. **6** (1957) 371.
[312] J.C. Taylor, *Ward identities and charge renormalization of the Yang-Mills field*, Nucl. Phys. **B33** (1971) 436.
[313] P.d. Teixeira, I.L. Shapiro and T.G. Ribeiro, *One-loop effective action: nonlocal form factors and renormalization group,* arXiv:2003.04503, to be published in Gravitation and Cosmology **26** (2020).
[314] Ya.P. Terletskii, *Paradoxes in the theory of Relativity*, (Plenum Press, NY, 1968).
[315] E. Tomboulis, *$1/N$ expansion and renormalization in quantum gravity*, Phys. Lett. **B70** (1977) 361; *Renormalizability and asymptotic freedom in quantum gravity*, Phys. Lett. **B97** (1980) 77; *Unitarity in Higher derivative quantum gravity*, Phys. Rev. Lett. **52** (1984) 1173.
[316] E.T. Tomboulis, *Superrenormalizable gauge and gravitational theories,* hep-th/9702146; *Nonlocal and quasilocal field theories,* Phys. Rev. **D92** (2015) 125037, arXiv:1507.00981.
[317] D.J. Toms and L. Parker, *Quantum field theory in curved spacetime: quantized fields and gravity,* (Cambridge Univ. Press, Cambridge, 2014).
[318] D.J. Toms, *Effective action for the Yukawa model in curved spacetime,* JHEP **1805** (2018) 139, arXiv:1804.08350.
[319] A.A. Tseytlin, *On singularities of spherically symmetric backgrounds in string theory.* Phys. Lett. **B363** (1995) 223, hep-th/9509050.
[320] Wu-Ki Tung, *Group theory for physicists,* (World Scientific, 1985).
[321] I.V. Tyutin, *Gauge invariance in field theory and statistical mecanics,* (Lebedev Physical Institute Preprint, no 39, 1975).
[322] R. Utiyama and B.S. DeWitt, *Renormalization of a classical gravitational field interacting with quantized matter fields,* J. Math. Phys. **3** (1962) 608.
[323] A.N. Vasiliev, *Functional methods in Quantum Field Theory and Statistical Physics,* (CRS Press, 1998).
[324] D.V. Vassilevich, *Heat kernel expansion: User's manual,* Phys. Rept. **388** (2003) 279, hep-th/0306138.
[325] M.J.G. Veltman, *Unitarity and causality in a renormalizable field theory with unstable particles*, Physica **29** (1963) 186.
[326] A.E.M. van de Ven, *Two loop quantum gravity*, Nucl. Phys. **B378** (1992) 309.

[327] A. Vilenkin, *Classical and quantum cosmology of the Starobinsky inflationary model*, Phys. Rev. **D32** (1985) 2511.

[328] G.A. Vilkovisky, *The unique effective action in quantum field theory*, Nucl. Phys. **B234** (1984) 125; *Effective action in quantum gravity*, Class. Quant. Grav. **9** (1992) 895; *Expectation values and vacuum currents of quantum fields*, Lect. Notes Phys. **737** (2008) 729; arXiv:0712.3379.

[329] C. Vishveshwara, *Stability of the Schwarzschild metric*, Phys. Rev. **D1** (1970) 2870; L. Edelstein and C. Vishveshwara, *Differential equations for perturbations on the Schwarzschild metric*, Phys. Rev. **D1** (1970) 3514.

[330] M. Visser, *Sakharov's induced gravity: A Modern perspective*, Mod. Phys. Lett. **A17** (2002) 977, gr-qc/0204062.

[331] A.A. Vladimirov and D.V. Shirkov, *The renormalization group and ultraviolet asymptotics*, Sov. Phys. Usp. **22** (1979) 860.

[332] M.B. Voloshin and K.A. Ter-Martirosyan, *Gauge theory of particle interaction*, (Nauka, 1981, in Russian).

[333] B.L. Voronov and I.V. Tyutin, *Models of asymptotically free massive fields*, Sov. Journ. Nucl. Phys. [Yad. Fiz.] **23** (1976) 664.

[334] B.L. Voronov, P.M. Lavrov and I.V. Tyutin, *Canonical transformations and the gauge dependence in general gauge theories*, Sov. J. Nucl. Phys. **36** (1982) 498.

[335] B.L. Voronov and I.V. Tyutin, *Formulation of gauge theories of general form. I.*, Theor. Math. Phys. **50** (1982) 218.

[336] B.L. Voronov and I.V. Tyutin, *Formulation of gauge theories of general form. II. Gauge-invariant renormalizability and renormalization structure*, Theor. Math. Phys. **52** (1982) 628.

[337] B.L. Voronov and I.V. Tyutin, *On renormalization of R^2 gravitation*, Yad. Fiz. (Sov. Journ. Nucl. Phys.) **39** (1984) 998.

[338] J.C. Ward, *An identity in Quantum Electrodynamics*, Phys. Rev. **78** (1950) 182.

[339] S. Weinberg, *High-energy behavior in quantum field theory*, Phys. Rev. **118** (1960) 838.

[340] S. Weinberg, *Gravitation and Cosmology: Principles and Applications of the General Theory of Relativity*, (John Wiley & Sons, New York, 1972).

[341] S. Weinberg, *Ultraviolet divergences in quantum theories of gravitation*, (In S. W. Hawking; W. Israel (eds.). General Relativity: An Einstein centenary survey. Cambridge University Press, 1979).

[342] S. Weinberg, *Effective gauge theories*, Phys. Lett. **B91** (1980) 51.

[343] S. Weinberg, *Anthropic bound on the cosmological constant*, Phys. Rev. Lett. **59** (1987) 2607.

[344] S. Weinberg, *The cosmological constant problem*, Rev. Mod. Phys. **61** (1989) 1.

[345] S. Weinberg, *The quantum theory of fields, Vol I. Foundations, Vol II. Modern applications, Vol II. I Supersymmetry*, (Cambridge University Press, Cambridge, 1995, 1996, 2000).

[346] C. Wetterich, *Average action and the renormalization group equations*, Nuc.

Phys. **B352** (1991) 529;*Exact evolution equation for the effective potential,* Phys. Lett. **B301** (1993) 90,arXiv:1710.0581.
[347] B. Whitt, *The stability of Schwarzschild black holes in fourth order gravity,* Phys. Rev. **D32** (1985) 379.
[348] C.M. Will, *Theory and experiment in gravitational physics,* (Cambridge University Press, Cambridge, UK, 2-nd Edition, 1993).
[349] A. Wipf, *Statistical approach to quantum field theory: an introduction,* Springer Lect. Notes in Physics, **864** 2013.
[350] R.P. Woodard, *Avoiding dark energy with $1/r$ modifications of gravity,* Lect. Notes Phys. **720** (2007) 403, astro-ph/0601672; *Ostrogradsky's theorem on Hamiltonian instability,* Scholarpedia **10** (2015) 32243, arXiv:1506.02210.
[351] Y. Yoon and Y. Yoon, *Asymptotic conformal invariance of $SU(2)$ and standard models in curved space-time,* Int. J. Mod. Phys. **A12** (1997) 2903, hep-th/9612001.
[352] O.I. Zavialov, *Renormalized Quantum Field Theory,* (Springer, 1990).
[353] Ya.B. Zel'dovich, *The Cosmological constant and the theory of elementary particles,* Letters to ZjETP **6** (1967) 1233; Sov. Phys. Usp. **11** (1968) 381; Gen. Rel. Grav. **40** (2008) 1557.
[354] F.J. Zerilli, *Effective potential for even parity Regge-Wheeler gravitational perturbation equations,* Phys. Rev. Lett. **24** (1970) 737.
[355] J. Zinn-Justin, *Renormalization of gauge theories,* (Trends in Elementary Particle Theory, Lecture Notes in Physics, **37**, Eds. H.Rollnik and K.Dietz, Springer, Berlin, 1975).
[356] J. Zinn-Justin, *Quantum field theory and critical phenomena,* (Oxford University Press, 1994).
[357] B. Zwiebach, *Curvature squared terms and string theories,* Phys. Lett. **B156** (1985) 315.

Index

The manufacturer's authorised representative in the EU for product safety is Oxford University Press España S.A. of el Parque Empresarial San Fernando de Henares, Avenida de Castilla, 2 – 28830 Madrid (www.oup.es/en or product.safety@oup.com). OUP España S.A. also acts as importer into Spain of products made by the manufacturer.

www.ingramcontent.com/pod-product-compliance
Ingram Content Group UK Ltd.
Pitfield, Milton Keynes, MK11 3LW, UK
UKHW051224160926
475595UK00022B/155

* 9 7 8 0 1 9 8 8 3 8 3 1 9 *